4th International Conference on Environment, Sustainability Issues, and Community Development (INCRID)

IOP Conference Series: Earth and Environmental Science
Volume 1098

Semarang, Indonesia
1 September 2022

ISBN: 978-1-7138-6442-4
ISSN: 1755-1307

Printed from e-media with permission by:

Curran Associates, Inc.
57 Morehouse Lane
Red Hook, NY 12571

proceedings.com

Some format issues inherent in the e-media version may also appear in this print version.

Institute of Physics
Dirac House, Temple Back
Bristol BS1 6BE UK

Phone: 44 1 17 929 7481
Fax: 44 1 17 920 0979

techtracking@iop.org

Additional copies of this publication are available from:

Curran Associates, Inc.
57 Morehouse Lane
Red Hook, NY 12571 USA
Phone: 845-758-0400
Fax: 845-758-2633
Email: curran@proceedings.com
Web: www.proceedings.com

TABLE OF CONTENTS

ENVIRONMENT, HEALTH AND SAFETY

ENVIRONMENTAL SCIENCE, TECHNOLOGY AND EDUCATION

GREEN INFRASTRUCTURE

ENERGY CONSERVATION AND EFFICIENCY

IOP Conf. Series: Earth and Environmental Science 1098 (2022) 011001 doi:10.1088/1755-1315/1098/1/011001

THE 4th INTERNATIONAL CONFERENCE ON ENVIRONMENT, SUSTAINABILITY ISSUES AND COMMUNITY DEVELOPMENT (INCRID)

Environmental Engineering Diponegoro University,
Jl. Prof. H. Soedarto, S.H. – Tembalang, Semarang, Indonesia
Telp. 081229323709, 085700924611
e-mail : incrid@live.undip.ac.id ; web : incrid.lingkungan.ft.undip.ac.id

---

On behalf of the committee, I would like to express my sincere gratitude to all colleagues, professors, lecturers, researchers, and welcome you all to the 4th INCRID 2022 "Supporting the Realization of Zero Carbon Environment by Implementing Circular Economy". This conference provides a great opportunity for researchers, students, industries, and governments to communicate their research results on the fundamentals and application of sustainability issues and community development.

INCRID 2022 was held on 1 September 2022 in the online system and the theater room 5th floor of the Faculty of Engineering Diponegoro University, Semarang (hybrid). Keynote and invited speakers were Prof. Ramaraj Boopathy Nicholls State University, USA, Dr. Premakumara Jagath (Institute for Global Environmental Strategies, Japan), Prof. Pau Loke Show (University of Nottingham Malaysia), Prof. Sudharto P. Hadi, MES, Ph.D. (Universitas Diponegoro, Indonesia), Prof. Ir. Tjandra Setiadi, M.Eng., Ph.D. (Institut Teknologi Bandung, Indonesia), Prof. Dr. Soraya Heuss-Aßbichler (Ludwig-Maximilians-Universität München, Germany).

At this conference we have contributions from seven countries. We received about 118 submissions of papers for presentation at this meeting. Each paper was evaluated by a reviewer and about 85 of these are accepted for presentation, divided into 12 parallel presentation sessions. The topics of this conference include Environmental, Health and safety; environmental science, technology, and education; Green infrastructure; and Energy Conservation and Efficiency. It is my hope that the 4th INCRID 2022 will be able to achieve its objective of creating an international forum for researchers, students, industries, and governments to communicate their research results, to share and exchange ideas on the fundamentals and application of environmental, sustainability issues, and community development.

By bringing up this theme, the Department of Environmental Engineering and the INCRID 2022 Committee want to support the efforts of the World to achieve the goal of emission reduction and Net Zero Emission (Carbon neutrality). Furthermore, this activity is expected to support efforts to implement the concept of the circular economy.

Last but not least, my deepest gratitude goes to the Advisory Board, Organizing Committee, International Scientific Committee, institutions, companies, and volunteers who have directly and indirectly supported the success of this conference. Although we try our best to be professional, on behalf of the committee, we request you to accept our sincere apologies for any inconvenience.

Dr. Yustina Metanoia Pusparizkita, S.T., M.T.
Chairman of the 4th INCRID 2022

IOP Conf. Series: Earth and Environmental Science 1098 (2022) 011001 doi:10.1088/1755-1315/1098/1/011001

NAME OF THE EVENT

The 4th International Conference on Environment, Sustainability Issues and Community Development 2022 (4th INCRID 2022)

THEME

"Supporting the Realization of Zero Carbon Environment by Implementing Circular Economy."

The topics of the conference are as follows.

A. Environment, Health, & Safety

- Environment, health, and safety system
- Environmental modeling and computation
- Risk analysis

B. Environmental Science, Technology, and Education

- Waste management and treatment
- Water and wastewater engineering
- Environmental education

C. Green Infrastructure

- Life cycle assessment
- Green building and technology option

D. Energy Conservation and Efficiency

- Clean and renewable energy
- Climate change and global warming

OBJECTIVES OF THE EVENT

The objections of INCRID 2022 are as follows:

- To create an international forum for researchers, students, industries, and governments to communicate their research results on the fundamentals and application of environment, sustainability issues, and community development.
- To share and exchange ideas, thoughts, and discussions on all aspects of the environment, sustainability issues, and community development.
- Facilitate the formation of networks among participants to enhance the quality and benefits of research and development.

PARTICIPANTS

This international conference is open to academicians, researchers, students, and professionals worldwide.

SPEAKERS

Keynote Speakers

No	Name	Institution and Country
1	Prof. Ir. Tjandra Setiadi, M.Eng., Ph.D	Institut Teknologi Bandung, Indonesia
2	Prof. Sudharto P. Hadi, MES, Ph.D	Universitas Diponegoro, Indonesia
3	Prof. Dr. Soraya Heuss-Aßbichler	Ludwig-Maximilians-Universität München, Germany
4	Prof. Pau Loke Show	University of Nottingham, Malaysia
5	Dr. Ramaraj Boopathy	Nicholls State University, USA
6	Dr. Premakumara Jagath	Institute for Global Environmental Strategies, Japan

SCHEDULE

1st Round Submission

Deadline for Abstract Submission	7 June 2022
Notification of Abstract Submission	9 June 2022
Deadline for Full Paper Submission	23 June 2022
Review Result	1 July 2022
Revised Paper Submission	9 July 2022
Deadline for Registration and Payment	23 June 2022

IOP Conf. Series: Earth and Environmental Science 1098 (2022) 011001 doi:10.1088/1755-1315/1098/1/011001

2nd Round Submission

Deadline for Abstract Submission	10 July 2022
Notification of Abstract Submission	12 July 2022
Deadline for Full-Paper Submission	26 July 2022
Review Result	2 August 2022
Revised Paper Submission	10 August 2022
Deadline for Registration and Payment	26 July 2022

Presentation File Submission

(28th August 2022)

Conference Day

(1st September 2022)

VENUE

Online and 5th floor Faculty of Engineering Universitas Diponegoro

Jl. Prof. H. Soedarto, S.H., Semarang, Indonesia

ADVISORY BOARD

1. Rector of Diponegoro University
2. Dean of Engineering Faculty, Diponegoro University
3. Head of Research and Community Service (LPPM), Diponegoro University

STEERING COMMITTEE

1. Prof. Dr. rer. nat. Heru Susanto, M.M., M.T. (Diponegoro University, Indonesia)
2. Prof. Ir. M Agung Wibowo, MM, M.Sc., Ph.D (Universitas Diponegoro, Indonesia)
3. Prof. Ir. Syafrudin, CES., M.T. (Universitas Diponegoro, Indonesia)
4. M. Arief Budihardjo, S.T., M. Eng.Sc., Ph.D. (Universitas Diponegoro, Indonesia)
5. Dr. Badrus Zaman, S.T., M.T. (Universitas Diponegoro, Indonesia)

IOP Conf. Series: Earth and Environmental Science 1098 (2022) 011001 doi:10.1088/1755-1315/1098/1/011001

ORGANIZING COMMITTEE

1. Dr. Ing. Sudarno, S.T., M.Sc (Head of Departement)
2. Dr. Yustina Metanoia Pusparizkita, S.T., M.T. (Chairman)
2. Dr. Ir. Haryono Setiyo Huboyo, S.T., M.T., IPM. (Vice Chairman)
3. Ir. Nurandani Hardyanti, S.T., M.T., IPM (Secretary)
4. Ir. Pertiwi Andarani, S.T., M.T., M.Eng., Ph.D., IPP (Publication and Indexing)

IOP Conf. Series: Earth and Environmental Science 1098 (2022) 011002 doi:10.1088/1755-1315/1098/1/011002

Peer Review Statement

All papers published in this volume have been reviewed through processes administered by the Editors. Reviews were conducted by expert referees to the professional and scientific standards expected of a proceedings journal published by IOP Publishing Publishing.

- **Type of peer review:** Single Anonymous
- **Conference submission management system:** Morressier
- **Number of submissions received:** 118
- **Number of submissions sent for review:** 98
- **Number of submissions accepted:** 85
- **Acceptance Rate (Submissions Accepted / Submissions Received × 100):** 72
- **Average number of reviews per paper:** 2
- **Total number of reviewers involved:** 14
- **Contact person for queries:**
  Name: Yustina Metanoia Pusparizkita
  Email: ympusparizkita@lecturer.undip.ac.id
  Affiliation: Departement of Environmental Engineering, Diponegoro University, Semarang, Indonesia

Environmental management system: The internal and external impact of ISO 14001 implementation on the manufacturing companies

A K Widiatami[1*], L K Pitaloka[1], A Nurkhin[1]

[1]Economic Education Department, Universitas Negeri Semarang, Indonesia
kania@mail.unnes.ac.id

Abstract. The activities of manufacturing companies are closely related to the environmental damage that occurs. It gives an imprint on the environmental damage from the production process until the community consumes the product . However, many manufacturing companies have tried to reduce the damage to nature by implementing an environmental management system. Being certified to ISO 14001 is the actual action of a manufacturing company in implementing an environmental management system. This study explores the impact of implementing an environmental management system that is ISO 14001 certified to the company internally and externally. Using a purposive sampling technique to the manufacturing companies listed on the IDX from 2017 – to 2020, 327 manufacturing companies were obtained as a sample of this study. The results of hypothesis testing using Panel Data Regression shows that the environmental management system has a significant positive effect on manufacturing cost, raw material cost, the number of outstanding shares, market value-added, and sales growth. Furthermore, an environmental management system is proven to significantly impact the company because it increases the number of outstanding shares, provides company value, and increases the company's sales growth.

1. Introduction

The government is now intensively informing business principles based on Environmental, Social, and Governance (ESG), so implementing ESG-based business is one of the most important things. It is due to the issue of global warming, which is becoming more and more widespread, caused mainly by the company's business activities—starting from the use of paper for office operations to the waste that arises as a result of the production process. However, few companies have tried reducing natural damage by implementing an environmental management system (EMS). ISO 14001 certification is the company's real action in implementing EMS. Quoted from the page [1], ISO 14001 is an international standard that provides a framework for organizations to protect the environment and respond to changing natural environmental conditions. ISO 14001 focuses on process management aimed at improving the prevention of environmental damage rather than environmental outcomes [2]. ISO 14001 is a voluntary environmental management system. So not all companies have this certification.

Implementing EMS is one of the company's implementations in reducing environmental damage. The company can obtain ISO 14001 if the company has developed an environmental management system that meets ISO requirements. However, it does not stop there. Companies ISO 14001 certified must conduct regular evaluations to improve the company's environmental management system. According to the ISO 14001 standard, there are six essential elements in the implementation of EMS, including environmental policy, planning, implementation, monitoring and improvement, management review, and continuous improvement. Therefore, investors see ISO 14001-certified companies as a signal that the company has international standard procedures for improving environmental management [3].

Globally, many companies in the world have implemented EMS. In 2020 survey results [4] showed that companies certified to ISO 14001 increased by 12% from the previous year. The three most countries that have ISO 14001 certified companies are China (48.25%), Japan (5.11%), then Italy (4.84%). Indonesia is ranked twenty-second globally and second after Thailand in Southeast Asia; as many as 0.69% of companies have been certified ISO 14001. The type of industry companies that have implemented the most EMS is Construction (14.07%), Wholesale & retail trade, repair of motor vehicles, motorcycles, personal & household items (9.31%), and electrical and optical equipment (7.53%).

Some studies have researched companies' motivation for implementing environmental management systems (EMS). Companies' main motivation for implementing it is to use it as a strategic tool for achieving specific company goals. Companies use it as a key to participate in government-funded projects and gain access to the international market [5]. EMS is believed to be able to open market access because it provides community legitimacy as a better company [6]. Other than that, companies use it to meet customer, community, and employee satisfaction. ISO 14001 certification means that the company has fulfilled the health and safety standard in the workplace. Also, employees could increase their skills and development related to the environment [7].

The following reasons for EMS implementation are improving the operational efficiency and quality of processes and products [8]. Implementing ISO 14001 proved to cut production costs, increase productivity and maximize profit [5]. Furthermore, the benefits of EMS implementation are improving the efficiency of fixed assets, raising the productivity of the company's workforce, increasing operational cycles and ROA, and reducing manufacturing costs [6].

The benefits of implementing EMS to reduce production costs and raw materials are still being debated. ISO 14001 certification does not increase the financial or turnover objectives for the company [7]. [9] stated that EMS certification had a significant positive effect on the company's operating costs. In addition to increasing the cost of raw materials in manufacturing companies, implementing ISO 14001 also increases labor costs and depreciation of fixed assets. The company's operational costs increase because in implementing EMS, companies must replace old equipment with new equipment with environmentally friendly technology and also replace new raw materials that do not damage the environment much [10]. Companies that have ISO 14001 certificates also do not increase sales growth [11]. As a result, ISO 14001 does not affect the evaluation of investors, both long-term and short-term [12]. Investors prefer companies that have high profits to companies that implement EMS as a strategy to prevent environmental damage [13].

The inconsistency of previous studies encourages researchers to explore the impact of implementing an environmental management system proxied by ISO 14001 certification on manufacturing companies listed on the Indonesian stock exchange. However, research on some of the impacts of implementing EMS has never been done in Indonesia. So, the research model of the influence of EMS as an independent variable on several dependent variables becomes the novelty of this study. The dependent variables in this study include internal impacts (i.e., production costs and raw material costs) and external impacts, namely the number of outstanding shares, market value-added, and sales growth. This research is expected to contribute in the form of knowledge of the impact of EMS implementation on manufacturing companies internally and externally. In addition, the research results are also expected to evaluate government policies in providing standards for companies to implement environmental management systems.

EMS is one of the company's responsibilities to stakeholders. Stakeholder theory describes to which party the company is responsible [14]. Stakeholders can be said to be a crucial component of the company and need to be considered because the existence of an organization is very dependent on stakeholders in the organization's environment. Communities are the stakeholders most affected and benefited from companies implementing EMS. By implementing EMS, it is hoped that there will be a reduction in pollution and waste that can pollute the environment around the company.

ISO specifically explains that the standards set can be used by any company that wants to make resource efficiency, save costs, and minimize waste, emissions, and environmental pollution [15]. Therefore, companies implementing EMS need sustainable practices to reduce the use of resources and pollution prevention practices so that companies can plan to reduce both the use of raw materials and energy [6]. EMS can reduce financial risk because it allows companies to diagnose critical problems and can improve operational processes [13]. Reduced of waste generated during production and savings from using raw materials and energy has been proven to improve the company's financial condition [16]. Carrying out recycling practices, using secondary raw materials, and implementing energy-saving policies have an impact on increasing the efficiency of business activities and reducing production costs [5]. Therefore, our hypothesis is as follows:

H_1: ISO 14001-certified companies can reduce manufacturing costs

H_2: ISO 14001-certified companies can reduce raw material costs

Besides having an internal impact, the implementation of EMS can also impact the outer side of the company. First, ISO 14001 certification increases customer satisfaction [7],[8]. Second, ISO 14001 certification will provide community legitimacy, thereby increasing the company's reputation [6]. Third, implementing ISO 14001 embeds positive routines, where companies can implicitly and uniquely improve their capabilities to have a competitive advantage [13]. Finally, investors see a company certified to ISO 14001 as a signal that the company has international standard procedures for improving environmental management [3]. Therefore, our hypothesis is as follows:

H_3: ISO 14001-certified companies can increase the number of outstanding shares

H_4: ISO 14001-certified companies can raise the market value-added

H_5: ISO 14001-certified companies can increase a sales growth

2. Methodology

This study uses secondary data from financial statements and annual reports of manufacturing companies listed on the IDX from 2017 to 2020. Manufacturing companies were chosen because they produce their products, so business activities are closely related to environmental damage. The selection of research samples by applying a purposive sampling technique with the criteria of manufacturing companies listed on the BEI during 2017 – 2020, not being delisted, issuing financial and annual reports during the year of observation, and using the conventional currency of the rupiah. Based on these criteria, we obtained as many as 327 samples of companies in this study. This research data is panel data, a combination of time-series and cross-section data. Therefore, this study uses panel data regression analysis to test the hypothesis. The panel data regression model formed in this study is as follows:

$$MC_{it} = \alpha + \beta 1EMS_{it} + \beta 2SENS_{it} + \beta 3AGE_{it} + \beta 4SIZE_{it} + \beta 5LIQ_{it} + \varepsilon_{it} \quad (1)$$
$$RMC_{it} = \alpha + \beta 6EMS_{it} + \beta 7SENS_{it} + \beta 8AGE_{it} + \beta 9SIZE_{it} + \beta 10LIQ_{it} + \varepsilon_{it} \quad (2)$$
$$LS_{it} = \alpha + \beta 11EMS_{it} + \beta 12SENS_{it} + \beta 13AGE_{it} + \beta 14SIZE_{it} + \beta 15LIQ_{it} + \varepsilon_{it} \quad (3)$$
$$MVA_{it} = \alpha + \beta 16EMS_{it} + \beta 17SENS_{it} + \beta 18AGE_{it} + \beta 19SIZE_{it} + \beta 20LIQ_{it} + \varepsilon_{it} \quad (4)$$
$$SG_{it} = \alpha + \beta 21EMS_{it} + \beta 22SENS_{it} + \beta 23AGE_{it} + \beta 24SIZE_{it} + \beta 25LIQ_{it} + \varepsilon_{it} \quad (5)$$

Where:

MC = Manufacturing Cost
RMC = Raw Material Cost
LS = Listed Share
MVA = Market Value Added
SG = Sales Growth
EMS = Environmental Management System
SENS = Environmental Sensitivity

AGE = Firm Age
SIZE = Firm Size
LIQ = Liquidity
α = Constant
β = Regression Coefficient
ε = Residual Error

This study uses five dependent variables (manufacturing cost, raw material cost, listed share, market value-added, and sales growth) with one independent variable (environmental management system) and four control variables (environmental sensitivity, age, size, and liquidity). In addition, this study uses a control variable to prevent bias in the study results due to the influence of other factors not examined. The definitions and measurements of each variable in this study are summarized in the table 1:

Table 1. Definition and measurement of variables.

Variable	Definition	Measurement
Dependent Variables		
Manufacturing Cost (MC)	The costs incurred by manufacturing companies consist of direct costs and indirect costs to produce a product.	Raw Material Cost + Direct Labour + Factory Overhead Cost
Raw Material Cost (RMC)	Costs incurred by manufacturing companies to purchase raw materials	Ln total raw material cost
Listed Share (LS)	The number of company shares owned by individual shareholders, institutions, or the government.	Number of outstanding shares reported by the company
Market Value-Added (MVA)	A benchmark can measure the company's market performance in maximizing company value.	(Total shares outstanding X Share price) – Total common stockholders' equity
Sales Growth (SG)	Growth of company sales in one accounting period	The percentage of the difference between $sales_t$ and $sales_{t-1}$ is divided by $sales_{t-1}$.
Independent Variable		
Environmental Management System (EMS)	Management implemented systems to reduce environmental damage	Binary variable, value 1 if it is ISO 14001-certified, value 0 otherwise
Control Variables		
Environmental Sensitivity (SENS)	Types of companies that are closely related to environmental damage (pulp & paper industry, chemicals, oil and gas, metals, and the like, as well as utilities)	Binary variable, value 1 if the company is a type of industry closely related to environmental damage, value 0 otherwise.
Age (AGE)	Firm Age since listed in IDX	The total age of the company from being listed on the IDX until the year of observation
Size (SIZE)	The size of the company in business	Natural logarithm of the company's total assets
Liquidity (LIQ)	Company's level of financial liquidity	The ratio of current assets to current liabilities

3. Results and discussion

The first step in panel data regression is determining the appropriate research model. The suitability test carried out is the Lagrange multiplier test to test the common or random effect, which is more suitable.

If the result of LM count > LM Table, then the random effect is more appropriate. On the other hand, if LM Count < LM Table, then the common effect is more in line with the research model. The summary results of the Lagrange multiplier test of this research model are summarized in table 2.

Table 2. Results of the research model suitability test.

Dependent Variable	Lagrange Multiplier Test		Result
	LM Count	LM Table	
MC	1.49529E-54	350.8743125	Common Effect
RMC	8.093E-111	350.8743125	Common Effect
LS	6.49501E-90	350.8743125	Common Effect
MVA	9.09424E-06	350.8743125	Common Effect
SG	7.99515E-76	350.8743125	Common Effect

It can be seen in Table 2 that the results of all the Lagrange multiplier tests show LM count < LM table, so the most suitable model for this research is the common effect model. The next test is the classical assumption test. In panel data regression with the Ordinary Least Squared (OLS) approach, not all classical assumption tests need to be performed. Only multicollinearity and heteroscedasticity are needed. All research models have carried out classical assumptions and have been free from problems of multicollinearity and heteroscedasticity.

Table 3. Results of hypothesis testing on the impact of EMS on company internals.

	MC			RMC		
	Coeff.	Prob.	Result	Coeff.	Prob.	Result
EMS	2.92E+12	0.0004	**Rejected**	1.61E+12	0.0068	**Rejected**
SENS	-2.30E+12	0.0283		-1.38E+12	0.0738	
AGE	1.55E+11	0.0001		1.01E+11	0.0004	
SIZE	-2125.740	0.1677		-1345.108	0.2314	
LIQ	-2.45E+10	0.5055		-1.60E+10	0.5514	
Adj. R^2		8.67%			5.79%	

The next test is to test the hypothesis that has been proposed. Table 3 shows the results of hypothesis testing on the impact of EMS implementation on the company's internals, namely manufacturing costs (MC) and raw material costs (RMC). The coefficient of the EMS variable on MC is 2.92E+12, with a probability value of 0.0004. At the same time, the coefficient of the EMS variable on RMC results in 1.61E+12 with a probability value of 0.0068. It means that the EMS variable has a positive effect significantly on the MC and RMC variables.

Companies with ISO 14001-certified are proven to increase manufacturing costs and raw materials. This result is in line with previous research, which stated that EMS significantly positively affected the company's operating costs, labour costs, and depreciation of fixed assets [9],[17]. Companies that implement EMS will manage the use of resources as efficiently as possible and reduce the production waste generated. So the company must replace all ordinary raw materials with eco-friendly raw materials and replace equipment that is more energy-efficient, which results in more costs for the company's operations [10]. The results of this study do not support previous research, which states that EMS can reduce production costs and use of raw materials and energy to improve financial conditions [5],[6],[16].

Tabel 4. Results of hypothesis testing on the impact of EMS on external companies.

	LS			MVA			SG		
	Coeff.	Prob.	Result	Coeff.	Prob.	Result	Coeff.	Prob.	Result
EMS	5.20E+09	0.0002	**Accepted**	0.100192	0.0351	**Accepted**	18112946	0.5555	**Accepted**
SENS	-2978609	0.1025		-0.42441	0.0000		-7771540	0.0508	
AGE	78687667	0.2460		-0.00397	0.0826		-5786915	0.6961	
SIZE	-1.26346	0.6358		2.59E-10	0.0042		0.002191	0.9700	
LIQ	6931052.	0.9136		-0.00201	0.3508		715886.4	0.6083	
Adj. R^2		4.12%			14.63%			0%	

Testing the impact of the implementation of EMS on external companies can be seen in table 4. The effect of the EMS variable on the LS variable obtains a regression coefficient of 5.20E+09 with a probability value of 0.0002. The effect of the EMS variable on the MVA variable obtained a regression coefficient of 0.100192 with a probability value of 0.0351. It shows that the EMS variable significantly has a positive affects the LS and MVA variables. In comparison, the effect of the EMS variable on the SG variable obtains a regression coefficient of 18112946 with a probability value of 0.5555, which means that the EMS variable has a positive but insignificant effect on the SG variable.

The results of this study support previous research, which states that implementing EMS is proven to increase community legitimacy [6]. The company has a competitive advantage because the implementation of EMS enhances the company's reputation [13]. Customers also have satisfaction with companies that implement EMS [7],[8]. In other words, obtaining ISO 14001 certification is proven to positively impact the company because it provides added value. EMS also increases shareholder confidence, which is indicated by increased outstanding shares. On the consumer side, consumers are increasingly proud and loyal to the company, so it impacts increasing sales growth.

4. Conclusion

This study examines the impact of EMS implementation on manufacturing companies listed on the IDX from 2017 to 2020. A total of 327 companies as samples in this study were tested using panel data regression. The results of this study prove that companies that have implemented EMS have an internal impact on the company, namely increasing manufacturing costs and raw material costs. It happens because ISO 14001-certified companies will manage the use of resources as efficiently as possible and reduce the production waste generated. So the company will replace ordinary raw materials with eco-friendly raw materials. In addition, the company must also replace production equipment that can save energy. Hence, it increases costs, both on raw materials and the company's production costs, to reduce waste in the production process.

The implementation of EMS also positively impacts the company's externals. Being certified to ISO 14001 will provide a competitive advantage and add value to the company. In addition, stakeholders prefer and are satisfied with companies that implement EMS. On the shareholder's side, shareholders believe that ISO 14001 certified companies have implemented international standard environmental damage prevention and repair procedures, thus impacting outstanding shares. Meanwhile, on the consumer side, consumers prefer companies that are aware of environmental damage through prevention from the production process to the repair of natural damage. It raises customer loyalty to buy products and impacts the company's sales growth.

This research contributes to providing insight to practitioners in implementing environmental management systems. The impact of implementing EMS is beneficial not only in the prevention and repair of environmental damage that is getting out of control but also in benefits that will be obtained for itself. Currently, ISO 14001 certification is still voluntary, so implementing EMS requires support from policymakers and the surrounding community. With an umbrella law from policymakers and demands from the surrounding community, it is hoped that more companies will implement EMS, one of which is by following ISO 14001 certification.

References

[1] Https://www.iso.org/home.html, "Introduction to ISO 14001:2015," 2015.
[2] Baek K 2017 *J. Bus. Ethics* **145** 325
[3] Sartor M, Orzes G, Touboulic A, Culot G, and Nassimbeni G 2019 *Qual. Manag. J.* **26** 32
[4] Https://www.iso.org/home.html, "ISO Survey 2020 Result : Number of Certificate and Sites Per Country and The Number of Sector Overall."
[5] Iatridis K and Kesidou E 2018 *J. Bus. Ethics* **148** 859
[6] Treacy R, Humphreys P, McIvor R and Lo C 2018 *Int. J. Prod. Econ.* **208** 319
[7] Chiarini A 2017 *Bus. Strateg. Environ.* **26** 844
[8] Tuppura A, Toppinen A, and Puumalainen K 2016 *Bus. Strateg. Environ.* **25** 355

[9] Wang J and Mao Y 2020 *Bus. Strateg. Environ.* **29** 2092
[10] Mao Y and Wang J 2019 *Int. J. Prod. Res.* **57** 7235
[11] Yusoff R, Yusoff H, Abd Rahman S A, and Darus F 2019 *J. Asia-Pacific Bus.* **20** 302
[12] Riaz H and Saeed A 2020 *Corp. Soc. Responsib. Environ. Manag.* **27** 681
[13] Ye Y, Yeung A C L, and Huo B 2020 *Int. J. Oper. Prod. Manag.* **40** 1829
[14] Freeman R E E and McVea J. McVea 2001 *Electron. J.*
[15] Camilleri M A 2022 *Corp. Soc. Responsib. Environ. Manag.* 1
[16] Sorooshian S, Qi L C, and Li Fei L 2018 *Environ. Qual. Manag.* **27** 97
[17] He W, Liu C, Lu J, and Cao J 2015 *China Econ. Rev.* **32** 43

Good Environmental Governance Model in Domestic Waste Management in Batang Arau, West Sumatera

A Frinaldi[1,2], B Saputra[1,2*], A Mubarak[1,2], Jumiati[1,2], I Renaldi[1], H Humaida[1]

[1]Department of Public Administration, Faculty of Social Science, Universitas Negeri Padang, Jl Prof Dr. Hamka Padang, 25132, Indonesia
[2]Policy, Law and Political Research Center, Universitas Negeri Padang, Jl Prof Dr. Hamka Padang, 25132, Indonesia

bonisaputra@fis.unp.ac.id

Abstract. River damage is often caused by human activities, one of which is massive domestic waste pollution in river flows. Using the Good Environmental Governance model, this study analyzes domestic waste management in Batang Arau, Padang City. This type of qualitative research with data analysis techniques Manual Data Analysis Procedure (MDAP). The results show that the domestic waste management of the Batang Arau City of Padang has not yet implemented the Good Environmental Governance model. Field data illustrates several pillars of Good Environmental Governance, such as low community participation, access to information on domestic waste management is not available, and Transparency and Accountability of domestic waste management in Batang Arau are constrained due to the difficulty of collecting domestic community data. Waste. Domestic waste processing in Padang City has not been decentralized and is still centralized. Environmental management institutions in Padang City have not been well integrated because there are still sectoral egos between institutions. The obstacle to managing Batang Arau domestic waste is that the Wastewater Treatment Plant in the city of Padang has not appropriately been facilitated in the city of Padang.

1. Introduction

River damage is often caused by human activities, one of the impacts is the massive domestic waste pollution that occurs in river flows. A waste disposal study recorded by the World Bank in 2021 explains that the waste records of 500 cities and provinces in Indonesia generate 7.8 tons of waste yearly, with 4.9 million tons of plastic waste not being appropriately managed [1]. This waste that is not appropriately managed will flow into various drainage channels and eventually end up in the Watershed (DAS).

Meanwhile, the phenomenon of river flow management is often not directly proportional to the needs and the economy. For developing countries, economic growth is an important thing to achieve for the progress of the country and the welfare of society. However, achieving this requires significant capital resources. Developing countries are still oriented to the production of natural raw materials, and natural resources must be sacrificed so that investment runs smoothly and the foreign exchange turnover of the country can increase and generate profits for the country even at the expense of the preservation of existing natural resources.

Economic development must have a solid basis for environmental sustainability because the costs incurred in economic development include all inherent aspects such as human resources, investment, and the environment. However, the environment has always been one aspect that stakeholders neglect in economic development. According to research by [2], regarding the relationship between environmental pollution and economic development, it is said that in developing countries, pollution management is prolonged to be implemented due to incomplete facilities and lack of use of technology. Apply technology in pollution management in developing countries, and it takes a long time to restore the environment. Moreover, weak environmental laws are an obstacle to balancing economic development and the environment. [3], Explained that the weakness of the existing environmental laws in Indonesia, known as the Environmental Impact Analysis, is also the cause of inefficient environmental management and decision-making so that environmental pollution and destruction continue to occur.

As a developing country in terms of environmental management of water resources, Indonesia still has much homework. As reported from Republika.co.id (2019), the Director of Forest and Freshwater of the World Wide Fund for Nature (WWF), Indonesia, Irwan Gunawan, explained: "*...Watershed (DAS) in Indonesia used for productivity activities are mostly in a worrying condition because they are polluted by household and industrial waste*."

The watershed is the parent of tributaries that cross land with integrated upstream and downstream landscapes. This study will analyze how to manage Domestic Waste in the Batang Arau River Basin, Padang City. The Batang Arau Watershed has an area of 174.67 km, with an upstream catchment area of only 30.90 km [4].

[5] revealed that pollution in the Batang Arau watershed is caused by industrial and domestic waste. The upstream point is a cement raw material mining area, and the middle point is a rubber factory and oil industry (CPO). While downstream, there are hospitals, workshop businesses, hotels, markets, and boat rest areas.

Initial observations made with the Head of the Law Enforcement and Environmental Section of the Padang City Environmental Service, Mr. Fuad Syukri, also confirmed that: "*...In this case, the Batang Arau watershed has indeed found incidents of industrial liquid waste disposal outside the established wastewater quality standards, as well as domestic waste such as detergents, food scraps from restaurants/communities, and public toilets, so it is not surprising that the Batang Arau watershed is known to be polluted by sewage. The liquid is disposed of from Industry and Households in the City of Padang*".

The last investigation by sampling conducted in May 2022 by the Nusantara River Expeditionary Team of West Sumatra found 420 microplastic particles in 100 liters of water in the Batang Arau River. Plastic microparticles are 1,000-2,500 microns in size with fiber and filament types [6]. To analyze the management of the Batang Arau watershed, the author will use the Good Environmental Governance Model (GEG), where it is said that the concept of Governance in Good Environmental Governance is defined as the framework of the state and society together to preserve environmental protection properly [7]. The Good Environmental Governance model used in this study is the [8] model, where the authors adopt elements such as Participation, Transparency, Accountability, Decentralization, Institutions, and Agencies. Meanwhile, in the discussion chapter, the obstacles faced in managing domestic waste in the Batang Arau watershed will be added. Thus, the theme raised in this study is a Good Environmental Governance Model in Domestic Waste Management in Batang Arau, Padang City.

2. Methodology

The research site is located in the Batang Arau River Basin in Padang, West Sumatra. This study uses a qualitative method with the withdrawal of informants using a purposive sampling technique. In this study, the data on the characteristics of the informants will be used as initials P1-P9 to facilitate coding, and the data of research informants shown in the following table:

Table 1. Informant Characteristics

Participant	Positions/Institute
P1	Head of Environmental Law Arrangement and Enforcement Division.
P2	Operational Technical Executor I County Hall Sungai V Padang.
P3	Senior Trainer Functional Planning of the Regional Planning and Development Agency (BAPPEDA) Padang City
P4	Head of Operational Responsibility for Padang City Sanitation Activities
P5	Staff HRD Padang City Public Works and Spatial Planning Office.
P6	Head of Sub-district Batang Arau
P7	West Sumatra Regional Executive Director of the Indonesian Forum for the Environment
P8	Chairman of the Batang Arau River Care Community
P9	Citizen of Sub-district Batang Arau

Collecting data using interviews, observation, and documentation. The interview results from the interviewees were made in the form of transcripts and analyzed using the Manual Data Analysis Procedure (MDAP) technique. [9] The Manual Data Analysis Procedure technique is a qualitative data processing technique carried out manually and consists of several stages: diaries, transcripts, Coding, categories, themes, and memos/preposition. The process stages are: The interview is listened to, read, and reread to ensure that the transcription matches the interview. Hard-copy transcripts were micro-analyzed by breaking them down into meaning units on a sentence-by-sentence basis through simple coding. Code is written into the margins of the transcript. Then the transcripts were micro-analyzed by breaking them down into meaning units on a sentence-by-sentence basis through simple electronic coding. Highlighted the relevant statement by dragging and clicking on the unit meaning. Codename-ing the highlighted text segments through the introduction of free nodes, i.e. simple code. Organized the simple code into complex and more encompassing categories by separating the relevant meaning units into groups, placing them in the same category, and creating a hierarchical representation of the umbrella themes, higher and lower level themes through the use of note cards, and paper-pencil lines.

3. Results and Discussion

3.1. Research Data Results

The study's preliminary results were obtained from structured and unstructured interviews where interviews were conducted with the authorities and observers of the Batang Arau watershed. The results of the research transcript were analyzed using the Coding stage, which from the Open Coding results resulted in the concept code #11, which would be brought to the Axial Coding analysis to arrive at the core problems of domestic waste management in Batang Arau, the following data are presented in the table:

Table 2. Axial Coding Analysis.

Concept	Subcategory
Concept #1 **Concept #2**	**Subcategory #1** Low public awareness in managing domestic waste in the Batang Arau watershed
Concept # 3 **Concept #4**	**Subcategory #2** Is the difficulty of controlling transparency and public responsibility in disposing of domestic waste
Concept #5 **Concept #6**	**Subcategory #3** Waste management and the Batang Arau watershed are still centralized in one agency
Concept #7 **Concept #8** **Concept #9** **Concept #10** **Concept #11**	**Subcategory #4** There is no synergy between agencies, and there is still sectoral ego between stakeholders **Subcategory #5** Domestic waste disposal containers are not available properly

Furthermore, research results will be obtained from the Axial Coding analysis, which is inputted into Selective Coding to obtain each of the Principles of the Good Environmental Governance Model. The following data is presented in the table:

Table 3. Selective Coding Analysis.

Subcategory #1	**Category #1 Participation** Low participation from elements of society
Subcategory #2	**Category #2 Transparency and Accountability** Closed and difficult to collect data on the community's domestic waste
Subcategory #3	**Category #3 Decentralization** Management is still centralized
Subcategory #4	**Category #4 Institutions and Agencies** There is still a sectoral ego
Subcategory #5	**Category #5 Management Constraints** The Wastewater Treatment Plant in Padang City has not been appropriately facilitated Padang City

3.2. Memos Visualization

Next, to proceed to the Manual Data Analysis Procedure stage, the coding analysis will be visualized here in a Memo; the following data are presented in the image:

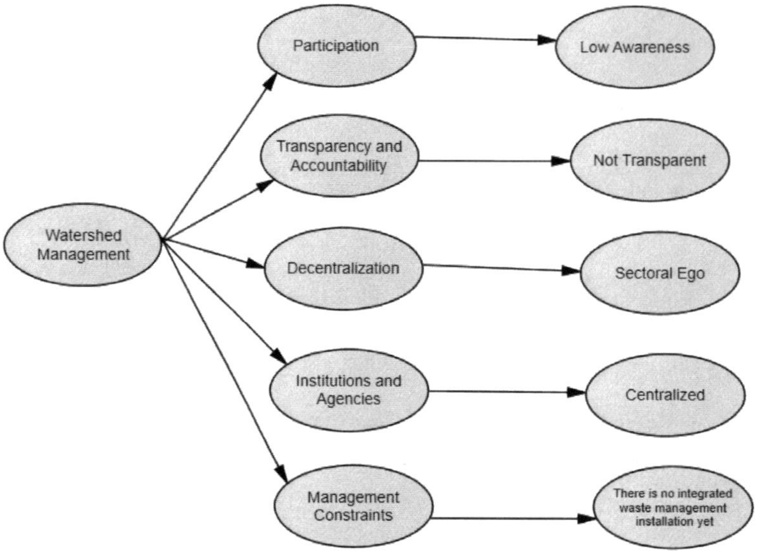

Figure 1. Visualization of Research Data Memos.

3.2.1. Participation

The results showed that community participation was low in maintaining and managing domestic waste. Many still did not heed the regulations not to litter and dispose of domestic waste such as detergents and leftovers from bathing, washing, and household latrines into the Batang Arau drainage channel. Participation is essential to the success of government administration and a program. As stated by [8], community participation has several significant impacts, such as increasing public awareness of social, political, and environmental concerns. Increase tolerance, empathy, and understanding of pluralism and Raise awareness of the implications of individual actions in the broader environmental and social context [10] [11] [12].

In Good Environmental Governance, a participatory framework is needed to protect the environment, so there is no information gap between the community and the Government. The purpose of environmental programs by the Government can run well if community participation, namely involving the community in implementing government programs. The low level of participation can cause a low level of environmental supervision, which results in the implementation of the law prohibiting throwing garbage from being ignored by the community. Meanwhile, supervisory activism can be streamlined by involving the community in environmental monitoring. [13].

3.2.2. Transparency and Accountability

Transparency is needed, so that information disclosure is conveyed to all elements of society. If we want to know environmental information, clear transparency is the key to open use and disposal activities that impact the environment [14]. If you look at the results of this study, the transparency and accountability of the community's domestic waste are not open. This is due to the difficulty of collecting data on the community's domestic waste. Community data collection is constrained by the difficulty of government supervision of domestic waste disposal activities from Micro, Small, and Medium Enterprises and houses in slum settlements in Padang City. Communities with slum settlements cannot be controlled because of the awareness of the community itself in disposing of garbage and domestic waste, which is directly dumped on the riverbanks. Meanwhile, disputes between the community and the Government over their place of residence have become public awareness in the programs launched by the Government for structuring slum settlements. Not only slums in the Batang Arau watershed, Micro, Small, and Medium Enterprises that dispose of their domestic waste, such as food scraps, and household-scale industrial production that disposes of the waste directly without prior processing can still be found.

Institutional pressure and firmness must be applied to the community to create good environmental hygiene [14]. Community institutions such as non-governmental organizations and the political role of the Government must unite in a common framework to manage domestic waste in Batang Arau. Moreover, lack of transparency and accountability is one of the elements of the Good Environmental Governance model. Within the framework of the Good Environmental Governance model, provisions related to accountability must be included in the main points of the state constitution. In addition to ensuring accountability, it is essential to make provisions and include environmental regulations and technical details about what the duties and functions of public officials are to the public [8].

3.2.3. Decentralization

The study results indicate no decentralization or delegation of authority in Domestic Waste Management in the Batang Arau watershed. Waste and solid waste management are still focused on one agency authorized to clean the city, the Padang City Environmental Service. Based on the Open Coding transcription, the Environmental Service gave an exciting discourse: managing waste in the city of Padang requires a large budget, especially if it is accompanied by the main tasks and functions of managing river cleanliness. The concept of decentralization is helpful for efficient waste management by combining economic efficiency with the environment [15]. In addition, decentralization is needed to encourage lower-level governments (local governments) to compete and show the best performance in management, and granting authority to lower-level governments (local governments) also reduces investment from the Government in economic and environmental development [16] [17]. An essential element of Good Environmental Governance, according to [8], is the principle of subsidiarity. Practice says that humans live in and around environmental resources, meaning there needs to be a subsidiary closest to the source of the existing problem or resource. So to realize the decentralization of domestic waste that pollutes Batang Arau, it is necessary to carry out a comprehensive decentralization involving all stakeholders throughout the Batang Arau watershed. The concept of decentralization is implemented by giving authority to clean up the river environment to sub-districts along Batang Arau by providing personnel and facilities for controlling river cleanliness that is integrated with the Environmental Service so that this decentralization can reduce the burden on the environment-serving in the supervision and cleaning of garbage and domestic waste in Batang Arau.

3.2.4. Institutions and Agencies

An essential element of governance is the existence of institutions, organizations, and agencies. In this case, these elements regulate the norms and rules that apply to all state elements. Their existence determines the success or failure of management, especially environmental management [8]. Findings indicate that domestic waste management institutions have not been well integrated. What is meant here is that stakeholders have sectoral egos in each agency which results in ineffective and well-coordinated management. According to the narrative of each agency in the interview transcripts, the first and second levels did not show the same spirit and vision to manage and overcome domestic waste in Batang Arau. As a policy maker, the Government as an authorized institution in environmental management needs to have good coordination, especially institutions that function in realizing a healthy and pleasing environment by making policies based on environmental management [18] [19], Especially in managing domestic waste in the Batang Arau watershed, Padang City.

3.2.5. Management Constraints

The study's results found that the obstacle to overcoming domestic waste in the Batang Arau watershed, Padang City, was the absence of an integrated Wastewater Treatment Plant facility in Padang City. The manager who is authorized and responsible in the field of City Sanitation, a stakeholder in sanitation management in the City of Padang in the Open Coding results, explains that due to the absence of this integrated Waste Water Treatment Plant, household waste is not managed correctly. The community freely drains their domestic waste into existing drainage channels. For waste related to bathing, washing, and latrines, there are still people who throw their garbage into the sewer without an integrated shelter.

With the integrated Wastewater Treatment Plant, the flow of all domestic waste, especially the types of bathing, washing, and latrines, can be appropriately managed in one integrated place. However, to build these facilities, the City Government complained about the large budget and time in constructing this integrated Wastewater Treatment Plant.

4. Conclusion

Based on the research conducted, it was found that public awareness participation is still low in domestic waste management. It is challenging to control transparency and public responsibility in disposing of domestic waste. After that, domestic waste management is still charged to one agency and has not been authorized by an excellent decentralized system. Synergies between agencies have not yet occurred, and there is still a sectoral ego between stakeholders in the environmental management of the Batang Arau watershed. The unavailability of proper domestic waste disposal containers has resulted in bathing, washing, latrines, and domestic waste still being dumped and flowing in the Batang Arau watershed. To realize the Good Environmental Governance model, all stakeholders must create an environmental framework that involves all elements in domestic waste management to reduce pollution levels in Batang Arau, Padang City.

References

[1] L. Boediono, "New Data-driven Approach Maps Plastic Pollution from Source to Sea in Indonesia, Offers Improvement Strategies," *www.worldbank.org*, 2021.

[2] Y. Yu *et al.*, "Relationship between environmental pollution and economic development in late-developing regions shows an inverted V," *Sci. Total Environ.*, vol. 838, p. 156295, 2022.

[3] D. Mitchell, Bruce, *Pengelolaan Sumberdaya Dan Lingkungan*. Yogyakarta: Gadjah Mada University Press, 2010.

[4] M. Yudha Ozman, M. Fauzi, and L. Darmayanti, "Analisis Kebutuhan Air Irigasi DAS Batang Arau Kota Padang (Studi Kasus : DAS Batang Arau Kota Padang)," *Sainstek (e-Journal)*, vol. 8, no. 2, pp. 56–60, 2020.

[5] R. Setiawan *et al.*, "Identifikasi Pencemaran Air Di Sepanjang Aliran Sungai Utama DAS Batang Arau Kota Padang," *J. Kapita Sel. Geogr.*, vol. 2, no. 5, pp. 29–34, 2019.

[6] Padek, "Koalisi Masyarakat Minta Bersihkan Batang Arau, Tindak Pembuang Sampah," *padek.jawapos.com*, Sumatera Barat, 2022.

[7] Purniawati dkk, "Good Environmental Governance in Indonesia (Perspective of Environmental Protection and Management)," *Indones. J. Int. Clin. Leg. Educ.*, vol. 2, no. 1, pp. 43–56, 2020.

[8] N. Belbase, "Environmental Good Governance in the Future Constitution of Nepal," *Cmsdata.Iucn.Org*, no. August, 2010.

[9] B. Bungin, *Post - Qualitative Social Research Methods*, 2nd ed. Jakarta: Kencana, 2021.

[10] S. Müllera, L. Hucka, and J. Markovab, "Sustainable community-based tourism in cambodia and tourists' willingness to pay," *Austrian J. South-East Asian Stud.*, vol. 13, no. 1, pp. 81–101, 2020.

[11] H. Malau, H. Fajri, P. Yuanjaya, B. Saputra, and K. D. Maani, "Knowledge of Local Communities Affected by the Development of Geothermal Energy," in *IOP Conference Series: Earth and Environmental Science*, 2020, vol. 448, no. 1.

[12] H. Fajri, P. Nurhabibi, B. Saputra, and P. Yuanjaya, "Community Engagement in a Social Movement: A Case of Geothermal Energy Development in Gunung Talang - Bukit Kili area," in *IOP Conference Series: Earth and Environmental Science*, 2020, vol. 448, no. 1.

[13] Z. Chu, C. Bian, and J. Yang, "How can public participation improve environmental governance in China? A policy simulation approach with multi-player evolutionary game," *Environ. Impact Assess. Rev.*, vol. 95, p. 106782, 2022.

[14] D. Sun, S. Zeng, H. Chen, X. Meng, and Z. Jin, "Monitoring effect of transparency: How does government environmental disclosure facilitate corporate environmentalism?," *Bus. Strateg.*

Environ., vol. 28, no. 8, pp. 1594–1607, 2019.
[15] S. Estévez, G. Feijoo, and M. T. Moreira, "Environmental synergies in decentralized wastewater treatment at a hotel resort," *J. Environ. Manage.*, vol. 317, no. May, p. 115392, 2022.
[16] X. Yang, J. Yan, K. Tian, Z. Yu, R. Yu Li, and S. Xia, "Centralization or decentralization? the impact of different distributions of authority on China's environmental regulation," *Technol. Forecast. Soc. Change*, vol. 173, p. 121172, 2021.
[17] H. Fajri, A. D. Akmal, B. Saputra, N. Wahyuni, Q. P. Ilham, and K. D. Maani, "Integration of renewable energy policies between central and regional governments," in *IOP Conference Series: Earth and Environmental Science*, 2021.
[18] H. Fajri, A. D. Akmal, B. Saputra, Q. P. Ilham, and N. Wahyuni, "Renewable energy policy and governance in West Sumatera Province: an overview," in *IOP Conference Series: Earth and Environmental Science*, 2021.
[19] E. van der Werff, L. Steg, and A. Ruepert, "My company is green, so am I: the relationship between perceived environmental responsibility of organisations and government, environmental self-identity, and pro-environmental behaviours," *Energy Effic.*, vol. 14, no. 5, pp. 1–21, 2021.

Assessment Of Health Service Facility Ash Waste Based On Policy (Case Study Of Moewardi Hospital Surakarta)

Siti Rachmawati [1,4*], Syafrudin Syafrudin [2], Budiyono Budiyono [3]

[1]Students Department of Doctoral Environmental Science, Faculty of Postgraduate, Diponegoro University, Semarang 50275, Indonesia
[2]Department of Environmental Engineering, Faculty of Engineering, Universitas Diponegoro, Semarang 50275, Indonesia
[3]Department of Chemical Engineering, Faculty of Engineering, Universitas Diponegoro, Semarang 50275, Indonesia
[4]Department of Environmental Science, Faculty of Mathematics and Natural Science, Universitas Sebelas Maret, Surakarta 57126, Indonesia

sitirachmawati1103@students.undip.ac.id

Abstract. The thermal process is one of the processes used in the treatment of solid waste generated from the activities of health service facilities (hospitals, health centers, clinics). The risk in the incineration process is the ash produced, the ash produced contains hazardous and toxic compounds so management needs to be carried out according to Government Regulations Ministry of Health RI number 56 of year 2015 and Regulation of the Minister of Health No. 9 of 2019. The purpose of this study was to find out how to handle health service facilities' waste ash following applicable standards and friendly to the environment. This research is qualitative descriptive research. The research was conducted at the Moewardi Hospital Installation section. Sources of data in this study came from primary and secondary data. Data collection techniques were carried out by collecting data through interviews with sanitation employees and incinerator operators, reviewing documents, and field observations. Waste that can be treated with an incinerator is waste generated in health service activities such as syringes, bottles/ampoules, infusion needles. Management of incinerator combustion ash is carried out based on applicable regulations. The utilization of incinerator ash needed to be an environmentally friendly and sustainable product.

1. Introduction

The increasing incidence of the COVID-19 pandemic for approximately 2 years has caused an increase in the waste generated by Health Service Facilities [1,2]. Health Service Facilities are all activities related to treatment and health recovery that that generate waste in the form of solid, air, and liquid [3]. The types of waste that have increased during the pandemic include waste contaminated with patient body fluids (infectious), sharp object waste, pharmaceutical waste (drugs), and personal protective equipment waste such as face masks, and gloves [4,5]. The average medical waste produced by health service facilities in Indonesia based on data from the Ministry of Environment and Forestry (KemenLHK) is estimated to reach ±242 tons/day from 2,813 with an average medical waste generation of 87 kg/day/hospital [6], then in India produces waste per day around 484 tons [7].

The principle of waste management produced by health service facilities consists of identifying the source and type of waste, receptacle, collection, packaging and storage according to the characteristics of the waste, transportation, processing, and utilization of waste [8]. Medical waste produced by health service facilities is hazardous and toxic waste (B3) which has infectious characteristics so effective waste management must be carried out. The good management of waste produced by health service facilities, especially medical waste, can prevent environmental pollution, occupational diseases, and work accidents [9].

The thermal process is one of the processes used in the processing of solid waste generated from household activities, offices, health service facilities (hospitals, health centers, clinics), and so on the waste that meets thermal requirements. Incineration using incinerators is often used in thermal processes. The incinerator is a treatment used in handling medical waste during the COVID-19 pandemic in China [10]. The incinerator is a combustion device that uses high temperatures in the waste combustion process [11]. The incineration process using an incinerator can remove the content of pathogenic compounds with a temperature range of 1100-1200^{0}C with a decrease in the volume of waste by 90%. The temperature and time are important factors in the process of burning waste, some find that there are still syringes and some bottles of medicine that have not been completely burned [12,13,14].

The negative impact on combustion produces toxic air emissions and ash from combustion, both fly ash and bottom ash. Ash from combustion contains heavy metals that can harm the environment [15,16]. Most of the ash produced by the health service facilities will be handed over to a third party for landfill (sanitary landfill), the heavy metal content in the ash causes it to be used sparingly while in the regulations it is allowed to be utilize B3 waste following certain requirements.

Moewardi Hospital is a Type A Hospital located in Surakarta City and is a reference for all hospitals around the city. There are outpatient, inpatient, sanitation, education, and research installations that generate waste. Waste is generated in the form of medical and medical. Based on Figure 1, the highest peak of producing medical waste in 2020 and 2021 because the COVID-19 pandemic occurred in early 2020, affecting the medical waste produced. This increase causes an increase in the combustion ash produced. Moewardi Hospital spends a lot of money on the management of medical waste, both in the treatment of the incinerator and the management of the ash produced. The disposal of ash with abundant availability causes reduced land availability, high costs in processing ash, and strict government regulations in licensing [17].

Based on the above background, research is needed on the management of health service facilities ash waste so that they can find out how to handle health service facilities waste ash by applicable standards and are friendly to the environment.

2. Methodology

This research is qualitative descriptive research. Descriptive research is a research method carried out with the main aim of making a picture or description of a situation objectively [18]. The research was conducted at Moewardi Hospital in the Sanitation Installation section which is located on Jl. Colonel Sutarto No. 132, Kec. Jebres, Surakarta City, Central Java from February to April 2022. The data sources in this study came from primary and secondary data. Data collection techniques were carried out by collecting data through interviews with sanitation employees and incinerator operators, reviewing documents containing information on medical waste, and field observations. The object of this research is the ashes of burning medical waste produced by Moewardi Hospital and its management includes identification of the waste produced so that it becomes ash, the content of chemical compounds contained in the ash, packaging, transportation, and utilization of the ash.

3. Results and Discussion

Incineration using an incinerator is one method that has proven to be effective in treating medical waste. Several countries use this system to overcome the problem of medical waste caused by the COVID-19 virus, such as China, Indonesia, and Italy [19, 20]. Moewardi Hospital has an incinerator type MS-2PSI with a capacity of 200 kg/hour and a chimney height (from ground level) of 17.4 meters. This is by Ministry of Health RI number 56 of year 2015 which explains that the height of the chimney in the incinerator is at least 14 meters. The fuel used is a diesel with an air pollution control device, namely a wet scrub. Waste is burned 2 times a day with a temperature of approximately 1000°C. Waste weight ±90 kg with a long burning time of 30 minutes or ±160 kg with a burning time of 120 minutes. The operation of the incinerator has an impact on health and the surrounding environment. The health effects given during combustion are the cause of cancer, neonatal disorders, reproductive and skin disorders, endocrine disorders, and suppression of the immune system due to the emissions produced [21].

3.1 Identification of Health Service Facilities Waste

Waste identification is carried out for the segregation of waste to be processed by incineration or to be handed over to third parties because not all waste generated can be processed using incinerators such as TL lamps, used batteries, used batteries, and others. The identifying waste is an effective management step for waste generated by health service activities [22].

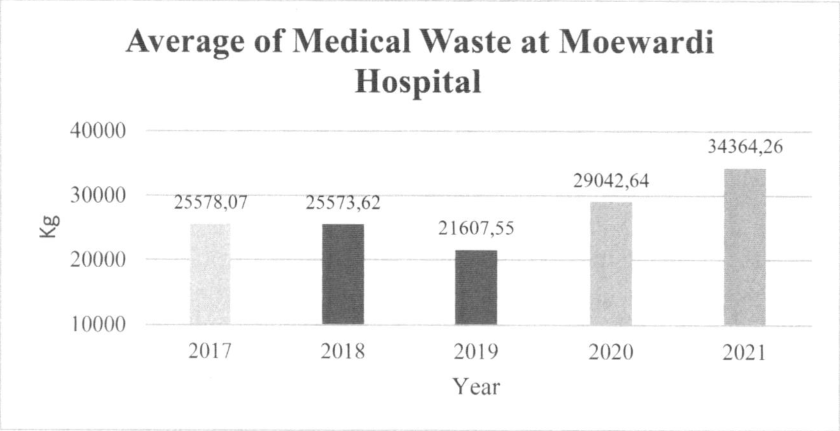

Figure 1. Average of medical waste at Moewardi Hospital
Source: Moewardi Hospital, 2022

Outpatient and inpatient installations contribute to the dominant medical waste generated by Moewardi Hospital. Based on figure 1, in 2020 and 2021 medical waste has experienced a significant increase, Moewardi Hospital in addition to burning waste using an incinerator also collaborates with third parties to carry out medical waste treatment because the hospital is overwhelmed by burning because the waste generated is increasing rapidly due to the pandemic COVID-19. Burning is done every day two times of burning. The type of waste burned in Moewardi Hospital consists of clinical waste originating from outpatient and inpatient activities, sludge wastewater treatment plant, and used cloth. Electronic waste, expired pharmaceutical products, hazardous and toxic waste packaging, expired chemicals, and hazardous and toxic waste contaminated laboratory equipment is handed over to third parties who have processing permits. This is following Minister of Health Regulation No. 7 of 2019 concerning Hospital Environmental Health, namely radioactive waste, high-pressure waste, and waste containing high metals, it is forbidden to burn using an incinerator.

3.2 Characteristics of Health Service Facilities Ash Waste

Fly ash is a residue that does not burn from the combustion process of medical waste and has volatile properties with the gas produced in the combustion process, which is then separated by several particle catching devices such as wet scrubbers, electrostatic separators, and so on. Bottom ash is a residue that falls to the bottom and does not evaporate along with the gas [23].

Based on figure 2, ash produced by Moewardi Hospital noted that the highest generation occurred in 2020 with an average figure of 4.419,13 kg/year, followed in 2021 with an average of 4.066,69 kg/year by the medical waste generated which showed an increase in the last two years. There has never been a characteristic test on the ash content produced by the incinerator, but a weighing has been carried out to determine the weight of the ash before it is put into the temporary storage of hazardous and toxic waste. After the combustion process, the resulting ash is in the form of bottom ash and fly ash. Fly ash is controlled with a wet scrubber, an air pollution control device mounted on the incinerator. While the base ash produced is not all ash but there are still ampoule bottles, syringes, wood charcoal, and lumps of food.

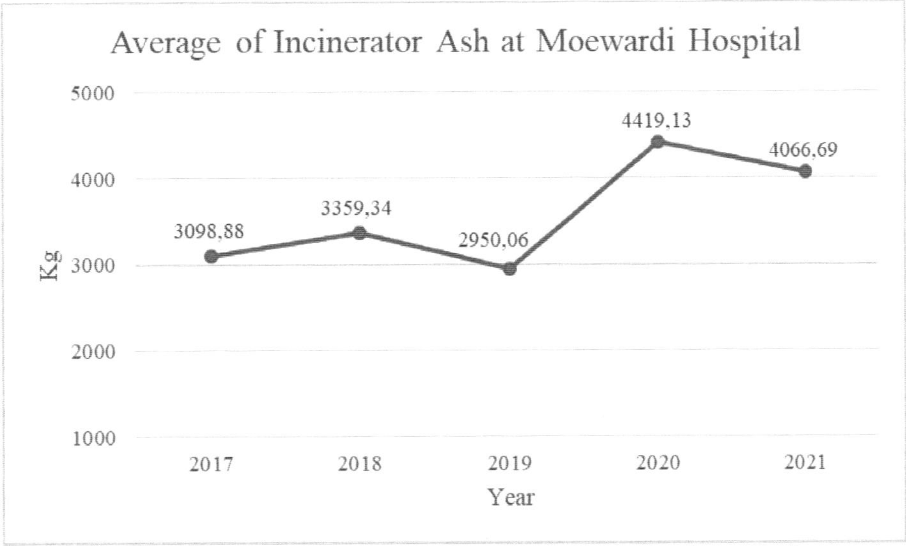

Figure 2. Incinerator ash generation at Moewardi Hospital
Source: Moewardi Hospital, 2022

Based on research, there are heavy metal content in fly ash and bottom ash from incinerator combustion in health care facilities. The type of waste burned will affect the heavy metal content in the combustion residue. The chemical properties contained in the ash consisted mostly of CaO (38.534%), Cl (%30.75), Al_2O_3 (6.9%), Fe_2O_3 (1.097%), and SiO_2 (7.963%) [24]. This is in line with research conducted by J. Miao et al. and Krishnamurthy and Chella that CaO compounds have the largest composition in bottom ash [19,25]. Ca compounds contributed the largest to bottom ash because the type of waste burned mostly came from health equipment such as protective clothing, face masks, and goggles made of polypropylene (PP) [26,27]. The heavy metals contained in the combustion residue are polycyclic aromatic hydrocarbons (PAH), Ag, As, Ba, Bi, Cu, Cd, Cr, Ni, Ti, Sb, Sn, Pb, and Zn which are harmful to the environment [16,28,29,30]. The compounds contained in hospital ashes in Pakistan are Pb, Zn, Cd, Cr, Ni, Cu [31].

3.3 Management of Health Service Facilities Ash Waste

Management of residues resulting from incinerator combustion must be carried out because they still contain hazardous and toxic compounds. E. Gidarakos et al., with his research that the fly ash and bottom ash produced were treated according to the applicable provisions in packaging, transportation, and stockpiling to utilization to avoid adverse effects on health and the environment [32].

Moewardi Hospital treats the incinerator ash well, the ash produced in the combustion process is packaged using a yellow plastic bag and labeled as infectious waste. Before entering the temporary storage of hazardous and toxic waste, a weighing is carried out to determine the weight of the ash produced. Temporary storage of hazardous and toxic waste at Moewardi Hospital is a building that has a roof equipped with standard operational procedure, good ventilation, there is a pallet for waste trays, equipped with label symbol and hazardous and toxic waste logbook, available emergency response system equipment such as fire extinguishers, first aid kits, sinks, and sawdust. This is following the provisions that apply to the Regulation of the Minister of Health Number No. 7 of 2019 concerning Hospital Environmental Health and PermenLHK P 56 of 2015 concerning Procedures and Technical Requirements for the Management of Hazardous and Toxic Waste from Health Service Facilities.

Ash transportation is carried out every three months by a third party who has a permit from the Government with special transportation equipped with a transportation permit, symbol, and label for the characteristics of hazardous and toxic waste. This is by PermenLHK P 56 of 2015 concerning Procedures and Technical Requirements for Hazardous And Toxic Waste Management from Health Service Facilities [33]. At the time of transportation, third parties provide drums which will later be filled with incinerator ash packages to be disposed of. This is followed by packaging using yellow plastic bags with infectious labels after they are put into drums with a capacity of 200 liters which are then solidified and then handed over to third parties.

Encapsulation and inertization are techniques used in the landfilling of hazardous and toxic waste, in general, these techniques use the solidification principle. The solidification/stabilization technique is one of the techniques that is often used to treat incinerator combustion residues so that the chemical compound content is more stable physically and chemically and is a means of preventing leaching caused by heavy metals [34]. The solidification technique required a binder that aims to bind toxic hazardous compounds in the ash, cement is often used as a binder in the solidification process [35,36].

Personnel in burning medical waste at Moewardi Hospital consists of two operators who are divided into two shifts. The use of Personal Protective Equipment (PPE) in the form of masks, head coverings, safety shoes, glasses, and clothes accompanied by aprons is PPE that is used every day in processing LB3. The potential dangers in handling incinerator burning ash are being punctured by wires, syringes, and exposed to broken glass that has not been completely burned. After burning the ash, it is flushed with water so that it is easy to pack and transport to the storage temporary of hazardous and toxic waste. The compliance of workers in using Personal Protective Equipment (PPE) or carrying it out following standard operating procedures at work can minimize the occurrence of work accidents and occupational diseases. Ramli explains that humans can be a source of danger at work [37]. PermenLHK P 56 of 2015 explains that B3 waste management personnel must be equipped with PPE, immunizations or vitamins, extra food, and regular health checks [33].

3.4 Utilization of Health Service Facilities Waste Ash

Most of the disposal of incinerator residue is carried out by stockpiling and most of it is utilized. In the health facility that owns an incinerator, the ash produced is handed over to a third party for stockpiling or burial, and minimal use of the residue is carried out. J. Peng M.M. et al., stated that the burial system was carried out after medical solid waste was heated at high temperatures [38].

Utilization of incinerator residue needs to be done by reducing risks to health and the environment, one of which is to produce materials with environmentally friendly uses. Some of the potential fly ash used include raw materials for the production of geopolymers with a strength of 2-8 MPa [39], building construction materials like fine aggregate [17]; substitute for cement in the manufacture of concrete [40]; a mixture of brick products [35]; synthetic glass-ceramic mixture modified by glass recycling [41]; in some European countries, ash is reused for the manufacture of pavements, bridges and structural masonry [32]; the use of solid waste as much as 30% replacing fine aggregate (sand) with a mixed composition of 1:3 using Portland cement type I in the manufacture of paving blocks will produce a compressive strength of 17.19 MPa. [42,43]; mortar mixture by providing a cement mixer as a binder [44]. Utilization is carried out to meet the requirements of the TCLP and LD50 tests to produce environmentally friendly and sustainable products. Based on PP 22 of 2021 concerning the Implementation of Environmental Protection and Management that the use of hazardous and toxic waste can be carried out by any hazardous and toxic waste producer, if the hazardous and toxic waste producer cannot use it himself, it can be submitted to a third party for utilization. The utilization of hazardous and toxic waste can be used as a substitute for raw materials, as a substitute for energy materials, and raw materials and can be utilized following the development of science and technology.

4. Conclusions and suggestions

The outpatient and inpatient installations at Moewardi Hospital provide the dominant contribution to medical waste in incinerator combustion. The ash produced in the combustion process using an incinerator is still classified as hazardous and toxic waste, which must be managed following applicable government regulations. Packaging, transportation, and disposal of ashes at Moewardi Hospital adapted to its characteristics based on the Regulation of the Minister of Health No. 9 of 2019 and PermenLHK P 56 of 2015. More extensive use needs to be made on fly ash and bottom ash by fulfilling the TCLP and LD50 quality standard tests to produce environmentally friendly and sustainable products.

References

[1] Abu-Qdais, H.A., Al-Ghazo, M.A., and Al-Ghazo, E.M. (2020). Statistical Analysis and Characteristics of Hospital Medical Waste Under Novel Coronavirus Outbreak. *Global Journal of Environmental Science and Management.* **6** (Special Issue (Covid-19)): 21–30.

[2] Morfi, C.W., Junaidi, A., Elsesmita, E., Asrini, D.N., Lestari, D.M., Medison, I., Russilawati, R., Fauzar, F., Kurniati, R. and Yani, F.F. (2020). Kajian Terkini Coronavirus Disease 2019 (COVID-19). *Jurnal Ilmu Kesehatan Indonesia*. **1(1)**.

[3] Adisasmito, Wiku. (2009). Sistem Manajemen *Lingkungan Rumah Sakit*. Jakarta: PT. Raja Grafindo Persada.

[4] Hendri, S. dan Fitriana, M. (2020). Tinjauan: Pengelolaan Limbah Medis untuk Covid19. *Jurnal Kesehatan Lingkungan*. **12 (1)**: 104–120. ISSN: 1829–7285.

[5] Sharma, H.B., K.R. Vanapalli, V.S. Cheela, V.P. Ranjan, A.K. Jaglan, B. Dubey, S. Goel, and J. Bhattacharya. (2020). Challenges, Opportunities, and Innovations for Effective Solid Waste Management During and Post COVID-19 Pandemic, Resour. *Conserv. Recycl*. 162: 105052. DOI: https://doi.org/10.1016/j.resconrec.2020.105052

[6] Ronald, T., Jootje, M.L., Umboh., Woodford, B.S., dan Joseph. (2018). Pengelolaan Limbah Medis Padat Bahan Berbahaya Beracun (B3) Di Rumah Sakit Umum Daerah (RSUD) Piru Kabupaten Seram Bagian Barat, Propinsi Maluku Pada Tahun 2018. Jurnal KESMAS, Vol. **7 No. 5**

[7] CPCB. (2016). In: Biomed. *In: Waste Manag. Cent. Pollut. Control Board.* Rules http://cpcb.nic.in/bio-medical-waste-rules/

[8] Ministry of Health RI. (2019). Regulation of the Minister of Health of the Republic of Indonesia Number 7 of 2019 concerning Hospital Environmental Health. Jakarta.

[9] Pertiwi, V. (2017) Evaluasi pengelolaan limbah Bahan Berbahaya Dan Beracun (B3) di Rumah Sakit Roemani Muhammadiyah Semarang. Jurnal Kesehatan Masyarakat **5(3)**, ISSN: 23P.56-3346.

[10] Ma, Y., Lin, X., Wu, A., Huang, Q., Li, X., and Yan, J. (2020). Suggested Guidelines for Emergency Treatment of Medical Waste During COVID-19: Chinese Experience. *Waste Dispos Sustain Energy*. **2 (2)**: 81–4. DOI: https://doi.org/10.1007/s42768-020-00039-8

[11] Achmad, S.N., Nugroho, R.A., Mardliyah, I., dan Oktavia, N. (2017). Rancang Bangun Insinerator Limbah Medis Berteknologi Plasma sebagai Filter Udara Hasil Pembakaran Limbah Medis. *Industrial Research Workshop and National Seminar*. 575–579. DOI: https://doi.org/10.35313/irwns.v8i3.652

[12] Kougemitrou, I., A. Godelitsas, C. Tsabaris, V. Stathopoulos, A. Papandreou, P. Gamaletsos, G. Economou, and D. Papadopoulos. 2011. Characterisation and Management of Ash Produced in the Hospital Waste Incinerator of Athens, Greece. *J. Hazard. Mater*. 187 (1-3): 421–432.

[13] Aneeta Mary Joseph, Ruben Snellings, and Philip Van den Heede. (2018). "The use of municipal solidwaste incineration ash in various building materials: A Belgian point of view," *Materials*

(Basel)., Vol. 11, No. 1, doi: 10.3390/ma11010141

[14] Rizal, A.M. dan Nurhayati, I. (2017). Pengolahan Limbah Bahan Berbahaya dan Beracun (B3) dengan Insinerator Tipe Reciprocating Grate Incinerator. *Jurnal Teknik WAKTU.* 15 (02): 21–27.

[15] Ramadhansyah, M.F., Nugroho, A., dan Nur, E.W. (2021). Policy Brief : Pengolahan Abu Hasil Pembakaran Limbah Medis sebagai Alternatif Pengganti Bahan Bangunan yang Berasal dari Fasilitas Pelayanan Kesehatan. *Serambi Engineering.* VI **(4):** 2354–2362.

[16] Rozumova, L., O. Motyka, K. Cabanova, and J. Seidlerova. (2015). Stabilization of Waste Bottom Ash Generated from Hazardous Waste Incinerators. *Journal of Environmental Chemical Engineering.* **3 (1)**: 1–9. DOI: http://dx.doi.org/10.1016/j.jece.2014.11.006

[17] Anamul, M.H., Rahman, J., and Tanvir, M., (2012). Zn and Ni of Bottom Ash as a Potential Diffuse Pollutant and Their Application as ''Fine Aggregate''. *J. Civ. Eng. Res.* **2**: 64–72.

[18] Notoatmodjo, Soekidjo. (2005). *Metodologi Penelitian Kesehatan*. Jakarta: Rineka Cipta.

[19] Miao, J., J. Li, F. Wang, X. Xia, S. Deng, and S. Zhang. (2022). Characterization and Evaluation of the Leachability of Bottom Ash from a Mobile Emergency Incinerator of COVID-19 Medical Waste: A Case Study in Huoshenshan Hospital, Wuhan, China. *Journal of Environmental Management.* **303**: 114161. DOI: https://doi.org/10.1016/j.jenvman.2021.114161

[20] Di Maria, F., E. Beccaloni, L. Bonadonna, C. Cini, E. Confalonieri, G. La Rosa, M. R. Milana, E. Testai, and F. Scaini. (2020). Minimization of Spreading of SARS-CoV-2 via Household Waste Produced by Subjects Affected by COVID-19 or in Quarantine. *Total Environ.* **743**: 140803. DOI: https://doi.org/10.1016/j.scitotenv.2020.140803

[21] Muduli, K. and Barve, A. (2012). Chalengges to Waste Management Practices in India Health Care Sector. International *Conference on Environment Science and Engineering*. IPCBEE, IACSIT Press, Singapore (Vol. 32).

[22] Su, E.C.Y. and Chen, Y.T. (2018). Policy or Income to Affect the Generation of Medical Wastes: An *Application* of Environmental Kuznets Curve by Using Taiwan as an Example. *J. Clean. Prod.* **188**: 489–496. DOI: https://doi.org/10.1016/j.jclepro.2018.04.011

[23] Antoni A., Klarens K., Indranata M., Al-Jamali L., and Hardjito D. (2017). The Use of Bottom Ash for Replacing Fine Aggregate in Concrete Paving Blocks. **138**: 01005.

[24] Akyıldız, A., E.T. Kose, and A. Yıldız. (2017). Compressive Strength and Heavy Metal Leaching of Concrete Containing Medical Waste Incineration Ash. *Construction and Building Materials.* **138**: 326–332.

[25] Krishnamurthy Masadi Patel1 & Chella Purushothaman Devatha. (2019). Investigation on leaching behaviour of toxic metals from biomedical ash and its controlling mechanism. *Environmental Science and Pollution Research* **26**:6191–6198

[26] Pechyen, C. and Ummartyotin, S. (2017). Development of Isotactic Polypropylene and Stearic Acid-Modified Calcium Carbonate Composite: A Promising Material for Microwavable Packaging. *Polymer Bulletin.* **74 (2)**: 431–444. DOI: https://doi.org/10.1007/s00289-016- 1722-3

[27] Gao, Y., Ren, K., Ning, N., Fu, Q., Wang, K., and Zhang, Q. (2012). Stretching-Induced Interfacial Crystalline Structures and Relevant Mechanical Properties in Melt-Spun Polypropylene/Whisker Composite Fibers. *Polymer.* **53** **(13)**: 2792–2801. DOI: https://doi.org/10.1016/j.polymer.2012.04.020

[28] Lo, H.M. and Liao, Y.L. (2007). The Metal-Leaching and Acid-Neutralizing Capacity of MSW Incinerator Ash CO-Disposed with MSW in Landfill Sites. *J. Hazard. Mater.* **142 (1-2)**: 512–519.

[29] Xie, Y. and Zhu, J. (2012). The Detoxification of Medical Waste Incineration Fly Ash Using Self-

Propagating Reaction. *Procedia Environmental Sciences*. **16**: 222–228.
[30] Zhao, L., F.S. Zhang, K. Wang, and J. Zhu. (2008). Chemical Properties of Heavy Metals in Typical Hospital Waste Incinerator Ashes in China. *J Waste Management*. **29 (3)**: 1114–1121.
[31] Sabiha-Javied TM, Khalid S .2008. Heavy metal pollution from medical waste incineration at Islamabad and Rawalpindi, Pakistan. Microchem J 90:77–81
[32] E. Gidarakos., Maria. P., Kalliopi., Anastasiadoua., Karl-W. S. (2009). Characterization And Hazard Evaluation Of Bottom Ash Produced From Incinerated Hospital Waste. *Journal of Hazardous Materials* **172**, 935–942. doi:10.1016/j.jhazmat.2009.07.080
[33] Kementerian Kesehatan RI. (2015). Peraturan Menteri Lingkungan Hidup dan Kehutanan Nomor 56 Tahun 2015 tentang Tata Cara dan Persyaratan Teknis Pengelolaan Limbah Bahan Berbahaya dan Beracun dari Fasilitas Pelayanan Kesehatan. Jakarta. p.35.
[34] Trihadiningrum, Y. (2016). *Pengelolaan Limbah Bahan Berbahaya dan Beracun*. Yogyakarta: Teknosain.
[35] Bunga, F. G., Syaiful, B., dan Elvi Y. (2017). Pemanfaatan Limbah Medis Padat Infeksius RSUD Arifin Achmad Pekanbaru dengan Teknik Solidifikasi sebagai Campuran Batako. *Jom F Teknik*. **4 (1)**.
[36] Chen, Q.Y., Tyrer, M., Hills, C.D., Yang, X.M., and Carey, P. (2009). Immobilisation of Heavy Metal in Cement-Based Solidification/Stabilisation: A Review. *Waste Management*. **29 (1)**: 390–403. DOI: http://dx.doi.org/10.1016/j.wasman.2008.01.019. 18367391.
[37] Ramli, Soehatman. (2010). Manajemen *Kebakaran*. Jakarta: Dian Rakyat.
[38] J. Peng MM, X. Wu, R. Wang MM, C. Li MM, Q. Zhang BM, D. Wei MD. (2020). Medical Waste Management Practice During The 2019-2020 Novel Coronavirus Pandemic: Experience In A General Hospital. American *Journal of Infection Control* **48** page 918–921. https://doi.org/10.1016/j.ajic.2020.05.035
[39] Sobiecka, E., A. Obraniak, and B. Antizar Ladislao. (2014). Influence of Mixture Ratio and pH to Solidification/Stabilization Process of Hospital Solid Waste Incineration Ash in Portland Cement. *Chemosphere*. **111**: 18–23. DOI: http://dx.doi.org/10.1016/j.chemosphere.2014.03.057
[40] Singh, A.K., V. Srivastav, and U. Kumar. (2016). Biomedical Waste Ash in Concrete: An Experimental Investigation. *International Journal of Innovative Research in Science, Engineering, and Technology*. **5 (6)**: 11547–11552.
[41] S. Papamarkou, C. Sifaki, P.E. Tsakiridis, G. Bartzas, K. Tsakalakis. (2018). Synthetic Wollastonitic Glass-Ceramics Derived from Recycled Glass and Medical Waste Incinerator Fly Ash. *Journal of Environmental Chemical Engineering*. **6 (5)**: 5812–5819.
[42] Haque, M. A, M.A Hoque, S. Saha, dan Hadiuzzaman. Md. (2014). Immobilization of Heavy Metals from Paving Block Constructed with Cement and Sand-Solid Waste Matrix. *Asian Journal of Applied Sciences*. **7 (3)**: 150–57.
[43] Sulistyowati, N.A. (2018). Compression Strength and Heavy Metal Leaching of Paving Block of Hazardous Material Waste (Metal Processing) Related to the Indonesia National Standard. *Jurnal Permukiman*. **13 (2)**: 69–76.
[44] Ababneh, A.R, Al-Rousan, W. Gharaibeh, and M.A. Dalo. (2020). Recycling of Pre-Treated Medical Waste Fly Ash in Mortar Mixtures. *Journal of Material Cycles and Waste Management*. **22**: 207–220. DOI: https://doi.org/10.1007/s10163-019-00928-z

Analysis of Ranjo Batu Community Perception of Unlicensed Gold Mining Activities on Environmental Quality

A Mubarak[1,2], B Saputra[1,2*], A Frinaldi[1,2], A Triono[1]

[1]Department of Public Administration, Faculty of Social Science, Universitas Negeri Padang, Jl Prof Dr. Hamka Padang, 25132, Indonesia
[2]Policy, Law and Political Research Center, Universitas Negeri Padang, Jl Prof Dr. Hamka Padang, 25132, Indonesia

bonisaputra@fis.unp.ac.id

Abstract. Since 2008, the activity of gold mining without a permit in Ranjo Batu Village has always been a polemic. There have been pros and cons in the community. Some people get various benefits, but others feel disadvantaged, giving rise to various perceptions. In this article, we focus more on the environmental aspect. The most apparent impact of mining activities is a decrease in environmental quality, but some people seem to ignore it because a pragmatic view is maintained. This study aims to comprehensively determine the publics perception of the environmental aspects of gold mining activities without a permit in Ranjo Batu Village. The research method uses descriptive and quantitative data. Data were obtained by distributing valid and reliable questionnaires to 95 respondents from a total population of 1,736 people. The descriptive analysis method will use the descriptive analysis method a percentage with the category of respondent achievement level (TCR). The results showed that the publics perception of environmental aspects was included in the unfavorable category, 53.61%. This perception score indicates that illegal gold mining from a community perspective hurts the environment, both in the water, on land, and in the air.

1. Introduction

The Indonesian government is currently very aggressive in distributing infrastructure development to revive the national economy. Indonesia is an archipelagic country that is rich in potential natural resources, one of which is mineral resources. By managing natural resources well, the Indonesian people can improve their economy. One of the concrete forms taken by the community in revitalizing the economy is opening up mining areas. The reference for the management of the mining sector itself has been explained in the 1945 Constitution Article 33 paragraph 3 that the earth, water, and natural resources contained therein are controlled by the state and used for the greatest prosperity of the people.

Departing from the definition of mining, according to Law Number 3 of 2020 concerning mineral and coal mining, it is stated that mining is "part or all of the stages of activities in the framework, management, and exploitation of minerals or coal." In mining management, the Indonesian government has complete control over mineral resources' management, management, and supervision, including granting mining permits to small-scale mining areas and limited investment. In Article 66, paragraph 2 of Law Number 3 of 2020, the applicant must apply to the Minister to obtain a People's Mining Permit. This application is to support justice for all Indonesian people to participate in enjoying the results of Indonesia's natural wealth.

In Ranjo Batu Village, Muara Sipongi District, Mandailing Natal Regency, the community, individually and in groups, manages Indonesia's natural wealth illegally or gold mining without a permit. According to [1][2], "Unlicensed Gold Mining is a mining activity that is prohibited by the state because it does not have a permit and it is feared that it will damage the environment." Unlicensed Gold Mining activities impact the community's economy, society, and environment. In economic development, the mining process must also be environmentally oriented; in this case, the government and society must maintain the balance of nature in every development activity to realize the principles of sustainable development [3][4]. Since 2008, Unlicensed Gold Mining has been carried out in Ranjo Batu Village, Muarasipongi District.

Figure 1. Location of Mining Activities in Ranjo Batu Village.

This activity is located on a residential hill only about 500 meters away. The community only uses simple traditional tools such as hoes, chisels, crowbars, and steel. Economic factors are the background of the rampant gold mining without permits in Ranjo Batu Village. There is a view from the community that mining gold without a permit is more productive than having to work as a farmer. The direct impact of this illegal mining activity is a decrease in environmental quality, as evidenced by the contamination of river water due to mining production waste containing mercury flowing into the river.

Seeing the impact of gold mining activities without permits creates perceptions or views among the community and triggers polemics. Some people are pro, and not a few are against this activity. The pros view that mining activity positively impacts the economy of the people of Ranjo Batu Village. Its existence is considered to open up jobs and community businesses. However, the people are against it, and they think that gold mining activities without a permit can cause a decrease in the quality of the environment. The mining process does not meet environmentally friendly standards and harms public health. Environmental issues have become a daily conversation in Ranjo Batu Village due to illegal gold mining activities.

According to [5][6], perception is a process in which messages and sensations received by a person will be selected and sorted, compiled, and finally interpreted. Every individual has a different perception. Differences in each individual's perception are influenced by various factors such as point of view, knowledge, experience, and others.

Departing from environmental problems due to illegal gold mining activities in Ranjo Batu Village, this activity is also touted as the cause of environmental damage. According to [2], the impact of Unlicensed Gold Mining activities seen from the environmental aspect is environmental damage and environmental degradation. What is meant by environmental degradation is the destruction of environmental resources such as water, air, and soil, in addition to the destruction of wildlife habitat ecosystems.

This mining is a source of income for the people of Ranjo Batu Village, which can support the regional economy if managed correctly and following environmental damage prevention instruments. Because these two things are essential for the progress of the life of the people of Ranjo Batu Village and its surroundings, all economic activities of the community, especially mining, are expected to consider not only the current short-term interests but also future generations and how their fate will be due to environmental damage. So it can be said that it is not following the principles of sustainable development.

The purpose of this research is to see how the people of Ranjo Village perceive the activities of Unlicensed Gold Mining on the environment and how the community perceives the environmental impact of this mining activity. Research on public perceptions is fundamental because knowing perceptions can prevent negative impacts on the community, and it is hoped that the results of this research can later be considered for the government in making policies related to regional problems.

2. Methodology

In this study, the researcher used a descriptive type of research with a quantitative approach. According to [7], this descriptive research is research that describes the symptoms of events and events that are happening now. This research was conducted in Ranjo Batu Village, Muara Sipongi District, Mandailing Natal Regency, with the research subject being the people of Ranjo Batu Village. This study uses a single variable, namely the public's perception of gold mining, without a permit from the environmental aspect. The population used in this study was the people of Ranjo Batu Village, amounting to 1736 people. This study uses a cluster sampling technique (cluster area) with a sample of 95 people determined using the slovin formula [8]. Data collection techniques used questionnaires as primary data and documentation as secondary data. Then the data is processed using the TCR formula (respondent's level of achievement) and classified as public perception. [9][10] explains that the percentage of TCR values is divided into three groups of criteria, namely: Good (76% - 100%), Fairly Good (56% - 75%), and Less good (< 56%).

3. Results and Discussion

The perception of the Ranjo Batu community about Unlicensed Gold Mining from an environmental perspective is to see the environmental impact of this mining activity. According to [11][12], perception is an interrelated activity. Then everything that exists in the individual, such as feelings, experiences, thinking abilities, frames of reference, and other aspects of the individual community, will play a role in that perception. Environmental problems here are considered as the gap between reality and actual expectations. According to [13][14][15], the emergence of environmental impacts due to mining activities generally decreases environmental quality with various forms of pollution and land damage. The following are some indicators in analyzing the perception of the people of Ranjo Batu Village seen from the environmental aspect based on a questionnaire that has been distributed to 95 respondents.

3.1. Public Perception Index Model Summary

The results of the recapitulation of the Community Perception Index of Unlicensed Gold Mining in Ranjo Batu Village seen from the environmental aspect can be seen in the table below:

Table 1. Community Perception Index Recapitulation of Unlicensed Gold Mining From Environmental Aspects.

No	Question	Average Respondents ' Answers	Respondent's Level of Achievement (Perception Index)	Category
1.	Does this unlicensed gold mining activity have any impact on water quality?	1,59	39,7	Not Good
2.	Do you think that unlicensed gold mining activities cause river pollution?	1,49	37,4	Not Good
3.	Do you think illegal gold mining activities cause land subsidence around the mine?	2,36	58,9	Pretty Good

No	Question	Average Respondents ' Answers	Respondent's Level of Achievement (Perception Index)	Category
4.	In your opinion, do illegal gold mining activities impact soil fertility?	2,41	60,3	Pretty Good
5.	Do you think mining activities cause noise?	2,54	63,4	Pretty Good
6.	Do you think mining activities impact decreasing air quality (air pollution)?	2,43	60,8	Pretty Good
7.	Do you think illegal gold mining activities have an impact on wildlife?	2,26	56,6	Pretty Good
8.	Do you think that the existence of this Unlicensed Gold Mining activity endangers living things (Flora and Fauna) around the mining area?	2,07	51,8	Not Good
	Average Amount	**2,14**	**53,61**	**Not Good**

3.2. Waste that Impacts Water Quality

Based on the results of field data processing related to the opinion of respondents regarding gold mining without a permit impacts water quality. As many as 58.9% of respondents said it was a very "impact," and 29.5% said "impact." Moreover, only 11.6% of respondents stated that the mining process had a "small impact" and "no impact" on water quality in Ranjo Batu Village, Muara Sipongi District, Mandailing Natal Regency. The average answer obtained from 95 respondents in this sub-indicator is 1.59. The public perception index in this sub-indicator is 1.59/4 x 100% = 39.7%, so the level of achievement or interpretation of respondents to the value is classified as Low or Less Good in a Negative sense. This means that most of these people are aware that this illegal mining process has an impact on the environment, especially the impact on the quality of water sourced from mining waste.

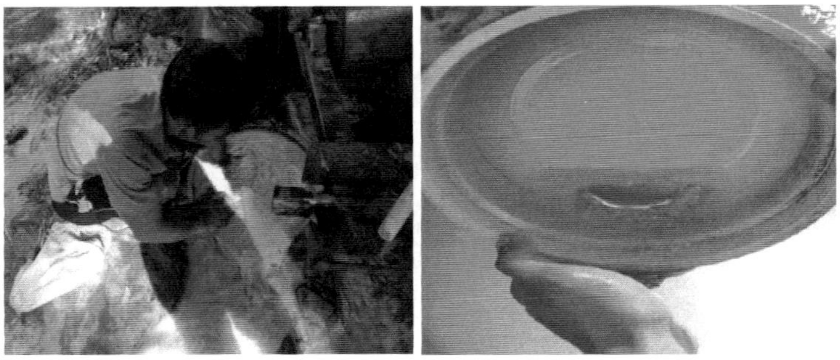

Figure 2. The Process of Milling Stone in the River Using Mercury.

According to the community, the existence of gold mining activities without a permit in Ranjo Batu Village has an impact on producing waste that affects water quality in Ranjo Batu Village. The existence of gold mining activities without a permit causes the river water in Ranjo Batu village to become cloudy. Gold mining is carried out in the river, so waste treatment using mercury in separating gold from gold-carrying stones causes water quality to decline and threatens public health. It is common to find people affected by the disease—skin (itching) due to the chemical mercury.

3.3. River Pollution

Based on the results of field data processing related to opinions about illegal gold mining have an impact on river pollution. As many as 60% of respondents stated "Highly Polluted River," and 30.5% of respondents stated "Polluted River." Moreover, the remaining 9.5% of respondents stated that the mining process "slightly pollutes the river," and none answered that this process has no impact on river pollution in Ranjo Batu Village. The average answer obtained from 95 respondents to this question is 1.49. Perception Index: The public response to this question is: 1.49/4 x 100% = 37.4%, so the respondents' level of achievement or interpretation is classified as Low/Poor in a negative sense. This means that most people agree that this illegal mining process has an impact on the environment, especially the impact of river pollution.

Figure 3. River Pollution Condition.

The community views and realizes that illegal gold mining activities in Ranjo Batu Village cause river pollution. The rivers in Ranjo Batu Village have been polluted by waste from illegal gold mining activities. The community is reluctant to use the river for daily activities because it is polluted and unfit.

3.4. Land Subsidence Around the Mine

Based on the results of processing field data related to public opinion about land subsidence around the mine, that is respondents' perception that as many as 15.8% think "very often" and 41.1% think "often" gold mining activities without a permit in Ranjo Batu Village cause land subsidence around the mine. In comparison, as much as 34.7% of others stated "rare," and the remaining 8.4% thought "none" was land subsidence around the mining area. The average answer obtained from 95 respondents to this question is 2.36. The public perception index of the question is 2.36/4 x 100% = 58.9%, so the level of achievement or respondents' interpretation of the value is classified as Medium/Quite good in a positive sense.

Figure 4. Miners are removing stones from the mine pit.

Descriptive analysis of the results of respondents' answers stated that the community believed that gold mining activities without a permit in Ranjo Batu Village did not cause or rarely occurred land subsidence around the mining area. According to residents, the mining pits dug by the miners are not very deep, and the mining site is surrounded by pine trees and smallholder rubber plantations so that it can help prevent landslides in the area. During gold mining activities, subsidence is rare.

3.5. Soil Fertility

Based on the results of processing field data related to respondents' opinions about gold mining without a permit have an impact on soil fertility there are 20% of respondents say mining activities cause the soil to be "very infertile," and 40% of respondents say "infertile" on soil fertility, 18.9% answered, "Slightly infertile." The remaining 21.1% answered mining activities "No Effect on Soil Fertility" in Ranjo Batu Village. The average answer obtained from 95 respondents to this question is 2.41. The index of public perception of the question is 2.41/4 x 100% = 60.3%, so the level of achievement or interpretation of respondents to the value is classified as moderate/good enough in a positive sense.

The community considers that the existence of gold mining activities without a permit in Ranjo Batu Village has no significant effect on soil fertility around the mining area in Ranjo Batu Village. However, the community rarely farms because most of their plantation area has been used as mining land. The conversion of agricultural land to mining land.

3.6. Generating Noise

Based on the results of processing field data related to respondents' opinions about gold mining without a permit have an impact on soil fertility, is that 17.9% of respondents stated that gold mining activities without a permit in Ranjo Batu Village were "boisterous," 24.2% of respondents stated that gold mining activities without permits were "noisy," 44.2% of respondents said "quite noisy. ". and 13.7% of respondents stated "no noise." The average answer obtained from 95 respondents to this question is 2.54. The index of public perception of the question is 2.54/4 x 100% = 63.4%, so the respondents' level of achievement or interpretation is classified as Medium or Good Enough in a positive sense.

The people of Ranjo Batu Village believe that Gold Mining activities without a permit do not cause noise that disturbs the comfort of the people of Ranjo Batu Village. The community also stated that although the processing uses a stone grinding machine that produces sound, the community does not feel disturbed.

3.7. Decreasing Air Quality (Air Pollution)

Based on the results of processing field data related to opinion respondents' about on the impact of gold mining without a permit on air pollution indicators. A total of 23.2% of respondents stated that mining activities impact air quality which is "highly polluted" (Air Pollution). 21.1% of respondents said the air was "polluted," and 45.3% said mining activities had an impact on air quality "quite polluted" (air pollution). The remaining 10.5% of respondents stated that gold mining activities without permits do not impact air quality (air pollution) in Ranjo Batu village. The average answer obtained from 95 respondents to this question is 2.43. The public perception index of the question is 2.43/4 x 100% = 60.8%, so the respondents' level of achievement or interpretation is classified as moderate or quite good in a positive sense.

The interpretation of these figures means that people believe illegal gold mining activities in Ranjo Batu Village will not impact air quality (Air Pollution). The community considers that extensive forests still surround the Ranjo Batu area and rubber plantations, and this mining activity does not cause air pollution in the village.

3.8. Impact on the Presence of Wild Animals

Based on the results of processing field data related to opinion respondents' about on the impact on the presence of wild animals, it is known that 4.25% of respondents answered that gold mining activities without a permit "have a major impact on the presence of wildlife" around the mine site, 33.7% stated "impact on the presence of wildlife," 30.5% of respondents said mining activities "enough to have an impact on the presence of wildlife" and the remaining 10.5% of respondents stated that gold mining activities without a permit in Ranjo Batu Village "have no impact on the presence of wildlife" around the mining site. The average answer obtained from 95 respondents for this question is 2.26. The index of public perception of the question is 2.26/4 x 100% = 56.6%, so the level of achievement or interpretation of respondents are classified as moderate or quite good in the sense of positive.

This means that the existence of illegal gold mining activities in Ranjo Batu Village does not have much impact on the presence of wildlife around the mining area. Wild animals, such as tigers, and forest people, live in nature. According to the community, gold mining activities without permits do not affect wildlife because gold mining activities are not too far from community settlements. In contrast, wild animals are usually found deep in the forest.

3.9. Threatening the Presence of Living Things (Flora and Fauna) Around the Mine

Based on the results of processing field data related to opinion respondents' about on the impact on the threatening the presence of living things (flora and fauna) around the mine, it can be concluded that as many as 22.1% of respondents stated that gold mining activities without a permit in Ranjo Batu Village are "hazardous" for living things (Flora and Fauna) around the mining area, and 57.9% of respondents stated that these activities mining "harms" living things (Flora and Fauna) around illegal mining. By comparison, 10.5% said it was "pretty dangerous." Only 9.6% of respondents stated that gold mining activities without a permit in Ranjo Batu Village had no impact and "no harm" to living things (Flora and Fauna) around the illegal mining sites in Ranjo Batu Village. The average answer obtained from 95 respondents to this question is 2.07. The public perception index of the question is 2.07/4 x 100% = 51.8%, so the level of achievement or interpretation of respondents to the value is classified as low/less in a negative sense.

This means the community believes that gold mining activities without a permit in Ranjo Batu Village can harm living things (Flora and Fauna) around illegal mining in Ranjo Batu Village. The flora is the type of plant or plant, while the fauna is the animals that live around the village of Ranjo Batu. According to the community in producing gold mining using mercury and disposing of river waste, a case was found that caused fish in the river Ranjo Batu village to die.

Based on the questionnaire distributed to 95 respondents, it was found that the public perception of Unlicensed Gold Mining viewed from the environmental aspect in Ranjo Batu Village, Muara Sipongi District, Mandailing Natal Regency has a low perception (index with a percentage of 53, 61%). So it can be concluded that the public perception of the Unlicensed Gold Mine from the environmental aspect is proven to damage the environment. Because according to the community, the production of this Unlicensed Gold Mine uses mercury/chemical substances that produce waste, which impacts water quality and can cause itching on the skin of the surrounding community. In addition, it also causes pollution of the rivers in Ranjo Batu Village. The river becomes cloudy and dirty, mixed with garbage so that it can threaten the lives of living things (Flora and Fauna) around the mining area, such as the case of fish deaths in the river due to waste containing mercury around the mining site.

4. Conclusion

Based on research and discussions on public perceptions of Unlicensed Gold Mining in terms of environmental aspects, public perceptions of this matter are included in the unfavorable category, which is 53.61 percent. This perception score shows that Unlicensed Gold Mining hurts the environment in Ranjo Batu Village, Muara Sipongi District, Mandailing Natal Regency because these activities use mercury or chemicals that produce waste that impacts water pollution, decreasing water quality and causing river pollution. So the advice to be given to the local government is to provide socialization to the community about the negative impact of illegal mining on the environment and socialization related to legal mining licensing issues so that official gold mines can be controlled through preventive procedures against environmental damage to suppress the spread of gold mines illegal activities that can damage the environment and threaten people's lives in the long term.

References

[1] Alisjahbana, *Sisi Gelap Pertambangan Emas di Indonesia*. Yogyakarta: Laksbang Pressindo, 2015.
[2] D. M. Sandi, "Aktivitas Penambang Emas Tanpa Izin (PETI) di Kecamatan Batang Natal Kabupaten Mandailing Natal (2004-2017)," *J. Educ. Dev.*, vol. 4, no. 1, 2018.
[3] R. Tambunan, "Perilaku Konservasi Pada Masyarakat Tradisional," *J. Harmon. Sos.*, vol. 2, no. 2,

2008.
[4] A. Luthfi and A. Wijaya, “Persepsi Masyarakat Sekaran Tentang Konservasi Lingkungan,” *J. Komunitas*, vol. 3, no. 1, 2011.
[5] B. Saputra, H. Fajri, A. D. Akma, N. Wahyuni, and H. S. Halawa, "Agree or disagree local youth's perception of renewable energy development," in *IOP Conference Series: Earth and Environmental Science*, 2021.
[6] R. Prasetijo and J. J. O. . Ihalauw, *Perilaku Konsumen*. Yogyakarta: Andi, 2005.
[7] N. Sudjana, *Penilaian Hasil Proses Belajar Mengajar*. Bandung: Remaja Rosda Karya, 1999.
[8] B. Prasetyo and L. M. Jannah, *Metode Penelitian Kuantitatif Teori dan Aplikasi*. Jakarta: Raja Garfindo Persada, 2014.
[9] M. Singarimbun and S. Effendi, *Metode Penelitian Survei*. Jakarta: LP3ES, 1982.
[10] Sugiyono, *Metode Penelitian Pendidikan Pendekatan Kuantitatif, kualitatif, dan R&D*. Bandung: Alfabeta, 2010.
[11] Bimo Walgito, *Bimbingan dan Konseling (Studi dan karier)*. Yogyakarta: Andi Offset, 2000.
[12] A. Agustin and K. Khuzaini, “Persepsi Masyarakat terhadap Penggunaan Transportasi Online (Go-Jek) di Surabaya,” *J. Ilmu dan Ris. Manaj.*, vol. 6, no. 9, 2017.
[13] E. Wahyudi and Slameto, “Dampak Sosial Penambangan Emas Tanpa Izin (PETI) terhadap Keberlanjutan Usaha Tani Padi di Kabupaten Merangin Provinsi Jambi,” in *Prosiding Seminar Nasional Agroinovasi Spesifik Lokasi Untuk Ketahanan Pangan Pada Era Masyarakat Ekonomi ASEAN*, 2014.
[14] S. Müllera, L. Hucka, and J. Markovab, "Sustainable community-based tourism in Cambodia and tourists' willingness to pay," *Austrian J. South-East Asian Stud.*, vol. 13, no. 1, pp. 81–101, 2020.
[15] H. Malau, H. Fajri, P. Yuanjaya, B. Saputra, and K. D. Maani, “Knowledge of Local Communities Affected by the Development of Geothermal Energy,” in *IOP Conference Series: Earth and Environmental Science*, 2020, vol. 448, no. 1.
[16] Undang-Undang Nomor 3 Tahun 2020 tentang Perubahan Atas Undang-Undang Nomor 4 Tahun 2009 tentang Pertambangan Mineral dan Batubara.

IOP Conf. Series: Earth and Environmental Science 1098 (2022) 012005 doi:10.1088/1755-1315/1098/1/012005

Cleanliness assessment of Lake Toba tourist attractions from stakeholders' perspective

H Khair[1,2,3*], R Utami[1], G Lordye[1], U Sari[1], and E Pasaribu[1]

[1]Environmental Engineering Department, Universitas Sumatera Utara, Medan, Indonesia 20155
[2]Toba Lake & Sustainable Tourism Working Group, Universitas Sumatera Utara, Medan, Indonesia, 20155
[3]Monitoring Center for Sustainable Tourism of Universitas Sumatera Utara, Medan, Indonesia, 20155

hafizhul@usu.ac.id

Abstract. Lake Toba area is one of the priorities for developing tourist destinations in Indonesia. As a strategic national tourism area, it is necessary to identify aspects of the tourism area development that are not yet optimal to increase the number of tourist visits to that location. This study aims to assess the cleanliness of tourist sites in Lake Toba based on the assessment of tourism site stakeholders, namely visitors, local businesses, and cleanliness managers. The tourist sites in Lake Toba studied were Pasir Putih and Bukit Holbung. This study found that Pasir Putih and Bukit Holbung were considered dirty, with a cleanliness index of 2.63 and 2.60. The assessment shows that access to the cleaning facilities in the two tourist sites is difficult to reach. However, regarding the completeness of the cleaning facilities, Pasir Putih is considered better with an average index of 2.94 compared to the cleaning facilities in Bukit Holbung, which are considered inadequate with an average index of 2.43. The number of tourists visiting a tourist site is directly proportional to the waste generated, and the habits of tourists who visit also affect the cleanliness of a tourist site.

1. Introduction

Sustainable tourism development ensures that optimal benefits will be obtained sustainably, which can only be realized with a comprehensive and integrated approach (policy) [1]. Article 2 of the Law of the Republic of Indonesia No. 10 of 2009 explained the principles of tourism development. Tourism development is achieved through the implementation of tourism development plans considering the diversity, uniqueness, and uniqueness of culture, nature, and human needs for tourism [2].

Tourism development will also affect a tourist area, especially under rapidly changing environmental conditions in terms of appearance. These changes are generally reflected in changes in land use, the development of tourist facilities and infrastructure, and even waste scattered due to tourist arrivals [3]. This problem is a form of participation of tourists who do not maintain cleanliness, litter, or improper habits. As a result, much waste has piled up and scattered in the tourist area [4].

The United Nations Environment Program (UNEP) states that tourists can generate six times as much waste on vacation. The volume of waste increases as the number of tourist visits increases. Usually, the amount of waste generated from an area is influenced by the socio-economic level. The higher the economy, the higher the consumption level, which will impact the amount of waste generated. The waste problem harms the environment. Waste can cause unpleasant odors around tourist sites. Furthermore, it can affect the satisfaction of other visitors, thus reducing the number of visitors and reducing the economic value [5].

Lake Toba is the world's largest caldera tectonic lake and the largest lake in Indonesia. Lake Toba is in the north Sumatra Province, 176 km west of Medan City as the provincial capital. Lake Toba has a length of 87 km in a northwest-southeast direction with a width of 27 km, a height of 904 meters above sea level, and the deepest lake depth is 505 meters. The Lake Toba area is a priority for developing tourist destinations in Indonesia that have a national and international service scale. Therefore, Lake Toba, as a development priority, should be able to contribute and have more adequate conditions for tourism aspects compared to other tourist areas. Therefore, it is necessary to identify problematic aspects of the development of tourist areas so that they can formulate efforts to increase the number of tourist visits in the tourist area of Lake Toba [6].

However, increasing tourist visits can cause problems with respect to waste generated by tourist activities [7]. The more visitors, the more waste generated from tourist activity [8]. Suppose that no changes are made in handling both technical and political policies. In that case, it is predicted that it will result in significant environmental pollution throughout the region, both directly and indirectly.

This study assesses the cleanliness of tourist attractions in the Lake Toba Tourism Area. It is hoped that management can be helped and know the condition of the tourist attractions they manage so that appropriate actions are produced in managing waste in tourist areas. This study takes the characteristics of tourist sites crowded with visitors and becomes an initial assessment of environmental conditions and pollution in the Lake Toba area.

2. Methods

The Lake Toba tourist area has at least ten popular tourist areas that offer lake views. This study was carried out in the Lake Toba tourist area with tourist characteristics classified as crowded with visitors and tourism classified as moderate by crowds of selected visitors based on the most reviews from Google Maps and TripAdvisor. The selected locations were Pantai Pasir Putih (2663 reviews) [9] and Bukit Holbung (1293 reviews) [10].

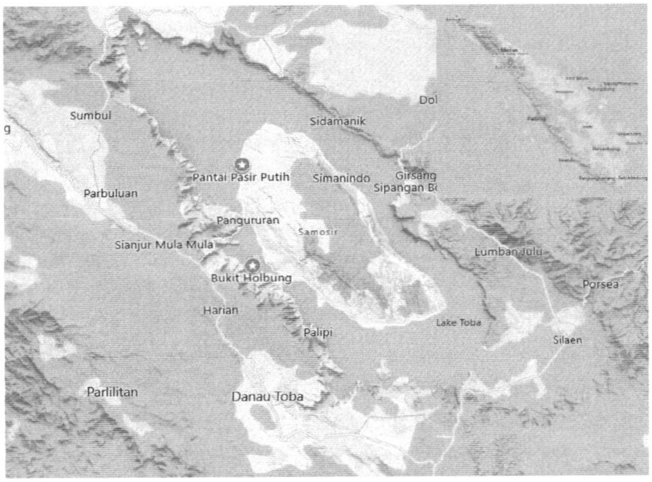

Figure 1. Sampling locations

This study used questionnaires as the primary data collection method. These questionnaires are intended to obtain information about the opinions of visitors or the surrounding community, which is used to measure or score the cleanliness of the study area location. This study assessed the opinions on the following topics:

1. Cleanliness of tourist sites
2. Access cleaning facilities at tourist sites
3. Completeness of cleaning facilities
4. Crowded tourist sites
5. A habit of carrying plastic bags

Respondents were asked to provide answers with a qualitative assessment of their experience regarding the conditions of each tourist site they visited. Additionally, this study also collects information related to the respondent profiles. The total number of respondents in this study was 135 respondents consisting of 100 respondents for the Pasir Putih Beach tourist spot and 35 for the Bukit Holbung tourist spot.

3. Result and discussion

Respondents from the two tourist attractions consisted of various backgrounds. It can be seen from the distribution of gender, respondent status, education, and regional origin. The complete profile and distribution of the respondents in this study are presented in Table 1.

Table 1. Profile of the respondent

Profile of Respondent		Number of Respondents (people)	
		Pantai Pasir Putih	**Bukit Holbung**
Gender	Male	67	13
	Female	33	22
Status	Visitors	69	28
	Local Business Actors	30	4
	Cleanness Manager	1	3
Education	Primary school	3	0
	Junior High School	5	1
	High School	72	28
	Higher Education	20	6
Origin	Lake Toba Area	57	24
	Outside the Lake Toba Area	41	9
	Other	2	2

The profile is important because it will affect an individual's waste disposal habit. These habits are influenced by several factors: predisposing factors (sex, education, knowledge, attitudes), supporting factors (infrastructure), reinforcing factors, and attitude factors. In addition, littering can affect the behavior of the surrounding environment in carrying out similar actions. The selection of respondents also needs to involve managers and business actors, because both parties contribute to maintaining the cleanliness of tourist sites. The results of the response to the assessment of the conditions of each location are presented in Figure 2.

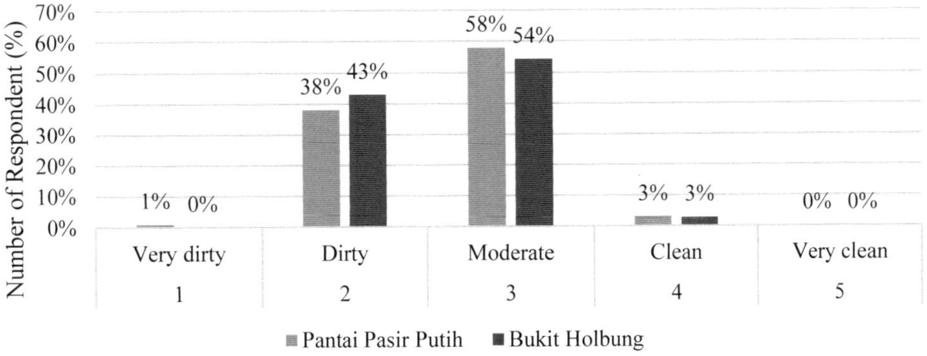

Figure 2. Responses to the assessment of the cleanliness of tourist sites

Based on the results of the questionnaire on the cleanliness assessment of the Cleanliness of Tourist Locations, the tourist location of cleanliness of the Pasir Putih Beach has an average value of 2.63, indicating that the level of cleanliness of the Pasir Putih Beach during this study was in the gross to moderate. Meanwhile, the average value for Bukit Holbung is 2.60, with the level of cleanliness in the dirty to moderate range. The average value is obtained by multiplying the index by the number of respondents who chose the index and then dividing by the total number of respondents.

Figure 3. Some examples of waste found at research sites

Based on the status of the respondents, tourism actors and visitors rate the cleanliness at Pasir Putih Beach to be in the dirty - moderate range, with an average index of 2.6. Meanwhile, the cleanliness manager considers that the cleanliness condition on the Pantai Pasir Putih is in the spring range with an index value of 3. Although the results of the cleanliness assessment are different from that in Bukit Holbung, the average index of cleanliness assessment from tourism actors is the highest, which is 3, by cleaning managers and tourism actors with an average index value of 3 and 2.5, respectively. Based on this research, it can be said that tourist visitors tend to have the lowest ratings from business actors and cleaning managers. This is possible due to the experience and expectations of visitors to these tourist attractions. The comparison of the average index value in the two locations is presented in Figure 4.

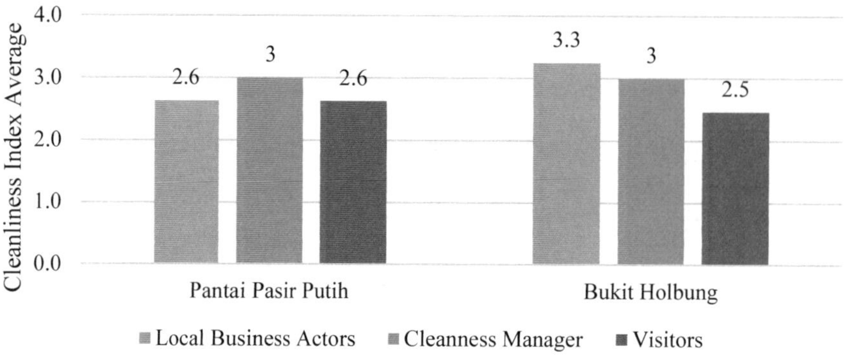

Figure 4. Cleanliness index average

The respondents' assessment of access to cleaning facilities showed that at Pasir Putih Beach, 53% rated access to cleaning facilities as moderate and 39% chose far. Meanwhile, in Bukit Holbung, the majority of the respondents, namely 57%, considered the access to cleaning facilities to be very far 9% considered it very far, and 29% considered it moderate. These data show that both places have minimal access to cleaning facilities, so tourists have difficulty reaching them. The results of the survey of access to cleaning facilities are presented in Figure 5. The access to cleaning facilities must be considered because the location of trash cans also affects the cleanliness of tourist sites [11]. The easier the trash can for visitors to reach, the more likely visitors are to dispose of trash in its place. It can happen because the location of the trash can will affect the level of awareness of visitors who enter it. However, it is difficult to reach if the trash can is far away. As a result, visitors will likely allow their trash to be placed where it should not be.

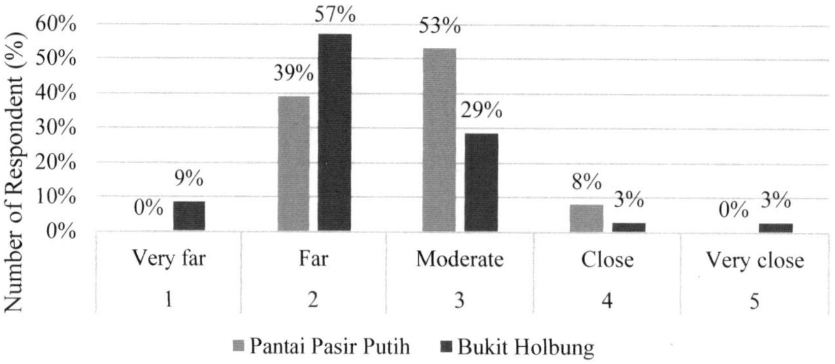

Figure 5. Access to cleaning facilities

This study also evaluates the completeness of cleaning facilities in tourist attractions based on the assessment of tourism actors. At Pasir Putih Beach, 40% of the respondents rated the cleaning facilities in the medium category, 33% chose incomplete, and 27% chose completely. Meanwhile, in the Bukit Holbung tourist location, 51% of the respondents chose incomplete, 37% chose moderate, and 6% chose very incompletely. Therefore, the cleanliness facilities at Pasir Putih Beach are considered to be better, with an average index of 2.94 compared to the cleaning facilities at Bukit Holbung, which are considered incomplete with an average index of 2.43. Adequate facilities and infrastructure must support cleanliness. The completeness of the facilities can affect the cleanliness of tourist sites because the more complete the cleaning facilities, the easier it is for visitors to dispose of garbage in its place [12]. The results of the response to the completeness of the the cleanliness of tourist sites are presented in Figure 6.

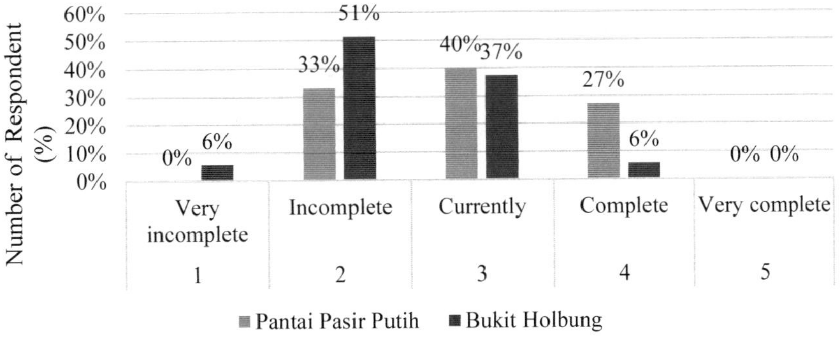

Figure 6. Completeness of cleaning facilities

When viewed from the crowds of tourist sites, as shown in Figure 7, for Pantai Pasir Putih, 49% of respondents said it was very crowded and 25% chose crowded. Meanwhile, at the Bukit Holbung tourist location, 63% of the respondents said it was quiet and 34% chose moderate/quite crowded. The higher the interest of tourists, the more waste generated [13]. This result follows data from the Google Maps and TripAdvisor applications, which show that Pasir Putih Beach is more crowded with visitors than Bukit Holbung.

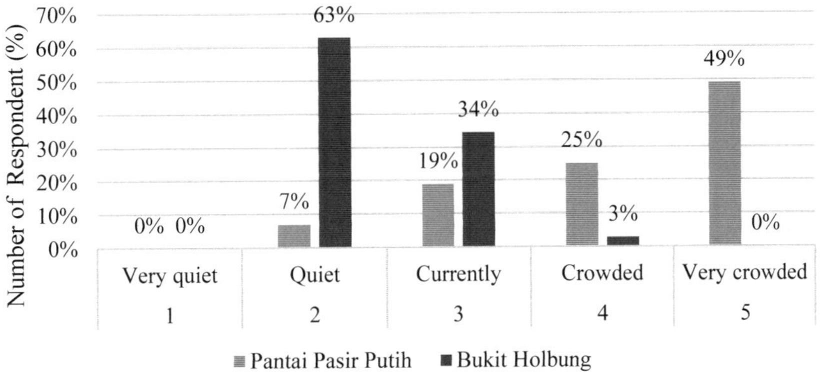

Figure 7. Crowd level of tourist locations

Waste and tourism are related because they are influenced by several factors [14]. An influencing factor is visitors' habits [4]. The results of the questionnaire support this statement at each tourist location regarding the visitors' habits to bring plastic bags as personal trash cans, as shown in Figure 8.

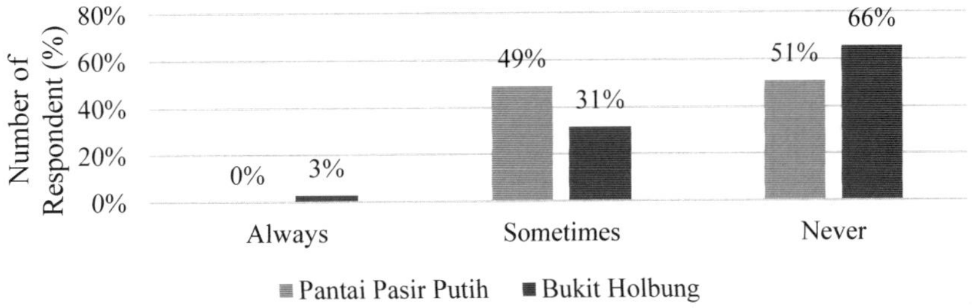

Figure 8. A habit of carrying plastic bags

At the Pasir Putih Beach tourist location, 51% of the respondents never brought personal trash bags and 49% chose sometimes. Meanwhile, at the Bukit Holbung tourist location, 66% of the respondents said that they never brought it, 49% of the respondents chose sometimes, and only 3% of the respondents always carried a garbage bag. Based on this percentage, it can be concluded that most of the visitors to the two tourist sites tend never to carry plastic bags as personal trash cans. It is a bad habit of visitors that can affect the cleanliness of a tourist site. Humans around tourist sites have a very important role in the cleanliness of tourist sites [15]. Therefore, visitors must begin to adapt to new habits by bringing plastic bags as personal trash cans to maintain the cleanliness of tourist sites.

4. Conclusion

In this study, it can be concluded that the cleanliness of Pasir Putih Beach and Bukit Holbung Beach is in moderate condition and tends to be dirty. At the Pasir Putih location with a busy to very crowded percentage of visits, the condition that tends to be dirty is caused by access, and the number of cleaning facilities considered moderate tends to be far away. Similarly, in the tourist destination of Bukit Holbung, with a low percentage of visitors or quiet, the resulting clean conditions also tend to be dirty. Furthermore, the level of awareness of visitors in both locations about bringing personal garbage bags is still low. Therefore, it also causes a level of cleanliness at tourist sites.

Increasing the number and access to cleaning facilities is expected to improve the hygiene conditions at tourist sites. Tourism actors, such as visitors, managers and business actors at tourist sites, have the same responsibility to maintain cleanliness. Of course, the government can provide assistance and guidance to tourism actors to realize sustainable tourism goals.

References

[1] A. Setijawan, "Pembangunan Pariwisata Berkelanjutan dalam Perspektif Sosial Ekonomi," *J. Planoearth*, 2018.

[2] N. Suhastini, "Strategi Kebijakan Pembangunan Berkelanjutan Wisata Pantai Di Kawasan Lombok Timur," *Komunitas*, 2019.

[3] Y. Prastika and I. N. Sunarta, "Studi Perkembangan Pariwisata Dan Pengaruhnya Pada Lingkungan Fisik Di Pantai Balangan, Desa Ungasan, Jimbaran," *J. Destin. PARIWISATA*, 2018.

[4] N. Prastiwi, Z. Shaluhiyah, and F. Agushybana, "Pengaruh Pemasangan Poster Anjuran Buang Sampah terhadap Pemanfaatan Tempat Sampah di Tempat Wisata," *J. Promosi Kesehat. Indones.*, vol. 14, no. 2, p. 149, Sep. 2019.

[5] H. Hu, J. Zhang, G. Chu, J. Yang, and P. Yu, "Factors influencing tourists' litter management behavior in mountainous tourism areas in China," *Waste Manag.*, 2018.

[6] R. A. Siregar, H. W. Wiranegara, and H. Hermantoro, "Pengembangan Kawasan Pariwisata Danau Toba, Kabupaten Toba Samosir," *TATALOKA*, vol. 20, no. 2, p. 100, May 2018.

[7] I. A. Suarinastuti and I. G. A. O. Mahagangga, "Pengelolaan Sampah di Daya Tarik Wisata Wanara Wana/ Monkey Forest, Desa Padangtegal, Ubud," *J. Destin. Pariwisata*, 2016.

[8] O. T. Utomo and N. Hilal, "Tinjauan Pengelolaan Sampah di Objek Wisata Dream Land Water Park Ajibarang Kabupaten Banyumas Tahun 2016," *Bul. Keslingmas*, 2017.

[9] Google, "Pantai Pasir Putih," *Google Map*, 2021. [Online]. Available: https://goo.gl/maps/CCqj8AeYHM2hPdh58. [Accessed: 29-Sep-2021].

[10] Google, "Bukit Holbung Samosir," *Google Map*, 2021. [Online]. Available: https://goo.gl/maps/6K5gN1424T2YJ9E66. [Accessed: 29-Sep-2021].

[11] Y. A. Marchita and A. W. Lestari, "Hubungan Pembangunan Tempat Pengolahan Sampah Terpadu Terhadap Tingkat Partisipasi Masyarakat Dibidang Kebersihan Lingkungan," *JISIP J. Ilmu Sos. dan Ilmu Polit.*, 2016.

[12] S. P. Setyowiniji and others, "Faktor--Faktor yang Berpengaruh Terhadap Kebiasaan Membuang Sampah Pada Mahasiswa Keperawatan STIKES Muhammadiyah Gombong," STIKES MUHAMMADIYAH GOMBONG, 2017.

[13] A. Fauzan and others, "Analisis Timbulan Dan Komposisi Sampah Di Kawasan Wisata Taman Pintar Dan Sindu Kusuma Edupark DI Yogyakarta," 2018.
[14] S. P. Wilson and K. M. Verlis, "The ugly face of tourism: Marine debris pollution linked to visitation in the southern Great Barrier Reef, Australia," *Mar. Pollut. Bull.*, vol. 117, no. 1–2, pp. 239–246, Apr. 2017.
[15] S. Suratinoyo, F. Lengkong, and V. Londa, "Partisipasi Masyarakat dalam Penanganan Kebersihan Pantai di Kecamatan Malalayang Kota Manado," *J. Adm. Publik*, vol. 3, no. 046, 2017.

Application of Hazard and Operability Study Methods (HAZOP) to asses and control hazard risk in spinning department using at textile industrial

Novie Susanto[1], Faradhina Azzahra[1], Alldo Haswandana Putra[1]

[1]Industrial Engineering Department, Faculty of Engineering, Diponegoro University Prof. Soedarto, SH, Tembalang – Semarang 50272

novie.susanto@ft.undip.ac.id

Abstract. PT. XYZ is a textile industry company located in Sukoharjo, Central Java. PT. XYZ produces 4 products, namely, yarn, greige, dyed and printed fabric, uniform and apparel. Based on initial observations, sources of danger were identified with a low hazard percentage of 11.7%, medium danger of 29.4%, high danger of 47.05%, and extreme danger of 11.7%. In effort to minimize workplace accidents, work accident analysis can be carried out using the Hazard and Operability Study (HAZOP) method through ranking priorities. The study results found 17 hazards in the spinning division and can be categorized into four hazard classes. From the results of the risk level assessment, there are two hazard sources classified as extreme, eight hazard sources classified as high, five hazard sources classified as moderate risk, and two hazard sources classified as low risk. This study provides recommendations based on control hierarchy with examples of inspection form, signs or posters regarding the use and need for PPE, supervision (punishment system and reward), prohibition of entering the work area without wearing PPE or work cards, and engineering controls.

1. Introduction

PT XYZ is a private textile company in Indonesia that was founded in 1966. It has two main factories and several subsidiaries that are spread and integrated in a group. Products from PT XYZ are in the form of spools with various diameter sizes. The product is made in a department called the Spinning Department. The yarn spools are made from raw cotton balls in the form of blocks that come from within and outside the country. Usually the products from this department are sold in rolls or can be processed by PT. XYZ. The processed products produced are in the form of various kinds of clothing, such as government uniforms and army uniforms of various countries in Europe.

In the production process, workers are in direct contact with heavy machinery. The machine rotates at high speed and there are processes that are done manually so that it has the potential to cause work accidents. In the manufacture of yarn spools, a fairly long process is needed and is directly related to the machine. The process requires regular monitoring of the machine and needs to be cleaned while the machine is operating. At the time of cleaning, workers are required to use sticks as personal protective equipment to pick up waste cotton, but there are still workers who ignore this and take it directly by hand. This is certainly very dangerous for workers because workers' fingers can be run over by the machine, such as in the case of an accident in the dust filter where the worker's finger was broken because it was run over by a machine. The second stage of the yarn winding process uses a speed frame machine, in which the yarn is rolled in a larger cops size. The machine winds the yarn at a high rotational speed. During operation the machine must also be supervised to ensure that no threads are entangled, not in position and to keep the thread spool free from waste. The work is supposed to be done when the engine stops spinning, but there are some workers who do the job without turning off the engine first. The machine should also be protected by a more closed cover to make it safer from the reach of the hands of workers who accidentally cross the machine's safe limit.

Judging from the production process, PT. XYZ will not be separated from the risk of accidents due to work. With the number of employees reaching 16,000 employees, the risk of work accidents can occur at any time when workers are doing their jobs. Field conditions show that there are still work accidents that occur at least one case every year in the factory area, especially those that occur in the carding machine, speed frame, ring frame, and dust filter. Based on an interview with the K3 section of the Spinning Department, the majority of workers in the carding department often lack concentration, especially employees who work on the night shift. Workers also do not use PPE when picking up waste so that it can cause their fingers to break off from being crushed by the machine. Workers do not use a stick when cleaning the rollers in the ring frame so that the fingers can be caught by the machine which rotates quite quickly and must lose the knuckles. In speed frame and ring spinning work stations that have noise levels up to more than 85 Db, workers also do not use PPE. With a room temperature above 30 degrees Celsius because the engine rotates quite fast and the micro cotton flying around, workers don't wear masks and earplugs. This can cause hearing problems to deafness and can cause respiratory problems due to the hot air and the accumulation of micro-cottons that enter the workers' lungs if left for a long time. Training on work accidents is only carried out once at the time of acceptance or training of new workers. What is often conveyed to workers when supervisors are around is only an appeal to always be careful when working and use the recommended PPE. Spinning Department has the highest number of accidents when compared to other departments. Judging from the high accidents at PT. XYZ, it is necessary to analyze the hazards contained in the Spinning Department of PT. XYZ and provide suggestions for improvement. This needs to be done in the Spinning Department because in that department workers are directly involved with large machines and high concentration is required when working. After repairs have been made and have produced positive results, improvements can be made in the entire spinning Department with the Spinning X Department as an example because all machines in the Spinning Department have the same number and type of machines and the same type of accident, namely fingers pinched on the machine even up to broken. In carrying out repairs, it is necessary to identify potential hazards first which is carried out using the HAZOP method.

HAZOP (Hazard and Operability) is a technique to identify potential occupational hazards by defining possible hazards, evaluating risks that occur through risk assessment using a risk assessment matrix. In addition, the HAZOP method also studies the flow of the machine operation that will be studied by interviewing field workers who work on the machine, it is hoped that more detailed potential hazards can be found in order to minimize and prevent accidents. This method is part of risk management and can determine the direction of the implementation of K3 in the company [1]. Several studies using HAZOP were carried out in different operating areas such as boilers [2], safety glass production [3], paper industry [4], garment companies [5], gas industry [6-7] and fertilizer factories [8]. The theoretical form of criticism of HAZOP is also discussed by [9-17].

Machines in the garment industry, especially the Spinning Department, have potential hazards and risks to workers in them. As a machine that requires regular direct supervision at the time of its operation, it causes the machine to be hazardous. This causes the need for a risk assessment to identify potential sources of danger and prevent the risk of accidents, even the loss of a person's life. So first, the hazard identification in the system is carried out using the Hazard and Operatibility Study (HAZOP) analysis. Through HAZOP, each part of the process can be tested to determine the possibility of deviations from the predetermined design state, and understand the causes and consequences. In addition, the HAZOP method was chosen because it has advantages, among others, it can be applied to all sequences of operations and can identify exactly what critical deviations occur and their causes. Hazard and Operability Study, otherwise known as HAZOP analysis is a standard technique used in the preparation of security establishments in new or modified systems against potential hazards or problems. Operate by identifying and evaluating hazards in planned or existing processes and operating in the most effective, economical and timely manner when all relevant considerations and constraints are considered. HAZOP can be used simultaneously in the hazard identification process and also in continuous operating systems. The goals of this HAZOP are to identify risks associated with system operation and maintenance. It aims at identifying at potential operability problems and causes of operational disruptions as well as possible deviations in the product that lead to product non-conformance as well. Other methods such as checklists, Fault Tree Analysis, Event Tree analysis, Failure Mode Effect Analysis, etc. provide detailed hazard risk assessments for machinery, product development and events but do not provide solutions to reduce hazard risks in work systems.

Based on the problem formulation that has been explained, the objectives to be achieved in the implementation of this research are to identify the potential hazards in production activities in the spinning department, analyze the risk control of work accidents, especially in activities that have a high level of risk and provide recommendations in handling the highest risks so as to minimize the risk. accident at PT. XYZ.

2. Methods

In this study, a risk assessment of the production floor in the Spinning Department will be carried out. The assessment was carried out using the Hazard and Operability Study (HAZOP) method. The data needed in this study are primary data and secondary data. Primary data consists of data on employee instructions, production flow, and the physical work environment. Secondary data obtained from the company include SOP, work accident data, and input from the company. Primary data can be obtained by means of direct interviews and field observations. In this study, the authors conducted interviews and discussions with Spinning Supervisor as many as one person, male field workers in Spinning as many as seven people (raw material, carding, drawing breaker, drawing finisher, speed frame, ring spinning, and winding, 1 person supervisor), and the head of production as much as one person who has worked for four years at PT. XYZ. The interviews conducted were unstructured interviews. The participatory method is also used in this study, especially in making decisions that involve the role of stakeholders in the company. The company stakeholders participating in the research are operators, supervisors, and managers. The roles of each stakeholder are:

- Operator: as a direct or main source of interviews, fill out the hazop worksheet and provide further explanations about the machine.
- Supervisor: as a discussion mentor for field conditions, validating the results of worker interviews, and providing likelihood and severity scores.
- Manager: as a discussion mentor for field conditions and providing recommendations for improvement, because the manager knows the resources he has in the department he manages.

The data that has been collected will be processed so that a descriptive result will be obtained. The analysis aims to determine the ranking of jobs that have the highest to lowest risk so that they can arrange the necessary improvements. Preparation of improvement methods using a control hierarchy involving the company. In preparing the recommendations, the company is asked to provide input on whether the proposed improvements can be implemented or need to be reviewed.
The research materials used in this study are as follows:

- Open interviews to collect data in the form of machine performance, risks that may occur, physical conditions in the field such as noise and room temperature.
- HAZOP Worksheet to identify risks and conduct hazard assessments.
- UNSW Health and Safety 2016 as a reference in providing a risk assessment.

Data processing is carried out based on data that has been previously collected in the HAZOP study and refers to the International IEC Standard 61882. The HAZOP procedure has 4 stages, namely definition, preparation, examination, as well as documentation and follow-up. The implementation of the HAZOP method starts from understanding the sequence of the production process, identifying potential hazards, completing the criteria in the HAZOP worksheet, ranking potentials by considering likelihood and consequences and making a risk matrix.

3. Results and Discussion

To identify any potential hazards contained in the production process of PT. XYZ, it is necessary to know the flow of the production process. The yarn production process of PT.XYZ Spinning Department is as follows: (1) Selection of raw materials to be used in the production process; (2) Mixing raw materials (from the same vendor) to reduce the unevenness of the yarn produced; (3) Blowing to distribute the raw material to the next process; (4) Carding to align the fibers to be made into yarn; (5) Drawing Breaker to align and add duplicates to the sliver; (6) Drawing Finisher to produce more even and parallel sliver; (7) Speed Frame to twist the sliver and result roving; (8) Ring Spinning to give the roving a twist until the yarn size (Ne) is in accordance with the desired yarn number; (9) Winding to roll the yarn in a cone and (10) Packing to pack the cones produced into cardboard and then stored in the finished product storage room. After that, direct field observations and interviews with trusted sources were carried out to obtain hazard findings. After that, ranking is carried out by taking into account the criteria for severity or risk ranking as follows:

- Likelihood (L) is the probability of an accident (table 1).
- Severity or Consequences (C) is the severity of injuries and lost working days (table 2)

Table 1. Likelihood Criteria.

Level	Likelihood Rate	Information
5	Almost Sure	More than 1 time per month
4	Most likely	More than 1 time per year to 1 time per month
3	Possible	1 time per 5 years to 1 time per year
2	Small chance	Happens once per 10 years
1	Seldom	Less than 1 time per 10 years

Table 2. Criteria Consequences.

Level	Likelihood Rate	Information
1	Not significant	Injury does not require first aid
2	Small	First aid needed
3	Medium	Medical treatment required
4	Heavy	Need to go to hospital
5	Disaster	Death, permanent disability

After determining the likelihood and consequences of each potential hazard, the next step is to multiply the likelihood and consequences values so that the risk level is obtained in the risk matrix which will be used in ranking the potential hazard sources that will be used as a reference. as a recommendation for improvement in accordance with the existing problems. The risk assessment itself is carried out using a risk matrix as shown in Table 3. From the risk matrix, it can then be calculated risk scores and priorities for taking corrective actions. Based on analysis and processing data, following results which could taken:

- High risk, namely in some work areas, including on *carding machines*, *ring spinning*, *speedframes*.
- Risk medium occur on two area work that ison draw *frame* and *winding*.
- And for risk low occur on machine *blowing*.

Table 3. Risk Matrix (UNSW Health and Safety, 2016).

Scale		Consequences				
		1	2	3	4	5
Likelihood	5	5	10	15	20	25
	4	4	8	12	16	20
	3	3	6	9	12	15
	2	2	4	6	8	10
	1	1	2	3	4	5

According to the risk matrix guidelines of PT. XYZ, the limit of the acceptable area is at the moderate level and the risk in the HAZOP result is the highest risk is extreme. In the attachment, it is known that there are several nodes that have the highest risk, the first is the Carding machine, the second is the Speed frame machine, and the third is the Ring Spinning machine. Based on the ranking, the above three machines will be prioritized for repair. As for the other machines, repairs will still be made but not as prioritized as the three machines above. This data does not directly provide an overview to PT. XYZ to pay attention to the operation process and to mitigate especially the risks that are of extreme value, namely Carding machines, Speed Frames, and Ring Spinning because they have the potential to harm workers if left unchecked. In accordance with the risk matrix guidelines of PT. XYZ risks with extreme to moderate levels of short-term risk mitigation plans must be approved by the line manager/equivalent.

Based on the results of the risk assessment, the highest deviation was caused by human factors with a percentage of 69%. These human factors are caused by the workers themselves who do not comply with regulations or lack of awareness of occupational safety and health. Furthermore, the second highest deviation was caused by miscellaneous and high temperature (28-30 Celcius degree) with a percentage of 12.5% and it causes a decrease in the level of worker and results in production defects to work accidents. Previous research on the spinning department has also been conducted at PT. Apac Core Corpora. Based on this research, the highest risk is based on the work environment and employee attitudes, while the lowest risk comes from OHC. The work environment and worker attitudes lead to the human factor. This is directly proportional to the research conducted by PT. XYZ, where the biggest risk occurs in the carding machine, namely pinched hands because workers do not use PPE in the form of sticks. It can be concluded that the main cause of extreme risk in the textile industry, both at PT. XYZ and PT. Apac Inti Corpora originates from the same thing, namely the human factor [18].

Based on the inspection process, the potential hazards that have been summarized in the HAZOP worksheet have their respective risk levels. This risk level will be translated into risk categories according to the risk matrix guidelines. Based on these guidelines, the classification of risk levels in the HAZOP study as follows: low (2 risks), medium (5 risks), high (8 risks) and extreme (2 risks). Each of these levels requires different actions. At a "low" level, it means that the area is still generally accepted where risk management is continuously embedded in the company's operations and there are no recorded incidents. Risk mitigation that is made must obtain approval from the highest field leader/supervisor/site controller. The "medium" risk level indicates the area is tolerable where short-term risk mitigation must be made appropriately with the approval of the line/equivalent manager to manage the risk. A "high" risk level indicates the danger is intolerable. A series of long-term and short-term risk mitigation plans must be carried out immediately to reduce risks with certain requirements, namely postponing work until the risk mitigation plan is carried out and the risk becomes medium, the short-term risk program must be approved by the HSE and the long-term mitigation plan is implemented and monitored. The "extreme" risk level is almost the same as the high risk level, except that continuing work and short-term risk program plans must be approved by HSE.

Recommendations are generated from evaluations related to deviations and causes that have the potential to occur. The focus of the recommendations is to complete safeguards in the form of instruments, operating procedures, mechanical tools, and static equipment. This recommendation is

IOP Conf. Series: Earth and Environmental Science 1098 (2022) 012006 doi:10.1088/1755-1315/1098/1/012006

expected to help the risk management program, especially in the carding area. It is known beforehand that there is a risk that has a high value at the extreme risk level with a value of 16. This needs to be given attention because if there is no proper mitigation or action, it can cause more victims. The list of recommendations made is not only related to the installation of safety instruments, but also preventive actions on certain tools or equipment. Recommendations for improvement are made based on the control hierarchy starting from elimination to the use of personal protective equipment. The next stage is filling out the HAZOP worksheet which contains information on deviation, causes, consequences, risk level, and recommendation (Table 4).

4. Conclusion and further study

The HAZOP study conducted in the Spinning Department area resulted in the fact that the sources of hazards and risks came from human factors and equipment or machines where the main focus of improvement was based on the top 3 hazard rankings according to the results of the risk assessment. The equipment is Carding, Ring Spinning, and Speed frame. The highest risk level found in the HAZOP study falls into the high category with the highest risk, namely the Carding machine and short-term risk mitigation must be approved by the line manager/equivalent. The priority recommendations for repairs on the Carding machine are providing a protective cover for the cotton roller gear, making a double safety lock on the machine cover, and making a bypass for the emergency button.

Some of the further study material that can be done is to revalidate HAZOP with a period of 2 to 4 years to update the potential hazards and risks that exist in production activities for the Carding machine area and in general for all facilities in the Spinning Department, follow up on mitigation plans risks, both short and long term (such as installation of engineering controls and preparation of SOPs) based on the list of recommendations that have been attached, taking into account company factors (costs, resources, etc.) in the research.

5. Reference

[1] Ramli S 2010. Pedoman Praktis Manajemen Risiko Dalam Perspektif K3 OHS Risk Management (Dian Agung: Jakarta)
[2] Euis J, Yadi Y H, Umyati A 2016 Jurnal Teknik Industri 35 3
[3] Restuputri D P dan Sari R P D 2015 Jurnal Ilmial Teknik Industri 14 1 Juni 2015 p 24-35
[4] Pujiono B N, Tama I P, Efranto R Y 2013 Jurnal Rekayasa dan Manajemen Industri p 253-264
[5] Ratri M and Widharto Y 2019. Accesed online: July 18th, 2022 https://ejournal3.undip.ac.id/index.php/ieoj/article/viewFile/23057/21071
[6] Alaei R, Mansoori S A A, Moghaddam A H, Mansoori S M, Mansoori N 2014 Journal of Natural Gas Science and Engineering 20 p 271-284
[7] Shafaghi, Ahmad and F. Bert Cook F B1988 IEEE Transactions on Relability 37 2
[8] Muthukumar K and Mohan M S 2018 IJRTE 7 4S
[9] Kotek L and Tabas M 2012 Procedia Engineering 42 4 p 808815
[10] Dunjo J, Fthenakis V, Vilchez J A, Arnaldos J 2009 Hazardous Materials 173 1 p 19 – 32
[11] Johnson R W 2010 Journal of Loss Prevention in the Process Industries 23 p 727-733
[12] Wang F, Gao J J, and Wang H 2012 Journal of Loss Prevention in the Process Industries 25 p 636-642
[13] Baybutt P 2015 Journal of Loss Prevention in the Pricess Industries 33 p 52-58
[14] Giardina M and Morale M 2015 Journal of Loss Prevention in the Process Industries 35 p 35-45
[15] Fattor M V and Vieira M G A 2019 Journal of Environmental Management 246 p 247-258
[16] Duisings L P M, Til S, Magielsen A J, Ronden D M S, Elzendoorn B S Q, Heemskerk C JM 2013 Fusion Engineering and Design 88 p 2688-2693
[17] Wieckol-Ryk, Krzemien A, Zawartka P, Glodniok M 2019 Process safety progress 39 1
[18] Sabrina M R W and Widharto Y 2016 Analisis Potensi Bahaya dengan Metode Hazard and Operability Study melalui Perangkingan Risk Assessment Studi Kasus: Divisi Spinning Unit 4 Ring Yarn PT APAC INTI CORPORA (Departemen Teknik Industri, Fakultas Teknik, Universitas Diponegoro: Semarang)

Table 4. HAZOP Worksheet

No	Node	Guideword	Deviation	Causes	Consequences	Safeguard	L	C	RR
1	Raw material	HumanFactors	Humanfactors	Moveraw materials heavy with tool manual	fatigue, injury muscle	No	4	2	8 (high)
			Humanfactors	Stack raw material which tall	hit by rawmaterial	No	3	2	6 (medium)
			Humanfactors	Workers don't usePPE: shoe	unprotected in accident	No	3	3	9 (high)
2	Mixing	High	Human factors	Worker does not use PPE: mask	congested breath	Mask	3	2	6 (medium)
			High temperature	Air stuffy	Decrease concentration	No	5	2	10 (high)
			Humanfactors	Cleanwaste cotton without use PPE: shoe and mask	slipped andcongested breath	Mask	2	3	6 (medium)
3	Blowing	-	-	Automatic	-	-			
4	Carding	Human Factors	Human factors	Taking waste in the machine without using PPE: stick	Machine run over	Sticks (PPE)	4	4	16 (extreme)
				Workers don't use PPE: mask	Cause shortness of breath	Mask	3	3	9 (high)
				Night shift workers lack of concentration	Lack of focus increases risk accident	Arrangement of lighting	3	3	9 (high)
5	Drawing	Human Factors	Human factors	Workers don't use PPE: mask	Cause shortness of breath	Mask	3	2	6 (medium)
6	Speedframes	More	More speed	The engine speed is fast and sometimes the cover doesn't close	Fingers or hands can get caught in the machine	Protective cover	3	4	12 (extreme)
			Miscellaneous	Noisy work environment	Hearing diminished and lowered working comfort	The engine area has been given a silencer	5	2	10 (high)
			Human factors	Workers not wearing PPE: masks and earplugs	Impaired breathing and hearing	Mask	3	2	6 (medium)

7	Ring spinning	High	Human factors	Workers not wearing PPE: masks and earplugs	Impaired breathing and hearing	Mask	2	2	4 (low)
			High temperature	Stuffy work environment	Concentration less workers	No	5	2	10 (high)
			Miscellaneous	engine sound noisy	Convenience interrupted work	Machine area has been dampened	5	2	10 (high)
8	Winding	Human Factors	Human factors	Workers do not use PPE: mask	Breathing can disturbed	The company provides mask	2	2	4 (low)

Integrating fuzzy logic and house risk method for operational risk management in automotive industry

Nia Budi Puspitasari[1], Muhammad Raihananda Ashafy Yuwono[1]

[1]Department of Industrial Engineering, Faculty of Engineering, Diponegoro University, Tembalang 50275, Semarang, Central Java, Indonesia

niabudipuspitasari@lecturer.undip.ac.id

Abstract. Risk management is very important for all organizations and nowadays companies need to be aware of the risks that endanger their operational activities. The object of this research is a motorcycle component manufacturing company that has not yet fully identified its operational risks and designed risk management even though these risks affect the company's performance. Therefore, the purpose of this study is to identify operational risks and design preventive actions. This study combines fuzzy logic with the house of risk (HOR) model which consists of two phases. The first phase determines the dominant risk agents and the second phase determines the effective actions to deal with them. From the results, there are 22 risk events and 28 risk agents with fifteen dominant risk agents such as human error, data misinformation, and others. Twelve preventive actions were also obtained, such as improving communication both internally and externally and others.

1. Introduction

In the current era of globalization, companies are trying to increase their competitive selling power to be able to compete with a focus on supply chain management. The competitive climate of the business world has shifted from competing between companies to competing between supply chains. If supply chain operations consistently run smoothly, the company can be said to be a good company [1].

Risk is the possibility of the realization of unintended consequences with undesirable outcomes such as harm or loss [2]. Risk management in every process is a challenge for companies today, where every company activity can't be separated from the uncertainty that can affect the flow of components in the supply chain [3]. This is in line with the definition of operational risk which is a risk that arises due to malfunctioning of internal processes, human error or fraud, system failures, and external factors [4].

The object of this research is a manufacturing company that produces motorcycle components in Karawang, West Java, Indonesia. Some components have a demand that reaches 20,000 pieces per month. With a large number of requests, the activities in the company's supply chain operational activities become increasingly complex so uncertainty can arise in these activities.

Currently, the company does not have structured risk management to cope with the risks that may occur in the company's operations. These risks prevent the achievement of the planned production targets and can result in delays in product delivery to customers. This can be seen from the percentage of sales of the two products with the highest demand, which from May to August 2021 was only 70 percent on average. This means that there is a large difference between the sum of orders and the sum of delivery which indicates that the company has not carried out good internal operational control, even though the risks that have occurred are inseparable from organizational processes.

In light of the assertions above, it is important to distinguish chances and their causes completely in the organization's activities and foster techniques so these dangers can be forestalled. This study utilizes House of Risk (HOR) model as an answer for the issues that were found in the examination object. The HOR model is an improvement of The Failure Mode and Effect Analysis (FMEA) model and the house of risk (HOQ) model by Pujawan and Geraldin [5] as a proactive gamble the board model. The HOR model spotlights on planning systems for the counteraction, decrease, and treatment of a few gamble specialists that can possibly cause more than one gamble. This model is utilized to focus on which risk specialist ought to be addressed first and to choose the best activity to diminish the gamble that the gamble specialist might actually present [5]. A few examinations have utilized this model to break down the dangers intrinsic in a paper printing organization [6] and a dairy-based item organization [7]. It is likewise used to deal with the store network dangers of a piece elastic organization [8] and a little family electrical machines organization [9]. Some additionally join the HOR model with different techniques, for example, the Analytical Network Process (ANP) to examine production network risk in the oil palm industry [10] and the wooden toy industry [11].

Fuzzy logic is additionally generally utilized in past exploration on risk the board. Ganguly and Kumar [12] and Salehi et al. [13] in their review consolidated fuzzy logic with Analytical Hierarchy Process (AHP) in store network risk the executives. Mangla et al. [14] and Immawan et al. [15] additionally join Fuzzy logic with Failure Mode and Effect Analysis (FMEA) for use in risk appraisal. Fuzzy logic in this study assumes a part in handling the seriousness, event, and relationship values to compute the aggregate risk potential (ARP) worth of a gamble specialist so it very well may be focused on. In past examinations, customarily, risk occasion seriousness and chance specialist event rates utilized a ten-point scale framework, with 10 addressing the most extreme for the seriousness level and the most regular for the event level [5]. Nonetheless, setting the seriousness esteem is an emotional subjective choice so a Fuzzy articulation is proposed for its portrayal in handling the ARP worth to decide the prevailing gamble specialist [8].

This study intends to plan functional gamble the board on the object of examination. Functional dangers that have happened and may happen and the gamble specialists that cause them are distinguished and surveyed involving fuzzy logic in the HOR model to decide the need of hazard specialists that should be dealt with. The consequences of the need of these gamble specialists are utilized as a kind of perspective for planning activities to forestall or diminish these gamble specialists. These activities are then evaluated and focused on to realize which activities the organization needs to take first.

2. Methodology

This exploration was led at an assembling organization participated in the auto area. In this review, there are two kinds of information utilized, in particular essential information and auxiliary information. Essential information was acquired through perception, direct meetings with parties straightforwardly engaged with the organization's activities, and finishing up polls. There are two kinds of polls as per the two periods of the HOR model. Poll A will be a survey to evaluate the degree of seriousness of the gamble occasion, the degree of event of the gamble specialist, and the degree of relationship between each chance occasion and the gamble specialist. Poll B is a survey that evaluates the level of trouble of every preventive activity and the degree of relationship between preventive activities and focused on risk specialists. The organizations that are utilized as respondents in these two polls are specialists in the field of organization functional exercises, adding up to three individuals who have worked in the organization for over five years. Optional information is gotten from reports and information from organizations connected with the organization's functional exercises, for example, material obtainment information, creation reports, completed item conveyance information, and others.

The plan of functional gamble the board in this study utilizes the HOR strategy. As a rule, HOR has two stages, where the main stage is to focus on risk specialists as indicated by their importance and afterward fills in as the reason for the second stage to decide proactive activities. Then, at that point, the initial step is to distinguish the gamble of the occasion and the gamble specialist. The distinguishing proof cycle is helped by the Supply Chain Operation Reference (SCOR) model as in the concentrate by

Pujawan and Geraldin [5], Ratnasari et al. [6], Ma and Wong [8], and Jiang et al. [16]. Then information handling starts by working out the Fuzzy Aggregate Risk Potential (FARP) worth of each hazard specialist through a Fuzzy cycle. The contribution of the Fuzzy interaction is an evaluation of the seriousness, event, and connection factors by research respondents from poll A. The request for need of hazard specialists as indicated by the FARP esteem is the result of the main period of HOR. For the subsequent stage, the worth of the Effectiveness to Difficulty Ratio (ETD) for every preventive activity is determined. The info is acquired from the worth of the level of trouble of each activity and the degree of connection with the gamble specialist in survey B. The request for need of preventive activities as indicated by the ETD esteem is the result.

3. Results and discussion

3.1. House of risk phase I

3.1.1. Identification and assessment of risk events, risk agents, and correlations

Table 1 shows the identified risk events and the appropriate severity level obtained in the aspects of the plan, source, make, delivery, and return. Each process is further explored to identify risk events in each sub-process. The respondent's assessment uses a scale of one to ten adapted from Shahin's 2004 study [17] where a score of one means almost no impact if the associated risk event occurs while a score of ten means a hazardous impact.

Table 1. Summary of Risk Events and Severity

Process	Sub-process	Risk events	Severity	Code
Plan	Production planning	A sudden change in production planning	Five	E1
	Supply chain alignment with financial planning	Budget planning for the fulfillment of raw materials is not accurate	Six	E2
	Inventory control for raw materials	The amount of material in the system with the amount of material in actual conditions does not match	Five	E3
Source	Contract agreement with the supplier	Disagreements in the payment method	Six	E4
		The desired material specifications with those on the market do not match	Six	E5
	Procurement process	Delivery of raw materials is not on time	Eight	E6
		Quality and material specifications from suppliers are not up to standard	Seven	E7
		The wrong amount of material was sent by the supplier	Six	E8
		Have not implemented supplier performance evaluation procedures	Six	E9
	Availability of production tools	Insufficient production tools	Five	E10
	Storage of goods	Limited material storage area	Four	E11
Make	Process of production activities	Damage to machinery or equipment	Seven	E12
		Set-up time is too long	Six	E13
	Production control	Changes in the product design when the production process has started	Seven	E14
		Delay in the production process	Seven	E15
		Work accident occurs	Six	E16
	Quality testing	The discrepancy between the drawing and the product that has been made (incorrect size/dimensions)	Seven	E17
Delivery	Packaging process	The wrong amount of product packaging	Four	E18

Process	Sub-process	Risk events	Severity	Code
	Finished product delivery process	Late delivery of products to customers	Five	E19
		Product damage occurred during delivery	Five	E20
Return	Handling returns to suppliers	Delay in the process of returning rejected raw materials to the supplier	Four	E21
	Handling product returns from customers	Delays in handling product returns from customers	Three	E22

The risk agent is considered the probable cause of the above-identified risk events, and it is assumed that one risk agent can cause more than one risk event and one event can be caused by more than one agent. Risk agents are identified along with their occurrence levels which are summarized in Table 2. Occurrence assessment uses a scale of one to ten where a score of one means that it rarely occurs and a score of ten means that it is almost certain to occur [15]. The correlations between risk events and risk agents are summarized in Table 3. Values of zero, one, three, or nine are used with the meanings respectively being no correlation, low correlation, moderate correlation, and high correlation [6].

Table 2. Identification of Risk Agent and Occurrence

Risk Agent	Occurrence	Code
Late arrival of materials	Five	A1
Sudden product demand from customers	Three	A2
Complete another project before the due date	Five	A3
There is a product design update from the previous design	Two	A4
Inaccurate price reference	Four	A5
Data not updated immediately	Five	A6
Human error	Six	A7
The difference between the supplier's desire and the company's capabilities	Four	A8
Limited raw materials in the market	Three	A9
No monitoring of supplier	Four	A10
Depends on only one supplier	Five	A11
Delay in making a purchasing requisition (PR)	Five	A12
Long delivery distance	Five	A13
There was a natural disaster or an accident during delivery	Three	A14
The supplier sends the material, not by the PR	Four	A15
Low material quality by suppliers in meeting demand	Three	A16
Wrong data information	Seven	A17
There is no supplier performance evaluation procedure	Six	A18
Limited company capital	Four	A19
Limited area for storage	Four	A20
Lack of regular checks on machines or equipment	Three	A21
Limited machines that can be used	Six	A22
There is a decrease in engine performance or engine damage	Three	A23
Disrupted power supply	Four	A24
Limited transportation for delivery	Three	A25
There is a problem during the product delivery process	Two	A26
The supplier's response time on the return of raw materials is too long	Three	A27
Lack of communication between the company and the customer	Four	A28

Table 3. Correlation Between Risk Events and Risk Agents

		Risk Agent																											
Proses	Risk Event	A1	A2	A3	A4	A5	A6	A7	A8	A9	A10	A11	A12	A13	A14	A15	A16	A17	A18	A19	A20	A21	A22	A23	A24	A25	A26	A27	A28
Plan	E1	9	3	3	1																								
	E2					9																							
	E3						9	3																					
Source	E4								3																				
	E5									3																			
	E6		1								9	9	3	9	9														
	E7															9	9												
	E8										3							3											
	E9																		9										
	E10																			3									
	E11																				9								
Make	E12																					9							
	E13																						3						
	E14				9																								
	E15	9																						3	9				
	E16							9														3							
	E17							9																3					
Delivery	E18							3										9											
	E19														9			3								9			
	E20																										9		
Return	E21																											9	
	E22																												9

3.1.2 Classifications of the Highest Risk Agents

After the defuzzification stage, fuzzy ARP (FARP) values can be obtained for each risk agent and prioritized from the largest to the smallest, as shown in Figure 1. It can be seen that there are 14 risk agents which together contribute 70 percent of the total FARP. As such, it is considered a high-risk agent (A7, A17, A1, A10, A14, A21, A23, A18, A2, A6, A11, A13, A12, A22) and requires special attention from the company. Among these risk agents, A7, the "human error" risk agent, is the most significant with a FARP value of 2.002. This result is following previous studies contained in the paper [18]. The paper states that human error is a risk agent with the highest score.

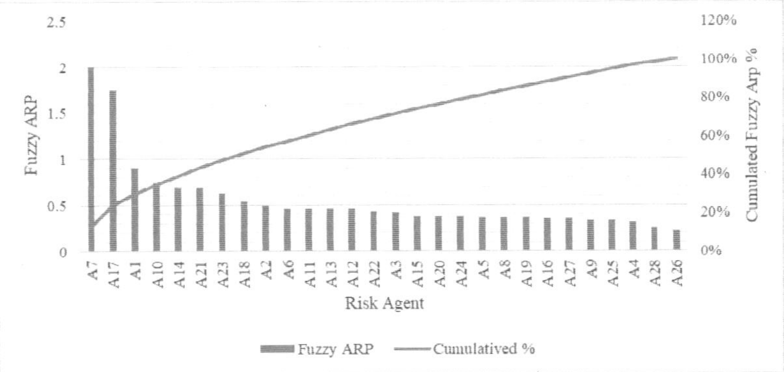

Figure 1. Pareto Diagram of Fuzzy Aggregate Risk Potentials of All Risk Agents

A7 correlates with many risk events in many different processes namely E3 and E16-E18 with different correlation values. A17 was rated as the second high-risk agent, associated with "data misinformation" where this finding is consistent with the discussion of Ma and Wong [8] followed by A1, which is associated with "late arrival of materials".

3.2 House of Risk Phase 2

3.2.1 Identification and Prioritizing Preventive Actions

The Pareto chart above shows that the level of importance of reducing the potential occurrence of each risk agent is different. Naturally, companies must prioritize risk agents that have a large aggregate risk potential. The second stage of the HOR model is useful for identifying and prioritizing mitigation actions that companies must take to maximize the effectiveness of efforts with existing financial commitments and acceptable resources [5]. Of the 14 risk agents that are prioritized as a basis, there are 12 proposed mitigation actions as described in Table 4.

Table 4. Proposed Mitigation Actions for Prioritized Risk Agents

Code	*Preventive Action*
PA1	Giving reward, punishment, and work motivation to all employees
PA2	Added training routines for tool and machine operators
PA3	Improve communication both internally and externally
PA4	Creating an integrated information system
PA5	Information sharing between departments via email notification
PA6	Conduct supplier performance evaluation
PA7	Collaborate with other suppliers
PA8	Mapping the characteristics of the procurement of raw materials (crucial level)
PA9	Reviewing safety stock materials with repetitive needs

Code	*Preventive Action*
PA10	Monitoring the delivery process
PA11	Carry out regular machine checks
PA12	Improve the quality of machine maintenance

The assessment of the difficulty of performing each action (Dk) follows the traditional HOR model which is classified into three categories: low with a score of three, moderate with a score of four, and high with a score of five. The level of difficulty should also reflect the money and other resources required to act. Therefore, the ratio will show the cost-effectiveness of each action.

The order of priority for each action is obtained based on the value of the ratio of effectiveness to the difficulty of action k (ETDk). The higher the ratio, the more cost-effective the proposed action is. From Table 5 it can be seen that the most cost-effective action is to improve communication both internally and externally followed by information sharing between departments via email notifications and so on.

Table 5. HOR Phase 2 of the Case Company

	Preventive Action												
Risk Agent	PA1	PA2	PA3	PA4	PA5	PA6	PA7	PA8	PA9	PA10	PA11	PA12	FARP
A7	9	3											2.00
A17			9	3	9								1.47
A1			9			3	9	9	9				0.91
A10			9		3								0.74
A14										9			0.69
A21		9									9		0.69
A23											9	9	0.64
A18						9							0.54
A2			9		9								0.50
A6			9	9									0.46
A11							9						0.46
A13			3							9			0.46
A22			9								3		0.43
A12			9	3									0.46
Tek	18.02	12.18	46.08	9.91	19.90	7.60	12.36	8.22	8.22	10.31	13.19	5.72	
Dk	4	4	4	5	3	3	4	3	4	3	3	4	
ETDk	4.505	3.045	11.521	1.982	6.634	2.533	3.089	2.739	2.054	3.438	4.397	1.431	
Rank	3	7	1	11	2	9	6	8	10	5	4	12	

4. Conclusion

In this study, the HOR model with fuzzy logic integration is used to implement operational risk management on the research subject. In this method, risk agents are selected based on the total potential risk, starting with the highest value. The second stage of the HOR model prioritizes preventive actions based on the overall effectiveness-to-difficulty ratio.

While this model is easy to implement in practice, it requires extensive data collection and brainstorming within your organization. References to previous work in relevant industries are useful in the brainstorming process to help cross-functional teams identify risk events and risk agents, their severity and magnitude, and the correlation between each risk agent and each risk event. The input data for this survey was collected from internal managers as survey respondents. However, as Internet of Things and Industry 4.0 become more common and mature, input data can be collected from various advanced technologies such as cloud databases to increase efficiency.

This study uses the HOR model, which includes the identification of risks and the determination of mitigation strategies, but does not consider the dependencies between risk events and thus the potential risks arising from dependence on risk events. This dependence can be taken into account in future studies.

References

[1] Trkman P, McCormack K. Supply chain risk in a turbulent environments-A conceptual model for managing supply chain network risk. *International Journal of Production Economics*. 2009; Vol. 119, No. 2, pp.247-258.
[2] Schlegel GL, Trent RJ. Supply Chain Risk Management: An Emerging Discipline (1st ed.). CRC Press; 2015.
[3] Svensson L. Open economy inflation targetting. Journal of International Economics. 2000; 50(1): 155-3.
[4] Djohanputro B. Manajemen Risiko Korporat. Jakarta: PPM; 2008.
[5] Pujawan I.N, Geraldin LH. House of Risk: a model for proactive supply chain risk management. Supply Chain Management: Business Process Management Journal. 2009; Vol. 5, No. 6, pp. 953-967.
[6] Ratnasari S, Hisjam M, Sutopo W. Supply chain risk management in newspaper company: House of risk approach. AIP Conference Proceedings. 2018.
[7] Wahyudin, Nur E, Santoso I. Modelling of risk management for product development of yogurt drink using house of risk (HOR) method. Asian Journal of Technology Management. 2016; Vol. 9, No. 2, pp. 98-108,
[8] Immawan T, Putri D. House of risk approach for assessing supply chain risk management strategies: A case study in Crumb Rubber Company Ltd. MATEC Web of Conferences. 2018.
[9] Ma H-L, Wong W-HC. A fuzzy-based House of Risk assessment method for manufacturers in global supply chains. Industrial Management & Data Systems. 2018; Vol. 118, No. 7, pp. 1463-1476.
[10] Kurniawan S, Marzuky D, Ryanto R, Agustine V. Risk and supply chain mitigation analysis using house of risk method and analytical network process. The Winners. 2021; 22 (2), 123-136.
[11] Tanjung WN, Khodijah RS, Hidayat S, Ripmiatin E, Atikah SA, Asti, SS. Supply chain risk management on wooden toys industries by using house of risk (HOR) and analytical network process (ANP) method. IOP Conference Series: Materials Science and Engineering. 2019.
[12] Ganguly K, Kumar G. Supply chain risk assessment: A fuzzy AHP approach. Operations and Supply Chain Management: An International Journal. 2019; 12(1), 1-13.
[13] Salehi HS, Khanbabaei M, Sabzehparvar M. A model for supply chain risk management in the automotive industry using fuzzy analytic hierarchy process and fuzzy TOPSIS. Benchmarking: An International Journal. 2018; Vol. 25 No. 9, pp. 3831-3857.
[14] Mangla SK, Luthra S, Jakhar S. Benchmarking the risk assessment in green supply chain using fuzzy approach to FMEA: Insights from an Indian case study. Benchmarking: An International Journal. 2018; Vol. 25 No. 8, pp. 2660-2687.
[15] Immawan T, Sutrisno W, Rachman A. Operational risk analysis with fuzzy FMEA (Failure Mode and Effect Analysis) approach (Case study: Optimus Creative Bandung). MATEC Web of Conferences. 2018.
[16] Jiang B, Li J, Shen S. Supply chain risk assessment and control of port enterprises: Qingdao381 port as case study. The Asian Journal of Shipping and Logistics. 2018; Volume 34, Issue 3, Pages 198-208.
[17] Shahin A. Integration of FMEA and the Kano model: An exploratory examination. Emerald Group Publishing Limited; 2004.
[18] Astutik D.W, Santoso P.B, Sumantri Y. Risk Management Strategy In The Supply Chain Of Organic Fertilizer By Using Fuzzy Analytical Hierarchy Process (FAHP) (Case Study in PT Tiara Kurnia, Malang). Jurnal Rekayasa dan Manajemen Sistem Industri. 2015; VOL. 3 NO. 3.

Define indicators for health protocol assessment at Prambanan Temple Tourism Park Yogyakarta

Ratna Purwaningsih[1*], Linda Anggita Leony [2], Ekawati [3], Ary Arvianto [4], Parwa Oryzanti [5]

[1,2,4] Industrial Engineering Department, Engineering Faculty, Diponegoro University. Jl. Prof. Soedharto, SH, Tembalang, Semarang, Central Java, Indonesia, 50275
[3] Public Health Faculty, Diponegoro University, Jl. Prof. Soedharto, SH, Tembalang, Semarang, Central Java, Indonesia, 50275
[5] Research Center for Ecology and Ethnobiology, Research Organization for Life Sciences and Environment, National Research and Innovation Agency, Jl. Raya Jakarta Bogor Km. 46, Cibinong, Kabupaten Bogor 16911

[a)] Corresponding author: ratna.tiundip@gmail.com

Abstract, Health protocols to control the Covid-19 outbreak must be applied in tourist attractions to maintain the health and safety of visitors. Ministry of Tourism and Creative Economy of the Republic of Indonesia also develop the CHSE (Clean, Health, Safety, Environment) standard. This study aims to develop indicators for the assessment of the health protocol. Literature study of various regulations issued done to identify indicators and then validated by experts. The selection of indicators is carried out by a cut-off point method and tested by implementing the assessment of CHSE to Prambanan Temple park. The Health Protocol Assessment Indicators for tourist attractions consist of 9 criteria, 38 sub-criteria, and 136 Indicators. The focus of the indicator is on the availability of various facilities for visitors consisting of hygiene, health, convenience, magnificence facilities, and other facilities that support tourism sustainability.

Keywords: facilities, safety, health, environment, tourism

1. Introduction

The existence of the Covid-19 Virus, which is a global health crisis due to its highly contagious nature, is a disaster around the world [1]. World Health Organization (WHO) urge all countries to restrict the individual movement for tourist attraction and suspend public events and business activities [2]. Proactive control of the Covid-19 outbreak has harmed the tourism industry [3]. The United Nations World Tourism Organization (UNWTO) predicts that the decline in international tourist arrivals due to the Covid pandemic is 60–80% [4].

Tourism is a social, cultural, and economic phenomenon involving people's movement to a country or place outside their usual environment for personal or business/professional purposes [4]. The Oxford English Dictionary 1811 defines tourism as the travel &tourism industry that contributes significantly to the global Gross Domestic Product (GDP) and creates jobs in the tourism sector [5]. In 2019 the tourism sector's contribution to the GDP was 4.80%. This contribution has increased by 0.30 points compared to 2018, 4.50% (ministry of tourism). In 2020, the spread of the coronavirus caused panic, anxiety, public health crisis, and widespread economic and social disruption. According to [6], the pandemic caused enormous losses in the tourism and hospitality sector [7]. The pandemic created a sense of fear and uncertainty, reflected in equity markets [8], [9]. Researchers explore public sentiment using socially mediated crisis communication models [10]. The world is witnessing a sharp surge in work-from-home activity and widespread unemployment [11], [12]. In 2019, 1,461 million people traveled globally, while only 381 million people traveled globally in 2020, with changes plummeting by 74% and a negative decrease of 1080 million compared to 2019 due to the pandemic [13]. Foreign and domestic tourists who came to Yogyakarta experienced a significant decline starting from the comparison in 2019 to 2021.

Table 1. Foreign and Domestic Tourists in Yogyakarta

Tourists	2019	2020	2021
Domestic Tourists	3.999.982	1.316.733	418.546
Foreign Tourists	461.873	56.867	653

Prambanan Park is the most favorite tour in Yogyakarta. The number of visitors to Prambanan Park decreased significantly, around 80% [26]. The impact of COVID-19 was able to reduce tourist visits by up to 90%. The European Union (EU) has stated that all countries must follow the rules of health and safety protocol. The Ministry of Tourism and Creative Economy of the Republic of Indonesia declared the Implementation of Standards and Guidelines in the prevention of Covid-19 by implementing CHSE in the tourism sector. The implementation of CHSE is quite good in increasing tourist visits.

Table 2. Foreign and Domestic Tourists in Prambanan Park

Tourists	YTD 2019	YTD 2020	YTD 2021
Domestic Tourists	2.326.122	666.903	261.757
Foreign Tourists	183.533	21.425	377

Research on tourist perceptions and attitudes towards a tourist attraction finds that (1) landscape beauty, (2) hospitality and behavior of locals, (3) climate and weather, and (4) safety are essential things for visitors [14]. Safety is the most crucial thing that tourists consider in traveling. Therefore, to restore tourist interest during this pandemic, efforts to ensure the visitors are safe are needed, including implementing health protocols.

Preparing standard indicators for health protocol facilities at tourist attractions is needed because the COVID-19 virus continues to be around us like the flu. As the number of infections increases, new variants will always appear. Although there are decreases in the death rate of COVID-19 from 1-2% to 0.25%, this death rate is still 2.5 times higher than the death rate of flu. Prevention and control measures must be done by the tourism management

After tourism object managers have sought various facilities to support the implementation of health protocols, an assessment standard is needed to assess how ready a tourist object to re-open. The Ministry of Tourism and Creative Economy Republic of Indonesia has issued regulations Number 13 of 2020 as a regulation for the Covid-19 Prevention and Control Protocol in the Tourism Sector. However, the indicators used are still lack of detail and need clarification by conducting a literature study. Assessment indicators focus only on management/governance, human resource readiness, and the involvement of visitors. The rules have not yet discussed the facilities and infrastructure for facilities and services at tourist sites. Identification of indicators conducted with literature studies and field observations. The literature study was carried out using references to the tourism theory (James J. Spillane, 1994 [15]) and regards health protocol standards from the Decree of the Minister of Health at year 2020, containing health protocols for the public in public places and facilities and decree of the Minister of Health, concerning Guidelines for Prevention and Control of Corona Virus Disease 2019 (Covid-19) in Office and Industrial Workplaces in supporting business continuity in a pandemic situation. Field observations were conducted at Prambanan Temple Park by seeing what facilities should be provided at a tourist location and conducting an interview with the management of Prambanan temple park.

Facilities are physical resources that must exist for consumer service [16]. Facilities are all physical equipment provided by service providers to support the convenience of tourists [17]. This study aims to (1) formulate standards for health protocol facilities and (2) design a method to assess the readiness of a tourist destination to implement health protocols, and (3) conduct a trial by conducting an assessment in Prambanan Temple Park. The research output is a standard of health protocol facility for tourist sites and a method for assessing the readiness of health protocol facilities to decide on whether the tourist location is suitable for open. If the health protocol facilities of a tourist location have not met the standards, then the next step are to identifying what facilities that must be completed for the health and safety of visitors.

The results of the design of the health protocol facility standards for tourist sites focus on health protocol indicators and on sustainable tourism. According to [18] baseline aspects for sustainable tourism indicators, one of which is socio-cultural effects of tourism on the host community, Local public safety, conservation of cultural heritage, and Tourism facilities on offer – provision of a variety of experiences.

The assessment is very technical, asks about the existence of health protocol facilities, the availability, and the conditions. Tourist facilities is facilities and infrastructure provided by management to fulfill customers' needs. The facilities divided into three parts [19]: (1) The main facilities are facilities and infrastructure needed by visitors while in a tourist attraction. The main facilities are Hygiene, Convenience, and Landscape Beauty, (2) Supporting facilities are facilities and infrastructure that function as a complement to the main facilities. It consists of a restaurant, accommodation (lodging), shopping facilities, and others, (3) Complete facilities are facilities and infrastructure that very needed while tourists are in a tourist attraction. Complete facilities consist of the basic facilities such as public toilets, Places of Worship / Mosque, Tour guides, Information centers, and Parking.

In pandemic conditions, facilities related to health and safety for the end of disease transmission get the top priority. The standard used is CHSE (Clean, Health, Safety, and Environment Sustainability), the implementation of health protocols based on hygiene, health, safety, and environmental sustainability. The Ministry of Tourism and Creative Economy, an agency in charge of tourism and the creative economy, strengthens the CHSE Health protocol so that community business entities can implement it. The program will provide a CHSE certification for tourism and creative economy business actors. The certification will show that business actors have implemented and improved health protocols in their respective businesses. The criteria for CHSE-based health protocols are [20]:

- Hygiene: Business actors must ensure that their place of business is always clean from viruses, bacteria, or germs. The availability of hand washing stations equipped with soap and hand sanitizers for visitors and spraying the location with disinfectants are also the main requirements to meet this aspect.
- Health: Management are required to maintain and ensure the health of workers and visitors by requiring the wearing of masks, checking body temperature, and implementing social restrictions.
- Safety: Business actors must have rescue procedures in the event of a disaster or unwanted emergency conditions to maintain the safety of the people at the location.
- Environment Sustainability: Business actors must ensure that their business implements environmentally friendly conditions by supporting water-saving and electricity-saving, using environmentally friendly materials.

Aggregation of indicators was done by combining indicators, their sub-indicators, and their components. It can be seen in the following illustration [21]:

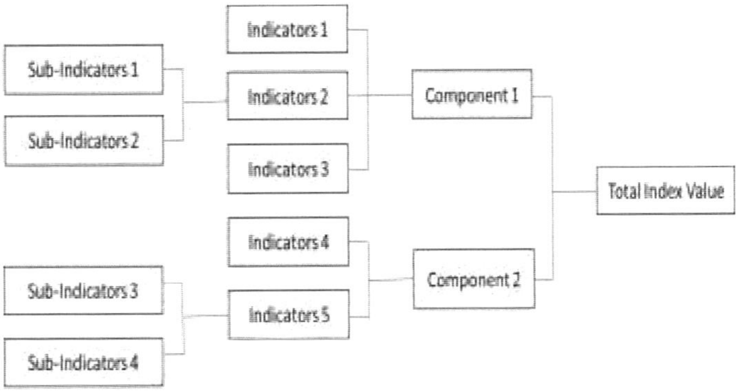

Figure 1. Indicator Aggregation Process

Similar studies that discuss the preparation of CHSE indicators include [22], [23], [24], [25], [26]. Research that discusses the assessment of CHSE health protocol standards among them is [27], [28], [29]. This research discussed the preparation of CHSE indicators and sustainability tourism and done assessment using this standard of health protocols at tourist facilities with trials at Prambanan Park.

2. Research Method

The research conducted in three steps, (1) the preparation of research indicators, indicators validation, and indicators testing for assessment in Prambanan park. Figure 2. explains the steps of the study.

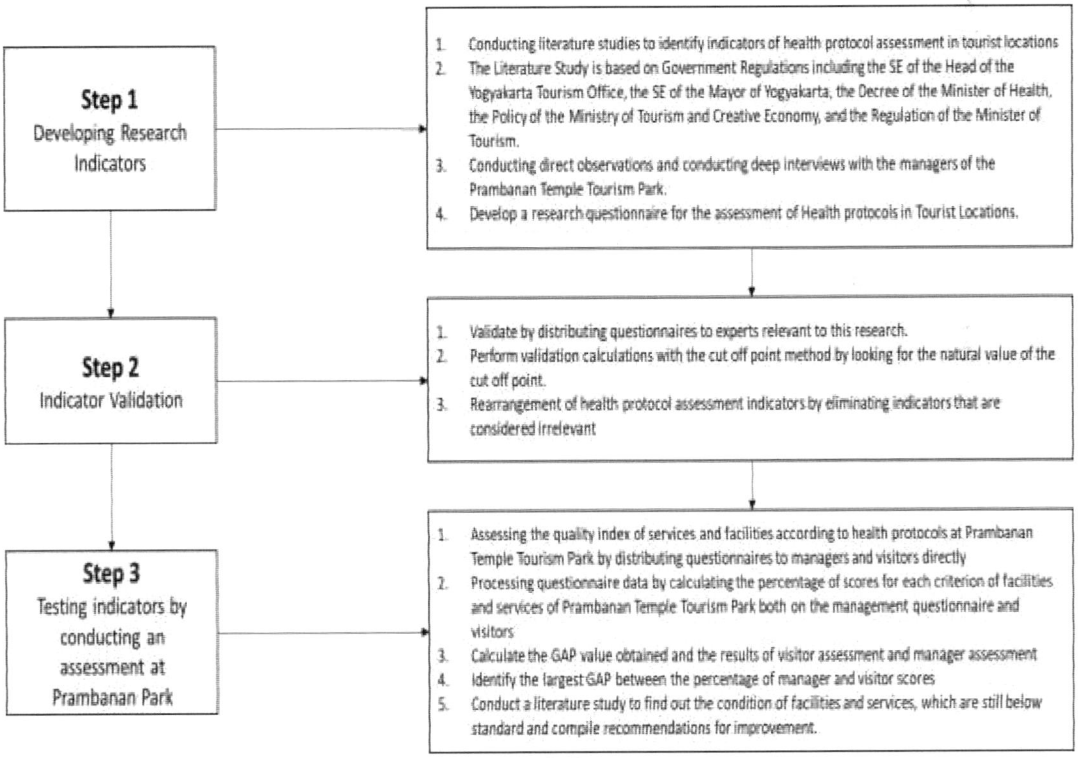

Figure 2. Research Steps

Validation of indicators done by distributing questionnaires to relevant experts and lecturers from the Ergonomic laboratory Industrial Engineering Department and Public Health Department. Indicator validation done by changing the answers to the questionnaire results. A score of 1 indicates the indicator is "Important," and a score of 0 means the indicator is "Not Important." Validation uses the cut-of-point method by looking for the natural cut-of-point value. Verification done by following ways:

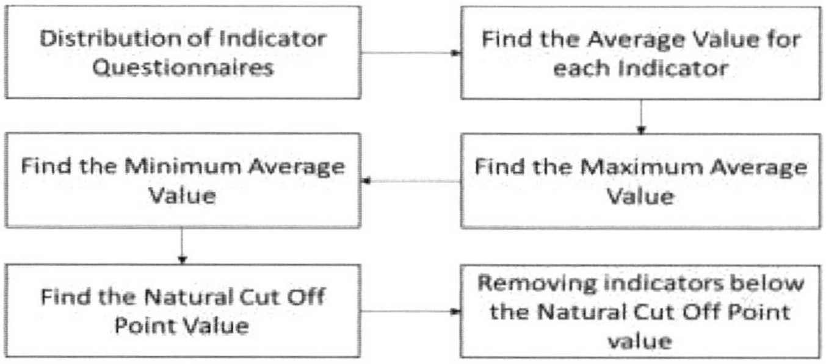

Figure 3. Validation Steps

Validation calculation using the formula:

Average Value $= \frac{R1+R2+R3+\cdots.+Rn}{n}$

Maximum Value = Largest Average Value

Minimum value = Least Average Value

The natural cut of point $= \frac{(Maximum\ Value + Minimum\ Value)}{2}$

The indicators omitted if the score is below the natural cut of point value. After validation of the indicator, an indicator test conducted by conducting an assessment at Prambanan park. The evaluation is conducted by looking for the percentage of the score of each criterion. The score percentage results classification is shown in table 3.

Table 3. Score Classification

Category	Value
Outstanding	>85%
Excellent	70% - 84 %
Very Good	55% - 69%
Good	45%- 54%
Pass	30% - 44%
Unclassified	< 30%

3. Result And Discussion

The identification of indicators of health facilities consists of hygiene criteria, health criteria, safety criteria, environmental sustainability criteria, convenience criteria, landscape beauty criteria, accommodation criteria, restaurant criteria, and essential facility criteria. Each criterion consists of several sub-indicators, and each sub-indicator is composed of several indicators. The literature review results provide nine criteria, 39 sub-indicators, and 131 indicators shown in Table 4.

Table 4. Research Indicators

Criteria	**Sub Criteria**	**Regulation & Reference**
(A) Hygiene	(A1) Washing hands using soap or a hand sanitizer	[30], [31], [32], [33], [34], [35]
	(A2) Availability of handwashing facilities with soap	[30], [31], [32], [33], [34], [35]
	(A3) Safe and appropriate cleaning of spaces and public goods with other disinfectants/cleaning fluids	[30], [31], [32], [33], [34], [35]
	(A4) Vectors and disease-carrying animals free	[30], [31], [32], [35]
	(A5) Cleaning and fitting of public toilets	[31], [32], [35], [36]
	(A6) Clean trash can	[30], [31], [35], [36]
(B) Health	(B1) Avoiding physical contact, setting a safe distance, preventing crowding	[30], [31], [32], [33], [34], [35]
	(B1) Avoiding physical contact, setting a safe distance, preventing crowding	[30], [31], [32], [33], [34], [35]

Criteria	Sub Criteria	Regulation & Reference
	(B2) Does not touch parts of the face, especially the eyes, nose, mouth	[30], [31], [32], [35]
	(B3) Body temperature check	[30], [31], [32], [33], [35]
	(B4) Wear the necessary personal protective equipment (PPE)	[30], [31], [32], [33], [34], [35]
	(B5) Simple medical equipment and supplies	[31], [32], [35]
	(B6) Public spaces and workspaces with good air circulation	[14], [30], [31], [32], [33], [35]
	(B7) Handling for visitors with health problems when doing activities at the location	[14], [30], [31], [32], [35]
(C) Safety	(C1) Disaster self-rescue procedure	[30], [31], [35]
	(C2) Avail-ability of First Aid kits in Accidents	[31], [35]
	(C3) Availability of fire extinguishers	[31], [35]
	(C4) Availability of assembly points and evacuation routes	[31], [35]
	(C5) Media and communication mechanisms for handling emergency conditions	[14], [30], [31], [35]
(D) Environmental Sustainability	(D1) The use of environmentally friendly fixtures and materials	[31], [35], [37]
	(D2) Efficient and healthy utilization of water and energy sources in order to maintain ecosystem balance	[14], [31], [37]
	(D3) Waste and liquid waste treatment is carried out thoroughly, healthy, and environmentally friendly	[31], [35]
	(D4) The surrounding environmental conditions are beautiful and comfortable, either naturally or by technical engineering	[14], [30], [31], [33], [34], [35]
	(D5) Monitoring and evaluating the implementation of guidelines and SOPs for the Implementation of Cleanliness, Health, Safety, and Environmental Sustainability	[30], [31], [32], [33], [34], [35]
(E) Convenience	(E1) Parking Lot	[38], [39]
	(E2) Disability Guidance	[38], [39], [42]
	(E3) *Rest area* (chairs and shelters)	[38], [39]
(F) Beauty Landscape	(F1) Geological	[40], [41]
	(F2) Educational	[40], [41]
	(F3) Tourist Attractions	[41]
(G) Accommodation	(G1) Hotel/Inn	[38], [39], [41]
	(G2) Transportation	[38], [39]
	(G3) Shuttle Bus/Train	[38], [39]
(H) Restaurant	(H1) Clean and hygienic food and beverage management	[31], [32], [35]
	(H2) Services provided	[38], [39], [41]
	(H3) Price	[38], [39], [41]
(I) Shopping Facilities	(I1) There are Shopping Facilities	[38], [39], [41]
(J) Basic Facilities	(J1) Worship place	[38], [39], [41]
	(J2) Tour Guide	[39]
	(J3) Information Center	[39], [41]
	(J4) Security Post	[38], [39]
	(J5) Availability of directional signs and location map displays.	[30], [31], [35], [39], [41]

The validation calculation results obtained a natural cut-off point value of 0.5. After validation, 12 indicators are considered irrelevant. To test the validity of grouping indicators the indicator validation can be done use several methods such as the Delphi method, expert opinion, survey, Dynamic Composite Indicator (DCI), or Structural Equation Modelling [43]

The assessment results of the quality index of services and facilities in the Prambanan Temple Park was done for each criterion. Visitors assess the indicators condition by giving rating for each indicator in 4 scale, there were 35 visitors who done the assessment while their tour. The recapitulation of the visitor assessment result shown in Table 5. The maximum value is the number of indicators multiply by 4. The comparison between the total value of each sub-indicator and the maximum value in percentage became the score for sub indicators.

Table 5. Visitor Assessment Recapitulation

No.	Criteria	Sub-indicators	Indicators	Maximum Value	Score	% Score	Classification
1	Hygiene Facilities	6	23	92	83,77	91,06%	Outstanding
2	Health Facilities	7	32	128	116,37	90,92%	Outstanding
3	Safety Facilities	5	15	60	57,46	95,76%	Outstanding
4	Environmental Sustainability Facilities	5	20	80	77,14	96,43%	Outstanding
5	Convenience Facilities	3	11	44	42,86	97,40%	Outstanding
6	Landscape Beauty Facilities	2	3	12	11,74	97,86%	Outstanding
7	Accommodation Facilities	3	8	32	31,31	97,86%	Outstanding
8	Restaurant Facilities	2	8	32	30,80	96,25%	Outstanding
9	Basic Facilities	5	16	64	63,37	99,02%	Outstanding
	Total	38	136	544	515	94,64%	Outstanding

Table 5 shows that all criteria are category as outstanding, means that all facilities are ready for tourism operation. All criteria have a value above 90%, the highest value is basic facilities, and the lowest value is health facilities. There are new indicators for health facilities which need to be completed due to pandemic era such as hand wash facilities and poster for social distancing, using masker, and cleaning service standard operating procedure. Not only completed the facilities, safety and security are associated with the existence of a system to monitor, prevent, inform, report and handle issues related to safety and security [44].

Even though all the late facilities are in outstanding condition, repairs still need to be done. The formulation of alternative recommendations is made to solve problems that occur in each dimension by submitting suggestions for improvements, Purwaningsih [45] define the recommendation to maintain the sustainability level for Borobudur temple and found that developing complain handling services is important to know the visitors need.

4. Conclusion

The identification of indicators of health facilities consists of nine criteria's: hygiene, health, safety, environmental sustainability, convenience, landscape beauty, accommodation, restaurant, and essential facilities. The preparation of this health protocol includes a sustainability indicator as an effort to preserve the environment as one of the measurement indicators. The4 scale and the data processing are easy to use. The check point was tested by using it in assessment of Clean, Health, Safety, Environment at Prambanan temple park. The result of assessment done by visitors shows that all criteria have outstanding value, means that all facilities are ready for tourism operational.

References

[1] Chan J, Yuan S, Kok K, To K, Chu H, Yang J, Xing F, Liu J, Yip C, Poon R, Tsoi H 2020 *Lancet.* **395** 514–523
[2] Ioannides D and Gyim´othy S 2020 *Tour. Geo.* **22** 624–632
[3] Crossley E 2020, *Tour. Geo.* **22** 536–546
[4] UNWTO 2020 *Annual report* (Madrid: World Tourism Organization)
[5] Obadic A and Maric I 2009 *Eco. Tho. Prac* **18** 93
[6] Yadav H, Kar A K, and Kashiramka S 2021 *EIM.* **35** 160–178
[7] Mahdikhani M 2022 *IJIM. Dat. Ins.* **2**
[8] Ghosh I and Sanyal M K 2021 *IJIM. Dat. Ins.* **1**
[9] Sarkar B D, Shankar R and Kar A K 2021 *Mod. Manag* **10**
[10] Obembe D, Kolade O, Obembe F, Owoseni A, and Mafimisebi O 2021 *IJIM. Dat. Ins.* **1**
[11] Chakraborty A and Kar A K 2021 IJILT. **38** 273–282
[12] Koch J, Plattfaut R, and Kregel I 2021 *IJIM. Dat. Ins.* **1**
[13] Verma S, Warrier L, Bolia B, and Mehta S 2022 *IJIM. Dat. Ins* **2**

[14] Jangra R, Kaushik S P, and Saini S S 2021 *Geo. Sustain.* **2** 48–58
[15] Spillane J 1994 *Pariwisata* (Yogyakarta: Kanisius)
[16] Tjiptono F 2002 *Manajemen Jasa* (Yogyakarta : Andy Offset)
[17] Kotler P 2005 *Manajemen Pemasaran* (Jakarta: PT Indeks Kelompok Media)
[18] Lozano-Oyola M, Blancas F J, González M, and Caballero R 2012 *Ecol. Indic.* **18** 659–675
[19] Akrom M 2014 (Semarang : Universitas Diponegoro)
[20] Kementrian Pariwisata dan Ekonomi Kreatif 2021 *Outlook Pariwisata & Ekonomi Kreatif Indonesia* (Jakarta: Kementrian Pariwisata dan Ekonomi Kreatif)
[21] Rumakat and Abdul A I J 2021 *Inst. Tec. Nasio.* **9** 23-33
[22] Lien M O, Dianawati N, Yulia I N S and Maemunah I 2021 *JNP.*
[23] Suwarni S, Siti K U L, and Rachmat H 2021 *Sosiohumaniora.* **23** 97
[24] Bagus G, Utama R, Agung A, Wirateja B, and Turker S B 2021 IJCSRR. **4** 11-12
[25] Achsa A, Destiningsih R, Islami F S, Novitaningtyas I, and Rahardjo B 2022 *Environ. Manag. Tour* **13** 346–353
[26] Supeno I, Gede M I, Gst Nym S M N, Benoa C A T, and Benoa T 2021 *Eelec.*
[27] Sinaga E K, Nurwitasari A, Kurniati C, and Nur K 2021 *Tour. Hosp.* **13** 64-74
[28] Putu I, Parma G, Yudha A A N, Mahardika M, Armawan I K, Yurika P, and Novia S 2021 *Econ. Bussin. Manag.* **197** 34-42
[29] Nurrahma H, Hakim L, and Parmawati R 2021 *Sumb. Akua. Indopas.* **5** 87-94
[30] SE Kepala Dinas Pariwisata Nomor : 556/266/SE/VIII/2020 (Yogyakarta: Pariwisata Kota Yogyakarta)
[31] Permenparekraf Nomor13 Tahun 2020 *Kreatif sebagai panduan bagi para pengusahan dan/atau pengelola, karyawan, serta pemandu wisata local di lokasi Daya Tarik Wisata dalam adaptasi kebiasan baru* (Jakarta: Kementrian Pariwisata dan Ekonomi Kreatif)
[32] Keputusan Menteri Kesehatan No. HK.01,07/MENKES/382/2020 *Protokol Kesehatan Bagi Masyarakat di Tempat dan Fasilitas Umum* (Jakarta: Kementrian Kesehatan)
[33] Untaru E N and Han H 2021 *Retail. Consum. Serv.* **60**
[34] Joshi U, Khan A, Muke S, Choubey S, Tugnawat D, Naslund J A, and Bhan A 2021 *Psychiatry.* **63** 102750
[35] Hafiddin L I, Kurniawan L R R, Assauri L S, Ammar K, Pujiana I M, and Nuriadi N 2021 *Jurnal Peng. Magis. Pend.* **4** 189-196
[36] Susetyarini O and Jussac M M 2018 *JNP.*
[37] Lagan P 2021 *Enviro.* **201** 111648
[38] Suryadana M L and O v 2015 *Bandung.* **30** 21–25
[39] Susetyarini O and Jussac M M 2018 *JNP.*
[40] Dowling R K and Newsome D 2006 Geo*tourism* (Australia: Foundation Professor of Tourism, School of Marketing,Tourism & Leisure, Edith Cowan University)
[41] Sasmita A 2018 *Chem. Info. Mod* **21** 1689–1699
[42] Kenyo A, Budiatiningsih M, Rahayu A, Surya A, and Rahayu N 2015 *JNP.*
[43] Purwaningsih R, Agusti F, Pranomo S N W, Susanty A and Purwanggono B 2020 *Assessment Sustainable Tourism* (Solo: IEOM Society International)
[44] Purwaningsih R, Santoso H, and Khasanah U 2020 *Rap-Tourism Method* (Bristol: IOP Publishing)
[45] Purwaningsih R, Ameliafidhoh Z, Susanty A, Pranomo S N W, Agusti F 2021 *Sustainability Status Assessment* (Semarang: ICENIS)

New risk assessment and prioritization failure modes based approach in a gas turbine system

A Chakhrit[1]*, M Bougofa[2], I H M Guetarni[1], N Nehal[1], A Bouafia[3], F Z Ghazli[1], R Kharzi[2], and M Chennoufi[1]

[1]Institut de Maintenance et de Sécurité Industrielle, Laboratoire de l'Ingénierie de la Sécurité Industrielle et du Développement Durable, Université Mohamed Ben Ahmed Oran 2, Sécurité Industrielle et Environnement, Oran, Algérie
[2]Laboratory of Transportation Engineering and Environment, Frères Mentouri Constantine 1 University, BP 325 Constantine, 25017, Algeria.
[3]Chemical Engineering and Environment Laboratory, Skikda University, Skikda, 21000, Algeria

*chakhritammar@gmail.com

Abstract. The dependability occupies a strong place in the performance achievement of the system. It describes the mechanisms that lead to failures of systems. Failure mode and effects, analysis (FMEA) is a classical safety technique widely used in several safety critical industries. This method uses the risk priority number (RPN) to assess the criticality value and prioritize failure modes. However, it suffers from some drawbacks regarding the situation where the in-formation provided is ambiguous or uncertain. Thus, in this work, a fuzzy criticality assessment based approach is carried out to evaluate the failure modes of the relevant system and gives an alternate prioritizing to that obtained by the conventional method. In addition, a novel hybrid approach is proposed that combines the grey relational approach (GRA) and fuzzy analytic hierarchy process. This approach offers a new ranking of failure modes by solving the shortcoming concerning the lack of established rules of inference system which necessitate a lot of experience and shows the weightage or importance to the three parameters severity, detection, and frequency, which are considered to have equal importance in the traditional method. A real case study from a gas turbine system provides encouraging results regarding the risk evaluation and prioritizing failures mode with handling different forms of ambiguity, uncertainty, and divergent judgments of experts.

1. Introduction

Failure mode and effect analysis (FMEA) is vastly employed as an analytical methodology for recognizing , ranking, and reducing different failures modes[1]. For such failure mode, three criticality factors: severity (S), non-detection (ND), and frequency (F) are assessed, and a risk priority number (RPN) is computed by multiplying these factors to evaluate the risk value[2] [3, 4].

Moreover, it demonstrates in different of applications that the FMEA still has many flaws. First, different integration of severity, detection, and frequency factors can provide an equal RPN value. While, the risk assessment for the different criticality can be wildly different. Second, in the calculation of the risk priority number, the potential importance of criticality parameters is not considered. Other disadvantage of the traditional FMEA methodology is the specific assessment of criticality factors concerning each failure mode. Nevertheless, because of restricted data and time pressure, criticality factors cannot correctly evaluating and the risk assessment process can be imprecise or uncertain [5-7].

To solve the flaws, several ameliorated FMEA method was suggested by different works. Ilangkumaran et al [8] suggested a risk assessment model based on analytic hierarchy method to assess the risk level of important failures in the paper industry. For the criticality assessment of risk in a 2- dimensional ambiguous linguistic context, Liu et al [9] utilized the interval 2-tuple hybrid weighted distance (ITHWD) measure; then Liu et al [10] employed the alternative queuing approach in order to classify the criticality of the different failure modes. Certa et al [11] gave an ELECTRE TRI-based approach to find risk level group of failure modes based on criticality factors. However, they still show some shortcomings due to the complication and ambiguity of the conventional technique. For this case, many researchers have utilized fuzzy system to resolve the issues mentioned above [12]. Wang et al [13] utilized the criticality parameters as fuzzy values for risk assessment and ranking of failure modes in FMEA. Liu et al [14] arranged and ranked the risk evaluation approach in FMEA to represent the links between risk parameters and riskiness[15,16]. More recently, a novel integrated decision-making model that combined the fuzzy analytic hierarchy process and grey relation analysis method were suggested by Chakhrit and Chennoufi[17]. They provided an alternative ranking for failure modes which reduce the shortcoming of insufficient constructed inference rules, which require a lot of experience.

In light of the research problem previously mentioned, the originality and innovations of this research is based to reduce the ambiguity and minimize the judgments uncertainty, for that a novel hybridized approach that integrates the GRA method and fuzzy AHP method may resolve this issue. This model provides an alternative ranking for the criticality and enables reducing the flaws regarding the deficiency of constructed inference rules that requires a lot of experience, and give the weightage for the different factors, which are estimated to be equally important in the conventional FMEA technique.

2. Fuzzy risk evaluation model

As being the FMEA method has considerable shortcomings in the calculation process and risk interpretation for these reasons a fuzzy risk evaluation based approach is given in section 2. The notion of fuzzy sets that was developed by Zadeh is the foundation for the fuzzy risk assessment model. It gives a more reliable to assess risk related with various failure modes [21], where the different factors utilized in the conventional technique will be fuzzified by using a proper membership function that employs knowledge rules IF - THEN that come from expert opinion, where the relation is represented as:

$$\text{criticality} = \text{frequency} \times \text{severity} \times \text{detection} \tag{1}$$

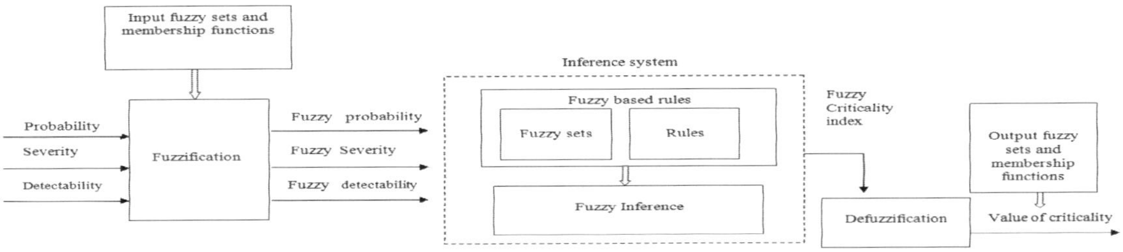

Figure 1. Fuzzy risk model

3. Proposed approach

A new model that integrates the GRA approach and fuzzy AHP is proposed to resolve flaws of the traditional technique. The proposed model can be acquired by the subsequent phases (figure 2).

3.1. Recognizing comparative series

The comparative series is a data set that contains values for the relevant components. The comparison series, which includes the three factors mentioned above, is as follows:

$$r_i(m) = (r_i(1), r_i(2), r_i(3), \ldots\ldots r_i(m)) \in R, \ldots i = 1,2,3, \ldots n \quad (2)$$

Where m denotes the risk parameters number and n is the failure modes number. r_i(m) represents the mth parameters of r_i and the n comparative series is given as follows:

$$r = \begin{bmatrix} r_1 \\ r_2 \\ \cdot \\ \cdot \\ r_n \end{bmatrix} \begin{bmatrix} r_1(1) & r_1(2) & \ldots & r_i(m) \\ r_2(1) & r_2(2) & \ldots & r_2(m) \\ \cdot & & & \\ \cdot & & & \\ r_n(1) & r_n(2) & \ldots & r_n(m) \end{bmatrix} \quad (3)$$

3.2. Standard series specification

Finding the degree of relation is the goal of recognizing the standard series; it displays the ideal level of all deciding elements. The standard series are as follows:

$$r_0(m) = (r_0(1), r_0(2), \ldots, r_0(m)) = (1,1, \ldots.,1) \quad (4)$$

3.3. Comparative and standard Series differences

$$\Delta_{0i}(m) = \begin{bmatrix} \Delta_{01}(1) & \Delta_{01}(2) & \ldots & \Delta_{01}(m) \\ \Delta_{02}(1) & \Delta_{02}(2) & \ldots & \Delta_{02}(m) \\ \cdot & & & \\ \cdot & & & \\ \Delta_{0n}(1) & \Delta_{0n}(2) & \ldots & \Delta_{0n}(m) \end{bmatrix} \quad (5)$$

Where $r_0(m)$ is the standard series, $r_i(m)$ is the comparative series, and$\Delta_{0i}(m) = |r_0(m) - r_i(m)|$

3.4. the Grey Relationship Coefficient

The standard series is examined with three risk variables. The following formula is used to calculate the grey relationship coefficient for frequency, non-detection, and severity factors [22]:

$$\gamma(r_0(m), r_i(m)) = \frac{\Delta_{\min} + \zeta\Delta_{\max}}{\Delta_{0i}(m) + \zeta\Delta_{\max}} \quad (6)$$

ζIs a predetermined coefficient with a typical value of 0.5.

3.5. the Degree of relation determination

Equation 7 is used to estimate the relative value of each criticality factor if they are all of equal importance.

$$\tau_i(m) = \frac{1}{n}\sum_{m=1}^{n} \Delta_i(m) \quad (7)$$

If the risk factors have various importances:

$$\tau_i(m) = \sum_{m=1}^{n} \Delta_i(m)\, \beta(m) .. and \sum_{m=1}^{n} \beta(m) = 1 \quad (8)$$

3.5.1. Fuzzy analytic hierarchy method

Saaty developed the fuzzy analytic hierarchy. It is an excellent method for resolving decision-making issues. The Fuzzy AHP technique is shown below[23,24]:

Step 1: As shown in equation 9[25], a pair-wise comparison matrix is created. The expert is asked to express linguistic terms in pair-wise comparisons across all criteria using triangular fuzzy numbers utilizing expert questionnaires.

$$A=\begin{bmatrix} (111) & j_{12}h_{12}k_{12} & j_{1n}h_{1n}k_{1n} \\ j_{21}h_{21}k_{21} & (111) & j_{2n}h_{2n}k_{2n} \\ j_{n1}h_{n1}k_{n1} & j_{n2}h_{n2}k_{n2} & (111) \end{bmatrix} \quad (9)$$

A_{ij} is a fuzzy number (j, h, and k) for reciprocal $A^{-1} = (j, h, k)^{-1} = (\frac{1}{k}, \frac{1}{h}, \frac{1}{j})$

Step 2: Compute the fuzzy geometric mean for each criterion as indicated in equation. 10.

$$r_i: A1 \otimes A2 \otimes An = (j1, \hbar1, k1) \otimes (j2, \hbar2, k2)(jn, \hbar n, kn) = (j1 * j2 * \ldots . jn, \hbar1 * \hbar2 * \ldots . \hbar n, k1 * k2 * \ldots . kn)^{1/n} \quad (10)$$

Step 3: The fuzzy weights are calculated using normalization. To determine the fuzzy weight of the ith criterion, apply the following equation:

$$w_i = r_i \otimes (r1 \oplus r2 \oplus \ldots \oplus rn)^{-1} \quad (11)$$

3.6. Criticality Ranking

The failure modes are prioritized according to how closely they are related to each other. Priority is given to the failure modes with the lowest degree of grey relation.

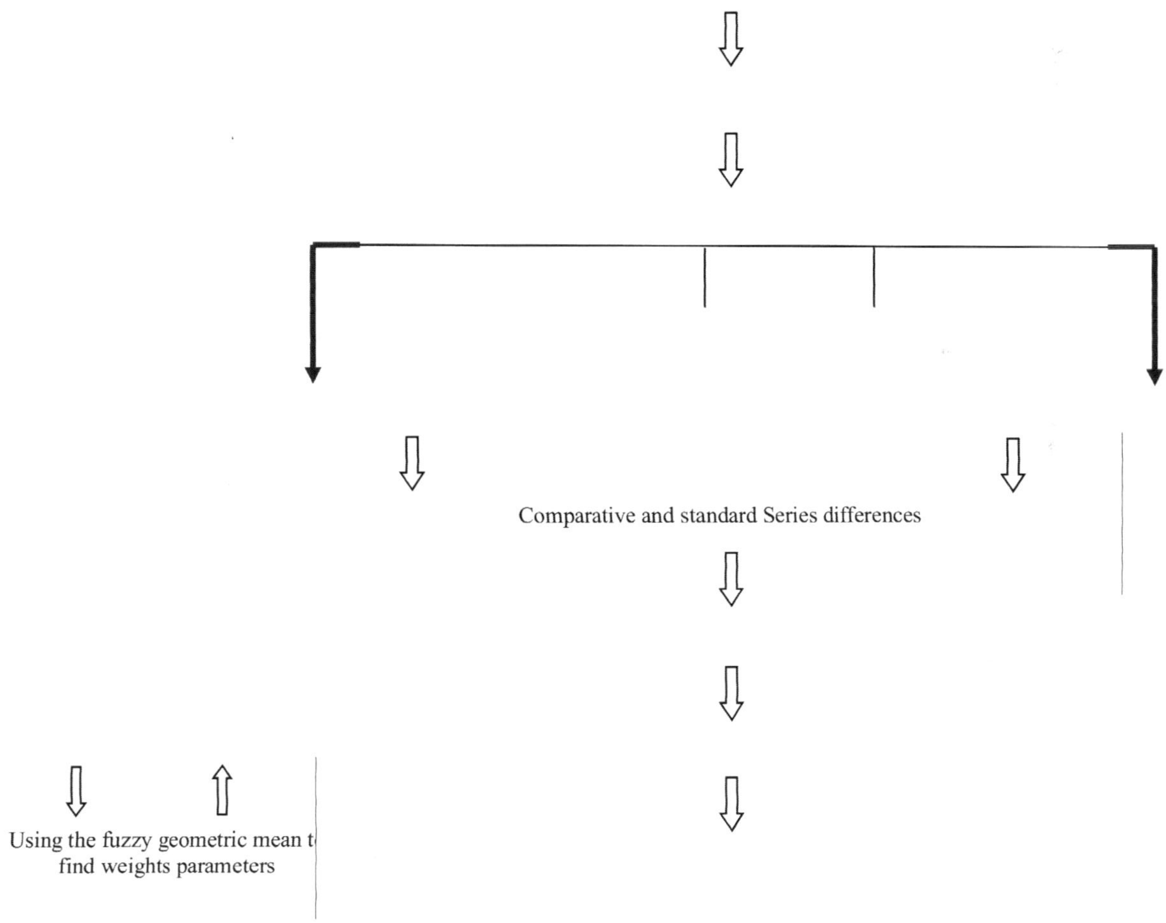

Figure 2. Grey proposed model

4. Case study

The system selected for this study is a gas turbine system, In Algeria, the spread of these systems are clearly seen. It is a combustion engine that can transform natural gas into mechanical energy. After that, a generator powered by this energy produces electricity. This system is shown in figure 3:

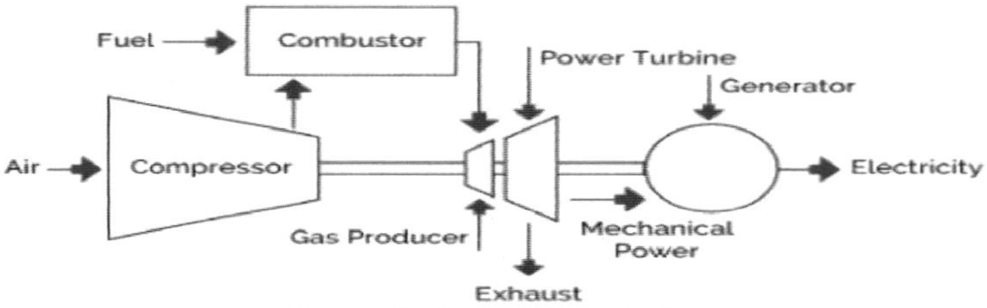

Figure 3. Gas turbine design

4.1. Proposed approach

FMEA data of the table 1 is employed in this part, and the grey relation method is used with different situations (risk factors with various and similar weights). The FAHP method is called to determine the risk parameters weights based on expert opinion.

The first step is to construct comparative series regarding to the various parameters by the matrix as follows.

$$\begin{bmatrix} r_1(1) & r_1(2) & r_1(3) \\ r_2(1) & r_2(2) & r_2(3) \\ & & \\ r_8(1) & r_8(2) & r_8(3) \\ & & \\ r_{14}(1) & r_{14}(2) & r_{14}(3) \\ r_{15}(1) & r_{15}(2) & r_{15}(3) \end{bmatrix} = \begin{bmatrix} 2 & 10 & 6 \\ 2 & 5 & 3 \\ & & \\ 5 & 6 & 6 \\ & & \\ 8 & 9 & 7 \\ 7 & 9 & 6 \end{bmatrix} \quad (12)$$

The difference between the standard and comparative series is given by Equation 5, which is shown in the matrix below.

$$\begin{bmatrix} \gamma_{01}(1) & \gamma_{01}(2) & \gamma_{01}(3) \\ \gamma_{02}(1) & \gamma_{02}(2) & \gamma_{02}(3) \\ \cdot & \cdot & \cdot \\ \cdot & \cdot & \cdot \\ \gamma_{08}(1) & \gamma_{08}(2) & \gamma_{08}(3) \\ \cdot & \cdot & \cdot \\ \gamma_{014}(1) & \gamma_{014}(2) & \gamma_{014}(3) \\ \gamma_{015}(1) & \gamma_{015}(2) & \gamma_{015}(3) \end{bmatrix} = \begin{bmatrix} 1 & 0.4 & 0.58 \\ 1 & 0.64 & 0.85 \\ \cdot & \cdot & \cdot \\ \cdot & \cdot & \cdot \\ 0.64 & 0.58 & 0.58 \\ \cdot & \cdot & \cdot \\ 0.48 & 0.44 & 0.52 \\ 0.52 & 0.44 & 0.58 \end{bmatrix} \quad (13)$$

As described previously, the grey relation coefficient is derived by equation 6.

$$\begin{bmatrix} \gamma_{01}(1) & \gamma_{01}(2) & \gamma_{01}(3) \\ \gamma_{02}(1) & \gamma_{02}(2) & \gamma_{02}(3) \\ \cdot & \cdot & \cdot \\ \cdot & \cdot & \cdot \\ \gamma_{08}(1) & \gamma_{08}(2) & \gamma_{08}(3) \\ \cdot & \cdot & \cdot \\ \gamma_{014}(1) & \gamma_{014}(2) & \gamma_{014}(3) \\ \gamma_{015}(1) & \gamma_{015}(2) & \gamma_{015}(3) \end{bmatrix} = \begin{bmatrix} 1 & 0.4 & 0.58 \\ 1 & 0.64 & 0.85 \\ \cdot & \cdot & \cdot \\ \cdot & \cdot & \cdot \\ 0.64 & 0.58 & 0.58 \\ \cdot & \cdot & \cdot \\ 0.48 & 0.44 & 0.52 \\ 0.52 & 0.44 & 0.58 \end{bmatrix} \quad (14)$$

If all risk factors have similar weights, equation 7 is used to compute the relation degree. For instance; the first failure mode is deduced as;

$$\tau_i(k) = \frac{1}{3}(\gamma_{01}(1) + \gamma_{01}(2) + \gamma_{01}(3)) = \frac{1}{3}(1 + 0.4 + 0.58) = 0.66$$

The second situation where the parameters with various weights; the fuzzy AHP technique is used to assign weights to the risk factors. The significance of the risk parameters has been compared by experts to triangular fuzzy values. The overall results are presented in table 2.

Table 1. Fuzzy weights calculation

	Frequency	**Non-detection**	**Severity**	Fuzzy geometric mean value r_i	Fuzzy weights w_i
Frequency	(1,1,1)	$(\frac{1}{7},\frac{1}{6},\frac{1}{5})$	$(\frac{1}{9},\frac{1}{9},\frac{1}{8})$	(0.251 0.264 0.292)	0.05
Non-detection	(5,6,7)	(1,1,1)	$(\frac{1}{9},\frac{1}{8},\frac{1}{7})$	(0.822 0.908 1)	0.173
Severity	(8,9,9)	(7,8,9)	(1,1,1)	(3.825 4.16 4.32)	0.777

Table 2. Ranking of conventional, fuzzy and GRA suggested model

Failure mode Sequence N°	Failure Mode	Cause	F	S	D	Conventional RPN	Ranking	Fuzzy RPN	ranking	GRA with similar weighs	ranking	GRA with different weight	ranking
1	Vibration	Incorrect signal	2	10	6	120	7	0.457	7	0.660	8	0.461	2
2		bearings not working	2	5	3	30	13	0.20	15	0.831	14	0.622	11
3	Over-temperature	soiled compressor rotor	3	5	4	60	10	0.250	13	0.737	11	0.666	14
4	Stall	changeable stator vanes Binding	3	6	4	72	9	0.274	12	0.718	10	0.619	10
5		Foreign objects degrade	3	4	5	60	10	0.230	14	0.737	11	0.600	9
6	Flame-out	Energy nozzles blockage	4	6	2	48	12	0.358	10	0.771	13	0.660	13
7	Hot spots on flame tube	Uneven flame dispersion around a malfunctioning flame tube.	4	7	2	56	11	0.370	9	0.751	12	0.649	12
8	Vibration	Defective vibration indication	5	6	6	180	5	0.457	6	0.567	3	0.583	8
9		Defective bearings	5	8	3	120	7	0.447	8	0.653	7	0.544	7
10	Over-speed	Excessive fuel flow	6	7	5	210	4	0.5	5	0.580	4	0.544	7
11	No start	particles, water, or air in fuel lines	2	9	5	90	8	0.308	11	0.693	9	0.500	4
12	Stall	irregular fuel pressure	6	8	5	240	3	0.531	4	0.600	5	0.512	5

13	Wrong temperature indication	short circuit in thermocouple circuit	6	9	3	162	6	0.572	3	0.620	6	0.516	6
14	Not reaching idle speed	Low electrical power	8	9	7	504	1	0.758	1	0.480	1	0.456	1
15	Defective speed indication	Internal tachometer failure	7	9	6	378	2	0.667	2	0.513	2	0.468	3

5. Results and discussion

Table 1 represents the findings of the different approach for gas turbine system. As given in the Table that the prioritization of failures mode obtained from the traditional FMEA is classified as FM14, FM15 FM12, FM10, FM8, FM13, (FM9, FM1), FM11, FM4, (FM5, FM3), FM7, FM6, FM2, respectively. However, after utilizing the fuzzy model provided a novel ranking for the criticality. For example, in the traditional methodology, FM13 was classified in the sixth rank. While, it ranked at the third classification in the fuzzy proposed model. Simultaneously, in both the methods, FM14 was the most critical mode.

By comparing the traditional finding of the fuzzy model with FMEA method, the flaws combined with conventional FMEA can evidently shown; the most significant flaw of the conventional methodology is that the various combinations of the different factors give an equal RPN value; however, the risk interpretations can be different, For instance, FM3and FM5 have an equal risk priority number of 60, while the risk outcome cannot exactly be the same, but the proposed fuzzy model differs in those. The second shortcoming of the traditional methodology disregards the importance between the different parameters. The three inputs are suggested to be equally important, but in actual implementations, the inputs' relative relevance is still present; for instance, the failure mode 13 with a moderate frequency, low detection, and very high severity (6, 3, 9) with a lower risk priority number of 162 than one with all factors moderate as the failure mode 8 with risk priority number of 180; on the contrary with the fuzzy model can distinctly observed that FM13 has a higher value than FM8with values 0.572, 0.457, and hence will be given more attention for preventive and corrective actions..

Concerning the suggested grey analysis model, there has been an observed variation in the ranking order with the previous approach. For example FM14, FM15 have higher priority in all model (the most significant modes of failure). Meanwhile, the failure mode 13 has lower priority in the suggested model and a higher classification in other models. The important cause is explicated by different risk assessment models and prioritization processes utilized in each approach.

Table 1 also represented the weightage of the different factors through the grey suggested model. When different weights parameters are utilized, it is clearly shown that is a marked rearranging in the classification for the failure modes, showing their importance. In the case of requests coming from modifications or variations, changing the weights can be a simple way to adjust the strategy. The GRA methodology is called when there are no particular inference rules that require a great deal of experience.

6. Conclusions

For identifying and rating potential failure modes in products and processes, FMEA has been broadly acknowledged as a standard technology technique. The traditional FMEA methodology received different criticism for its shortcomings, especially in the evaluation process and risk priority computation. In this work, a novel risk prioritization approaches to assess the criticality of the different failures in FMEA is proposed.

Compared to the traditional methodology, the advantage of fuzzy model evaluation permits experts to more objectives combining of the different factors by utilizing their experience to reduce the flaws creating in performing of the classical FMEA technique. This study represented the efficiency of the fuzzy criticality approach to control various kind of ambiguities in the failure evaluation technique, such as, fuzziness, incompleteness and vagueness

Fuzzy analytic hierarchy and grey relation analysis approach are utilized in this study to assess and prioritize the failure modes more factual and dynamic. The results can give important outcomes in the decision-making process. The results represent in a real case study of risk ranking of failures modes that the combination of the fuzzy analytic hierarchy and grey relation analysis methods can give a more acceptable risk prioritization order. In addition, the proposed models can be used when the predetermined inference rules are inadequate and require a lot of expertise.

References

[1] H.-C. Liu, J.-X. You, M.-M. Shan, and Q. Su 2019 *Total. Qual. Manag. Bus. Excell* **30** 537-564
[2] A. Chakhrit and M. Chennoufi 2021 *J. Intell. Fuzzy Syst.* 1-16
[3] D. Panchal, A. K. Singh, P. Chatterjee, E. K. Zavadskas, and M. Keshavarz-Ghorabaee 2019 *Appl. Soft Comput.* **74** 242-254
[4] I. H. Guetarni, N. Aissani, E. Châtelet, and Z. Lounis 2019 *Process Saf. Prog.* **38** e12011
[5] H.-C. Liu, L. Liu, and Q.-L. Lin 2013 *IEEE Trans Reliab* **62** 23-36
[6] M. Bougofa, M. Taleb-Berrouane, A. Bouafia, A. Baziz, R. Kharzi, and A. Bellaouar 2021 *Process Saf Environ Prot* **153** 486-499
[7] M. Bougofa, A. Bouafia, and A. Bellaouar 2020 *Int. J. Perform. Eng,* **16** 354-366
[8] M. Ilangkumaran, P. Shanmugam, G. Sakthivel, and K. Visagavel 2014 *Int. J. Product. Qual.,* **14** 296-313
[9] H.-C. Liu, J.-X. You, and X.-Y. You 2014 *Comput. Ind. Eng.* **78** 249-258
[10] H.-C. Liu, Y.-P. Hu, J.-J. Wang, and M. Sun 2018 *IEEE Trans. Reliab.* **68** 554-565
[11] A. Certa, M. Enea, G. M. Galante, and C. M. La Fata 2017 *Comput. Ind. Eng* 108 100-110
[12] L. A. Zadeh, G. J. Klir, and B. Yuan 1996 *Fuzzy sets, fuzzy logic, and fuzzy systems: selected papers* vol. 6: World Scientific
[13] Y.-M. Wang, K.-S. Chin, G. K. K. Poon, and J.-B. Yang 2009 *Expert Syst. Appl.* **36** 1195-1207
[14] H.-C. Liu, L. Liu, and N. Liu 2013 *Expert Syst. Appl.* **40** 828-838
[15] R. Fattahi and M. Khalilzadeh 2018 *Saf. Sci.* **102** 290-300
[16] G. F. Can 2018 *Hum. Factors Ergon. Manuf.* **28** 130-147.
[17] A. Chakhrit and M. Chennoufi 2021 J. Eng. Des. Technol *10.1108/JEDT-07-2021-0395*
[18] I. A. Elbadawi, M. A. Ashmawy, W. A. Yusmawiza, I. A. Chaudhry, N. B. Ali, and A. Ahmad, 2018 *Eng. Technol. Appl. Sci. Res* **8** 3023-3027
[19] N. M. Thoppil, V. Vasu, and C. Rao 2019 *J. Fail. Anal. Prev.* **19** 1153-1157
[20] A. S. Relkar 2021 *J. Fail. Anal. Prev.* **21** 793-805
[21] S. Y. Far, R. Mirzaei, M. B. Katrini, M. Haghshenas, and Z. Sayahi 2018 *Eng. Appl. Sci. Res,* **8** 2689-2692
[22] F. Sarraf and S. H. Nejad 2020 *Eval. Program Plann.* **79** 101762
[23] T. Bani-Mustafa, N. Pedroni, E. Zio, D. Vasseur, and F. Beaudouin 2020 *Proc Inst Mech Eng O J Risk Reliab* **234** 748-763
[24] P. T. Harker and L. G. Vargas 1987 *Manag. Sci.* **33** 1383-1403
[25] J. J. Buckley, "Fuzzy hierarchical analysis," *Fuzzy Sets Syst.* **17** 233-247

The effectiveness of EM4 and Local Micro-organisms (LOM) Activators in Organic Waste Processing in Brikama Market West Coast Region, The Gambia

Babucarr Jassey[1,2*], **Syafrudin**[1], **Badrus Zaman**[1], **Kemo Ceesay**[2], **Ibrahim Touray**[2], **Juma Ngum**[2], **and Habibi Prakoso**[1]

[1] Department of Environmental Engineering, Diponegoro University, Semarang Indonesia
[2] Public Health Directorate, Department of Public Health Services, Quadrangle, Banjul, Ministry of Health, The Gambia.

*baxjas@gmail.com, baxjas@alumni.undip.ac.id

Abstract. Compost is the result of decaying process of organic materials due to the interaction between decomposing microorganisms found within. Compost is a type of organic fertilizer that originates from decomposed organic materials. Observations at Brikama Market indicated 20 vegetable vendors with at least 5 pieces of vegetables decaying, market sales decline by approximately 5 kg for every transaction. This study aims to determine the effectiveness of organic waste treatment using EM4 and LOM. This research is purely experimental, uses Posttest Group design and Independent Sample T-Test. This research was conducted in groups, with different forms of waste-treatments which included the use of EM4 and local microorganism activators and observed as it decays. The results obtained differences in the average processing of organic waste using EM4 and LOM activators. The average temperature of EM4 activator 29.89°C while LOM 29.97°C. The average humidity 48.67 for EM4 activator while LOM 49.64 and the average pH of EM4 activator 5.96 while LOM 5.43. The research revealed significant changes in EM4 and LOM activators seen from the measurement of temperature, humidity and pH. It is recommended to the community to participate in managing waste, especially organic waste by making compost using EM4 and LOM activators.

1. Introduction

Waste according to Jassey et al., is something that is not used, not intended to be used, disliked, intended to be thrown away or something that is thrown away that comes from human activities and does not happen by itself [1]. The research conducted by Wang et al. [2] on renewable hydrogen production from municipal solid waste. The results of the study concluded that composting of various types organic waste has shown differently the effectiveness of the composting process. Composting is purely conducted on organic waste matter. It has been proven significantly reduce the volume inside country and provide solutions for agriculture as a substitute for fertilizer chemical fertilizers. This also has a similar findings with that of [3].

Other studies conducted by Kohlstock and Kraft [4] in the European Union found that organic waste that was converted into compost could be very beneficial for farmers. The process of recycling organic waste provides economic benefits for the community to mobilize unused waste.

Data from the environment unit of Brikama Area Council states that the estimated comparison data from 2017-2021, lagging waste produced in various places, namely Brikama Town 314,225.00m³/day,

Farato Village 177,004.00 m^3/day, Busumballa with 37,403.55 m^3/Day, Kitty Village with a waste volume of 10,115.50 m^3/day, and Kembujeh area with a waste volume of 2,901.00 m^3/day [5). Waste from Brikama town is produced by many housing as many as 218.90 m^3 and from market waste as much as 72.96 m^3 from 446.27 m^3 of waste/day, the average domestic waste generation of Brikama Town is 1.49 liters/person/day [5].

Waste management by sorting waste according to its type is actually going well, but the lack of community knowledge of organic waste management makes this waste management not to proceed effectively so that there is a lot of build-up of organic waste [1,6].

Organic waste can be processed into fertilizer using fermentation process. Organic fertilizer made using the fermentation process is called compost. Making compost in a conventional way takes a long time so it is less effective for overcoming the problem of build-up of organic waste. In order to solve the issue, it is required to identify alternative or more efficient composting techniques. [4,7,8].

Compost is the result of weathering process of organic materials due to the interaction between decomposing microorganisms that work in it. Compost is also one type of organic fertilizer because it comes from weathered organic matter.

LOM is a fermented microorganism of the material in the surrounding environment and is easy to obtain. The use of raw materials is adjusted to the potential in a region. LOM contains elements of micro and macro nutrients and microbes, the LOM solution has the potential to overhaul the organic matter, stimulant growth and disease control agents and plant pests [8]. LOM that works requires carbohydrate intake as a source of energy. Carbohydrates can be obtained from rice laundry water, where on the skin of rice there is a high carbohydrate content. When washing, the carbohydrate will dissolve in the washing water. The carbohydrate content contained in rice washing water can be used as a source of food suppliers for microorganisms [7,8].

LOM can be used as organic fertilizers, as decomposer or the culprit of making compost fertilizer and as natural occurring pesticides. According to the results of the study, giving LOM as organic fertilizer/LOM is able to fertilize red spinach plants [3,9].

EM4 can accelerate the composting process; the addition of EM4 can eliminate the odour that arises during the composting process if the composting process takes place smoothly EM4 solution is a bioactivator used to make compost in solid form called bokashi. Organic materials commonly composted are vegetables, grass, husks, sawdust [2]. Compost activator that is easily found is EM4. According to review of data, [5,6,10] the advantage of the EM4 included the composting time of 4-7 days, the resulting compost is not hot, it does not smell foul and does not contain pests and diseases [11,12]. The increasing level of compost waste that could have been used as manure because of its high content of micro and macro nutrients important for plant growth is not fully utilized. The basis of this study is to determine the effectiveness of organic waste treatment using EM4 and LOM. Which of the two methods is more effective and saves time while causing lesser harm to the environment.

2. Methodology

This study is purely experimental research where the researcher aimed at looking for the effectiveness of both EM4 and LOM activators in organic waste processing. This study was conducted in ten groups, with different forms of treatment.

The 10 groups were selected by the researchers at every 20 meters of the market. Waste samples were collected, processed and treated. The results of the samples were observed for composting in the time frame of 4-7 days.

The resulting temperatures were also read to see if the temperature increases or not, see of it has developed some form of foul smell as it should under normal circumstances. It was also observed for the presence of pests and substances that might disturb the growth pattern of the plant. This study used a posttest group design. This design allowed the researchers to measure the effect of waste treatment (intervention) on the experimental group by comparing the results of the various groups. The research was conducted in the Brikama market in the west coast region of The Gambia from February-April 2022.

3. Result and Discussion

This study examines the effectiveness of using EM4 and LOM in organic waste processing in Brikama market.

3.1. Univariate Analysis

Univariate analysis was carried out to describe the distribution of each research variable:

3.1.1. Average Composting Temperature by Using Activator EM4. Based on table 1 the average composting temperature using EM4 activator is 29.89˚C with standard deviation 1.31. The researcher's assumptions on compost temperature using EM4 activator is that I was supposed to be close to soil temperature, which ranges from 28-50˚C where the average EM4 compost temperature 29.89˚C. Temperature at the end of compost according to the Gambian National Standards (GNS) should not exceed 30˚C.

Table 1. Average Composting Temperature By Using Activator EM4

Variable	Mean	SD	Min Temp (˚C)	Max Temp (˚C)
Temperature of EM4	29.89	1.31	24	31

*SD: Standard Deviation, Min: Minimum, Max: Maximum

3.1.2. Average Composting Temperature by Using LOM. Based on table 2 it is known that the average composting temperature using LOM is 29.97 with a standard deviation of 1.24. According to the researcher's assumptions, on compost waste temperature comparing to the results of LOM already approaching the temperature of ground soil that ranges from 28-50˚C where the average LOM temperature is 29.97˚C. Final temperature of compost according to GNS should not exceed 30˚C.

Table 2. Average Composting Temperature Using LOM

Variable	Mean	SD	Min Temp (˚C)	Max Temp (˚C)
Temperature of LOM	29.97	1.24	25	31

*SD: Standard Deviation, Min: Minimum, Max: Maximum

3.1.3. Average humidity of compost EM4 activator is used. Based on Table 3 it is known average composting humidity by using activator EM4 is 48.67 by standard deviation of 2.41. According to the researcher's assumptions, on the results of moisture research compost with EM4 activator where is the mean humidity EM4 48.67 while according to GNS 22-1012-2014 max limit compost moisture is 50% so the results of the terms of composting meet specifications compost from organic matter.

Table. 3 Average humidity of compost using EM4 as the activator

Variable	Mean	SD	Min Humidity	Max Humidity
Humidity EM4	48.67	2.41	44	53

*SD: Standard Deviation, Min: Minimum, Max: Maximum

3.1.4. Average Humidity Composting Using LOM. Based on Table 4 it is known average composting humidity using LOM is 49.64 with a standard deviation of 2.53. According to the researcher's assumptions, on the results of moisture research compost with LOM where the average humidity LOM 49.64 while according GNS 22-1012-2014 maximum humidity limit of compost is 50% so the result of the condition composting meets the requirements Compost specifications of materials organic.

Table 4. Average Humidity Composting With Using LOM

Variable	Mean	SD	Min Humidity	Max Humidity
Humidity LOM	49.64	2.53	43	52

*SD: Standard Deviation, Min: Minimum, Max: Maximum

3.1.5. Average pH of Composting using EM4 Activators. Based on Table 5 it is known average pH of composting with using EM4 activator 5.96 with a standard deviation of 0.71. According to the researcher's assumptions, on the results of research on compost pH with EM4 and LOM activators where the average is 5.96. pH that meet the requirements, namely the pH is close to neutral pH.

Table 5. Average pH Of Compost using EM4 Activators

Variable	Mean	SD	Min pH	Max pH
pH EM4	5.96	0.71	4	6

*SD: Standard Deviation, Min: Minimum, Max: Maximum

3.1.6. Average pH of composting using LOM. Based on table 6, it is known that the average pH of composting with using a LOM of 5.43 with a standard deviation of 0.85. According to the researcher's assumptions, the results of the research on the pH of compost using LOM where the average LOM pH is 5.43. LOM pH is smaller than EM4 pH. The pH that meets the requirements is a pH that is close to neutral pH.

Table 6. Average humidity of composting using LOM

Variable	Mean	SD	Min pH	Max pH
pH LOM	5.43	0.85	4	8

*SD: Standard Deviation, Min: Minimum, Max: Maximum

3.2. Bivariate Analysis

Based on bivariate analysis that the researchers conducted with the title the effectiveness of organic waste processing using EM4 and LOM activators, using the Independent sample t-test statistical test as follows:

3.2.1. Composting temperature effectiveness using EM4 and LOM activators. Based on Table 7, it is known that the average composting temperature using EM4 activator was 29.89°C with a standard deviation of 1.31 while the average composting temperature using LOM is 29.97°C with a standard deviation of 1.24. The difference between the two variables is 0.50. Statistical test results obtained p value 0.39 at (α) 0.05 meaning Hypothesis (o) is accepted which means that there is no difference between the EM4 temperature and the LOM temperature. According to the research, the results showed that the temperature of the compost with EM4 activator and LOM was close to the soil temperature, which ranged from 28-50°C, where the average temperature of EM4 compost was 29.89°C, and the LOM temperature was 29.97°C. The final temperature of compost according to GNS should not exceed 30°C. Compost temperatures EM4 and LOM are equally effective, both composts are equally good and both composts can be used [12,13]. So waste processing is more effective using EM4 activator (14). The length of time for composting using EM4 activator is 15 days and the length of time for composting using LOM is. The length of time for composting is faster than using EM4 [14,15].

3.2.2. Humidity effectiveness using EM4 and LOM activators. Based on table 8, it is known that the average humidity of composting using EM4 activator is 48.67 with a standard deviation of 2.41 while the average humidity of composting using LOM is 49.64 with a standard deviation of 2.53. There is no significant difference between mean deviations of the two variables which is 0.851 and 0.852. Statistical test results obtained p value of 0.37 at 0.05 (α) means Ho is accepted which means that there is no difference between EM4 humidity and LOM humidity. According to the researcher's assumptions [13,14,15], the results of the research on compost moisture using EM4 and LOM activators where the average humidity of EM4 is 48.79 and LOM humidity is 49.64. The more effective humidity is LOM humidity because it is close to the maximum moisture limit in compost and humidity at EM4 is also good [6,16]. Meanwhile, according to GNS 22-1012-2014 the maximum limit of compost moisture is 50%, so the results of the composting requirements meet the specifications for compost from organic matter. The length of time for composting using EM4 activator is 15 days and the length of time for composting using LOM is 19. The length of time for composting based on this experiment is faster than using EM4.

Table 8. Humidity effectiveness of composting with EM4 and LOM activators

Variable	Mean	SD	MD	P-Value
EM4 Humidity	48.67	2.41	0.85	0.37
LOM Humidity	49.64	2.53	0.85	0.38

*SD: Standard Deviation, MD: Mean Deviation.

3.2.3. Effectiveness of pH using EM4 and LOM activator. Based on Table 9, it is known that the average pH of composting using EM4 activator is 5.96 with a score of standard deviation of 0.71 while the average pH of composting using LOM is 5.43 with a standard deviation of 0.85. The difference between the two variables is just 0.53. Statistical test results obtained p value 0.50 at 0.05 (α) means Ho is accepted which means that there is no difference between pH, EM4 and pH LOM. According to the researcher's assumptions, the results of the research are the pH of compost with EM4 and LOM activators where the average is 5.96 and the pH using LOM is 5.43. A more effective pH is the pH of composting using EM4 because the pH is closer to normal pH. The pH that meets the requirements is a pH that is close to neutral pH, so pH EM4 is better than pH LOM so it is more effective to use EM4 and both are equally effective and good for use as compost [13,17]. The length of time for composting using EM4 activator is 15 days and the length of time for composting using LOM is 19. The length of time for composting is faster than using EM4 based on this experiment. This finding is on contrast with the literature review conducted by Anderson et al in 2018 and Puccetti et al in 2021 [18,19] which stated otherwise using PM4 to control the and minimise the pH of organic waste in healthcare setting. When the same test was conducted using vegetables by Mirwandono and co [20], the results match that of the experiment

Table 9. Effectiveness of pH using EM4 and LOM

Variable	Mean	SD	MD	P-Value
EM4 pH	5.96	0.71	0.50	0.10
LOM pH	5.43	0.85	0.51	0.11

*SD: Standard Deviation, MD: Mean Deviation.

4. Conclusions

The average temperature of compost using EM4 activator is 29.89°C. The average temperature of composting using LOM is 29.97°C. The average moisture content of compost using EM4 activator is

48.67. The average humidity of composting using LOM is 49.64. The average pH of compost using EM4 activator is 5.96. The average pH of composting using LOM is 5.43.

The effectiveness of temperature using EM4 activator is 29.89°C and LOM 29.97°C. Meaning the most effective among the two is EM4 in terms of temperature. The effectiveness of humidity using EM4 is 48.67 and using LOM is 49.64. Therefore, it can be concluded that LOM is more effective in terms of moisture accumulation. The effectiveness of pH using EM4 activator is 5.93 and that using LOM is 5.43. The most effective among the two in terms of pH is EM4.

This research finally concludes that local organic microorganisms are very good in moisture keeping in compost. The acidity level and the alkalinity level are best controlled using the EM4. Therefore, it is the most ideal processing method.

Acknowledgments

The writers would like to give a special thanks to the department of public health services, the director of public health services, the management and staff of Briama market. We would also like to thank the *Institut Teknologi Nasional* Yogyakarta and *Direktorat Riset dan Pengabdian Masyarakat, Deputi Bidang Penguatan Riset dan Pengembangan, Kementerian Riset, Teknologi/Badan Riset dan Inovasi Nasional Republik* Indonesia for the support research funding.

References

[1] Jassey B, Zaman B, and Syafruddin 2021 Dynamic model of municipal solid waste management from households in Sukuta Nema, The Gambia *IOP Conf. Ser.: Earth Environ. Sci.* **896** 012071

[2] Wang S, Yang H, Shi Z, Nuran I, and Wen Y 2022 Renewable hydrogen production from the organic fraction of municipal solid waste through a novel carbon-negative process concept. *Renew. Nat. Energy* **252** 124056

[3] Kadir A A, Azhari N W, and Jamaludin S N 2016 An Overview of Organic Waste in Composting. *MATEC Web Conf.* **47** 0502525

[4] Meyer-kohlstock D, Schmitz T, and Kraft E 2015 Organic Waste for Compost and Biochar in the EU: Mobilizing the Potential *Resources* **4**(3) 457-475

[5] BAC. Profile of Waste Generation at Dumpsite in Tambana by differen Communities within the Catchment Area. Ministry of Local Government The Gambia; 2021 p. 203–75.

[6] Hendriani N, Juliastuti S R, and Masetya H N 2017 Composting of Corn By-Product using EM4 and Microorganism *Azotobacter sp.* as Composting Organism *KnE Life Sci.* **3**(5) 158-166

[7] Vlachokostas C, Achillas C, Diamantis V, Michailidou A V, Baginetas K, and Aidonis D 2021 Supporting decision making to achieve circularity via a biodegradable waste-to-bioenergy and compost facility *J. Environ. Manag.* **285** 112215

[8] Sufficiency E, Qamar S A, Fernando L, Ferreira R, Franco M, Iqbal H M N, and Bilal M 2022 Emerging biotechnological strategies for food waste management: A green leap towards achieving high-value products and environmental abatement *Energy Nexus* **6**(16) 100077

[9] Rahman MA, Khan MN, Akter S, Rahman A, Alam MM, Khan MA, Rahman M M 2020 Determinants of Faster Decomposition using EM4; Evidence from Nationally Representative Survey Data. PLoS One [Internet]. 2020;15(7):1–14. Available from: http://dx.doi.org/10.1371/journal.pone.0236080

[10] Composting domestic waste [Internet]. [cited 2022 May 1]. Available from: https://www.citizensinformation.ie/en/environment/waste_and_recycling/composting_domestic_waste.html

[11] Suryawan I W K, Prajati G, Afifah A S, Apritama M R, and Adicita Y 2019 Continuous piggery wastewater treatment with anaerobic baffled reactor (ABR) by bio-activator effective microorganisms (EM4) *Indonesian J. Urban Environ. Tech.* **3**(1) 1-12

[12] Mirwandono E, Sitepu M, Wahyuni T H, Ginting N, Siregar G A and Sembiring I 2018 Nutrition quality test of fermented waste vegetables by bioactivator local microorganisms (MOL) and effective microorganism (EM4) *IOP Conf. Ser.: Earth Environ. Sci.* **122** 012127

[13] Naqvi Z H, Mushtaq-ul-Hassan M, Chaudry Z, Akram I M, and Ahmad R 2020 Effect of effective microorganisms (EM4) on health of layers *Pak. J. Biol. Sci.* **3**(9) 1516-1518

[14] Indrawati S, Anggoro D, Sukamto H, Puspitasari N, Indarto B, Prajitno G, Arvitrida N. and Yuwana L 2020 The Effectiveness of the Addition of EM4 and Molasses in Increasing Levels of N, P and K in Environmentally Friendly Liquid Fertilizers Made From Banana Pseudostem. In *International Joint Conference on Science and Engineering (IJCSE 2020)* 365-369. Atlantis Press.

[15] Basri A B, Azis A, and Iswoyo H 2021 Application of coffee husk compost and EM4 on growth and yield of chili pepper (*Capsicum Frutescens L.*) *IOP Conf. Ser.: Earth Environ. Sci.* **807** 042040

[16] Ali F and Sari T I 2019 Utilization of PT. Hok Tong liquid waste rubber industry in making of liquid organic fertilizer with addition of eceng gondok and EM4 (Effective Microorganism 4). *IOP Conf. Ser.: Mater. Sci. Eng.* **620** 012013

[17] Rahmi A, Jannah N, and Apriyanto R 2022 Effect of Planting Media Composition and Probiotics of Effective Microorganisme 4 Application on the Production of Oyster Mushroom (Pleurotus Ostreatus) *Budapest Intl Res. Exact Sci. (BirEx) J.*, **4**(1) 99-104.

[18] Anderson D J, Moehring R W, Weber D J, Lewis S S, Chen L F, Schwab J C, Becherer P, Blocker M, Triplett P F, Knelson L P, and Lokhnygina Y 2018 Effectiveness of targeted enhanced terminal room disinfection on hospital-wide acquisition and infection with multidrug-resistant organisms and Clostridium difficile: a secondary analysis of a multicentre cluster randomised controlled trial with crossover design (BETR Disinfection) *Lancet Infect. Dis.* **18**(8) 845-853

[19] Puccetti M, Pariano M, Renga G, Santarelli I, D'Onofrio F, Bellet M M, Stincardini C, Bartoli A, Costantini C, Romani L, and Ricci M 2021 Targeted drug delivery technologies potentiate the overall therapeutic efficacy of an indole derivative in a mouse cystic fibrosis setting *Cells* **10**(7) 1601

[20] Mirwandono E, Sitepu M, Wahyuni T H, Ginting N, Siregar G A and Sembiring I 2018 Nutrition quality test of fermented waste vegetables by bioactivator local microorganisms (MOL) and effective microorganism (EM4) *IOP Conf. Ser.: Earth Environ. Sci.* **122** 012127

Environmental impact assessment approach to dynamic safety evaluation : a liquefied natural gas plant case study

Abderraouf Bouafia[1*], Mohammed Bougofa[2], Bilal Zeroauli[1], Rabeh Kharzi[3], Ammar Chakhrit[4], Amin Baziz[2], Salah Aberkane[5], Mohamed Salah Medjram[1], Ahmed Mebarki[6]

[1]University 20 Aout 1955 Skikda, Laboratory Genie Chimique et Environnement de Skikda, Faculty of technology, Department of Process engineering, 26 Road El Hadaiek, 21000, Skikda, Algeria.
[2]University Freres Mentouri Constantine I, Laboratory Ingenierie de Transport et Environnement, 325 Road Ain El Bey, 25017, Constantine, Algeria.
[3]University M Hamed Bougara of Boumerdes.
[4]University of Oran 2, Institute of maintenance and industrial safety, Oran, 31000, Algeria.
[5]University of Khenchela, Department of Psychology, BP 1252 Road of Batna Khenchela, 40004, Algeria.
[6]University Gustave Eiffel, Laboratory Multi Scale Modeling and Simulation (UMR 8208 CNRS/UPEC/U.Eiffel), 5 Bd Descartes, 77454, Marne-La-Vallee, France.
* abderraouf.bouafia@gmail.com

Abstract. Industrial discharges pose global ecological risks. This study investigates Algeria's GL1K (gas liquification complex of Skikda) plant's hazardous effluents. The impact assessment evaluates the environmental repercussions of the researched facilities, including the effects on populations and their way of life. Thus, it is possible to stress suggestions to improve facility design to remove or restrict negative effects and to minimize or compensate for the current facility's unfavorable repercussions. This technique is consistent with establishing and monitoring the industrial plant's environmental management system. The method utilized to determine impacts in this EIA can be used to evaluate the EMS's significant aspects and effects and provide improvement options. An evaluation matrix can rate impacts, the grid and criteria are based not regulation but on "good practices" for this type of assessment, and results obtained from modelling the effects using PHAST software. Therefore, they can be modified to fit the facility's activity. Based on examining activities and identifying elements likely to interact with the environment, environmental aspects are identified using the proposed grid and criteria. This study describes a section-by-section approach. Each determined environmental impact can be graded based on environmental factor criteria.

1. Introduction

Skikda is one of the most important industrial ports in Algeria. The Port Enterprise of Skikda as well as EGZIK (Skikda industrial zone management company) keep records of all cases of major accidental pollution [1]. Several accidents have been recorded, mainly, hydrocarbon pollution are the primary sources of pollutants generated by routine operations. Pollutant discharge rules impose stringent limitations on the concentrations of contaminants that can be released [2]. There may be additive or synergistic harmful effects of industrial effluents on humans, animals, and marine life [3]. Air, soil, rivers, and marine life poisoned by such effluents have a disastrous effect on the entire community [4]. During operation, accidental spills of oil and chemicals can occur. In 2012, 122 minor occurrences with a total oil discharge of 16 m^3 were reported. In the previous decade, there have been between 100 and 150 occurrences of acute chemical spills per year. Prior to the mid-1990s, the primary source of environmental pollution was crude hydrocarbons released by the petroleum industry. One of the primary purposes of environmental monitoring is to determine if discharge rules are protective enough. The past of sediment monitoring has revealed that the detection of unanticipated biological consequences alone has resulted in tougher discharge laws. In the field of environmental protection, sustainable development and the preservation of biodiversity, Algeria has adhered to many international protocols and conventions. Among the most important conventions is the one for the protection of the Mediterranean Sea against pollution, the Barcelona Convention, which was adopted in 1976 and came into force in 1978. It was amended in 1995 and entitled "Convention for the Protection of the Marine Environment and the Coastal Region of the Mediterranean"[5]. The Convention and its six Protocols constitute the legal framework of the Mediterranean Action Plan (MAP). The Barcelona Convention and the Protocols drawn up under its aim to reduce pollution in the Mediterranean Sea area and to protect and improve the marine environment in this area with a view to contributing to its sustainable development. Algeria adheres to this convention by the decree n° 80-14 of January 26, 1980[6]. The toxic release inventory (TRI) is a key component of the 1986 Emergency Protection and Community Right-to-Know Act (EPCRA) [7]. Today the TRI is an important national database of toxic chemical releases that is widely used by economists, public interest groups, journalists, policy makers, consultants, and investors to assess trends in the environmental commitment industry [8]. The TRI has been widely lauded as the best source of data nationwide on toxic chemical releases[9]. Various countries have implemented similar pollutant release and transfer registries. The environmental impact assessment (EIA) is a study required by the regulations relating to classified facilities subject to authorization. This work represents a decision support tool on the choice of the variants of a project by taking into account in addition to the socio-economic criteria and the environmental criteria [2]. The EIA must take into account the impacts on the environment at all stages of the project, from its design to its operation, including its closure, and makes it possible to design a project that is more sensitive to the receiving environment, without jeopardizing its technical and economic feasibility [10]. The impact assessment takes into account all the components of the biophysical and human environments likely to be affected by the project. It makes it possible to analyse and interpret the relationships and interactions between the factors influencing ecosystems, resources and the quality of life of individuals and communities. The public may express opinions and reservations about the proposed developments or works [11]. This paper aims to assess environmental impact based on the MEEDDM methodology (ministry of ecology, energy, sustainable development and the sea) for the industrial complex of Skikda. The second section is reserved for the modelling results of the environmental measurement of O_3, NO_2, SO_2 on the atmosphere, soil and marine quality. The third one presents the threshold effects by inhalation of gases such as nitrogen dioxide, carbon monoxide and particulate matter (as a physical agent: PM10) and the relationship between self-consumption and pollution, hence, its necessary to discuss the obtained results, while the last section is dedicated to a conclusion.

2. Methodology

The methodology used in this study refers to the MEEDDM methodological guide (see figure 1) "Evaluation of health risks in impact studies of facilities classified for environmental protection " (2002)[12]. We use an approach to map the impact of air emissions over a long period of time in order to obtain results that can be used for health risk assessment, which focuses on the effects of exposures of potentially exposed populations over long periods of time (chronic exposure). Samples were collected from databases provided by the company's laboratory utilizing highly developed sensors and detectors deployed in the various facilities for atmospheric emissions. Additionally, sea and soil samples were collected, analyzed, and entered into their data base for subsequent studies. The modelling tool PHAST used correspond to the recommendations of the US-EPA and the Institut de Veille Sanitaire (InVS) for the study of the health impact of atmospheric emissions from fixed sources [13], [14]. PHAST is a software tool that has been developed and maintained by DNV software [15] to assess the consequences of gas leaks, fires, explosions, toxicity and other technological hazards related to various industries. PHAST simulates the evolution of an accidental release of a toxic and/or flammable product, from the initial leak to the atmospheric dispersion in the far field.

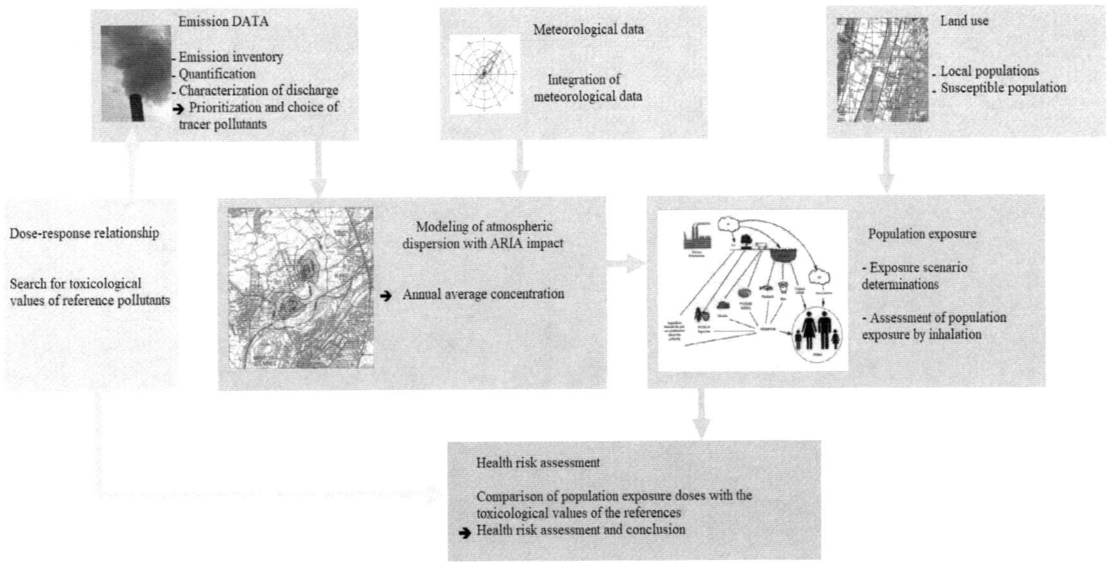

Figure 1. MEEDDM methodological guide [12]

In order to rank the impacts, an evaluation grid can be used. The grid and criteria we propose are based on regulations. They are also directly transposable to the rating method required by paragraph 4.3.1 of the ISO 14001-2004 standard [16]. They can therefore be modified to better adapt the grid to the activities of the complex. The identification of environmental aspects is based on the analysis of activities and the identification of elements that may interact with the environment. This is the approach that has been carried out and that is presented in this study. Each environmental impact identified can be rated according to the criteria that we have associated with each of the environmental aspects.

3. Uncertainties

This EIA was carried out using the methods and data recommended by the (InVS)[14], MEEDDM[12],US-EPA[13] and the WHO [17]. Nevertheless, the EIA process is necessarily accompanied by a degree of uncertainties due to gaps or imprecision in the data collected and the established assumptions [18]. The main sources of uncertainty are:

- The extrapolation of toxicological data from epidemiological studies and animal experiments and the assimilation of a mixture of substances to a penalising substance of the same family,
- Uncertainties in the quantification of emissions and therefore in the choice of tracer pollutants, including the nature of the substances emitted,
- Uncertainties related to the atmospheric dispersion model used,
- Uncertainties about population exposure and human variability to different factors.

The purpose of this work is to present the main uncertainties, some of which overestimate the risk, others underestimate it. The EIA should be read as an estimate of the potential environmental impact based on knowledge at the time of the study and on a number of assumptions.

4. Presentation of the modelling results

All the concentrations modelled and discussed in our study are attributable to the studied emissions of the GL1K complex in Skikda. The results of the study are given in the form of maps and tables. These results only concern the contribution of the studied discharges. The maps are made of colored areas representing each a concentration interval. The tables present the most significant situation in terms of impact, insofar as the values used are the maximums for the entire study area.

4.1 Environmental measures

4.1.1 Air quality

Figure 2. Location of ambient air (yellow) and soil (red) measurement points

The following table 1 shows the measurement results of passive O_3, NO_2 and SO_2 tubes from Skikda:

Table 1. Measurement results of O_3, NO_2, SO_2

Points	Premises	Date of installation	Drop off date	Concentration ($\mu g/m^3$)					
				O_3	O_3 limit	NO_2	NO_2 limit	SO_2	SO_2 limit
P1	GL1K fence	06/05/2022	09/05/2022	350	200	45	200	<0.3	350
P2	Guard post N1	06/05/2022	09/05/2022	343	200	39	200	<0.3	350
P3	GL1K city	06/05/2022	09/05/2022	311	200	52	200	<0.3	350

The measured O_3 concentrations (for 3 days) are above the limit values (annual average). Continuous measurements must be carried out to compare them with the limit values, alert thresholds and air quality objectives of Executive Decree 06-02, hence the need to install an air quality measurement station in the industrial zone of Skikda.

4.1.2 Soil quality

The choice of piezometer locations was based on hydrogeological knowledge of the area. A groundwater monitoring network was set up with a piezometer placed upstream of the site (PzA) and two others placed downstream of the site (PzB1 & PzB2). The soil test points were selected from among the zones representing a potential source of pollution for the environment. The locations of the piezometers and soil borings are indicated on the plan as stars (see figure 2). No significant organic pollution on the measured parameters (C5-C40 hydrocarbons, BTEX, PAHs, VOCs) was detected on the studied site. Concerning metals, abnormal lead and zinc levels were found in the surface soils of borings S1 and S2. All the levels found correspond either to the ranges of values commonly observed in ordinary soils or to the ranges of values observed in the case of moderate anomalies.

4.1.3 Quality of the marine

To determine the quality of the marine environment in the vicinity of the GL1K complex, seawater and sediment sampling points were taken. They are distributed in 03 areas of the shoreline in front of the GL1K complex in the bay of Skikda (see figure 3).

Figure 3. Location of seawater and sediment sampling points

A first estimation of the current state of the physico-chemical quality of the sea water at 2 m depth in the GL1K complex, shows:

- of significant mineralization
- the presence of a high organic load (BOD5) for samples B & C
- for sample A, of a slightly high organic load.

This ecological state becomes critical during the summer period due to the lack of water supply (absence of dilution), which favors the accumulation of organic matter at the bottom of the water bodies. The results of physico-chemical analyses are summarized in the following table 2:

Table 2. Results of physico-chemical analyses of sediments

Parameters	A	B	C
Total hydrocarbons (mg/kg)	0.92	8.18	9.07
Hg (mg/kg)	0.01	0.01	0.01
Pb (mg/kg)	88	125	130
Zn (mg/kg)	162	198	199
Cd (mg/kg)	10	12	38
Cu (mg/kg)	0.9	1	1.3
PH	8.3	8.36	9.24
T°C	19.3	19.4	19.4
Ni (mg/kg)	98	112	161
Cr (mg/kg)	190	200	211
As (mg/kg)	38	41	68

Below level L1	
Between level L1 and L2	
Beyond level L 2	

Table 2 shows the level of mercury, zinc and copper is below the L1 level for all samples. The high levels of various pollutants in the sediment make it difficult to manage. As the dumped sediments are fine, it is also possible that wave action has contributed to dispersing the sediments during and after disposal operations at sea and transporting them away from the disposal site to accumulate in depressions on the sea floor. Dumping and landfilling pose significant environmental problems.

4.2 Relationship between self-consumption and pollution:

The main sources of air pollution generated by the GL1K natural gas liquefaction complex are boilers, flares, chemical pits, fire school drills, burners. The greatest pollution comes from the first two sources: boilers and flares, as these emit pollutants continuously. To better assess the degree of air pollution resulting from natural gas liquefaction complexes, we have a quantity called self-consumption which represents the amount of NG required to be consumed to ensure the production process of LNG, butane, propane and gasoline. Self-consumption is defined as the difference, at a given time, between the incoming and outgoing thermies in the form of finished products (LNG, propane, butane, gasoline, etc.). It is expressed in percent (%). Self-consumption thus appears to be a valuable indicator of both the performance of the complex and the degree of pollution. The greater the self-consumption, the greater the pollution because the complexes emit large quantities of smoke from the combustion of natural gas in the boilers and flares. The fumes are generally composed of CO_2, CO, NOx, SO_2, at a very high temperature.

4.2.1 Results of the health risk assessment

The following maps (see figures 4,5,6) are given for the pollutants selected in this study using PHAST software [15]. The study area is a square of 10 km side.

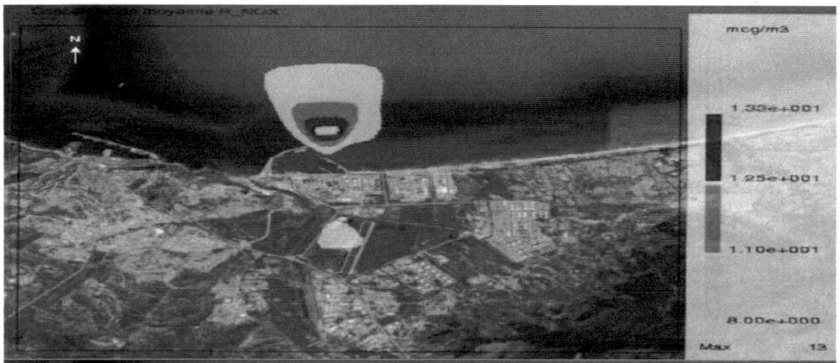

Figure 4. Map of modelled annual average air concentrations for nitrogen oxides assimilated to NO_2

Figure 5. Map of modelled annual average air concentrations for carbon monoxide (CO)

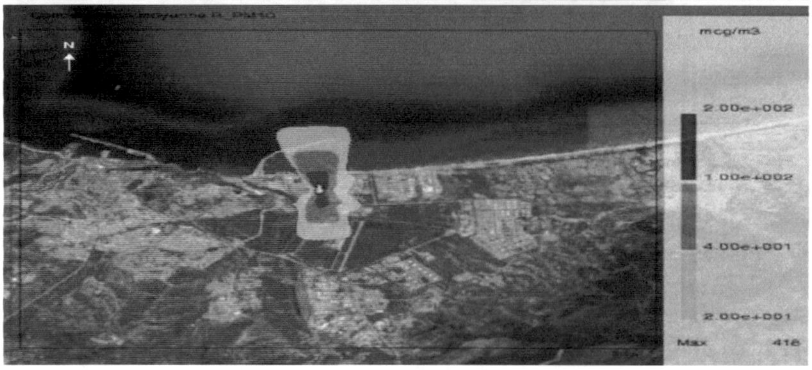

Figure 6. Map of modelled annual average air concentrations for particulate matter (similar to PM10)

4.2.2 Threshold effects by inhalation

4.2.2.1 Results

The maps presenting the results of the atmospheric dispersion modelling show that the maximum atmospheric concentrations modelled as an annual average:

- For nitrogen dioxide, they are located at sea (no targets in the sense of the Health Risk Assessment, which evaluates the exposure of local populations).
- For carbon monoxide and particles (PM10), they are located within the property limits of the GL1K complex. On the complex, the people present are workers: they are not targets in the sense of the Health Risk Assessment, which evaluates the exposure of neighboring populations. These concentrations are indicated in the tables below for information.

For the health risk assessment, the maximum annual average modeled air concentrations at the point of potential exposure of targets (approximately 500 m from the western boundary of the GL1K complex) were used.

4.2.2.2 Combustion

- gases Nitrogen dioxide

Table 3. The modelled annual average concentration at the worst-case point

Annual average concentration (µg/m³)		TRI for chronic exposure
Concentration modeled at the most penalizing point (at sea: no targets)	13.5	No TRI, otherwise ✓ WHO guideline =40 µg/m³ ✓ Air quality objective (directive n 06-02) = 135 µg/m³
Modeled concentration at the point where potential targets are likely to be most exposed	1.9	

Table 3 shows the modelled annual average concentration at the worst-case point is below the WHO guideline and the air quality objective.

- Carbon monoxide

Table 4. The modelled annual average concentration at the worst-case point

Annual average concentration (µg/m³)		TRI for chronic exposure
Concentration modeled at the most penalizing point (on the right-of-way of the complex)	700	No TRI, otherwise ✓ WHO recommendation for the maximum daily moving average over 8 hours = 10 µg/m³
Modeled concentration at the point where potential targets are likely to be most exposed	17.5	

Table 4 shows the modelled annual average concentration at the most penalizing point is below the WHO recommendation.

- Particulate Matter (as a Physical Agent: PM10)

Table 5. The modeled annual average concentration at the worst-case point

Annual average concentration (µg/m³)		TRI for chronic exposure
Concentration modeled at the most penalizing point (on the right-of-way of the complex)	418	No TRI, otherwise ✓ WHO guideline =20 µg/m³ ✓ Air quality objective (directive n 06-02) = 50 µg/m³
Modeled concentration at the point where potential targets are likely to be most exposed	9.6	

Table 5 shows the modelled annual average concentration at the most critical point where targets are likely to be present is below the WHO guideline and the air quality objective [17].

4.3. Discussion

The impact assessment was carried out for the emissions of discharges attributable to the installations of the GL1K complex operated on the Skikda platform by SONATRACH. The maximum atmospheric concentrations modelled as an annual average are located either at sea for nitrogen dioxide or within the property limits of the GL1K complex. In terms of the Health Risk Assessment, which examines the exposure of nearby populations, the occupants of the complex are not targets. The maximum atmospheric concentrations modelled as an annual average at the point where targets are potentially exposed (approximately 500 m from the western limit of the GL1K complex) make it possible to respect the values recommended by the WHO [17] and the air quality objectives of the Executive Decree n°06-02 [19]. On the observed parameters (C5-C40 hydrocarbons, BTEX, PAH, and VOC), the site under study did not exhibit any major organic pollution. On the surface soils of drill holes S1 and S2, abnormal quantities of lead and zinc were discovered. All levels obtained correspond to either the ranges of values frequently recorded in ordinary soils or the ranges of values observed for moderate anomalies. PZA and PZB2 piezometers produce water with mild salinities, whereas PZB1 piezometer water is quite brackish. The water studies suggest that groundwater is contaminated with metals (As and Pb) and toluene, indicative of a pollution background of the water table at the level of the Skikda industrial zone. Due to the nature of the water cycle and its consequences (dissolution, leaching, concentration), the marine environment contains the most soluble salts (such as NaCl) in a wide variety of concentrations.

5. Conclusion

Public concern has been raised regarding operational discharges from the petrochemical industry because they represent a very substantial, continuous intake of contaminants from numerous widely scattered point sources. In addition, it is notoriously difficult to examine the impacts of discharges on population structure and ecological system performance. All evidence indicates that the impacts of current discharges are local and generally confined within 1 km of an outlet, both in the water and on the land, and that the risk of discharges having a wider influence is low. It is crucial to note, however, that aside from studies on the effects of drilling waste on sediment, there is almost no published evidence on the effects on populations or communities. Consequently, one must not disregard the possibility of subtle, cumulative effects from operational discharges that we are currently unable to measure, despite risk assessments indicating that this will not be the case. It is disheartening that there appear to be no alternatives to risk-based modeling for estimating potentially substantial effects of pollution discharges at the human and natural system levels. The published research has not yet been able to validate with certainty or empirically confirm that the discharges have only local impacts. We believe that research addressing this challenge should emphasize, first, the development of effects methods and endpoints that can be used with confidence in health screening of organisms at a scale large enough to reflect population health conditions, and, second, effects studies that encompass natural species interaction at the ecological system level.

Rererences

[1] A. Bouafia, M. Bougofa, M. Rouainia, and M. S. Medjram, "Safety Risk Analysis and Accidents Modeling of a Major Gasoline Release in Petrochemical Plant," *J Fail. Anal. and Preven.*, vol. 20, no. 2, pp. 358–369, Apr. 2020, doi: 10.1007/s11668-020-00826-9.

[2] T. Bakke, J. Klungsøyr, and S. Sanni, "Environmental impacts of produced water and drilling waste discharges from the Norwegian offshore petroleum industry," *Marine Environmental Research*, vol. 92, pp. 154–169, Dec. 2013, doi: 10.1016/j.marenvres.2013.09.012.

[3] R. Avagyan, G. Luongo, G. Thorsén, and C. Östman, "Benzothiazole, benzotriazole, and their derivates in clothing textiles—a potential source of environmental pollutants and human

exposure," *Environ Sci Pollut Res*, vol. 22, no. 8, pp. 5842–5849, Apr. 2015, doi: 10.1007/s11356-014-3691-0.

[4] J. Rovira, M. Nadal, M. Schuhmacher, and J. L. Domingo, "Human exposure to trace elements through the skin by direct contact with clothing: Risk assessment," *Environmental Research*, vol. 140, pp. 308–316, Jul. 2015, doi: 10.1016/j.envres.2015.03.032.

[5] "Convention for the Protection of the Marine Environment and the Coastal Region of the Mediterranean — European Environment Agency." https://www.eea.europa.eu/policy-documents/convention-for-the-p...f-the-mediterranean (accessed Jul. 22, 2022).

[6] "JOURNAL OFFICIEL DE LA RÉPUBLIQUE ALGERIENNE." gov algérie. Accessed: Jul. 22, 2022. [Online]. Available: https://www.joradp.dz/FTP/jo-francais/1985/F1985002.PDF

[7] O. US EPA, "Emergency Planning and Community Right-to-Know Act (EPCRA)," Jul. 24, 2013. https://www.epa.gov/epcra (accessed Aug. 07, 2022).

[8] D. A. Koehler and J. D. Spengler, "The toxic release inventory: Fact or fiction? A case study of the primary aluminum industry," *Journal of Environmental Management*, vol. 85, no. 2, pp. 296–307, Oct. 2007, doi: 10.1016/j.jenvman.2006.09.025.

[9] "Peer Reviewed: Are Toxics Release Inventory Reductions Real?," *Environ. Sci. Technol.*, vol. 32, no. 15, pp. 368A-374A, Aug. 1998, doi: 10.1021/es983672m.

[10] B. Mohammed, B. Abderraouf, and B. Ahmed, "An Integrated Quantitative Bayesian Network in Risk Management for Complex Systems," *Int J Performability Eng*, vol. 16, no. 3, pp. 354–366, 2020, doi: 10.23940/ijpe.20.03.p4.354366.

[11] R. Slootweg, F. Vanclay, and M. van Schooten, "Function evaluation as a framework for the integration of social and environmental impact assessment," *Impact Assessment and Project Appraisal*, vol. 19, no. 1, pp. 19–28, Mar. 2001, doi: 10.3152/147154601781767186.

[12] C. Boudet, "Evaluation du risque sanitaire dans l'étude d'impact des installations classées pour la protection de l'environnement (ICPE). Principes méthodologiques et retour d'expérience," *Environnement, Risques & Santé*, vol. 1, no. 2, pp. 101–7, Aug. 2002.

[13] "U.S. Environmental Protection Agency | US EPA." https://www.epa.gov/ (accessed Jul. 22, 2022).

[14] "Institut de veille sanitaire," *Haute Autorité de Santé*. https://www.has-sante.fr/jcms/c_436647/fr/institut-de-veille-sanitaire (accessed Jul. 22, 2022).

[15] DNV, "PHAST Tutorial Manual." DNV Software, London, 2016.

[16] 14:00-17:00, "ISO 14001:2004," *ISO*. https://www.iso.org/cms/render/live/en/sites/isoorg/contents/data/standard/03/18/31807.html (accessed Jul. 22, 2022).

[17] World Health Organization. Regional Office for Europe, *Air quality guidelines: global update 2005: particulate matter, ozone, nitrogen dioxide and sulfur dioxide*. World Health Organization. Regional Office for Europe, 2006. Accessed: Jul. 22, 2022. [Online]. Available: https://apps.who.int/iris/handle/10665/107823

[18] M. Bougofa, A. Bouafia, and and A. Bellaouar, "Availability Assessment of Complex Systems under Parameter Uncertainty using Dynamic Evidential Networks," *International Journal of Performability Engineering*, vol. 16, no. 4, p. 510, Apr. 2020, doi: 10.23940/ijpe.20.04.p2.510519.

[19] "Executive Decree n°06-02 of 7 Dhou El Hidja 1426 corresponding to January 7, 2006." gov algérie. [Online]. Available: https://onedd.org/storage/app/media/uploaded-files/decret-executif-n006-02.pdf

Review of seepage behaviour on concrete facing rockfill dam based on FEM simulation and chemical aspect

G Pamungkas[1*], T T Putranto[2] and Suharyanto[3]

[1] Master's Degree Student of Department of Civil Engineering, Diponegoro University, Jalan Prof Sudharto SH, Semarang, Indonesia
[2] Lecturer at Department of Geological Engineering Diponegoro University, Jalan Prof Sudharto SH, Semarang, Indonesia
[3] Professor at Department of Civil Engineering, Diponegoro University, Jalan Prof Sudharto SH, Semarang, Indonesia

*insinyur.goji@gmail.com

Abstract. Concrete Facing Rockfill Dam (CFRD) is a popular type of dam for hydropower plant due to its relatively good stability than others. However, CFRD have a potential technical issue concerning seepage and leakage as it has a thin impermeable layer. In this study, dam stability against seepage was evaluated by finite element simulation. Concrete facing layer and dam foundation were assumed to undergo permeability coefficient (K) increasing within 3 stages and would be simulated within 15 scenarios. Furthermore, seepage chemistry analysis was conducted to evaluate seepage water interaction with the dam foundation. FEM simulation shows that increasing of permeability coefficient on concrete facing < 10^{-5} m/second tends to cause the exceedance of allowable seepage discharge ($q_{-simulation} > q_{-allowable}$). On other hand, the increase of permeability coefficient on the dam foundation tends to cause the dam failure against seepage force. According to seepage chemistry analysis, calcium ion Ca^{2+} concentration increase 1.8 times from 13.6 meq/L on 2019 sampling to 24.8 meq/L on 2020 sampling. The seepage chemistry analysis shows the sign of carbonate rock interaction as dam foundation with seepage water.

1. Introduction

Many experiences on CFRD, the dam performance can be evaluated based on dam deformation, dam crest settlement and seepage or leakage of the dam [1]. Seepage analysis on CFRD is rarely evaluated than its deformation and slope safety due to the high strength of rockfill material. Though it is rarely to be conducted, seepage safety evaluation is suggested for dams as the dam aging process could affect dam seepage [2]. In some case, seepage flow through CFRD on first filling stage can cause collapse settlement as result of reduction on rockfill shear strength. Furthermore, significant seepage flow at operational stage can cause dam instability if the dam does not have proper drainage system [3].

The Guhou Dam was one of famous CFRD dam which has stability failure related to seepage flow. The dam failure started after first filling to the maximum reservoir level. Investigation shows that seepage water flows through the construction joint of concrete slab. In addition, the concrete slab does not lie directly above the transition filter. Some defects in the construction process also decrease the dam's stability. The sand and gravel or alluvium layer which lies below the dam structure increase the seepage discharge [4]. In other hands, a crack on concrete slab can be formed because of high water pressure as happened in Barra Grande dam, Campos Navos dam, Aguamilpa dam, Khao Laem dam and Xingo dam [5]. The seepage may flow through the crack on concrete slab.

The dam foundation can experience instability issue according to its geological feature. The dam with high seepage energy can cause mineral flushing and dissolution especially on sandstone and limestone formation. At initial condition, the seepage water flow through the rock foundation crack or joints. Continuous seepage water flow will flush and dissolve rock minerals, which can cause the wider crack. This phenomenon can be evaluated based on seepage chemistry characteristic [6]. The chemical analysis of seepage water can be used to find the interaction between seepage water and rock foundation of the dam [7]. Furthermore, seepage chemistry analysis is also used to investigate the integrity of cement grouting layer on the upstream of the dam [8].

The objective of this research is to study whether the dam is designed to withstand againts exsessive seepage flow. Series seepage simulation will be conducted to investigate whether if the concrete slab is degraded, the seepage flow will decrease the dam safety. The change of seepage chemistry may indicate the increase of foundation permeability because of mineral flushing and dissolution of the rock foundation. Considering the seepage chemistry aspect, this research is also intended to study whether the increase of rock foundation permeability will affect the dam safety against uplift forces.

2. Research Method

This section explains CFRD seepage evaluation method. Firstly, the study was started with seepage chemistry analysis of the seepage water sample, which was obtained from the dam location. From the seepage chemical analysis, it might be determined whether the dam foundation experienced mineral flushing and dissolution or not. After that, the study would be continued to conduct numerical simulation for seepage analysis. The simulation would consider the presence of crack on concrete slab and the increase of coefficient permeability of the dam foundation.

2.1. Dam Location

A concrete facing rockfill dam (CFRD) usually consist of compacted rockfill as dam body which lies on rock foundation. Dam zoning consist of concrete face protection layer, concrete supporting layer and rockfill [9]. The CFRD uses thick concrete slab on the upstream face of the dam as impermeable layer [10]. The study will use one of CFRD in Indonesia ("X" Dam) which located in V-shaped valley of mountainous area. The "X" dam was built from breccia rockfill with 3 transition and filter layer between the concrete facing and the rockfill layer. The dam height is about 126 meters, and it has 1 meter thickness of concrete facing. The dam section is shown on Figure 1.

The dam foundation consists of volcanic product, sandstone, and claystone alternation. The claystone divided into sandy claystone and calcareous claystone. Carbonate mineral is found on the rock foundation such as marl limestone and calcite. The volcanic product on the dam foundation consists of volcanic breccia, andesite, and tuffaceous conglomerate.

2.2. Seepage Chemistry Analysis

Seepage chemistry analysis should be conducted as qualitative method to evaluate the seepage behaviour of the dam [6]. The analysis was conducted by comparing the chemical content between water sample from reservoir and seepage water, which had been taken in 2019 and 2020. The chemical content difference between reservoir and seepage water may indicate mineral flushing and dissolution of dam foundation. Seepage chemistry should be analysed based on major cation (Na^+, K^+, Mg^{2+}, Ca^2), major anion (Cl^-, HCO_3, SO_4^{2-}, NO_3^-), electric conductivity, and pH [11]. The increase of certain ion concentration might lead to mineral dissolution or ion exchange.

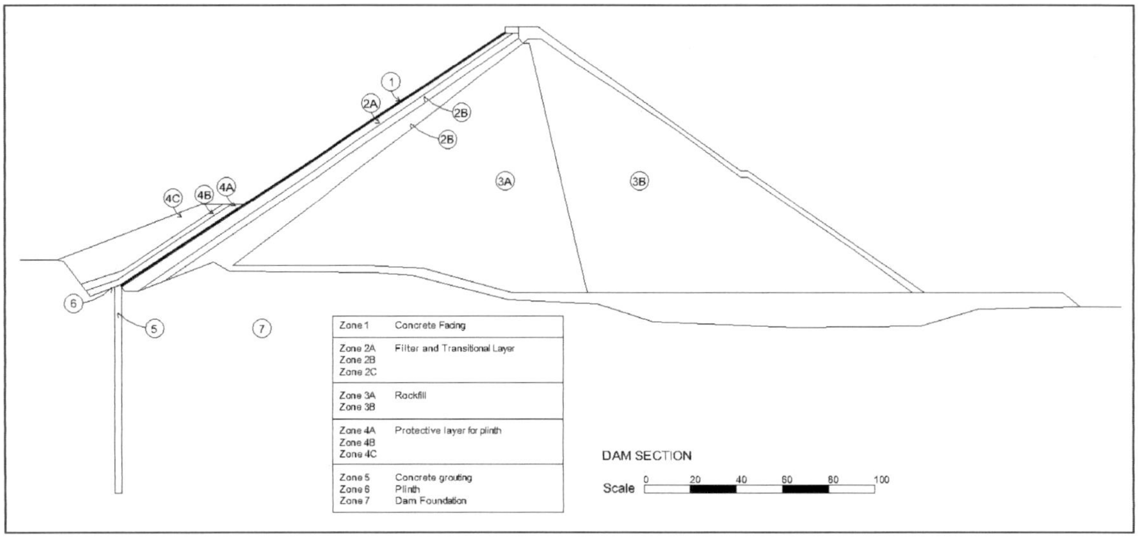

Figure 1. The section of "X" concrete facing rockfill dam

2.3. Numerical Seepage Simulation

Seepage simulation was conducted by increasing coefficient of permeability of the concrete facing and the dam foundation. The simulation would be assumed as steady-state seepage and calculated by using finite element method. The calculation result should be evaluated based on critical hydraulic gradient and allowable seepage discharge. Critical hydraulic gradient could be calculated as in equation (1) and equation (2) [12], while the allowable seepage discharge could be obtained based on dam height [13]. Hydraulic conductivity parameter for seepage simulation is listed on Table 1.

$$i_{critical} = \frac{\gamma'}{\gamma_w} \quad (1)$$

$$i_{critical} = \frac{G_s - 1}{1+e} \quad (2)$$

Table 1. Hydraulic conductivity and material parameters for seepage simulation

Dam Zone	Zone	G_s	γ-m kN/m³	γsat kN/m³	e	K_x (m/s)	K_y (m/s)	I_{cr}
Concrete Facing	1	3	21.58	23.54	0.43	10^{-10}	10^{-10}	2.40
Cushion	2A	2.65	20.40	22.56	0.27	10^{-4}	10^{-4}	2.30
Transition	2B	2.65	20.40	22.56	0.27	5×10^{-4}	5×10^{-4}	2.30
Filter	2C	2.30	20.40	22.56	0.00	10^{-3}	10^{-3}	2.30
Rockfill A	3A	2.80	19.62	21.58	0.50	10^{-2}	10^{-2}	2.20
Rockfill B	3B	2.80	19.62	21.58	0.50	2×10^{-2}	2×10^{-2}	2.20
Protective layer	4A	2.65	20.40	22.56	0.27	10^{-4}	10^{-4}	2.30
	4B	2.65	20.40	22.56	0.27	5×10^{-4}	5×10^{-4}	2.30
	4C	2,3	20.40	22.56	0.00	10^{-3}	10^{-3}	2.30
Plinth	1	3	21.58	23.54	0.43	10^{-10}	10^{-10}	2.40
Grouting	1	3	21.58	23.54	0.43	10^{-10}	10^{-10}	2.40
Foundation	7	2.80	19.62	21.58	0.50	5×10^{-6}	5×10^{-6}	2.20

Concrete facing layer and dam foundation were assumed to undergo permeability coefficient (K) increasing within 3 stages and would be simulated within 15 scenarios after the existing simulation based on Table 1. On the concrete facing, the seepage was simulated by assuming 1-meter width of the damaged area (code C_dam). The dam geometry and finite element discretization is shown on Figure 2. On the dam foundation, the increase of coefficient of permeability was simulated due to rock foundation dissolution phenomena (code F_dam). Hydraulic conductivity parameter for seepage simulation scenario is listed on Table 2.

Table 2. Hydraulic conductivity for 15 scenarios of seepage simulation

Simulation Scenario	Scenario Number	Coefficient of permeability Concrete Facing (m/s)	Coefficient of permeability Dam Foundation (m/s)
Initial Condition	G0	10^{-10}	5×10^{-6}
C_dam 1	G1	10^{-8}	5×10^{-6}
C_dam 2	G2	10^{-5}	5×10^{-6}
C_dam 3	G3	10^{-3}	5×10^{-6}
F_dam 1	G4	10^{-10}	5×10^{-5}
F_dam 2	G5	10^{-10}	5×10^{-3}
F_dam 3	G6	10^{-10}	5×10^{-2}
C_dam 1 + F_dam 1	G7	10^{-8}	5×10^{-5}
C_dam 1 + F_dam 2	G8	10^{-8}	5×10^{-3}
C_dam 1 + F_dam 3	G9	10^{-8}	5×10^{-2}
C_dam 2 + F_dam 1	G10	10^{-5}	5×10^{-5}
C_dam 2 + F_dam 2	G11	10^{-5}	5×10^{-3}
C_dam 2 + F_dam 3	G12	10^{-5}	5×10^{-2}
C_dam 3 + F_dam 1	G13	10^{-3}	5×10^{-5}
C_dam 3 + F_dam 2	G14	10^{-3}	5×10^{-3}
C_dam 3 + F_dam 3	G15	10^{-3}	5×10^{-2}

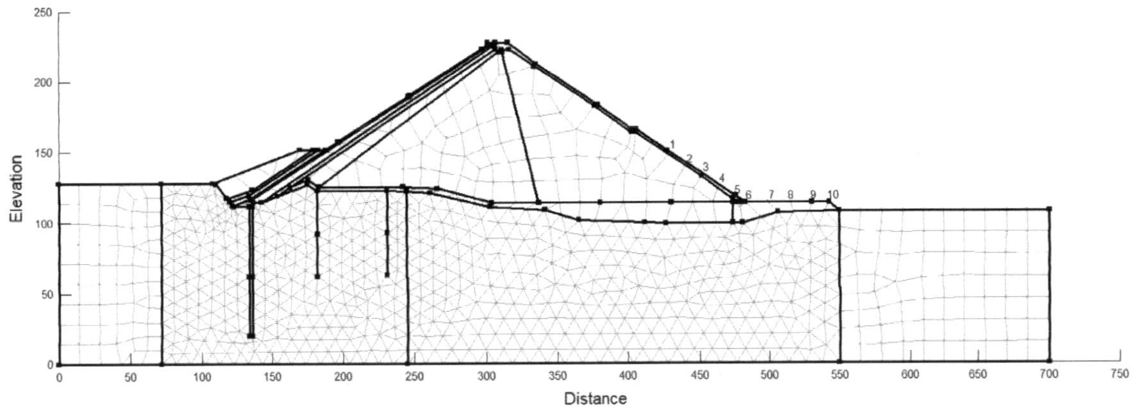

Figure 2. The finite element discretization of "X" dam for seepage simulation

3. Results and Discussion

3.1. Chemical analysis of seepage water

Based on chemical test of water samples, major ion, and electric conductivity (EC) have different result between reservoir sample and seepage water sample. The increase of EC on water sample may also indicate the increase of ion concentration. According to laboratory test result of 2019 sample, the EC of reservoir water is 208 μS/cm while seepage water EC is 2299 μS/cm. The EC of 2020 water sample is similar with laboratory test result of 2019 sample, in which the seepage water has greater EC than the reservoir sample.

In line with EC, ion concentration of the seepage water is greater than reservoir water sample as shown on Figure 3. Ion Ca^{2+} is the dominant major ion that is found on seepage water sample, while HCO_3^- ion is the dominant major ion on reservoir water sample. Ion Ca^{2+} concentration of 2019 test result on the reservoir water sample is 1.11 meq/L, while concentration on the seepage water sample is 13.7 meq/L. Furthermore, the ion Ca^{2+} concentration on the seepage water of 2020 test is greater than concentration of 2019 test. The concentration increases almost twofold, from 13.7 meq/L to 24,84 meq/L.

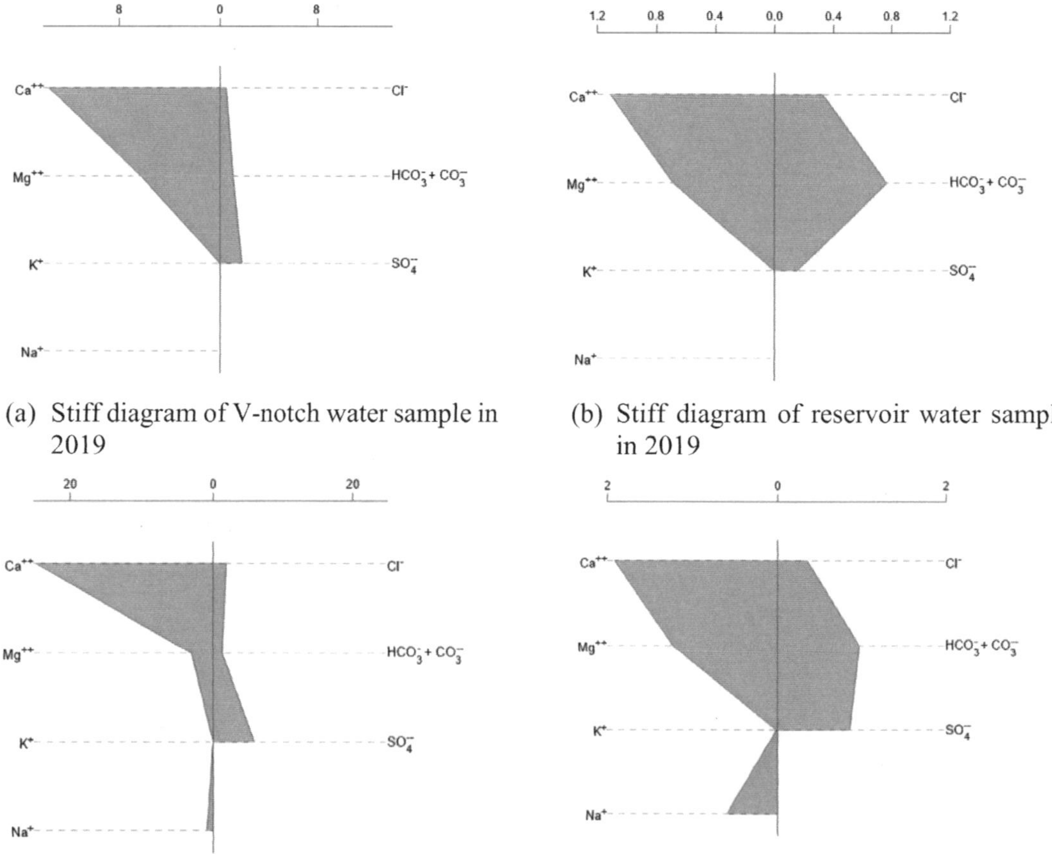

(a) Stiff diagram of V-notch water sample in 2019

(b) Stiff diagram of reservoir water sample in 2019

(c) Stiff diagram of V-notch water sample in 2020

(d) Stiff diagram of reservoir water sample in 2020

Figure 3. The stiff diagram of major ion which is obtained from laboratory test (meq/L)

The increase of carbonate ion on seepage water may indicate dissolution process related to carbonate or calcareous rock of the dam foundation. The increase of carbonate ion is found on Tannur Dam which may lead to grouting dissolution if the alkalinity of seepage water also increases (pH > 8) [7]. Extreme dissolution on grouting and rock foundation is also found on Xin'an Jiang Hydropower where the increase of carbonate ion is accompanied by the increase of water alkalinity (pH > 11). According to the laboratory test, seepage water of "X" dam increases from pH 6.93 on 2019 to pH 7,41 on 2020. It may be concluded that the carbonate rock of "X" dam foundation is started to interact with the seepage flow from the dam reservoir.

3.2. Dam safety based on Numerical Simulation

According to simulation scenario as listed on Table 1, the increase of coefficient of permeability also increases seepage quantity and hydraulic gradient of "X" dam. The simulation result is listed on Table 3. Simulation G3 where the coefficient of permeability if the concrete facing increases to 10^{-3} m/s, the seepage quantitiy prediction increases into 2,000 L/s. Though the seepage quantity exceeds the minimum criteria, the dam body is on safe condition according to safety againts seepage force.

Table 3. Numerical seepage simulation result of "X" CFRD

Scenario	Scenario Number	Critical Hydraulic Gradient	Hydraulic Gradient Exit	Seepage Force FoS	Remarks FoS (min 2.00)	Seepage quantity (L/s)	Max. Seepage 925 L/s by Look (2014)
Initial Condition	G0	2.20	0.0027	815	Safe	13.6	Safe
C_dam 1	G1	2.20	0.003	733.33	Safe	13.7	Safe
C_dam 2	G2	2.20	0.023	95.65	Safe	110	Safe
C_dam 3	G3	2.20	0.43	5.12	Safe	2,120	Not safe
F_dam 1	G4	2.20	0.023	95.65	Safe	110	Safe
F_dam 2	G5	2.20	1.13	1.95	Not safe	5,100	Not Safe
F_dam 3	G6	2.20	2.73	0.81	Not safe	11,300	Not Safe
C_dam 1 + F_dam 1	G7	2.20	0.023	95.65	Safe	110	Safe
C_dam 1 + F_dam 2	G8	2.20	1.13	1.95	Not safe	5,100	Not Safe
C_dam 1 + F_dam 3	G9	2.20	2.73	0.81	Not safe	11,370	Not Safe
C_dam 2 + F_dam 1	G10	2.20	0.04	55.00	Safe	220	Safe
C_dam 2 + F_dam 2	G11	2.20	1.14	1.93	Not safe	5,170	Not Safe
C_dam 2 + F_dam 3	G12	2.20	2.74	0.80	Not safe	11,380	Not Safe
C_dam 3 + F_dam 1	G13	2.20	0.44	5.00	Safe	2,200	Not Safe
C_dam 3 + F_dam 2	G14	2.20	1.32	1.67	Not safe	6,000	Not Safe
C_dam 3 + F_dam 3	G15	2.20	2.78	0.79	Not safe	11,600	Not Safe

According to Table 3, if the simulation considers the increase of coefficient of permeability, it increases the seepage quantity and the safety against seepage force to the failure condition. Simulation G6 where the coefficient of permeability if the dam foundation increases to 5 x 10^{-2} m/s, the seepage quantity prediction increases into 11,600 L/s. In addition, the safety against seepage force decreases to 0.79 (below the acceptable criteria FoS > 2.00). If the simulation considers combining the increase of concrete facing and dam foundation coefficient of permeability, simulation G15 will be on the dam failure condition.

The increase of concrete facing coefficient of permeability will not lead to dam failure [14, 15]. The CFRD which is built from high strenght rockfill is relatively stable againts seepage force. If the CFRD is designed properly, high rate of seepage quantity because of crack on concrete facing will not endanger the dam stability, although the seepage quantity is greater than minimum criteria as suggested [9, 13]. It is clear that the increase of dam foundation coefficient of permeability may lead to seepage problem and decrease the dam stability.

4. Conclusion

This paper studied the dam stability against seepage by using finite element simulation and chemical aspect of seepage water. Based on numerical simulation, the increase of concrete facing coefficient of permeability will not lead to dam failure. Based on chemical test of water samples, major ion, and electric conductivity (EC) have different result between reservoir sample and seepage water sample. It is found that the ion Ca^{2+} concentration and water alkalinity in 2020 is greater than 2019 water sample. The chemical test of seepage water may lead to the increase of foundation coefficient of permeability due to carbonate rock dissolution. According to simulation result, The increase of dam foundation coefficient of permeability may lead to the dam instability.

References

[1] ICODS 2015 Interagency Committee on Dam Safety, Evaluation and Monitoring of Seepage and Internal Erosion Oklahoma: Federal Emergency Management Agency

[2] Shi Z, Gu C, Zhao E, and Xu B 2020 A Novel Seepage Safety Monitoring Model of CFRD with Slab Cracks Using Monitoring Data *Mathematical Problems in Engineering* **2020** 1-13

[3] Wen L, Junrui C, Xu Z, Qin Y, and Li Y 2017 A statistical review of the behaviour of concrete face rockfill dams based on case histories *Géotechnique* **68**(9) 749-771

[4] Zhang L.M and Chen Q 2006 Seepage Failure Mechanism of The Guhou Rockfill Dam During Reservoir Water Infiltration *Soil and Foundation Japanese Geotechnical Society* **46**(5) 557-568

[5] Escobar C and Posada A M 2008 Recent Experience on Design, Construction and Performance of CFRD Dams *International Conference on Case History in Geotechnical Engineering*, Arlington

[6] Craft D 2005 *Seepage Chemistry Manual* (Denver: US Bureau of Reclamation)

[7] El Naqa A and Al Kusisi M 2004 Hydrogeochemical modeling of the water seepages through Tannur Dam Southern Jordan *Environ. Geol.* **45** 1087-1100

[8] Han-Zhou S. and Jian-fu H 2006 Evolution of Groundwater Quality Around Dam-site and Its Significance *Dams and Reservoirs, Societies and Environment in 21st Century* **2** 485-489

[9] ICOLD 2010 *Concrete face rockfill dams: Concepts for design and construction,* No.141 (Paris: International Commission on Large Dams)

[10] US Bureau of Reclamation 2014 *Design Standard No.13 Embankment Dam* (Washington DC: US Department of Interior)

[11] Appelo C and Postma D 2005 *Geochemistry, Groundwater and Pollution* (Amsterdam: A.A. Balkema)

[12] Das B M and Sobhan K 2014 *Principle of Geotechnical Engineering* (Connecticut: Cengage Learning)

[13] Look B G 2014 *Handbook of Geotechnical Investigation and Design Second Edition* (London: Taylor & Francis)

[14] Gan L, Shen X, and Qing W 2017 Seepage Analysis of Concrete Fece Rockfill Dam for Material Damage of Water Stop Structure *Key Eng. Mater.* **729** 93-98

[15] Haselsteiner R and Ersoy B 2011 Seepage Control Of Concrete Faced Dams With Respect to Surface Slab Cracking *6th International Conference On Dam Engineering,* Lisbon

Analysis of Water Quality Change in Tourist Attractions to Reduce Ecological Destruction: Perspectives on Science and Environmental Theology

Zainul Abas[1], Dita Purwinda Anggrella[2],Fathurrohman Husen[3], Purwono Purwono[4]

[1]Ushuluddin and Dawah Faculty, UIN Raden Mas Said Surakarta, Jl. Pandawa, Pucangan, Kartasura 57168, Indonesia
[2]Center for Halal Research, UIN Raden Mas Said Surakarta, Jl. Pandawa, Pucangan, Kartasura 57168, Indonesia
[3]Dakwah Manajement, UIN Raden Mas Said Surakarta, Jl. Pandawa, Pucangan, Kartasura 57168, Indonesia
[4]Center for Science and Technology, UIN Raden Mas Said Surakarta, Jl. Pandawa, Pucangan, Kartasura 57168, Indonesia

zainul.abas@staff.uinsaid.ac.id

Abstract
This study aims to determine changes in water quality due to water tourism activities in Umbul Ponggok and Jumog Waterfall, Central Java, Indonesia. This research uses a mixed-method approach that combines science (laboratory tests) and environmental theology study. Water samples were taken from tourism locations and then analyzed in the laboratory. Environmental theology studies were conducted by interviewing managers & visitors to find out about the care and utilization of water sources. Based on statistical tests, the pH, Temperature, TDS, and conductivity values of the water at the research location did not change significantly; both the quality of the water from the source, the water when it was used and the water after it was used for water tourism activities. The water pH at the Umbul Ponggok tourist and Jumog Waterfall is between 8.7-9.1 and 8-8.1, respectively. Water temperature ranges between 20-21°C and 24-25°C. TDS levels ranged from 67-121 mg/L, and water in the Umbul Ponggok ranged from 126-129 mg/L. Jumog waterfall's water conductivity is 123-217 mg/L and Umbul Ponggok 235-240 mg/L. The implementation of environmental theology was shown by the existence of preventive and curative actions against water pollution that water tourism managers have carried out.
Keywords: water tourism; water quality; theology; pH; TDS

1. Introduction

The phenomenon of environmental damage is currently a global issue [1]. Indonesia, as a developing country, has abundant natural resources and has the potential to be very vulnerable to environmental damage. The availability of water from water sources provides benefits for humans and the surrounding environment [3]. Generally, the decrease in water quality and quantity is caused by human activities [4], such as industrial activities and changing habitats into tourist attractions [5,6]. This phenomenon is supported by research [7] which states that the decline in water quality is caused by outdoor activities such as tourist attractions/water recreation [8]. Therefore, that water-based outdoor recreation would impair water quality. On the other hand, the large number of pollutants that enter water bodies can limit the use of water for tourists [9].

Research in Klaten Regency shows the quality of river water sources on the parameter pH, TDS, EC, BOD, COD, and phenol do not meet quality standards and the level of pollution increases from upstream to downstream due to industrial waste [11]. The low quality of water sources also occurs in Karanganyar. The low water quality in Karanganyar and Klaten is interesting to explore regarding water tourism that uses water sources as a tourist destination. Jumog is one of the tourism destinations in Karanganyar, which utilizes springs like waterfalls and swimming pools as its tourism advantages [13]. While in Klaten, there is the Umbul Ponggok tour, whose swimming pool is designed to contain a variety of freshwater fish, so tourists can swim and snorkel in it [14]. Both of these tours utilize water sources that can trigger environmental damage.

Tourism activities have the potential to cause environmental damage if they are exploitative-pragmatic. According to [15], preventive efforts must include green ideology, green theology, and green mentality. Environmental theology examines the consequences of human actions on the environment from a religious approach [16]. Environmental theology focuses on the interests of ecological preservation, environmental awareness, and human opportunities and obstacles in caring for God's creation [17]. The human attitude that exploits nature arbitrarily shows the fading value of theophany or sacralization of nature. The value of theophany is the harmonization of humans, nature, and God. Therefore, modern critical thinking is needed to integrate the metaphysical dimension of divine reality with the development of science [18]. Therefore, researchers are interested in researching changes in water quality (physical and chemical properties) from the perspective of environmental theology due to outdoor recreation in water tourism in Karanganyar and Klaten. The implication of this research is in the form of essential information for water tourism managers to scientifically and theologically convince about water quality protection to meet the needs of tourists.

2. Method

2. 1. Type and location of research

This type of research uses a mixed method approach, namely a qualitative and quantitative approach that connects science (changes in water quality from physical and chemical properties) and environmental theology. The object of this research is water quality in water tourism sites, and the subject is tourism managers & tourists. The research locations were conducted in Central Java, namely Umbul Ponggok water tourism, Klaten, Central Java, Indonesia, and Jumog water tourism, Karanganyar, Central Java, Indonesia. The selection of research sites was based on the following criteria: 1) having a water source located in the lowlands or highlands; 2) having swimming pool facilities; 3) having tourist visit activities, and 4) deliberately managed tourism objects.

2. 2. Water sampling methods

Water sampling at two points: water sources, water used for tourism activities, and water after being used for tourism. The sampling method refers to SNI No. 06-2413-1991 [19]. The parameters of pH, TDS, Temperature, and electrical conductivity were measured based on that standard [20]. A total of 5 L of water samples were transferred to the Balai laboratory at the science and technology laboratory of UIN Raden Mas Said Surakarta and the Health and Medical Device Testing Laboratory, Semarang, Central Java, Indonesia, to determine the chemical and physical quality of the water. The need for testing other Physico-chemical parameters is carried out by preservation by storing them at a temperature of $\pm 4^0$C using a cooling box.

2. 3. Qualitative methods

Observations are carried out to observe water conditions, management of water tourism, and water utilization. In addition, researchers will observe water sources used at the research site, water distribution networks, and chemicals used such as chlorine, floor cleaners, detergents, and other chemicals that affect water quality. Interviews were conducted with tourism managers to find out information about water conditions, water treatment, and utilization of water sources in tourist attractions, such as water distribution networks, chemicals used such as chlorine, floor cleaners, detergents, and other chemicals. Interviews were also conducted with visiting tourists to find out the experience of using water in tourist attractions used at the research site. The criteria for informants from tourism managers are those who have the authority to make policies and implement specialized care and use of water sources. The requirements for tourist informants were chosen randomly

(randomly) with snowball sampling. Determination of samples and respondents was done by purposive sampling. Respondents are tourists and managers, with the criteria of knowing tourism management and having travelled to that tourism.

2. 4. Data Analysis Techniques

The data analysis technique on water quality used in this study is the metal analysis method using AAS (Buck Scientific VGP 210, Buck Scientific, Inc.). The stages of analysis follow the model of Robert K. Yin. which consists of 1) pattern matching by comparing patterns from observations, interviews, and documents (secondary data and laboratory test results of water sample quality); 2) data explanation (explanation building), explanation of data on data matching and generalizing it; 3) analysis of the series (time series), investigation of events under study against patterns of information from time to time (development of events/events studied). In addition, the water treatment and utilization analysis uses the perspective of environmental theology, which includes efforts to preserve the environment, manage a friendly environment, protect water quality, and use water proportionally.

3. Result and Discussion

3.1 Water quality Change at the Umbul Ponggok tourist site and Jumog waterfall

Based on observations, the water quality in the swimming pool from the water's surface is clear, tasteless, and has no odour. Surface water pH values 6-9 can be used for tourism activities [21]. The measurement of pH uses the principle of electrometry. The main component that regulates pH in natural waters is carbonate, which consists of CO_2, H_2CO_3 dan HCO_3 [20].

The results of the research on the pH value at the Umbul Ponggok tourist site and Jumog Waterfall are shown in Fig.1.The pH of the water in Jumog waterfall is 9; 9.1; 8.7. The pH of the water in Umbul Ponggok tourism 8; 8.1; 8 respectively for the location of water sources, water used for tourism activities, and water after being used for tourism. Based on statistical tests, it is known that the pH value in Jumog and Umbul Ponggok waterfalls does not change significantly (sig.>0,05) (α 0,48775>0,05; 0,94213>0,05) due to the use of water in the tour. Based on the Regulation of the Minister of Health of the Republic of Indonesia No. 32 of 2017, the standard pH for swimming pools is 7-7.8 when using chlorine and 7-8 when using bromine as a disinfectant.

Research from [22] shows that changing water pH affects the content of C02, which is soluble in water; the higher the CO2 level, the more acidic a water solution. Water-soluble CO2 comes from the air through the diffusion process, carried by rainwater, the result of the respiration process of microorganisms, and from the decomposition of organic substances by microorganisms [23]. Therefore, the high pH in Jumog and Umbul Ponggok water sources indicates the low content of CO2 soluble in water, indicating that the microorganisms in Jumog water sources are low. Therefore, based on PP No. 82 of 2001 (class II), Jumog's water sources are safe and suitable for tourism activities and are not polluted because they do not exceed water quality standards [21]. Based on the analysis results before and after being used for swimming activities, the pH of the water in Jumog and Umbul Ponggok did not change. The tourists' swimming activities do not affect the CO2 level of the pool water because disinfectants affect the pH value (chlorine, bromine) and microorganisms. [24,25].

The water temperature at Jumog waterfall ranges from 20^0C-21^0C and Ponggok 24^0C -25^0C. Changes in water temperature due to tourism activities are shown in Fig.1. It is known that the temperature of the Jumog and Umbul Ponggok waterfalls does not change significantly (sig.>0,05) (α 0,33333>0,05; 0,33333>0,05) as a result of the use of water in the tourism. Based on the Minister of Health Regulation No. 32 of 2017, the temperature parameters for the water quality of swimming pools are around 16^0C - 40^0C [26], so the water temperature of the Jumog and Umbul Ponggok swimming pools has met the threshold determined by the Minister of Health. Thus the temperature in the Jumog and Umbul Ponggok swimming pools is safe to use for swimming.

Changes in TDS concentration that exceed the quality standard can be dangerous because they can cause changes in ion composition, salinity, and toxicity [27]. The water TDS (Total Dissolve Solid) at Jumog waterfall ranges from 67-121 mg/L, and Ponggok ranges from 126-129 mg/L.

Conditions in water TDS due to tourism activities are shown in Fig.1. It is known that the temperature of the Jumog and Umbul Ponggok waterfalls does not change significantly (sig.>0,05) (α 0,9897>0,05; 0,33333>0,05) due to the use of water in the tour. The amount of TDS is one of the water quality parameters; the more turbid the waters, the higher the TDS value. According to PP No.22 of 2021, the TDS parameters for class 2 (water recreation facilities) are 1.000mg/L, so the TDS value for the Jumog and Umbul Ponggok swimming pools before and after swimming activities have met the quality standards that have been set. This condition indicates that Jumog swimming pool water is safe for swimming.

The conductivity of the water in the Jumog waterfall ranges from 123-217 mg/L, and Ponggok ranges from 235-240 mg/L. Changes in water conductivity due to tourism activities are shown in Fig.1. It is known that the temperature of the Jumog and Umbul Ponggok waterfalls does not change significantly (sig.>0,05) (α 0,97604>0,05; 0,73985>0,05) due to the use of water in the tour. This condition is supported by research [28] that adding organic substances affects water quality, turbidity, and the number of water ions. Based on the study's results, there was no significant difference in the value of the electrical conductivity of water before and after activities in the Jumog and Umbul Ponggok swimming pools, proving that tourist activities in swimming pool water did not affect the number of ions in the swimming pool.

The results of this study indicate that tourism activities in Umbul Ponggok and Jumog waterfalls do not worsen water quality based on pH, Temperature, TDS, and conductivity values.

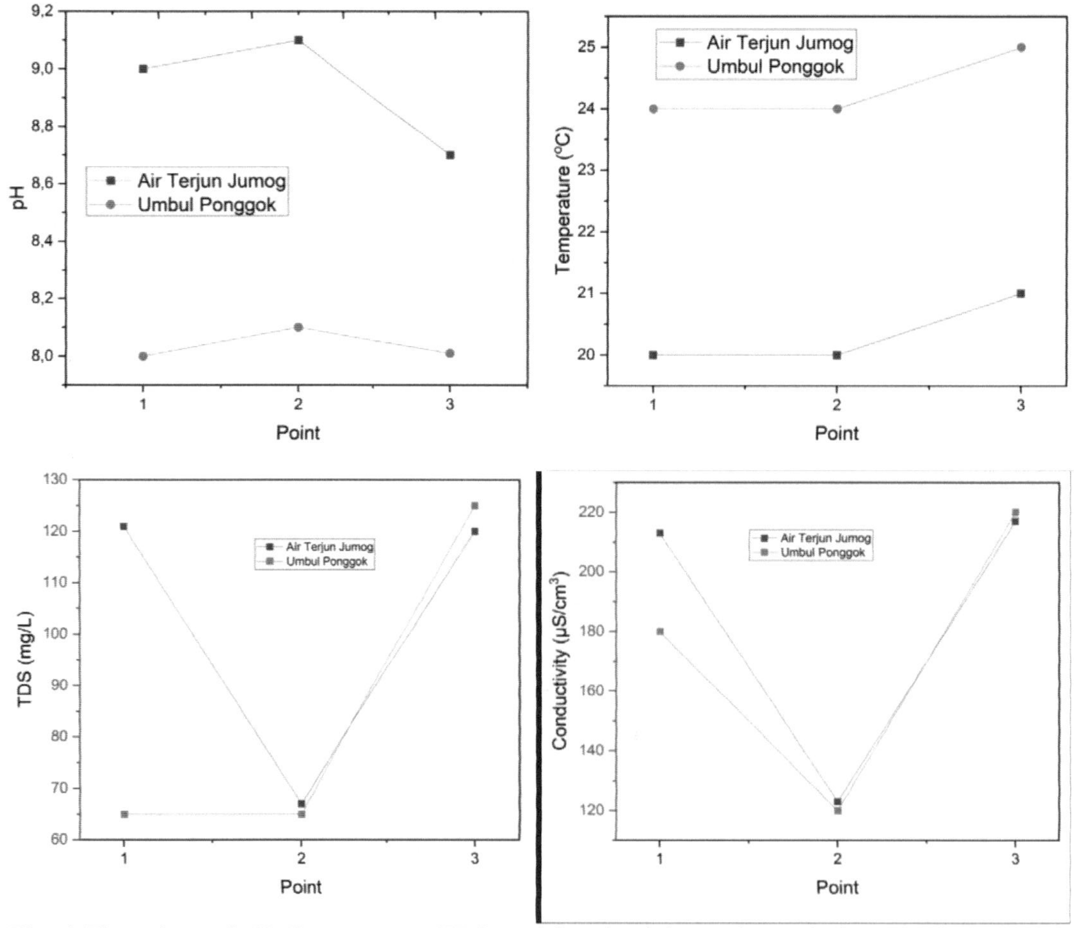

Fig. 1 The values of pH, Temperature, TDS, and Conductivity at the Umbul Ponggok tourist location and Curug Jumog in three places, namely (1) water sources, (2) water when used for tourism activities, and (3) water after being used for tourism

3.2 Management of water tourism

Based on the results of interviews with the manager of the Umbul Ponggok tourist spot, the bottom of the swimming pool is still natural in the form of rocks. The manager does not use chlorine to clean the swimming pool. The clean water supply system for Umbul Ponggok tourism from the underground Clean Water Treatment Plant and Umbul Ponggok springs uses an up feed distribution system and a down feed system. The distribution network of swimming pool water uses an overflow system in the pool. In this system, the water is sucked in by the pump from the Balancing Tank and then sent to the swimming pool and rides through a filtration process in the filter. Water that enters the pool through the Inlet will overflow. It is made to overflow and spill into the Outter or channel that is made as a reservoir for the overflow, and then through the Outter Drain, the water returns to the Balancing Tank, where it will be sucked back in by the Circulation Pump. [29] Stated that the reservoir of water sources is one of the factors that cause contamination with microorganisms because it has never been opened and cleaned by the manager. Hence, it needs regular water hygiene management by tourism managers. The distribution of dirty swimming pool water, namely dirty water from the swimming pool, is channelled to the water treatment and then accommodated in the water tank so that it can be used to water plants or garden ponds.

As for Jumog tourism, the tourist water comes from the springs flowing from the Jumog waterfall, and once a week, cleaning will be carried out using a water machine with high pressure to clean dirt such as leaves and other garbage that is carried by the current when it rains and settles on the riverbed. For cleaning, the swimming pool is drained every two days, and cleaning the walls of the swimming pool is every two weeks. The manager does not use chlorine in the swimming pool because the water is pure from the source directly. The manager only carries out maintenance through regular cleaning. The water distribution network is located at several points of water sources, such as the primary source as a tourist attraction. Toren has been provided and is used to collect water from other sources, which are then used or channelled to stalls, swimming pools, prayer rooms, bathrooms, and drinking water. Swimming pool wastewater will have flowed back into the river, but a different place is provided for the waste of stalls and bathrooms. In addition to showing the beauty of the waterfall, the manager also maintains the plant vegetation around the waterfall, such as moss, ferns, and trees, so the tour feels cool.

The explanation above shows that tourism managers only depend on natural water sources to maintain water quality. There is no monitoring and evaluation of water quality. Praja W et al. [30] revealed that maintaining water quality requires organizing, managing, monitoring, and evaluating swimming pool management. According to [31], regular water quality monitoring will improve swimming pool sanitation hygiene. Based on the regulations stipulated by the Minister of Health No. 32 of 2017, to maintain the quality of swimming pool water, the manager must; 1) provide a small pool near the pool to wash feet before swimming; 2) conduct periodic inspections of pH and chlorine following environmental health quality standards; 3) provide information on the prohibition of using swimming pools for those who have infectious diseases; 4) there must be water recirculation. However, based on the results of interviews with several tourists in the two locations, there has never been an impact of pool water on the health of tourists [26].

3.3 Analysis of environmental theology perspectives on water tourism management

Water management at the Jumog waterfall attraction is the local BUMDES. Data from interviews with managers show that the water used for the swimming pool has a different path from the water flowing in the waterfall, even though the water source is the same. This separation aims to maintain the clarity of the water enjoyed by tourists. Maintenance activities for waterfall attractions are carried out routinely, such as river combing to clean garbage and water spray machine technology to clean dirt that settles in the water. Maintaining water clarity at the tourist attraction does not use chlorine, but the intensity of water treatment by cleaning dirt and regulating the flow of water in the tourist attraction according to needs. For example, trash cans are provided at several locations to prevent visitors from littering.

The addition of facilities such as swimming pools, prayer rooms, and food stalls slightly changes the contours of the land but still pays attention to water channels and maintains the existing rocks. Meanwhile, the manager routinely drains the swimming pool every two days and cleans the pool walls every two weeks. Rejuvenation of Bogor trees around the river is also routinely carried out. The tools used to maintain the facility include cleaning equipment such as brooms, brushes, and sickles. In addition, the manager uses special tools such as Alkon (water pump) and Karcher (jet cleaner) to clean the river area to reduce moss on the rocks. For conditional use of Karcher when the moss is thick.

In terms of environmental theology, Jumog tourism objects have met the principles of integrating metaphysics of divine reality with the development of science. Additionally, the human attitude not to exploit nature arbitrarily becomes a benchmark for implementing environmental theology. As a result, the creation of harmonization between humans (as caliph fil ard), nature, and God. This theological concept was formulated by Sayyed Hossein Nasr [18].

Being careful in utilizing natural products indicates an attitude of trust and responsibility in the afterlife (commitment), resulting in an environmental crisis. Abdul Quddus explains these indicators as objects of Islamic-constructive theology [32]. In the Judeo-Christian tradition (Jonah 3:1-10), Biblical teachings can play a role in preserving the natural environment. [33]. In Kenya, young Pentecostal Christians are the driving force for environmental awareness. Loreen Maseno explains her research results that PBC (Parklands Baptist Church) implements its vision for up to 2040. The concept is to preserve the environment based on anthropocentrism and religious communities' role in creating prosperity [34].

In the concept of environmental theology, it is legal to preserve customs that support the preservation of nature and maintain sharia principles (*maqasid sharia*). In his research, Suyatman (2018) concluded that the traditions that exist in Sundanese society, such as the *seren taun* ceremony (a ritual of gratitude preceded by reforestation), the *ngahutu Kala* system (zig-zag irrigation pattern), *pamali* (sacralization of specific water sources), and *leuweung kolot* (forest Baduy area conservation).

Based on this analysis, water management at Jumog tourism object has complied with the principles of environmental theology. These principles are implemented to use water sources as a tourist attraction without damaging the environment. The manager's responsibility as *khalifah fil ard* (leader on earth) is shown by treating water by cleaning the water flow and draining the swimming pool that was built. The use of special tools such as Alkon (water pump) and Karcher (jet cleaner) to clean the moss is done carefully and sufficiently so as not to damage the originality of nature. Therefore, even without customs, the implementation of *maqashid sharia* to maintain visitors' safety and preserve nature has been fulfilled. This action is indicated by maintaining the freshness of natural water and the absence of cases of disease caused to visitors.

4. Conclusion

This study aimed to determine changes in water quality (physical and chemical properties) from the sciences and environmental theology perspectives due to outdoor recreation in water tourism in Karanganyar and Klaten. The study results show that tourism activities in Umbul Ponggok and Jumog waterfalls do not worsen water quality based on pH, Temperature, TDS, and conductivity values. There is no influence of tourist activities before and after entering the swimming pool on the quality of the pool water. The Jumog and Umbul Ponggok managers maintain natural authenticity by using springs as water distribution in the swimming pool so that managers do not use chlorine to maintain water quality. However, there is still no monitoring and evaluation of water quality in the swimming pool. In Jumog tourism, in addition to utilizing water sources for swimming pools, the manager also maintains the authenticity of plant vegetation, such as ferns and mosses. In Umbul Ponggok, the manager takes care of and cultivates freshwater fish to take underwater photos. Based on the study of environmental theology, the manager has fulfilled the principles of ecological theology by utilizing water sources without damaging the environment.

5. The acknowledgements

Thank you to Litapdimas Ministry of Religion and Institute for Research and Community Service (LP2M) UIN Raden Mas Said Surakarta, who have fully funded this research for the 2022 budget year. Thank you also to Jumog waterfall tourism and Umbul Ponggok managers for providing data related to tourism water management.

6. Reference

[1] Wagianto, Hamdani D M, Edison E F, Adam S, Umar I and Barlian E 2021 Ecological Theory: Anthropocentrism as an Evolving Phenomenon in the Global Environment *Cent. ASIAN J. Theor. Appl. Sci.* **2** 278–82
[2] Prastiyo S E, Irham, Hardyastuti S and Jamhari 2020 How agriculture, manufacture, and urbanization induced carbon emission? The case of Indonesia *Environ. Sci. Pollut. Res.* **27** 42092–103
[3] Westall F and Brack A 2018 The Importance of Water for Life *Space Sci. Rev.* **214** 1–23
[4] Sun B, Feng Y, Mo X, Zheng P, Wang Q, Li P, Peng P, Liu X, Chen Z, Huang H, Zhang F, Luo W, Niu X, Hu P, Wang L, Peng H, Huang Z, Feng L, Li F, Zhang F, Li F, Zhong N and Chen L 2020 Kinetics of SARS-CoV-2 specific IgM and IgG responses in COVID-19 patients *Emerg. Microbes Infect.*
[5] Effendi H 2003 Telaah Kualitas Air Bagi Pengelolaan Sumber Daya dan Lingkungan Perairan, Edisi 5, Penerbit Kanisius, Yogyakarta. *Kanisius*
[6] Giri S 2021 Water quality prospective in Twenty First Century: Status of water quality in major river basins, contemporary strategies and impediments: A review *Environ. Pollut.* **271**
[7] Shi P, Zhang Y, Li Z, Li P and Xu G 2017 Influence of land use and land cover patterns on seasonal water quality at multi-spatial scales *Catena* **151** 182–90
[8] Barnett M J, Jackson-Smith D and Haeffner M 2018 Influence of recreational activity on water quality perceptions and concerns in Utah: A replicated analysis *J. Outdoor Recreat. Tour.* **22** 26–36
[9] Wibisono Y, Sucipto S, Perdani C G, Astuti R and Dahlan M 2018 Halal Compliance on Drinking Water Industries: A Future Perspective *Proc. 3rd Int. Halal Conf. (INHAC 2016)* 555–64
[10] Dinas Kepemudaan dan Pariwisata Provinsi Jawa Tengah O 2020 Pariwisata Jawa Tengah Dalam Angka *J. Chem. Inf. Model.* **53** 1689–99
[11] Budianta W and Sutrisno 2021 Kajian kualitas air sungai dengkeng di kecamatan bayat, kabupaten klaten, jawa tengah. *KURVATEK Vol.* **6** 153–64
[12] Patil. P.N Deshmukh. R.N S D V 2012 Physico-chemical parameters for testing of water, a review paper *Int. J. Chem. Stud.* **3** 1194–207
[13] Hafiyyan F N 2018 Eksotisme Air Terjun Jumog Sebagai Destinasi Wisata Alam Kabupaten Karanganyar *Domest. Case Study* 1–8
[14] Zakiyah U and Idrus I A 2017 Srategi Pengelolaan Sumber Daya Alam Desa Ponggok *JIP (Jurnal Ilmu Pemerintahan) Kaji. Ilmu Pemerintah. dan Polit. Drh.* **2** 84–95
[15] Manshur S 2019 Bersahabat dengan Bumi: Modal Simbolik dalam Mitigasi Bencana Alam *TALENTA Conference Series: Local Wisdom, Social, and Arts (LWSA)* (Indonesia: TALENTA Publisher Universitas Sumatera Utara) pp 310–7
[16] Ituma E A 2013 Christocentric Ecotheology and Climate Change *Open J. Philos.* **03**
[17] Kroeck T 2021 Environmental awareness of protestant youth in germany: Perspectives from an empirical exploration *HTS Teol. Stud. / Theol. Stud.* **77** 1–7
[18] Maftukhin M 2016 Teologi Lingkungan Perspektif Seyyed Hossein Nasr *Din. Penelit. Media*

Komun. Penelit. Sos. Keagamaan **16** 337–52
[19] Badan Standardisasi Nasional 2008 Metoda pengambilan contoh air permukaan *Standar Nasional Indonesia SNI Air dan Air Limbah* (Jakarta: Badan Standardisasi Nasional) pp 1–20
[20] APHA 2017 *Standard Methods for the Examination of Water and Wastewater* (Washington)
[21] Pemerintah Republik Indonesia 2001 *Peraturan Pemerintah Republik Indonesia Nomor 82 Tahun 2001 Tentang Pengelolaan Kualitas Air Dan Pengendalian Pencemaran Air Presiden Republik Indonesia* (Jakarta: Sekretariat Negara)
[22] Mukarromah R, Yulianti I and Sunarno 2016 Analisis Sifat Fisis Kualitas Air Di Mata Air Sumber Asem Dusun Kalijeruk, Desa Siwuran, Kecamatan Garung, Kabupaten Wonosobo *Unnes Phys. J.* **5** 40–5
[23] Susana T 1988 Karbon dioksida *Oseana* **12** 1–11
[24] Elmia Kursani Rika Aqrianti B Y 2019 Analisis Kadar Sisa Klorin Dan Ph Air Di Kolam Renang Umum Kota Pekanbaru *J. Kesehat. Al-Irsyad* **12** 11–22
[25] Wahyuningtias I, Sunarko B and Rustanti EW I 2019 KEBERADAAN BAKTERI Escherichia coli PADA AIR KOLAM RENANG UMUM *Gema Lingkung. Kesehat.* **17** 87–91
[26] Indonesia M K R 2017 Peraturan Menteri Kesehatan Republik Indonesia Nomor 32 Tahun 2017 Tentang Standar Baku Mutu Kesehatan Lingkungan Dan Persyaratan Kesehatan Air Untuk Keperluan Higiene Sanitasi, Kolam Renang, Solus Per Aqua dan Pemandian Umum *Peratur. Menteri Kesehat. Republik Indones.* 1–20
[27] Weber-Scannell P K and Duffy L K 2007 Effects of total dissolved solids on aquatic organisms: A review of literature and recommendation for salmonid species *Am. J. Environ. Sci.* **3** 1–6
[28] Agustina E 2022 Pengaruh Eco-Enzyme pada Air Baku Sungai Borang Palembang Terhadap Nilai Parameter Conductivity, Total Dissolved Solid (TDS), dan Zat Organik *J. Kolaboratif Sains* **5** 284–9
[29] Talita S, Nurjazuli and Dangiran H L 2016 Studi kualitas bakteriologis air kolam renang dan faktor – faktor yang mempengaruhinya di kolam renang kota semarang *J. Kesehat. Masy.* **4** 196–203
[30] Praja W, Imansyah F, Jasmani S P and Selatan S 2022 Analisis manajemen pengelolaan kolam renang lumban tirta di provinsi sumatera selatan 1,2,3 *J. Penjaskesrek* **9** 11–25
[31] El-Salam M M A 2012 Assessment of water quality of some swimming pools: A case study in Alexandria, Egypt *Environ. Monit. Assess.* **184**
[32] Quddus A 2017 Ecotheology Islam: Teologi Konstruktif Atasi Krisis Lingkungan *Ulumuna J. Stud. Keislam.* **16** 311–46
[33] Redelinghuys N 2019 Negotiating an Eco-conscious Translation of the Hebrew Bible: Jonah 3:1–10 as Test Case *Old Testam. Essays* **32** 821–45
[34] Maseno L and Mamati K 2021 An appraisal of the pentecostal eco-theology and environmental consciousness among youths in parklands baptist church, Kenya *HTS Teol. Stud. / Theol. Stud.* **77** 1–7
[35] Suyatman U 2018 Teologi Lingkungan dalam Kearifan Lokal Masyarakat Sunda *Al-Tsaqafa J. Ilm. Perad. Islam* **15** 77–88

Review of the Circular Economy of Plastic Waste in Various Countries and Potential Applications in Indonesia

Z Murti*, Dharmawan, Siswanto, D Soedjati, A Barkah, P Rahardjo

Research Center for Sustainable Production System and Life Cycle Assessment, Research Organization for Energy and Manufacture, National Research and Innovation Agency, Puspiptek, South Tangerang, Banten

*Email : zulwelly.m@brin.go.id

Abstract. One of the main environmental problems faced by the world, including Indonesia, is plastic waste. Plastic waste has polluted the soil, rivers, and the sea. The nature of plastic that is not easily biodegradable, takes hundreds of years to decompose naturally. Plastic waste management which is believed to be effective, in addition to efforts to reduce use is by recycling plastic waste. In many countries, recycling plastic waste has led to a circular economy. This circular system allows plastic waste to be recycled into new products. In fact, this concept is claimed to be able to encourage environmentally friendly economic growth. This study uses a literature study to identify the application of a circular economy of plastic waste in various countries. As a result, 21 papers were collected and analyzed based on year of publication, geographical location and research methodology. Furthermore, this study highlights the obstacles faced by many countries towards a circular economy followed by a comparison of their plastic waste recycling rate achievements. This research also discusses the state of plastic waste management and the potential for adoption of circular economy best practices in Indonesia.

1. Introduction

Plastic waste is one of the environmental polluter issues, both as a pollutant of the soil and the sea. The nature of plastic waste that is not easily decomposed or takes hundreds of years to decompose, as well as processing processes that are toxic and carcinogenic, make it a worrying environmental pollutant. So that the existence of plastic which was originally intended to help preserve the environment due to the exploitation of paper as a wrapping material, in its development became a challenge, especially in terms of waste management.

The United States (US) is by far the largest contributor to plastic waste in the world, generating 35.7 million tons of plastic waste in 2018, more than the 29 million tons of European countries and 26.7 million tons of China[1].

The circular economy is a concept that promotes the repair, improvement and redistribution of second-hand goods. Circular economy is basically an idea, which drives the change of a linear process into a circular process, which involves reduction, reuse, recycling, arrangement, assembly and circulation[2]. The principle of circular economy is theoretically divided into 3Rs, namely "Reduce, Reuse, Recycle"[3]. Through the circulation economy model, it is hoped that it will be able to extend the life of the waste into something useful to be reused. Plastic waste as an alternative raw material or recycled into new products, so it can save production costs or become new products that sell well. Various policies have been adopted by various countries to achieve this circular economy.

To move towards a more circular economy, in 2015 the European Commission proposed the first Circular Economy Package and continued with the Second Circular Economy Package in 2018[4]. The plan sets out a concrete program of action and measures to stimulate Europe's transition to a circular economy. In Indonesia, plastic waste accounts for 15% of the total waste and is still not managed properly even though various policies have been issued[5].

In this study, a literature review was conducted on the application of the circular economy in various countries in the world. Various articles related to the circular economy were collected and then identified various obstacles faced, policies and methods used to increase the recycling rate. The results of this compilation are used as adoption for the Indonesian state which methods and methods are suitable and suitable to be applied. The first part tells about the introduction related to plastic waste. The second part describes the research methods used. The third section describes the results of data processing. The fourth part is a discussion that tells how to handle plastic waste in Indonesia and the adoption of a circular economy and the last part contains the conclusions of this research.

2. Methodology

To study how the circular economy concept is actively applied in various countries, a literature review of existing scientific publications is carried out. This methodology section describes how data was collected, analyzed and presented in this article. The first stage is to identify the search criteria. The second stage is to search for suitable papers based on the title and abstract. The third stage is to determine the paper that will be taken more deeply, including reading the contents and conclusions until the final number is obtained. The final step is to describe the collected papers with the parameters of the year of publication, research methodology, geographic location, barriers, methods and policies as well as the recycling rate in the country.

Papers that are included in the search criteria are published in the 2018-2022 range from the Scopus database which is in English with the subject area of environmental science. The keywords used are circular economy, plastic waste and the state. The combination of this search resulted in 145 papers with a filter process 2 times, the first filter process by reading the title and abstract of the paper resulted in 67 papers, the second filter by reading the contents and conclusions resulted in 36 papers and the last filter by adjusting the parameters relevant to the scope and The research objective was to produce 21 final papers (Table 1). Papers related to the condition of plastic waste management in Indonesia apart from all the selected papers.

Table 1. Keywords for Paper Search

Year	Keyword	Scopus	Filter 1	Filter 2	Final
2018-2022	"Circular Economy" AND "Plastic Waste" AND "Country"	145	67	36	21

3. Result

This section analyzes descriptively from 21 selected papers related to the year of publication, research methodology and geographical location which are presented in the form of Table 2. From the table it can be seen that since 2018 the issue of the circular economy has been in high demand and has been studied, especially in the last two years. Based on research methodology, analysis is the most widely used method followed by article review. The analysis that is often done is the Material Flow Analysis (MFA) model. MFA is a tool that is widely used in the environmental field for decision making. MFA results can contribute to saving natural resources and minimizing environmental impacts, as well as pursuing sustainable development goals[6][7]. Then from the geographical side, it is divided into 5 continents and 10 countries are in Asia. This shows that countries in Asia are currently running quite fast towards a circular economy. The countries involved in the review and being discussed in the selected paper are Sub-Saharan Africa[8], European Union[4], South Africa[7], Denmark[6], Qatar[9], Singapore[10], Taiwan[11][12], Brazil[13], India[1], Italy[3], Kenya[14], Malaysia[15], Sweden[16], USA[17], Netherlands[18], South Korea[2][19], China[20] and Australia[21].

Table 2. Compilation of Year, Methodology and Geographical Location of 21 Selected Papers

Year		Research Methodology		Geographic Location	
2018	3	Review	5	Africa	3
2019	0	Analysis	7	Europe	6
2020	7	Survey	3	Asia	10
2021	7	Comparative	2	America	1
2022	4	Descriptive	3	Australia	1

4. Discussion

This section discusses the main factors of this research related to the scope of research, namely the circular economy of plastic waste.

4.1. Best Practice dari Ekonomi Circular

The process towards a circular economy has been proclaimed by several countries, there are countries that are successful in implementation and there are also countries that have difficulties. In this research, it was found that there are several reasons that hinder the low progress towards this. The obstacles faced by many countries can be briefly seen in Figure 1.

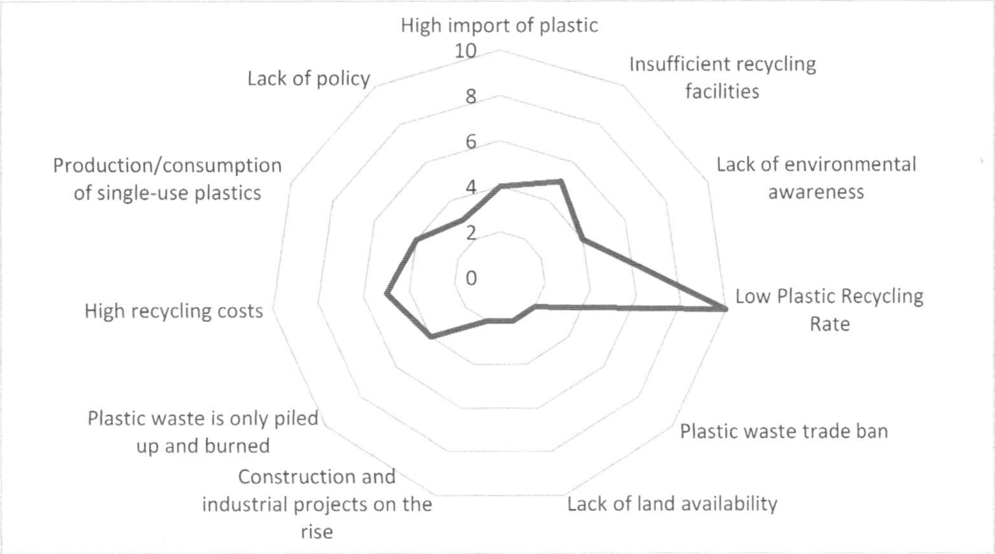

Figure 1. Circular Economy Barriers in various countries

In Figure 1 we can see that the main problem with the low level of the circular economy of plastic waste in several countries is the low recycling rate of plastic waste. The trigger for this occurrence is due to the lack of recycling facilities and the high costs that must be incurred for this process. This is the case in many developing countries.

A comparison of the recycling rates of plastic waste between countries can be seen in Figure 2. This figure shows which countries have high and low levels of plastic waste recycling. South Korea (60%)[19], the Netherlands (57%)[18] and Taiwan (51%)[11] are examples of best practice countries that have succeeded in recycling their plastic waste. On the other hand, Singapore (6%)[10], USA[17],

Australia[21] and Qatar[9] (9%) respectively are countries with low plastic waste recycling rates. The reasons for the success of the circular plastic waste economy program in the three countries can be seen in Table 3.

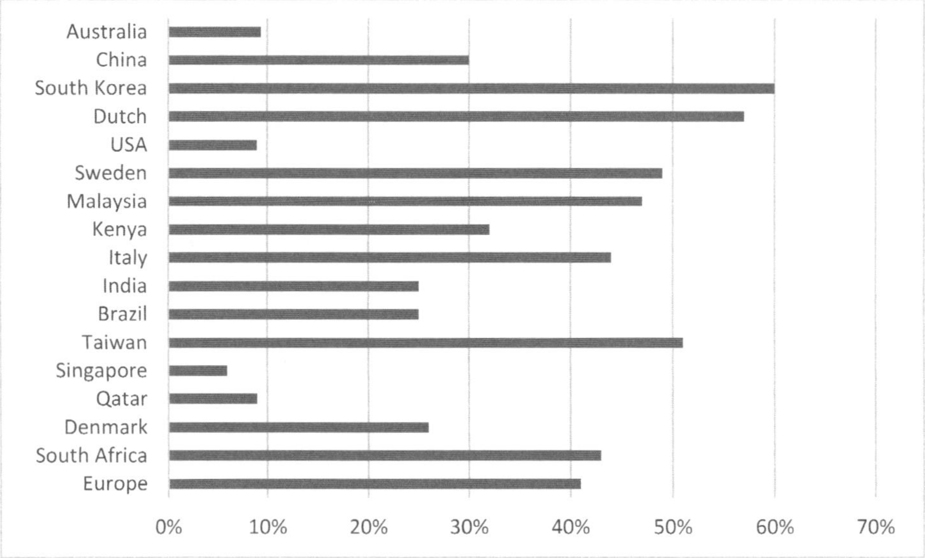

Figure 2. Recycling Rate

4.2. Plastic Waste Management in Indonesia

Indonesia will produce 28.60 million tons of waste in 2021. This number decreased by 12.80% compared to the previous year which was 32.82 million tons. As much as 15 percent of which is plastic waste[5]. A global study, Indonesia is the second largest country in the world after China which contributes to the generation of plastic waste into the ocean[22].

The amount of plastic waste generation in Indonesia tends to increase from year to year along with the increase in population. Types of plastic found in urban waste include Light Density Poly Ethylene (LDPE), PP (Poly Propylene), High Density Poly Ethylene (HDPE), Poly Vinyl Chloride (PVC), Polyethylene Terephthalate (PET), Styrofoam, etc. Most of them are PP and HDPE. A lot of plastic bag waste does not reach the landfill and only a little is recycled, so a lot of plastic bag waste ends up in waterways, rivers, and finally into the sea. A study conducted by the Directorate General of Pollution Control and Environmental Damage (PPKL) of the Ministry of Environment and Forestry shows that the most common type of waste found in marine waters in Indonesia is plastic (42%)[23].

In response, the Government of Indonesia has stated a target of reducing plastic waste by up to 70% by 2025. As a strategic guide for ministries/institutions and a reference for the community and business actors to accelerate the handling of plastic waste until 2025, the National Action Plan (RAN) for Marine Plastic Waste for 2018-2025 is established. The RAN for marine plastic waste consists of five main strategies as follows: (1) national movement to increase awareness of all stakeholders, (2) management of plastic waste from (land) sources, (3) management of plastic waste in coastal and marine areas, (4) mechanisms to increase institutional strength, funding, and law enforcement, and (5) research and development.

The government is also in the 2020-2024 RPJMN targeting a reduction in the percentage of waste thrown into the sea in 2024 by 60% of the amount of plastic waste wasted in 2019. Several local governments have also banned the use of single-use plastics (straws and plastic bags). in shopping centers.

In formal regulation, Indonesia has issued quite some policies related to waste management in general, ranging from laws, government regulations, presidential regulations, ministerial regulations, to regional head regulations, but the problem of solid waste is still not resolved. This may be due to the minimal budget, inadequate facilities and especially the lack of public awareness and the low responsibility and commitment of producers to recycle.

4.3. The Potential for Adoption of the Circular Economy

The circular economy can contribute to three important aspects [61]. First, a circular economy can increase productivity and increase economic growth. Second, it can improve the quality and quantity of labor. Third, a circular economy can help save lives by reducing environmental impacts such as air and water pollution and climate change. The three countries with the highest recycling rates (Figure 1) can be good examples in this study for adoption.

South Korea is the best example that can be adopted by Indonesia in moving toward a Circular Economy. South Korea divides the circulation system into four stages, namely, production, consumption, disposal and recycling, with 3 main objectives: reducing waste generation, increasing the recycling rate and reducing the final disposal rate (this is following the 3R principle).

Table 3. The Circular Economy of Plastic Waste in South Korea, Netherland and Taiwan

Country	Method of CE	Policy
South Korea[2][19]	Recycling from the household separated from the source Implementing a pay-as-you-throw system Reduce production and consumption of single-use plastic packaging Waste mart Implementing Extended Producer Responsibility (EPR) methods for recycling	Safe disposal - waste policy in Korea from 1980 (waste Control Act) Recycling 1992-2003 RCA (Resource Circulation Act) 2016 RCMP (Resource Circulation Master Plan) 2018-2027 PWCP (Plastic Waste Control Plan) 2018-2039 Target - Reduction PW 50%, Recaycling PW 70%
Netherland[18]	Have a strong plastic waste management system Recycle plastic waste 57% Energy recovery 42%	Become 100% circular by 2050 and its Circular Economy Action Plan focuses on plastics 23 policies towards a circular plastic economy
Taiwan[11][12]	Recycling plastic containers, Disposable plastic restrictions Green design Source reduction Recycling enhancements Circulation and regeneration	Develop a roadmap in 1997 to reduce the generation of plastic waste (Phase I 1997-2018) to a circular economy (Phase II 2019 - present) Targeting Single-Use Plastic Packaging (recycled contents) 2025: 25%

The database of 21 reviewed papers can be accessed at the following link: https://s.id/1cIVg

Some of the methods applied there include recycling from the household separated from the source, implementing a pay-as-you-throw system, reducing the production and consumption of single-use plastic packaging, most of the plastic waste is processed by energy recovery (waste to energy), waste mart (where to exchange recycled plastic for goods/cash/points) and apply EPR methods for recycling. With the implementation of these methods, South Korea can increase the recycling rate of plastic waste by up to 60%, the highest of several countries reviewed in this paper (Figure 2).

The Netherlands is also a concern. They have a strong and adequate waste handling infrastructure so that they can recycle large amounts of waste and apply waste to energy. The reason is that the Dutch Economy Action Plan focuses on handling plastic waste, establishes an institution that ensures producers are responsible for recycling their products (similar to EPR) and is complemented by the policies issued. Likewise, Taiwan has prepared a roadmap for handling plastic waste since 1997 in order to transition to a circular economy. They applied various successful methods and the process was divided into 2 phases as shown in Table 3.

EPR regulation of South Korea, producers must annually collect and recycle the specified amount according to the mandatory recycling rate notified by the Government, those who fail to meet their recycling obligations must pay a fine of more than the recycling fee. The recycling contribution is calculated by multiplying the market in the volume of each product (including imports) by the mandatory recycling target rate per product and the unit cost. After the EPR policy was launched in

2003, increased recycling rate by 32.3%, compared to the period before the EPR system was implemented. Also, it is estimated that 2,264.3 billion won (2.05 billion USD) of economic benefits and 4,260 new jobs have been created in the 5 years[24]. The benefits of EPR to the environment are equally great. With the enforcement of EPR (2001-2008) total energy saving approximately 2.6 trillion won and reduction 15 million tons of CO2[25]. This EPR recycling scheme has succeeded in increasing the recycling rate of plastic in South Korea and also in line with the circular economy objectives as discussed at the beginning of this sub-chapter. This method has also applied in India, Netherlands, Sweden, Italy and many other European countries.

The policy regarding EPR in Indonesia has actually been mandated in Law Number 18 of 2008 concerning Waste Management, especially in Article 1 which reads: "Producers are obliged to manage the packaging and/or goods they produce that cannot or are difficult to decompose by natural processes". Although this article does not textually mention EPR, the presence of this law is the legal basis for demanding the roles and responsibilities of producers in efforts to reduce and manage waste[26].

However, even though it has been mandated in law, it was revealed through Government Regulation Number 81 of 2012[27] and Ministry of Environment and Forestry Regulation Number 75 of 2019[28] that the implementation of EPR has not become mandatory. EPR is still implemented voluntarily. According to producers, the application of EPR will increase the cost of producing a product, plus the use of biodegradable products, which are not cheap and therefore economically inefficient. Therefore, it is necessary for the government to be serious in implementing the EPR method, for example by establishing institutional and technical regulations related to this matter, stronger law enforcement by providing incentives to industry players who have implemented and providing punishment for industries that do not implement EPR. Likewise with other circular economy models that carry the concept of reduce, reuse and recycle. The regulations already exist in Indonesia but their implementation has not been going well.

5. Conclusion

Plastic waste production continues to grow along with industrial growth and population growth. Plastic waste that is difficult to decompose causes damage to the environment, especially the sea. The circular economy agenda changes the community's paradigm of waste management including plastic. The transition to circularity has been implemented in many countries in the world by issuing various policies and methods. Low recycling rates are a major obstacle in many countries. Due to the lack of facilities and high costs for it. South Korea is one of the leading countries in the process towards a circular economy. With a well-thought-out strategic plan and a systematic policy, this country will continue to achieve the targets they want, one of which is a high recycling rate with their EPR system. In addition, several countries can be good examples to be adopted in Indonesia, such as the Netherlands and Taiwan. Indonesia has formally issued quite a several policies in the field of waste but it has not been implemented properly.

References

[1] R. Hossain *et al.*, "Plastic Waste Management in India: Challenges, Opportunities, and Roadmap for Circular Economy," *Sustain.*, vol. 14, no. 8, 2022, doi: 10.3390/su14084425.

[2] Y. C. Jang, G. Lee, Y. Kwon, J. hong Lim, and J. hyun Jeong, "Recycling and management practices of plastic packaging waste towards a circular economy in South Korea," *Resour. Conserv. Recycl.*, vol. 158, no. March, p. 104798, 2020, doi: 10.1016/j.resconrec.2020.104798.

[3] M. Lombardi, R. Rana, and J. Fellner, "Material flow analysis and sustainability of the Italian plastic packaging management," *J. Clean. Prod.*, vol. 287, p. 125573, 2021, doi: 10.1016/j.jclepro.2020.125573.

[4] M. Robaina, K. Murillo, E. Rocha, and J. Villar, "Circular economy in plastic waste - Efficiency analysis of European countries," *Sci. Total Environ.*, vol. 730, 2020, doi: 10.1016/j.scitotenv.2020.139038.

[5] M. of E. and F. of the R. of Indonesia, "Waste Management Performance Achievements," 2021.

https://sipsn.menlhk.go.id/sipsn/ (accessed Jul. 22, 2022).
[6] E. Vingwe, E. Towa, and A. Remmen, "Danish plastic mass flows analysis," *Sustain.*, vol. 12, no. 22, pp. 1–22, 2020, doi: 10.3390/su12229639.
[7] K. I. Olatayo, P. T. Mativenga, and A. L. Marnewick, "Comprehensive evaluation of plastic flows and stocks in South Africa," *Resour. Conserv. Recycl.*, vol. 170, no. October 2020, 2021, doi: 10.1016/j.resconrec.2021.105567.
[8] O. O. Ayeleru *et al.*, "Challenges of plastic waste generation and management in sub-Saharan Africa: A review," *Waste Manag.*, vol. 110, pp. 24–42, 2020, doi: 10.1016/j.wasman.2020.04.017.
[9] J. N. Hahladakis and H. M. S. J. Aljabri, "Delineating the plastic waste status in the State of Qatar: Potential opportunities, recovery and recycling routes," *Sci. Total Environ.*, vol. 653, pp. 294–299, 2019, doi: 10.1016/j.scitotenv.2018.10.390.
[10]Z. Chen and A. Tan, "Exploring the circular supply chain to reduce plastic waste in singapore," *Logforum*, vol. 17, no. 2, pp. 271–286, 2021, doi: 10.17270/J.LOG.2021.564.
[11]C. Y. Wu, M. C. Hu, and F. C. Ni, "Supporting a circular economy: Insights from Taiwan's plastic waste sector and lessons for developing countries," *Sustain. Prod. Consum.*, vol. 26, pp. 228–238, 2021, doi: 10.1016/j.spc.2020.10.009.
[12]Y. Y. Lai and Y. M. Lee, "Management strategy of plastic wastes in Taiwan," *Sustain. Environ. Res.*, vol. 32, no. 1, 2022, doi: 10.1186/s42834-022-00123-0.
[13]L. R. Lima, R. F. Gutierrez, and S. A. Cruz, "Challenges in the context of single-use plastics and bioplastics in Brazil: A legislative review," *Waste Manag. Res.*, vol. 40, no. 7, pp. 998–1006, 2022, doi: 10.1177/0734242X211055548.
[14]B. Horvath, E. Mallinguh, and C. Fogarassy, "Designing Business Solutions for plastic waste management to enhance circular transitions in Kenya," *Sustain.*, vol. 10, no. 5, 2018, doi: 10.3390/su10051664.
[15]H. L. Chen, T. K. Nath, S. Chong, V. Foo, C. Gibbins, and A. M. Lechner, "The plastic waste problem in Malaysia: management, recycling and disposal of local and global plastic waste," *SN Appl. Sci.*, vol. 3, no. 4, pp. 1–15, 2021, doi: 10.1007/s42452-021-04234-y.
[16]L. Milios, A. E. Davani, and Y. Yu, "Sustainability impact assessment of increased plastic recycling and future pathways of plastic waste management in Sweden," *Recycling*, vol. 3, no. 3, 2018, doi: 10.3390/recycling3030033.
[17]J. Di, B. K. Reck, A. Miatto, and T. E. Graedel, "United States plastics: Large flows, short lifetimes, and negligible recycling," *Resour. Conserv. Recycl.*, vol. 167, no. October 2020, p. 105440, 2021, doi: 10.1016/j.resconrec.2021.105440.
[18]M. Calisto Friant, D. Lakerveld, W. J. V. Vermeulen, and R. Salomone, "Transition to a sustainable circular plastics economy in the netherlands: Discourse and policy analysis," *Sustain.*, vol. 14, no. 1, pp. 1–32, 2022, doi: 10.3390/su14010190.
[19]S. K. Shin, N. Um, Y. J. Kim, N. H. Cho, and T. W. Jeon, "New policy framework with plastic waste control plan for effective plastic waste management," *Sustain.*, vol. 12, no. 15, 2020, doi: 10.3390/su12156049.
[20]A. Yoshida, "China's ban of imported recyclable waste and its impact on the waste plastic recycling industry in China and Taiwan," *J. Mater. Cycles Waste Manag.*, vol. 24, no. 1, pp. 73–82, 2022, doi: 10.1007/s10163-021-01297-2.
[21]A. W. M. Ng, S. Ly, N. Muttil, and C. N. Nguyen, "Issues and challenges confronting the achievement of zero plasticwaste in Victoria, Australia," *Recycling*, vol. 6, no. 1, pp. 1–24, 2021, doi: 10.3390/recycling6010009.
[22]A. A. Phelan, H. Ross, N. A. Setianto, K. Fielding, and L. Pradipta, "Ocean plastic crisis—Mental models of plastic pollution from remote Indonesian coastal communities," *PLoS One*, vol. 15, no. 7 July, pp. 1–29, 2020, doi: 10.1371/journal.pone.0236149.
[23]Ministry of Environment and Forestry of the Republic of Indonesia, "Status of Environment and Forestry 2020," pp. 14–50, 2020.

[24]J. H. Kim, "Extended producer responsibility (EPR) and job creation in Korea," *WIT Trans. Ecol. Environ.*, vol. 163, pp. 75–82, 2012, doi: 10.2495/WM120071.
[25]K. E. Institute, *Korea Environmental Policy Bulletin*, vol. IX, no. 1. 2010.
[26]I. President, "Law of the Republic of Indonesia Number 18 of 2008 regarding Waste Management," pp. 61–64, 2008.
[27]I. President, "Government Regulation of the Republic of Indonesia No. 81 of 2012 Management of Household Waste and Types of Household Waste," 2012.
[28]Ministry of Environment and Forestry of the Republic of Indonesia, "Regulation of the Minister of Environment and Forestry of the Republic of Indonesia Number 75 of 2019 concerning Roadmaps for Reducing Waste by Producers," vol. 53, no. 9, pp. 1689–1699, 2019.

Dynamic Availability Assessment Using Dynamic Evidential Network: Water Deluge System Case Study

Mohammed Bougofa[1*], Abderraouf Bouafia[2], Ammar Chakhrit[3], I H M Guetarni[3], Amin Baziz[1], Salah Aberkane[4], Bilal Zerouali[5], Rabeh Kharzi[1], and Ahmed Bellaouar[1]

[1] Laboratory of Transportation Engineering and Environment, Frères Mentouri Constantine 1 University, BP 325 Constantine, 25017, Algeria
[2] Department of Process Engineering, Chemical Engineering and Environment Laboratory, Skikda University, Skikda, 21000, Algeria
[3] Institute of maintenance and industrial safety, Oran 2 University, Oran, 31000, Algeria
[4] Department of Psychology, Khenchela University, BP 1252 Road of Batna Khenchela, 40004, Algeria
[5] Department of Process Engineering, Skikda University, Skikda, 21000, Algeria

*mohamed.bogoffa@umc.edu.dz

Abstract. Probabilistic modeling is widespread in engineering practices, mainly to evaluate the safety, risk analysis, and reliability of complex systems. However, insufficient data makes it difficult to estimate the state probability of components or the global system in dynamic complex systems. Furthermore, conventional methods for dependability analysis typically have little capacity to cope with dependence, failure behavior, epistemic uncertainty, and common cause failure simultaneously. This paper presents the application of an extended discrete-time dynamic evidential network (DEN) model to assess the availability of complex systems. The model application combines Dempster-Shafer's theory to treat epistemic uncertainty over a new state-space reconstruction of components and the dynamic Bayesian network to present multi-state system dependability. This model is demonstrated in a real case study of a water deluge system installed as a safety barrier from Algeria's oil and gas plant. The results show the significant influence of these factors on the system's availability. The goal of this modeling is to assure the high availability of a safety barrier in a volatile setting by providing a decision-making tool to prioritize maintenance tasks, preventing the failure of complicated redundant systems, and recommending alterations to the design.

1. Introduction

The increasing complexity of technology, design methodologies, processes, manufacturing systems, and maintenance has presented engineers with significant challenges throughout the past century. High dependability values are required to develop systems with better performance and safety. More probabilistic methodologies, such as fault tree analysis (FTA) [1], Markov chain (MC) [2], Petri nets (PN) [3], Bayesian networks (BN) [4], and Monte Carlo simulation (MCS) [5], have been developed for studying the reliability and availability of complex systems. The majority of these methods were intended to address aleatory and epistemic uncertainty by incorporating numerous theories, such as possibility theory, evidence theory, and fuzzy set theory [6,7].

For complex systems, the BN has grown increasingly prominent in dependability, safety, and risk analysis [1,8,9]. The paper [10] examines a recent statistical overview of BN applications to ensure the dependability of safety-critical systems. Because of its flexibility to manage multi-state, dependence, dynamic failure behavior, and uncertainty, BN is an effective modeling tool for complex systems. In the last decades, researchers have used BN to address engineering issues, particularly in dependability and decision-making. Bobbio et al. [11] described using FTA for constructing static BN for reliability assessment. To predict system reliability using the BN model, Lee et al. [12] established a predictive maintenance strategy for complex systems. After a series of studies devoted to static models, Boudali et al. [13] Introduced a method for converting a dynamic fault tree (DFT) to a dynamic Bayesian network (DBN) for predicting reliability by incorporating the dynamic feature. Numerous researchers worked on BN based on the theory of evidence to solve the problem of epistemic uncertainty in reliability evaluation. Weber and Simon [14] presented the evidentiary network (EN), a synthesis of evidence theory and BN, to evaluate the reliability of non-repairable systems. Zuo et al. [15] showed how a static evidential network could be used to improve the reliability of systems with CCF based on the β-factor model and uncertainty of parameters. They constructed an EN depending on a static fault tree and described the technique for the transformation of traditional static logic gates (i.e., AND-gate and OR-gate). In addition, they represented the CCF as a constant-valued node, which is illogical. However, all prior research has focused on the reliability of non-repairable systems using EN but has missed the availability evaluation of complex repairable systems via EN. Bougofa et al. [16] suggested an evidential network-based model for assessing system availability under parameter uncertainty. Another paper [9] focuses on extending DEN to determine the availability based on transforming dynamic spare gates and CCF using the β-factor model on the same model. This study presents an application of DEN to assess the availability of a water deluge system based on a CCF and epistemic uncertainty in an oil storage area.

This present paper is based on DEN to evaluate complex system availability. Part two is dedicated to the basics of evidence theory. The third part describes constructing a DEN from DFT, and the last part demonstrates the proposed method's application.

2. Evidential networks (EN)

The EN is a directed acyclic graph that is comparable to the BN. Mass functions of conditional beliefs show how various parts are dependent on one another by drawing arcs between them. In a graphical probabilistic model, P(x) denotes each parent node's marginal belief mass distribution X, while P(E|pa(E)) is the CBMT for each child node E.

The CBMT is given in a matrix form like follows:

$$M(E/pa(E))=\begin{bmatrix} m(A_1^Y/A_1^X) & \dots\dots. & m(A_P^Y/A_1^X) \\ \dots & \dots\dots. & \dots \\ m(A_1^Y/A_Q^X) & \dots\dots\dots & m(A_P^Y/A_Q^X) \end{bmatrix} \quad (1)$$

The junction tree approach was used by Simon et al. [19] to calculate each node's P(x). The functions of m(A) can be presented by experts' judgments (A).

2.1 DEN for spare gates

Industrial systems' increasing complexity requires technical requirements in terms of reliability, such as redundancy. To explain the complicated dynamic properties of redundant systems, such as failure behavior, logic connections have already been extended from the standard logic gates to spare gates (hot spare gate (HSP), cold spare gate (CSP), and warm spare gate (WSP)). In addition, numerous publications have studied the CBMT for conventional logic gates [9,20].

Figure 1 depicts two components connected with a spare logic gate, and the dormancy factor is α. There are three m(A) (up, down, up-down) assignments for each component in the framework of DST. The CBMT for X at step (k + 1) is shown in Table 1. The CBMT at step (k + 1) for node Y, depending on the type of spare gate, is displayed in Tables (2, 3 and 4). The reference [9] provides additional information on this modeling and validate all algorithm presented in this paper with Markov chain. Following this modeling, the CBMT for the spare gate node is modeled as a normal AND-gate, as illustrated in Table 5.

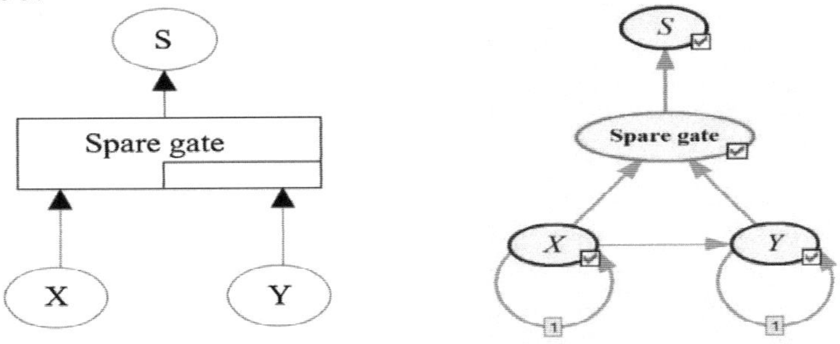

(a) (b)

Figure 1. (a) Spare gate using DFT, (b) Spare gate using DEN [9]

Table 1. CBMT at (k+1) for node X

Step (k+1)	Up (k)	Down (k)	Up-down (k)
Up	$e^{-\lambda_{upper}}$	$1 - e^{-\mu_{low}.\Delta t}$	$1 - e^{-\mu_{low}.\Delta t}$
Down	$1 - e^{-\lambda_{low}}$	$e^{-\mu_{upper}.\Delta t}$	$1 - e^{-\lambda_{low}}$
Up-down	$e^{-\lambda_{low}} - e^{-\lambda_{upper}}$	$e^{-\mu_{low}.\Delta t} - e^{-\mu_{upper}.\Delta t}$	$e^{-\lambda_{low}.\Delta t} + e^{-\mu_{low}.\Delta t} - 1$

Table 2. CBMT at (k+1) for node Y using WSP [9]

X (k+1)	Up			Down			Up-down		
Y (k)	Up	Down	Up-down	Up	Down	Up-down	Up	Down	Up-down
Up (k+1)	$e^{-\alpha\lambda_{upper}\Delta t}$	$1 - e^{-\mu_{low}\Delta t}$	$1 - e^{-\mu_{low}\Delta t}$	$e^{-\lambda_{upper}\Delta t}$	$1 - e^{-\mu_{low}\Delta t}$	$1 - e^{-\mu_{low}\Delta t}$	$e^{-\lambda_{upper}\Delta t}$	$1 - e^{-\mu_{low}\Delta t}$	$1 - e^{-\mu_{low}\Delta t}$
Down (k+1)	$1 - e^{-\alpha\lambda_{low}\Delta t}$	$e^{-\mu_{upper}\Delta t}$	$1 - e^{-\alpha\lambda_{low}\Delta t}$	$1 - e^{-\lambda_{low}\Delta t}$	$e^{-\mu_{upper}\Delta t}$	$1 - e^{-\lambda_{low}\Delta t}$	$1 - e^{-\lambda_{low}\Delta t}$	$e^{-\mu_{upper}\Delta t}$	$1 - e^{-\lambda_{low}\Delta t}$
Up-down (k+1)	$e^{-\alpha\lambda_{low}\Delta t} - e^{-\alpha\lambda_{upper}\Delta}$	$e^{-\mu_{low}\Delta t} - e^{-upper\,\Delta t}$	$e^{-\alpha\lambda_{low}\Delta t} + e^{-\mu_{low}\Delta t} - 1$	$e^{-\lambda_{low}\Delta t} - e^{-\lambda_{upper}\Delta t}$	$e^{-\mu_{low}\Delta t} - e^{-\mu_{upper}\Delta t}$	$e^{-\lambda_{low}\Delta t} + e^{-\mu_{inf}\Delta t} - 1$	$e^{-\lambda_{low}\Delta t} - e^{-\lambda_{upper}\Delta}$	$e^{-\mu_{low}\Delta t} - e^{-\mu_{upper}\Delta}$	$e^{-\lambda_{low}\Delta t} + e^{-\mu_{low}\Delta t} - 1$

Table 3. CBMT at (k+1) for node Y using CSP [9]

X (k+1)	Up			Down			Up-down		
Y (k)	Up	Down	Up-down	Up	Down	Up-down	Up	Down	Up-down
Up (k+1)	1	1	1	$e^{-\lambda_{upper}\Delta t}$	$1 - e^{-\mu_{low}\Delta t}$	$1 - e^{-\mu_{low}\Delta t}$	$e^{-\lambda_{upper}\Delta t}$	$1 - e^{-\mu_{low}\Delta t}$	$1 - e^{-\mu_{low}\Delta t}$
Down (k+1)	0	0	0	$1 - e^{-\lambda_{low}\Delta t}$	$e^{-\mu_{upper}\Delta t}$	$1 - e^{-\lambda_{low}\Delta t}$	$1 - e^{-\lambda_{low}\Delta t}$	$e^{-\mu_{upper}\Delta t}$	$1 - e^{-\lambda_{low}\Delta t}$
Up-down (k+1)	0	0	0	$e^{-\lambda_{low}\Delta t} - e^{-\lambda_{upper}\Delta t}$	$e^{-\mu_{low}\Delta t} - e^{-\mu_{upper}\Delta t}$	$e^{-\lambda_{low}\Delta t} + e^{-\mu_{low}\Delta t} - 1$	$e^{-\lambda_{low}\Delta t} - e^{-\lambda_{upper}\Delta t}$	$e^{-\mu_{low}\Delta t} - e^{-\mu_{upper}\Delta t}$	$e^{-\lambda_{low}\Delta t} + e^{-\mu_{low}\Delta t} - 1$

Table 4. CBMT at (k+1) for node Y using HSP [9]

X (k+1)	**Up**			**Down**			**Up-down**		
Y (k)	**Up**	**Down**	**Up-down**	**Up**	**Down**	**Up-down**	**Up**	**Down**	**Up-down**
Up (k+1)	$e^{-\lambda_{upper}\,\Delta t}$	$1 - e^{-\mu_{low}\,\Delta t}$	$1 - e^{-\mu_{low}\,\Delta t}$	$e^{-\lambda_{upper}\,\Delta t}$	$1 - e^{-\mu_{low}\,\Delta t}$	$1 - e^{-\mu_{low}\,\Delta t}$	$e^{-\lambda_{upper}\,\Delta t}$	$1 - e^{-\mu_{low}\,\Delta t}$	$1 - e^{-\mu_{low}\,\Delta t}$
Down (k+1)	$1 - e^{-\lambda_{low}}$	$e^{-\mu_{upper}\,\Delta t}$	$1 - e^{-\lambda_{low}\,\Delta t}$	$1 - e^{-\lambda_{low}\,\Delta t}$	$e^{-\mu_{upper}\,\Delta t}$	$1 - e^{-\lambda_{low}\,\Delta t}$	$1 - e^{-\lambda_{low}\,\Delta t}$	$e^{-\mu_{upper}\,\Delta t}$	$1 - e^{-\lambda_{low}\,\Delta t}$
Up-down (k+1)	$e^{-\lambda_{low}\,\Delta t} - e^{-\lambda_{uppe}}$	$e^{-\mu_{low}\,\Delta t} - e^{-\mu_{upper}\,\Delta t}$	$e^{-\lambda_{low}\,\Delta t} + e^{-\mu_{low}\,\Delta t} - 1$	$e^{-\lambda_{low}\,\Delta t} - e^{-\lambda_{upper}\,\Delta t}$	$e^{-\mu_{low}\,\Delta t} - e^{-\mu_{upper}\,\Delta t}$	$e^{-\lambda_{low}\,\Delta t} + e^{-\mu_{low}\,\Delta t} - 1$	$e^{-\lambda_{low}\,\Delta t} - e^{-\lambda_{upper}\,\Delta t}$	$e^{-\mu_{low}\,\Delta t} - e^{-\mu_{upper}\,\Delta t}$	$e^{-\lambda_{low}\,\Delta t} + e^{-\mu_{low}\,\Delta t} - 1$

Table 5. CBMT of the "AND" gate

		AND-gate		
X	**Y**	**Up**	**Down**	**Up-down**
Up	Up	1	0	0
Up	Down	1	0	0
Up	Up-down	1	0	1
Down	Up	1	0	0
Down	Down	0	1	0
Down	Up-down	0	0	1
Up-down	Up	1	0	0
Up-down	Down	0	0	1
Up-down	Up-down	0	0	1

2.2 Modeling CCF with DEN

The node CCF has two possible states, "occur" (indicating CCF exists) and "not occur" (indicating CCF does not exist)., as shown in Figure 2.

At time zero, node CCF is in the "not occur" state, and Table 6 shows the CBMT for that node. Table 7 lists the C1 and C2 CBMTs at. Following this section's logic gates, system availability can be determined (static or spare gates).

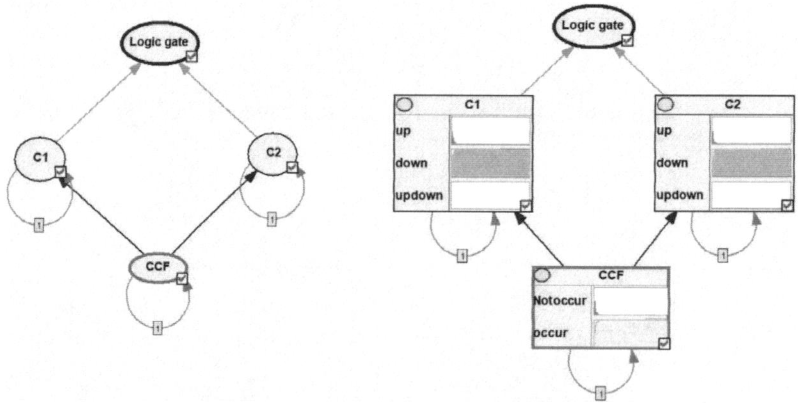

Figure 2. DEN for CCF and Spare gate

Table 6. The CBMT for CCF at T = k+1

CCF (k)	Not occur	Occur
Not occur	$e^{-\lambda_{CCF}.\Delta t}$	0
Occur	$1 - e^{-\lambda_{CCF}.\Delta t}$	1

Table 7. The CBMT for C1/C2 at T = k+1

CCF (k+1)	Not occur			Occur		
C (k)	**Up**	**Down**	**Up-down**	**Up**	**Down**	**Up-down**
Up (k+1)	$e^{-\lambda_{upper}\ \Delta t}$	$1 - e^{-\mu_{low}\ \Delta t}$	$1 - +e^{-\mu_{low}\ \Delta t}$	0	0	0
Down (k+1)	$1 - e^{-\lambda_{low}\ \Delta t}$	$e^{-\mu_{upper}\ \Delta t}$	$1 - e^{-\lambda_{low}\ \Delta t}$	1	1	1
Up-down (k+1)	$e^{-\lambda_{low}\ \Delta t} - e^{-\lambda_{upper}\ \Delta t}$	$e^{-\mu_{low}\ \Delta t} - e^{-\mu_{upper}\ \Delta t}$	$e^{-\lambda_{low}\ \Delta t} + e^{-\mu_{low}\ \Delta t} - 1$	0	0	0

3. Case study

In this part, modeling a water deluge system has been established to evaluate dependability (availability) performance. This system is installed as a safety barrier at the crude oil storage area of the CSC project in Rhourd Nouss /Algeria. The storage area of the CSC project has three oil storage tanks. Two centrifugal pumps send the treated oil to the pumping station outside the complex.

Storage areas are where the most catastrophic scenarios occur, including a finished product destined for marketing. This is why the study of the availability of the water deluge system is significant to ensure the effectiveness of the safety barrier at the level of the centrifugal shipping pumps.

3.1 Description of the water deluge system

The CSC plant has a deluge system comprising a water storage tank, a fire pump system, and a fire-fighting water distribution network. The fire pump system includes several redundant pumps calculated for the worst-case scenario. The dimensioning is considered for the water demand corresponding to a fire at the level of the complex. This fire scenario requires a water flow of 680 m^3/h. To ensure this water flow {two jockey pumps 2*35 m^3/h at 7 bar, an electric pump 680 m^3/h at 12 bar, and a diesel pump 680 m^3/h at 12 bar. The first jockey pump runs full time to keep the network under pressure. The other pumps intervene automatically by the fire and gas system (Detronex and Triconex) if a fire detection is confirmed by 2 of the three fire detectors. At the same time, the fire and gas system sends the signal to the pumps to work, and it also activates the deluge valve (solenoid valve) so that the water can pass to the nozzles of the deluge system (sprinkler).

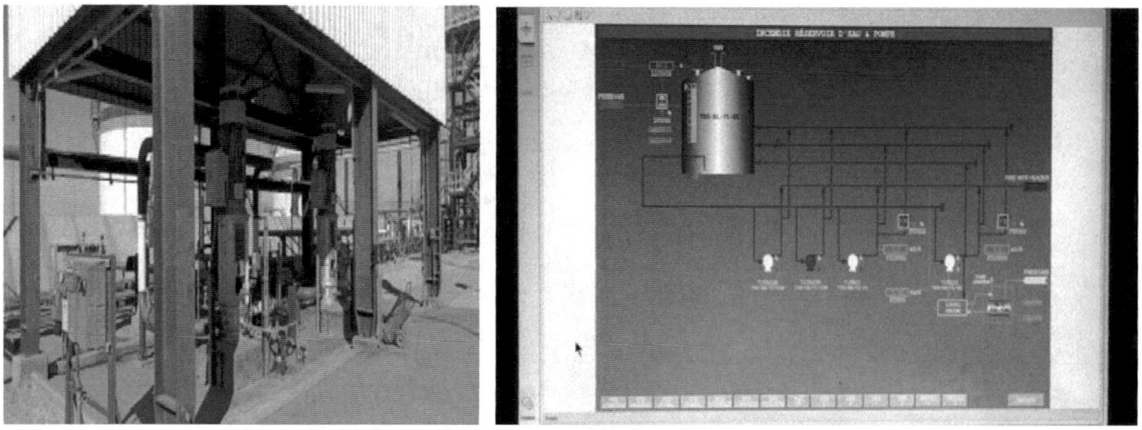

Figure 3. (a) Expedition pumps, (b) fire-fighting water system

3.2 DEN modeling for case study

Table 8 displays the estimated failure and repair rates of all components as a function of time, based on both statistical generalizability and expert judgment [21]. The CCF for the components (X10, X11, and X12) is β=0.05, which relies on the loss of electrical energy. The dormancy factor α for the WSP gate in the jockey pump subsystem is assumed to be 0.3. Figure 5 shows a DEN model for deluge system availability derived from DFT analysis using the methods outlined in Section 3 (Figure 4). When time began (t = 0), everything was in a state of perfection.

Table 8. Failure and repair rates of components

Symbol	Component and their mode of failure	Failure Rate ($Week^{-1}$)	Repair rate ($Week^{-1}$)
X1	The solenoid valve does not open on demand	[3E-4, 4.5E-4]	[7E-3, 8E-3]
X2	Nozzle blockage or reduced flow	Failure probability = 1E-6	/
X3/X4/X5	Flame detector	[8.5E-5, 1E-4]	[3E-3, 4E-3]
X6/X7/X8/X9	Failure of the jockey/electric/diesel pump	[2.2E-4, 3.5E-4]	[2.5E-3, 3.5E-3]
X10/X11/X12	Electric motor failure	[2.4E-4, 3.1E-4]	[2.75E-3, 3.42E-3]
X13/X14/X15/X16	Failure of the pressure sensor (pressure switch)	[7.7E-4, 9.2E-4]	[7E-3, 8E-3]
X17	Diesel engine failure	[2.1E-3, 2.9E-3]	[1.6E-2, 2.1E-2]

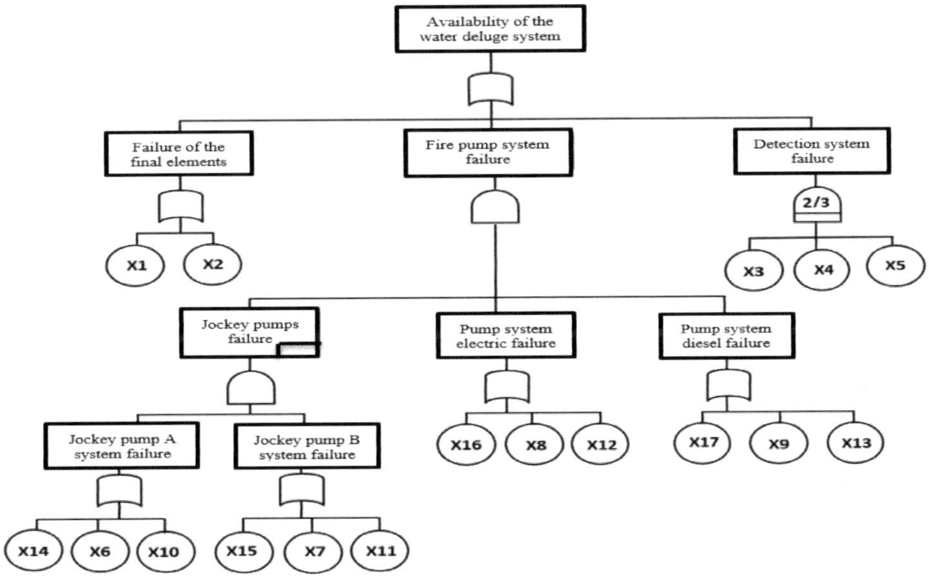

Figure 4. DFT analysis for system availability

3.3 Results and discussion

The values of the availability of the studied system are presented in table 9 for 13 years of service. Complete system availability curves under various conditions, including those where uncertainty and CCF are considered, either included or not, are presented in Figure 6.

According to the results, the availability of the water deluge system is reduced over time but remains very effective after 13 years of service. In addition, it is found that CCF factors and epistemic uncertainty significantly impact availability. To preserve these results, a well-established maintenance plan should be followed.

One of the advantages of DEN is the updating of results after obtaining new information. In the system history, the detector X4 failed after ten years of service and has remained in this condition until now. Due to a lack of spare parts, the manager neglected this technical defect and the repair priority. The new information about X4 in the DEN model is updated, and the results are presented in Table 10. Unfortunately, the availability has continued to drop more than last.

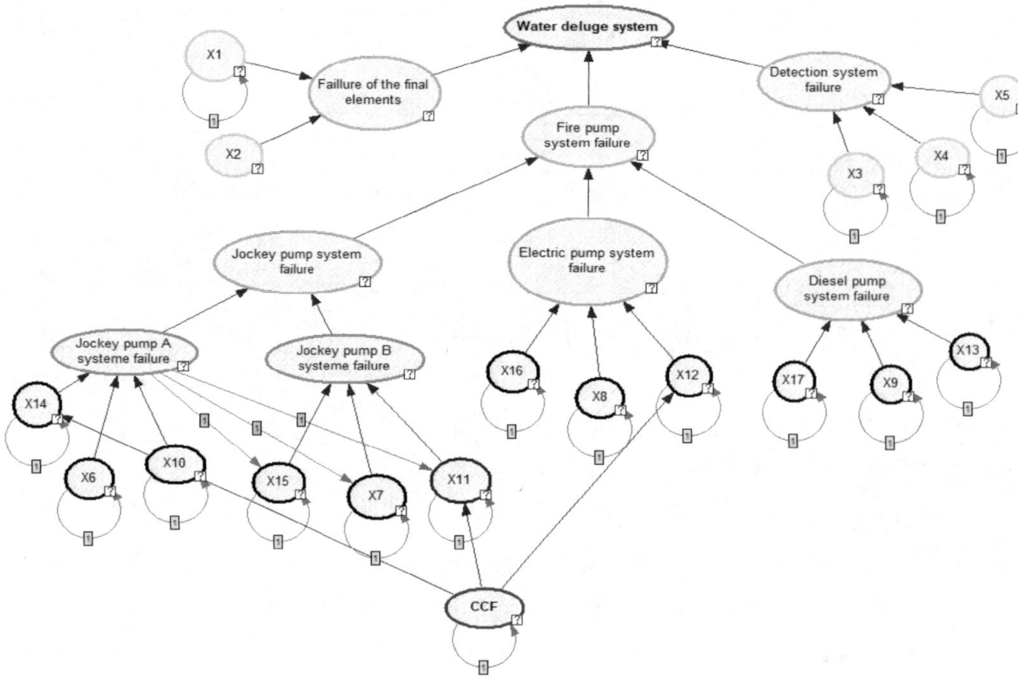

Figure 5. DEN converted using DFT

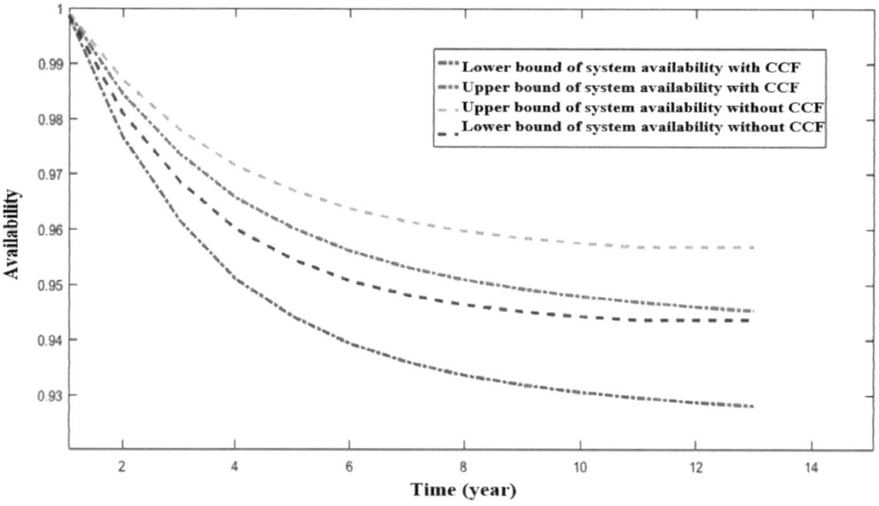

Figure 6. Water deluge system availability

Table 9. System availability after 13 years

System	States	With CCF	Without CCF
Availability at	Bel	0.9282	0.9436
T= 13 years	Pls	0.9452	0.9568

Table 10. Updated system availability after 13 years

System	States	With CCF
Availability at	Bel	0.8841
T= 13 years	Pls	0.8996

3.4 Sensitivity analysis

Another advantage of DEN is the sensitivity analysis or diagnosis using Bayesian inference given a posteriori likelihood, which allowed us to establish a decision support tool. Several methods, such as the ratio of variance (RoV), determine the latter. For example, in any failure and sensitivity study of a system, the RoV in Equation (2) can be relied upon to include a reliable proportion of importance. [4,8].

$$RoV(Xi) = \frac{\pi(Xi) - \theta(Xi)}{\theta(Xi)} \tag{2}$$

Where $\pi(Xi)$ and $\theta(Xi)$ signify Xi's posterior and prior probabilities, respectively.

In this case, the following evidence is updated in this DEN model:

- The unavailability of the water deluge system is 100%
- X4 flame detector failure is 100%
- The unavailability of the jockey pump system is 100%: if the deluge system is operated only by jockey pumps, the system is insufficient to act as a safety barrier due to the low water flow.

According to figure 7, the results show that the probabilities of X3, X5, X1, and X2 represent the highest percentage increases in RoV and therefore represent the most critical basic events contributing to the unavailability of the water deluge system. In this case, installing a preventive maintenance plan is necessary to avoid an untimely opening (slow opening) of the solenoid valve due to salt grains being permanently in the water. These grains interfere with the opening of the solenoid valve. In the same way, it is necessary to avoid the clogging of the nozzles by limestone and salt too. In addition, the detector X4 is faulty. In this simulation, the most critical event is the failure of one of two flame detectors, X3 or X5, due to the 2 out of 3 voting architectures. The decision-maker must act in priority to repair the detector X4.

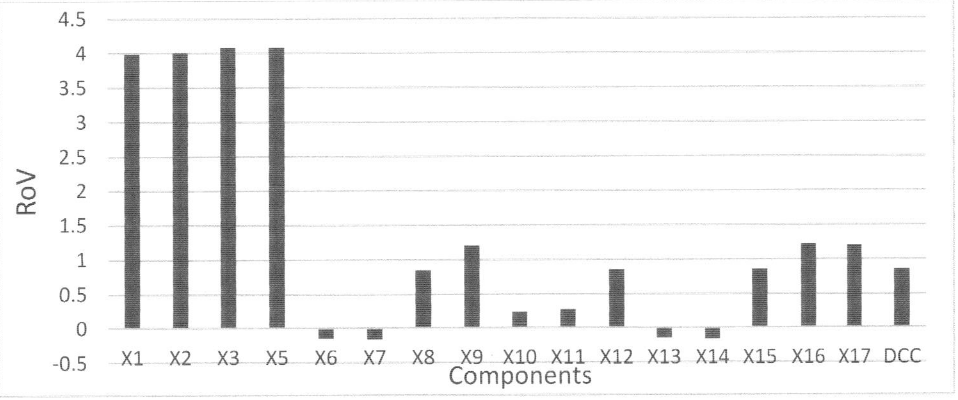

Figure 7. Sensitivity analysis results

4. Conclusion

According to this study, the spare gate, including HSP, WSP, and CSP gates with CCF, can be considered when evaluating a multi-state system's availability. The introduced method takes the epistemic uncertainty and CCF into account because they are important in the dependability analysis. Lack of data leads to many difficult issues, including state space explosion, reliance, dynamic failure behavior, CCF, and parameter ambiguity, can be handled with the approach described here. This study examines the CBMT for static gates as well as all of the spare gates. It is used to reduce the level of model uncertainty, show the interdependence of the many model components, and compute the system's mass distribution. Finally, a real-world example illustrates how this approach can be used (a water deluge system). According to these findings, system availability is impacted significantly by factors like CCF and epistemic uncertainty, which may be modeled simultaneously in a single model. Furthermore, the model illustrates DEN's ability to incorporate new information and revise all previous findings. This model, which uses sensitivity analysis, could serve as a decision-making tool for enhancing availability and averting undesirable phenomena.

References

[1] Kabir S, Taleb-Berrouane M and Papadopoulos Y 2019 Dynamic reliability assessment of flare systems by combining fault tree analysis and Bayesian networks *Energy Sources A: Recovery Util. Environ. Eff.* 1–18

[2] Talebberrouane M, Khan F and Lounis Z 2016 Availability analysis of safety critical systems using advanced fault tree and stochastic Petri net formalisms *J. Loss Prev. Process Ind.* **44** 193–203

[3] Taleb-Berrouane M, Khan F and Amyotte P 2020 Bayesian Stochastic Petri Nets (BSPN) - A new modelling tool for dynamic safety and reliability analysis *Reliab. Eng. Syst. Saf.* **193** 106587

[4] Bougofa M, Bouafia A, Baziz A, Aberkane S, Kharzi R and Bellaouar A 2021 Risk analysis-based reliability assessment approach under epistemic uncertainty using a dynamic evidential network *IOP Conf. Ser.: Earth Environ. Sci.* **896** 012035

[5] Wang W, Maio F D and Zio E 2017 Three-loop Monte Carlo simulation approach to Multi-State Physics Modeling for system reliability assessment *Reliab. Eng. Syst. Saf.* **167** 276–89

[6] Buckley J J 2005 *Fuzzy Probabilities: New Approach and Applications* (Berlin Heidelberg: Springer-Verlag)

[7] Ben Yaghlane A links open overlay panelBoutheina and Mellouli K 2008 Inference in directed evidential networks based on the transferable belief model *Int. J. Approx. Reason.* **48** 399–418

[8] Bougofa M, Bouafia A and Bellaouar A 2020 An Integrated Quantitative Bayesian Network in Risk Management for Complex Systems *Int. J. Perform. Eng.* **16** 354-366

[9] Bougofa M, Taleb-Berrouane M, Bouafia A, Baziz A, Kharzi R and Bellaouar A 2021 Dynamic availability analysis using dynamic Bayesian and evidential networks *Process Saf. Environ. Prot.* **153** 486–499

[10] Kabir S and Papadopoulos Y 2019 Applications of Bayesian networks and Petri nets in safety, reliability, and risk assessments: A review *Saf. Sci.* **115** 154–75

[11] Bobbio A, Portinale L, Minichino M and Ciancamerla E 2001 Improving the analysis of dependable systems by mapping fault trees into Bayesian networks *Reliab. Eng. Syst. Saf.* **71** 249–60

[12] Lee D and Pan R 2017 Predictive maintenance of complex system with multi-level reliability structure *Int. J. Prod. Res.* **55** 4785–801

[13] Boudali H and Dugan J B 2005 A discrete-time Bayesian network reliability modeling and analysis framework *Reliab. Eng. Syst. Saf.* **87** 337–49

[14] Weber P and Simon C 2008 Dynamic evidential networks in system reliability analysis: A Dempster Shafer approach (IEEE) 2008 16th Mediterranean Conference on Control and Automation 603–608

[15] Zuo L, Xiahou T and Liu Y 2019 Evidential network-based failure analysis for systems suffering common cause failure and model parameter uncertainty *Proc. Inst. Mech. Eng.* C *J. Mech. Eng. Sci.***233** 2225–2235

[16] Bougofa M, Bouafia A and Bellaouar A 2020 Availability Assessment of Complex Systems under Parameter Uncertainty using Dynamic Evidential Networks *Int. J. Perform. Eng.* **16** 510-519

[17] Dempster A P 1967 Upper and Lower Probabilities Induced by a Multivalued Mapping *Ann. Math. Statist.* **38** 325–339

[18] Jensen F V 1996 *An introduction to Bayesian networks* (London: UCL Press)

[19] Simon C, Weber P and Evsukoff A 2008 Bayesian networks inference algorithm to implement Dempster Shafer theory in reliability analysis *Reliab. Eng. Syst. Saf.***93** 950–63

[20] Mi J, Cheng Y, Song Y, Bai L and Chen K 2019 Application of dynamic evidential networks in reliability analysis of complex systems with epistemic uncertainty and multiple life distributions *Ann. Oper. Res.* **311** 311-333

[21] Guetarni Islam Hadj M 2019 *Analyse Quantitative des Risques : Application sur les Bacs de Stockage* Thesis (Université d'Oran 2 Mohamed Ben Ahmed)

IOP Conf. Series: Earth and Environmental Science 1098 (2022) 012016 doi:10.1088/1755-1315/1098/1/012016

The impact of disposable mask waste pollution in peat soil

Alvi Eka Mentari [1], Adian Khoironi *[1,3] and Hadiyanto [2,3]

[1,2] Environmental Health Department, Health science Faculty of Dian Nuswantoro University
[2]Chemical Engineering Department, Diponegoro University, Semarang
[3]Centre of Biomass and Renewable Energy, Diponegoro University, Semarang

*adian.khoironi@dsn.dinus.ac.id

Abstract. The abundance of plastic waste in Indonesia has increased due to the contribution of disposable masks waste during the COVID-19 pandemic. The improper waste management causes the habit of dumping waste in the soil system tostill frequently occur. The secondary microplastics from disposable mask waste has the potential to damage soil quality and soil capillary water. Therefore, this study was conducted with the aim of understanding the effects of disposable mask waste in peat soil. The research was conducted by setting up 3 reactors containing peat soil with disposable mask waste buried in it and some plants being grown on it. For 45 days, observations and samples were taken, both filtrate water and soil for analysis. The analysis results showed significant changes in soil properties withchanges in the concentration of inorganic matter, organic matter, porosity, bulk density, and water content. Further research needs to be done to see the effects of disposable mask waste in the soil system in real conditions where several confounding variables cannot be controlled.

1. Introduction

Between 2020 and mid of 2021, about 4,088,166 Covid-19 SARS-CoV-2 positive cases were counted in Indonesia with total death cases reached 128,252 [1]. To curb the inclusion of positive cases coronavirus increasingly and more widespread, Indoneisa Government released a decree by Ministry of Health No 0107/MENKES/382/2020 about Public Health Protocols in Public Places and Facilities for the Prevention and Control of Coronavirus Disease 2019 or Covid-19. This regulation strict the health protocols must be implemented by the community, such as washing hands maintaining distance or avoiding congestion, and increasing self-immune due to a balanced consumption of nutrition and practice healthy lifestyle and not forget to wear personal protective equipment likes mask that acts as a PPE outside your home

Increasing public awareness to use masks as one of the personal protective equipment from the spread of the covid-19 virus resulted in the community's need for disposable masks to increase significantly [2]. The increase in the consumption of single-use masks has trigger a new problem, namely the emergence of medical waste in the form of single-use masks in large quantities. In Indonesia, the total amount of disposable mask waste from the beginning of the pandemic until mid-2021 reached 21.768 tons [3]. Another study even noted that around 538,40-807,59 million pieces/day of disposable mask were generated [4]. This very large number is increasing along with the management of medical waste in Indonesia which is still inadequate, Aragaw (2020) [5] explained that the lack of facilities and in the management of disposable mask waste resulted in open dumping in the ground being an alternative in managing disposable mask waste. Furthermore, this condition will deteriorate environmental health [6] due to decrease of the soil quality where disposable mask waste is buried and certainly will have an impact on the final quality of the soil [7].

Disposable mask waste is a secondary source of microplastics, this is because the basic material used to make disposable masks is polypropylene fiber which is a type of polymer used as a plastic raw material [8]. The fiber structure of polypropylene in disposable masks produces fiber-shaped microplastics that have greater distribution capabilities compared to fragments and filaments of microplastics [9]. In addition, in their research, Guo et al., 2020 [7] and Auta et al., 2017 [10] proved that disposable mask waste buried in the soil will interact with variousnatural phenomena such as exposure to ultraviolet light, hydrolysis process by rainwater, temperature changes and so on which result in disposable mask waste will be fragmented into microplastics.

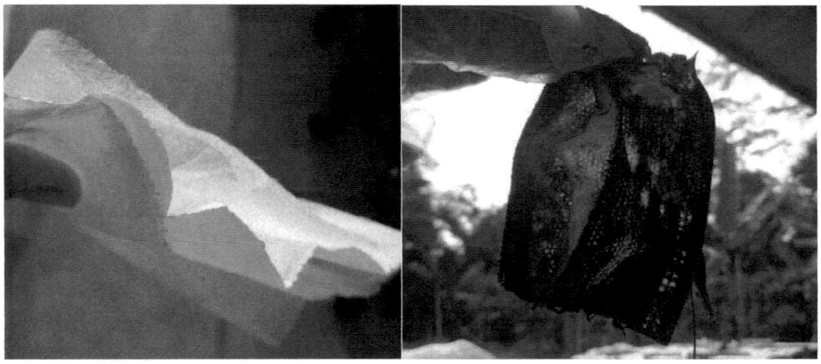

Figure 1. The Impact of Disspoable Mask Waste

From previous studies, it was found that there was a significant effect of exposure to microplastics in the soil on changes in soil properties, groundwater quality and further impact on plant growth [11]. Microplastics can enter the terrestrial ecosystem due to the leching process. In soil, microplastic contamination affected the increase of soil porosity, adsorption capacity of pollutants in soil solution, activity of enzyme, and more over the capacity of microplastic surface to adsorb pathogenic. Soil with more organic matter content has a largest adsorption value which is easily desorb Zn^{2+} contains in microplastic therefore increase the bioavailability of Zn^{2+}, Even though particle size, specific surface area and surface characteristics take important role in adsorption process between microplastic and metal ion [12].

The impact of microplastics on the soil system is strongly influenced by the characteristics of the soil itself) [13]. Indonesia is a country consisting of islands with different topography including diverse soil characteristics. The islands of Sumatra and Kalimantan are islands that have soil characteristics are dominated by organic matter elements so that many types of peat soil are found [14]. Peat soil is soil that is rich in organic material because it is formed from the accumulation of dead plant residues [15]. Peat soil has several levels of maturity called fibric, namely raw peat soil, hemic, or semi- cooked peat soil, and sapric which is mature peat soil which is usually found on the top or surface of the soil [14,16]. The type of organic material that composes peat soil will affect the characteristics of peat soil, especially for water holding capacity [17].

In a study conducted by Zhou et al, 2021 [18] on loamy soils, it was found that microplastics reduced bulk density concentrations and changed soil pH which also affected the microbial activity of the soil itself. In addition, research from Leed and Smithsoon, 2019 [12] also states that the influence of microplastic contamination on soil ecosystems can directly affect soil physical and chemical properties, but the impact of microplastics is also influenced by various environmental factors. Therefore, due to various kinds of research resultsobtained previously while there is still little researchon how the influence of microplastics on changes in the characteristics of peat soils, this research was carried out.

2. Methods and material

2.1. Preparation sample
The reactor used consists of one control reactor and three treatment reactors, each of which hasa height of 28 cm with a diameter of 25 cm, and the height of the seat for the reactor is 30 cm. Other equipment needed is a measuring cup used in the watering process, a container to accommodate infiltration water resulting from watering, chili plants, NPK fertilizer, a ruler to measure the growth of chili plant height, as well as PET bottles, 1000 ml plastic boxes, and plastic clips for saving sample.

The peat soil used in the study was taken from Kumpai Batu Bawah Village, South Arut District, West Kotawaringin Regency, Central Kalimantan (2°46'33.3"S 111°31'24.9"E) on August 6, 2021 at 09:25 WIB. Peat soil samples were taken at a depth of approximately 30 cm from the surface and were classified as capric peat soils. This peatland extraction site is a peatland that is managed into agricultural

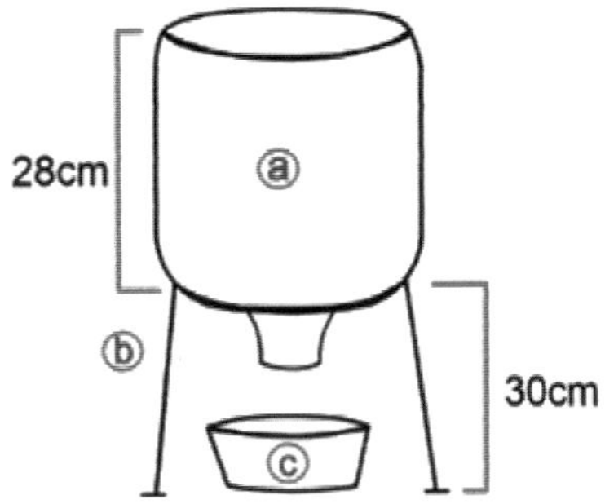

Figure 2. Size of the reactor

2.2. Analysis
Peat soil samples were analyzed for heavy metals and soil minerals with XRF WDXRF Rigaku Supermini200 and Visible Spectrophotometer. First, the soil sample is weighed and then put into a test tube and then added with reagents, then compress the volume with distilled water. The test tube was then allowed to stand for 24 hours. The same treatment for the glucose standard series. The sign that reaction has been successful is green. Then the absorbance was measured. As for the analysis and identification of microplastics in the soil, the soil is dried using a vacuum oven, after grinding the soil dry until the lumps are destroyed, then as much as 3 grams of soil is dissolved in NaCl with a concentration of 60 grams/100 mL. Soil solution with NaCl is then added to centrifuge with a speed of 2000 rpm for 15 minutes. After finishing the centrifuge, separate the floating part with filter paper, then the material contained in the filter paper is dissolved with 70% ethanol with the aim of eliminating bacteria [19].

Furthermore, after the soil extraction process is complete, add peat soil into all reactors and then proceed with burying disposable medical mask in gallons of treatment reactors (R1, R2, R3). While the control reactor does not have a buried mask because it will be used as a comparison before and after treatment. In addition, all reactors received treatment inthe form of routine watering every day with 200 ml of well water for 45 days where the capillary water from this watering process will be collected as shown in Figure 2.

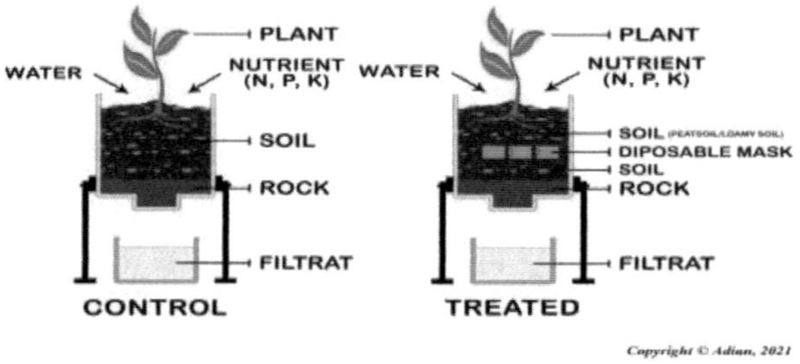

Figure 3. Reactors set up

After that, the filter again and identify microplastics from the last filtered solution with an NSZ 606 microscope.

3. Result and discussion

3.1. The impact of disposable mask contamination to the soil quality

Soil quality can be determined from changes in soil characteristics that greatly affect the ability of the soil to perform its functions. One of the functions of the soil is to support the life of living things that exist on the surface of the soil or in the soil body. A number of studies, one of which was conducted by de Souza Machado [13], proved the impact of microplastics on soil function in supporting the growth of onion blub plants [20], several types of plastic polymers have been shown to inhibit plant growth, especially in the roots. The ability of microplastics to inhibit plant growth is influenced by the shape of the microplastic, the size of the microplastic and the length of exposure time [18]. With very small sizes ranging from 1-5 mm, microplastics can be absorbed by the roots. Disruption of root growth due to microplastics in the soil is also due to the nature of the microplastics themselves which can absorb contaminants including heavy metals [21] (Hahladakis et al., 2018) thus disrupting the life of organisms and microbial activity that helps plant growth. Soil.

Table 1. Assessment for the change of mineral content in peat soil

Sample	Elements	Consentration (%)		
		Blank	Control	Treated
Peat soil	Al	1,33	3,81	3,48
	Si	2,60	7,71	7,11
	P	0,09	0,08	0,06
	S	0,39	0,27	0,22
	Cl	0,07	0,01	0,01
	K	0,55	0,38	0,37
	Ti	0,50	0,35	0,33
	Fe	1,67	1,03	0,73
	Zr	0,02	0,01	0,01
	Nb	0,0075	Unidentified	Unidentified
	Mo	0,0087	Unidentified	Unidentified
	Ru	0,23	Unidentified	Unidentified
	Cd	0,03	0,006	0,01
	Te	0,09	Unidentified	Unidentified
	I	0,10	Unidentified	Unidentified
	Cs	0,10	Unidentified	Unidentified
	Cr	0,01	0,055	0,003
	Ni	0,01	0,002	0,0005

Mn	0,01	0,009	0,01
Cu	0,007	0,002	0,003
Zn	0,005	0,005	0,004
As	Unidentified	Unidentified	Unidentified
Hg	Unidentified	Unidentified	0,00105
Pb	0,007	0,004	0,0038
Mg	Unidentified	0,087	Unidentified
Ca	Unidentified	0,128	0,102
Sr	Unidentified	0,002	0,003
Rb	Unidentified	Unidentified	0,002
Co	Unidentified	Unidentified	0,004

Micronutrients in the form of Cl, Fe, Mn, Cu, B, Mo, and Zn are needed in smaller amounts but their presence is important in the formation of enzymes and hormones in plants. After 45 days of experiment, the blank, the control and soil exposure to disposable mask were analyzed using the XRF instrument. The results of the XRF analysis (Table 1) showed a very compared to control soils and soils treated with disposable masks. Overall, there was a decrease in macro and micronutrients in the control and quality, fertility level, and plant growth are strongly influenced by soil nutrients. Macronutrient elements in the form of C, H, O, P, S, Ca, N, K, and Mgare used by plants for respiration and photosynthesis processes so that macronutrients are needed in large quantities.

Treatment soil compared the blank soil. One of the reasons for the decrease in the concentration of inorganic matter in the control and treated soils was the use of inorganic elements as a source of macro micronutrients for the growth of chili plants that were cultivated in a reactor with control and treated soils. As we can see in Table 1, the concentration of macronutrients P and K in the blanks was 0.09% and 0.55%, decreased in the control to 0.08% and 0.38% while in the treated soil it decreased by 0.06% and 0.37%. This condition is understandable, changes in soil nutrient concentrations can occur due to the presence of microplastics and plastic additives that affect the enzymatic activity of microbes in the soil.

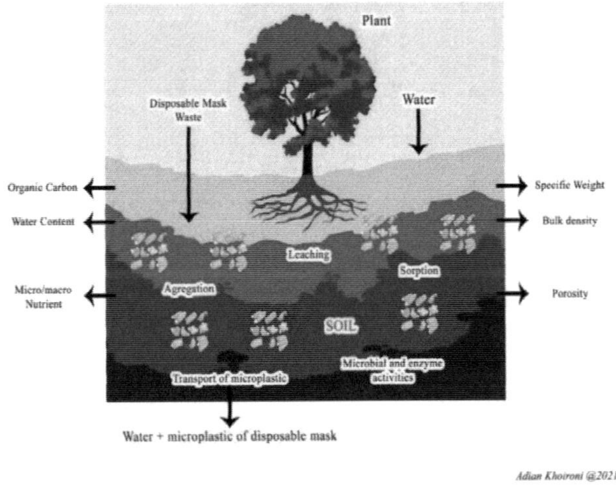

Figure 4. The Impact of Disposable Mask Waste to Soil Characteristic

The enzymatic activity of microbes in the soil greatly affects the availability of nutrients in the soil. In the research of Wang et al., 2016 showed the ability of plastic residues to significantly reduce the diversity of microbial communities and reduce carbon and nitrogen concentrations. The change of the inorganic composition in soil could be caused the HDPE particles altered the cation exchange in the soil and allowed free exchange of protons in the soil water, given their relatively large, possibly reactive, surface area.

The next characteristic change caused by microplastic contamination is a change in the physical properties of the soil. Referring to several previous studies conducted on loamy soils, the presence of microplastics in the soil resulted in a decrease in bulk density, one of which was because plastic had a lower density than the density of soil colloidal particles.

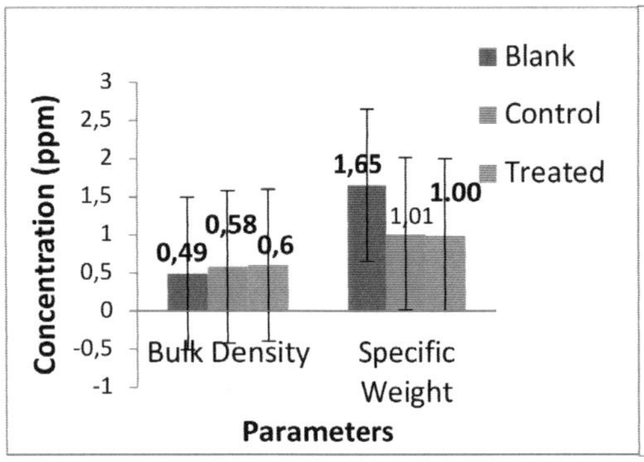

Figure 5. Bulk Density and Specific Weight

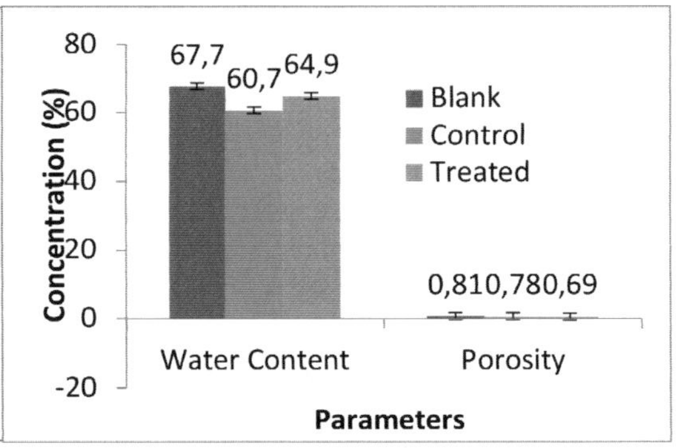

Figure 6. Water Content & Porosity

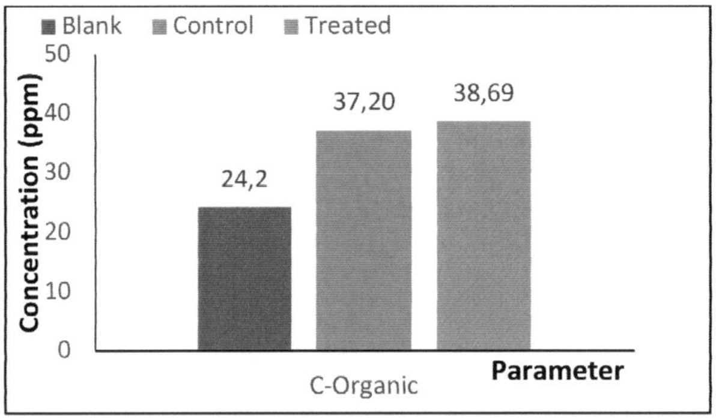

Figure 7. C-Organic

From the result of the study, bulk density measurements showed an increase in the bulk density of the soil along with a decrease in the specific gravity of the soil on the soil that was treated with disposable This condition occurs because the density of peat soil particles is lower than the density of polypropylene. The density of polypropylene is 0.9 – 0.91 ppm [19] while the particle density of hemic peat soil is only 0.075 – 0.0195 ppm mas waste. In this research using peat soil treated by polypropylene film of disposable mask, changes of some physical soil properties shown the increase using peat soil treated by polypropylene film of disposable mask, changes of some physical soil properties shown the increased of bulk density 0,11 ppm from 0,49 ppm in blank peat soil to 0,6 ppm in reactor treated (Figure 5) [19]. As a result, the increase in bulk density in the soil is caused by the increase in soil mass by microplastics so that by decreasing the specific gravity of the soil it does not decrease the bulk density of the. In this study, a result shows a close relationship to the increase in bulk density due to the presence of microplastics resulting from the fragmentation of disposable masks with changes in soil porosity. The decrease in porosity in the treated soil (0.69%) compared to the blank soil (0.81%) and control (0.78%) indicated the presence of microplastic filling of the pores in the soil body. Bulk density has negative correlation with porosity since bulk density influenced by the total mass filled in soil body. Filling the soil pores with microplastics can increase the density of the soil mass and reduce the presence of empty voids in the soil. This increase occurred in the treated soil (64.9%) when compared to the control soil (60.7%) where both became chili plant growth media.

C-organic in the soil is an important factor in determining soil properties. The C-organic content in the soil is able to improve soil aggregates, where high soil aggregates will increase macro porosity and increase microbial activity [22]. Liu, H., et al., 2017 [23] have shown that the addition of polypropylene microplastics to the soil at 28% w/w altered microbial activity as measured by the hydrolysis of fluorescein diacetate and nutrient content of the dissolved organic matter was higher than when no microplastics were added. Figure 7 shows an increase in C-Organic by 38.69 ppm in the treated soil, while in the control soil the C-organic content is slightly lower at 37-2 ppm and significantly different in the blank soil by 24.2 ppm. The increase in C-organic in the soil that became the growth medium for chili plants (control soil and treated soil) was caused by the formation of a life cycle mechanism caused by the plants that used the soil. The presence of plants stimulates the life of microorganisms and stimulates the activity of microorganisms. The increase in C-organic in the treated soil when compared to the control soil was due to the carbon contained in the microplastics which manipulated the natural C-organic content of the soil. C-organic in the soil is an important factor in determining soil properties. The C-organic content in the soil is able to improve soil aggregates, where high soil aggregates will increase macro porosity and identification of microplastics in the soil is one process to provide evidence of the ability of plastics degradation process in the soil. From the identification process of microplastics in the soil can show results as in Table 2 below.

Table 2 and Figure 8 show the presence of microplastics in the soil that received single-use mask treatment in reactors 1,2 and 3 with an average number of microplastic particles of 3.8 particles in each ml of the soil extracted sample. Each identified microplastic particle has varying sizes but has the same shape, namely fiber. From the results of the identification of microplastics, there is evidence that the practice of open dumping of disposable medical masks in peat soils contributes to microplastic pollution, where microplastics have a greater ability to transport to groundwater systems than macro plastics. This amount of microplastic concentration will have the potential to increase with the increasing residence time of single- use mask waste in the soil.

Table 2. Characteristic of Microplastic After Treated During 45 Days in Peat soil

Reactors	Days 0	Days 45		
		Particle (part/ml)	Type	Particle Size (mm)
Control	Unidentified	Unidentified	Unidentified	Unidentified
1	Unidentified	4,5±0,5	Fiber	1,036±1,93
2	Unidentified	3±0	Fiber	1,971±2,029
3	Unidentified	3,9±0,5	Fiber	0,999±1,916

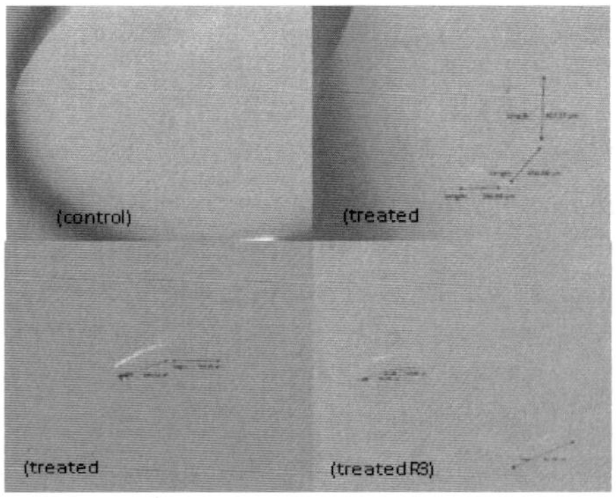

Figure 8. Identification of Microplastic in Peat Soil

4. Conclusion

Disposable medical masks that are disposed of in open dumping on peat soils have contributed to microplastic contamination and have the potential to release additives which, when entered into the soil system through abiotic processes such as precipitation, adsorption and leaching, will have an effect on changes in soil and groundwater characteristics. Changes in the physical characteristics of peat soil due to microplastic contamination from disposable masks can be evidenced by an increase in bulk density, an increase in specific weight, an increase in the ability of the soil to support groundwater and a decrease in soil porosity. On the other hand, changes in the chemical characteristics of peat soils are indicated by a decrease in the concentration of macro-micro nutrients in peat soils and changes in the concentration of C-organic soil.

The research that has been carried out by researchers is still on a prototype scale, so it is necessary to carry out similar research in real conditions in the field where the variables involved are in uncontrollable conditions.

Acknowledgments

The authors appreciate Inggar Dian Ratri and Falvocha Alifsmara Joelyna from UPT Laboratory Diponegoro University for helping water quality analysis using, for identification microplastics, and analysis peat soil minerals. This research was supported by the Centre of Biomass and Renewable Energy (C- BIORE) Research Group, Diponegoro University under grant of World Class Research (WRC) Program 2021 by Diponegoro University.

References

[1] Kemenkes, R. (2021) 'Infeksi Emerging Kementerian Kesehatan RI', *Infeksi Emerging*. Available at: https://infeksiemerging.kemkes.go.id/dashboard/covid-19.

[2] Tedjomurti, F. and Suyanto, B. (2021) 'Behavior Trend of Buying Mask as A New Lifestyle for Women in The City of Surabaya During Pandemic', *The Journal of Society and Media*, 5(1), p. 199. doi: 10.26740/jsm.v5n1.p199-217.

[3] Hamdani, A., and, A. H.-J. of G. S. and 2021, undefined (2021) 'The Face Mask Waste Recycling Generated During Covid-19 Pandemic In Indonesia', *Journal.Unpad.Ac.Id*, 5(2), pp. 31–34. Available at: http://journal.unpad.ac.id/gsag/article/view/34996.

[4] Sari, G. L. *et al.* (2021) 'Infectious Waste Management as the Effects of Covid-19 Pandemic in Indonesia', *Asian Journal of Social Science and Management Technology*, 3(2), pp. 62–75. Available at: https://www.researchgate.net/publication/351276744.

[5] Aragaw, T. A. (2020) 'Surgical face masks as a potential source for microplastic pollution in the

COVID-19 scenario', *Marine Pollution Bulletin*, 159(August), p. 111517. doi: 10.1016/j.marpolbul.2020.111517.

[6] De Sousa, F. D. B. (2020) 'Pros and cons of plastic during the COVID-19 pandemic', *Recycling*, 5(4), pp. 1–17. doi: 10.3390/recycling5040027 [20] Kementrian Pariwisata dan Ekonomi Kreatif 2021 *Outlook Pariwisata & Ekonomi Kreatif Indonesia* (Jakarta: Kementrian Pariwisata dan Ekonomi Kreatif) [in Indonesian Language].

[7] Guo, J. J. *et al.* (2020) 'Source, migration and toxicology of microplastics in soil', *Environment International*. Elsevier, p. 105263. doi: 10.1016/j.envint.2019.105263.

[8] Huang, J. T. and Huang, V. J. (2007) 'Evaluation of the efficiency of medical masks and the creation of new medical masks', *Journal of International Medical Research*, 35(2), pp. 213–223. doi: 10.1177/147323000703500205.

[9] Shim, W. J., Hong, S. H. and Eo, S. (2018) *Marine microplastics: Abundance, distribution, and composition*, *Microplastic Contamination in Aquatic Environments: An Emerging Matter of Environmental Urgency*. Elsevier Inc. doi: 10.1016/B978-0-12-813747-5.00001-1.

[10] Auta, H. S., Emenike, C. U. and Fauziah, S. H. (2017) 'Distribution and importance of microplastics in the marine environment A review of the sources, fate, effects, and potential solutions', *Environment International*, 102, pp. 165–176. doi: 10.1016/j.envint.2017.02.013.

[11] Yu, M. *et al.* (2020) 'Effects of microplastics and earthworm burrows on soil macropore water flow within a laboratory soil column setup', *Vadose Zone Journal*, 19(1). doi: 10.1002/vzj2.20059.

[12] Leed, R. and Smithson, M. (2019) 'Ecological Effects of Soil Microplastic Pollution', *Science Insights*, 30(3), pp. 70–84. doi: 10.15354/si.19.re102.

[13] De Souza Machado, A. A. *et al.* (2019) 'Microplastics Can Change Soil Properties and Affect Plant Performance', *Environmental Science and Technology*, 53(10), pp. 6044–6052. doi: 10.1021/acs.est.9b01339.

[14] Dariah, A., Maftuah, E. and Maswar (2013) 'Karakteristik Lahan Gambut', *Panduan Pengelolaan Berkelanjutan Lahan Gambut Terdegradasi*, 6(2), pp. 16–29 [in Indonesian Language].

[15] Rezanezhad, F. *et al.* (2016) 'Structure of peat soils and implications for water storage, flow and solute transport: A review update for geochemists', *Chemical Geology*, 429, pp. 75–84. doi: 10.1016/j.chemgeo.2016.03.010.

[16] Boelter, D. H. (no date) 'DIVISION S-6 — SOIL AND WATER MANAGEMENT Physical Properties of Peats as Related to Degree of Decomposition 1', *Water*, pp. 606–609.

[17] Caron, J., Price, J. S. and Rochefort, L. (2015) 'Physical Properties of Organic Soil: Adapting Mineral Soil Concepts to Horticultural Growing Media and Histosol Characterization', *Vadose Zone Journal*, 14(6), p. vzj2014.10.0146. doi: 10.2136/vzj2014.10.0146.

[18] Zhao, T., Lozano, Y. M. and Rillig, M. C. (2021) 'Microplastics Increase Soil pH and Decrease Microbial Activities as a Function of Microplastic Shape, Polymer Type, and Exposure Time', Frontiers in Environmental Science, 9(June), pp. 1–14. doi: 10.3389/fenvs.2021.675803

[19] Hidalgo-Ruz, V. *et al.* (2012) 'Microplastics in the marine environment: A review of the methods used for identification and quantification', *Environmental Science and Technology*, 46(6), pp. 3060–3075. doi: 10.1021/es2031505.

[20] Rillig, M. C. *et al.* (2019) 'Microplastic effects on plants', *New Phytologist*, 223(3), pp. 1066–1070. doi: 10.1111/nph.15794.

[21] Hahladakis, J. N. *et al.* (2018) 'An overview of chemical additives present in plastics: Migration, release, fate and environmental impact during their use, disposal and recycling', *Journal of Hazardous Materials*, 344, pp. 179–199. doi: 10.1016/j.jhazmat.2017.10.014. [9] Sarkar B D, Shankar R and Kar A K 2021 *Mod. Manag* **10**

[22] Blanco-Canqui, H. *et al.* (2013) 'Soil organic carbon: The value to soil properties', *Journal of Soil and Water Conservation*, 68(5), pp. 129–134. doi: 10.2489/jswc.68.5.129A [4] UNWTO 2020 *Annual report* (Madrid: World Tourism Organization)

[23] Xu, B. *et al.* (2020) 'Microplastics in the soil environment: Occurrence, risks, interactions and fate–A review', *Critical Reviews in Environmental Science and Technology*, 50(21), pp. 2175–2222. doi: 10.1080/10643389.2019.1694822.

[24] Jiang, X. J. *et al.* (2017) 'Residual plastic mulch fragments effects on soil physical properties and water flow behavior in the Minqin Oasis, northwestern China', *Soil and Tillage Research*, 166, pp. 100–107. doi: 10.1016/j.still.2016.10.011.

Stability Evaluation of Bener Dam Diversion Tunnel During Construction

MSH Danuartha[1,3] **, K W Sadono**[1] **and T T Putranto**[2]
[1] Civil Engineering Department Diponegoro University
[2] Geological Engineering Department Diponegoro University
[3] Virama Karya Engineering Consultant

E-mail : satyahimawan@gmail.com

Abstract. Bener Dam Diversion Tunnel is located in two rock units such as andesite and andesite breccia. Based on the engineering geology investigation, the subsurface tunnel will penetrate andesite breccia with good rock mass quality (fresh-moderately weathered), while surrounding the tunnel consist of andesite breccias with moderate and poor rock mass quality. Geological and engineering geology conditions are used as the basis for stability analysis of excavation methods and determining primary tunnel support systems. Geological Strength Index is used for determining rock mass parameters. The stability analysis of the openings and tunnel supports was carried out using an empirical and numerical method approach. The results of the empirical analysis of support system stability were then evaluated using the finite element method. The most suitable excavation method is applied by blasting. Evaluation of the support system based on rock mass rating with the finite element method modeling shows that the lowest safety score in the tunnel geomechanical zone is in zone 2 with a safety factor value at the final support is 1.786, however in general they are in a good condition which are safe from the initial excavation until the installation of supports stage.

Keywords : diversion tunnel, geological strength index, support system, stability, finite element method modeling, safety factor

1. Introduction

Bener Dam diversion tunnel is a part of building at Bener Dam site which are located in Kemiri Village, Kepil District, Wonosobo Regency and Guntur Village, Bener District, Wonosobo Regency. Bener dam is concrete face rockfill dam which has 168 m height from the depest foundation until crest dam, Bener Dam will embankt Bogowonto River.

The Bener Dam Diversion Tunnel based on the final design has a horseshoe section, with rock bolts, shotcretes, and wire mesh as the lining reinforcements. The dimension of excavation will be 9 m high and 9 m wide, and has a length of ±851.63 m [1]. Parameters calculated using the empirical method in the form of geological strength index (GSI) are used as input to the finite element method [2]. Geological studies both field and laboratory investigations were carried out, the tunnel route was divided into four geotechnical zones based on the results of this study. Then for each zone, the support capacity is evaluated using geological strength index and finite element methods. Combining between empirical method and numerical method can obtain the result more reliable[3]. The previous research in same field did not give and explain the safety factor of tunnel construction stages, the geomechanical zones, and the GSI reduction parameter for fault [4]. To evaluate the design, knowing the safety factor and fault reduction parameter is a must. Therefore the main objectives are to divide the geomechanical zones, determine supports design, assess safety factor of tunnel during each stage of construction, and measure effect of fault toward installed support. The minimum safety factor (SF) of constructions are based on SNI code , it must be greater than 1.5 [5].

2. Methodology

Tunnel excavation designs are based on the empirical method by assessing GSI and using finite element method analysis. The modeling processes of diversion tunnel excavation and support are began by determining rock mass, support system, and geometry as input parameters. Rock mass input parameters must be calculated by some equations (1 until 12), support configurations are based on equation (13), overburden load is calculated by equation (14). Finally the result of running can be obtained after all parameters had been calculated and input properly as shown in Figure 1.

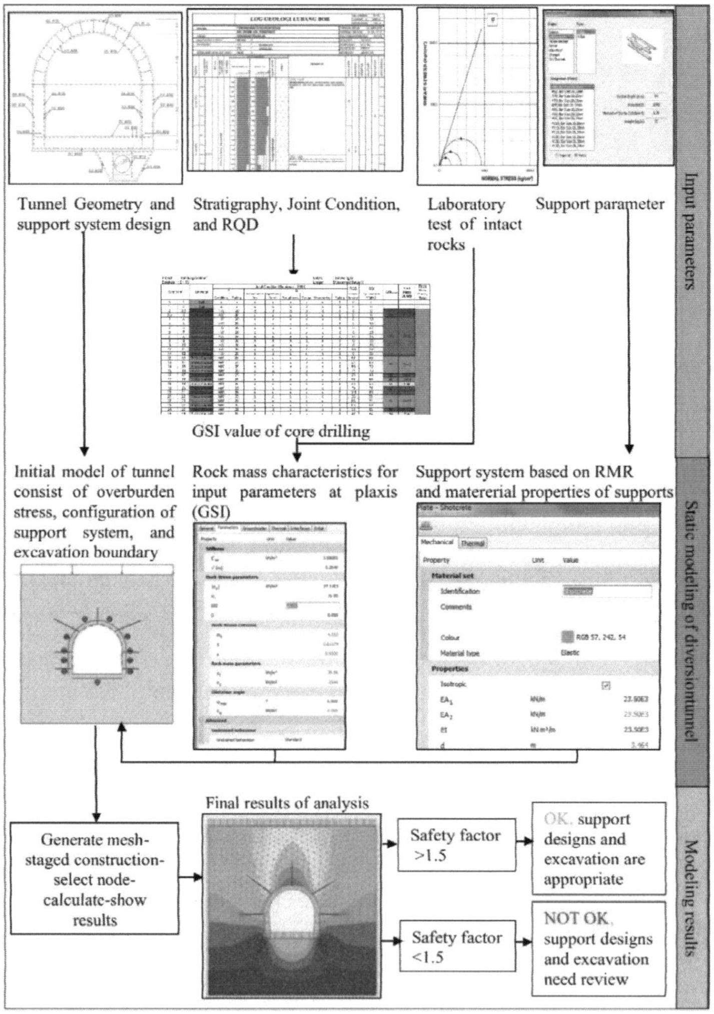

Figure 1. Overall research and modelling process of diversion tunnel

There are 5 boreholes drilled in the diversion tunnel to determine lithostratigraphy, then some rock samples are used for laboratory testing such as unconfined compression strength (UCS), Triaxial (cohesion, internal friction, confining pressure), Ultrasonic Velocity (compression wave and shear wave), and index properties of rock (porosity, natural water content, saturation, dry density, and natural density).

The scanline-survey was used for mapping and core drills description in order to determine geological strength index (GSI) values. GSI values are considered both structure condition and surface conditions [6]. Structure conditions in this research area mostly in very blocky condition. The surface conditions are determined by persistence, orientation, spacing, roughness, opening, and degree of weathering and filling of discontinuities in the rock masses. Surface mapping of GSI is determined and can result in an engineering geological map as shown in Figure 6.GSI calculations for subsurface datas use equation (1) as quantification approach method [6].

$$GSI = \sigma 1.5JCond_{89} + \frac{RQD}{2} \quad (1)$$

where Jcond is Joint condition and RQD is rock quality designation [6 and 7]. The GSI calculation will supply equations (2-4,7-10, and 13) and will be used as software input. *mb*, *s*, and *a* are the rock mass material constants, are given by equations (2, 3, and 4).

$$m_b = m_i \exp\left(\frac{GSI\text{-}100}{28\text{-}14D}\right) \quad (2)$$

$$s = \exp\left(\frac{GSI\text{-}100}{9\text{-}3D}\right) \quad (3)$$

$$a = \frac{1}{2} + \frac{1}{6}\left(e^{\frac{-GSI}{15}} - e^{\frac{-20}{3}}\right) \quad (4)$$

where *mi* is a material constant of the intact rock, D is the disturbance factor. The relationship between the compressive to tensile strength ratio, and the Hoek-Brown parameter *mi* as shown in equation (5)[6].

$$\sigma_{ci}/|\sigma_t| = 0.81\ m_i + 7 \quad (5)$$

where σ_t is rock tensile strength. The rock mass modulus young value obtained from the conversion of the intact rock young modulus value which is based on GSI value as in equation (6) [8].

$$E_{rm} = E_i \left\{0.02 + \frac{1\text{-}D/2}{1+\exp[(60+15D\text{-}GSI)/11]}\right\} \quad (6)$$

Residual hoek brown parameters of fault are derived from the zone 1 parameters as shown in Table 4. For reducing parameters, it will use residual GSI and Hoek Brown parameters were calculated using following equations suggested from equations (7-10)[9] as below :

$$GSI = GSI\ e^{-0.0134GSI} \quad (7)$$

$$m_{br} = m_i e^{\left(\frac{GSI_r\text{-}100}{28}\right)} \quad (8)$$

$$s_r = e^{\left(\frac{GSI_r\text{-}100}{9}\right)} \quad (9)$$

$$a_r = \frac{1}{2} + \frac{1}{6}\left(e^{GSI_r/5} - e^{-20/3}\right) \quad (10)$$

where “r”means residual. For reducing uniaxial compression strength using equation (11), the limitation for rock masses must be characterized by a thin infilling or slightly weathered to unweathered joint walls [10].

$$(\sigma_i)_r = 0.2\ \sigma_i \quad (11)$$

dilatancy angles can be obtained from the most widely available parameters for rocks and rock masses based on Equation 12[11].

$$\psi_{peak} = \frac{\varphi}{1+\log_{10}\sigma_{ci}} \log_{10}\frac{\sigma_{ci}}{\sigma_3+0.1} \quad (12)$$

Where ψ_{peak} = dilatancy peak angle. Correlation of GSI and RMR_{89} use equation (13) [12]. Tunnel support in all zones are considered by rock mass rating (RMR_{89}) classification for tunnel support shown in Table 1 [13].

$$RMR_{89} = GSI + 5 \quad (13)$$

Equation (14) has been proposed for a rock tunnel, taking the tunnel depth and tunnel radius as two additional parameters with RMR for non squeezing conditions [14].

$$P = (2.32\text{-}0.035.RMR_{89}+0.001.H+0.03.\ a)\ MPa \quad (14)$$

where P is rock mass load toward tunnel support, "H" is the tunnel depth (m) and "a" is the radius of the tunnel (m).

Table 1. Tunnel support based on rock mass rating RMR_{89}

No	Rock Mass Class	Excavation	Support		
			Rock bolts (20 mm dia.,fully bonded)	Shotcrete	Steel sets
1	Good Rock RMR : 61-80	Full Face : 1,0-1,5 m advance, Complete support 20 m from face	Locally bolts in crown 3 m long, spaced 2,5 m with occasional wire mesh	50 mm in crown where required	None
2	Fair Rock RMR : 41-60	Top heading and bench : 1,5-3 m advance in top heading : commerce support after each blast ; complete support 10 m from face	Systematic bolts 4 m long, spaced 1,5 m-2 m in crown and walls with wire mesh in crown	50-100 mm in crown and 30 mm in sides	None

3. Results and Discussions

Engineering geological map and cross section has already created and shown in Figure 4 consist of four geomechanical zones and four rock mass class units such as good-fair andesite with fresh-moderate weathering, good-very good andesite breccia with fresh-moderate weathering, fair andesite breccia with fresh-moderate weathering,fair andesite breccia with highly weathering. Rock mass input parameters in the model can be shown in Table 2, whereas support parameters are shown in Table 3 and Figure 3.

Table 2. Input parameters summary in finite element software modeling

Property	Zone 1	Fault	Zone 2	Zone 3.a	Zone 3.b	Zone 4.a	Zone 4.b
Lithology	Breccia	Breccia	Breccia	Breccia	Breccia	Andesite	Breccia
E_{rm} (KPa)	9.08E6	1.405E6	7.338E6	7.338E6	1.179E7	1.957E7	1.389E6
E_i (Kpa)	17.46E6	1.726E7	9.692E3	9.692E3	1.932E7	3.462E7	4.26E6
Poisson's ratio	0.264	0.279	0.272	0.272	0.269	0.222	0.444
UCS (KPa)	2.751E4	5502*	9.69E4	9.69E4	1176	4.516E4	3.43E4
GSI	60	31*	67	67	64	62	51
	Good	Poor	Good	Good	Good	Good	Fair
mb parameter	4.553	0.3928*	2.805	2.805	5.529	6.435	1.082
S parameter	0.0117	0.0004*	0.00678	0.00678	0.01832	0.0147	0.0059
a parameter	0.502	0.522*	0.5017	0.5017	0.5021	0.502	0.5054
γ unsat	25	21	21	21	25	26	26
γ sat	26	22	23	23	26	31	31
ψ	37.5		41.6	41.6	9.8	41	40.7
Thickness (m)	58	5	100	78	22	5	15
Load (KPa)	303	1685	310	743	197	325	820
Disturbance	0.8	-	0.8	0.8	0.8	-	0.5

Notes : E_{rm} = Rock mass modulus ; E_i = Intact rock modulus ; γ unsat = Unsaturated density; γ sat = Saturated density; *Parameters have already reduced

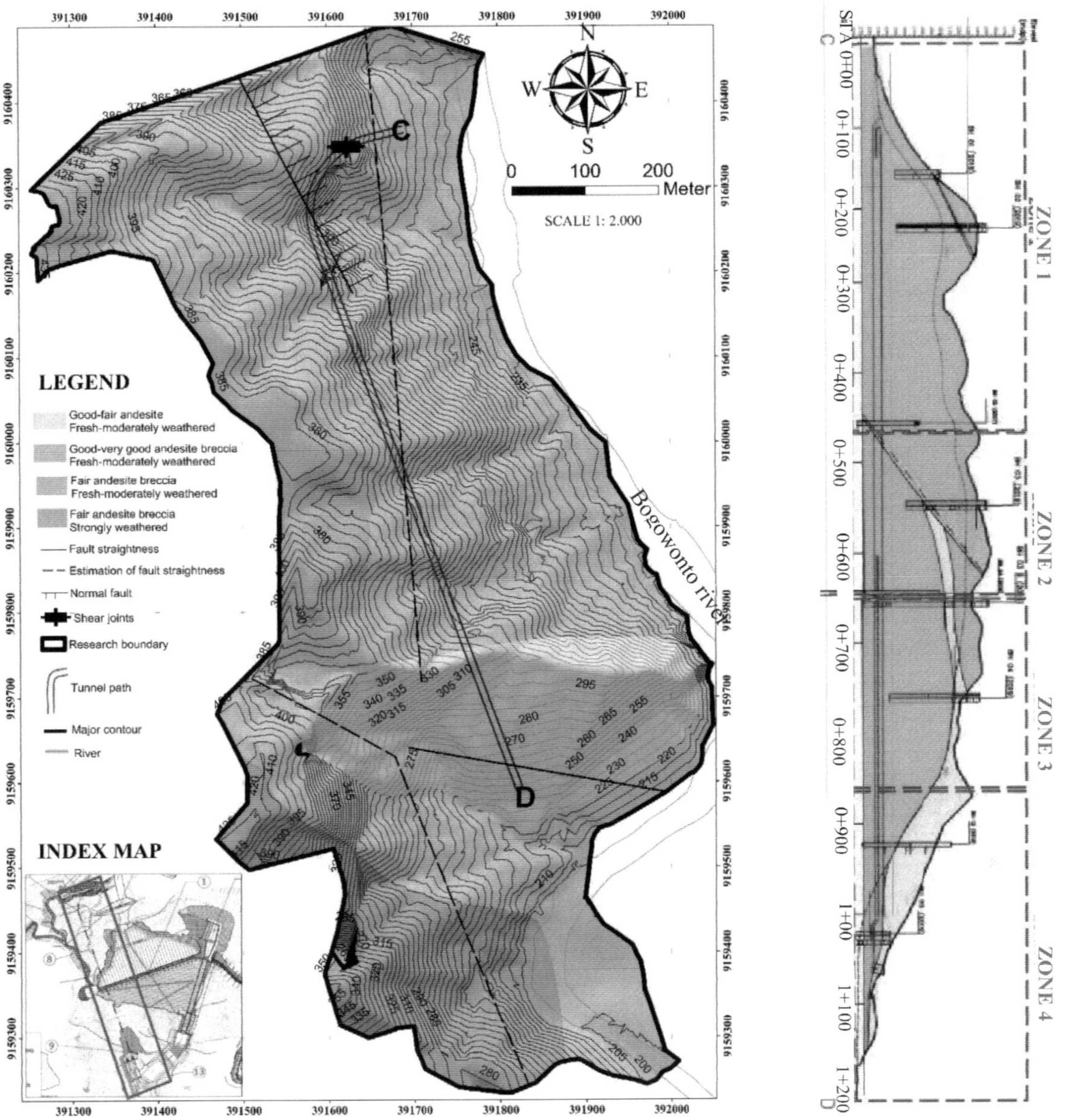

(a) (b)

Figure 2. Engineering geological map of diversion tunnel (a) and cross section of diversion tunnel (b)

Table 3. Support system parameters which are applied in model

Property	Shotcrete	Rock Bolt	Wire mesh
E modulus (GPa)	23.5	200	200
Poisson's ratio	0.2	-	0.25
UCS (MPa)	25	-	400
Type		Ø 25 mm	Ø 8 mm 100x 100 mm

The value of GSI at four geomechanical zones are shown in Figure 2 respectively 62-63 (good); 63-70 (good); 64-66 (good); 50-61 (fair-good). Figure 3 shows the resistance forces are influenced by the dilatancy angle of rock mass at the surrounding of the tunnel. Equation (12) shows the important data which are important for measuring dilatancy angle such as UCS and Triaxial datas (internal friction and confining pressure) [15]. There is common assumption that the intact specimens are homogeneous and isotropic and that the values of the unconfined compressive strength σ_{ci} and the constant m_i are representative of the intact rock in the blocks of the rock mass. In fact, this assumption is not always valid since in many rock masses, defects such as veins, micro-fractures, and weathered or altered components can reduce the intact rock strength as shown in Table 4[8].

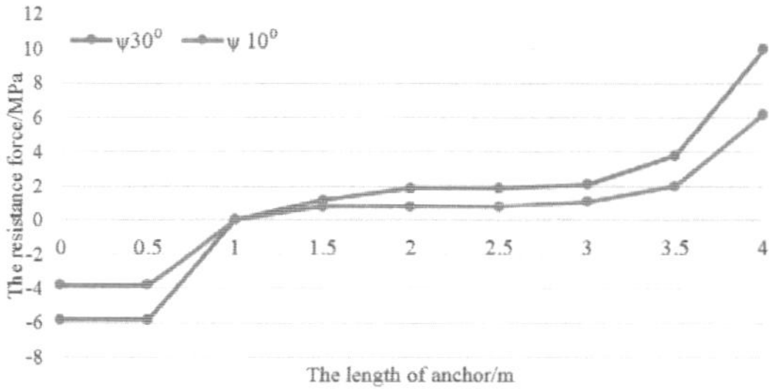

Figure 3. The curves of resitance force for a rockbolt with different dilatancy angle [15]

Table 4. Intact rock samples properties for designing rock bolt

Code	σ_3 (MPa)	φ (...°)	σ_{ci} (MPa)	ψ (...°)	Lithology	Core samples
BH 1/ Zone 1	0.98	65.5	27.5	37.5	Andesite Breccia (Isotropic)	
BH 2	0.49	46.0	0.8	6.74	Andesite Breccia (Anisotropy)	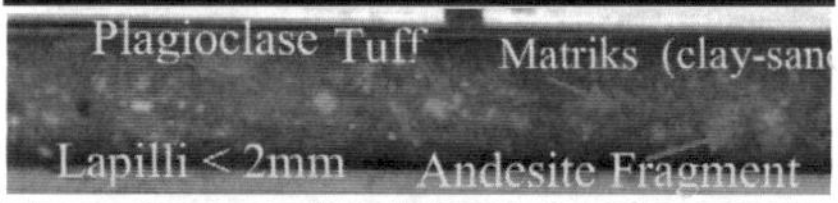
BH 3/ Zone 3.a	0.49	56.0	96.9	41.6	Andesite Breccia (Anisotropy)	
BH 4/ Zone 3.b	0.49	35.0	1.176	9.8	Andesite Breccia (Anisotropy)	
BH 05/ Zone 4.a	0.49	59	34	41.0	Andesite breccia (Anisotropy)	
BH 05/ Zone 4.b	0.58	59.3	45.16	40.7	Andesite (Isotropic)	

Notes : σ_3= Confining pressure ; φ = Internal friction ; σ_{ci} = unconfined compressive strength
Ψ = dilatancy angle ; BH is abrreviation of borehole

Numerical analysis for example finite element methods have been used in many geotechnical projects by many researchers [2,3,4]. Safety factors in this discussion are conducted in every stage of tunnel construction such as excavation, rock bolt installation, and shotcrete+wire mesh installation.

Overall every stage already fulfilled the minimum requirements (SF>1.5) as shown in Figure 4 (a). There are some differences between the effects of rock bolt installation. Both in zone 1 and zone 4 give significant safety factors, however in zone 2 and zone 3 there are no significant value between excavation and rock bolt installation stages. Although composite of shotcrete and wire mesh gives significant safety factor in zone 3 and zone 4.

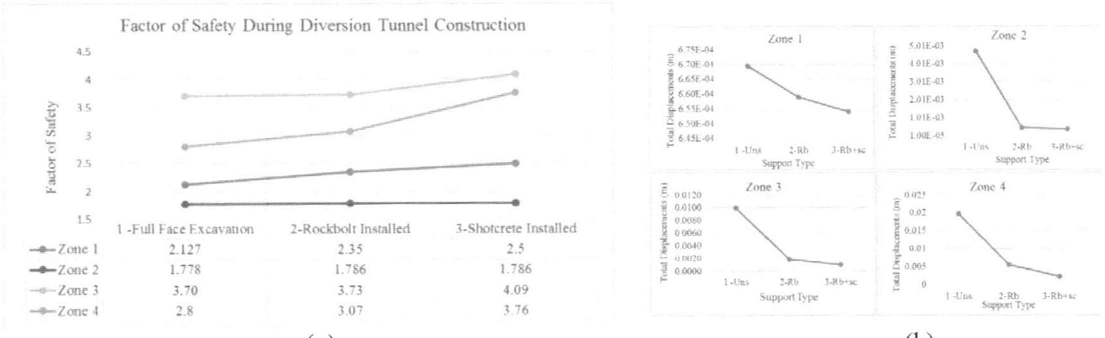

(a) (b)

Notes: Uns = Unsupported; Rb = Rock bolt; Rb+sc = Rock bolt+Shotcrete+Wiremesh

Figure 4. Safety factor evolution (a) and total displacements evolution of diversion tunnel (b)

Total displacement evolution of diversion tunnel are shown in Figure 4(b). From Figure 4 (b) can be learned that supports decrease total displacement of tunnel excavation. The total displacement contour in zone 2 and zone 3 are decreasing significantly after installation of support as shown in Figure 4. Hence, from the biggest until the smallest total displacements are zone 4, zone 3, zone 2 and zone 1 respectively, then the model results of total displacements are shown in Figure 5.

For determining support, almost all zones are referred to RMR_{89} as mentioned in Table 1, except for zone 2. The justification is set up in zone 2 which contains the fault zone below the tunnel floor. Rock bolts are designed with both closer space and longer arms than the RMR_{89} classification for good rock mass excavation. In this modification, the rock bolts have 2 m space and 4 m length, and their configurations are systematic at the wall, roof, and floor.

Potential failure in the zone 1 is crown fall and in Zone 2 is floor heave or buckling as shown in Figure 5, however in zone 3 and zone 4 there is no potential failure could happen.

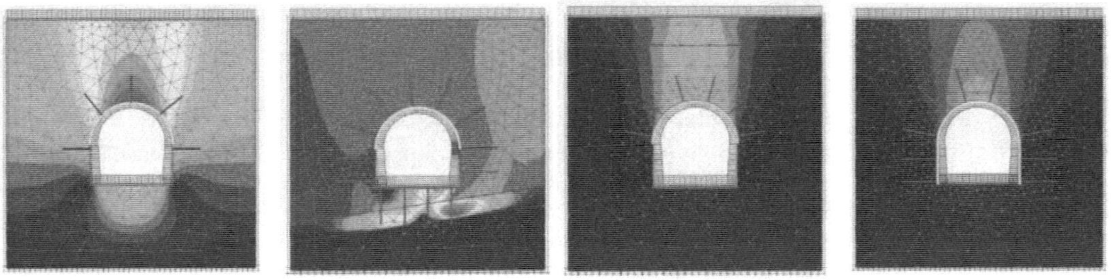

(a) (b) (c) (d)

Figure 5. Final stage results in Zone 1 (a) in Zone 2 (b) in Zone 3 (c) in Zone 4 (d)

4. Conclusions

In this study, GSI were used to evaluate the rock mass in the Bener Dam diversion tunnel. RMR_{89} is obtained by making correlation with GSI value. RMR_{89} was used as empirical methods to determine support system configuration. For support parameters, especially rock bolts which are affected significantly by dilatancy angle of rock mass, it must be calculated to get precise rock bolts resistance force parameters. In future, radial displacement of the rockbolt must be tested in order to get preciser rock bolts friction resistance parameters. Each site must be tested since the difference of rock mass characteristics.

The finite element analysis shows every stage had different safety factors and total displacement values. Every stage of tunnel construction in all zones (4 zones in total) have already fulfilled the minimum requirements (SF>1.5). Zone 2 is more critical than other zones in every stage because of the fault zone below the tunnel floor, the peak value after final support is 1.786 or slightly greater than minimum safety factor standard (>1.5). Zone 1 has a smaller safety factor than zone 3 and zone 4, the peak value after final support is 2.5.

From this study we know some findings of fault zone effects toward non squeezing deep tunnels. The fault zone will give critical damage to the supports if it is found below and near the lining, however if it is found far above the tunnel lining, the fault zone will just give additional rock load value toward supports which had been already installed. Hence, the RMR_{89} support system classification needs some updates in order to cover reduction of GSI parameters because of fault or another poor quality rock mass below the tunnel lining for example in zone 2.

Acknowledgements

Thank you to the supervision team of Virama Karya Corporation for handling the Bener Dam Project for escorting and collecting geotechnical investigation data and Ministry of Public Works and Housing (BBWS Serayu Opak) as owner who permit this research to be carried out and published.

References

[1] Virama Karya Co. Engineering Consultant 2018 *Design Certification of Bener Dam Report*
[2] Gurocak Z 2011 *Analysis of Stability and Support Design for a Diversion Tunnel at The Kapikaya Dam Site, Turkey* (*Bulletin Engineering Geology Environment* vol 70) (Berlin:Springer) pp 41-52
[3] Rasouli M 2009 *Engineering geological studies of the diversion tunnel, focusing on stabilization analysis and support design, Iran.* (*Engineering Geology* vol 108) (Amsterdam : Elsevier) pp 208-224
[4] C A Bagio, A Rifa'i, and S Ismanti 2021 *Tunnel stability in fault zone base on 3D numerical simulation.* In IOP Conference Series: Earth and Environmental Science (Vol. 871, No. 1, p. 012052). IOP Publishing.
[5] Indonesian National Standard Institution 2017 SNI 8460:2017 *Geotechnical Design Standard*
[6] Hoek E, Carter TG, and Diederichs M S 2013 *Quantification of the geological strength index chart.* In 47th US rock mechanics/geomechanics symposium. OnePetro.
[7] Deere, Don U , and Don W Deere 1989 *Rock Quality Designation (RQD) After Twenty Years.* Washington: US Army Corps of Engineers
[8] Hoek E and Brown E T 2019. *The Hoek–Brown failure criterion and GSI–2018 edition.* (*Journal of Rock Mechanics and Geotechnical Engineering* vol 11)(Elsevier: Amsterdam) pp 445-463
[9] Cai M, Kaiser P K , Tasaka Y., and Minami M, 2007.*Determination of residual strength parameters of jointed rock masses using the GSI system.* (International Journal of Rock Mechanics and Mining Sciences, 44(2), pp.247-265.
[10] Ribacchi R 2000 *Mechanical tests on pervasively jointed rock material: insight into rock mass behaviour* (*Rock Mechanics and Rock Engineering* vol 33) (Wien: Springer) pp 243-246
[11] Alejano L R and Alonso E 2005. *Considerations of the dilatancy angle in rocks and rock masses.* (*International Journal of Rock Mechanics and Mining Sciences* vol 42)(Elsevier :Amsterdam) p 481-507
[12] Ceballos F, Olalla C and Jiménez R 2014 *Relationship between RMRb and GSI based on in situ data. In ISRM regional symposium-EUROCK 2014.* OnePetro.
[13] Bieniawski Z T 1989. *Engineering Rock Mass Clasification Mining and Mineral Resources Research Institute.* Pennsylvania State University.
[14] Rehman H , Ali W , Naji, A M , Kim J J , Abdullah R A, and Yoo H K 2018 *Review of rock-mass rating and tunneling quality index systems for tunnel design: Development, refinement, application and limitation* (*Applied sciences* vol 8)(MDPI:Basel) p 1250-1277
[15] Zhu X G 2013 *The Studying of Reinforcement Mechanism of Fully Grouting Rock Bolt in Brittle Wall Rock Mass* (*Advanced Materials Research* Vol. 690) Trans Tech Publications Ltd P 786-792

The Effect of Laptop Usage Behaviors on Eye Fatigue on Work-from-Home (WFH) Lecturers during The Covid-19 Pandemic

T L Wardani[1], L Retnowati[1], R Fajariani[1], I Qadrijati[1], Sumardiyono[1], S Rinawati[1], and T B Atmojo[1]

[1]Diploma 4 Occupational Health and Safety, Vocational School, Sebelas Maret University, Surakarta, Central of Java, Indonesia

*tyasliliaw@staff.uns.ac.id

Abstract. The Covid-19 pandemic in Indonesia has encouraged changes in society, especially in the aspect of education. Learning is conducted online and educators carry out activities from home. However, this has a negative impact on physical health, which may occur due to WFH behavior, which is Eye Fatigue. Laptop usage behavior depend on the size of the laptop, the duration and frequency of laptop usage, the visual distance, the position of the monitor and eye rest. This study intends to determine the effect of laptop usage behavior on Eye Fatigue. This study uses an observational design with a cross-sectional design. The sample of this study is 175 lecturers and use a simple random sampling technique. The used instruments are laptop usage behavior and visual fatigue index questionnaires distributed through Google Forms. Bivariate analysis was performed using the Somers'd correlation statistical test. The results show a significant level of laptop size with eye fatigue with a p-value = 0.005, while the other variables were not related. The behavior of using a laptop has affected eye fatigue especially because of the size of the laptop. They use standard laptop size and use laptop support when operating laptops.

1. Introduction

Covid-19 is a disease caused by infection of the SARS-CoV-2 virus. Covid-19 is a new virus attacking the respiratory system with symptoms including fever, cough, runny nose, shortness of breath, fatigue, anosmia, and ageusia that is able to death cause in vulnerable individuals [1]. The incubation period of the virus is approximately 2 weeks and able to spread through droplets [2]. The status of Covid-19 as a global pandemic was first set by WHO in mid-March of 2020 due to its contagiousness and prompt spread to countries around the world [2]. The COVID-19 pandemic in Indonesia began to occur in March 2020. Indonesia was one of the countries with the fastest transmission of the Covid-19 virus with more than 100 new cases every day [3]. Based on the Covid-19 global case report, on June 18, 2021, there were 177,108,695 cases of Covid-19 and 3,840,223 deaths in 222 countries affected by Covid-19 and 149 countries where local transmission of Covid-19 occurred. In Indonesia as of June 18, 2021, there have been 1,963,266 confirmed cases of Covid-19 from 18,438,025 specimens examined, and 54,043 cases of death have occurred [4].

This pandemic has changed people's lifestyles in various aspects, one of which is education. Since the beginning of the COVID-19 pandemic in Indonesia, government strategies have consisted of conducting physical distancing and enforcing online activities or work from home, both in office and educational institutions [3]. In Indonesia, around 80% of the population performed various activities, both working and studying from home because of the pandemic. The use of laptops was very necessary to support activities of work from home (WFH) so there was an increasing number of laptops used in the Indonesian population [5]. Online learning and working from home for lecturers change to sustain the learning process. Besides changing the learning system, Work from Home on lecturers have also changed their lifestyle: increasing the frequency of laptop use, expanding sitting time rather than standing, and replacing physical activities with digital activities.

The implementation of the WFH policy increases the intensity of time in laptop usage. Most people have not been able to balance the duration of work from home and homework so they do not realize that their laptop use has exceeded the normal duration of work [6]. According to Rosenfield's research, staring at a computer screen for more than 4 hours every day will increase symptoms of eye strain [7]. This condition is caused by prolonged use of computers, tablets, e-readers, and cell phones and causes symptoms including eye fatigue, headaches, neck, shoulder, eye discomfort, dry eyes, and blurred vision [8]. Based on a survey conducted by the American Optometric Association/AOA (2016) in the United States regarding Most Americans Experience Digital Eye Strain from Overexposure to Computers According to Survey, the average working time spent working with computers is 5.8 hours or 69 % of the total 8 hours of work. Excessive use of computers often results in an increased risk of work disorders, one of which is eye health problems due to the continuous use of computer monitors [9].

Eye fatigue is eye strain caused by an impaired vision for a long period which is usually accompanied by uncomfortable viewing conditions [10]. Eye fatigue is known as eye strain or asthenopia, i.e. ocular fatigue or strain on the visual organs in which eye problems and headaches are associated with intensive eyewear [11]. Eye fatigue can occur as deep or intensive stress on eye function such as working accommodation muscles requiring careful observation or on the retina as a result of inaccurate contrast/lighting [12]. Muscle and nerve fatigue in the eye occurs as a result of constant tension. However, it does not cause permanent eye damage, but increases the workload, fatigue, loss of working hours, frequent breaks, reduced job satisfaction, decreased production quality, increased error frequency, impaired concentration, and reduced work productivity [10]. Complaints of eye fatigue include throbbing or pain around the eyes and behind the eyeballs; double vision, blurred vision, and difficulty focusing vision; stinging, redness, eye pain, and watering; headache; sometimes accompanied by nausea, aches, and easily emotional [13].

A study conducted (by Jovanovic, et al., 2021) found that 54.4% of office workers who use computers working more than 4 hours per day experience vision problems. Improper use and long-term use of computers have decreased health such as muscle pain, trauma, and abnormalities in the limbs, back, and neck [14]. Based on Riskesdas (2018) that 94.6% of workers were injured due to negligence / unintentional. Eye fatigue in office workers can increase the risk of injury due to negligence/accident [15]. A lecturer has a job that possesses a risk of eye fatigue because, during the Covid-19 pandemic, they conducted the learning process and their jobs online, so their daily activities revolved around laptops.

Based on these problems, the researchers conducted a study on The Effect of Laptop Use Behaviors on Eye Fatigue on Work-from-Home (WFH) Lecturers during the Covid-19 Pandemic. The study aims to analyze the effect of laptop use on Eye Fatigue and provide solutions to minimize the negative impacts. The benefit of this research is that workers, especially lecturers, can understand the importance of proper laptop use behaviors to minimize the risk of Eye Fatigue.

2. Methodology

2.1. Design study
This study uses an analytic observational design. Based on research conducted cross-sectional the cause/effect variable and the cause/effect are related or collected at the same time and carried out at the same time.

2.2 Population, sample, and sampling method
This study was conducted in June 2021 in Surakarta City, Central Java Province at higher education conducting WFH. The population of this research was lecturers who performed WFH, both in State Higher Education and Private Higher Education Institutions in Surakarta. A random sampling technique was used to determine its samples. They consisted of 175 respondents. The independent variable in this study was the laptop usage behaviors (laptop size, duration of laptop use, frequency of laptop use, visual distance during laptop use, monitor position, and eye rest) while the dependent variable was eye fatigue.

2.3 Data collection method
The instruments in this study are the laptop usage behavior questionnaire and the Visual Fatigue Index questionnaire. The questionnaires used had been tested for validity and reliability before their applications. The questionnaires were distributed to respondents via online Google Form.

2.4. Analysis of data
Univariate analysis was conducted to observe the distribution of frequency and percentage of respondents' characteristics, laptop use behaviors (laptop size, duration of laptop use, frequency of laptop use, visual distance during laptop use, monitor position, and eye rest), and eye fatigue. The bivariate analysis applied Somers'd test.

3. Results

3.1 Univariate analysis
The univariate analysis of this study included respondent characteristics, laptop use behaviors, and eye fatigue.

3.1.1 Characteristics of respondents. Characteristics of respondents in this study included gender, age, years of service, and laptop location during WFH. Based on the study on 175 lecturers, the results are described in Table 1.

Table 1 shows that the number of lecturers in the study was dominated by women as many as 109 respondents (62.3%). Most respondents, as many as 105 people (60%), had an age of 36 years. Most respondents, as many as 93 people (53.1%), had years of service of fewer than 5 years. Most respondents, as many as 131 people (74.9%), placed laptops during WFH at work desks with chairs with backrests.

Table 1. Characteristics of Respondents

Characteristics	n	%
Gender		
Man	66	37.7
Woman	109	62.3
Age		
≤ 36	105	60.0
>36	70	40.0
Years of service		
≤ 5	93	53.1
>5	82	46.9
Laptop Location and User Position		
A little table and the user sitting on the floor	26	14.9
A work desk and the user sitting in a chair with a backrest	131	74.9
A bed and the user sitting	7	4.0
A work desk and the user sitting in the chair without a backrest	3	1.7
Not fixed	6	3.4
User's lap and the user sitting on a sofa or chair	1	0.6
The user lies prone	1	0.6
Total	175	100.0

3.1.2 Laptop use behaviors. In this study, the parameters in the laptop usage behaviors are the laptop size, duration of laptop use, frequency of laptop use, and user's body posture during laptop usage. Based on the study conducted on 175 lecturers, the following results were obtained:

Table 2. Laptop use behaviors

Laptop Use Behaviors	n	%
Laptop Size		
Small (< 14 inches)	66	37.7
Big (≥ 14 inches)	109	62.3
Laptop Use Duration		
Low (< 2 hours each time using a laptop)	20	11.4
High (> 2 hours each time using a laptop)	155	88.6
Laptop Use Frequency		
Low (≤ 5 days a week)	36	20.6
High (> 5 days a week)	139	79.4
Visual distance during laptop use		
Low risk (≥ 50 cm)	16	9.1
High risk (< 50 cm)	159	90.9
Monitor position		
No risk (monitor at eye level)	80	45.7
At risk (monitor is not at eye level)	95	54.3
Eye rest		
< 15 minutes	59	33.7
> 15 minutes	78	44.6
Not rest	38	21.7
Total	175	100.0

Table 2 shows that the majority of respondents, as many as 109 people (62.3%), used a large-size laptop (≥ 14 inches). The majority of respondents, as many as 155 people (88.6%), had a high duration of laptop use, namely > 2 hours in each use. Most respondents, as many as 139 people (79.4%), had a high frequency of laptop use, namely > 5 days a week. The majority of respondents, as many as 159 people (90.9%), had a visual distance to the laptop monitor screen during laptop use with a high-risk category. The majority of respondents, as many as 95 people (54.3%), respondents with the position at the top of the monitor at risk (monitor not at eye level). Most respondents, as many as 78 people (44.6%), do eye rest for > 15 minutes.

3.1.3 Eye fatigue. Based on research conducted on 175 lecturers, the following results were obtained in Table 3.

Table 3. Eye fatigue

Eye Fatigue	n	%
Not Fatigue	103	58.9
Fatigue	72	41.1
Total	175	100.0

Table 3 shows that the majority of respondents, as many as 103 people (58.9%), had no eye fatigue.

3.2 Bivariate analysis

The results of the bivariate analysis are shown in the table of Somers'd correlation test between laptop use behaviors and Eye fatigue, the following results were obtained:

Table 4. Somers'd Correlation

Variable	**Eye Fatigue**				
	Not Fatigue	Fatigue	Total	*P value*	r
Laptop size					
Small	30	36	66		-0,215
Big	73	36	109	**0,005**	
Laptop use duration					
Low	12	8	20	0,912	0,013
High	91	64	155		
Laptop use frequency					
Low	22	14	36	0,756	0,028
High	81	58	139		
Visual distance during laptop use					
Low risk	8	8	16	0,462	-0,097
High risk	95	64	159		
Monitor position					
No risk	48	32	80	0,778	0,021
At risk	55	40	95		
Eye rest					
< 15 minutes	31	28	59		
>15 minutes	53	25	78	0,850	-0,012
Not rest	19	19	38		

Table 4 shows that the variable related to Eye fatigue was the size of the laptop with a p-value of 0.005. The relationship between the two variables was very weak with a correlation coefficient value of 0.215 in a negative direction. Thus, the smaller the size of the laptop was, the higher the risk of affected Eye fatigue.

4. Discussions

Based on the results of this study, respondents who used a large laptop size (≥ 14 inches) did not experience eye fatigue when using a laptop as many as 73 respondents. Based on the results of the Somers'd correlation test, it was found that there is a significant relationship between laptop size and eye fatigue with a p value of 0.005 (<0.05) while for the strength of the correlation r = - 0.215 which is a weak correlation with a negative correlation direction which means that the smaller the size of the laptop used, the higher the risk of eye fatigue. Octaviani (2007) in Puspita (2012) states that in determining the size of the laptop used, whether large or small, it is determined based on the statement in "Ergonomic For Laptop Users" that a laptop with a size of 14 or 15 inches can work better for all applications, with such a size then the user can more easily adjust the height of the monitor to adapt eye level. If the monitor size is smaller, then more settings are needed. Otherwise, it will lead to a bent neck posture forward [16].

The design of a laptop that is so portable further adds to the potential for ergonomic risks for long-term use (Mozes, 2010) in Amanda, Zulkarnain, 2012). Some of the main ergonomic issues on laptops are related to the keyboard, monitor, and pointing device where the laptop body construction blends between the screen and the keyboard close to each other. If the monitor screen is too low or too high, the neck muscles have to work continuously to hold the head in a viewing position which can result in fatigue and discomfort (Computer Ergonimics, 2004 in Amanda, Zulkarnain, 2012 [17]. In Saito's research, he found that the combination of the screen and keyboard in the form of a notebook PC causes bad work postures. So that the notebook PC user operator adopts a tilted head posture in front of the screen and a short viewing distance. This can lead to visual and musculoskeletal disorders. The smaller the laptop we have, the smaller the size of the keyboard and monitor, therefore we must pay attention to whether the laptop used is for our needs and also make sure we can see and type comfortably with the size of the laptop used [18].

Research respondents who use laptops equipped with laptop support feel comfortable when typing with a laptop. Through its function, laptop support helps increase the viewing angle on the laptop screen to match the eye level of laptop users (Amanda, Zulkarnain, 2012). In setting the workstation when using a laptop, you should adjust the height of the laptop screen so that the viewing angle ranges from 10-20°, the top line of the laptop screen does not exceed eye level, and is not below 20°. The position of the laptop screen which is equivalent to the eye's visibility will reduce physical tension in all visual organs so that laptop users are comfortable at work (Nurmianto in Amanda, Zulkarnain, 2012) [17].

A laptop stands or often called a laptop table is a functional tool that is portable and helps maintain a good sitting position. So, your body will not get sore quickly. Currently, many laptop stands are equipped with a cooling system. Some products are compact. There are three materials commonly used to make laptop stands. Each material has its advantages. Wood looks artistic. The wood grain and minimalist design make the wooden laptop stand to look stunning. Not only that, the wood used is generally thick so it has good resistance. Unfortunately, wooden laptop stands usually can't be folded so they can't be taken anywhere. In addition, the tilt angle cannot be adjusted. Plastic, the price is economical. Besides being pocket-friendly, plastic laptop stands are also easy to maintain. You only need to wipe it with water when it gets dirty. However, a stand made of plastic cannot support a heavy laptop because the material is not sturdy enough. Aluminum is lightweight and has excellent durability. In addition to being rust-resistant, aluminum laptop stands are also not easily broken so they are ready for long-term use. Interestingly, this material can reduce heat from the laptop. You can also adjust the height. The most important thing in choosing a laptop stand is choosing a size that fits the size of your laptop. Generally, there are three sizes of laptop stands, namely A4 (14-15.6 inches), B5 (11.6-14 inches), and mobile (less than 11.6 inches). The right size of the stand will certainly add to your comfort when using a laptop. A laptop stand that tilts and cannot be adjusted in height makes the sitting position uncomfortable. This has the potential to cause the body to quickly feel tired and achy, it can even make your head dizzy. Therefore, we recommend that you choose a laptop stand that can be tilted and height-changed. Unlike PCs, laptops are designed to be compact and lightweight so you can carry them with you easily. For that, you should also choose a laptop stand that is lightweight and easy to carry anywhere. You can choose a laptop stand that is thin or can be folded so that its size is more compact and does not hinder mobility. Products made of aluminum or plastic are suitable to be chosen because they are light and can be folded. So, you can carry it in your bag without any major difficulties [19].

The results of this study indicate that respondents who use laptops with a high duration of > 2 hours each time they use a laptop tend not to experience eye fatigue when using a laptop as many as 91 respondents. Based on the results of the Somers'd correlation test, it was found that there was no relationship between the duration of laptop use and eye fatigue with a p value of 0.912 (> 0.05). The researcher assumes this is because respondents have maintained a healthy lifestyle such as eating healthy foods and always taking vitamins to maintain eye health and lighting according to standards. In addition, respondents did not experience eye fatigue because the intensity of the monitor's lighting was good. Lighting that is too bright causes a glare effect so that the characters of letters or images on the monitor screen become blurry and there are some points where the lights are damaged/off so that the distribution of light is uneven and causes a lack of lighting. Eye fatigue can occur in poor lighting quality, for example when lighting in the visual task area is much brighter than the surrounding area. This affected the eyes to make frequent adjustments (adaptation and accommodation) when the view moves from the bright part to the dark part. From the dark part to the bright part repeatedly.

The results of this study indicate that respondents who use laptops with a high frequency of > 5 days a week tend not to experience eye fatigue when using laptops as many as 81 respondents. Based on the results of the Somers'd correlation test, it was found that there was no relationship between the frequency of laptop use and eye fatigue with a p value of 0.756 (> 0.05). The researcher's assumption is this can happen because the lighting in the room used by the respondents is quite good, but this cannot be analyzed because of the limitations of the study.

The results of this study indicate that respondents who have a high risk of distance vision tend not to experience eye fatigue when using a laptop as many as 95 respondents. Based on the results of the Somers'd correlation test, it was found that there was no relationship between visual distance when using a laptop and eye fatigue with a p value of 0.462 (> 0.05). OSHA explains that the recommended distance for viewing a computer monitor is 18-24 inches (45.72 – 60.96 cm), and the average distance is 50.80 cm. Shantakumari stated that the ideal distance for viewing a computer is 50-70 cm. Kanitkar et al explained that at a distance of 50-70 cm the eye is in a physiological condition, in which the eye will rest from the accommodation. The researcher assumes that this can happen because respondents always take care of their health and also blink often. This rule of 20-20-20 can be implemented using socialization through counseling, reminding each other between lecturers, and placing posters on the wall or the desk.

The results of this study indicate that respondents who use a risky monitor position that is higher or lower than the horizontal eye level tend not to experience eye fatigue when using a laptop as many as 55 respondents. Based on the results of the Somers'd correlation test, it was found that there was no relationship between the position of the monitor when using a laptop and eye fatigue with a p value of 0.778 (> 0.05). The researcher assumes that the respondent is still 35 years old so they have not felt the effect of eye fatigue due to laptop use. Physiologically increasing age results in a decrease in the function of the eye organs resulting in decreased vision [20]. Workers who are over 40 years old will be more susceptible to vision, in line with the process of physiological changes and aging in the eye. Increasing age causes the elasticity of the eye to decrease and in old age the elasticity will be lost, causing a decrease in the ability of the eye lens to focus on objects on the retina, causing discomfort in the eyes and accelerating the occurrence of eye fatigue.

The results of this study indicate that respondents who do eye rest for > 15 minutes tend not to experience eye fatigue when using laptops as many as 53 respondents. Based on the results of the Somers'd correlation test, it was found that there was no relationship between eye rest when using a laptop and eye fatigue with a p value of 0.850 (> 0.05). The researcher assumes that this happens because the majority of respondents do not experience eye refraction. Eye fatigue is felt more quickly in individuals with refractive errors.

5. Conclusions

The laptop usage behaviors have affected Eye fatigue because of the size of the laptop. Therefore, it is recommended to use a standard laptop size (≥ 14 inches) and use laptop support when operating a laptop.

Laptop support can be used like a laptop stand. The most important thing in choosing a laptop stand is choosing a size that fits the size of your laptop, choose a laptop stand that can be tilted and its height can be changed, choose a laptop stand that is thin or can be folded so that its size is more compact and does not hinder mobility. Products made of aluminum or plastic are suitable to choose from because they are light and can be folded.

References

[1] WHO. (2020). Q&A on coronaviruses (Covid-19). Diakses Pada Tanggal 5 Juli2021.https://www.who.int/emergencies/diseases/novelcoronavirus2019/question-and-answershub/q-a-detail/q-acoronaviruses#:~:tex
[2] Hartati R, Imbiri M, and Setiani D 2020 Mengukur Pengetahuan Mahasiswa Poltekkes Kemenkes Jayapura Terhadap Infeksi Covid-19 Selama Pembelajaran Daring *Gema Kesehatan* **12**(1) 7-13 [in Indonesian Language]
[3] Ardianti N, Bramanti A, Mohanty P, Narayan K, and Saputro A 2020 COVID19 Impact On Indonesian Attitudes & Behaviours: Learning For Brands www.kantarworldpanel.com. (Accessed September 22, 2022)
[4] Kemenkes RI 2021 Situasi Terkini Perkembangan Coronavirus Disease (COVID-19) Laporan Kasus COVID-19.https://covid19.kemkes.go.id/situasi-infeksi-emerging/info-corona-virus/situasi-terkiniperkembangan-coronavirus-disease-covid-19-21-september-20 https://infeksiemerging.kemkes.go.id/situasi-infeksi-emerging/situasi-terkini-perkembangan-coronavirus-disease-covid-19-19-juni-2021 (Accessed August 22, 2022) [in Indonesian Language]
[5] Dampati P S, Chrismayanti NKSD, Veronica E 2020 Pengaruh Penggunaan Smartphone dan Laptop terhadap Muskuloskeletal Penduduk Indonesia pada Pandemi COVID19 *Gema Kesehatan* **12**(2) 57-67 [in Indonesian Language]
[6] Marimuthu P and Vasudevan H 2020 The Psychological Impact Of Working From Home During Coronavirus (COVID 19) Pandemic: A Case Study *CnR's International Journal of Social & Scientific Research* **6**(1) 18-29
[7] Rosenfield M 2011 Computer Vision Syndrome: A review of Ocular Causes and Potential Treatments *Ophthalmic and Physiological Optics* **31**(5) 502–515
[8] American Optometric Association 2020 Computer Vision Syndrome. Retrieved from https://www.aoa.org/healthy-eyes/eye-and-vision-conditions/computer-visionsyndrome?sso=y (Accessed August 22, 2022)
[9] America Optometric Assosiation 2016 Most american Experience digital Eye strain from overexposure to computers according to survey. Retrieved from https://www.aoa.org/about-the-aoa/press-room/press-releases/most-americans-experience-digital-eye-strain-from-overexposure-to-computers-according-to-survey?sso=y
[10] Pheasant S 1991 *Ergonomics, Work and Health* (London: Macmillan Press)
[11] Hanum I F 2008 Efektifitas Pengguna Screen Pada Monitor Komputer Untuk Mengurangi Kelelahan Mata Pada Pekerja Call Center di PT. Indosat NSR Medan. *Master Thesis Universitas Sumatera Utara* [in Indonesian Language]
[12] Suma'mur P K 2013 *Higiene Perusahaan dan Kesehatan Kerja (Hiperkes),* Second Edition (Jakarta: CV. Agung Seto) [in Indonesian Language]
[13] Yulia Y, Rizyana N P, and Rahmi A 2021 Determinan Kelelahan Mata Pada Pekerja Kantor yang Work From Home (WFH) di Kota Padang *Jurnal Ilmu Kesehatan* **5**(2) 247-254 [in Indonesian Language]
[14] Jovanovic J, Sarac I, Jovanovic J J and Jovanovic S 2021 Work Related Health Disorders And Work Ability Among Computer-Operating Workers. *IPSI BgD Internet Research Society* **17**(1) 22-28
[15] Kementerian Kesehatan RI 2018 *Riset Kesehatan Dasar* (Jakarta: Badan Penelitian dan Pengembangan Kesehatan) [in Indonesian Language]

[16] Puspita A 2012 Hubungan Antara Perilaku Penggunaan Laptop Dan Keluhan Kesehatan Akibat Penggunaan Laptop Pada Mahasiswa Sarjana Reguler Fakultas Ilmu Komputer Universitas Indonesia. Depok. Universitas Indonesia *Undergraduate Thesis Universitas Indonesia* [in Indonesian Language]
[17] Amanda Z 2012 Peranan Laptop Support Dalam Mengurangi Kelelahan Pada Pengguna Laptop. *INSAN* **14(**2) [in Indonesian Language]
[18] Hedge A 2011 10 Tips for using a computer mouse. Cornell University. http://ergo.human.cornell.edu/cumousetips.html (Accessed August 22, 2022)
[19] 10 Stand Laptop Terbaik. https://my-best.id/49277 (Accessed August 22, 2022)
[20] Tarwaka, H. (2015) Ergonomi Industri Dasar-Dasar Pengetahuan Ergonomi Dan Aplikasi Di Tempat Kerja Edisi II. Surakarta: Harapan Press [in Indonesian Language]

Development of Agritourism Potential as an Alternative Form to Preservation of Rural Environment in Samboja East Kalimantan

Virgilia Anna Gustiniani Pakalla[a*], Achmad Ghozali[b]

[1]Urban and Regional Planning Departement, Institut Teknologi Kalimantan, Balikpapan, 76127, Indoensia
[2]Urban and Regional Planning Departement, Institut Teknologi Kalimantan, Balikpapan, 76127, Indoensia

[*]08181081@student.itk.ac.id

Abstract. The relocation of the capital city of Republic Indonesia to East Kalimantan, which is called Nusantara, will certainly increase resident migration to East Kalimantan. This relocation will affect the tourism activity needs. Although, it can also threaten the existence of water system, flora and fauna, pollution and other environmental damage. Samboja, one of Subdistrict of Kutai Kartanegara, is the buffer zones of Nusantara. Samboja has a great potential for developing agritourism because it has the largest land area and amount of agricultural production in Kutai Kartanegara. The combination of agriculture and tourism or called agritourism is a form of creative economic development in the agricultural sector that provides added value to agribusiness and also can contribute to the rural environment preservation. Therefore, this research was conducted to determine the agritourism development in Samboja by identify the agritourism potential using scoring analysis with the 4A approach, namely Attraction. Accessibility, Amenity and Ancillary. The results of this research showed that the potential attractions in Samboja Subdistrict are agricultural land, various agricultural commodities, and agricultural activities carried out in the forms of planting, management, and harvesting. Then, the villages with high potential classification for agritourism development are Handil Baru, Sanipah and Karya Jaya.

1. Introduction

Tourism is one of the sectors pursued by the Indonesian government because it is very influential in Indonesia's development, especially as the largest contributor to international trade from the service sector [4]. Indonesia has a large agricultural area, natural wealth and biodiversity so that it becomes a potential source of economics in the fields of agriculture and tourism. The combination of tourism and what is commonly called agritourism is a form of creative economic development in the agricultural sector that provides added value to agribusiness businesses in order to improve the welfare of farmers in Indonesia [3]. There are five development priorities in the Kalimantan region, one of which is a village/rural area development project in the form of accelerating the development of tourist villages and social, cultural, economic, infrastructure transformation, as well as superior products in villages/rural areas [2]. This is in line with the achievements of East Kalimantan Province which received the Abdi Bakti Tani award in 2021 as Rank II for the province with the highest commodity exports organized by the Agriculture Ministry of the Republic of Indonesia [2]. The relocation of the capital city of Republic of Indonesia to East Kalimantan, which was named Nusantara, was located in part of Penajam Paser Utara District and part of Kutai Kartanegara District. This relocation will certainly increase the population in East Kalimantan. Tourism has become a trend and part of the basic needs of the community. Therefore, the increase of the population in Nusantara area will affect the development of existing tourism involve Samboja Subdistrict.

Samboja Subdistrict is one of the Nusantara buffer zones. Based on the Urban Land Use Plan of Kutai Kartanegara in 2013 – 2033, Samboja was directed as a service center function which includes agricultural development and tourism services [6]. This is in line with the tourism strategic plan of Kutai Kartanegara for 2016 – 2025, that Samboja was designated as a strategic tourism area with one of its attractions being agritourism [5]. Samboja has great potential for agritourism development. In 2020, Samboja Subdistrict is the largest harvested area and production of vegetable crops in Kutai Kartanegara, each 1,195 hectares (31.23%) and 92,594 tons (27.7%) [1]. The plantation area in Samboja Subdistrict is 53,417 hectares or 14.9% of the plantation area in Kutai Kartanegara. Then, in the fisheries sector, Samboja has capture fisheries production of 10,246 tons (13.5%) and aquaculture production of 12,034 tons (9.6%) [1]. Samboja also has a natural beauty, as well as cool air which can be a attraction in supporting the agritourism development. The average temperature in these two villages is around 27.8°C, and even has the lowest temperature of 23.5°C so it can create a sense of comfort [1]. This is a potential of agritourism development in this region, but it is also a challenge to have sustainable agricultural products so as to create an ideal agritourism. In addition, agritourism is a relatively new type of tourism in Kutai Kartanegara District. So, the agritourism development in Samboja has the potential for interregional tourist visits, such as people from outside Samboja who visit to relieve daily boredom, and as edutourism in the agricultural sector.

2. Methodology

This research was carried out in Samboja Subdistrict in Kutai Kartanegara District, East Kalimantan, Indonesia. The analysis used in this research is a scoring analysis in every villages. There are 13 villages in Samboja Subdistrict. This analysis identifies the potential of agritourism in Samboja based on the criteria for assessing agritourism potential. After that, the analysis is continued with the desciptive analysis to find out the strategies of agritourism development in Samboja. The picture below is the administrative map of the Samboja Subdistrict which is the location of this research study.

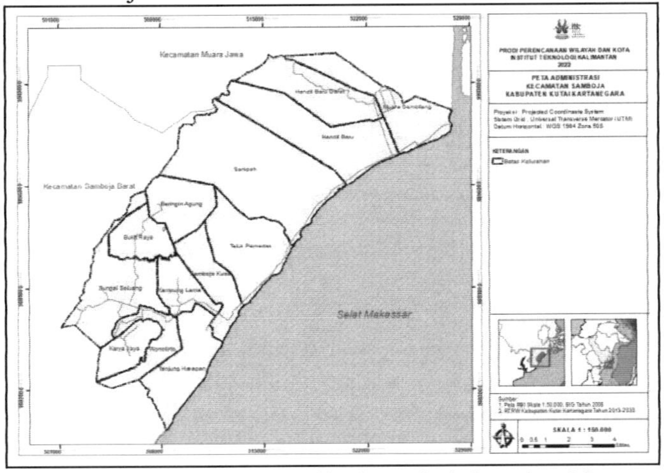

Figure 1. Samboja Subdistrict Administration Map

2.1. Identification of Agritourism Potential in Samboja Subdistrict

Table 1. Research Variables.

Variables	Value	Subvariables
Attraction	0.43	Rare Attraction, Natural Attraction, Unique Attraction, Variety Of Agricultural Activities, Sanitation, and Safety
Accessibility	0.27	Road Conditions, Road Distance From The District Capital, Road Distance From The Provincial Capital, and Public Transportation
Amenity	0.20	Hostelry, Facilities and Infrastructure
Ancillary	0.10	Institutions And Community Groups, and Involvement Of Local Workforce

The scoring analysis steps in this study are as follows :

- consider the score by using the formula:

$$S = C \times V \quad (1)$$

Description:
S : Score
C : Criteria score
V : Value

- Classify potential and analyze agritourism potential based on scoring results
- Perform potential class intervals, with the formula.

$$Interval = \frac{maximum\ score\ of\ criteria\ i - maximum\ score\ of\ criteria\ i}{3\ (the\ number\ of\ potential\ classes\ includes\ High, Medium, and\ Low)} \quad (2)$$

- Classifying potential based on the interval value of each criterion. The following table shows the classification of potential criteria for agritourism potential:

Table 2. Determination of the Interval of Each Criteria

Criteria	Minimum Score	Maximum Score	Interval
Attraction	77	26	17
Accessibility	27	11	5
Amenity	24	8	5
Ancillary	6	2	1
Total	134	47	29

Table 3. Classification/Class of Potential Each Criteria

Criteria	High	Medium	Low
Attraction	77 – 60	59 – 42	<41
Accessibility	27 – 22	21 – 16	<15
Amenity	24 – 19	18 – 13	<12
Ancillary	6 – 5	4 – 3	<2
Total	134 – 105	104 – 75	<74

2.2. Analysis of Agritourism Development Strategies in Samboja Subdistrict

Based on the results of the scoring analysis in each village in Samboja Subdistrict, it is continued with a qualitative descriptive analysis to determine strategies for developing agritourism in Samboja Subdistrict.

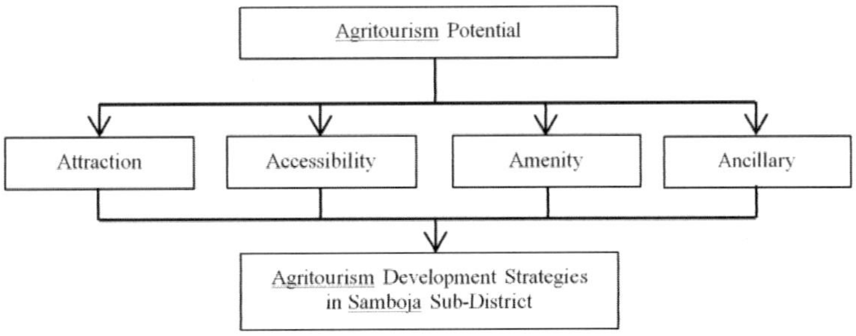

Figure 2. Research framework

3. Results and Discussion

3.1. *Identification of Potential Agritourism in Samboja*

The analysis carried out is a scoring analysis. The method of data collection for the first stage analysis are questionnaires to collecting information in each village and observations to validate the previously obtained information. At this stage the researchers conducted an analysis related to the potential of agritourism in Samboja Subdistrict based on the criteria of agritourism attractions, accessibilities, amenities, and ancillaries. After the analysis process of identifying agritourism potential through scoring analysis has been carried out, then the next step is to determine the class of agritourism potential (high, medium, and low) on each agritourism potential variable.

Table 4. Scoring Analysis of Attractions in Samboja Subdistrict

Number	Village	Attractions			Potential Classification
		C	V	S	
1	Handil Baru	165	0.43	70.95	High
2	Karya Jaya	160	0.43	68.8	High
3	Sanipah	155	0.43	66.65	High
4	Muara Sembilang	145	0.43	62.35	High
5	Samboja Kuala	140	0.43	60.2	High
6	Wonotirto	140	0.43	60.2	High
7	Beringin Agung	140	0.43	60.2	High
8	Teluk Pemedas	135	0.43	58.05	Medium
9	Bukit Raya	135	0.43	58.05	Medium
10	Handil Baru Darat	120	0.43	51.6	Medium
11	Kampung Lama	115	0.43	49.45	Medium
12	Tanjung Harapan	115	0.43	49.45	Medium
13	Sungai Seluang	105	0.43	45.15	Medium

This analysis was carried out using the variables of scarcity, naturalness, uniqueness, variety of agricultural activities, sanitation and safety. Based on the results of the scoring analysis carried out, it was found that the village/village with the high score for potential agritourism development was Handil Baru, then, followed by Karya Jaya, Sanipah, Muara Sembilang, Kuala Samboja, Wonotirto and Beringin Agung.

Figure 3. Agritourism Potential in Samboja Subdistrict

Figure 4. Agritourism Potential in Samboja Subdistrict

The agricultural activities that have the potential to develop agritourism are preparing plant seeds, planting, maintaining and harvesting, and also producing agricultural products. Fishery in the Samboja area is a commodity with great potential as an agritourism area development. The agricultural activities that can be carried out in the fishery area are fishing, shrimp, and crab breeding, then maintenance and harvesting. The potential of cattle in Samboja Subdistrict also has the opportunity to support the development of agritourism in Samboja Subdistrict. Agricultural activities carried out in Samboja Subdistrict are the maintenance and harvesting of cattle in the form of cows, and chickens.

3.1.1. Accessibility

Table 5. Scoring Analysis of Accessibilities in Samboja Subdistrict

Number	Village	Accessibilities			Potential Classification
		C	V	S	
1	Muara Sembilang	75	0.27	20.25	Medium
2	Handil Baru Darat	70	0.27	18.9	Medium
3	Handil Baru	70	0.27	18.9	Medium
4	Kampung Lama	70	0.27	18.9	Medium
5	Sanipah	70	0.27	18.9	Medium
6	Samboja Kuala	70	0.27	18.9	Medium
7	Teluk Pemedas	70	0.27	18.9	Medium
8	Wonotirto	70	0.27	18.9	Medium
9	Sungai Seluang	70	0.27	18.9	Medium
10	Karya Jaya	70	0.27	18.9	Medium
11	Bukit Raya	65	0.27	17.55	Medium
12	Beringin Agung	65	0.27	17.55	Medium
13	Tanjung Harapan	60	0.27	16.2	Medium

Based on the results of the analysis above, it is known that Muara Sembilang is the village with the largest score, which is 20.25. This is due to the proximity factor to the provincial and district capitals.

Meanwhile, the road conditions in each village are relatively same that having a road width of ±3 meters and asphalt pavement or cementation.

3.1.2. Amenity

Table 6. Scoring Analysis of Amenities in Samboja Subdistrict

Number	Village	Amenities			Potential Classification
		C	V	S	
1	Sungai Seluang	110	0.2	22	High
2	Sanipah	105	0.2	21	High
3	Teluk Pemedas	95	0.2	19	High
4	Handil Baru	90	0.2	18	Medium
5	Kampung Lama	90	0.2	18	Medium
6	Tanjung Harapan	85	0.2	17	Medium
7	Samboja Kuala	75	0.2	15	Medium
8	Handil Baru Darat	70	0.2	14	Medium
9	Wonotirto	70	0.2	14	Medium
10	Muara Sembilang	65	0.2	13	Medium
11	Bukit Raya	65	0.2	13	Medium
12	Beringin Agung	65	0.2	13	Medium
13	Karya Jaya	60	0.2	12	Low

Based on the results of the analysis above, it is known that the largest score in amenity is Sungai Seluang Village, followed by Sanipah Village and Teluk Pemedas Village. This is influenced by the quantity of accommodations and facilities and infrastructure that can support tourism activities. Accommodation in the form of lodging are only found in Kampung Lama, Sanipah, and Sungai Seluang.

3.1.3. Ancillary

Table 7. Scoring Analysis of Ancillaries in Samboja Subdistrict

Number	Village	Amenities			Potential Classification
		C	V	S	
1	Handil Baru	60	0.1	6	High
2	Sanipah	60	0.1	6	High
3	Wonotirto	60	0.1	6	High
4	Bukit Raya	60	0.1	6	High
5	Beringin Agung	60	0.1	6	High
6	Karya Jaya	60	0.1	6	High
7	Muara Sembilang	55	0.1	5.5	High
8	Handil Baru Darat	50	0.1	5	High
9	Kampung Lama	45	0.1	4.5	Medium
10	Samboja Kuala	45	0.1	4.5	Medium
11	Teluk Pemedas	45	0.1	4.5	Medium
12	Sungai Seluang	45	0.1	4.5	Medium
13	Tanjung Harapan	35	0.1	3.5	Medium

Based on the results of the analysis above, it is known that the high potential of ancillary in Samboja Subdistrict is in Handil Baru Village, followed by Sanipah Village, Wonotirto Village, Bukit Raya Village, Beringin Agung Village, Karya Jaya Village, Muara Sembilang Village, and Handil Baru Darat Village. In this case, the existence of farmer groups in Samboja District has great potential because of the large community and they also active in operations.

3.1.4. Identification and Classification of All Potential Components of Agritourism Development in Samboja Subdistrict

Based on the results of identification and scoring analysis on attractions, accessibilities, amenities and ancillaries variables, the total score and potential classification (high, medium, low) in each village based on agritourism potential are as follows.

Table 8. Results of the Scoring Analysis of Agritourism Potential in Samboja Subdistrict

Village	Attractions			Accessibilities			Amenities			Ancillaries			Total Skor	Potential Classification
	C1	V1	S1	C2	V2	S2	C3	V3	S3	C4	V4	S4	S1+S2+S3+S4	
Handil Baru	165	0.43	70.95	70	0.27	18.9	90	0.2	18	60	0.1	6	113.85	High
Sanipah	155	0.43	66.65	70	0.27	18.9	105	0.2	21	60	0.1	6	112.55	High
Karya Jaya	160	0.43	68.8	70	0.27	18.9	60	0.2	12	60	0.1	6	105.7	High
Muara Sembilang	145	0.43	62.35	75	0.27	20.25	65	0.2	13	55	0.1	5.5	101.1	Medium
Teluk Pemedas	135	0.43	58.05	70	0.27	18.9	95	0.2	19	45	0.1	4.5	100.45	Medium
Wonotirto	140	0.43	60.2	70	0.27	18.9	70	0.2	14	60	0.1	6	99.1	Medium
Samboja Kuala	140	0.43	60.2	70	0.27	18.9	75	0.2	15	45	0.1	4.5	98.6	Medium
Beringin Agung	140	0.43	60.2	65	0.27	17.55	65	0.2	13	60	0.1	6	96.75	Medium
Bukit Raya	135	0.43	58.05	65	0.27	17.55	65	0.2	13	60	0.1	6	94.6	Medium
Kampung Lama	115	0.43	49.45	70	0.27	18.9	90	0.2	18	45	0.1	4.5	90.85	Medium
Sungai Seluang	105	0.43	45.15	70	0.27	18.9	110	0.2	22	45	0.1	4.5	90.55	Medium
Handil Baru Darat	120	0.43	51.6	70	0.27	18.9	70	0.2	14	50	0.1	5	89.5	Medium
Tanjung Harapan	115	0.43	49.45	60	0.27	16.2	85	0.2	17	35	0.1	3.5	86.15	Medium

Based on the results of the analysis above, it is known that the villages with high potential classification for agritourism development are Handil Baru, Sanipah and Karya Jaya.

3.2. Agritourism Development Strategy in Samboja Subdistrict

Strategies for developing agritourism in Samboja Subdistrict is as follows:

- Optimization of the agritourism development on agricultural potential in the form of agritourism of rice fields, food crops, horticulture, plantations, and fruits, with the attractions are: beautiful and natural agricultural panoramic views, agricultural activities by farmers include planting/seeding, maintenance, harvesting and producing agricultural products into food/drink/crafts.
- Optimization of the agritourism development on fishery commodities with attractions are: agricultural activities by farmers including seeding, maintenance, harvesting and producing of agricultural products into food/drinks/crafts.
- Optimization of the agritourism development on cattle commodities with attractions are: agricultural activities by farmers including breeding, maintenance, harvesting and producing of agricultural products into food/drink/crafts.
- Increase the availability and services of amenities in all villages in Samboja Subdistrict.
- Application of new concepts with technology to make the agricultural product innovation to increase the attractions and become the characteristic of agritourism in Samboja Subdistrict.
- Supervision of land use change in Samboja Subdistrict to maintain environmental sustainability, and guard against pollution. Then, the development of idle lands into productive agricultural land as agritourism potential. as an Alternative Form of Environmental Conservation in Samboja Subdistrict.

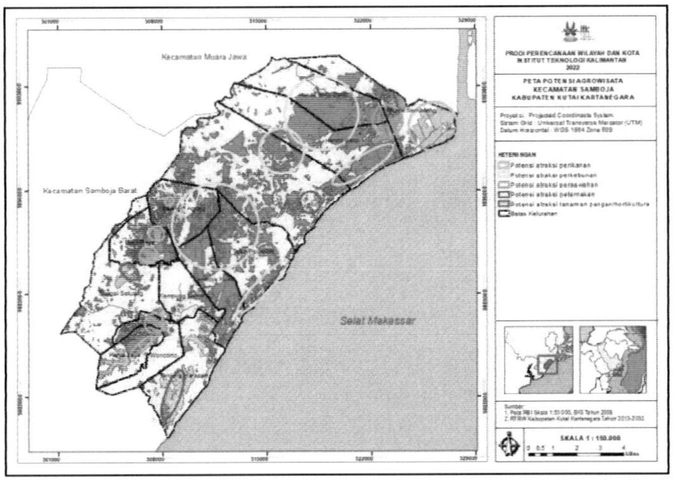

Figure 5. Potential Areas for Agritourism Development in Samboja Subdistrict

4. Conclusion

Based on the results of the identification and analysis that has been carried out for research with the title 'Agritourism Potential Development Strategy in Samboja Subdistrict', the conclusions that can be drawn is Samboja Subdistrict has the potential for agritourism development, especially in Handil Baru, Sanipah and Karya Jaya. The potential attractions in Samboja Subdistrict were agricultural land, various agricultural commodities, and agricultural activities carried out in the forms of planting/seeding, maintenance/management, and harvesting.

References

[1] BPS. 2020. Kabupaten Kutai Kartanegara Dalam Angka 2021. BPS Kabupaten Kutai Kartanegara
[2] Central government. 2020. Peraturan Presiden (PERPRES) tentang Rencana Pembangunan Jangka Menengah Nasional Tahun 2020 –2024, Jakarta
[3] Directorate of Business Development and Investment. 2012. Pedoman Umum Agriwisata. Agriculture Ministry of the Republic Indonesia. 2012, Jakarta
[4] Nafisa, Lina Rizqi. 2016. Arahan Pengembangan Kawasan Wisata Pantai Pidakan di Kecamatan Tulakan, Kabupaten Pacitan Berbasis Partisipasi Masyarakat. Surabaya.
[5] Regional Government of Kutai Kartanegara Regency. 2016. Peraturan Daerah Kabupaten Kutai Kartanegara No. 19 Tahun 2016 Tentang Rencana Induk Pembangunan Kepariwisataan Kabupaten Kutai Kartanegara Tahun 2016 – 2025, Kutai Kartanegara
[6] Regional Government of Kutai Kartanegara Regency. 2013. Peraturan Daerah Kabupaten Kutai Kartangeara No 9 tahun 2013 tentang Rencana Tata Ruang Wilayah Kabupaten Kutai Kartanegara Tahun 2013 – 2033, Kutai Kartanegara
[7] Said, Keliwar, Eman S., Riska R, A. Rinto D.A.. 2021. Analysis of the Potential Development of Tourism Villages in Jembayan Tengah East Kalimantan Using AHP Method. Advances in Social Science, Education and Humanities Research, volume 647 : 407 - 413

A Time-Series Analysis on the Covid-19 Mortality, PM2.5 Levels, and Weather Variables in Denpasar City, Indonesia

Amari Dihaning Suhendarto[1], Yeny Dhokhikah[1*], and Abdur Rohman[1]

[1]Study Program of Environmental Engineering, Dept. of Civil Engineering Faculty of Engineering, Universitas Jember
Jalan Kalimantan 37, Jember, Jawa Timur, Indonesia 68121

yeny.teknik@unej.ac.id

Abstract. The objective of this study is to explore the relationship between Covid-19 mortality cases and environmental variables, namely PM2.5 concentration and weather variables, in Denpasar City, Indonesia. Regression models were used. The response variable was the monthly Covid-19 mortality from March 2020 to December 2021 and the predictor variables were the mean concentration of PM2.5, temperature, wind speed, rainfall and duration of sunshine. All data analyzed were provided by the Indonesian Government. Simple linear regression (SLR) and dynamic regression with ARIMA error models were used. Further, of the 22 monthly data, the first 19 months data were used to train the models and the remaining data were used as the test data. It is found that both wind speed and the interaction between PM2.5 concentration and wind speed have statistically significant relationships with Covid-19 mortality. The estimates of SLR and ARIMA (0,1,1) with interaction models show that on average, in case of 0.5 m/s wind speed, an increase of 1 μg/m3 in the monthly mean of daily PM2.5 concentrations associates with 17.4 and 16.3 increase in the monthly Covid-19 mortality case, respectively. Although this study is observational, its findings suggest the importance of controlling PM2.5 concentration.

1. Introduction

The Covid-19 pandemic was first declared by the WHO (World Health Organisation) on 12 March 2020 after the death cases from Covid-19 increased significantly. This pandemic started in Wuhan, Hubei Province of China on 12th December 2019, which caused 80 deaths of 2,794 laboratory-confirmed infections by 26 January 2020 [1][2]. Covid-19 is officially named by WHO to describe the disease caused by the new coronavirus, whilst the International Committee for Taxonomy of Viruses (ICTV) established that the official name of the disease is Severe Acute Respiratory Syndrome Coronavirus 2 (SARS-CoV-2). *Our World in Data* reported that the cumulative confirmed Covid-19 cases were found in almost all countries, with the five highest cumulative cases per 30 December 2021 were recorded in the United States, India, United Kingdom, France, and Germany [3]. Infected persons might have different responses to Covid-19. Most people who are exposed to this virus will experience mild to moderate symptoms and will recover without needing to be hospitalized. However, some people will experience severe pain and require medical assistance. Covid-19 can spread directly or indirectly, such as through respiratory droplets of an infected person when she speaks, sings, breathes, coughs, or sneezes. The droplets can be coarse aerosols with aerodynamic diameter greater than 5 µm or fine-particle aerosols with aerodynamic diameter less than 5 µm [4]. Some climate conditions (weather, air pollution, temperature, and humidity level) contribute to the increase in confirmed cases of Covid-19 and total death in Milan, Italy [5]. A similar study conducted in Turkey analyzed the correlation between weather (temperature, dew point, humidity, and wind speed), population, and the number cases of Covid-19. The authors stated that high wind speed and low temperature correlate with the number of Covid-19 cases [5]. To examine the correlation between weather and the number of cases or deaths of Covid-19, some models are applied, namely Spearman's correlation analysis [6][7], multiple regression model [8], Pearson's correlation coefficient, linear and non-linear regression, LOESS, two-way ANOVA, and others [9].

This study aims to discover the correlation between the deaths from Covid-19, the concentration of PM2.5, and the weather variables in Denpasar, Indonesia using simple linear regression and ARIMA models.

2. Methods

The area of study in this research is Denpasar, the capital city of Bali province, Indonesia. Denpasar is geographically located between 8°35'31"S to 8°44'49"S and 115°00'23"E to 115°16'27"E, with a total area of 127,78 square kilometers. In 2021 the population of Denpasar was 726,599 and its population density was 5,686 per square kilometer [10].

2.1. Datasets

The datasets of this study were provided by three separate institutions of the Government of Indonesia. The monthly mortality cases data due to Covid-19 were provided by the Public Health Department of Denpasar City. Regarding the predictor variables, the daily concentration of PM2.5 (in micrograms per m3), the daily speed of the wind (in microgram/m^3), the daily wind speed (in m/s), and the daily temperature (^{0}C) were provided by the Bali and Nusa Tenggara Ecoregional Control and Development Directorate, Ministry of Environment and Forestry. The daily precipitation records (in mm) and the daily duration of the sunshine (in hour/day) were provided by BMKG (Meteorological, Climatological, and Geophysical Agency). Because mortality case data were available only in monthly intervals, not daily ones, all daily data of the predictor variables were summarised to obtain a mean value for each month. Missing values were excluded from the summarization step. In total, there were 22 data points for each variable, representing the corresponding 22 months of observations (March 2020 to December 2021). The datasets were then divided into two groups, namely, training group and testing group. The training group consisted of data points from March 2020 to September 2021 and was used to build the models. The testing group consisted of data points from October to December 2021 and was used to measure and compare the accuracy of the models built on the training data. The summary of the 22-monthly datasets is presented in Table 1.

Table 1. The data summaries of the variables used in this study

Variable	Min.	1st Qu.	Median	Mean	3rd Qu.	Max.	SD
Mortality case (people)	0.00	4.25	21.00	44.91	52.00	357.00	78.76
PM2.5 concentration (μg/m^3)	7.73	9.91	12.16	12.33	14.20	18.70	3.13
Air Temperature (°C)	26.21	26.87	27.14	27.19	27.56	28.32	0.58
Wind speed (m/s)	0.30	0.45	0.67	0.82	1.07	2.04	0.47
Rainfall (mm)	0.03	2.87	6.45	8.20	10.90	28.56	7.93
Sunshine duration (hour/day)	3.31	5.71	7.03	6.50	7.40	8.41	1.39

For the mortality case, the mean value was 45 cases/month, but the number of cases varied widely, as indicated by the large standard deviation. The highest mortality case occurred in August 2021 with 357 cases. The significant rise occurred shortly after the outbreak of the *delta* variant of the Covid-19 virus in July 2021. The mean daily concentration of PM2.5 varied in the range of 8 to 19 μg/m^3. The mean daily temperature over the months was relatively stable in the range of 26 to 28 degree Celsius. The mean daily wind speed was low, mostly lower than 2 m/s. A closely similar time series pattern was found between the wind speed data and the mortality case from Covid-19. Because Indonesia has a dry season and a rainy season, the rainfall varied as expected with a mean daily value of 8.20 mm and the highest values occurred during the usual rainy season period, namely around November to January each year. The duration of the daily sunshine varied slightly, with a mean value of 6.5 hours/day and a standard deviation of 1.4 hours/day. This value was also expected because Denpasar is located relatively close to the Equator.

2.2 Simple Linear Regression

In a simple linear regression, the response variable is modelled as a linear function of the predictor variables with an error term assumed to be independent and identically distributed random variables that follow a normal distribution with constant variance, as presented in equation (1)

$$y = \beta_1 x_1 + \beta_2 x_2 + \ldots + \beta_n x_n + \varepsilon \quad (1)$$

where y denotes the response variable, x is the response variable, β denotes the coefficient of the predictor variables. The values of the coefficients are estimated from the observation data using the least-squares estimation. Each coefficient represents the change in the mean response variable for every unit increase in the corresponding predictor variable when all the other predictors are kept constant. The independence of the errors is evaluated in this study by testing the serial correlation of the residuals of the model because the presence of serial correlation implies that the residuals, and hence the errors, are not independently distributed. The variables in the regression are assumed to be independent of each other. The response variable in the Simple Linear Regression is the monthly mortality case due to Covid-19, and the predictor variables are the other five variables presented in Table 1.

2.3. Dynamic Regression Model with Autoregressive Integrated Moving Average (ARIMA)

An ARIMA model is constructed from the combination of differencing, autoregression, and a moving average model. Integration is the inverse of differencing. The entire model can be written as in equation (2)

$$y_{t,d} = \mu + \alpha_1 x_{1,d} + \alpha_2 x_{2,d} + \ldots + \alpha_n x_{n,d} + \beta_1 y_{t-1} + \beta_2 y_{t-2,d} + \ldots + \beta_p y_{t-p,d}$$

$$+\epsilon_t + \lambda_1 \epsilon_{t-1} + \lambda_2 \epsilon_{t-2} + \ldots + \lambda_q \epsilon_{t-q} \quad (2)$$

where $y_{t,d,p}$ denotes the response variable at timestamp t, lagged with p order, and differenced with d order, μ denotes the intercept, $x_{i,d}$ denotes the predictor variable i that is differenced with d order, and α, β, and λ denote coefficients of the predictor variables. Their values are estimated from the observation data by using least squares estimation, ϵ_{t-q} denotes the error lagged with q order. This is known as an ARIMA(p,d,q) model.

3. **Results and Discussion**

3.1. Serial Correlation in Each Variable Data

The presence of serial correlation, or, in another term, autocorrelation, in each variable data was inspected through the ACF (Autocorrelation Function) plot and PACF (Partial Autocorrelation) plots. It is observed that a significant serial correlation was found in each variable. In addition, in all variables the lag = 1 shows significant serial correlation, indicating that ARIMA models with d = 1 differencing are highly probably appropriate to represent the serially correlated data.

3.2. Goodness-of-Fit of the Models

Based on the aforementioned ACF and PACF plots, we examined ARIMA (0,1,0) and several models with relatively close specifications. The AIC (Akaike Information Criterion) and the corrected AIC (AICc) were used to assess the goodness of fit of each model with the observation data. The corrected AIC is also used in this study because the sample size (n = 22) can be considered as a small sample size, in which AICc is recommended because it takes the number of parameters into consideration [11]. Table 2 shows that ARIMA (0,1,1) is the model with the best-fit based on AIC values. However, based on the corrected AIC, ARIMA (0,1,0) is the best model. Both ARIMA models are better suited than the SLR model. Therefore, we further analyse the three models, namely the SLR, ARIMA (0,1,0) and ARIMA(0,1,1).

Table 2. Estimates, Standard Errors, and p-values in the SLR, ARIMA (0,1,0), and ARIMA(0,1,1) models without interaction.

Term	Simple Linear Regression			ARIMA (0,1,0)			ARIMA (0,1,1)		
	estimate	std. error	p.val	estimate	std. error	p.val	estimate	std. error	p.val
Intercept	775.48	844.72	0.375	-5.03	13.65	0.717	0.58	3.01	0.849
MA(1)	undefined			undefined			-1.00	0.17	0.000
Temperature	-32.08	32.12	0.336	3.74	37.02	0.921	-30.61	28.33	0.294
PM2.5	9.56	6.39	0.158	1.25	7.80	0.875	9.21	5.73	0.125
WindSpeed	181.21	40.89	0.000	231.97	50.37	0.000	175.35	46.09	0.001
Rainfall	2.10	3.73	0.582	–0.24	2.99	0.935	1.90	3.33	0.575
SunshineDur	-21.46	22.06	0.348	-32.09	22.47	0.170	-21.15	18.80	0.275
AIC	213.11			210.27			208.52		
Corrected AIC	223.30			221.47			224.52		

Our analyses consist of two steps. First, we analysed the three models without any interaction between the predictor variables. The results are shown in Table 3. It is found that in the three models, the mean daily wind speed is the only variable that is statistically highly significant (with p-values far lower than 0.05). It is not reasonable that wind speed alone is associated with mortality due to Covid-19 because the disease is caused by a virus. Based on [11], we suspect that the concentration of PM2.5 probably interacts with the wind speed during virus transmission.

Therefore, in the second step, we analysed the models including the interaction term between PM2.5 concentration and wind speed. The results show that all models with interaction terms are better suited than those without an interaction term, as seen in Table 3. Based on the AIC values, when the interaction term is included, the best-fit model is ARIMA(0,1,1), but based on the corrected AIC values, the SLR model is the best-fit model.

Table 3. Estimates, Standard Errors, and P-values in the SLR, ARIMA (0,1,0), and ARIMA(0,1,1) with Interaction Terms

Term	Simple Linear Regression			ARIMA (0,1,0)			ARIMA (0,1,1)		
	estimate	std. error	p.val	estimate	std. error	p.val	estimate	std. error	p.val
Intercept	1068.39	675.37	0.139	-1.88	10.79	0.864	4.11	2.25	0.085
MA(1)	undefined			undefined			-1.00	0.14	0.000
Temperature	-54.48	26.51	0.062	-25.54	30.46	0.413	-48.02	20.24	0.028
PM2.5	32.24	9.17	0.004	29.97	10.61	0.011	33.77	6.93	0.000
WindSpeed	502.42	113.11	0.001	580.39	112.19	0.000	518.27	85.30	0.000
Rainfall	3.63	2.99	0.248	1.80	2.44	0.469	2.50	0.05	0.298
SunshineDur	-14.73	17.60	0.418	-4.74	19.52	0.811	-11.31	13.33	0.407
PM2.5 : WindSpeed	-29.68	10.01	0.011	-36.45	10.98	0.003	-34.98	8.06	0.000
	Goodness-of-Fit Metrics for the Models								
AIC	204.68			203.67			197.62		
Corrected AIC	219.08			219.67			220.12		

Table 3 shows that in the three models, each concentration of PM2.5, the speed of the wind and the interaction of the two variables have a statistically significant linear relationship with the mortality case due to Covid-19 with slightly different slopes between the models. Due to the presence of the interaction, to interpret the relationship, the coefficients of the interaction term must be accounted for. For example, the SLR models implies that on average, in case of 0.5 m/s wind speed, and other variables are held constant, an increase of 1 μg/m3 in mean daily PM2.5 concentration in a month associates with 32.24 - (0.5) (29.68) = 17.4 increase in mortality case due to Covid-19 per month. As for ARIMA (0,1,0) and ARIMA (0,1,1), the models implies association between 1 μg/m3 in mean daily PM2.5 concentration in a month with an 11.7 and 16.3 increase in mortality case due to Covid-19 per month. Likewise, the SLR implies that, in case of PM 2.5 concentration is 10 μg/m3 and other variables are held constant, an increase in daily mean wind speed of 1 m/s associates with 502.42 - (10) (32.24) = 180 increase in mortality case due to Covid-19 per month, on average.

The temperature is found to be statistically significant in the ARIMA (0, 1,1) model, but not statistically significant in other models. In all models, the coefficient of the temperature variable is negative, which implies that the increase in temperature is associated with a decrease in mortality due to Covid-19. The rainfall variable and the duration of sunshine are not statistically significant in all models. The fitness of the models in each month data are visually presented in Figure 1.

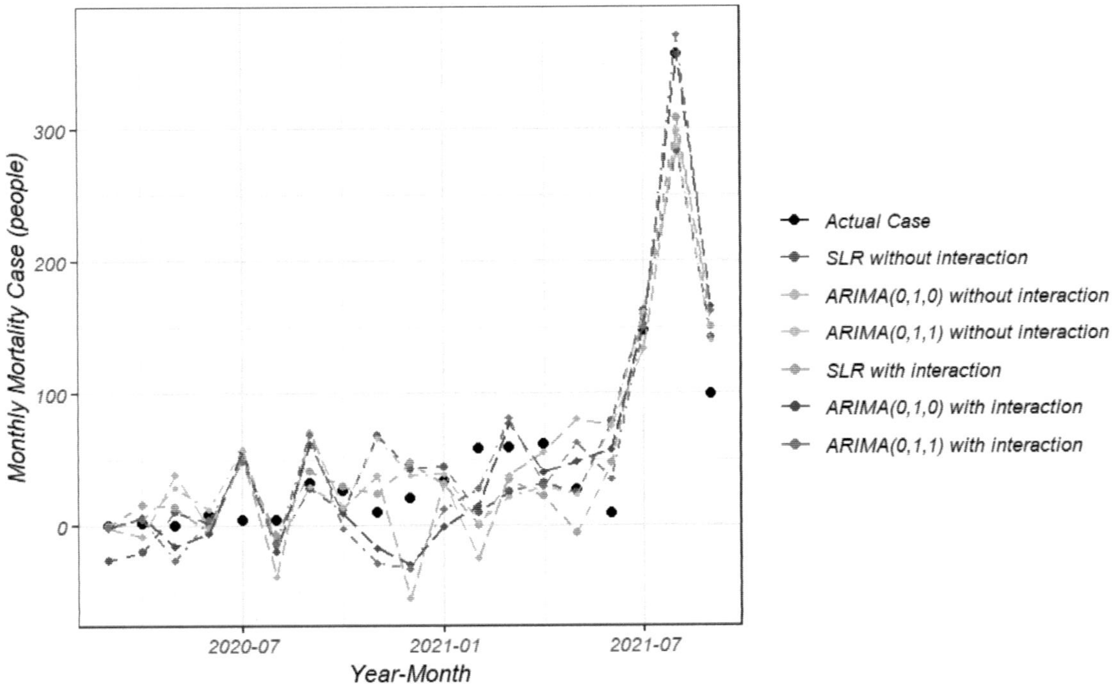

Figure 1. Actual values of monthly mortality cases and the corresponding fitted values of the models

3.3. Accuracy of the Models

After the goodness-of-fits of the six models are evaluated based on their AIC and the corrected AIC values, the accuracies of the models are compared. The comparison is conducted by applying each model to the test data of the predictor variables to generate the predicted mortality case for the last three months of 2021. The accuracy is measured by using the Mean Absolute Error (MAE). The comparison results are presented in Table 4.

Table 4. Predicted mortality cases due to Covid-19 of SLR, ARIMA (0,1,0), and ARIMA (0,1,1) models

		without interaction			with interaction		
Month	Actual Case	SLR	ARIMA (0,1,0)	ARIMA (0,1,1)	SLR	ARIMA (0,1,0)	ARIMA (0,1,1)
2021 Oct	21	269	157	266	210	108	58
2021 Nov	5	49	-86	41	56	-106	-154
2021 Dec	2	79	-30	70	46	-86	-126
Mean Absolute Error		123.0	86.3	116.3	94.7	95.3	108.0

Table 4 shows that each model results in a much larger mean absolute error than the actual case. This poor accuracy can be attributed to the small sample size of the model and the highly uncertain nature of Covid-19 development. Table 4 also shows that the best-fit model in the training phase does not always show the best accuracy in the testing phase.

3.4. Comparison with previous studies

This study found that each of the concentration of PM2.5, the wind speed, and the interaction of the two variables have a significant relationship with the case of mortality from Covid-19. The relationship between particulate matter concentration and Covid-19 has been strongly supported by the finding in [13] in which SARS-CoV-2 RNA was found in PM10 samples in Bergamo, Italy. In another study [14], it was found that other things being equal, a municipality in the Netherlands with 1 $\mu g/m^3$ more PM2.5 concentrations will have 9.4 more Covid-19 cases, 3.0 more hospital admissions, and 2.3 more deaths. These findings are clearly in agreement with the findings of our study.

Since our study is observational, no causal relationship can be inferred from the results. Nevertheless, our findings are in close agreement with the findings in [15] in which it is found that mean wind speed has a significant correlation with Covid-19 confirmed cases in Bandung City, Indonesia. However, in [15] it is found that the mean temperature has a significant correlation with the Covid-19 confirmed case, while in our study, the relationship between the mean temperature and the Covid-19 mortality case is not significant in most models.

4. Conclusion

The Covid-19 pandemic spread throughout the world in a few months since it exploded in Wuhan Province of China in late 2019. This study analysed six models consisting of Simple Linear Regression (SLR) and ARIMA models. Weather conditions and air pollution are found to have contributed to the increasing case of the Covid-19 mortality rate. Each of the PM2.5 concentration, the wind speed and the interaction between the PM2.5 concentration and the wind speed have a statistically significant relationship with the Covid-19 mortality cases. Temperature is not statistically significant in most models. The SLR, ARIMA (0,1,0) and ARIMA(0,1,1) with interaction models implies that on average, in case of 0.5 m/s wind speed, and other variables are held constant, an increase of 1 μg/m3 in mean daily PM2.5 concentration in a month associates with 17.4, 11.7, and 16.3 increase in mortality case due to Covid-19 per month, respectively. Our study is observational, therefore, the relationships found between Covid-19 mortality case and PM2.5 concentration and wind speed cannot be claimed to be causal. However, our findings are in agreement with previous studies in Italy, the Netherlands, and Indonesia. These findings suggest the importance of PM2.5 concentration control to hinder the spread of Covid-19.

References

[1] Zhou P, Yang XL, Wang XG, et al. 2020. A pneumonia outbreak associated with a new coronavirus of probable bat origin. Nature. 579 (7798):270–273. https://www.nature.com/articles/s41586-020-2012-7

[2] Ciotti M, Ciccozzi M, Terrinoni A, Jiang WC, Wang CB, Bernardini S. 2020. The Covid-19 pandemic. Critical Reviews in Clinical Laboratory Sciences. 57 (6): 365-388. https://www.tandfonline.com/doi/full/10.1080/10408363.2020.1783198

[3] Our World in Data. 2022. *Cumulative confirmed COVID-19 cases, Dec 30, 2021.* https://ourworldindata.org/coronavirus#coronavirus-country-profiles (Accessed: 18 September 2022).

[4] Leung NHL, Chu DKW, Shiu EYC, Chan KH, McDevitt JJ, Haul BJP, Yen HL, Li Y, Ip DKM, Peiris JSM, Setol WH, Leung GM, Milton DK and Cowling BJ. 2020. Respiratory virus shedding in exhaled breath and efficacy of face masks. Nature Medicine. 25 (5): 676-680. https://www.nature.com/articles/s41591-020-0843-2

[5] Zoran MA, Savastru RS, Savastru DM, Tautan MN. 2020. Assessing the relationship between surface levels of PM2.5 and PM10 particulate matter impact on COVID-19 in Milan, Italy. Science Total Environment. 738: 1-12. https://www.sciencedirect.com/science/article/pii/S0048969720333453

[6] Sahin M. 2020. Impact of weather on COVID-19 pandemic in Turkey. Science of The Total Environment. 728 (8): 1-5. https://www.sciencedirect.com/science/article/pii/S0048969720323275

[7] Tosepu R, Gunawan J, Effendy SD, Ahmad LOAI, Lestari H, Bahari H, Asfian P. 2020. Correlation between weather and Covid-19 pandemic in Jakarta, Indonesia. Science of The Total Environment. 725: 1-4. https://www.sciencedirect.com/science/article/pii/S0048969720319495

[8] Rath S, Tripathy A, Tripathy AR. 2020. Prediction of new active cases of coronavirus disease (COVID-19) pandemic using multiple linear regression model. Diabetes & Metabolic Syndrome: Clinical Research & Reviews. 14 (5): 1467-1474. https://www.sciencedirect.com/science/article/pii/S1871402120302939

[9] Ganegoda NC, Wijaya KP, Amadi M, Erandi KKWH & Aldila D. 2021. Interrelationship between daily COVID-19 cases and average temperature as well as relative humidity in Germany. Scientific Report. 11:11302. https://www.nature.com/articles/s41598-021-90873-5

[10] BPS-Statistics of Denpasar Municipality. 2022. Denpasar Municipality in Figures 2022. https://denpasarkota.bps.go.id/publication/2022/02/25/68f4c38625094b798b0471a6/kota-denpasar-dalam-angka-2022.html (Accessed: 18 September 2022).

[11] Dyer, O. 2021. Covid-19: Indonesia becomes Asia's new pandemic epicentre as delta variant spreads. BMJ (Clinical Research Ed). 16 July 2021. Available at https://www.bmj.com/content/bmj/374/bmj.n1815.full.pdf . Accessed on August 8, 2022.

[12] Hyndman, R.J., & Athanasopoulos, G. (2021) *Forecasting: principles and practice*, 3rd edition, OTexts: Melbourne, Australia. https://otexts.com/fpp3 . Accessed on 20 April 2022.

[13] Setti, L., Passarini, F., De Gennaro, G., Barbieri, P., Perrone, M. G., Borelli, M., Palmisani, J., Di Gilio, A., Torboli, V., Fontana, F., Clemente, L., Pallavicini, A., Ruscio, M., Piscitelli, P., & Miani, A. (2020). SARS-Cov-2RNA found on particulate matter of Bergamo in Northern Italy: First evidence. *Environmental Research*, *188*(May), 109754. https://doi.org/10.1016/j.envres.2020.109754

[14] Cole, M. A., Ozgen, C., & Strobl, E. (2020). Air Pollution Exposure and Covid-19 in Dutch Municipalities. *Environmental and Resource Economics*, *76*(4), 581–610. https://doi.org/10.1007/s10640-020-00491-4

[15] Azmy, A.R and Wulandari, R.A. 2022. Hubungan Cuaca dengan Kasus Terkonfirmasi Aktif COVID-19. Media Kesehatan. Available at https://journal.poltekkes-mks.ac.id/ojs2/index.php/mediakesehatan/article/view/2459

Precipitated calcium carbonate derived from green mussel shells extracted from magnesium chloride containing solution

M.A. Irfa'i[1,2*], A. Prihanto[3], S. Muryanto[4], J. Jamari[5], R. Ismail[5], A.P. Bayuseno[5]

[1]Doctoral Students at the Department of Mechanical Engineering, Diponegoro University, Tembalang, Semarang, Indonesia
[2]Mechanical Engineering Department, Universitas Negeri Surabaya, Surabaya, Indonesia
[3]Chemical Engineering Department, Catholic Polytechnic Mangun Wijaya, Semarang, Indonesia
[4]Department of Chemical Engineering, UNTAG University in Semarang, Indonesia
[5]Mechanical Engineering Department, Diponegoro University, Semarang, Indonesia

arifirfai@unesa.ac.id

Abstract. Green mussel shells are one of the calcium carbonate-rich biogenic sources. They contain 95.7 to 98.2 wt. % calcium. The high calcium content of green mussel shells was extracted and converted into calcium oxide material in the study by calcination at 900°C for 5 h. Later, at stirring times of 30, 60, 90, and 120 minutes, a solution medium containing magnesium chloride was used to dissolve the resulting calcination powder and subsequent carbonation for producing PCC (precipitated calcium carbonate). According to XRD, Rietveld, and FTIR analyses, the resulting PCC product with calcite-containing Mg was obtained primarily for a short stirring time of 30 minutes during the carbonation process, whereas calcite and aragonite were generated at longer stirring times (from 60 to 90 minutes). Instead, brucite was obtained after 120 minutes of stirring. The PCC product has a non-uniform polyhedrallic morphology of various small and large sizes as observed by SEM. This powder processing method allowed for faster PCC synthesis at a lower cost while requiring no chemical addition to increase basicity.

1. Introduction

Green mussel shells (*Perna Viridis*) are a type of biogenic waste that is commonly discarded in landfills. This type of waste, on the other hand, is high in calcium sources and could be extracted for use as a low-cost source of raw biomaterials. Processing methods such as grinding, calcination, and carbonation at specific temperatures and times are used to improve the properties of required biomaterials. Calcination followed by mineral carbonation, in particular, is an appealing route for preparing a powder precursor in CO_2 storage while also reducing basicity in the medium leaching agent and producing precipitated calcium carbonate with added value[1], [2]. Calcination could also be used to valorize unused and alkaline biogenic waste [3–5]. Significant effort has been invested in researching the potential biogenic wastes produced by the carbonation process, with a focus on developing processing routes that increase chemical reactivity with CO_2 while reducing energy requirements [6]-[8].

In practice, the main mineral carbonation routes are as follows: a) indirect carbonation, in which Ca and/or Mg are extracted from wastes while CO_2 is injected into the solids to precipitate carbonate; and

b) direct carbonation, in which solid wastes are dissolved into medium solutions and directly injected with CO_2 to precipitate carbonate and other mineral formations. In this condition, the presence of Ca and Mg elements in the solution may pose a significant barrier to influencing high conversion rates and CO_2 capture [9]. As a result, appropriate leaching agents with high extraction efficiencies have been proposed to address this issue. In fact, carbonates can precipitate when the pH changes to basicity because alkaline-earth elements have a lower affinity than carbonates [1]. Moreover, ammonium salts (NH_4Cl, CH_3COONH_4, NH_4NO_3) and acetic acid were proposed as leaching agents for Ca-containing solid wastes (CH_3COOH). Currently, it has been demonstrated that nitric acid (HNO_3) is effective in the synthesis of PCC from mussel shell powders [11], [12]. However, a strong base chemical (e.g., NH_4OH) was required to promote a neutralization reaction during carbonate precipitation, which could result in a high processing cost. Furthermore, ammonia (NH_3) production may result in other environmental issues.

On the other hand, magnesium chloride ($MgCl_2$) is a potentially effective approach to calcium leaching and promoting aragonite as a result of the mineral carbonation reaction [13], [14]. Aragonite, in particular, has the potential to form when seawater is used as a leaching agent [15]. Furthermore, using $MgCl_2$ as a leaching agent may cause aragonite-accompanied Mg^{2+} to be incorporated into the calcite crystal structure, slowing its growth. A decrease in pH may also result in a decrease in CO_3^{2-} supersaturated solutions. Using $MgCl_2$ as a leaching agent during the extraction of calcium from mussel shell powders may significantly increase calcium mobility, resulting in increased PCC production in mineral carbonation.

As a result, the current research looked into how calcium carbonate was precipitated from mussel shell powders that had been calcined and then carbonated in $MgCl_2$-containing solution. The goal of this research was to gain a thorough understanding of the efficacy of mineral carbonation in crystal-forming solutions containing $MgCl_2$. The effect of stirring times on PCC formation with desired phase compositions was also discussed. Furthermore, the PCC product was characterized using analytical techniques such as XRD, SEM, and FTIR. The findings could perhaps improve the powder processing strategy for recycling green mussel shells with magnesium chloride.

1.1. Theoretical background

In practice, using $MgCl_2$ to react with the calcium source of $Ca(OH)_2$ slurry is effective for making calcium soluble while also producing $Mg(OH)_2$ (Eq. 1). Furthermore, calcium carbonate precipitates from the solution as a result of the reaction that produces magnesium chloride (Eq. 2). Exothermic heat is required in this case to be responsible for both chemical reactions.

$$\begin{aligned} &Ca(OH)_{2(s)} + MgCl_{2(aq)} + H_2O_{(l)} \rightarrow Mg(OH)_{2(s)} + CaCl_{2(aq)} \\ &\Delta H = -19.4\ kJ/mol \end{aligned} \tag{1}$$

$$\begin{aligned} &CaCl_{2(aq)} + H_2CO_{3(aq)} + Mg(OH)_{2(s)} \rightarrow CaCO_{3(s)} + MgCl_{2(aq)} + 2H_2O_{(l)} \\ &\Delta H = -160.8\ kJ/mol \end{aligned} \tag{2}$$

Brucite [$Mg(OH)_2$] could also be formed during the carbonation reaction. When carbonic acid reacts with calcium dissolved in the slurry, the calcium carbonate polymorphs calcite and aragonite are formed[2]. The inhibition of brucite formation as the carbonate reaction progresses indicates the formation of $MgCl_2$. This condition is undesirable because it may result in additive loss and lower product purity. When the reaction is complete, however, only calcium carbonate can be formed.

In fact, $CaCl_2$ is less soluble than $MgCl_2$, resulting in a less soluble $CaCO_3$ product when compared to Mg-carbonates[3]. However, this condition may prevent the formation of Mg-carbonates in the solution, allowing the production of pure calcium carbonates. Because of the HCl generation from the solution, simultaneous generation of Ca- and Mg-carbonates may not occur unless the solution is neutralized with ammonia (NH_3)[4], [5]. Reusing magnesium chloride solution for frequent carbonation cycles, on the other hand, may not necessitate recrystallization and purification[6], [7]. However, precipitated $Mg(OH)_2$ could form in the Ca-rich solution during carbonation, limiting it to the direct carbonation system[8], [9]. As a result, it was proposed in the current study that $MgCl_2$ has the potential for carbonate mineral forming solution-derived green mussel shell powders[10]. Furthermore, magnetic stirring was a simple piece of equipment for assisting the calcined mussel shell powders in reacting with CO_2, which resulted in an intensification of the mineral carbonation reaction.

2. Material and methodology

2.1. Raw Material and Sample Preparation

Green mussel shells were collected from Kenjeran Beach in Surabaya. The mussel shells were cleaned with water, followed by drying in the sun. The mussel shells were then ground in a ball mill and sieved to pass 100 mesh. Calcination at 900°C for 5 hours was used to remove the organic compounds found in the resulting ground shell powders. Calcined powders emitted carbon dioxide (CO_2) to produce calcium oxide (CaO) compounds, which were then converted to $Ca(OH)_2$ by absorbing moisture during storage.

2.2. Carbonation Methods

The PCC product was made by preparing 0.6 M $MgCl_2$ (50 ml) of solution-rich $MgCl_2$. Following that, 2 g $Ca(OH)_2$ was added to the prepared solution and stirred for 0, 30, 60, 90, and 120 minutes on a magnetic stirrer at a constant speed of 300 rpm. Carbon dioxide was injected at a rate of 50 ml/min into this mixture solution and was stopped when the pH reached 7. The precipitated solid was separated using filter paper and then dried for 24 hours at 80°C. Before the subsequent material crystallization, the obtained powders were kept in a plastic bag.

2.3. Characterization of Solid Materials

The PCC product was ascertained and characterized using the following instruments. The chemical elements of raw green mussel shell powder were determined using X-ray Fluorescence (XRF) analysis. Moreover, x-ray powder diffraction (XRD) analysis was performed using a Bragg Brentano mode (Philips PW 3050/60) equipped with a Cu K (λ = 0.15406 nm) radiation source and measuring the XRD data (10-90° 2θ, 0.02° steps, and 30 s/step). Minerals were identified using a search-match program (QualX PC-based match software), which was later validated using XRD Rietveld profile refinements (Program X'Pert plus version 1.0-Philip Analytical B.V)[11], [12]. The AMCSD model (American Mineralogist Crystal Structure Database) was used in the refinement program[13]. The XRD-Rietveld method was used to determine the relative weighted fractions of each phase of the PCC, which includes that the sum of the weighted percent amounts of each crystalline phase be 100 % [14]. Instead, a Hitachi H-7100 SEM (scanning electron microscope) was used to examine the morphology of PCC products (20.0 kV; 1.00000 nA). Gold was coated onto the powder's surface. Furthermore, FTIR spectrometry (IR-Prestige88) assisted in the analysis of PCC product formation. Collection of all spectra was performed at 200 scans/s with a spectral resolution of 2 cm^{-1} in bands ranging from 400 to 4000 cm^{-1}.

3. Results and discussion

3.1. Raw Green Mussel Shell Composition

The results of XRF analysis of ground green mussel shells are presented in Table 1, confirming that the shells contain a rich calcium source (16.3 wt. percent) with minor elements less than 0.3 wt. percent. After converting wt. percent elements to 100%, the most abundant chemical element is clearly calcium (72.3496 wt. percent), which may be associated with the calcium carbonate mineral.

Table 1. The results of XRF analysis of Ground Green Mussel Shells.

Element	Wt.%	Wt.%
Mg	0.1490	0.6614
Al	0.0446	0.1980
Si	0.3220	1.4292
P	5.3300	23.6578
S	0.1450	0.6436
CI	0.0765	0.3396
K	0.0794	0.3524
Ca	16.3000	72.3493
Fe	0.0536	0.2379
Zn	0.0087	0.0386
Sr	0.0208	0.0923
Total	22.5296	100

The main composition of waste mussel shells observed in the current study, in addition to salt and protein, was calcium carbonate, implying that subsequent calcination was required to extract the Ca-content in the starting powder[15]. As a result, the ground mussel shell powder was calcined to convert it into calcium oxide while removing the shells' water, salt, mud, and meat. In this case, calcination temperatures of 900 ^{O}C for 5 h were thought to be a good strategy for converting into calcium oxide while emitting carbon dioxide[16].

3.2. XRD Characterization of Precipitated Calcium Carbonate

Figure 1 depicts the resulting PCC products at various stirring times as determined by the XRD Rietveld method. The carbonation process produced primarily calcite after only 30 minutes of stirring. There were no other peaks visible in the XRD spectra for that sample. However, when the time was increased from 60 to 90 minutes, aragonite appeared over calcite. Brucite, on the other hand, is the only mineral found in the sample obtained at a stirring time of 120 minutes, implying that a longer stirring time allowed magnesium to react with water according to Eq. 2. The synthesis factors of pH change from 11 to 7 and stirring time were determined in this study to contribute to the formation of calcium carbonate polymorphs.

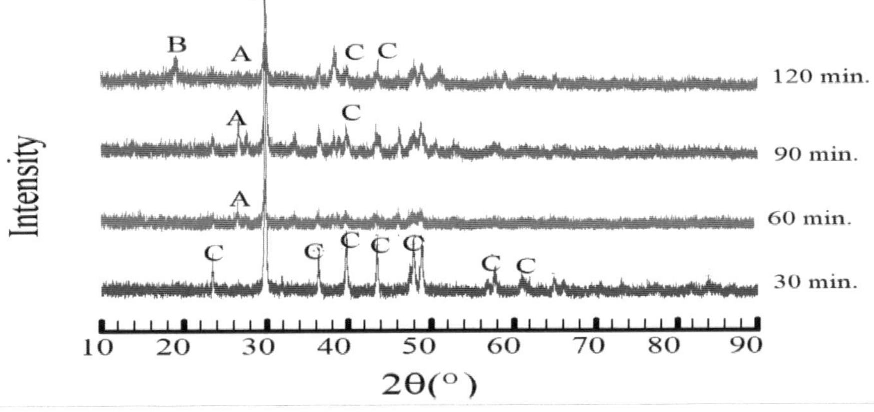

Figure 1. X-ray diffractograms of synthesized PCC at various stirring times. A is for aragonite, B is for brucite, and C is for calcite, respectively.

The phase abundance of PCC products obtained during the study is shown in Table 2. In the experiments, Portlandite [$Ca(OH)_2$] appeared to be converted primarily into calcite during the short aqueous crystallization process (30 min. stirring time), but minor aragonite other than major calcite appeared at longer subsequent stirring times (60-90 min). Instead of calcite and aragonite, brucite [$Mg(OH)_2$] could be produced abundantly after 120 minutes of stirring. These phases have been shown to be critical in the use of PCC for biomaterial synthesis.

Table 2. XRD phase composition analysis of PCC produced from varying stirring time duration

Sample at stirring time (min)	Mineral		
	Calcite	Aragonite	Brucite[Mg(OH)]$_2$
30	100		
60	67.1(7)	32.9 (8)	
90	75.4 (4)	24.6 (7)	
120	60.6 (6)	10.0 (3)	29.4 (4)

3.3. Morphology Analysis of the PCC Product

SEM was also used to investigate PCC powders derived from mussel shells. Figure 2 depicts the morphology of the PCC product, which was primarily rhombohedrallic calcite crystal aggregates produced after 30 minutes of stirring. As the stirring time (60-90 minutes) increased, rod-like aragonite precipitated alongside calcite. After 120 minutes of stirring, a SEM micrograph of brucite with aggregate on plate was also obtained. The current study's findings are consistent with previous reports[6], [7].

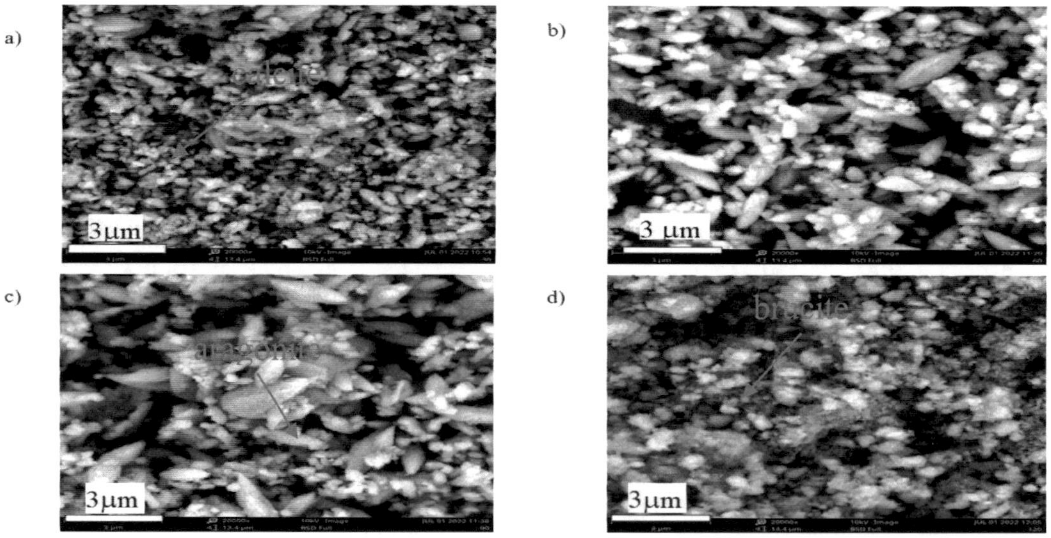

Figure 2. SEM micrographs of PCC powder products obtained at various stirring times of; a) 30 min; b) 60 min; c) 90 min; d) 120 min, magnification 10000x

3.4. Analysis of Using FTIR Method

Figures 3 & 4 depict the FTIR analysis results for PCC powders. Because of the presence of hydrogen bonding, the peaks had a positional shift in comparison to standard ones[17]. The FTIR spectra primarily showed calcium carbonate at 873.31 and 712.33 cm^{-1} bands. Furthermore, the Ca-O and C-O bonds at

1405.80 and 1795.07 cm^{-1} bands could be matched with the calcium carbonate spectra of the samples' FTIR standard.

FTIR analysis could distinguish the types of minerals formed in various PCC products. Aragonite corresponding bands of 700 to 900 cm^{-1} have been confirmed. Strong relative spectra relating to O-H bonds at around 3400-3600 cm^{-1} could be attributed to the presence of $Mg(OH)_2$ with a sharp band of 3354.09 cm^{-1}. Furthermore, the 711.11cm^{-1} band may be related to aragonite[17].

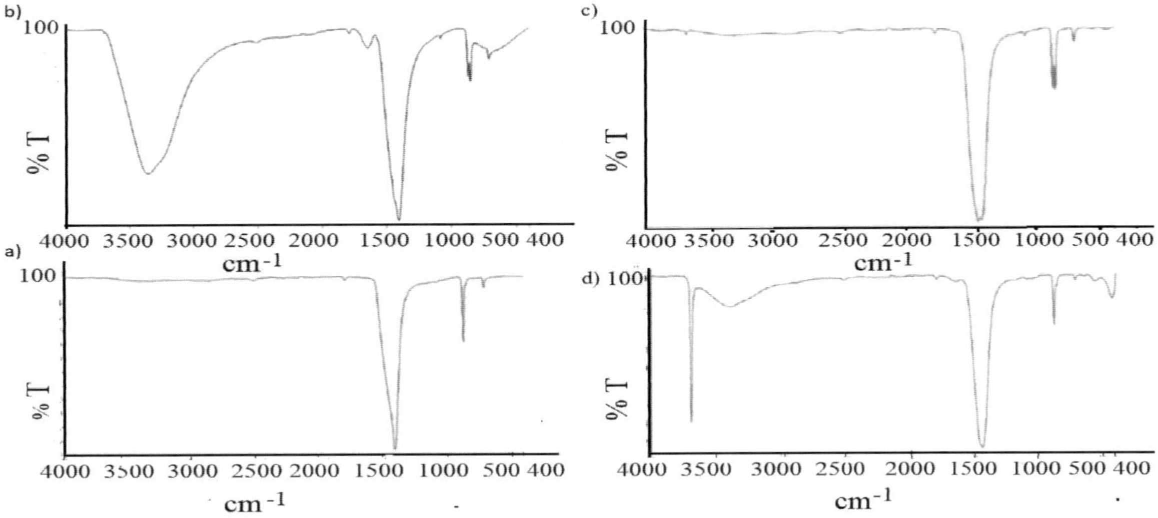

Figure 3. FTIR analysis of PCC powders synthesized at various stirring times of; a) 30 min; b) 60 min; c) 90 min; d) 120 min

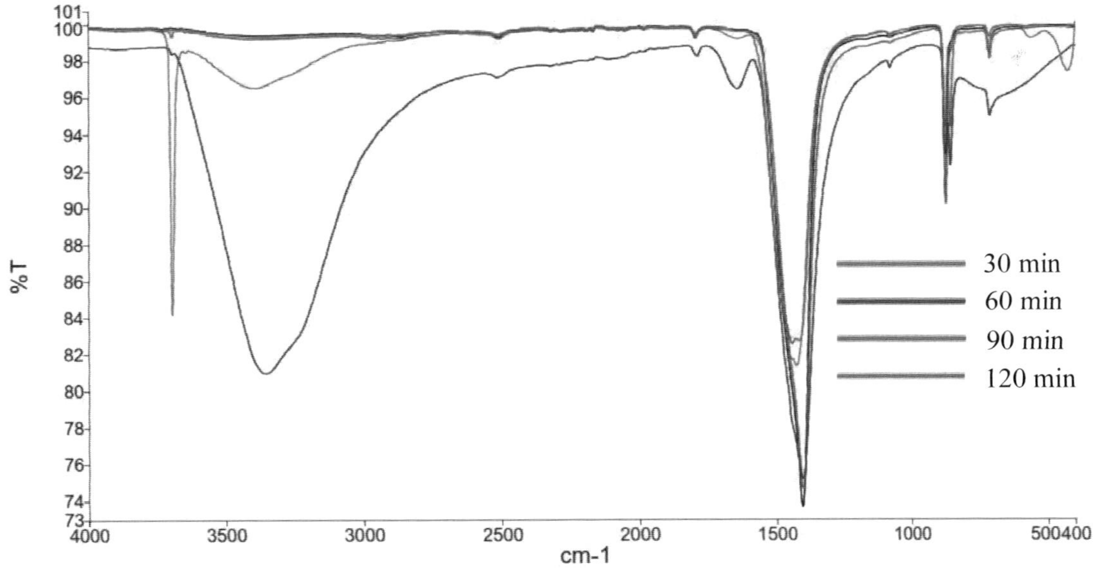

Figure 4. The FTIR analysis of all PCC powders obtained at various stirring times is combined into a single graph.

4. Conclusion

This present research demonstrated a powder processing method for recycling green mussel shells into PCC. The calcium carbonate source in mussel shells was extracted using calcination first, then carbonation. At short stirring times, calcite was the dominant phase, while major calcite and minor aragonite were obtained at stirring times ranging from 60 to 90 minutes. The calcined powder's longer carbonation process produced brucite after 120 minutes of stirring. The calcium carbonate polymorph derived from mussel shells via magnesium chloride leaching was also confirmed by FTIR analysis, which revealed the formation of aragonite, corresponding to a band of 700-900 cm^{-1} obtained after 60 to 90 minutes of stirring. Brucite, on the other hand, was observed in the 3400-3600 cm^{-1} range after 120 minutes of stirring. Calcite formation was primarily observed in the band ranges 873.31 and 712.33 cm^{-1} after 30 minutes of stirring. The findings would contribute to our understanding of how to recycle mussel shells by leaching calcium sources with magnesium chloride.

Acknowledgment

The authors would like to thank the Deputy for Strengthening Research and Development, the Ministry of Research and Technology, Higher Education, and the Republic of Indonesia for funding Ph.D. study grants under contract number: 345-36/UN7.6.1/PP/2022. Special thanks to Mangunwijaya Catholic Polytechnic for allowing the author to use the facilities in the Basic Chemistry Laboratory.

References

[1] R. Chang, S. Kim, S. Lee, S. Choi, M. Kim, and Y. Park, "Calcium Carbonate Precipitation for CO2 Storage and Utilization: A Review of the Carbonate Crystallization and Polymorphism," *Front Energy Res*, vol. 5, no. 17, pp. 1–12, Jul. 2017.

[2] B. Feng, A. K. Yong, and H. An, "Effect of various factors on the particle size of calcium carbonate formed in a precipitation process," *Materials Science and Engineering A*, vol. 445–446, 2007, doi: 10.1016/j.msea.2006.09.010.

[3] Z. S. Hu *et al.*, "Mg2+ content control of aragonite whisker synthesised from calcium hydroxide and carbon dioxide in presence of magnesium chloride," *Materials Science and Technology*, vol. 24, no. 12, 2008, doi: 10.1179/174328407X168865.

[4] P. L. Althoff, "Structural refinements of dolomite and a magnesian calcite and implications for dolomite formation in the marine environment," *American Mineralogist*, vol. 62, no. 7–8, pp. 772–783, Aug. 1977.

[5] V. Ferrini, C. de Vito, and S. Mignardi, "Synthesis of nesquehonite by reaction of gaseous CO2 with Mg chloride solution: Its potential role in the sequestration of carbon dioxide," *J Hazard Mater*, vol. 168, no. 2–3, 2009, doi: 10.1016/j.jhazmat.2009.02.103.

[6] J. Ma, H. Y. Liu, and Y. F. Chen, "Reuse of MgCl2 as crystal controlling agent in the synthesis process of needle-like calcium carbonate," *Wuji Cailiao Xuebao/Journal of Inorganic Materials*, vol. 26, no. 11, 2011, doi: 10.3724/SP.J.1077.2011.01199.

[7] R. M. Santos, M. Bodor, P. N. Dragomir, A. G. Vraciu, M. Vlad, and T. van Gerven, "Magnesium chloride as a leaching and aragonite-promoting self-regenerative additive for the mineral carbonation of calcium-rich materials," *Miner Eng*, vol. 59, 2014, doi: 10.1016/j.mineng.2013.07.020.

[8] F. Liendo, M. Arduino, F. A. Deorsola, and S. Bensaid, "Nucleation and growth kinetics of CaCO3 crystals in the presence of foreign monovalent ions," *J Cryst Growth*, vol. 578, 2022, doi: 10.1016/j.jcrysgro.2021.126406.

[9] W. Kurdowski and S. Duszak, "Changes of C-S-H gel in strong chloride solution," *Advances in Cement Research*, vol. 7, no. 28, 1995, doi: 10.1680/adcr.1995.7.28.143.

[10] L. Xiang, Y. Wen, Q. Wang, and Y. Jin, "Synthesis of dispersive CaCO3 in the presence of MgCl2," *Mater Chem Phys*, vol. 98, no. 2–3, 2006, doi: 10.1016/j.matchemphys.2005.09.015.

[11] P. Y. Mahieux, J. E. Aubert, M. Cyr, M. Coutand, and B. Husson, "Quantitative mineralogical composition of complex mineral wastes - Contribution of the Rietveld method," *Waste Management*, vol. 30, no. 3, 2010, doi: 10.1016/j.wasman.2009.10.023.
[12] R. S. Winburn, D. G. Grier, G. J. McCarthy, and R. B. Peterson, "Rietveld quantitative X-ray diffraction analysis of NIST fly ash standard reference materials," *Powder Diffr*, vol. 15, no. 3, 2000, doi: 10.1017/S0885715600011015.
[13] R. T. Downs and M. Hall-Wallace, "The American Mineralogist crystal structure database," *American Mineralogist*, vol. 88, no. 1, 2003, doi: 10.5860/choice.43sup-0302.
[14] R. J. Hill and C. J. Howard, "Quantitative phase analysis from neutron powder diffraction data using the Rietveld method," *J Appl Crystallogr*, vol. 20, no. 6, 1987, doi: 10.1107/S0021889887086199.
[15] M. C. Barros, P. M. Bello, M. Bao, and J. J. Torrado, "From waste to commodity: transforming shells into high purity calcium carbonate," *J Clean Prod*, vol. 17, no. 3, 2009, doi: 10.1016/j.jclepro.2008.08.013.
[16] A.P. Bayuseno, A.I. Prasetya, R. Ismail, B. Setiyana, and J.Jamari, "REUSE OF WASTE CRAB SHELLS FOR SYNTHESIS OF CALCIUM CARBONATE AS A CANDIDATE BIOMATERIAL," *Rasayan J. Chem*, vol. 15, no. 1, pp. 523–528, Mar. 2022.
[17] N. v. Vagenas, A. Gatsouli, and C. G. Kontoyannis, "Quantitative analysis of synthetic calcium carbonate polymorphs using FT-IR spectroscopy," *Talanta*, vol. 59, no. 4, 2003, doi: 10.1016/S0039-9140(02)00638-0.

Utilization of green mussel shell waste for calcium carbonate synthesis through the carbonation method with temperature variation

A Prihanto*[1], S Muryanto[2], R Ismail[3], J Jamari[3] and AP Bayuseno[3]

[1] Doctoral student at the Department of Mechanical Engineering, Diponegoro University, Tembalang, Semarang, Indonesia

[1] Chemical Engineering Vocational Program, Catholic Polytechnic Mangun Wijaya, Semarang, Indonesia

[2] Department of Chemical Engineering, UNTAG University in Semarang, Indonesia

[3] Department of Mechanical Engineering, Diponegoro University, Semarang, Indonesia

*antoniusprihanto@ymail.com

Abstract. In this study, PCC (precipitate calcium carbonate) was synthesized from green mussel shell waste via calcination and subsequent carbonation methods. Organic substances were removed from green mussel shell powder using a 5-hour calcination at 900 °C. Furthermore, the carbonation method was used in the Ca $(NO_3)_2$ solution at constant stirring speed with pH control by NH_4OH, followed by the injection of carbon dioxide at 50, 70, and 90 °C temperature variations to precipitate calcium as $CaCO_3$ (PCC). According to Rietveld's quantitative XRD analysis, PCC products at 50 °C, 70 °C, and 90 °C exhibited primarily calcite and aragonite phases, with a significant needle-like morphology of aragonite growth during synthesis. Aragonite growth appears to have increased with increasing temperature. The results show that a simple, low-cost approach to green recycling works.

1. Introduction

The management of solid and liquid industrial waste is a major concern for governments all over the world. Waste management efforts to reduce environmental impacts necessitate large sums of money. In this context, waste recycling is one of the most effective alternatives for reducing environmental impact while also adding socioeconomic value to the population. One type of waste that contributes to pollution is marine waste, such as Asian green mussel shells (*P.viridis*). This waste will have a negative impact on the environment if it is not properly managed. Shell waste management is a major issue in a number of countries. Clam shells, for example, are the main component of industrial waste in New Zealand, accounting for 90 percent of factory waste of around 100,800 m^3 per year [1].

In recent years, the introduction of new methods for producing desirable composite materials has significantly increased the recycling of various types of shell byproducts. Shell waste has been extensively tested as a substitute for aggregate in construction materials [2, 3]. Many studies have been conducted to investigate the incorporation of this shell waste into composites such as concrete [4, 5], cement mixtures [6, 7], artificial stone, and bricks [8, 9].

Shell waste recycling is also widely used in many aspects of daily life. Mussel shell waste is used in agriculture as a pH regulator, a bacterial agent with a high antibacterial effect. It is used as an adsorbent and in the synthesis of biodiesel in the chemical industry [10, 11]. Shellfish waste has been used in the environmental field to remove dyes and heavy metals from wastewater [12]. Shell ash also has a high adsorption capacity of more than 94% for metals such as Hg (II) and As (V) [13], metallic lead (Pb), and cadmium (Cd) from contaminated soils [14, 15]. By converting this material into a sustainable reagent for efficient phosphate removal from wastewater and setting up fertilizer plants for recycling oysters, Korean industry has solved the problem of water eutrophication [16, 17]. Peru has been testing the extraction of lime from seashells for building materials [18].

Further, Asian green mussel shells are a biological waste with a high carbonate mineral concentration [19, 20]. These shells account for approximately 31-33 percent of the total weight of mussels, which are typically discarded in landfills [21, 22]. Shells, on the other hand, are a potential composite biomaterial that contains 95-99 wt. percent calcium carbonate, with the remaining 1-5 wt. percent composed of organic compounds [23]. The shell can be ground into a powder and used to produce calcium carbonate, which can then be used for a variety of other applications.

Many scientists from various fields are interested in the reuse of mussel shells for the synthesis of calcium carbonate (PCC) [24] Various phases of calcium carbonate synthesized from green mussel shells can be reacted with phosphoric acid in an aqueous medium at room temperature in a short time [25,26] to form monocalcium phosphate monohydrate [$Ca(H_2PO_4)_2 \cdot H_2O$]. This is significant because calcium carbonate can be found in a variety of industrial and application [27].

Several synthesis routes and methods for producing calcium carbonate precipitates in the laboratory have been proposed in recent years, with or without a focus on specific applications or simply to conduct basic research on crystal precipitation processes [28]. Rhombic calcite, needle-like aragonite, and spherical vaterite are the three types of anhydrous calcium carbonate. Under normal conditions, calcite is known to be the most stable of these polymorphs [29]. The formation of one of these three polymorphs is highly dependent on several parameters, including solution temperature, saturation, and pH [30, 31]. However, little research has been conducted into the effect of temperature on the formation of specific calcium carbonate polymorphs and morphology.

The current study concentrated on the utilization of green mussel shell waste in the synthesis of calcium carbonate via temperature-varying calcination and carbonation processes. Calcination is a straightforward method for producing calcium oxide powder required for subsequent carbonation that involves dissolving CO_2 in nitric acid and passing it through calcined powder. This method is simple and inexpensive, but it has the potential to yield extremely high PCC yields. The effect of temperature as a process parameter on the polymorphic PCC product was investigated in this study, followed by material characterization using XRD (X-ray diffraction) and SEM (scanning electron microscopy), respectively. This powder processing strategy will provide depth understanding on recycling Asian green mussel shells.

2. Methodology

2.1. Powder processing of green mussel shells

Green mussel shells were collected as waste in the fishing village of Tambak Lorok Semarang. After that, the shells were washed, sun-dried, ground, and sieved through a 100-mesh sieve (Endecott, UK). Table 1 shows the elemental composition of the mussel shell powder obtained before calcining as determined by XRF. The shell powder was then calcined for 5 h at 900 ^{O}C in electric furnace (Thermolyne 1400). During the calcination of green mussel shell powder, the following reaction occurs, resulting in calcium oxide (Eq.1).

$$CaCO_3 \rightarrow CaO + CO_2 \quad (1)$$

Table 1. Elemental composition in green mussel shell

No	Element	wt.%
1	Al	0.0272
2	Si	0.257
3	P	0.0373
4	S	0.0797
5	Cl	0.0313
6	K	0.0429
7	Ca	32.8
8	Fe	0.0310
9	Cu	0.0085
10	Br	0,.0138
11	Sr	0.156

2.2. Powder synthesis

In the experimental study, 5 grams of calcined mussel shell powder were dissolved in a beaker glass with 100 ml of 2M HNO_3 made by diluting HNO_3 Pro Analysis (Merck Germany). By adding an ammonia or ammonium hydroxide solution, the pH of the resulting solution was raised to 11. (NH_4OH). The solution was filtered through filter paper to separate the precipitate formed (Whatman 42). The filtrate was heated to 50, 70, and 90 ^{o}C and flowed with CO_2. When the pH reached 7, the carbonate would be terminated. The calcium carbonate precipitate ($CaCO_3$) obtained was separated using a paper filter, washed, and dried for 2 h in a 110 °C oven. The reaction pathway of the carbonation process is

Dissolving for CaO : $CaO + HNO_3 \rightarrow Ca(NO_3)_2 + H_2O$ (2)

Ionization process : $Ca(NO_3)_2 \rightarrow Ca^{2+} + 2NO_3^{-}$ (3)

Dissolving for CO_2 : $CO_2 + H_2O \rightarrow H_2CO_3 \rightarrow 2H^{+} + CO_3^{-2}$ (4)

Precipitation : $Ca^{2+} + CO_3^{-2} \rightarrow CaCO_3$ (5)

2.3. Characterization of oyster shell powder and PCC

All powders were analyzed using XRD and SEM. The AMCSD crystal structure model (American mineralogist from the crystal structure database) was used to run the Rietveld program [32]. Furthermore, Rietveld XRD performed a quantitative phase analysis by determining the relative weighted fraction of each PCC polymorph [33, 34, 35]. Meanwhile, SEM analysis (Hitachi H-7100) with a voltage of 20.0 kV and a probe current of 1,000000 nA revealed morphology and particle size characteristics for all powder samples.

3. Result and Discussion

3.1. Calcined product derived from mussel shells

In this study, calcium is the most important element required for PCC synthesis. The calcium ion found in green mussel shells may be related to the mineral calcium carbonate (Table 1). As a result of removing water and other organic substances from the shells, shell powders were calcined to obtain pure calcium oxide. It was previously demonstrated that pure calcium oxide could be obtained by calcining mussel shells at 900 °C for 5 h, which was used in the current experiment [19]. After calcination in this condition, the composition of the powder mineralogy phase changed during the experiment. According to the Rietveld XRD method, the calcined shell powder and subsequent hydration under the influence of humidity resulted in portlandite [$Ca(OH)_2$]. As a result, the calcined powder mass (remaining mass) was 63.66 wt., which corresponded to approximately 30 % of the weight loss of mussel shells, as previously reported [19]. These results are also consistent with previous thermo-gravimetric analysis, which showed that calcium carbonate decomposes at 900 °C with a mass loss of nearly 30 % [37, 38].

3.2. Characterization of calcium carbonate precipitate

According to Eq. 5, PCC appears to be the result of calcium ions in the solution reacting with carbon dioxide. The synthesis factor of temperature variations of 50, 70, and 90 °C was regulated under constant conditions of CO_2 flow rate and pH range of 7 to 11. The x-ray diffraction spectra of all PCC products are shown in Figure 1, confirming that calcite and aragonite phases could be generated during the synthesis study.

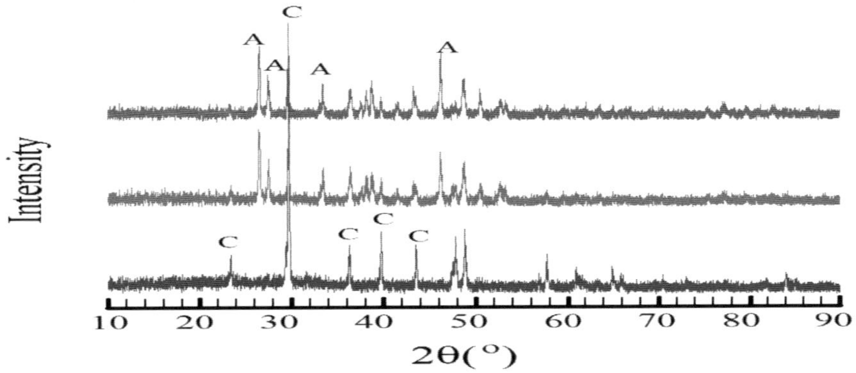

Figure 1. X-ray diffractograms of PCC products obtained at varying temperatures; 50 °C (blue), 70 °C(red) and 90 °C (green) with A = Aragonite and C = Calcite

Furthermore, according to Eq. 5, portlandite appears to be transformed into aragonite and calcite during CO_2 injection into the solution at various temperatures. Furthermore, the mineral composition of PCC powder was quantitatively determined for various set temperatures using the Rietveld XRD method, with the results shown in Table 2. It appears that increasing the carbonation temperature has the potential to increase aragonite production. Aragonite has grown in importance as a PCC in drug and biomedical applications [26]. The economic feasibility of PCC recycling product may be implied from the use of 5 g calcined powder in the experiment that approximately 6.461 g of PCC could be produced by a nearly 30 percent increase in production for a 0.5 h-batch process. This approach could be expanded for PCC product improvement by controlling process variables like CO_2 flow rate and temperature.

Table 2. XRD phase composition of PCC obtained at varying temperatures

Sample synthesized at temperature (°C)	Calcium carbonate polymorph	
	Aragonite	Calcite
50	23.4(9)	76.6(5)
70	70.6(2)	29.4(7)
90	73.1(5)	26.9(6)

The amount of each phase grown is the goal of cost-effective mussel shell recycling, and morphology is a key variable in calcium carbonate synthesis. Calcite has a cubic crystal morphology in SEM micrographs, whereas aragonite has a flower or needle-like morphology. According to the SEM analysis, when the PCC product was deposited at 50 °C, it produced crystals in the form of cubes of various sizes (Figure 2a). While carbonation at 70 °C produced crystals with a flower or needle shape and a more uniform particle size (Figure 2b), carbonation at 90 °C produced needle-shaped crystals with a uniform particle size (Figure 2c). As previously reported [39], the mean particle size ranged between 1 and 5 µm.

Figure 2. PCC morphology obtained at carbonation synthesis 50 °C (a), 70 °C (b) and 90 °C (c)

4. Conclusions

This study provides useful information on how to recycle Asian green mussel shells as a feedstock for the synthesis of precipitated calcium carbonate (PCC) with the desired phase and morphology. Calcium oxide was produced by extracting calcium sources from mussel shells using a quick and simple powder processing method. Carbonation at various temperatures produced PCC of varying morphology and particle size. The most common calcium carbonate phases found during the study were aragonite and calcite. The increasing aragonite product is primarily controlled by temperature increases. This discovery could help us better understand the powder processing strategy for preparing waste mussel shells for biomaterials.

Acknowledgment

The authors would like to express their heartfelt gratitude to Diponegoro University for awarding the world-class university research (WCUR) Grant-No: 118-32/UN7.6.1/PP/2022 to carry out this research.

References

[1] Barnaby C 2004 An investigation into the reuse of organic waste produced by the New Zealand Mussel Industry
[2] Yoon H, Park S, Lee K, Park J 2004 *Waste Manag. Res.* **22** 158-170
[3] Yoon G-L, Kim B-T, Kim B-O, Han S-H 2003 *Waste Manage.* **23** 825-834
[4] Olivia M, Mifshella AA, Darmayanti L 2015 *Procedia Eng.* **125** 760–764
[5] Erni S, Gagoek H, Purwanto K 2016 *J. Teknol.* **78** 203–207
[6] Seo JH, Park SM, Yang BJ, Jang JG 2019 Calcined oyster shell powder as an expansive additive in cement mortar *Materials* (Basel) **12** 8
[7] Chen D, Zhang P, Pan T, Liao Y, Zhao H 2019 *J. Clean. Prod.* **237** 117811
[8] Li G, Xu X, Chen E, Fan J, Xiong G *J. Clean. Prod.* **91** 279–287
[9] Silva TH, Mesquita-Guimarães J, Henriques B, Silva FS, Fredel MC 2019 *Resources* **8** 1

[10] Zheng H, Ye Y, Xia M, Yao Z, Li H, Chen T 2013 Bivalve shell: not an abundant useless waste but a functional and versatile biomaterial *Crit. Rev. Environ. Sci. Technol.* **44** 2502–2530
[11] Perea A, Kelly T, Hangun-Balkir Y 2016 *Green Chem. Lett. Rev.* **9** 1
[12] Papadimitriou CA, Krey G, Stamatis N, Kallianiotis A 2017 *J. Chem. Technol. Biotechnol.* **92** 8
[13] Seco-Reigosa JC, Peña-Rodríguez N, Nóvoa-Muñoz S et al 2013 *Environ. Sci. Pollut. Res.* **20** 2670– 2678
[14] Ok M, Oh YS, Ahmad S 2010 *Environ. Earth Sci.* **61** 1301–1308
[15] Hong J, Kim CO, Gutierrez SY et al 2010 *Biol. Fertil. Soils* **46** 491–498
[16] Cheon S-M, Song T-W. 2003 *J. Korean Ceram. Soc.* **40** 1235-1240
[17] Kwon H-B, Lee C-W, Jun B-S, Yun J-D, Weon S-Y, Koopman B 2004 *Resour. Conserv Recycl* **41** 75–82
[18] Villarrial RAM, Córdova MGF 2021 *Revista Ingeniería de Construcción RIC* **36** 380-388.
[19] Barros MC, Bello PM, Bao M, Torrado, JJ 2009 *J. Clean Prod.* **17** 400- 407
[20] Magnabosco G, Giuri D, Bisceglie APD, Scarpino F, Fermani S, Tomasini C, Falini G 2021 *ACS Sustainable Chem. Eng.* **9** 6203-6208
[21] Francis AA, Abdel Rahman MK 2016 *J. Clean Prod.* **137** 1432-1438
[22] Fukui K, Arimitsu N, Kidoguchi S, Yamamoto T, Yoshida H 2009 *J. Hazard Mater* **163** 391-395.
[23] Wan M-c, Qin W, Lei C, Li Q-h, Meng M, Fang M, Song W, Chen J-h, Tay F, Niu L-n 2021 *Bioact. Mater.* **6** 4255-4285
[24] Hamester MRR, Balzera PS, Becker D 2012 *Mater. Res.* **15** 204-208
[25] Ismail R, Fitriyana DF, Santosa YI, Nugroho S, Hakim AJ, Al Mulqi MS, Jamari J, Bayuseno AP 2021 *J. Cryst. Growth* **572** 126282
[26] Prihanto A, Fitriyan DF, Muryanto S, Masykur I, Ismail R, Jamari J, Bayuseno AP 2021 *J. Environ. Chem. Eng.* **9** 106913
[27] Mititelu M, Stanciu G, Drăgănescu D, Ionit AC, Neacs SM, Dinu M, van Staden R-IS, Morosan E 2022 *Mar. Drugs* **20** 25
[28] Mantilaka MMMGPG, Karunaratne DGGP, Rajapakse RMG, Pitawala HMTGA 2013 *Powder Technol.* **235** 628–632
[29] Knez S, Klinar D, Golob J 2006 *Chem. Eng. Sci.* **61** 5867–5880
[30] Ibrahim A-R, Vuningoma J, Huang Y, Wang H, Li J 2014 *Int. J. Mol. Sci.* **15** 11350
[31] Montes-Hernandez G, Daval D, Chiriac R, Renard F 2010 *Cryst. Growth Des.* ***10*** *4823–4830*
[32] Downs RT, Hall-Wallace M 2003 The American mineralogist crystal structure database *Am. Mineral* **88** 247
[33] Hill RJ, Howard CJ 1987 *J. Appl. Crystallogr.* **20** 467-474
[34] Winburn RS, Grier DG, McCarthy GJ, Peterson RB 2000 *Powder Diffr.* **15** 163-172
[35] Mahieux PY, Aubert JE, Cyr M, Coutand M, Husson B 2010 *Waste Manage.* **30** 378-388
[36] Krajewska, B 2018 *J. Adv. Res.* **13** 59–67
[37] Ren FZ, Wan XD, Ma ZH, Su J.H 2009 *Mater. Chem. Phys.* **114** 367-370
[38] Kamba AS, Ismail M, Ibrahim TAT, Zakaria ZAB 2013 *J. Nanomater.* 10.1155/2013/398357
[39] Feng B, Yong AK, An H 2007 *Mater. Sci. Eng.* **46** 170–179
[40] Goodwin AL, Marc Michel F, Brian L, Phillips BL, Keen DA, Martin T, Dove MT, Reeder RJ 2010 *Chem. Mater.* **2** 3197-3205
[41] Biradar S, Ravichandran P, Gopikrishnan R, Goornavar V, Hall JC, Ramesh V., Baluchamy S, Jeffers RB, Ramesh GT 2011 *J. Nanosci. Nanotechnol.* **11** 6868-6874
[42] Parakhonskiy B, Zyuzin MV, Yashchenok A, Carregal-Romero S, Rejman J, Möhwald H, Parak WJ, Skirtach AG 2015 *J. Nanobiotechnol.* **13** 1-13
[43] Svenskaya YI, Pavlov AM, Gorin DA, Gould DJ, Parakhonskiy BV, Sukhorukov G B 2016 *Colloids and Surf. B. Biointerfaces* **146** 171-179
[44] Chang R, Kim S, Lee S, Choi S, Kim M, Park Y 2017 *Front. Energy Res.* **5** 1-12
[45] Trushina DB, Bukreeva TV, Kovalchuk MV, Antipina MN 2014. *Mater. Sci. Eng.* C 45: 644- 658

[46] Kalaimani N, Muraleedharan V, Nair TSU, Joseph KG 1984 *Preparation of lime from mussel shell waste*. Fishery technology, India: Society of Fisheries Technologists. 21: 142

A Review on household water end-use consumption to mitigate the urban water crisis: insight from various cities in Indonesia

R N Azizah[1,2*], L F Sinaga[1,2], D Awfa[1,2], A D Imami[1,2], R Yanda[1,2], Z F Saraswati[3,4], Y Putri[1,2], E Setiawati[1,2]

[1]Department of Environmental Engineering, Institut Teknologi Sumatera, Jl. Terusan Ryacudu, Lampung Selatan, Indonesia 35365
[2]Center of Research and Innovation for Environment and Sanitation, Institut Teknologi Sumatera, Jl. Terusan Ryacudu, Lampung Selatan, Indonesia 35365
[3]Department of Regional and City Planning, Institut Teknologi Sumatera, Jl. Terusan Ryacudu, Lampung Selatan, Indonesia 35365
[4]Center of Research and Innovation for Sustainable Infrastructure, Institut Teknologi Sumatera, Jl. Terusan Ryacudu, Lampung Selatan, Indonesia 3536

rifka.azizah@tl.itera.ac.id

Abstract. The present paper reviews the recent perspectives of end-use studies as a part of urban water crisis mitigation within the past decade in various regions in Indonesia. The paper integrates a critical review of water-end use consumption from five major islands in Indonesia. Comparative methods were used to determine the domestic water consumption behaviour from the collected data from 2013 to 2020. Furthermore, the collected data were classified into two major groups (i.e., regions on Java Island and regions on non-Java Island). At the end, this review reflected a different pattern between domestic water consumption in various socio-economic regions in Indonesia and expected to provide strategies that could be used to mitigate the urban water crisis in the future.

1. Introduction

The current water crisis is a severe concern both globally and nationally. Due to population growth occurring every day, the need for water to supply the domestic, agricultural and industrial sectors is also increasing. In addition, the water crisis is also caused by water pollution, both surface water and groundwater. This problem makes it difficult for the community to access water in terms of quality and quantity as determined by WHO[1],[2]. Climate change also plays a role in causing water scarcity [3],[4]. The factors mentioned above make people look for alternative water resources with all efforts. If people's efforts go unchecked, the reckless use of water will bring catastrophic changes to the world. More than half of the world's population is projected to be chronically water-deficient by 2050 [5], and governments and policymakers worldwide are focusing on reducing water consumption [6].

For Indonesia, various efforts have been made to prevent water scarcity, such as reuse water, land use planning, minimizing erosion and reclamation, and peatland utilization [7]–[10]. In addition, to prevent other water scarcity, the water-energy nexus approach can also be carried out, namely in the form of mitigating the release of greenhouse gases by reducing household water end-uses [11]. Water end-use studies can separate the quantity and frequency of household-scale water use. This study is widely used as a reference for urban recycling plans and strategies.

Water end-use studies were undertaken in several parts of the world, indicating that parameters like geography, proximity to water sources, education, socio-economic development, environmental beliefs, population size, and climatic conditions govern water consumption patterns within a population [6]. Determinant socio-demographic factors such as income, gender, education and geographical conditions create accuracy for better urban water management.

The economic center is centralized on Java island, making Indonesia have different conditions from one region to another [12]. As a result of this condition, the use of water in Java and non-Java areas tend to be different. For example, water consumption for laundering and bathing differs among each region's income settings. This is made possible by how the community engages in activities such as bathing and laundering (using machines or tools) [6]. On this basis, it is necessary to know how differences in social demographic and economic conditions affect the quantity of end-use water to provide a regional policy approach.

2. Methodology

End-use data of water consumption collected is data from year 2013 to 2020. The data is classified into three regional groups: Java, Non-Java and Asia. For Java, data represent an area with a dense population, and non-Java data represent an area whose population is spread out. Meanwhile, the Asian group is used to compare water consumption conditions outside Indonesia. The group classification is also used to show different social, demographic and economic conditions, where the dominant condition of Regional Budget Receipts (PAD) is high in the Java area and other variations for each different social and demographic condition.

Furthermore, for water end-use consumption, the distribution consists of six types. The type of water end-use selected is the type that has the most significant nominal proportion of water end use per person per day. The types are drinking, cooking, laundry, bathing, and toilet flushing. The type of water end-use that is not written is referred to as any other uses.

3. Results and discussion

3.1. Socio-demographic characteristics

3.1.1. Regional budget receipts

Economic factors such as regional budget receipts (i.e., income) shown to be associated with the number of end use water consumptions [13], [14]. High number on end use water consumption tends to be affected by the higher income due to the improvement of life quality to obtain high quality food/drinks and health care [15]. It is clearly shown that the regional budget receipts in Java Island is relatively higher than Non-Java Island (Figure 1). High regional budget receipts could change the domestic, municipal, and industrial sectors in a city; which might cause higher urbanization rates followed by the acceleration to exploit natural resources [16]. On the other hand, the study conducted by Ostos [17] in Brazil found that the regional budget receipts could affect the end use water consumption only “at a certain level”. Long term evolution of water intensity in this study (1860-2011) clearly indicates that regional budget receipts did not have a direct effect on water consumption. Other economical aspect such as the amount of investment in the development of facilities, socio-cultural (i.e., how people spend their money), and political issues at stakeholders tend to be the most direct and significant factors. Therefore, exploring and develop the scenario to predict end use water consumption using the significant economical factor is an urgent need to obtain sustainable urban water metabolism planning.

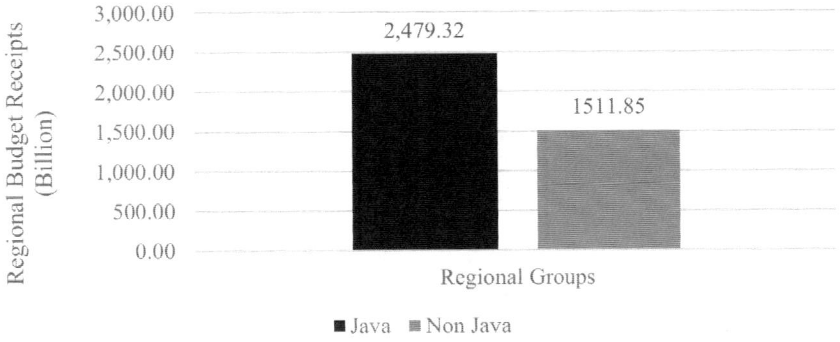

Figure.1 Regional budget receipts Java and Non-Java [18].

3.1.2. *Gender*
The gender proportion itself seen different in this scheme of region. Figure 2 and 3 show in java it almost balance proportion between female and male but in non-Java Region female has a bigger number than a male. The effect of this difference makes a two phenomenon in water consumption. There is finding when water consumption correlated with gender.

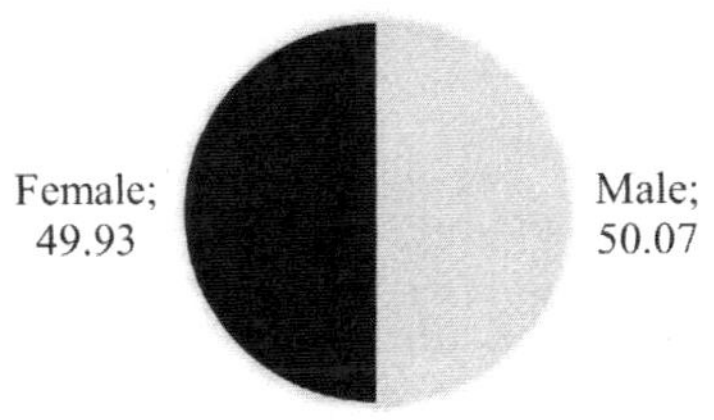

Figure 2. Gender proportion in Java Region

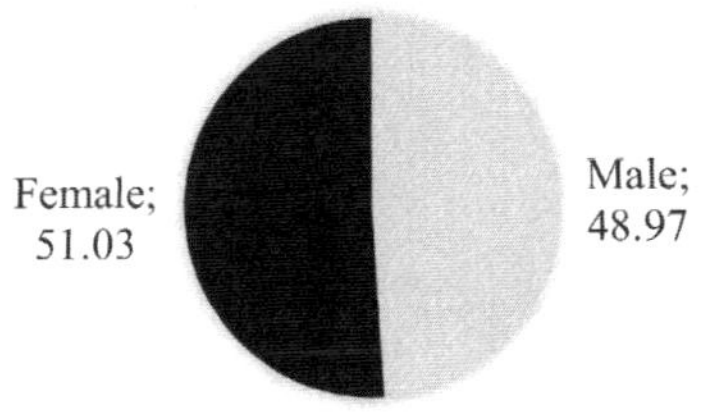

Figure 3. Gender proportion in non-Java Region

Female tend to be more conscious in water conservation [19], it stated because research shown Turkish women practices water saving behavior. It could be related with the assumption that female aware of the water bill they pay. Besides [20], emphasized that women contribute in teaching their children of how water supposed to be consumed. Because of this, gender will always correlate with people responsibility, right, and participation toward conservation.

3.1.3. *Region's Average Temperature*
Water consumption rates depend on the consumer and industrial demand, socio-demographic, water pricing, secondary water losses and many other factors including climatic variables. Climatic variables e.g. precipitation and air temperature classified as short term variables in water consumption particularly throughout the year [21]. Previous studies from [22] in Calgary, [23] in Seoul, and [21] in Belgrade show some association between the region's air temperature and both water production and water consumption. In Calgary, water consumption increases with increasing temperature especially when the temperature is above 15°C. The air temperature factor is believed to influence the in-door use, irrigation, sanitation, public use, and public park sprinkling [22]. Seasonal urban water consumption in Seoul can be explained by a regression model by 61% due to weather variables. The study shows, maximum daily temperatures is a good predictor of water consumption which show positive correlation with the seasonal

water use [23]. [21] conclude that the air temperature also influences water production especially in May and June period which shows by coefficient of determination (R^2) = 0.502 which means about 50% variation of water production can be predicted by the air temperature.

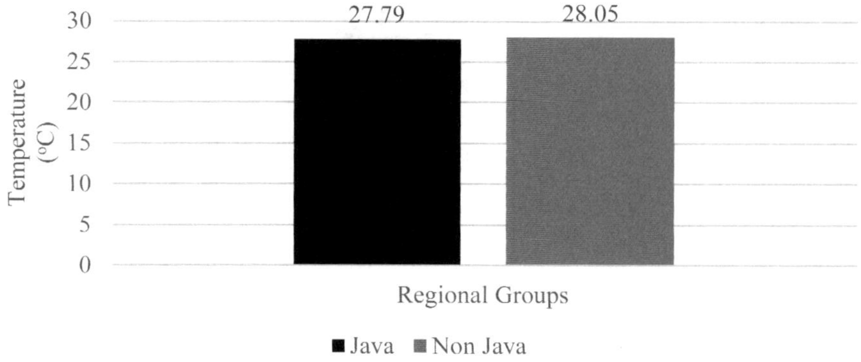

Figure 4. Region group temperature Java and Non-Java [24].

The climate in Indonesia is typically divided into three areas (excluding sea surface warm water) i.e. temperatures on coastal areas are about 28°C on average, 26°C in the inland area, and about 23°C for the higher mountain area [24]. In this study, the two regions (Java and Non-Java) show insignificant differences in average temperature ranging about 27.5°C and 28.5°C (Figure 4). It shows that both water production and water consumption in these regions seems to be similar based on average temperature variables. Other climatic variables can be identified for further research e.g., wind speed, relative humidity, and precipitation.

3.1.4. Higher level of Education (Complete 12 years of education)

Based on literature, socio and demographic factors appear to influence people in water consumption behavior. In Canada water end use research was carried out using 2 major factors; education level and presence of children [25]. Findings by [26] in Annapolis Valley and [25] in Canada notes education was a significant factor in how the people chose their drinking water sources (tap, filtered, or bottled water). Individuals who are educated beyond high school are more likely to drink their municipal tap water than are less educated individuals. Highly educated people are believed to have better ability to digest the scientific information that presents arguments in favor of water quality while less educated respondents will easily fall for health claims made by home filters manufacturers [25].

Relation between education level to water consumption in Poland and Greece [27], showing that all of respondents from Poland with higher education level (high school above) are more likely to have their own property. It influences their water end behaviors like shower time and shower length, gardening, and swimming activities. In Poland (with 100% higher educated respondent), respondent chose more efficient water saving strategies e.g. Insulate hot pipes to save energy and hot water use, buying new 2 button toilet, buying new water saving dishwasher, etc. [27]. It shows that education level also influences people in choosing strategies to saving water.

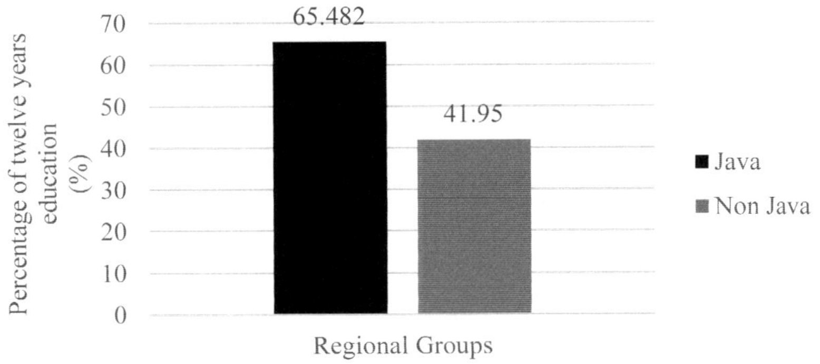

Figure 5. Twelve years education level Java and Non Java [28].

Based on previous research, more water was used by people with higher education due to their additional activities (showering, swimming, gardening, etc.). In this study, respondents with higher education (12 years above) are dissimilar between Java regional and non-Java regional groups. In Java, almost 70% (Figure 5) of respondents complete their high school education. It means there will be different approaches regarding policy and strategy to prevent the water crisis. For the Java region, subsidy or incentive in purchasing a water device will be more effective since the people with higher education backgrounds are willing to pay more for the saving water device. While in Non-Java region, less than a half of respondents are not complete their higher education (58,1%) (Figure 5). Some strategies that may be implemented are water price restructuring, increasing awareness about taking shorter shower time, and collecting rainwater for gardening. Those strategies are known as cost effective strategies that can be implied by people in non-Java Region.

3.2. Water end-use consumption

There is no significant difference for Java, Non-Java and Asia in terms of end use consumption. Bathing is the highest number of liters per person per day over all the categories (Table 1). Compared to any other region like Australia-Pacific, the same pattern appears when it comes to the highest residential end uses. Named showers, clothes washing, irrigation, toilet and tap use in sequence rank [29]–[31]. In the Java region Bathing/Shower takes about half the proportion of all uses in the amount of 55,2 Liter/person/day over 112,3 Liter/person/day in total.

Table.1 Water end-use consumption.

End-uses	Water consumption (Liter/person/day)		
	Java	Non-Java	Asia
Drinking	3.4	3.3	5.3
Cooking	11.3	5.6	12.2
Laundry	22.6	10.1	17.8
Bathing/Shower	55.2	25.6	29.7
Toilet Flushing	14.7	11.5	14.6
Any other use	15.2	24.8	31.2

Meanwhile, one per three of shower uses are shown in the non-Java region. It is less than Java, possibly affected by the environmental attitude of consumers and culture of daily bustle [4]. For a deeper dig of reason, the attitude socio-demographic had linked back and forth through the attitude that was shown in regional budget receipt and educational level (Figure 6). Summarizing the total water consumption over the end use as the data provided in Figure 6 [32]–[41] the same pattern shows that Java region water consumption per person per day was the higher overall region. About 122.3 liters/person/day the value is almost the same as the Department of Health and Public Works based on scale City (Metropolitan) of 126.9 liters/person/day, 150 liters/person/day and 170 liters/person/day. The number of metropolitan cities on the island of Java has almost reached more than 50 cities [42].

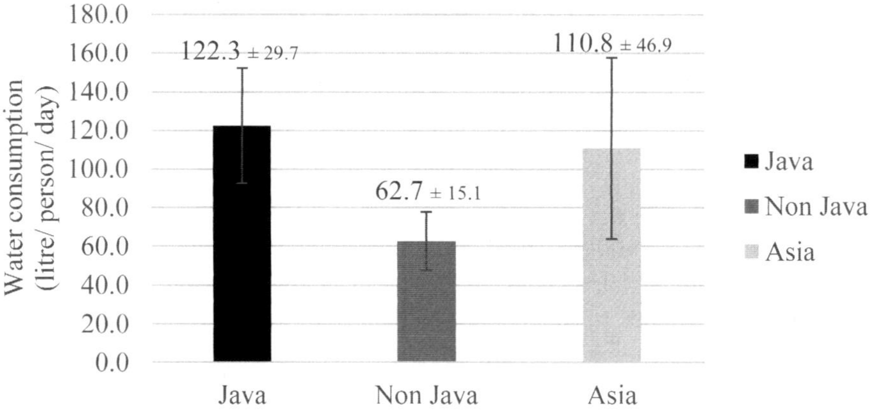

Figure 6. Total Water Consumption.

4. Result and Discussion

This review study concludes that the sociodemographic profile which is separated by centralized made the end-use water consumption has a significant gap. It is also concluded, between two group there is no number of total consumptions meet the government estimation. This study offers insight into the end-use behavior of water consumption at the household level based on region classification and also adopting the end-use proportion of each activity (bathing, cooking, toilet flushing) to predict better estimation for a mitigation effort.

Reference

[1] P. H. Gleick, "Basic water requirements for human activities: Meeting basic needs," *Water Int.*, vol. 21, no. 2, pp. 83–92, 1996, doi: 10.1080/02508069608686494.

[2] Howard Guy and Bartram Jamie, "Domestic Water Quantity, Service Level and Health," *World Heal. Organ.*, p. 38, 2003.

[3] C. Zhan-Ming and G. Q. Chen, "Virtual water accounting for the globalized world economy: National water footprint and international virtual water trade," *Ecol. Indic.*, vol. 28, pp. 142–149, 2013, doi: 10.1016/j.ecolind.2012.07.024.

[4] R. M. Willis, R. A. Stewart, K. Panuwatwanich, P. R. Williams, and A. L. Hollingsworth, "Quantifying the influence of environmental and water conservation attitudes on household end use water consumption," *J. Environ. Manage.*, vol. 92, no. 8, pp. 1996–2009, 2011, doi: 10.1016/j.jenvman.2011.03.023.

[5] A. L. Sønderlund, J. R. Smith, C. Hutton, and Z. Kapelan, "Using smart meters for household water consumption feedback: Knowns and unknowns," *Procedia Eng.*, vol. 89, pp. 990–997, 2014, doi: 10.1016/j.proeng.2014.11.216.

[6] M. Kumar, A. Sharma, N. Tabhani, and Y. Otaki, "Indoor water end-use pattern and its prospective determinants in the twin cities of Gujarat, India: Enabling targeted urban water management strategies," *J. Environ. Manage.*, vol. 288, no. 336, p. 112403, 2021, doi: 10.1016/j.jenvman.2021.112403.

[7] Kiswanto, Wintah, S. Sriwahyuni, and Nurdin, "Post-mining pond water suitability for fisheries culture in West Aceh, Indonesia," *AACL Bioflux*, vol. 15, no. 1, pp. 436–445, 2022.
[8] N. Ihsan Fawzi, A. N. Rahmasary, and I. Z. Qurani, "Minimizing carbon loss through integrated water resource management on peatland utilization in Pulau Burung, Riau, Indonesia," *E3S Web Conf.*, vol. 200, 2020, doi: 10.1051/e3sconf/202020002019.
[9] B. Muis Ariadi, "Land use planning for water resource conservation in the Krueng Aceh watershed, Indonesia," *AES Bioflux*, vol. 11, no. 2, pp. 67–74, 2019.
[10] R. Juniah, "Potential Economic Value of Water Resource Sustainability for Sustainable Environment : A Case Study in South Sumatra , Indonesia," pp. 1–23, 2021.
[11] R. P. Adriani, G. A. Kristanto, and M. A. Pratama, "Greenhouse gas emission estimations for Depok's (West Java, Indonesia) middle-class household water end-uses," *IOP Conf. Ser. Earth Environ. Sci.*, vol. 423, no. 1, 2020, doi: 10.1088/1755-1315/423/1/012042.
[12] N. A. Phelps, T. Bunnell, M. A. Miller, and J. Taylor, "Urban inter-referencing within and beyond a decentralized Indonesia," *Cities*, vol. 39, pp. 37–49, 2014, doi: 10.1016/j.cities.2014.02.004.
[13] K. Hubacek, D. Guan, J. Barrett, and T. Wiedmann, "Environmental implications of urbanization and lifestyle change in China: Ecological and Water Footprints," *J. Clean. Prod.*, vol. 17, no. 14, pp. 1241–1248, 2009, doi: 10.1016/j.jclepro.2009.03.011.
[14] Z. Pan, Z. Fang, J. Chen, J. Hong, Y. Xu, and S. Yang, "Driving factors of decoupling between economic development and water consumption in food and energy in north-west china—based on the tapio-lmdi method," *Water (Switzerland)*, vol. 13, no. 7, 2021, doi: 10.3390/w13070917.
[15] K. Hubacek, D. Guan, and A. Barua, "Changing lifestyles and consumption patterns in developing countries: A scenario analysis for China and India," *Futures*, vol. 39, no. 9, pp. 1084–1096, 2007, doi: 10.1016/j.futures.2007.03.010.
[16] ilibrary.org/environment/water-consumption-and-sustainable-water-resources-management.
[17] J. R. Ostos and E. Tello, "A long-term view of water consumption in Barcelona (1860–2011): from deprivation to abundance and eco-efficiency" *Water Int.*, vol. 39, no. 5, pp. 587–605, 2014, doi: 10.1080/02508060.2014.951252.
[18] "PAD," [Online]. Available: https://djpk.kemenkeu.go.id/portal/data/apbd.
[19] S. Hablemitoglu and E. Ozmete, "Sustainable water management: A case study on saving behaviour of Turkish women for domestic water usage," *Eur. J. Soc. Sci.*, vol. 12, no. 3, pp. 447–456, 2010.
[20] C. Ngai and V. Nitivattananon, "The Role of Gender in Domestic Water Conservation in Malaysia," *Malaysian J. Environ. Manag.*, vol. 8, pp. 109–129, 2007.
[21] D. Dimkić, "Temperature Impact on Drinking Water Consumption," p. 31, 2020, doi: 10.3390/environsciproc2020002031.
[22] A. Akuoko-Asibey, L. C. Nkemdirim, and D. L. Draper, "the Impacts of Climatic Variables on Seasonal Water Consumption in Calgary, Alberta," *Can. Water Resour. J.*, vol. 18, no. 2, pp. 107–116, 1993, doi: 10.4296/cwrj1802107.
[23] S. Praskievicz and H. Chang, "Identifying the relationships between urban water consumption and weather variables in Seoul, Korea," *Phys. Geogr.*, vol. 30, no. 4, pp. 324–337, 2009, doi: 10.2747/0272-3646.30.4.324.
[24] "BMKG. (2020)." .
[25] D. Dupont, W. L. Adamowicz, and A. Krupnick, "Differences in water consumption choices in Canada: The role of socio-demographics, experiences, and perceptions of health risks," *J. Water Health*, vol. 8, no. 4, pp. 671–686, 2010, doi: 10.2166/wh.2010.143.
[26] J. Janmaat, "Divergent drinking water perceptions in the Annapolis Valley," *Resour. J.*, vol. 32(2), no. Canadian Water, pp. 99–110, 2007.
[27] Y. Shan, L. Yang, K. Perren, and Y. Zhang, "Household water consumption: Insight from a survey in Greece and Poland," *Procedia Eng.*, vol. 119, no. 1, pp. 1409–1418, 2015, doi: 10.1016/j.proeng.2015.08.1001.

[28] "BPS 2020." https://www.bps.go.id.
[29] P. Loh, M., & Coghlan, "Domestic water use study in Perth, Western Australia, 1998-2001," no. Perth: Water Corporation., pp. 1–235, 2003.
[30] S. F. Barker, J. O'Toole, M. I. Sinclair, K. Leder, M. Malawaraarachchi, and A. J. Hamilton, "A probabilistic model of norovirus disease burden associated with greywater irrigation of home-produced lettuce in Melbourne, Australia," *Water Res.*, vol. 47, no. 3, pp. 1421–1432, 2013, doi: 10.1016/j.watres.2012.12.012.
[31] M. Heinrich, "Water end use and efficiency project (WEEP)," no. Branz, 2007.
[32] M. Akbar, "Pola konsymsi air pada perumahan teratur," vol. 12, no. 2, p. 12, 2018.
[33] M. M. D. Widiastuti and P. Mangera, "Studi Pola Konsumsi Air Domestik Masyarakat Multietnis Di Kelurahan Karang Indah Distrik Merauke," *Kritis*, vol. 27, no. 1, pp. 1–15, 2020, doi: 10.24246/kritis.v27i1p1-15.
[34] M. Naslilmuna, C. Muryani, S. Santoso, P. S. Studi, and P. Kependudukan dan Lingkungan Hidup, "Analisis Kualitas Air Tanah dan Pola Konsumsi Air masyarakat Sekitar Industri kertas PT. Jaya Kertas Kecamatan Kertosono Kabupaten Nganjuk," *J. GeoEco*, vol. 4, no. 1, pp. 51–58, 2018.
[35] R. A. Lomi, J. J. Messakh, and P. G. Tamelan, "Pemanfaatan Air Bersih Untuk Kebutuhan Rumah Tangga Dari Mata Air Oelnaisanam Di Kelurahan Bakunase II, Kota Kupang," *J. Batakarang*, vol. 2, no. 1, pp. 32–38, 2020.
[36] Laksono, N. B., & Suprayogi, S. (2019). Analisis Ketersediaan Air Hujan untuk Kebutuhan Air Domestik di DAS Jono, Kecamatan Piyungan, Kabupaten Bantul. *Jurnal Bumi Indonesia*, *8*(3).
[37] Kusumo, B. J., & Priyana, Y. (2016). Pola Konsumsi Air Untuk Kebutuhan Domestik Pada Perusahaan Daerah Air Minum (PDAM) Di Kecamatan Manyaran Kabupaten Wonogiri (Doctoral dissertation, Universitas Muhammadiyah Surakarta).
[38] Kurniasari, Ratna. Distribusi Kuantitas Dan Kualitas Mataair Untuk Memenuhi Kebutuhan Air Domestik Di Kecamatan Jatinom Dan Karanganom, Kabupaten Klaten, Jawa Tengah. pp.1-11, 2019.
[39] M. Al–Amin, K. Mahmud, S. Hosen, and M. A. Islam, "Domestic water consumption patterns in a village in Bangladesh," *4th Annu. Pap. Meet 1st Civ. Eng. Congr.*, no. 4, pp. 83–85, 2011, [Online]. Available: http://www.iebconferences.info/356.pdf.
[40] O. Singh and S. Turkiya, "A survey of household domestic water consumption patterns in rural semi-arid village, India," *GeoJournal*, vol. 78, no. 5, pp. 777–790, 2013, doi: 10.1007/s10708-012-9465-7.
[41] B. PASAKHALA, H. HARADA, S. FUJII, S. TANAKA, B. R. SHIVAKOTI, and S. SHRESTHA, "Household Coping Measures With Water Scarcity: a Case Study in Kathmandu, Nepal," *J. Japan Soc. Civ. Eng. Ser. G (Environmental Res.*, vol. 69, no. 7, p. III_73-III_81, 2013, doi: 10.2208/jscejer.69.iii_73.
[42] "DPPW," 2002. [Online]. Available: Departemen Permukiman dan Prasarana Wilayah.

IOP Conf. Series: Earth and Environmental Science 1098 (2022) 012024 doi:10.1088/1755-1315/1098/1/012024

Upcycle strategy on tree branches to improve eco-efficiency towards a circular economy using life cycle assessment

Sri Hartini*[1], Rani Rumita[1], and Muhammad Hafidz Al Huda[1]

[1]Department of Industrial Engineering, Diponegoro University, Prof Sudarto Tembalang, Semarang, Semarang, Indonesia

*srihartini@lecturer.undip.ac.id

Abstract. A circular economy (CE) is considered capable of being a solution to reduce the use of natural resources and reduce waste without harm. A circular economy system uses products and materials at maximum value and function. Various strategies were developed for this purpose. However, these efforts often require additional energy and materials to create more valuable products. Does the increase in value-creation mean more than the energy consumption and materials to make it happen? This study conducted a study of efforts to increase wood branch waste, which is usually only used for firewood, which is processed into hanger products with an upcycle strategy. This study uses the eco-efficiency method to determine the increase in value in the economic aspect compared to the environmental impact it causes. The cost of environmental impact is calculated using a life cycle assessment with Simapro software. The findings from this study are that upcycle efforts can increase product eco-efficiency. The eco-efficiency index (EEI) of hanger products is 4.512 and the EEI of firewood is 3.816.

1. Introduction

A circular economy (CE) differs from a linear economy (LE). The linear economy applies the economic model using "take, make, use and throw away". Post-use products are underutilized and pollute the environment [1]. LE utilizes extracted product resources as much as possible with efficiency and product life is not maximized [2]. Some argue that the circular economy (CE) can be a solution to reduce the use of natural resources and reduce waste without harm [3]; [4]. CE is a regenerative system that minimizes resources, inputs, waste emissions, and energy leakage by slowing, closing, and narrowing material and energy loops [5]. CE can be achieved through durable design, maintenance, repair, reconditioning, refurbishing and recycling. The CE system uses products and materials at their maximum value and functionality.

Much of the decline in environmental conditions from industrial activities spread globally. Global furniture production in 2016 reached more than 99 million m^3 from 50 countries [6]), but only 40 to 60% of the total volume of raw wood is effectively used, and the rest becomes waste [7]. The burden faced by the traditional wood furniture industry is exacerbated by its inefficient production process resulting in a lot of wood waste [8]. On the other hand, the furniture industry that uses tree branches is starting to develop. So far, efforts to use tree branches have only been used as fuel. This effort is known as an upcycle. Upcycle is a recycling concept to reuse discarded objects or materials without deterioration, resulting in a product that has a higher quality or value than the original [9].

Hanger is one of the products that are upcycled. The reuse of waste from wood tree branches supports the implementation of a circular economy. However, upcycle requires additional energy and materials to create more valuable products. Thus, on the one hand, upcycle efforts will increase added value and reduce waste, but on the other hand, it still requires materials and energy that impact the environment. It is interesting to study whether upcycle efforts in the case of utilizing teak waste will increase the value of eco-efficiency. This study intends to study wood waste's environmental impact and eco-efficiency if it is upcycled into a more valuable product and compare products when tree branch waste is burned.

2. Methodology

2.1. Case Study
PT X is a medium-scale furniture company in Central Java that produces various furniture products from waste teak tree branches. The reason for choosing teak branches as the primary raw material is the low price factor and the ease of availability and processing techniques. Hanger is a product most in demand by consumers whom Europe and America dominate. Product hangers function to hang hats and coats without painting. Finishing is done by leaching to produce a brighter and more natural wood color using white agent 250.

2.2. Circular Economy
Several studies on circular economy in the furniture industry have been carried out previously. There are investigations about Circular Economy practices, barriers, and drivers of CE [10]), the challenges and opportunities [11], the design of a business model [12], the relationship between varying ESCC practices on the CE-targeted performances [13], and the strategy to reduce wood waste [8]. Empirical studies of using tree waste as raw material for furniture products in a circular economy perspective are still rarely carried out. When CE is combined with 6R, a closed-loop material flow can be realized, thus relieving inherent challenges in sustainable manufacturing systems [14].

2.3. Life cycle Assessment
Several stages of LCA will be discussed in this subchapter, namely Goal and scope, Life Cycle Inventory, and Life Cycle Impact Assessment.

2.3.1. Goals and Scopes. This study aims to compare, identify the impact, and determine the environmental costs arising from the upcycle strategy of teak wood waste into a more valuable product, in this case, the hanger product. An analysis of the effect of upcycling on the value of eco-efficiency was also carried out. The scope of the study is Life Cycle Assessment, a hanger production system that includes stripping, material, sanding, assembly, treatment, and finishing. The production process flow chart of the hanger is shown in Fig 1.

2.3.2. Life Cycle Inventory (LCI). Life Cycle Inventory (LCI) shows the inputs (material, energy, and waste) and outputs (product output and non-product output) related to the product throughout the specified life cycle. The inputs include material requirements (kg) and electrical energy (kWh). While the output is in the form of product and non-product output in the form of waste generated, for example, scrap and pulp (kg). Energy and material consumption are needed to measure the environmental impact of a production process. Energy and material consumption can be seen in the following Table 1 and Table 2. The output of the hanger production process in the upcycle process can be seen in Table 3. Electrical energy consumption data is calculated based on the power of the production machine and the time of using the machine as measured by a stopwatch time study.

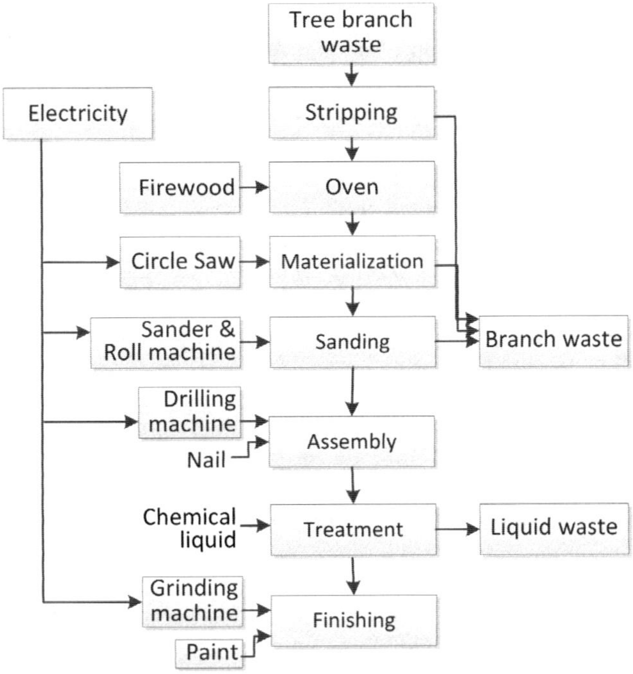

Figure 1. Hanger Production Process Flow Chart (Upcycle)

Table 1. Energy consumption

No	Machine	*Power (watt)*	*Duration (Hour)*	Consumption of energy (kWh)
1	Circular Saw	2200	36,281	79,818
2	Sander	520	21,78	11,326
3	Roll	1200	37,69	45,228
4	Boring	250	0,97	0,243
5	Grinding	450	16,77	7,547

Table 2. Material consumption

No	Material	Unit	Consumption	Function
1	Branches	Kg	666,25	Material
2	Firewood	Kg	450	Burning/oven
3	M8 x 1.5. Screws	Pcs	1400	Assembly
4	Kerosene	L	10	Treatment
5	Chemical material (DTM)	ml	50	Treatment
6	White Agent 250	L	2,5	Finishing

Table 3. The output of the production process

Category	Output		Total
Products Output (PO)	Hanger	Kg	387,5
Non Product Output (NPO)	Wood Waste	Output LCIA Indicator Simapro v 9.1.1.7	
	Electricity	kWh	144,161
	Liquid waste	Liter	0,635

2.3.3. Life Cycle Impact Assessment (LCIA) phase. This phase aims to evaluate how much environmental impact the system has on the product under study. The LCIA phase uses the Eco-cost 2017 version 1.5 method on the Simapro v.1.1.7 software. The advantage of eco-cost is that the output of this calculation is expressed in monetary value (€ or US$), which seems easy to understand directly. The LCIA phase is divided into several stages, namely characterization, normalization, weighting, and a single score. The indicators analyzed are climate change, human health, ecosystem quality, and resources. The units of analysis are 50 hanger units.

2.3.4 Interpretation. Researchers will discuss the results of the analysis at the interpretation stage and summarize them into a conclusion. At this stage, the researcher will usually provide recommendations on the objectives and scope of the research.

2.4. Eco-Efficiency

The results of the LCA measurement are information to measure the Eco-efficiency index as one of the parameters in a clean production strategy. Eco-efficiency index is the environmental cost effectiveness which is determined from the comparison of net value, production costs, and Eco-costs [17], [[18], [19]. Eco-efficiency is an indicator of the sustainability of a system in several studies. If the eco-efficiency value is negative, the production system is included in the non-affordable category. It is not affordable because, economically, it still suffers losses because production costs are still greater than profits. If the eco-efficiency value is between 0-1, then the system is in the affordable but not sustainable category. The system is included in the affordable category because it has already made a profit and is economically feasible. However, because the environmental costs are still greater than the profit earned, the system is in the unsustainable category. If the eco-efficiency value is more than one, it means that the system is included in the affordable and sustainable category. If the cost of environmental impact is low, the system can have a higher level of eco-efficiency if the net value of the system does not change [20]. The higher the ratio, the better the eco-efficiency and vice versa. If the cost of environmental impact is low, the system can have a higher level of eco-efficiency if the net value of the system does not change.

$$\text{Net Value} = \text{Selling Price} - \text{Production cost} \tag{1}$$

$$EEI = \frac{Net\ Value}{EcoCost} \tag{2}$$

$$EVR = \frac{Eco\ cost}{Net\ value} \tag{3}$$

$$ERR\ Rate = (1 - EVR)100\% \tag{4}$$

3. Result and Discussion

3.1. Impact of the tree branches as firewood

Impact Category of tree branches as firewood is climate change 8.28 (kg CO2 eq), acidification 0.154(kg SO2 eq), Photochemical Oxidant Formation 0.033 (Kg C2H2 eq), eutrophication 0.019 (kg PO4 eq), fine dust 0,28 (kg PM 2.5 eq), human toxicity 1.78E-7 (Cases), ecotoxicity 26600 (PAF.m3.day), metals scarcity 0.144 (Euro), water stress indicator 0.019 (WSI Factor), oil & gas depletion excl energy 0,19 (kg oil equ), land use 0 (Bio Factor), and waste 0 (MJ).

3.2. Life Cycle Assessment of the tree branches production process

Table 4 shows the impact characterization values of the hanger production process with a strategic upcycle. The environmental impact value is obtained from Simapro software calculation based on life cycle inventory. The characterization stage identifies the factors of production and groups them into several categories of environmental impacts.

Table 4. The result of the characterization stage

Impact Category	Firewood	Upcycle
Climate Change (kg CO2 eq)	8,28	127
Acidification (kg SO2 eq)	0,154	0,595
Eutrophication (kg PO4 eq)	0,0189	0,0365
Photochemical Oxidant Formation (kg C2H2 eq)	0,0327	0,021
Fine Dust (kg PM 2,5 eq)	0,276	0,149
Human Toxicity (Cases)	1,78E-7	0,000003
Ecotoxicity (PAF.m3.day)	26600	47600
Metals Scarcity (Euro)	0,144	6,28
Oil & Gas Depletion excl Energy (kg oil equ)	0,197	1,86
Waste (MJ)	0	0
Land Use (Bio Factor)	0	0
Water Stress Indicator (WSI Factor)	0,0185	0,289

To assess the contribution of an activity to the environmental impact, normalization is carried out. The normalization is done by multiplying the characterization and normalization values to have the same units. The method used to determine the normalization factor is Eco-Cost 2017 v1.5. Furthermore, weighting is carried out on each category of environmental impact. The weighting factor has different values depending on the method used and the level of importance of the impact category. In this study, all impact categories have a weight value of 1. Thus, the normalization and weighting stages have the same impact category value. The single score stage aims to classify the impact category values of each process or activity. Processes that have contributed to environmental damage can be identified from the results of a single score. Table 5 shows the results of a single score for the production process of upcycling tree branches into hangers. Activity assessed include stripping & oven, materialization, sanding, assembly, treatment, and finishing in 10 impact categories. Table 5 shows the single score after the euros' normalization and weighting stages. While Table 6 shows the result of a single score in rupiah. Figure 2 is the result of a single score in a bar chart.

Table 5. Single Score of Upcycle process (Euro)

Impact Category	Stripping &	Materializa	Sanding	Assembly	*Treatment*	*Finishin*
Climate Change	0,387	6,930	4,910	0,946	0,436	1,097
Acidification	0,543	2,080	1,480	0,266	0,300	0,534
Eutrophication	0,031	0,050	0,036	0,009	0,009	0,015
Photochemical	0,137	0,020	0,014	0,028	0,005	0,013
Fine Dust	3,780	0,613	0,435	0,136	0,025	0,086
Human Toxicity	0,066	0,493	0,349	1,732	0,022	0,097
Ecotoxicity	0,059	0,087	0,061	0,021	0,006	0,029
Metals Scarcity	0,058	0,238	0,168	1,071	0,033	4,723
Oil&Gas Depletion	0,063	0,741	0,525	0,033	0,009	0,117
Water Stress Indicator	0,007	0,108	0,076	0,061	0,008	0,029
TOTAL	5,133	11,361	8,055	4,303	0,853	6,741

The hanger from the upcycle process has an eco-cost of IDR. 12,559.06. This value states that the company must incur environmental impact mitigation costs for hanger (upcycle) product. 12,559.06 for one product. Firewood has a total eco-cost value of IDR 4.388,02. This value states the cost of mitigating the environmental impact of firewood for 1116.25 kg or equivalent to the number of wood branches used to make upcycle hanger products.

The value of the environmental impact on the hanger production process was greatest in the climate change category of 127 Kg CO_2 equivalent. Meanwhile, the impact of climate change on firewood is 8.28 Kg CO_2 equivalent. Climate change is a change in geothermal conditions caused by the increase in greenhouse gases, one of which is CO_2 in the troposphere. Greenhouse gases are like the greenhouse effect, which reflects radiation from the earth back to the earth [21]. This is due to the continuous use of fossil fuels and land. In the hanger production process with an upcycle, this CO_2 gas is produced from the oven process and is caused mainly by the use of electrical energy caused by the use of coal to produce electrical energy.

Table 6. Single Score of Upcycle process

Impact Category	Upcycle		Woodfire	
Climate Change	IDR	253.366,701	IDR	16.546,397
Acidification	IDR	89.798,674	IDR	23.268,371
Eutrophication	IDR	2.619,846	IDR	1.359,907
Photochemical Oxidant Formation	IDR	3.757,411	IDR	5.842,946
Fine Dust	IDR	87.385,658	IDR	161.499,727
Human Toxicity	IDR	47.570,891	IDR	2.826,676
Ecotoxicity	IDR	4.550,259	IDR	2.533,667
Metals Scarcity	IDR	108.241,012	IDR	2.481,960
Oil&Gas Depletion excl Energy	IDR	25.681,387	IDR	2.723,261
Waste	IDR	-	IDR	-
Land Use	IDR	-	IDR	-
Water Stress Indicator	IDR	4.981,155	IDR	318,863
Total/batch	IDR	627.952,994	IDR	219.401,774
Total/product	IDR	12.559,060	IDR	4.388,020

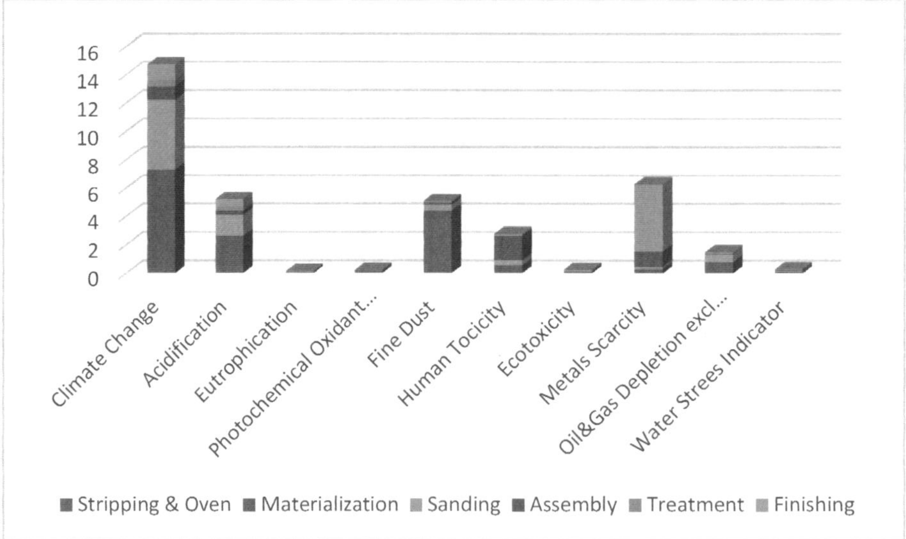

Figure 2. Single score

3.3. Eco-Efficiency Index (EEI)

The calculation of the eco-efficiency index aims to determine whether the hanger production (upcycle) is efficient from an economic and ecological perspective. This calculation refers to equation 2.

Firewood has an eco-efficiency index value of 3.816 and hanger products have an eco-efficiency index of 4.512. Firewood has less impact because it does not require materials and energy when disposed of. The upcycling process requires energy and other materials to become more valuable products. The calculation above found that firewood has a smaller eco-cost value than upcycle products. However, when firewood is upcycled, it can increase the value of the eco-efficiency index (EEI) even though it triggers a greater environmental impact due to energy and material consumption during production.

According to the eco-efficiency concept, the production process is included in the affordable and sustainable category [20].

Table 7. Eco Efficiency Performance

Indicator	Hanger (upcycle)	Firewood
Price (IDR)	6,750,000	837.187,5
Production cost	3,916,667.92	0
Net value (IDR)	2,833,332.08	837.187,5
Amount	50 unit	1116
Net (IDR) value/product	56.666.64	
Eco-cost (unit)	12,559.060	219.401,774
EEI	4.512	3,816
EVR	0.222	0.262
EER	77.83 %	73.8 %

Although based on the results of the EEI calculation, it is found that the hanger production process (upcycle & wood logs) does not harm the environment, but waste is still formed from the production process, which has an impact on the operation of the environmental ecosystem. The cost of environmental impact or eco-cost arising from the waste produced must be incurred by the company in order to overcome the impact that occurs on the environment. Hanger production has an EVR of 0.222 for upcycle products. The smaller the EVR, the better, and the hanger product is more feasible. The more efficient a production activity is, the lower the negative impact on the environment. Eco-efficiency ratio hanger rate is 77.837%. Efforts to increase the eco-efficiency rate can be made by increasing the net value or minimizing the contributors to production costs and environmental costs. Cost minimization can be done by minimizing the waste of production raw materials, using more environmentally friendly materials, and avoiding double or repeated processes in the production process.

Wood waste generated during production is better utilized in products with marketability value. The resulting sawdust can be recycled into fertilizer or sold to collectors to be used as particle board to generate profits for the company. In addition, sawdust can be used to make briquettes. If wood waste can be reduced, it will automatically reduce the environmental impact. The biggest impact of the hanger manufacturing process is in the manufacturing process caused by the use of electrical energy with fossil fuels. Therefore, based on LCA measurements, the improvement focus is on the improvement stage. More and more fossil fuels for power generation will produce hazardous chemicals and pollute the environment. Substitution towards renewable power generation needs to be done. Of the three renewable power plants, namely solar panels, biomass, and hydropower, it is known that solar panels impact climate change, carbon dioxide, methane, and nitrogen oxides which are 20 times lower than other renewable energies. A more in-depth study of solar energy as an energy source in the production process is needed.

4. Conclusion

Efforts to increase the lifespan of natural resources by using a circular economy often require additional energy and materials. In the upcycle process of tree branch waste into a hanger, it was found that the environmental impact of the production process was greater than if the waste was used as fuel. However, the upcycle effort has increased the net value so that the eco-efficiency index of upcycled products is more sustainable.

Waste recycling efforts should be carried out continuously to produce cleaner production. It can be done by reducing waste and production costs and environmental costs to improve the eco-efficiency index. Further research on efforts to substitute energy sources from fossils to solar panels needs to be carried out to realize an increasingly environmentally friendly industry.

References

[1] Ajwani-Ramchandani R, Figueira S, Torres de Oliveira R, and Jha S 2021 *Resour. Conserv. Recycl.* **168** 105468
[2] Okorie O, Salonitis K, Charnley F, Moreno M, Turner C, and Tiwari A 2018 *Energies* **11**(11) 3009
[3] Frig M and Sorsa V P 2018 *Bus. Soc.* **59**(6) 1151–1180
[4] Oskam I, Bossink B, and Pieter de Man A 2017 *Exploring a changing view on organiizing value creation : Developping New Business Models* **2511** 51–52.
[5] Geissdoerfer M, Pieroni M P P, Pigosso D C A, and Soufani K 2020 *J. Clean. Prod.* **277** 123741
.[6] Selvatti T D S, Cristina H, Soares C, De Souza N, Moreira L, and Junior C 2018 *Revista Árvore* **42**(5) 420–506
[7] Kouchaki-Penchah H, Sharifi M, Mousazadeh H, and Zarea-Hosseinabadi H 2016 *J. Clean. Prod.* **112** 351–358
[8] Hartini S, Wicaksono P, Rizal M, and Hamdi M 2021 *IOP Conf. Ser.: Mater. Sci. Eng.* **1072** 012067
[9] Kamleitner B, Thürridl C, and Martin B A S 2019 *J. Mark.* **83**(6) 76–92
[10] Barbaritano M, Bravi L, and Savelli E 2019 *Sustainability* **11**(11) 3089
[11] Oliveira G A, Tan K H, and Guedes B T 2018 *Int. J. Prod. Econ.* **205** 62–73
[12] Wicaksono P A, Hartini S, Sutrisno, and Nabila T Y 2020 *IOP Conf. Ser.: Earth Environ. Sci.* **448** 012061
[13] Susanty A, Tjahjono B, and Sulistyani R E 2020 *Prod. Plan. Control* **31** 1336-1348
[14] Bradley R, Jawahir I S, Badurdeen F, and Rouch K 2018 *Resour. Conserv. Recycl.***135** 141-149
[15] Rinawati D I, Sriyanto, Sari D P, and Prayodha A C 2018 *E3S Web Conf.* **31** 1–5
[16] Hartini S, Wicaksono P A, Prastawa H, Hadyan A F, and Sriyanto 2019 *IOP Conf. Ser.: Mater. Sci. Eng.* **598** 012078
[17] Hartini S, Ramadan B S, Purwaningsih R, Sumiyati S, and Kesuma M A A 2021 *IOP Conf. Ser.: Earth Environ. Sci.* 894 012004
[18] Hartini S, Sari D, Aisy N, and Widharto Y 2020 *E3S Web Conf* **202** 10004
[19] Hartini S, Widharto Y, Indarto S R, and Murdikaningrum G 2021 *IOP Conf. Ser.: Earth Environ. Sci* **896**(1) 012066
[20] Purwaningsih R, Simanjuntak C F, and Rosyada Z F 2020 *Jurnal Teknik Industri* **22**(1) 47–54
[21] Samiaji T 2011 *Berita Dirgantara LAPAN* **12**(2) 68–75

Bibliometric Analysis of Thermal Comfort and Sleep Quality Research Trends in Indonesia

Wiwik Budiawan[1], Kazuyo Tsuzuki[2], Heru Prastawa[3]

[1]Department of Industrial Engineering, Diponegoro University, Semarang, 50275, Indonesia
[2]Department of Architecture, Faculty of Environmental & Urban Engineering, Kansai University, Suita 564-8680, Japan
[3]Department of Industrial Engineering, Diponegoro University, Semarang, 50275, Indonesia

[1]wiwikbudiawan@ft.undip.ac.id

Abstract. Using a bibliometric analysis of the publication output associated with research on thermal comfort and sleep quality in Indonesia during the ten-year period 2012–2022, the purpose of this study was to describe current trends and future research areas. The database Scopus was queried for information covering the years 2012 to 2022. We analyzed selected documents containing "thermal comfort," "Indonesia," and "sleep quality" as part of the title, abstract, or key words and reported the following parameters: publication output trends, cooccurrence, author institution, author key words, and index key words. We utilized Visualization of Similarities (VOS) viewer to analyze the files of a bibliographic database in which five co-occurrences occurred. Air conditioning, ventilation, and the tropics accounted for three main cluster of thermal comfort in Indonesia. Meanwhile, in the sleep quality, the main cluster consisted of human, adult, and major clinical study. This study provides a bibliometric analysis demonstrating that, over the past 26 years, the annual number of publications pertaining to sleep quality in Indonesia has increased at a significantly faster rate than literature on thermal comfort. The latest keywords (trend) of thermal comfort are energy efficiency, temperature effect, and field measurement. In the sleep quality topic, the latest keywords are pandemic, controlled study, and Pittsburgh Sleep Quality Index (PSQI).

1. Introduction

Sleep is primarily a behavior. The identification and classification of sleep was based on changes in posture, continued behavior throughout sleep, and an enhanced arousal threshold [1]. A third of the day is devoted to sleep [2], which is an essential daily activity for maintaining good health [3]. A quality night's sleep allows the body to recuperate and prepares them for the day ahead. Sleep is necessary for relieving physical and psychological exhaustion [4], enhancing work performance, and preserving wellness at both school and work [5]. During a typical sleep time, a person has four to six sleep cycles [6]. A newborn requires 12–18 hours of sleep. Between the ages of 5 and 10, children have a 10-hour sleep demand. As teenagers need around 8 and 9 hours while adults need around 7 and 8 hours, the demand lowers further [7]. Quantity and quality of sleep are distinct concepts. Sleep quantity is defined by how long a person sleeps each night, whereas sleep quality is influenced by how thoroughly a person sleeps. Taking the effort to analyze your sleep quantity will allow you to quickly determine whether or not you are sleeping enough each night. As a general rule, high sleep quality is identified by the following properties: rapid onset of sleep, fewer awakenings, sleep duration commensurate with age recommendations [4], rapid re-entry into sleep, and morning clarity. Alternately, Rehman (2021) concluded that a number of factors could be leading to poor sleep quality [8].

Some research suggests that getting too little sleep can be harmful to your health, and that getting too much sleep might be just as harmful [9]. Thirty-five percent of American adults say they get less than seven hours of sleep per night on a regular basis [10]. Insomnia (30-48% of adults [11] ; women get a 40% higher mortality rate of insomnia than men [12]); obstructive sleep apnea or OSA (15-30% of males [13] and 10-30% of females [14] suffer OSA; 2-9% of adults in the USA suffer OSA); central sleep apnea or CSA; restless leg syndrome; narcolepsy; teeth grinding; talking in sleep; sleep walking Multiple medical effects, including high blood pressure, right ventricular hypertrophy, arrhythmia, and other irregular cardiac rhythms, as well as ischemic heart disease and cerebrovascular dementia [15], have been linked to sleep apnea [15].

Sleep deprivation, reduced attention and judgment, lower work efficiency, and diminished physical strength all diminish life quality and contribute to serious accidents. Moreover, sleep disruptions affect immunological function, biological defense, and maintenance processes [16]. The people may be ignorant of the loss in job productivity induced by sleepiness, which increases the likelihood of incidents and human mistake [5]. Numerous prior studies on indoor thermal conditions, particularly the microenvironment (such as the sleeping space or bedroom), have shown that maintaining a pleasant temperature is essential for sleep quality [17]–[19]. Seasonal variations in interior temperature may have variable effects on sleep measurements. During the summer, there was a considerable association between the inside and outdoor temperatures, however during the cooler months, the correlation was weaker [20].

Generally, thermal comfort is enhanced by complying to the relevant standard's recommendations. Nevertheless, thermal comfort requirements (ASHRAE 55, ISO 7730) are generally applied to all individuals, irrespective of ethnicity or nationality. In addition, physical elements including bedroom temperature, humidity, sound, light, and wind might be considered sleep quality factors. Physical elements resulting from variations in house design and physiological factors resulting from the living environment may influence the factors that affect sleep quality. For instance, there is a difference in the quantity of items in the room and the thermodynamic efficiency between the room of a person living alone in an apartment and the room of a person living alone in a house, and there may be a little difference in the setting.

In developed countries, comfort temperature studies have been extensive: comfort temperatures in workplaces [21], [22], in dwellings during the summer [23]–[25], and throughout the seasons [26]. In Indonesia, however, sleep thermal comfort research (review and empirical) is still scarce. For the purpose of describing this phenomenon, a bibliometric study was performed to determine the trend of research on thermal comfort as well as sleep quality within Indonesia. The current analysis relied on the Scopus database of scholarly literature (in last 10 years). The purpose of this study was to examine the trend of thermal comfort and sleep quality studies in Indonesia. In addition, opportunities for research topics could also be identified.

2. Method

Bibliometric analysis has been widely used in previous review-based research. However, in the thermal comfort and sleep quality is still limited. In the current study, we conducted Bibliometric Analysis to answer the research question (see Figure 1).

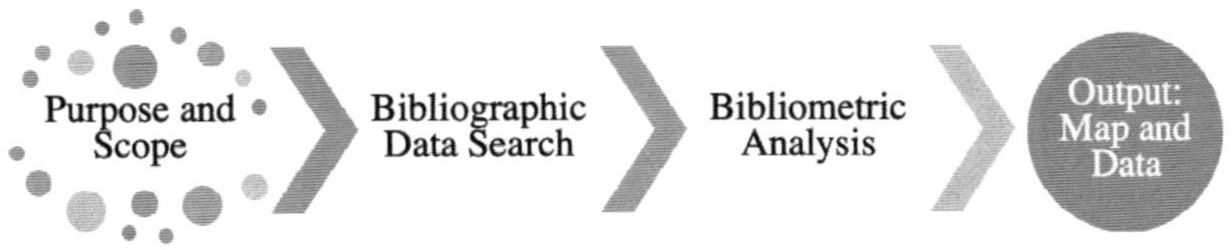

Figure 1. Bibliometric Analysis Steps

In the first step, we stated the purpose and scope of the research. The purpose was to explore the research trend of thermal comfort and sleep quality in Indonesia. Furthermore, two main scopes were approached (keywords and publication time). In the second step, we explored bibliographic data in the Scopus database. Three combinations keywords were applied in bibliographic data search, namely: (1) “thermal comfort” AND “Indonesia”; (2) “sleep quality” AND “Indonesia”; and (3) “thermal comfort” AND “sleep quality” AND “Indonesia.” Searching bibliographic data was conducted within article title, abstract, and keywords. Furthermore, the selection of article studies was carried out by examining the publication time on the development of thermal comfort and sleep quality in Indonesia between 2012 and 2022. In the third step, co-occurrence analysis was applied based on keyword(s) from two or more documents. Co-occurrence analysis is intended to analyze the content, patterns and trends (trends) of a collection of documents by measuring the strength of keywords/ terms [27]. Co-occurrence analysis is performed to determine the frequency with which three keyword combinations from a research document simultaneously appear in the article under consideration. In the second step, this keyword combination is determined beforehand. The greater the frequency with which comparable keywords appear in a group of papers, the stronger their association. VOS viewer version 1.6.18 (open-source license, Centre for Science and Technology Studies (CWTS) of Leiden University, Netherlands) was applied in the current step to explore data and visualization of network. Network visualization represent the cluster that produced by VOS viewer. Number of research publication each year was analyzed by Microsoft Excel for Mac version 16.59 (Microsoft 365 Subscription, Microsoft, Redmond-United State).

3. Results and Discussions

The growth of publications on the topic of thermal comfort and sleep quality in Indonesia has increased significantly in 2018-2021. As seen in table 1 and figure 1, the number of publications in 2021 reached 21% (out of 135 publications) on thermal comfort topic and 27% (out of 90 publications) on sleep quality topic. In the thermal comfort topic, the documents type is dominated by conference paper (47.4 %), article (39.3 %), and conference review (5.9 %). Institut Teknologi Bandung (17 documents) was the most published article in thermal comfort topic, followed by Universitas Indonesia (13 documents) and Institut Teknologi Sepuluh Nopember (12 documents). In the sleep quality topic, the documents type is dominated by article (92.2 %), conference paper (5.6 %), and conference review (1.1 %). Universitas Indonesia (46 documents) was the most published article in sleep quality topic, followed by Universitas Airlangga (9 documents) and Universitas Gajah Mada (5 documents).

Table 1. Number and percentage publication by year.

Year	TC (f)	TC (%)	SQ (f)	SQ (%)
2012	3	2	1	1
2013	3	2	0	0
2014	3	2	2	2
2015	10	7	1	1
2016	6	4	2	2
2017	13	10	2	2

2018	27	20	11	12
2019	16	12	21	23
2020	16	12	16	18
2021	28	21	24	27
2022	10	7	10	11
Total	135	100	90	100

Abbreviations: TC (f), frequency of thermal comfort; TC (%), percentage of thermal comfort; SQ (f), frequency of sleep quality; SQ (%), percentage of sleep quality.

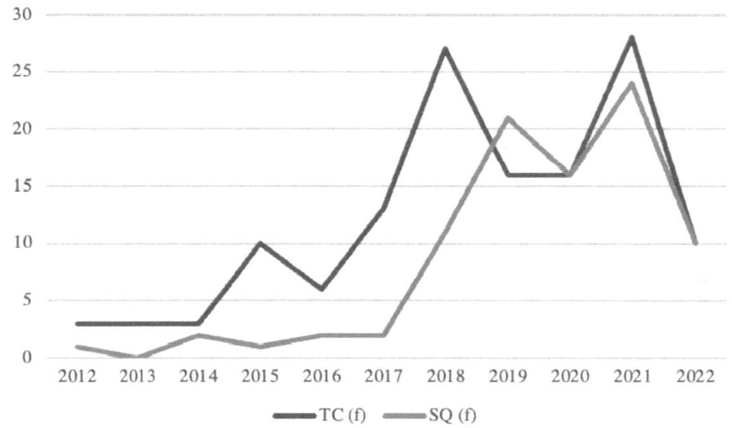

Figure 2. Trend of publication by year.

Abbreviations: TC (f), frequency of thermal comfort; SQ (f), frequency of sleep quality.

Figure 3(a) showed that based on keywords ("thermal comfort" AND "Indonesia"), co-occurrence map of thermal comfort consisted of three clusters. Cluster 1 with red color consisted of 16 keywords: air conditioning, air temperature, atmospheric temperature, buildings, comfort temperatures, buildings, comfort temperatures, cooling, hot-humid climate, indoor air temperature, indoor thermal environment, Jakarta, microclimate, office buildings, PMV, surveys, temperature effect, and thermal environment. Cluster 2 with green color consisted of 11 keywords: architectural design, energy conservation, energy efficiency, energy utilization, heating, houses, housing, residential building, sustainable development, tropical climates, and tropics. Cluster 3 with blue color consisted of 7 keywords: air, air quality, computational fluid dynamics, field measurement, natural ventilation, thermal condition, and ventilation. Figure 3(b) showed that based on keywords ("sleep quality" AND "Indonesia"), co-occurrence map of sleep quality consisted of four clusters. Cluster 1 with red color consisted of 19 keywords: adult, aged, complication, depression, disease severity, educational status, fatigue, health status, insomnia, middle aged, pain, priority journal, psychology, quality of life, sleep, sleep disorder, sleep wake disorders, survey and questionnaire, and young adult. Cluster 2 with green color consisted of 18 keywords: age, article, body mass, child, clinical article, cross-sectional study, female, human, human experiment, hypertension, male, obesity, observational study, physical activity, questionnaire, sleep time, smoking, and stress. Cluster 3 with blue color consisted of 13 keywords: adolescent, controlled study, coronavirus disease 2019, demography, major clinical study, mental health, pandemic, Pittsburgh sleep quality index, prevalence, psychometry, risk factor, social media, and social status. Cluster 4 with yellow color consisted of two keywords: anxiety, and sampling.

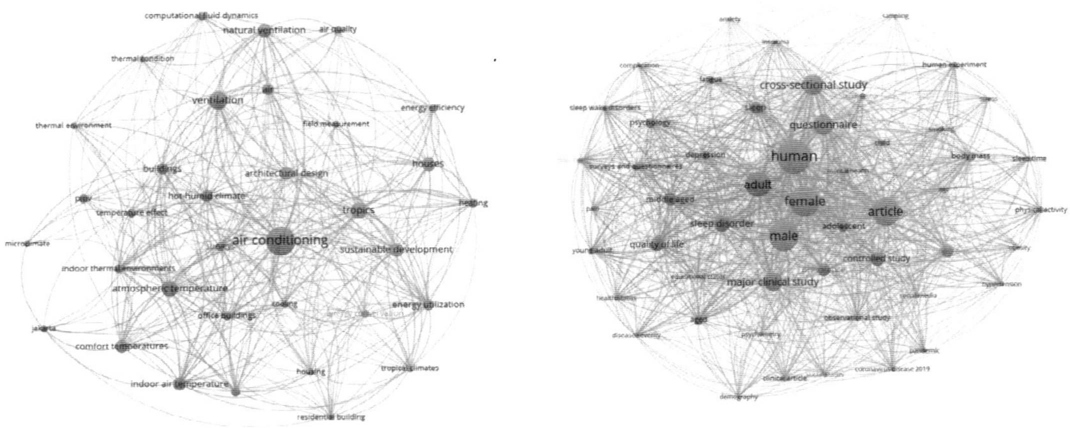

(a) (b)

Figure 3. Co-Occurrence Map of (a) Thermal Comfort; (b) Sleep Quality

Based on a map of two keywords combinations (Figure 3: "thermal comfort" AND "Indonesia", and "sleep quality" AND "Indonesia"), thermal comfort and sleep quality were not related to each other. Furthermore, by combining the two combinations of keywords ("thermal comfort" AND "sleep quality" AND "Indonesia"), the data search results were also nil. Even though previous research has stated that thermal comfort is essential to maintaining and improving sleep quality [17]–[19]. Indonesia, which is located on the equator line, has a tropical climate and with a dry and rainy season. Indonesia receives roughly 12 hours of sunlight per day with the average temperature same throughout year (21.4 - 28.7 °C) [28]. This may indicate that thermal comfort has not been sufficiently considered in relation to sleep quality. In the future, however, with the emergence of the phenomenon of global warming, the study of thermal comfort and sleep quality will be crucial.

4. Conclusion

The highest frequency publication between 2012 and 2022 was 28 (21%) publications for thermal comfort and 24 (27%) publications for sleep quality. The document type dominated by conference article for thermal comfort and article for sleep quality. Furthermore, analysis found the latest keywords (year: 2022) of thermal comfort and sleep quality topic. The latest keywords (trend) of thermal comfort are energy efficiency, temperature effect, and field measurement. In the sleep quality topic, the latest keywords are pandemic, controlled study, and Pittsburgh Sleep Quality Index (PSQI). In sleep quality studies, thermal comfort has not been accounted for. With the emergence of the phenomenon of global warming, it will be fascinating to further investigate the relationship between thermal comfort and sleep quality.

Acknowledgement

This study was funded by the International Publication Research scheme grant (No: 569-158/UN7.D2/PP/VII/2022) from Diponegoro University in 2022.

References

[1] R. D. Ogilvie, "The process of falling asleep," *Sleep Med. Rev.*, vol. 5, no. 3, pp. 247–270, 2001, doi: 10.1053/smrv.2001.0145.

[2] M. Fabbri, L. Tonetti, M. Martoni, and V. Natale, "Sleep and prospective memory," *Biol. Rhythm Res.*, vol. 45, no. 1, pp. 115–120, Jan. 2014, doi: 10.1080/09291016.2013.830510.

[3] K. A. Lee, "An overview of sleep and common sleep problems.," *ANNA J.*, vol. 24, no. 6, p. 86716, 1997.

[4] M. Hirshkowitz *et al.*, "National sleep foundation's sleep time duration recommendations: Methodology and results summary," *Sleep Heal.*, vol. 1, no. 1, pp. 40–43, 2015, doi:

10.1016/j.sleh.2014.12.010.

[5] Japan Ministry of Health Labour and Welfare, "Sleep guidelines for health promotion (in Japanese)," 2014, [Online]. Available: https://www.mhlw.go.jp/file/06-Seisakujouhou-10900000-Kenkoukyoku/0000047221.pdf

[6] A. K. Patel, V. Reddy, and J. F. Araujo, "Physiology, Sleep Stages.," Treasure Island (FL), 2021.

[7] K. Truong, "Sleep Statistics," *Sleep Foundaition*, 2021. https://www.sleepfoundation.org/how-sleep-works/sleep-facts-statistics (accessed Oct. 05, 2021).

[8] A. Rehman, "How To Determine Poor Sleep Quality," *Sleep Foundaition*, 2021. https://www.sleepfoundation.org/sleep-hygiene/how-to-determine-poor-quality-sleep (accessed Oct. 05, 2021).

[9] D. F. Kripke, L. Garfinkel, and D. L. Wingard, "Mortality Associated With Sleep Duration and Insomnia," vol. 59, 2002.

[10] CDC, "Short Sleep Duration Among US Adults," 2017. https://www.cdc.gov/sleep/data_statistics.html (accessed Oct. 05, 2021).

[11] D. Patel, J. Steinberg, and P. Patel, "Insomnia in the elderly: A review," *J. Clin. Sleep Med.*, vol. 14, no. 6, pp. 1017–1024, 2018, doi: 10.5664/jcsm.7172.

[12] J. A. Mong and D. M. Cusmano, "Sex differences in sleep: Impact of biological sex and sex steroids," *Philos. Trans. R. Soc. B Biol. Sci.*, vol. 371, no. 1688, 2016, doi: 10.1098/rstb.2015.0110.

[13] T. Young, M. Palta, J. Dempsey, P. E. Peppard, F. J. Nieto, and K. M. Hla, "Burden of sleep apnea: Rationale, design, and major findings of the Wisconsin sleep cohort study," *Wis. Med. J.*, vol. 108, no. 5, pp. 246–249, 2009.

[14] P. E. Peppard, T. Young, J. H. Barnet, M. Palta, E. W. Hagen, and K. M. Hla, "Increased prevalence of sleep-disordered breathing in adults," *Am. J. Epidemiol.*, vol. 177, no. 9, pp. 1006–1014, 2013, doi: 10.1093/aje/kws342.

[15] J. P. Kiley, N. Edelman, and S. Derderian, "Cardiopulmonary disorders of sleep," in *Wake Up America: A National Sleep Alert*, Vol.2 (Rep., U.S. Department Health and Human service, Ed. U.S. Department Health and Human service, 1994, pp. 10–75.

[16] D. F. Dinges, S. D. Douglas, S. Hamarman, L. Zaugg, and S. Kapoor, "Sleep deprivation and human immune function," *Adv. Neuroimmunol.*, vol. 5, pp. 97–110, 1995, doi: 10.1016/0960-5428(95)00002-j.

[17] G. Zheng, K. Li, and Y. Wang, "The effects of high-temperature weather on human sleep quality and appetite," *Int. J. Environ. Res. Public Health*, vol. 16, no. 2, pp. 1–13, 2019, doi: 10.3390/ijerph16020270.

[18] J. A. F. van Loenhout *et al.*, "The effect of high indoor temperatures on self-perceived health of elderly persons," *Environ. Res.*, vol. 146, pp. 27–34, 2016, doi: 10.1016/j.envres.2015.12.012.

[19] W. Budiawan and K. Tsuzuki, "Thermal Comfort and Sleep Quality of Indonesian Students Living in Japan during Summer and Winter," *Buildings*, vol. 11, no. 8, p. 326, Jul. 2021, doi: 10.3390/buildings11080326.

[20] J. L. Nguyen, J. Schwartz, and D. W. Dockery, "The relationship between indoor and outdoor temperature, apparent temperature, relative humidity, and absolute humidity," *Indoor Air*, vol. 24, no. 1, pp. 1–19, 2014, doi: 10.1111/ina.12052.The.

[21] S. A. Damiati, S. A. Zaki, H. B. Rijal, and S. Wonorahardjo, "Field study on adaptive thermal comfort in office buildings in Malaysia, Indonesia, Singapore, and Japan during hot and humid season," *Build. Environ.*, vol. 109, pp. 208–223, 2016, doi: 10.1016/j.buildenv.2016.09.024.

[22] M. S. Mustapa, S. A. Zaki, H. B. Rijal, A. Hagishima, and M. S. M. Ali, "Thermal comfort and occupant adaptive behaviour in Japanese university buildings with free running and cooling mode offices during summer," *Build. Environ.*, vol. 105, pp. 332–342, 2016, doi: 10.1016/j.buildenv.2016.06.014.

[23] H. B. Rijal, M. A. Humphreys, and J. F. Nicol, "Adaptive model and the adaptive mechanisms

for thermal comfort in Japanese dwellings," *Energy Build.*, vol. 202, no. August, p. 109371, 2019, doi: 10.1016/j.enbuild.2019.109371.

[24] H. B. Rijal, M. Honjo, R. Kobayashi, and T. Nakaya, "Investigation of comfort temperature, adaptive model and the window-opening behaviour in Japanese houses," *Archit. Sci. Rev.*, vol. 56, no. 1, pp. 54–69, 2013, doi: 10.1080/00038628.2012.744295.

[25] H. B. Rijal, M. A. Humphreys, and F. Nicol, "Study on adaptive model. Part 3. Development of the adaptive model for thermal comfort in Japanese houses," *Anu. Meet. Archit. Inst. Japan*, pp. 403–406, 2014.

[26] H. B. Rijal, M. A. Humphreys, and J. F. Nicol, "Towards an adaptive model for thermal comfort in Japanese offices," *Build. Res. Inf.*, vol. 45, no. 7, pp. 717–729, 2017, doi: 10.1080/09613218.2017.1288450.

[27] M. A. De Looze and J. Lemarié, "Corpus relevance through co-word analysis: An application to plant proteins," *Scientometrics*, vol. 39, no. 3, pp. 267–280, 1997, doi: 10.1007/BF02458530.

[28] BMKG, "Climate change extreem in Indonesia," *Meteorological, Climatological, and Geophysical Agency*, 2021. https://www.bmkg.go.id/iklim/?p=ekstrem-perubahan-iklim&lang=ID (accessed Jan. 20, 2021).

Increasing The Efficiency of Bone Charcoal in Water Defluoridation Using Eggshell Powder and Dry Banana Peel Medium (DBPM)

P W Gathere*[1], Sudarno[1], and A Sarminingsih[1]
[1]Department of Environmental Engineering, Faculty of Engineering, Diponegoro University, Semarang 50275, Indonesia

*peterwachiragm@gmail.com

Abstract. The world health organization (WHO) has set an upper limit of 1.5 mg/l fluoride concentration in water meant for human consumption. Ground water has high fluoride level which exceeds the value recommended by the WHO. Defluoridation is the only viable way of dealing with the issue of high fluoride levels in water. The most common defluoridation method is the use of bone charcoal. This research shows how adding dry banana peel medium (DBPM) and egg shell powder to bone charcoal increases its defluoridation capacity as well as the water defluoridation rate. Raw water samples were collected from different boreholes and their initial fluoride concentration measured. The groundwater was then subjected to different adsorbent combinations of bone charcoal, eggshell powder and DBPM and the rate of fluoride removal as well as the final concentration of fluoride in the water after two hours, were determined. Eggshell powder and DBPM were found to increase the fluoride removal capacity of bone charcoal by more than 20%. The highest fluoride removal was achieved when bone charcoal was combined compositely with eggshell powder and DBPM. Hence the addition of eggshell powder and DBPM was found to increase the defluoridation rate as well as the fluoride removal capacity of bone charcoal.

1. Introduction

Ground water is the main source of drinking water for more than half of the world's population. The quality of ground water is quite different between locations, for example in coastal areas, ground water can be polluted by sea water intrusion [1]. Ground water contains various minerals, some of which are essential to human beings while others are harmful. One important mineral in water is the fluoride ion, whose concentration in drinking water is of great concern. The fluoride ion in drinking water is essential in preventing tooth decay and cavities. However, the fluoride ions could become a contaminant in drinking water, if present in large quantities [6]. Excessively high fluoride levels in water causes pain and tenderness of bones ultimately leading to increased likelihood of fractures. This is called *skeletal fluorosis*. Additionally, high fluoride levels may cause browning of teeth, a condition called *dental fluorosis* [13].

Table 1. Effect of prolonged use of drinking water on human health, related to fluoride content [3]

F^{-1} concentration, mg/l	Health outcome
< 0.5	Dental caries
0.5 – 1.5	Optimum dental health
1.5 – 4.0	Dental fluorosis
4.0 – 10	Dental and skeletal fluorosis
> 10.0	Crippling fluorosis

High fluoride levels in drinking water from underground sources is a problem that affects large parts of Africa, specifically the east African rift (through Kenya). In Kenya, a survey conducted by the World Health Organization (WHO) in conjunction with the United Nations Environment Programme (UNEP) showed that Nairobi, the rift valley and some of the central regions of the country had the highest fluoride concentration in ground water [13]. On average, the fluoride content in the aforementioned regions was 1.1 to 6.5mg/l. The who has recommended an optimum fluoride level of between 0.5 to 1.5mg/l [15]. For this reason, the fluoride levels in ground water need to be reduced to the allowable levels that do not have any adverse effects to human life.

Various defluoridation methods have been employed to reduce the fluoride content of water, with one of the most effective and cheapest method being adsorption using bone char. However, bone char filters have been found to be less durable hence prompting frequent regeneration or exchange of filters. In Kenya, Nairobi County currently generates approximately 1900 tons of solid waste per day. It is projected that this amount could double by the year 2030 [9]. Natural waste such as banana peels and eggshells account for about 4% of the total solid waste generated. This translates to close to 28,000 tons of banana peels and eggshells waste per year [9]. Lack of proper disposal of solid leads to accumulation of such waste in the environment, causing pollution. Some of the adverse effects of pollution by such waste is deterioration of the natural beauty of the environment and blockage of drainage and sewerage systems.

Research has shown that increasing the calcium and pHospHate levels in bone charcoal water filters boosts their efficiency and durability. Eggshells and banana peels are known to be rich in calcium and pHospHate respectively [4]. This research investigates the effectiveness of using bone charcoal, eggshell powder and dry banana peels medium (DBPM) to reduce fluoride levels in ground water.

The science behind the beneficial and harmful effects of fluoride on the skeletal structure can be explained by ion exchange reactions between hydroxide ions in the calcium hydroxy-pHospHate (the main component of the skeletal structure) and fluoride ions [3]. The substitution of hydroxide ions with fluoride ions results in a structure that is more resistant to damage by acid. This compound that is resistant to attack by acid is called fluoroapatite.

$$Ca_5(PO_4)_3OH + F^- \longrightarrow Ca_5(PO_4)_3f + OH^-$$

Fluoroapatite is more resistant to acid attack compared to hydroxyapatite and hence it provides a protective layer to the tooth enamel that prevents attack of the enamel by the acid produced by bacteria in the mouth [3]. However, excessive intake of fluoride leads to the formation of calcium decafluoride.

$$Ca_5(PO_4)_3f + 9F^- \longrightarrow Ca_5F_{10} + 3PO_4o^{3-}$$

Calcium decafluoride is very hard and brittle, hence not suited for functions of the skeletal structure.

Additionally, fluoride in water helps to maintain good bone density [15]. As explained above, fluoride combines with calcium ions in the bones which ultimately ensures that bones remain strong.

The component of bone char that is responsible for adsorption of fluoride is calcium pHospHate also called hydroxyapatite [11]. When fluoride ion reacts with hydroxyapatite, fluorapatite and hydroxyl ions are formed [12]. The chemistry of defluoridation using bone char can be described as consisting of ion-exchange, adsorption, recrystallization and precipitation processes. Direct adsorption of fluoride takes place when the fluoride ions accumulate on the empty sites on the bone char surface [14]. The ionexchange process takes place when the fluoride ion exchanges position with the hydroxyl, carbonate, bicarbonate or pHospHate ion in the bone char. Then recrystallization takes place whereby the hydroxyapatite and bone minerals dissolve and precipitate with fluoride as fluoroapatite [11]. The major reactions in the defluoridation process are:

$$Ca_{10}(PO_4)_6(OH)_2 + 2F^- \longrightarrow Ca_{10}(PO_4)_6F_2 + 2OH^-$$

$$(Ca_3(PO_4)_2)_nCaCo_3 + 2F^- \longrightarrow (Ca_3(PO_4)_2)_nCaF_2 + CO_3^{2-}$$

Fluoride removal from groundwater using bone char is commonly used due to the relatively low cost of acquiring the adsorbent whose defluoridation efficiency is well above 65%. The efficiency of defluoridation using bone char is highly dependent on temperature, pH of the raw water and the contact time of the adsorbent and the water. In defluoridation plants, maximum defluoridation using bone char adsorbent requires a contact time of 9 hours. The pH of the raw water should be 7-7.5 since the adsorbent tends to dissolve at low pH. Defluoridation using bone char has an efficiency of up to 66% [4].

Eggshells, specifically the hen eggshell, are largely composed of inorganic materials (about 98%) and organic materials form only a small proportion of the eggshells. Eggshell is mainly composed of calcium carbonate (94.03%) and it also contains calcite and calcareous soil [5]. Eggshell has a cellulosic structure and contains amino acids, hence would be a good biosorbent. This composition makes the eggshell to have a very porous structure. On the inner lining of the eggshell, is the eggshell membrane which is largely composed of fibrous proteins [10]. For this reason, eggshells have emerged as non-toxic and efficient adsorbents. It is the eggshell porosity and the eggshell membrane that makes the eggshell a good adsorbent, capable of removing heavy metals and non-metallic ions such as fluorides from water. Recent studies have shown that eggshell powder can be effectively used as a low-cost adsorbent for defluoridation of groundwater [5]. Nevertheless, the effectiveness of defluoridation using eggshells is dependent upon parameters such as pH, adsorbent dose, contact time, initial concentration of fluoride and temperature [14]. It has been found, empirically, that maximum adsorption of fluoride from an aqueous solution by eggshell adsorbent, was obtained at a pH range of 4-7 [4].

Banana peel is a major waste that is found in almost all regions of Kenya. Experiments have shown that banana peels have excellent antioxidant potential and hence can be used in water purification [2]. Although there has not been extensive research on the ability of banana peel to remove fluoride from water, banana peel has been extensively used for removal of heavy metals such as lead and nickel from aqueous solutions. Banana peels contain nitrogen, sulfur, pHospHates and carboxylic acids [2]. The carboxylic acids that are usually present in large amounts in the banana peels, are responsible for binding toxic metals and non-metals present in water [6]. Due to the relatively high adsorptive properties of banana peels, they could be an economically viable and low-cost adsorbent of fluoride from groundwater. Just like in the case of eggshells, the effectiveness of the dry banana peel medium on defluoridation of groundwater depends on parameters such as pH, adsorbent dosage, temperature, contact time and the initial fluoride concentration of the water. Studies have shown that banana peels have an efficiency of up to 85% defluoridation when used at optimum conditions of pH, temperature and adsorbent dosage [6]. The major advantage of using banana peels as fluoride adsorbent is the fact that banana peels are readily available at a relatively low cost (Saranya and Anu, 2016). In the same vein, banana peels can be easily used in small scale water treatment plants [7]. Therefore, the use of banana peels could provide an alternative method for fluoride removal from groundwater.

This study will focus on determining the effect of adding eggshell powder and DBPM to bone charcoal, with regard to defluoridation capacity as well as fluoride removal efficiency. Due to the relatively high fluoride levels of untreated groundwater, this study will investigate the fluoride concentration of raw groundwater samples obtained from boreholes in Kibera area of Nairobi County. Hence Kibera area will be the area of study.

In a bid to assess the effect of adding eggshell powder and DBPM to bone charcoal, constant amounts of each of these adsorbents will be used. Different adsorbent combinations will be setup under the same ambient conditions of temperature, pH and contact time, for easy comparison of these different adsorbent combinations. The effect of adding eggshell powder and DBPM on other water quality parameters, save from fluoride, will not be considered in this study. In the same vein, the effect of varying the temperature, pH, contact time and adsorbent dosage on each of the different adsorbent combinations will not be considered, hence these conditions will be kept constant.

2. Methodology

Experimental research was used to determine the effect of adding Dry Banana Peel Medium (DBPM) and eggshell powder to bone charcoal in water defluoridation. Raw water samples for use in the study were collected from randomly selected boreholes and their initial fluoride concentrations measured using the Ion-selective electrode fluoride meter. Bone charcoal with particles measuring between 0.5mm and 3mm were was obtained for use as the primary defluoridation adsorbent. Eggshell powder was obtained by collecting eggshells, washing them using deionized water, oven-heating them at 60°c for 24 hours, grinded and sieved using a 60-100 mesh/0.25-0.104 mm. Dry Banana Peel Medium was made from washed fresh banana peels which were sun-dried and later oven-dried at 100°c for 36 hours. The dried banana peels were then crushed to produce the DBPM. The eggshell powder and DBPM were to be used as the biosorbents in the study.

The average initial fluoride concentration was determined based on the fluoride meter readings from the different raw water samples collected for use in the experiment. Different experimental setups were used to determine the fluoride removal efficiencies of different adsorbent combinations. Different adsorbent combinations were used to treat 200ml raw water samples as shown in the table below.

Table 2. Adsorbent_Dosages

Setup	Raw Water Sample	Adsorbent Dosage		
		Bone Charcoal (g)	Eggshell Powder (g)	DBPM (g)
1	200 ml	12.8	0	0
2	200 ml	12.8	5	0
3	200 ml	12.8	0	5
4	200 ml	12.8	5	5

The fluoride concentration in each setup was measured at intervals of 30 minutes for a total of two hours and the results recorded accordingly. The residual fluoride concentration in each setup was determined and would be critical in calculating the relative defluoridation efficiencies of the different adsorbent combinations.

3. Results and Discussion

The average fluoride concentration in raw ground water samples was calculated as 3.83 mg/l. In the first setup, bone charcoal reduced the fluoride concentration of the groundwater sample from an initial concentration of 3.83 mg/l to 1.4 mg/l in 120 minutes (two hours). Based on the initial and residual fluoride concentrations of water in the sample, the removal efficiency (re) as a percentage was calculated at 30-minute intervals using the following equation [14]:

$$RE\ (\%) = \frac{Co-Ce}{Co} \times 100 \qquad (1)$$

where, *co* is the initial fluoride concentration, *ce* is the residual fluoride concentration.

In the second setup, the combination of both bone charcoal and eggshell powder reduced the fluoride concentration in the groundwater sample from 3.825 mg/l to 0.92 mg/l in 120 minutes (2 hours). It follows that in setup 3 where both bone charcoal and dry banana peel medium were combined in the defluoridation of groundwater, it was observed that the fluoride concentration of the sample reduced from an initial of 3.83 mg/l to 0.54 mg/l in 120 minutes (2 hours). A combination of the three sorbents, bone charcoal, eggshell powder and DBPM, was found to reduce the fluoride concentration of the raw groundwater from 3.83 mg/l to 0.44 mg/l in 120 minutes (2 hours).

The following is a grapH depicting a comparison of all the above sorbent combinations used in defluoridating raw groundwater samples having the same initial fluoride concentration.

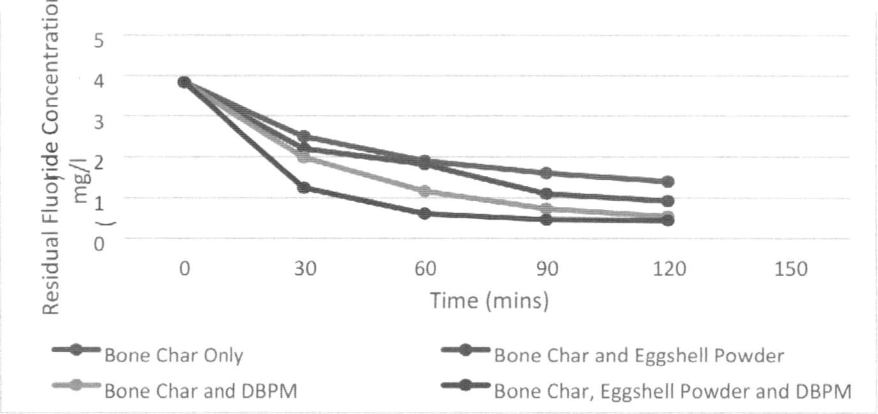

Figure 1. Variation of fluoride concentration with time for the different adsorbent combinations.

The percentage of fluoride removal efficiencies of the four adsorbent combinations are shown in the chart below.

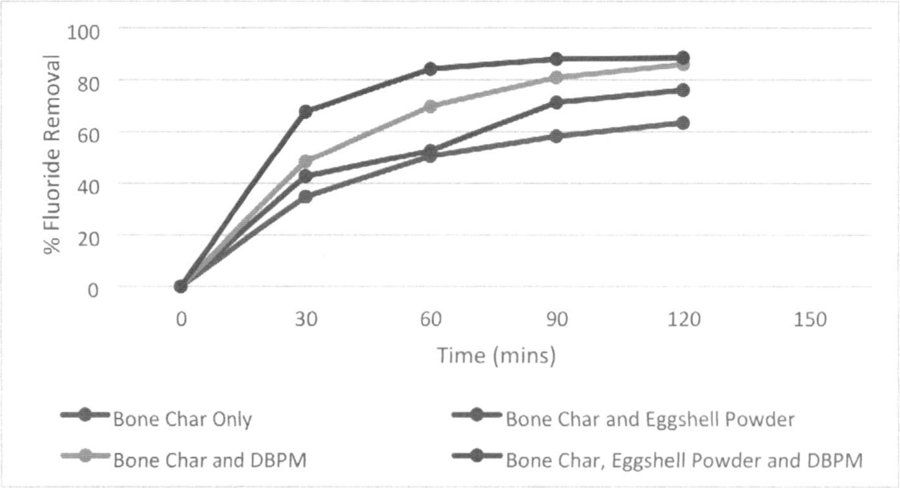

Figure 2. Fluoride removal curves for the different adsorbent combinations.

A composite combination of the three adsorbents- bone charcoal, eggshell powder and DBPM- reduced the fluoride concentration of the raw groundwater sample from 3.83 mg/l to 0.44 mg/l in two hours' time. The final fluoride concentration after the two-hour period was far much lower than the final concentration achieved when using the other sorbent combinations explained above. The composite combination of the three adsorbents gave a fluoride removal of 88.5%, representing a 25.1% increase in defluoridation efficiency from the 63.4% obtained when using bone charcoal only to defluoridate groundwater.

The use of bone charcoal and the other biosorbents (eggshell powder and DBPM) to defluoridate groundwater made the water to become turbid. However, the clarity of water improved through filtration. The impact of the adsorbents on other water quality parameters was not assessed in this research. Further research needs to be done to assess how the use of eggshell powder, DBPM and bone charcoal to defluoridate water, affects other water quality parameters in addition to redcing fluoride content in water.

4. Conclusions

The addition of eggshell powder and DBPM to bone charcoal was found to increase the fluoride removal of bone charcoal from 63.4% to 88.5%. Hence, the addition of the biosorbents- eggshell powder and DBPM- to bone charcoal increases the defluoridation capacity of bone charcoal as well as its fluoride removal efficiency. In light of the above, eggshell powder and dry banana peel medium (DBPM) can be added to bone charcoal as low-cost adsorbents to improve its fluoride removal efficiency. As such, this study may find application in the development of sustainable, low cost and eco-friendly household water treatment systems for removal of fluoride.

This research shows how banana peels and eggshells, which are solid wastes that are disposed in an unhygienic manner in the environment, can be transformed into useful material. This will significantly reduce the quantity of solid waste that is dumped in the environment. In the same vein, this research provides an effective way of incorporating natural material (in this case banana peels and eggshells) into bone charcoal filters in order to make them more durable and effective in water defluoridation. Incorporation of banana peels and eggshells into bone charcoal filters will reduce the need for regeneration of the filter medium, hence making such filters more economical.

References

[1] Susanto, Syafrudin, TT Putranto 2021, *IOP Conf. Ser.: Earth Environ. Sci.* **896** 012057
[2] Chaparadza A and Hossenloppl JM 2012 *Water Sci. Technol.* **65**(5) 940-947
[3] Malago J, Makoba E and Muzuka AN 2017 *Am. J. Water Resour.* **3**(1) 1-17
[4] Waghmare SS and Arfin T 2015 *Int. J. Innov. Res. Technol. Sci. Eng.* **4**(9) 8090-8102
[5] Bhaumik R, Mondal NK, Das B, Roy P, Pal KC, Das C and Baneerjee A 2012 *J. Chem.* **9**(3) 1457-1480
[6] Emaga TH, Andrianaivo RH, Wathelet B, Tchango JT and Paquot M 2007 *Food Chem.* **103**(2) 590-600
[7] Jamal P, Saheed OK and Alam Z 2012 *Asian J Biotech.* **4** 1-14
[8] Kaseva ME 2006 *Journal of water and health*, **4**(1) 139-147
[9] Kasozi A and Von Blottnitz H 2010 *Solid Waste Management in Nairobi: A Situation Analysis Technical Document Accompanying the Integrated Solid Waste Management Plan* (Cape Town: Scientific Research Publishing Inc.)
[10] Mohapatra M, Anand S, Mishra BK, Giles DE and Singh P 2009 *J. Environ. Manage.* **91**(1) 67-77
[11] Moreno-Piraján JC, Giraldo L and García-Cuello VS 2011 *Water Resour. Prot.* **3**(03) 176.
[12] HWTFR 2011 *Calgary, Alberta, Canada: CAWST (Centre for Affordable Water and Sanitation Technology).*
[13] Mumma A, Lane M, Kairu E, Tuinhof A and Hirji R 2011 *Kenya groundwater governance case study* (Washington DC: World Bank)
[14] Kashi G, Mehree A, Zaeimdar M, Khoshab F and Madaree AM 2015 *Bulg. Chem. Commun.* **47** 187-192

Medical waste management improvement in several private hospitals in Amanat Al-Asimah –Yemen

Adeeb Hayel Zaid[1,2], Sri Sumiyati[1], Haryono Setiyo Huboyo[1],

[1] Department of Environmental Engineering, Faculty of Engineering, University of Diponegoro, Indonesia
[2] Faculty of Engineering, Aden University, Aden, Yaman

Correspondence email: adeebhayel24@gmail.com

Abstract. Handling medical waste presents several environmental challenges in most Yemeni hospitals that need immediate attention before they get out of control. This study aims to assess current practices of medical waste management in some of the major private hospitals in Amanat AlAsimah-Yemen. This study relied on several methods of data collection, visits, observation, and interviews with health cadres, cleaners, and workers in the municipality, with the use of a checklist that contains several questions and then follow-up. The results of the study, which were conducted in the four largest private hospitals in Amanat Al-Asimah, which are the Science and Technology Hospital, Azal Hospital, European Modern Hospital, Dr. Abdul Qader Al Mutawakil Hospital showed that there were no independent and specialized administrative units within the hospitals studied how to manage, follow up on the collection, transport, and disposal of medical waste and the lack of laws and regulations on medical waste management and how to deal with them. Moreover, liquid medical waste is discharged through a sewage network that is treated only from organic matter regardless of harmful chemicals that in turn affect humans, the environment, and agriculture.

1. Introduction

Continuous progress in medical sciences and technology and expansion in the number of health institutions worldwide has been accompanied by increasing quantities of potentially hazardous medical waste. The risks include occupational exposure of health workers and waste handlers and environmental exposure to the public caused by medical waste mismanaged [1].

Medical waste is defined as the waste generated from healthcare establishments, such as hospitals, primary healthcare centers, private clinics, research facilities, and laboratories used in diagnosis, monitoring, and prevention activities (curative or palliative) drugs in the field of human and veterinary medicine including infectious hazardous materials and human body parts[2].

Although medical waste represents only a small fraction of urban municipal waste, there should be a greater consensus on its hazardous nature and how much of the waste generated is infectious or hazardous. Infectious or hazardous hospital waste represents only a small part of total medical waste; yet, because of ethical questions and potential health risks, it is a focal point of public interest. Most hazardous and toxic waste is coming from clinical and hospital waste. Only a small amount is coming from domestic or industrial sources[3].

In Yemen, as in many developing countries, solid wastes, in general, and medical and hazardous wastes, in particular, have not received sufficient attention in terms of studying the current practices and methods applied in their collection and disposal (Ministry of Water and Environment, 2008). Where medical waste is disposed of without any sorting process, it is all collected in plastic bags and then delivered to municipal vehicles that are transferred to landfills or by trying to burn in any area or bury it. The environmental consequences and the potential sound or standard environmental management plans that need to be applied nationwide are meant to be for alleviating potential disease means and sources and consequently improving medical and health services through proper waste management [4].

Unfortunately, up to date, medical wastes and hazardous material generated from hospitals in Sana'a city and other cities in Yemen, are still handled and disposed of together, mixed with domestic wastes. Hence, such practices have commonly created or are expected to create high risks for, or great impacts on, the health of municipal workers and the surrounding environment (Ministry of Water and Environment, 2008). As such, there are many different kinds of wastes that may be produced in hospitals, such as infectious, radioactive, and chemical ones, in addition to heavy metals and regular municipal wastes. Transporting or disposing of such mixed wastes in sewers or dumping sites in the surrounding environment has ultimately been found to cause soil and groundwater pollution in such environments and consequently cause health hazards for living species[5]. Therefore, medical waste management is of great importance due to its potential environmental hazards and public health risks [6]. Nevertheless, the management of medical wastes requires reliable information, regarding their generation rates, quantities, compositions, handling, and transportation i.e. from cradle to grave [7].

This study aims to evaluate the current practices of medical waste management in some of the major private hospitals in Amanat Al-Asimah, identifying environmental and health potential effects on the surrounding environment and people, and a comprehensive national plan for the management of medical waste in the future.

2. Methodology

This study assesses the current practices of medical waste management in four private hospitals in Amana Al-Asimah, namely Science and Technology Hospital, Azal Hospital, Dr. Abd Alkader Al-Mutawakel Hospital, and Modern European Hospital(fig.1). These hospitals are ranked in Group A, which is one of the best in terms of quality services and capabilities according to the Ministry of Public Health and Housing 2022. Where 80% of these hospitals were studied, these hospitals are one of the largest and most recent and include private hospitals in Amana Al-Asimah, and include many departments such as Operations Department, Maternity Department, Kidney Department, Cardiac Catheter Department, Lump Department, Laboratory Department, X-ray Department, etc. The methodology for this study involved the following steps (Fig. 2)

1. Interviews: containing several questions with several relevant officials in the management of medical waste in hospitals such as the Director of the Infection Department, the Director of Quality, the Director of Nursing, and the Director of Services and Measures to know the regulations, laws, policies, and possibilities in the management of medical waste.
2. Observation: spending enough time in the different hospital departments to record the observation through a Checklist containing several questions including sorting, collection, transportation, and treatment of medical waste.
3. Follow up in each hospital's departments to learn how to manage medical waste. Site visits were very useful in obtaining information on common waste management practices.
4. Data analysis was performed using Statistical Package for Social Science (SPSS) version 26, and Excel software programs. Descriptive statistics such as frequencies and percentages were computed.

The results obtained were discussed to ensure that medical waste is treated in the light of written policies and established international standards in this (WHO,1999). This study is purely a case and analyses mainly descriptive. The analysis data was extracted from personal interviews and observations and its checklist and follow-up by the researcher.

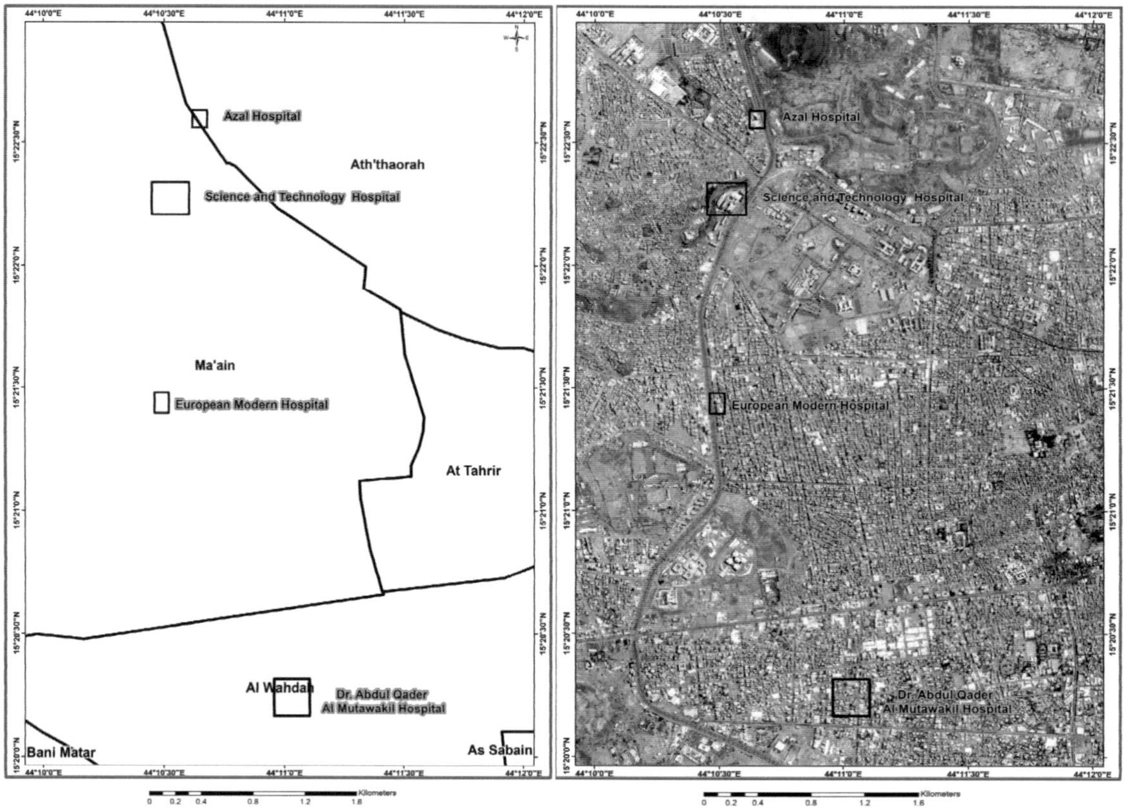

Figure 1. Map and an aerial photograph showing the location of the four hospitals under study

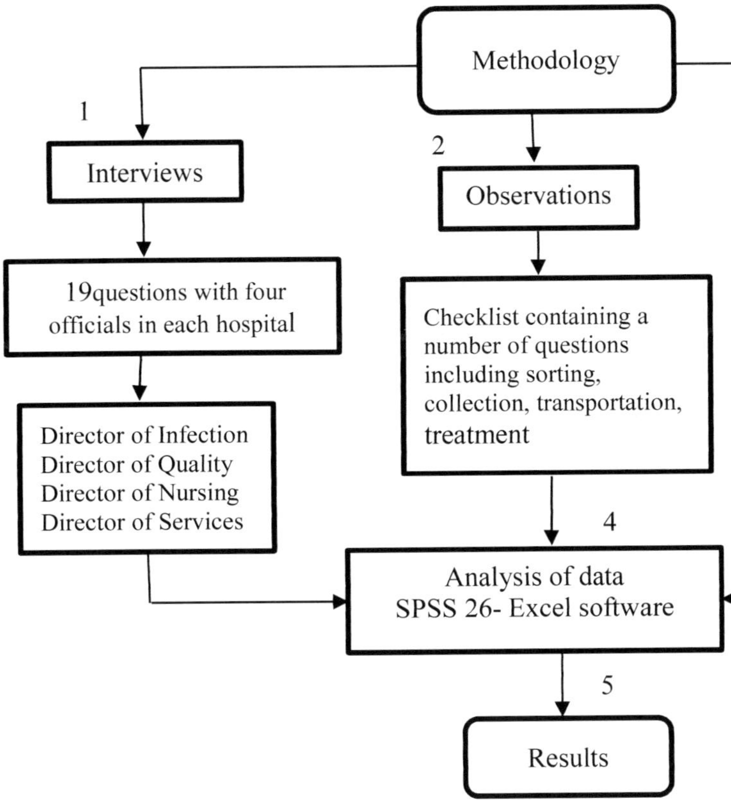

Figure 2. Methods of study.

3. Results

Waste Production: Waste is generated from the assorted activities performed within the hospitals. Wastes produced within hospitals include general and medical wastes. General waste produced at the hospitals is said to food preparation, administrative departments, and landscaping. This kind of waste is analogous to household and city waste. Within the four hospitals, different sorts of therapeutic procedures like Laparoscopic excision of benign, hematological, and malignant tumors, hemodialysis, surgery, childbirth and removal of damaged organs, etc., are meted out and end in the assembly of infectious wastes, contaminated sharps with patients' blood and secretions, radioactive wastes and chemical materials which are considered to be hazardous wastes(Prüss,1999). The number of waste generated in hospitals depends upon various factors like the number of beds, varieties of health services provided, economic, social, and cultural status of the patients, and therefore the general condition of the realm where the hospital is situated. As an example, in hospitals located in low socioeconomic areas of the cities, most of the waste consists of residues from fruits which are voluminous and abundant, whereas in those located in high socioeconomic areas of the city; most of the waste contains flowers, cans, and single-use containers for food[8]. However, during an interview with waste management staff within the hospitals, they may not tell the number of waste generated in the hospital daily. While they provided simple information that supported their observation concerning which departments generate the highest and lowest amounts of medical waste in the hospitals. Proper waste management will reduce waste at the source. For example, sorting organic and inorganic waste effectively reduces the amount of waste to more than 20% of the original waste generation [9].

Table 1. Hospitals Characteristics

Characteristics	Science and Technology Hospital	Azal Hospital	Modern European Hospital	Dr. Abd Alkader Al-Mutawakel Hospital
1. Class	general	general	general	general
2. Area	12900 m^2	3016 m^2	2438m^2	1700 m^2
3. Departments	25	25	14	22
4. Example of departments	Operations, Emergency, Laboratory, Heart, Kidney, Orthopedics, etc.	Operations, Emergency, Laboratory, Heart, Therapeutic nutrition, urology, oncology, children, etc.	Neurology, Internal Medicine, Cardiac Surgery, Ear, Nose, and Throat, Dermatology, etc.	Dental, ophthalmology, physiotherapy, central care, general surgery endoscopy, etc.
4. Beds	181	110	112	105
5. Doctors	87	75	65	34
6. Nurses	309	240	120	160
7. Administrators	110	77	70	35
8. Cleaners	154	80	45	30
9. Medical waste generation	Pathological, tissues, organs, blood, pus, and body parts and Fluids. Needles, syringes, broken glass, blades, placenta, Hazardous chemicals, Common medicines that are expired, etc.	Needles, syringes, broken glass, blades, Pathological, tissues, organs, blood, pus, body parts and Fluids, hematological and malignant tumors, hemodialysis, surgery, placenta, Common medicines that are expired, etc.	Needles, syringes, broken glass, blades, Pathological, tissues, organs, blood, pus, body parts and Fluids, placenta, Hazardous chemicals, Common medicines that are expired, etc.	Needles, syringes, broken glass, blades, Pathological, tissues, organs, blood, pus, body parts and Fluids, placenta, Hazardous chemicals, Common medicines that are expired, etc.

The results of this study, through observation and the checklist and follow-up, showed the existence of a real defect in the management of medical waste and recommended the necessity of urging decision-makers to take real measures to develop proper disposal processes (Table 2).

Table 2. Existing problems, their effects, and ways to solve them.

No	Problem	Effect	Solution
1	Medical waste is not properly separated and sorted from the source.	It causes damage to the cleaners who carry out the collection process.	Raising awareness among nurses to carry out the sorting process and provide red, yellow, and black bags.
2	The presence of open baskets in some sections.	Affect patients, nurses, and visitors.	Providing closed baskets and replacing damaged baskets.
3	No temporary collection rooms and waste collection in the bathroom corridor.	Ease of access by visitors and children, which leads to harm to them.	Providing temporary collection rooms in each section so that they are closed and that no one can open them except for the waste collection worker and with good ventilation.
4	Using public elevators to transport waste outside the hospital.	Leading to the transmission of infection to all users of these elevators.	Allocating special corridors and elevators to transport medical waste.
5	The fluids from the dialysis departments and others are discharged into the sewage network without any treatment.	Affect the environment and groundwater.	Using the chemical disinfection method to treat liquid medical waste resulting from laboratories, operations, and others.
6	Leaving medical waste in the collection room outside the hospital, mixed and exposed, for a long time.	Affects plastic collectors, animals, and people passing near them.	Creating a special room for hazardous waste and another room for non-hazardous waste are closed so that access to it is prevented, with the waste being transported daily to the treatment or disposal site.
7	Leaks occur from trucks transporting medical waste from hospitals to landfills, and these trucks are often exposed. These trucks follow the municipality.	Affect humans and the environment.	Allocating special trucks to hospitals to transport waste, which is tightly closed and impermeable to liquids.
8	All hazardous medical waste produced by the four hospitals is disposed of with normal waste in the Alzraqeen landfill.	It affects plastic collectors, animals, and people who pass by them, as well as soil, surface, and groundwater.	Establishing a central incinerator to treat medical waste away from populated areas and creating a special landfill for waste residues

4. Conclusions

Almost all the hospital facilities studied suffer from weak sorting of medical waste because there are insufficient awareness and attention from medical staff. The collection of medical waste inside hospitals is done in incorrect ways and by means that do not conform to quality specifications and standards. Hospitals' waste collection sites do not meet environmental specifications and precautions are taken to prevent liquid leakage from bags. Medical waste remains inside the collection site for prolonged periods and there is no final treatment of medical waste in some of the hospitals studied, even liquid waste is discharged through the sewage system. There is also no special transportation of waste from within hospitals to outside hospitals and is transported by municipal vehicles. There was poor awareness about the risks of medical waste and safe treatment procedures among hospital managers, and most hospitals did not differentiate between household and medical waste. Budgets were not allocated for waste management, resulting in a shortage of waste treatment equipment and supplies and the absence of training programmers for staff, resulting in poor knowledge and practices for waste workers, and the risk of hazardous waste exposure for staff and visitors.

5. References

[1] A. A. Al-Emad, 'Assessment of medical waste management in the main hospitals in Yemen اليمن , *East. Mediterr. Heal. J. La Rev.*, vol. 17, no. 10, 2011.
[2] E. Science, 'Medical Waste Management in Primary HealthCare Centers and Private Clinics : Jenin District as a Case Study Majdi Qasem Jameel Abu-Awwad Dr . Hssan Arafat', 2008.
[3] N. Akter, 'Medical Waste Management: A Review School of Environment, Resources and Development', 2000.
[4] C. E. Da Silva, A. E. Hoppe, M. M. Ravanello, and N. Mello, 'Medical wastes management in the south of Brazil', *Waste Manag.*, vol. 25, no. 6 SPEC. ISS., pp. 600–605, 2005, doi: 10.1016/j.wasman.2004.03.002.
[5] A. Z. Alagöz and G. Kocasoy, 'Determination of the best appropriate management methods for the health-care wastes in İstanbul', *Waste Manag.*, vol. 28, no. 7, pp. 1227–1235, 2008, doi: 10.1016/j.wasman.2007.05.018.
[6] M. H. Dehghani, K. Azam, F. Changani, and E. Dehghani Fard, 'Assessment of medical waste management in educational hospitals of Tehran University Medical Sciences', *Iran. J. Environ. Heal. Sci. Eng.*, vol. 5, no. 2, pp. 131–136, 2008.
[7] M. E. Birpinar, M. S. Bilgili, and T. Erdoğan, 'Medical waste management in Turkey: A case study of Istanbul', *Waste Manag.*, vol. 29, no. 1, pp. 445–448, 2009, doi: 10.1016/j.wasman.2008.03.015.
[8] M. Askarian, M. Vakili, and G. Kabir, 'Results of a hospital waste survey in private hospitals in Fars province, Iran', *Waste Manag.*, vol. 24, no. 4, pp. 347–352, 2004, doi: 10.1016/j.wasman.2003.09.008.
[9] R. A Prasasti, M A Budihardjo and B P Samadikum 2021, *IOP Conf. Ser.: Earth Environ. Sci.* 896 012067.

IOP Conf. Series: Earth and Environmental Science 1098 (2022) 012028 doi:10.1088/1755-1315/1098/1/012028

A study of work-culture changes at post Covid-19 pandemic in greater Jakarta

Surya Gunanta Tarigan[1,3*], Khalid Abdul Mannan[1,3], Nur Uddin[2,3]

[1]Department of Architecture, Universitas Pembangunan Jaya, Tangerang Selatan, Indonesia
[2]Department of Informatics, Universitas Pembangunan Jaya, Tangerang Selatan, Indonesia
[3]Center for Urban Studies, Universitas Pembangunan Jaya, Tangerang Selatan, Indonesia

surya.gunanta@upj.ac.id

Abstract. This paper presents an examination of culture change in working places due to the Covid-19 pandemic, with a study case in the Greater Jakarta. During the Covid-19 pandemic, most of companies around the world made policy for the employee to work from home (WFH). Implementation of the WFH policy has showed a great success not only in keeping the employee safety and health but also in maintaining employee performance. Now, the pandemic has been shifting to endemic such that working culture is going to change. Health, safety, and flexibility are being keywords of the new working culture. This study assessed the employee perspective about a trend of workplace changing and the needs of supporting facilities. The assessment was done through a survey and analysing the survey data. The result show different types of working system, where 66% work from office, 30% work in hybrid system, 2% work from home, and 2% work from anywhere.

1. Introduction

Working life has changed dramatically in the last two years due to the Covid-19 pandemic. It forcing us to rethink our perceptions of the workplace and the nature of work itself. So far, our understanding of work has been interwoven with concepts of place, which are associated with an office or workplace. Several studies explore how the Covid-19 pandemic affects to the workplace culture in the real world [1–3]. Those studies give a message about adaptation and preparation for success in future, especially in the face of unexpected disruptions like the Covid-19 pandemic. As [4] points out, the Covid-19 pandemic has shown many weaknesses of many systems that have been applied and implemented in daily life around the world. The pandemic has raised numerous questions for businesses such as about the working culture. Some current changes are expected to be temporary, while others will have long-term implications. Arguably, the pandemic has pushed several pre-existing trends in the office sector, such as flexibility, activity-based working, health and wellness, and the desire for better space utilization.

The pandemic has pushed most of the companies to adopt work from-home (WFH) since beginning of the Covid-19 pandemic. Surprisingly, the pandemic has shown that working from home is possible. Even though it is far from ideal, the absence of a commute and the ability to balance work and life are highlighted as significant advantages of remote working. In fact, remote working had been implemented in several countries before the pandemic [4–6]. However, the pandemic has enabled millions of employees to work from home worldwide. Arguably, with the advent of cloud storage and the shift to a paperless working culture, the need to work from an office has become obsolete, as work can be done anywhere.

This trend has also had an impact on Indonesia's working culture. Many companies are finally rethinking about the importance of workplace and how work should be done. In order to attract potential employees, many companies are becoming more flexible in their work environments. Employees are the most valuable and costly resource in any organisation, so employee needs must be prioritised [7]. Indonesia has a distinct work culture. Arguably, the differences between its work culture and other countries significantly impact how businesses adapt to the new normal post-Covid-19 pandemic. This study examines how the companies in Greater Jakarta have adapted to the new normal during the Covid-19 pandemic.

2. Method

This study is evaluating working culture change after the Covid-19 pandemic. Three aspects are considered in the evaluation: 1) rethinking about remote working, 2) acceleration of digital transformation, and 3) hybrid working and workplace culture. A survey about the working culture change is conducted to collect data from employees. The resulted survey data is used in the evaluation.

2.1 A Rethink of Remote Working

As of July 2022, most companies are already back to work from the office, as the public activity restrictions (PPKM) have been relaxed for more than five months. Since the beginning of the pandemic in 2020, the office must implement social distancing and a 50% reduction in office capacity to prevent workplace-related clusters. There are several reasons why remote working will be a viable option such as:

1. Many companies were left with cash-flow issues due to the pandemic, having to pay high rent on offices they were not using. This has been a distressing experience for them, and they will increasingly want to advocate for a more flexible model [4].
2. Previously held meetings and collaborations in meeting rooms are now being replaced by online platforms such as Zoom, Ms Team, Cisco Webex, Google Meet, and others. Because virtual spaces can be accessed from anywhere, the office function as a place for collaboration has shifted to virtual spaces with greater flexibility. Many companies of all sizes are opting for a more decentralised working model as more businesses embrace the concept of remote working. People no longer need to travel to a central office and can work from anywhere. As a result, companies must broaden the definition of" workplace".
3. Viability of working from home or the office has been accepted by both employers and employees. Finally, the current generation (Generation Y & Z) of workers desires more job satisfaction and control over how, where, and when they work. They also want places where they can collaborate, share ideas, and innovate [8]. Generation Y & Z are more digitally savvy and value flexibility than previous generations.

On the other hand, many managers from local and national companies may prefer to work in the office because it gives them more control and reduces the number of emails to which they must respond. Companies are also concerned about the remote working affecting the workforce. Many companies attract employees by emphasising values and cultural practices that are more visible in a formal office setting (for example, training, on-the-job learning, and mentoring). Maintaining these practices is critical for corporate culture, particularly as employees spend more time away from work. In addition, a study also highlighted that employees were concerned about missing out on conversations and office interaction in a situation of isolation [9]. Office is still required as it is a place for employee to have a social interaction.

2.2. Acceleration of Digital Transformation

The requirement to move work online has emphasised the significance of digital transformation. Undoubtedly, the Covid-19 pandemic has accelerated digital technology adoption by several years. The Covid-19 crisis has caused years of change in how businesses in all sectors and regions operate [10]. The crisis has vividly demonstrated the power of digital technologies, and some changes may now be too significant to postpone [11]. According to a McKinsey survey, respondents said their companies moved 40 times faster than expected before the remote working pandemic. Respondents previously stated that implementing the level of remote working during the crisis would have taken more than a year. It took an average of 11 days [12].

As email has altered how we work and communicate, technological advancements will continue to change our work culture and methods. Technology will play a more significant role in office development in the future. Many businesses expect to continue investing in technology such as data security, remote working, and artificial intelligence to maintain productivity. Furthermore, paperless working and collaboration via video conferencing and asynchronous training collaboration have become the new norms in almost all businesses. This is both an opportunity and a challenge for companies, especially given the lack of good infrastructures such as fast internet connections and cloud-based work systems in many developing countries such as Indonesia. Work, space, and information technology (IT) alignment are necessary for success.

2.3. Hybrid Working and Workplace Culture

According to some surveys, even during the Covid-19 pandemic, many employees reported a positive experience with remote working. Nonetheless, very few of them prefer to work in this manner all of the time, preferring instead to work in a hybrid work environment. According to a recent Hays survey, more than half of employees will shift to a hybrid work model [13]. In addition to traditional cubicles and factory spaces, a hybrid workforce is dispersed across multiple locations [9]. Flexibility and choice characterise the working model. Furthermore, according to another survey, hybrid working is driving the office footprint strategy [14].

It is worth noting that hybrid models emphasise activities that build and demonstrate trust. A diverse set of essential values that all employees respect and practice contribute to the positive and effective functioning of the workplace. Organisations can recruit talent better, achieve innovation, and create value for all stakeholders thanks to hybrid work models [15]. Furthermore, the absence of commuting means higher individual productivity in the case of work from-home (WFH) and hybrid model strategies [9]. Employees can choose between at-will employment, split weeks, shift work, and a flexible workweek.

Several studies have begun to look into the benefits and drawbacks of hybrid working. It has been discovered that hybrid working contributes to the reduction of toxic workplace cultures. According to a recent study, most employees would not consider joining a company with a higher salary but a poor work culture [16]. Cooperation and interaction, on the other hand, are more critical than ever in the remote working environment. Working remotely may make the employees to be isolated such that they will lack of conversations and office interaction [9]. Humans are social creatures, and social interaction at work can be difficult for many people, especially during Covid-19.

2.4. Survey to Employees in Greater Jakarta

Employee perspectives are essential in understanding how workplace culture has changed in the last two years during the Covid-19 pandemic. This viewpoint has become critical in the current situation in which businesses reflect their office requirements. A quantitative approach using questionnaires was chosen to understand changes in working culture in various companies and regions of Greater Jakarta. From June to July 2022, data was collected using a Google form via the WhatsApp application. Respondents worked in offices in and around Greater Jakarta for national and multinational corporations. The questionnaire respondents were 61 employees from various industries and locations throughout Greater Jakarta. The questionnaire uses a combination of multiple-choice and open-ended questions to provide a more detailed picture of how workplace culture is changing.

Results of the survey are applied in qualitative research to investigate how the pandemic affected the setting of working arrangements. Moreover, inductive analysis is done in analysis and generalisation. Objectives of using the inductive approach are to: (a) condense raw textual data into a brief summary format, (b) establish clear links between research objectives and the summary of findings derived from the raw data, and (c) develop a framework of the underlying structure of experiences or processes visible in the raw data [17]. This method is beneficial for studying phenomena that are still new and changing; the author recognised that pandemic cases would greatly influence responses during questionnaire distribution.

3. Results and Discussion

3.1 Rethinking business model

The presented result in this paper is at the early stage of the research. At this stage, the survey got 61 respondents who are employees in the Greater Jakarta area. The survey results show a changing pattern of working cultures across many companies in Greater Jakarta. According to the survey, nearly 30% of respondents now work in a hybrid model, 2% work entirely from home (WFH), 2% work from anywhere (WFA), and the rest return to the office. The survey result demonstrates various working models that companies in Greater Jakarta implemented even though office capacity is no longer restricted. Some businesses that return to the office also implement more flexible working hours to assist their employees in adapting to their individual needs. The findings also show that the working culture in Indonesia needs to be more adaptable to the possibility of working from home. Many businesses prefer their employees to present in the office, while others use" wait and see" approach by following the Covid-19 cases day to day. Nevertheless, working cultures in many companies appear to be rapidly changing in response to the new normal needs. According to the questionnaire, the readiness of companies' digital technology significantly impacts the choice of the working model. Leading companies have increased capital expenditures in digital technology and corporate strategies during this period of disruption. Indicators of success are not only strategy and leadership, but also the technology [18].

3.2. Transformation of working culture

What stands out in the data is that almost all respondents said they are now shifting from face-to-face meetings to online meetings; even among respondents who have returned to the office, online meetings are still the preferred option. In the possibility of Covid-19 cases, people likely feel more comfortable having meetings online than offline, and some respondents highlight that online meeting are far more effective. It is included that business travel is significantly reduced because most tasks can be completed through online meetings. Most respondents stated that they have not travelled for business in the last few months. The majority of respondents also want more flexibility in their current working arrangement, with respondents who have the option of working from home or anywhere expressing greater satisfaction with their current working situation. Some respondents also feel unsafe and tired of wearing masks all day in the office, highlighting the importance of health and safety for most employees. According to several reports, the office will be a place to build corporate culture, conduct training, socialise, meet clients, and attract potential talents [4, 18, 19].

3.3. The New Normal Post Covid-19 Pandemic

The covid-19 pandemic reinforced a fact of modern life: we live in a period of change and uncertainty, in which everything can change in a matter of weeks or days due to dramatic economic changes, natural disasters, or other disruptions. The workplace is evolving toward greater collaboration and teamwork. Employees are becoming more empowered; work/life balance is becoming more important; and there is a widespread desire for egalitarian and socially engaging workplaces. In order to respond to these trends, the workplace must be adaptable enough to accommodate a wide range of working styles.

When asked what they thought about the challenges in the new normal following the Covid-19 pandemic, the majority of respondents reflected on this. The constant fear of health when going to work, the fast-paced working culture, how to meet targets with economic uncertainty, and how to collaborate with other colleagues in the new working arrangements are some of the survey respondents' responses. Thus, understanding work and providing flexible workplaces that enable continuous improvement of that work are two things that organisations can do to help differentiate in the era of new normal post Covid-19 pandemic.

4. Conclusion

With advancements in Covid-19 treatment and vaccine rollout, countries are beginning to imagine a future in which the pandemic has no impact on employee and employer decisions. The Covid-19 pandemic will eventually end, and it will be up to the business sector to develop new ways of working in order to maintain and foster this positive shift in workplace management. One thing is certain: many of our offline meetings will be replaced by online meetings in the near future. Short-term and long-term changes in working culture will significantly impact how the office operates following the Covid-19 pandemic.

A hybrid form of work may predominate in terms of the working system: a model in which employees can work at the office, from home, or mix it up throughout the working week. While some workdays may require all employees to be physically present, others may necessitate a combination of physical and virtual presence. Depending on the nature of their tasks and their personal needs or preferences, employees and managers must find ways to work that combine the benefits of face-to face contact with the flexibility of remote work [20].

Technology must play a more significant role in increasing remote workers' productivity and making workplaces safer for returning employees. Employees must be flexible in order to benefit from both modes of operation. People anticipate a much more open and flexible work environment in the future. Furthermore, rather than simply reverting to the past, responsible companies will seize the opportunity in the new normal to reinvent their working model, fostering collaboration and productivity and creating a better human experience for their talents.

Acknowledgement

The authors are very grateful to the Directorate of Research, Technology, and Community Services at the Ministry of Education, Culture, Research, and Technology of the Republic of Indonesia for financial support through a research grant with contract number 156/E5/PG.02.00.PT/2022; 116/SP2H/RT-MONO/LL4/2022; 001/PER-P2M/UPJ-DIKTI/06.22.

References

[1] KevinMKniffin, Jayanth Narayanan, Frederik Anseel, John Antonakis, Susan P Ashford, Arnold B Bakker, Peter Bamberger, Hari Bapuji, Devasheesh P Bhave, Virginia K Choi, et al. Covid-19 and the workplace: Implications, issues, and insights for future research and action. American Psychologist, 76(1):63, 2021.

[2] Andr´e Spicer. Organizational culture and covid-19. Journal of Management Studies, 57(8):1737–1740, 2020.

[3] Antonio de Lucas Ancillo, Mar´ıa Teresa del Val N´u˜nez, and Sorin Gavrila Gavrila. Workplace change within the covid-19 context: a grounded theory approach. Economic Research-Ekonomska Istraˇzivanja, 34(1):2297–2316, 2021.

[4] James Finestone, James Ward, Marcus Morrell, Neil Rebeugeot, Paula Walsh, Rebecca Goldberg, Stu Manton, and Tristan More. Future of offices: in a post-pandemic world, Aug 2020.

[5] Ylenia Curzi, Tommaso Fabbri, and Barbara Pistoresi. The stressful implications of remote working: Evidence from europe. 2020.

[6] Irene Hardill and Anne Green. Remote working—altering the spatial contours of work and home in the new economy. New Technology, Work and Employment, 18(3):212–222, 2003.

[7] Barry P Haynes. Office productivity: a shift from cost reduction to human contribution. Facilities, 2007.
[8] Nicola Gillen. Future Office: Next-generation workplace design. Routledge, 2019.
[9] Michal Beno and Jozef Hvorecky. Data on an austrian company's productivity in the pre-covid-19 era, during the lockdown and after its easing: To work remotely or not? Frontiers in Communication, 6:641199, 2021.
[10] Kevin Sneader and Shubham Singhal. The next normal arrives: Trends that will define 2021—and beyond. McKinsey. com, January, 2021.
[11] Marco Tregua, Cristina Mele, Tiziana Russo-Spena, Maria Luisa Marzullo, and Adriana Carotenuto. Digital transformation in the era of covid-19. In International Conference on Applied Human Factors and Ergonomics, pages 97–105. Springer, 2021.
[12] Kevin Sneader and Shubham Singhal. Beyond coronavirus: The path to the next normal. McKinsey & Company, 5, 2020.
[13] Neil Franklin. Half of firms set to adopt flexible working, Jul 2020.
[14] US PWC. It's time to reimagine where and how work will get done. In PWC, 2021.
[15] Christoph Hilberath, Julie Kilmann, Deborah Lovich, Thalia Tzanetti, Allison Bailey, Stefanie Beck, Elizabeth Kaufman, Bharat Khandelwal, Felix Schuler, and Kristi Woolsey. Hybrid work is the new remote work. Boston consulting group. Viitattu, 8:2021, 2020.
[16] Donald Sull, Charles Sull, and Ben Zweig. Toxic culture is driving the great resignation. MIT Sloan Management Review, 63(2):1–9, 2022.
[17] David R Thomas. A general inductive approach for analyzing qualitative evaluation data. American journal of evaluation, 27(2):237–246, 2006.
[18] Laura LaBerge, Clayton O'Toole, Jeremy Schneider, and Kate Smaje, Oct 2020.
[19] Ian Brinkley, Ben Willmott, M Beatson, and G Davies. Embedding new ways of working: Implications for the post-pandemic workplace, 2020.
[20] Monika Grzegorczyk, Mario Mariniello, Laura Nurski, and Tom Schraepen. Blending the physical and virtual: a hybrid model for the future of work. Technical report, Bruegel Policy Contribution, 2021.

Study on the analysis of demographic data based on spatial information systems

A P Wijaya[1*], A L Nugraha[1], A Sukmono[1], H S Firdaus[1]

[1]Department of Geodetic Engineering, Diponegoro University
Prof. Sudarto Street, District of Tembalang, Semarang city, Central Java, Indonesia, 50277

[*]Corresponding author's e-mail: arwanputrawijaya@lecturer.undip.ac.id

Abstract. The demography database system should be very useful for various purposes in resolving socio-economic problems, such as the distribution of charity, especially for people affected by the Covid-19 pandemic. But unfortunately, the demography database system in Indonesia still has many problems. Things to consider are that there is no integration between databases at the central and local governments, and there is no synchronization of data between institutions. The next problem is that our demographic database system is not based on spatial information. Therefore, we need a demography database system that is integrated with the spatial information system. This study aims to integrate a demographic database system into a spatial information system. The demographic database system in principle always refers to certain administrative units, such as villages, subdistricts, districts/cities, and the state. The administrative unit can contain information about a geographic perspective, such as elevation, position, topography, etc. As a result, in addition to demographic policies, geographic perspectives can be considered, and the identification of asynchronous data between central and local governments becomes easier. In addition, there are impacts, which need to be considered in transforming tabular to spatial data, such as; scale factor, generalization, and classification.

1. Introduction

Rapid population growth, where every day there is always an increase in population from birth and arrival rates, of course, requires fast and accurate data recording as well. However, in reality, there are many problems with our Demographic data. For example, unemployment and poverty rates are out of sync between data on the central government and data on local governments. Especially since the Covid-19 pandemic, poverty and unemployment data have become very important because they are related to the distribution of aid that must be on target. Demographic data out of sync also occurs between agencies, such as; data on the Ministry of Social Affairs, the Central Statistics Agency, state ministries, or other agencies. [1] In addition to mitigating disasters, it is necessary to have a map of vulnerabilities in various sectors, one of which is social vulnerability. In the identification of social vulnerabilities, complete and accurate demographic data is required, including; internal factors (such as; gender, age, health, etc.), ability to recognize disaster (such as; education, experience, etc.), and ability to take preparedness (such as; employment, income, residence, etc.). [2]

The problem of asynchronous data can be caused by the absence of a clear reference to the demographic data. As an example; On the time dimension, one year's data can be multi-interpreted both between the central and the local as well as between agencies. Because each agency has its reference for one year's data, some agencies reference that the beginning of the year is January, and some refer to June or other months. The second problem with our Demographic data is that it is not based on spatial data, (such as location, distance, area, and various other spatial information) but is still tabular or graphic. The impact is that many government policies are related to population issues, only in terms of policies or regulations, but there are many problems in their implementation because the information used by the government is less oriented to spatial data. Therefore, to produce comprehensive population data, it is not only integrated into one reference but also oriented to spatial data.

The Special Region of Yogyakarta is one of the provinces in Java island, Indonesia, and consists of 4 districts and one city, each district being further divided into 78 subdistricts. [3] In carrying out the administration of demographic data, the provincial government also often faces problems, such as the asynchronous between data at the provincial government and data at the sub-district level. This also causes data confusion, for example, the unemployment data at the sub-district level if the total is one hundred people, but when compared to the provincial level data, the female population does not reach one hundred people. Therefore, it is necessary to consider a system that can be integrated, so that problems like this can be avoided.

Geographic information-based dashboards function as media that can present and analyze maps, graphs, and tables simultaneously.[4] For example, the female population in an administrative unit is only in the form of numbers, in fact, we do not only need numerical data but also need graphic information, and the distribution is tabular and spatial. Geographic information-based dashboards can also function to aggregate data. [5]. For example, data on the unemployment rate at the subdistrict level can be aggregated at the district level. Previous research has used Geographic information-based dashboards for various purposes, such as monitoring the performance of smart cities [6], mapping and Tracking Novel Coronavirus pandemics [7], and Decision-making support towards e-Governance [8] many impacts can be generated from the transformation of tabular data into spatial data or based on geographic information, including scale and generalization factors. [9]

This study aims to integrate demographic data, especially in the case study of the Special Region of Yogyakarta in a spatial information system, and as research material, indicators for demographic data are used, indicators of female male, and total population. This spatial information system is implemented on a dashboard based on geographic information. This geographic information-based dashboard will show how data from the subdistrict level is aggregated at the district level spatially and how it affects, especially scale and generalization factors. This aims to show that the implementation of spatial data needs to be considered in addition to data integration because there are impacts.

2. METHODOLOGY

2.1. Demography database systems

In making Geographic information-based dashboards, a database system framework is needed first. A demographic database system is implemented in a multidimensional database system represented on a cube. As a representation of a multidimensional data base system, a cube generalizes a tabular spreadsheet such that there can be any number of dimensions (and not just two as in a spreadsheet). [10] There are three important elements as a basis for building a data cube. The first element is the dimension. Dimensions are important ideas for building data cubes. This dimension aims to group data at the desired level of detail, and that level of detail can be arranged in a hierarchical system. For example, in the spatial dimension, there is a hierarchy from the lowest level, such as individuals, households, and buildings, to the highest level, such as country or region. The second element is a measure. The measurement in this data cube has two meanings. The first meaning is the value of a fact, for example, the selling price/profit of production. The second meaning can mean a formula that can be used to combine several indicators/variables/values into one, and the third element is a fact. Facts are objects that want to be analyzed. This object can be an event or a process. [11]

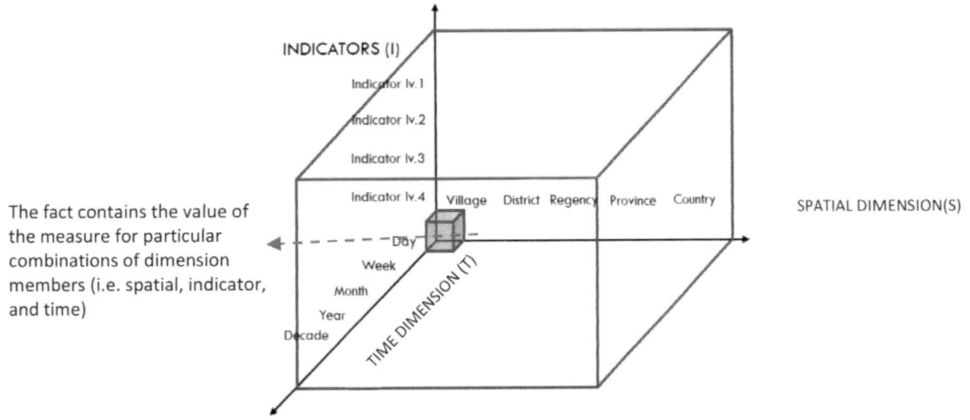

Figure 1. Demography database systems in the cube framework

In the demographic database system, several important elements need to be informed, such as; spatial information, variable/indicator information, and time information. These three elements are implemented into the cube, as shown in figure 1. Each of these elements is implemented as a dimension in the cube, so each becomes a spatial dimension, an indicator dimension, and a time dimension. The small cube in Figure 1 shows certain information or facts from the lowest hierarchical level of each dimension. The information or facts can be values or formulas. For example, the lowest level of the hierarchy from the spatial dimension is the village, from the indicator dimension it is the female population, and from the time dimension it is days. Then a small cube will show information on the number of female population in a certain village, on a certain day.

Information or facts on a group of small cubes can be aggregated into large cubes on the hierarchical level. For example, a group of small cubes from several villages in a subdistrict can be aggregated into a piece of information or fact on the large cube of a sub-district. This is called aggregation on one dimension only, namely the spatial dimension. [12] Likewise, if there are several sub-districts within the same district, the information can be aggregated into information or facts at the district level, and so on it can be aggregated up to the highest level of the hierarchy. The situation can be illustrated in Figure 2.

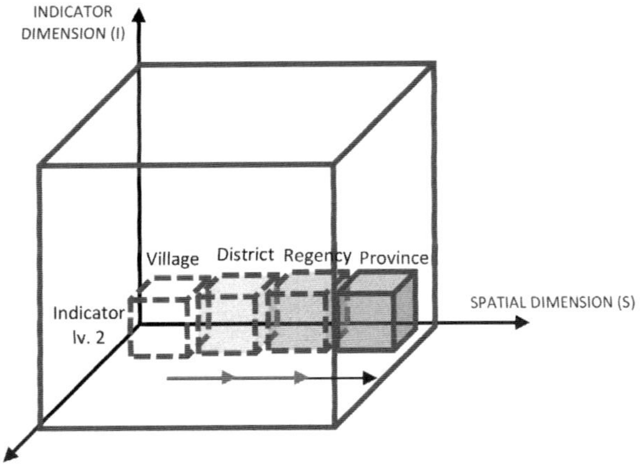

Figure 2. The aggregation process in spatial dimension

2.2. Spatial Information Systems

After the multidimensional database system is formed and the demographic data has been successfully entered, the next step is to join the demographic data in spatial data based on geographic information on a dashboard. There are several choices of elements on the dashboard that can be selected as elements to be displayed in the layout, where more than one element can be selected, according to the information that needs to be displayed. Among them, there are options such as; maps, graphics, and more.

In principle, a fact / small cube, which contains the spatial information shown in the spatial dimensions of the cube, represents an object/location on the earth's surface. [13] So that all facts / small cubes in a cube perspective contain spatial information as if it were a map that represented objects on the earth's surface. And various kinds of indicators and times are presented in the indicator and time dimensions, representing the number of layers with various themes on a map. For example, a collection of facts / small cubes from all villages that are members of the Purwosari district with the indicator used is Female population with 2019 observational data, which can be replaced with a village map in Purwosari district, with the theme female population in 2019. The situation can be illustrated in Figure 3.

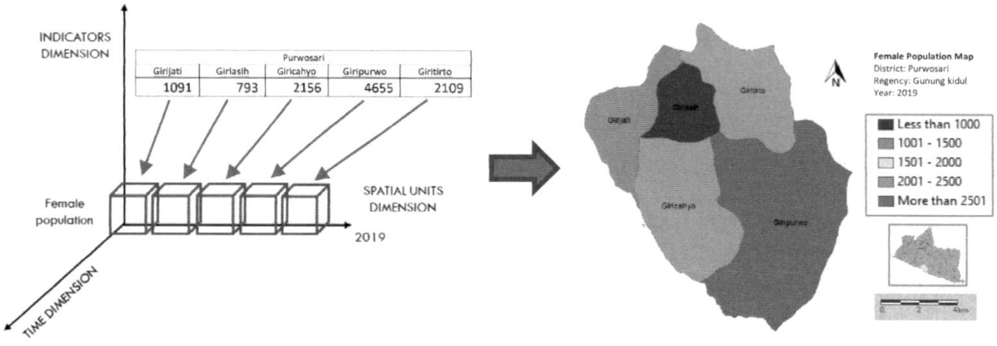

Figure 3. A map as a representation of all facts / small cubes in a cube perspective

Spatial information systems in Geographic information-based dashboards can be formed after all the small cubes in the demographic data are represented in the spatial data according to their respective locations on the spatial dimension. [14] While the value of the indicator dimensions and time in each small cube is represented as an attribute in the spatial data, it can be a piece of separate information or a theme for mapping. Furthermore, the dashboard design can be adjusted to the needs, mainly to represent the concept of integration of demographic data. [15]

Design layout for the dashboard that includes spatial information (regency level and district level) to represent the spatial dimension, time information (choice of year) to represent the time dimension, and indicator information (choice of indicators) to represent the indicator dimensions/ variable. The dashboard itself can be further divided into several pieces of information, such as; maps, pie charts, and bar charts. The map function to show spatial data, pie charts, and bar charts will show the value of the small cube on the demographic data, either partially or in aggregate.

3. RESULTS AND DISCUSSION

3.1. Layout result of the dashboard

Figure 4 shows that there are four layouts that we can access. The first layout shows a map with the theme of a particular indicator. For example; the female population. The map shows which location has the highest female population. The percentage at each location is shown through a pie chart in the second layout. Furthermore, the comparison between indicators in one theme is displayed in the third layout, in this case, for example, on the gender theme, the indicator is Male and female population. The results of the aggregation of the number of male and female populations from all regions on the map can be shown in the fourth layout. To get more detailed information on each region, you can enter the district-level menu. then the distribution of the female population in one regency will be described in each district.

In each district, there is similar information consisting of four accessible layouts. With this system, data can be cross-checked both at the district level and at the regency level. For example; the number of the female population in all districts in the same regency. should be the same as the number of the female population in that regency. On the contrary, the number of female population in one regency should be the same as the number of the female population in all districts in the same regency

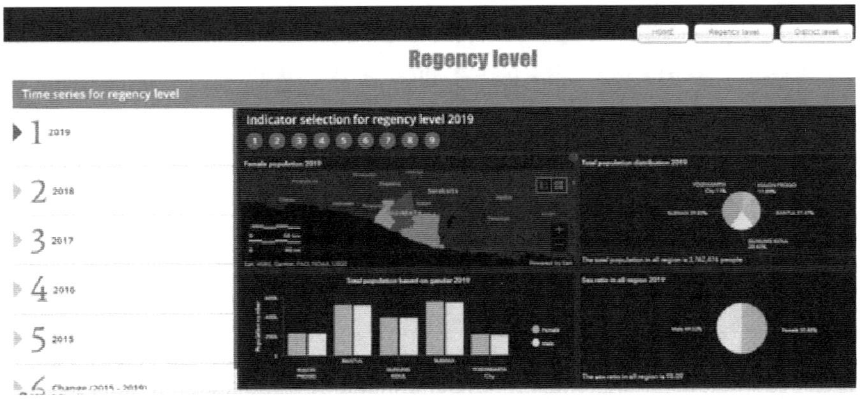

Figure 4. Geographic information-based dashboard layout results for demographic data

This condition, if applied to our real demographic data, will greatly facilitate the government in making policies related to demography, as well as the public or researchers in accessing demographic data. because it is integrated into one system. Each agency related to demographic data has its system; even the central and local governments have different systems for their demographic data.

3.2. Impact of spatially integrated demographic data

One of the advantages of maps in replacing data cubes to represent objects on the earth's surface is the visualization of data from the scaling, generalization, and classification process. [16] However, in a map, these changes affect not only the value but also affect the map scale, generalization, and classification processes. In a sense, if there is a transformation to the higher hierarchical level in the spatial dimension of the data cube (for example, from the village level to the district or regency level), then the transformation process map will have an impact on decreasing the scale (downscaling), resulting in a process of data generalization, and classification also becomes simpler. However, if there is a transformation to the lower hierarchical level in the spatial dimension of the data cube (for example, from the regency level to the district or village level), then the transformation process map will have an impact on increasing the scale (upscaling) so that the distribution of spatial data becomes more and the classification becomes more varied.

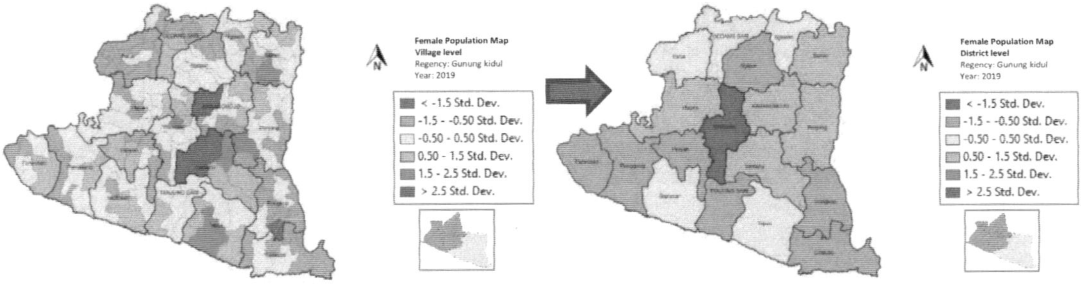

Figure 5. A transformation to the higher hierarchical level in the spatial dimension of the data cube in the map that presents

For example, the transformation to a higher hierarchical level from the village level to the district level, the 2019 female population indicator in the Gunung Kidul regency, and to be able to see the difference, the data were classified according to standard deviations. At the village level, data can be classified into six classes; however, after generalizing at the district level, the data are classified into only four classes. If we look in more detail at the comparison of each village in each district, in one district the spatial distribution is very diverse, it can be classified into at least two classes, three classes, or even four classes. The results of comparisons like this, of course, we cannot observe in the data tabular only. (Figure 5)

In the multi-dimensional level transformation process in the data cube (for example, spatial dimensions and indicator dimensions), if it is presented on a map, the transformation between levels in the indicator dimension will not affect the scaling and generalization process, because, in principle, the various indicators used and their changes are only changes in the theme or layer in a map. It will only affect the classification because each theme will certainly produce its classification on a map. For example, the transformation to a higher hierarchical level from the village level to the district level in the spatial dimension, and from male and female population to the total population in the indicator dimension especially in gender criteria. In each indicator, both female population, male population, and total population, we can see that there are differences in classification, although not many, both at the village level and at the district level. Meanwhile, the transformation from village to district, on all indicators (i.e. female population, male population, and total population), led to generalizations and changes in classification. (Figure 6)

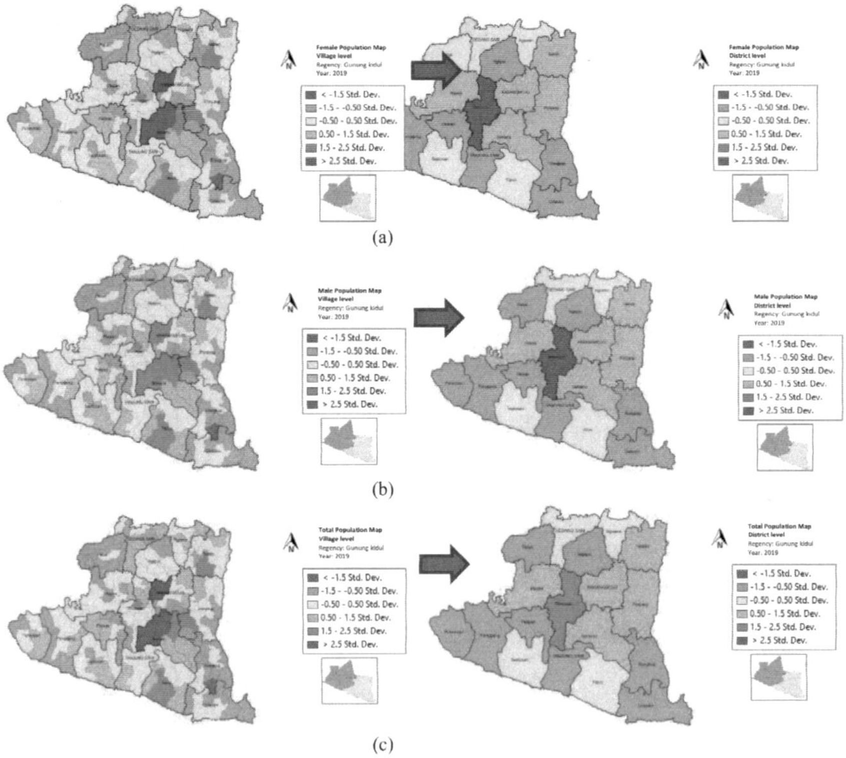

Figure 6. A transformation to the higher hierarchical multidimensional level in the spatial and indicator dimension of the data cube in the map presents: (a) Female population indicator at village and district level, (b) Male population indicator at village and district level, (c) Total population indicator in district and village level

The process of aggregation (generalization) and zoning (classification), especially in spatial dimensions, is very sensitive to MAUP (Modifiable Areal Unit Problem). [17] MAUP is a situation in which the boundary of the zone used in spatial analysis changes from the initial zone, the statistical inference and interpretation derived from the zone are also different. [18] This is indicated, in the transformation process from a lower spatial unit to a higher spatial unit, by a decrease in the scale and a reduction in the number of zones. In the reverse process, the transformation from a higher spatial unit to a lower spatial unit increases the scale and number of zones. For example, in the spatial unit transformation process from village to district in Gunung Kidul district in 2019, both in the female population indicator, male population, and total population, data aggregation and zoning processes occur, so that the number of zones becomes smaller, from five to six classes into four classes only. (Figure 6)

4. CONCLUSION

The implementation of a spatial information system for demographic data can have an impact on various aspects. The first aspect is to enter the raw data of each indicator on each small cube representing each fact in the cube that identifies the three dimensions in the cube (i.e., spatial, time, and indicator). The number of small cubes formed in the entire cube is the product of the number of members from the lowest hierarchy of each dimension. This becomes interesting for the dimension of the indicator because the hierarchy is not parallel like the spatial dimension, so it takes a lot of cubes to be able to implement every raw data that is on each small cube. It even takes a cube within a cube for indicators whose hierarchical members are not perfectly distributed at lower levels of the hierarchy. In principle, a fact / small cube, which contains the spatial information shown in the spatial dimensions of the cube, represents an object/location on the earth's surface. So, all facts / small cubes in a cube perspective contain spatial information in it as if it were a map that represented objects on the earth's surface. And various kinds of indicators and times are presented in the indicator and time dimensions, representing the number of layers with various themes on a map. One of the advantages of maps in replacing data cubes to represent objects on the earth's surface is the visualization of data from the scaling, generalization, and classification process. The transformation process between hierarchical levels in the cube dimension, in principle, is an operational process of value / raw data that is in fact / small cube. However, in a map, these changes not only affect the value, but also affect the map scale, generalization, and classification processes. However, unfortunately, the process of aggregation (generalization) and zoning (classification) especially in spatial dimensions is very sensitive to MAUP (Modifiable Areal Unit Problem). The second aspect is that demographic data at the central government can be controlled by local governments. This can be indicated by the demographic data in all districts in the same regency, which will have the same value as the demographic data in that regency.

Acknowledgments

Authors wishing to acknowledge assistance or encouragement from colleagues, the department of geodetic engineering, Diponegoro university, and financial support from the engineering faculty, Diponegoro University.

REFERENCES

[1] Ridwan 2021 Ego Sektoral, Penyebab Belum Terintegrasi Basis Data Kependudukan *dpr.go.id*

[2] Wijaya A P and Hong J-H 2018 Quantitative assessment of social vulnerability for landslide disaster risk reduction using gis approach (case study: Cilacap regency, province of central Java, Indonesia) *International Archives of the Photogrammetry, Remote Sensing and Spatial Information Sciences - ISPRS Archives* vol 42

[3] Afifah Z N and Sudaryatno 2019 EFFECTIVENESS OF CLASSIFICATION METHOD AND COLOR SYMBOL SCHEME ON CHOROPLETH MAP OF POPULATION DENSITY IN SPECIAL REGION OF YOGYAKARTA. *Int. Arch. Photogramm. Remote Sens. Spat. Inf. Sci.* **XLII-4/W16** 55–61

[4] Esri team 2010 What is a dashboard—ArcGIS Dashboards *arcgis.com*

[5] Djiroun R, Boukhalfa K, Alimazighi Z, Atigui F and Bimonte S 2016 A data cube design and construction methodology based on OLAP queries *Proc. IEEE/ACS 13th Int. Conf. Comput. Syst. Appl. AICCSA*

[6] Jing C, Du M, Li S and Liu S 2019 Geospatial Dashboards for Monitoring Smart City Performance *Sustainability* **11**

[7] Sulaiman N, Abid S K, Chan S W, Nazir U, Mahmud N P N, Abdul Latib S K K, Hafidz H F M, Shahlal S A B, Sapuan S H M and Fernando T 2020 Geospatial dashboards for mapping and tracking of novel coronavirus pandemic *Proc. Int. Conf. Ind. Eng. Oper. Manag.* **59** 2326–35

[8] Lacasandile A D, Abisado M B, Labanan R M and Abad L P 2020 Development of an Information-Based Dashboard: Automation of Barangay Information Profiling System (BIPS) for Decision Support towards e-Governance *2020 The 4th International Conference on E-Society, E-Education and E-Technology* ICSET'20 (New York, NY, USA: Association for Computing Machinery) pp 68–75

[9] Houston D 2014 Implications of the modifiable areal unit problem for assessing built environment correlates of moderate and vigorous physical activity *Appl. Geogr.*

[10] Datta A and Thomas H 1999 The cube data model: a conceptual model and algebra for on-line analytical processing in data warehouses1A related paper introducing this model was presented at the Workshop on Information and Technology (WITS), Atlanta, GA, December 1997.1 *Decis. Support Syst.* **27** 289–301

[11] Jensen C S, Pedersen T B and Thomsen C 2010 *Multidimensional Databases and Data Warehousing* vol 2, ed M T (University ofWaterloo) Özsu (Morgan & Claypool Publishers series)

[12] Xu Z, Lee J, Park D and Chung Y 2017 Multidimensional analysis model for highly pathogenic avian influenza using data cube and data mining techniques *Biosyst. Eng.* **157** 109–21

[13] Wolff M and Asche H 2009 Geovisualization Approaches for Spatio-temporal Crime Scene Analysis -- Towards 4D Crime Mapping *Computational Forensics* ed Z J M H Geradts, K Y Franke and C J Veenman (Berlin, Heidelberg: Springer Berlin Heidelberg) pp 78–89

[14] Takama Y and Yamada T 2009 Visualization Cube: Modeling Interaction for Exploratory Data Analysis of Spatiotemporal Trend Information *2009 IEEE/WIC/ACM International Joint Conference on Web Intelligence and Intelligent Agent Technology* (Milan, Italy: IEEE)

[15] Sedrakyan G, Mannens E and Verbert K 2019 Guiding the choice of learning dashboard visualizations: Linking dashboard design and data visualization concepts *J. Comput. Lang.* **50** 19–38

[16] Yelland P M 2013 *Rolling Up Random Variables in Data Cubes*

[17] Horner M W and Murray A T 2002 Excess Commuting and the Modifiable Areal Unit Problem *Urban Stud.* **39** 131–9

[18] Xu P, Huang H, Dong N and Abdel-Aty M 2014 Sensitivity analysis in the context of regional safety modeling: Identifying and assessing the modifiable areal unit problem *Accid. Anal. Prev.*

IOP Conf. Series: Earth and Environmental Science 1098 (2022) 012030 doi:10.1088/1755-1315/1098/1/012030

Climate Change Issue and Consumer Behavior in Purchasing Beauty Product

Lola Kurnia Pitaloka[1*] **and Anna Kania Widiatami**[1]

[1]Faculty of Economics, Universitas Negeri Semarang

*lolakp@mail.unnes.ac.id

Abstract. The purpose of this study is to see how consumer awareness, trust, and social influence can interest consumers buy green cosmetic products. This research is an exploratory quantitative analysis with Structural Equation Model as a data analysis with warpPLS as a tool analysis. The population is active users of cosmetic products in Indonesia with an age range of 15 to 35 years. The sampling technique used accidental sampling because there were no exact data and the sample are 100 respondents. Data collected using a close questionnaire. In this study, all variables affect consumer purchasing decisions as large as 0,03 and 0,001, but this study found that social influences weaken brand trust. It can be said that the green character of consumers is still difficult to form. Consumers, as a social individual, still depend on the decisions of others because they want to be accepted in their environment. Unfortunately, there are still many people in Indonesia who do not have a strong green character. Awareness of climate change still needs improvement, and it requires the commitment of environmental organizations to socialize the dangers of climate change and increase the green character in the community to influence consumer purchases.

1. Introduction

Climate change due to increased greenhouse gas emissions from human activities is a global problem today. The increase in greenhouse gas emissions has the impact of climate change that has been felt in various parts of the world. Some of the impacts that are often felt are an increase in the incidence of natural disasters due to an increase in very extreme weather. The increase in temperature and sea level has also begun to worry many parties, this has caused widespread flooding and several islands have been submerged [1]. Indonesia itself has a high level of vulnerability as an archipelagic country to the impacts of climate change. Since 1997, Indonesia has been trying to reduce national greenhouse gas emissions, especially in the sectors that contribute the highest emissions, namely forestry, energy, transportation, industry, and waste.

In a report issued to the Intergovernmental Panel on Climate Change (IPCC), global air temperatures have increased by 0.6 degrees Celsius (1 degree Fahrenheit) since 1861. Warming is mainly caused by human activities that add greenhouse gases to the atmosphere. In fact, the IPCC predicts that the increase in global average temperatures will increase by 1.1 to 6.4 degrees Celsius (2.0 to 11.5 degrees Fahrenheit) between 1990 and 2100. This condition causes the climate to continue to warm for a certain time due to emissions that previously released, and carbon dioxide remains in the atmosphere for hundreds of years or more before nature can absorb it again [2].

One of the causes of the increasingly hot climate is the greenhouse gas effect caused by industry production, especially large industries. Today, several large industries are mastered by the cosmetics industry. The cosmetics industry is growing rapidly due to the opening of the world market and the changing culture of generations. In 2019, the growth of the cosmetics industry grew by 7% and is expected to continue to increase according to Indonesian Ministry of Industry in 2020. This growth encourages entrepreneurs to make many innovations to maintain their existence in the industry. One of the innovations that cosmetic brands are doing these days is to create environmentally friendly cosmetic products. The growth of the market for eco-friendly cosmetics is huge and rapid. Especially for the millennial generation who grew up with the tagline of environmental conservation.

According to the research of Liobikiene et al. [3] at least the world has produced hundreds of millions of tons of waste every year and it is recorded that the cosmetics industry contributes 120 million tons of waste annually. Franca and Ueno [4] in their research produced a study on trends of the green cosmetics industry and saw that 30% of consumers in the United States today prefer to buy green cosmetics. However, behind that, there is a gap between what consumers talk and do. In his research, it is said that there are several factors that are a problem for consumers in the purchase decision of these green cosmetics. These obstacles include awareness of green cosmetic brands and trust in the brand.

Awareness can be interpreted as the knowledge that there is something or an understanding of a situation or subject now based on information or experience. The awareness of green brand in this study is interpreted as consumers' understanding of the content of ingredients, the production methods, and cosmetics impact of the production of cosmetics on the environment. Environmental issues are closely related to the awareness of the consumer of environmental issues. Raharja [5] explained that the growth of consumer awareness of the environment makes consumers aware of their response to live a sustainable life. Today's consumers clearly have such an awareness of environmental issues that most of them decide to consume green products, including green cosmetics. But the reality is that consumers in Indonesia themselves still have little action to shift consumption to green cosmetics. An initial survey conducted by researchers showed that 35% of consumers are willing to move to cosmetic brands that carry environmental sustainability. This is a separate gap between the theory of awareness and the reality that exists. According to Sharma [6], green brand awareness is one of the considerations for consumers when making decisions to buy green cosmetic products. This is based on the consumer's final choice when he realizes what he is doing, which will all have an impact on the environment.

Al-Haddad et al. [7] stated that some cosmetic brands take only the claim of green cosmetic sentences. But they themselves do not apply green cosmetics. The perception of green cosmetics is not only based on the ingredients used for the manufacture of cosmetics, but refers to the entire production process of these cosmetics, including the packaging that is used and after sales offered to the product packaging. Trust is one of the factors that influence buying decisions. Brand trust is defined as a sense of security owned by product users, in their interaction with a brand that is based on the perception that the brand can be trusted, in this research means that the brand can be trusted that it is truly producing according to the foundation of sustainability for the environment.

There are differences in the results in the research of Le Thanh et al. [8] and the research of Huang et al. [9], where Le Thanh states that brand trusts have no say in consumer's buying decisions for green cosmetic products. The use of cosmetics itself is an awareness of everyone, so the decision to buy everything is returned to the awareness of that individual in consuming. Therefore, brand trusts do not play a big role in purchasing decisions. Meanwhile, Huang explained that brand trusts have a great influence on the decision of green cosmetic products. People's growing perception makes their consumption also develop. The existence of the environmental care movement makes the younger generation of people feel ashamed if they do not participate in the campaign. The problem is that in Indonesia, Indonesians are still skeptical of the claims made by cosmetic manufacturers. So, the purchasing decision of the Indonesian people for green cosmetics is still low.

Decisions and behavior of others towards the environment can be caused by the social influences around them [10]. Individuals tend to adhere to what is the basis of their environment to be well received. Individuals with an environment that cares about the environment and health, will indirectly make them have the same character [11]. Social influence in this study became a moderation variable to see how social influence can strengthen both other factors in purchasing decisions. According to

Febrian and Fadly [12], the biggest influence in this era of social networks comes from influencers. In his research, it is mentioned that the factors influencing consumers' purchasing decisions increase in the presence of strong social influences.

2. Research Methods

This research method used an exploratory quantitative method, where this research wanted to look deeper into consumer purchasing decisions by suspecting other factors that could be influence dependent variable. This method is also used to dig deeper so that it can describe social phenomena related to the change in consumer consumption of green cosmetics to support the slowdown in climate change. This research was carried out in Semarang City, one of the metropolitan cities in Indonesia that has a large distribution chain of cosmetic products and has become one of the major markets for green cosmetics in Indonesia.

The population in this study was individuals who fall into the category of generation Y and generation Z, or people with an age range of 15 to 35 years in 2022. The research was carried out on 3 cosmetic brands known to the public as green cosmetic manufacturers, namely The Body Shop, Love Beauty and Planet, and Innisfree. The sampling technique used accidental sampling with a total study of 7 days. This sampling technique was taken because there was no exact number of populations. The saturated sample taken in this study was 100 respondents.

Data collection techniques used close questionnaires and analyzed used Structural Equation Model because this analysis could identify the dimensions of the construct and at the same time, is able to measure the influence or degree of relationship between factors for which the dimensions have been identified. This research analysis tool uses WarpPLS as a development of the PLS analysis tool.

3. Results and Discussion

3.1. SEM Measurement and Model Analysis

3.1.1. *Common Method Bias*

Common bias methods are commonly used to count for errors in measurements because the questionnaire method may cause bias. The common method bias test is seen from the value of full collinearity VIFs which is the result of full collinearity testing, including vertical and lateral multicolonierity. The criteria for the full collinearity of VIFs is < 3.3.

Table 1. Full Collinearity Value VIFs.

Variable	*Full Collinearity VIFs*
Green Brand Awareness (X1)	1.128
Brand Trust (X2)	1.159
Social Influence (X3)	1.417
Purchasing Decision (Y)	1.419
X3*X1	1.129
X3*X2	1.114

Source: data processed in 2022

As shown in the table above, the value of VIFs in all variables in this study has a value of < 3.3. So, this research model does not detect multicollinearity and it is certain that independent variables can affect dependent variable.

3.1.2. *Outer Model*

Evaluation of the outer model is carried out to construct each indicator of the existing variables to find if any errors have occurred. This evaluation includes assessing the convergence validity and composite reliability. The construction of the indicator value can be seen from the value of combined loads and cross loadings. The outer accepted model indicated by the loading factor of each indicator must be >

0.70. However, acceptance of the loading factor value < 0.70 and > 0.40 can still be considered for acceptance by looking at the AVE value. The result of this study still shows some indicators whose value < 0.40 and these indicators should be eliminated to avoid bias.

Table 2. Loading Factor P Value, AVE, Composite Reliability.

Variables/Items	Loading Factors	P-Values	AVE	Composite Reliability
Green Brand Awareness				
X1. 1	0.651	< 0.001		
X1. 2	0.555	< 0.001		
X1. 3	0.868	< 0.001		
X1. 6	0.933	< 0.001	0.616	0.916
X1. 7	0.863	< 0.001		
X1. 8	0.821	< 0.001		
X1. 9	0.731	< 0.001		
Brand Trust				
X2.1	0.893	< 0.001		
X2.2	0.853	< 0.001		
X2.3	0.698	< 0.001	0.710	0.924
X2.4	0.877	< 0.001		
X2.5	0.874	< 0.001		
Social Influence				
X3.1	0.833	< 0.001		
X3.2	0.872	< 0.001	0.672	0.891
X3.3	0.821	< 0.001		
X3.4	0.748	< 0.001		
Purchasing Decision				
Y.1	0.819	< 0.001		
Y.5	0.635	< 0.001	0.601	0.856
Y.6	0.842	< 0.001		
Y.8	0.790	< 0.001		

Source: data processed in 2022

The research model is acceptable if it can meet convergence validity and composite reliability. Convergent validity of the model seen from the value of the loading factor of each indicator and the AVE of each variable, while the composite reliability is seen in the composite reliability coefficients when the value > 0.70 then the questionnaire is reliable. The table above shows that the composite reliability coefficients have been qualified so that the outer model of this study can be accepted without changes in indicators.

Table 3. Correlations among Latent Variables and errors.

	Green Brand Awareness	Brand Trust	Social Influence	Purchasing Decision
Green Brand Awareness	0,785	0,213	0,108	0,255
Brand Trust	0,213	0,842	0,112	0,343
Social Influence	0,108	0,112	0,820	0,410
Purchasing Decision	0,255	0,343	0,410	0,775

Source: data processed in 2022

Based on the table above, there is a correlation between all variables in their diagonal values. All variables have a good correlation value with other variables. It can be described that the entire variable meets the discriminant validity. Therefore, based on the result of convergence validity, composite reliability, and discriminant validity, this research model can be analysed further.

Table 4. *Model Fit* and *Quality Indices.*

No.	***Model Fit* and *Quality Indices***	**Fit Criteria**	**Analysis Results**	**Note**
1	APC[1]	$p < 0.05$	0.228 p = 0.00 4	Accepted
2	ARS[2]	$p < 0.05$	0.384 P < 0.001	Not Accepted
3	AARS[3]	$p < 0.05$	0.351 P < 0.001	Accepted
4	AVIF[4]	*Acceptable if* <= 5, *ideally* <= 3.3	1.076	Ideal
5	AFVIF[5]	*Acceptable if* <= 5, *ideally* <= 3.3	1,228	Ideal
6	GoF[6]	*Small* >= 0.1, *medium* >= 0.25, *large* >= 0.36	0.542	Big
7	SPR[7]	*Acceptable if* >= 0.7, *ideally* =1	0,800	Accepted
8	RSCR[8]	*Acceptable if* >= 0.9, *ideally* = 1	0,974	Accepted
9	SSR[9]	*Acceptable if* >= 0.7	1,000	Accepted
10	NLBCDR[10]	*Acceptable if* >= 0.7	0,900	Accepted

[1]Average Path Coefficient
[2]Average R-Squared
[3]Average Adjusted R-Squared
[4]Average block VIF
[5]Average full collinearity VIF
[6]Tenenhaus GoF
[7]Sympson's Paradox Ratio
[8]R-Squared contribution Ratio
[9]Statistical Suppression Ratio
[10]Nonlinier Bivariate Causality Direction Ratio

Based on the preliminary research data above, this research model has met the construction requirements to be continued as a research model. The structures in the SEM model must be accepted by all to ensure that the model and variables do not encounter errors for hypothesis testing.

3.2. Hypothesis Test

Model testing will be seen from the results and values of the model that come out in the SEM analysis that has been carried out. The purpose of testing this model is to see the direction, relationship, and magnitude of the coefficients between variables.

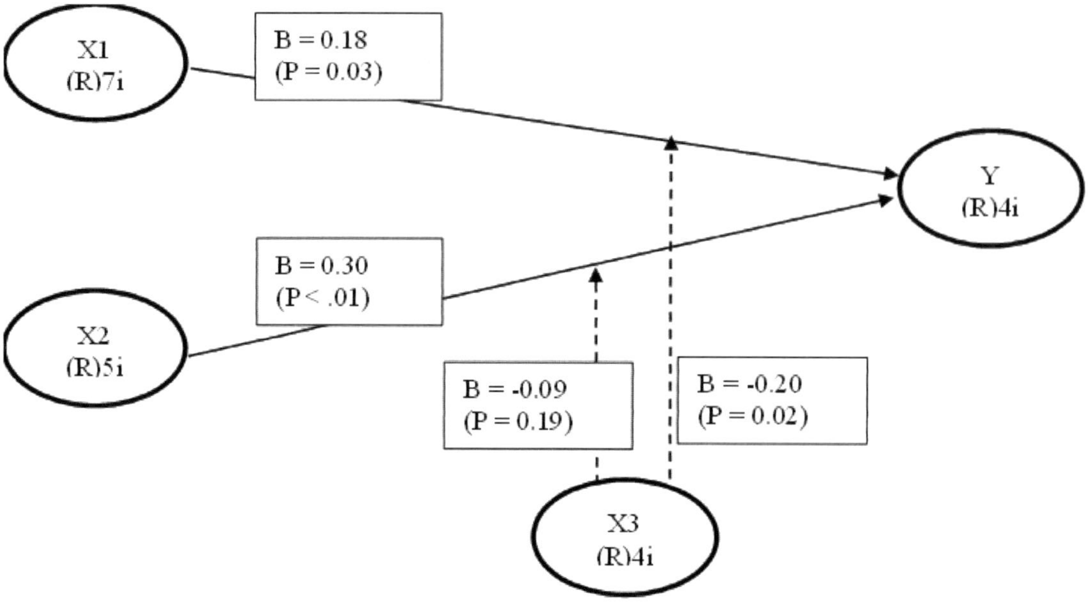

Figure 1. Research Model.

Table 5. Research Model Results.

No.	Line	Coefficient	P *Value*
1.	Green brand awareness affects the purchase decision of green cosmetic products	0.181	0.030
2.	Brand trust influences the purchase decision of green cosmetic products	0.296	< 0.001
3.	Social influence strengthens the influence of green brand awareness on the purchase decision of green cosmetic products	-0.205	0.016
4.	Social influence strengthens the influence of brand trust on buyer decisions green cosmetic products	-0.085	0.192

3.3. The Influence of Green Brand Awareness on the Purchasing Decision of Green Cosmetic Products

In the hypothesis testing table, the green brand awareness variable has a positive direction towards purchasing decisions with a coefficient of 0.181 and a p-value requirement of 0.030. Therefore, it can be concluded that awareness of the green brand has an influence on the purchase decision of green cosmetic products. Even so, green brand awareness has a small opportunity to influence consumers' purchasing decisions.

This study describes green brand awareness as a consumer's understanding of the content of ingredients, production methods, and their impact on the environment. It can also be interpreted as a collection of consumer perceptions related to environmental issues. Consumers in Semarang City apparently realize many things about the perception of health for their bodies. Some elements for this variable question emphasize consumers' concern for the impact on themselves by consuming green cosmetics. In addition, consumers in generation Y (age range 28-35 years) turned out to have the same reason as generation Z (age range 15-27 years) about awareness of green products. Today, it is easy for consumers to search for information about the dangers they face if they consume conventional cosmetics in the long term.

This research is supported by Mukaromah et al. [13] who state that awareness of green brands is one of the factors that can determine consumers' purchasing decisions for green products. There are many factors that can control consumer awareness of the environment, some of them are age, education, gender, and income [14]. Education affects consumer behavior, one of them is green behavior. This is supported by the result from Lee et al. [15], the higher the consumer education, the higher their pro-environment level.

Green cosmetic product is developing rapidly. Even today, green cosmetic products have already dominated the market. Cosmetic consumers in Indonesia with their awareness, are ready to move to green cosmetic products. This statement is reinforced by the research of Hussain et al. [16], which states that the higher the con-consumer awareness the production process of an item can damage the environment, the higher the decision consumer purchases of green products. However, on the other hand, there is a distortion in the green awareness, that is greenwashing. In Zhang et al. [17], greenwashing has a negative impact on consumers' green awareness, which has an impact on purchasing decisions. For some manufacturers, changing the direction of their products is a challenge because it requires large costs. So, they go the dirty way to be able to compete in the market. The practice of greenwashing becomes biased toward the consumers' green awareness. This distortion is the reason why awareness of the green brand has a weak influence on purchasing decision.

However, according to consumers, there are several other factors that they consider when deciding to buy green cosmetic products. The biggest factor is the price of such products. During this time, the market raised prices for green cosmetic products. This is certainly not without reason, of course, the price of green cosmetic production materials is more expensive than conventional cosmetic products. Therefore, green products have a higher selling price. Consumer income tends to be an obstacle due to high prices. In this study, awareness has an influence on consumers' purchasing decisions, but its influence is small. There are other factors that cause why consumers decide not to buy green cosmetics.

3.4. *The Effect of Brand Trust on the Purchasing Decision of Green Cosmetic Products*

In the hypothesis testing table, the brand trust variable has a positive direction with a coefficient of 0.296 and a p-value requirement of < 0.001, which means that brand trust has a very strong influence on buying decisions green cosmetic products. In this study, brand trust is defined as the trust of a consumer in a brand for their campaign to continue producing cosmetics with environmental sustainability rules.

Consumers of Semarang City apparently do not just buy a product, especially green cosmetics. They choose brands that claim to produce green cosmetics. For them, because the lifestyle of urban people today is really concerned about health and the environment, manufacturers look at it very sharply. Thus, many manufacturers began to claim that their products were environmentally friendly. Market expansion makes producers more aggressive to follow what the market wants to continue to exist in their fields.

Trust is a concept that is difficult to define clearly. This is because it concerns human psychology about what the brand builds and believes in consumers. However, trust is the most important thing in a business for a manufacturer. Even with consumers believing in the brand, consumers will not hesitate to continue making purchases and discuss them with others. This statement is supported by research by Huang and Guo [18] that states that brand trusts have the highest effect on consumer purchases. The more consumers believe in the brand, the greater the decision to buy and even promote it.

The results of interviews with consumers in Semarang City, stated that "The Body Shop" is a very trusted brand when consumers are talking about green cosmetic products. Apart from its early appearance, "The Body Shop" has claimed that it is a green cosmetic product that is environmentally friendly and without animal testing, The Body Shop does play an active role in the preservation of the environment and is committed to providing harmless ingredients both for consumers and the surrounding environment in their products. Most of the respondents when asked about cosmetic products that they believe really pay attention to environmental issues will answer The Body Shop. They believe and believe in the commitment of the product because there is already a lot of real evidence. From this we can see that the campaign built by The Body Shop brand has fostered consumer trust and it can be concluded that trust has a strong influence to purchasing decisions.

3.5. Social Influence Strengthens the Influence of Green Brand Awareness on Purchasing Decisions of Green Cosmetic Products

In this study, the social influence variable is a moderator variable for the green brand awareness variable on consumer purchasing decisions. Previous results have shown that awareness of green brand has a weak influence on the purchasing decision for green cosmetic products. According to the hypothesis testing table, the social influence variable in its role in strengthening the green brand awareness variable has a negative direction with a coefficient of -0.205 and a p value of 0.016. This means that social influence can be a moderator between awareness of purchasing decisions, but its nature tends to weaken.

The social status of an individual is certainly related to the family and the surrounding environment. In this study, social influence was focused on the environment within individuals' social media, including influencers. Today, we are finding a new job in the digital world, namely influencers. The social influence of the family environment, friends, and the surrounding environment has been a long-standing research topic where the three environments can affect the consumption behavior patterns of a person consumers. Consumers will tend to follow the consumption patterns of their social environment because the psychological nature of human beings is to want to continue to be accepted by the surrounding environment. Therefore, if the surrounding environment is aware of the consumption of green cosmetic products to maintain environmental sustainability, consumers will tend to follow this mindset and can strengthen the awareness of purchasing decisions.

According to Ambarwati et al. [19] the influence of digital influencers today is very strong on consumers' purchasing decisions. But unfortunately, nowadays we also find cases of many lies committed by digital influencers that cause influencers to have a bad image in society. This influences products that have been promoted or campaigned by the influencer. People start not believing what influencers say because they think influencers are the same as advertisements on television. This is what causes social influence to weaken the variable of green brand awareness in purchasing decisions.

3.6. Social Influence Strengthens Brand Trust's Influence on Purchasing Decisions of Green Cosmetic Products

Social influence is also a moderator variable for brand trust variables on consumers' purchasing decisions. Previous results have shown that brand trusts have a very strong influence on the purchasing decisions of green cosmetic products. In the hypothesis testing table, the social influence variable in its role to strengthen the brand trust variable has a negative direction with a coefficient of -0.085 and a p-value requirement of 0.192. It can be interpreted that social influence has no influence on the ability of brand trust to influence consumers' purchasing decisions.

Although awareness can be influenced by social influences, beliefs are abstract things that are difficult for others to influence. Consumer confidence in a brand will differ depending on the conditions that occur in them. Especially if we talk about cosmetic products, between the consumer and others will have different preferences because they have different skin types and needs. Moreover, consumer confidence in a brand on whether it is right to apply the principle of green cosmetics cannot even be built on just promotion through digital influencers.

According to interviews with key people, they have more confidence in brands that have succeeded in convincing them that their campaign is not just a campaign. According to respondents, trust is something that is difficult to describe and impossible to control by others, especially people they do not really know. Claims from digital influencers are often exaggerated, and they do so because they already have a cooperation contract that cannot be violated. So, for consumers, their social environment only helps introduce some brands that do have campaign claims about green cosmetics, but it is the trust issues that make consumers buy the brand from the internal side of the consumer.

4. Conclusion

The results of the existing discussion can be concluded that awareness of and trust in the green brand influence on the decision to buy green cosmetic products directly. However, the awareness of the green brand has a weak influence. This is due to several other factors that consumers think more about and the biggest one is the price issue. Furthermore, the issue of greenwashing in the community has reduced the awareness of green awareness. Social influence has no influence on the strengthening brand trust in consumers' purchasing decisions. However, in this study, we saw that social influence weakened the influence of green brand awareness on consumers' purchasing decisions. This can be seen in the indicators of influencers in the direction of social media. Due to the many cases that have happened to influencers, now people no longer trust digital influencers and apparently this makes green brand awareness weak for purchasing decisions.

References

[1] Masson-Delmotte V, Zhai P, Pirani A, Connors S L, Péan C, Berger S, ... - Zhou B 2021 Climate change 2021: the physical science basis *Contribution of working group I to the sixth assessment report of the intergovernmental panel on climate change* 2

[2] Arneth A, Shin Y J, Leadley P, Rondinini C, Bukvareva E, Kolb M, ... - Saito O 2020 Post-2020 biodiversity targets need to embrace climate change *Proc. Natl. Acad. Sci.* **117**(49) 30882-30891

[3] Liobikienė G and Bernatonienė J 2017 Why determinants of green purchase cannot be treated equally? The case of green cosmetics: Literature review *J. Clean. Prod.* ***162*** 109-120

[4] Franca C C V and Ueno H M 2020 Green cosmetics: perspectives and challenges in the context of green chemistry *Desenvolvimento e Meio Ambiente* **53** 133-150

[5] Raharja B S 2021 Content quality, religious consciousness, and brand awareness: The empirical evidence on islamic sharia product *J. Manag. Appl.* **19**(3) 572-584

[6] Sharma A P 2021 Consumers' purchase behaviour and green marketing: A synthesis, review and agenda *Int. J. Consum. Stud.* **45**(6) 1217-1238

[7] Al-Haddad S, Awad A, Albate D, Almashhadani I, and Dirani W 2020 Factors affecting green cosmetics purchase intention *J. Manag.. Inf. Decis. Sci.* **23**(4) 332-342

[8] Le Thanh T, Ngo H Q, and Aureliano-Silva L 2021 Contribution of corporate social responsibility on SMEs' performance in an emerging market–the mediating roles of brand trust and brand loyalty *Int. J. Emerg. Mark.* 10.1108/IJOEM-12-2020-1516

[9] Huang C and Guo R 2021 The effect of a green brand story on perceived brand authenticity and brand trust: the role of narrative rhetoric *J. Brand Manag.* **28**(1) 60-76

[10] Lazaric N, Le Guel, F, Belin J, Oltra V, Lavaud S, and Douai A 2020 Determinants of sustainable consumption in France: the importance of social influence and environmental values *J. Evol. Econ.* **30**(5) 1337-1366

[11] Rustam A, Wang Y, and Zameer H 2020 Environmental awareness, firm sustainability exposure and green consumption behaviors *J. Clean. Prod.* **268** 122016

[12] Febrian A and Fadly M 2021 Brand Trust As Celebrity Endorser Marketing Moderator'S Role *Journal of Management Applications* **19**(1) 207-216

[13] Mukaromah A L, Teja I G N A E, and Anggraini N P N 2019 The effect of green marketing, brand awareness and price perception on purchase decision *Int. J. Appl. Bus. Int. Manag.* **4**(3) 75-83

[14] Arttachariya P 2012 Environmentalism and green purchasing behavior: A study on graduate students in Bangkok, Thailand *BU Acad. Rev.* **11**(2) 1-11

[15] Lee Y K, Kim M S, Katz-Gerro T, and Kim Y 2019 Personal values, perceived consumer effectiveness and demographic effects on green purchasing behavior of Korean consumers *Environ. Eng. Manag. J.* **18**(2) 349-358

[16] Hussain M A, Khokhar M F, and Asad A 2014 Green Awareness Effects on Consumers Purchasing Decision: A Case of Pakistan *Glob. J. Manag. Bus. Res.* **14**(6) 8-15

[17] Zhang L, Li D, Cao C, and Huang S 2018 The influence of greenwashing perception on green purchasing intentions: The mediating role of green word-of-mouth and moderating role of green concern *J. Clean. Prod.* **187** 740-750
[18] Huang C and Guo R 2021 The effect of a green brand story on perceived brand authenticity and brand trust: the role of narrative rhetoric *J. of Brand Manag.* **28**(1) 60-76
[19] Ambarwati M F L, Damaryanti H, Prabowo H, and Hamsal M 2019 The Impact of a Digital Influencer to the Purchase Decision *Sci. Tech. J. of Proc. Ser.* **5** 220-224

Quality of surface water due to sand mining activity: a case study from the Progo River, Daerah Istimewa Yogyakarta Province, Indonesia

P T Trisnaning[1], A Zamroni[1,3*], O Sugarbo[1], H N E Prasetya[2], S T Sagala[2], M Y Hardiansyah[4,5]
[1]Department of Geological Engineering, Institut Teknologi Nasional Yogyakarta, Indonesia
[2] Department of Mining Engineering, Institut Teknologi Nasional Yogyakarta, Indonesia
[3]School of Environmental Science and Management, University of the Philipines, Los Baños, Laguna 4031, The Philippines
[4]Global Agriculture Technology and Genomic Science, International College, National Taiwan University, Taiwan
[5]Department of Agronomy, Agrotechnology Study Program, Faculty of Agriculture, Hasanuddin University, Indonesia

*) Corresponding author: akhmadzamroni@itny.ac.id

Abstract. Due to sand mining, the study objective is to assess water quality in the Progo River, Daerah Istimewa Yogyakarta Province. It is beneficial to evaluate the physicochemical parameters of the Progo River and to appraise water quality for drinking purposes based on the Indonesia Minister of Health regulation No: 492/ Menkes/Per/IV/2010. The physicochemical parameters used in this study included pH, EC, TDS, DO, SO_4^{2-}, NO^{3-}, and PO_4^{3-}. The pH decrease is thought to be due to the opening of sand deposits around the river so that minerals such as feldspar, hornblende, and quartz dissolve and mix in water. The TDS increase was estimated due to the increase in the riverbed and bank erosion associated with sand mining at the mining and downstream sites. In addition, sand mining activities affect higher DO because bubbles aerate flowing water in fast-moving water as it churns over rocks and falls into hundreds of tiny waterfalls. Meanwhile, other parameters, such as EC, SO_4^{2-}, NO^{3-}, and PO_4^{3-} are influenced by natural factors, such as seawater mixing and soil erosion. Almost all parameters are permissible for drinking purposes, except for Station 3 at the downstream which TDS and SO_4^{2-}- values are.

1. Introduction

The structure, productivity, abundance, and physiological conditions of water organisms are influenced by water's chemical and physical characteristics [1]. Rivers are essential to the survival of all lives, offering primary residential, industrial, and agricultural water supplies. Unfortunately, the indiscriminate dumping of sewage, toxic waste, and many human activities affecting their physicochemical parameters and microbiological efficiency pollute river waters [2]. Due to rapid urbanization, industrialization, and related developments, gravel and sand demand has increased exponentially over the past few decades [3]. Sand is raw material used in all kinds of development projects, including the construction of artificial islands, land reclamation, airports, dams, homes, bridges, pipelines, highways, roads, coastal stabilization, and septic tanks [3, 4]. Consequently, the sand needs would contribute to a rise in numerous sand mining operations in watersheds [5]. These ventures have economic and social advantages, especially in developing countries, but at the same time sand mining may caused environmental issues. Environmental concerns occur the moment gravel, sand, and other materials extraction exceeds the rate at which natural processes produce these materials [4, 6]. Sand mining is perceived as a disruption that increases sediment contact with the changes and flows of the river's morphology. The rugged nature of stream and canal flows causes the transportation of bed loads [7].

In the last few decades, demand for gravel and sand has increased dramatically in line with increasing economic development [8]. The rapid population rise, exponential urbanization, and growing economic growth have led to high and continual sand demand as a residential and general infrastructure building material. River channels, as the sand's primary source, provide high-quality material at a low cost [9]. Water quality means the key parameters that decide the water supply quality. Chemical, metabolic operations, growth, reproduction, feeding, physical, and biological water are parameters. In an aquatic environment, such as river bed erosion, sand mining operations can weaken bridge supports, pipelines, or other structures. Mining can displace or deplete marine biota or its food chain [10]. Alluvial extraction of sand and gravel often creates severe problems for surface water and groundwater sources. The high amount of suspended particles in the water column causes significant impairment of river habitats due to sand mining activities [11].

Sand mining may develop into an environmental concern as the sand demand in manufacturing and construction increases. Instream sand mining is one of the human operations to degrade the physicochemical parameters of river water. Instream mining activity leads to sediment transport interruptions and suspended solid particle loads that increase turbidity by creating higher concentrations of finer particles [12]. Mining and its related operations can be responsible for significant environmental degradation if this activity is conducted without sufficient guidance and can lead to vegetation clearance, soil erosion, and mass movement [13]. The weathering process of surface rock combines basic anions and cations into the river water caused by several anthropogenic and terrestrial activities. Deepening the channel bed contributes to a rise in turbidity by interrupting the stream's flow. Then, the turbidity phase deteriorates the water quality and decreases light penetration into the flow. Any changes in physicochemical parameters destroy the channel's aquatic environment. Mining effects not only be focused at mining sites, but also may spread across the stream's ecosystem [12]. Heavy metal contamination also tends to increase in mining areas [14]. River water is churned up to form clouds consist of fine organic and inorganic particles by the destruction caused by indiscriminate sand and gravel mining. High suspended sediment concentrations obstruct the respiratory systems of fish and other marine species, such as bivalves [11].

Evaluating the quality of river water is vital since river water is also used for everyday life. It is critical to determine the state of water quality and the physicochemical evolution of water as a result of mining. Hence, evaluating water quality is important for present use and for future consumption [15]. Some researchers have conducted some studies during this decade on the effect of sand mining on water quality [8, 12, 16, 17, 18, 19]. This study aims to assess the impact of sand mining on the water quality in the Progo River, Daerah Istimewa Yogyakarta Province, Indonesia. This study benefits in evaluating the physicochemical parameters of the Progo River and appraising its water quality for drinking purposes based on the regulation of the Indonesia Minister of Health [20] so that stakeholders can give more attention to conducting water treatment.

2. Method

The study area is located in the Progo River, Daerah Istimewa Yogyakarta Province, Indonesia (Figure 1.). The Progo River is a river that flows through the Indonesian province of Central Java and Daerah Istimewa Yogyakarta, which crosses to the southeast, and its estuary is the Indian Ocean [21]. Several construction structures are found along the Progo River, such as revetment, bridge, intake of irrigation, weir, and groundsill. In addition, human activities such as sand mining and opening up agricultural land were also reported along the Progo River [22]. The Progo River is one of the rivers with a volcanic and non-volcanic source of sediment [21]. People use the material since the quality of sediment accumulation is best for building materials. Using conventional machinery, local people perform sand mining operations, and mining companies use heavy industrial machines [21]. As a means of drinking water, irrigation, and sand mining operations, the Progo River is currently being used. To sustain regional economic activities, water quality and the production of water supplies are essential [23].

The water samples were taken during the dry season on 10 September 2020. The river was divided into three sites to test the effect of sand mining, based on the methodology used by [24], consisting of the upstream (Station 1), mining site (Station 2), and downstream (Station 3). The upstream of the mining site was believed to be undisturbed and was therefore regarded as a reference site. A region of active sand mining was the mining site. The downstream zone is the region without active mining but within the mining activity's effect. Station 1 has a river width of approximately 19 meters. Station 2 has a river width of approximately 63 meters. It has a distance of approximately 29.5 kilometers from Station 1. It is a mining location carried out by excavators and pump machines. Station 3 has a river width of roughly 558 meters. It has a distance of approximately 12.86 kilometers from Station 2 and 820 meters north of the Indian Ocean. The mining activities are still using pump machines one kilometer upstream from this location. One water sample was taken for each site. The water sampling locations are shown in Figure 2.

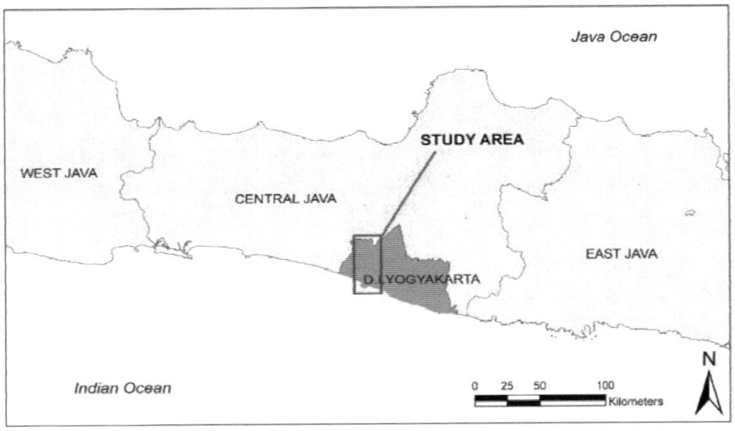

Figure 1. The study area

According to [10], who studied water quality due to sand mining, the physicochemical parameters used in this study included "pH, Electrical Conductivity (EC), Total Dissolved Solids (TDS), Dissolved Oxygen (DO), Sulphate (SO_4^{2-}) Nitrate (NO^{3-}), and Phosphate (PO_4^{3-})". Samples were taken in 1000 ml volume polyethylene bottles. Before filling with water samples, these bottles were rinsed to reduce the risk of any contamination. A pH meter was used to measure the pH. A conductivity meter was used to measure EC. A TDS meter was used to measure the TDS. A DO meter was used to measure the DO. A spectrophotometer was used to detect SO_4^{2}, NO^{3-} and PO_4^{3-} ions.

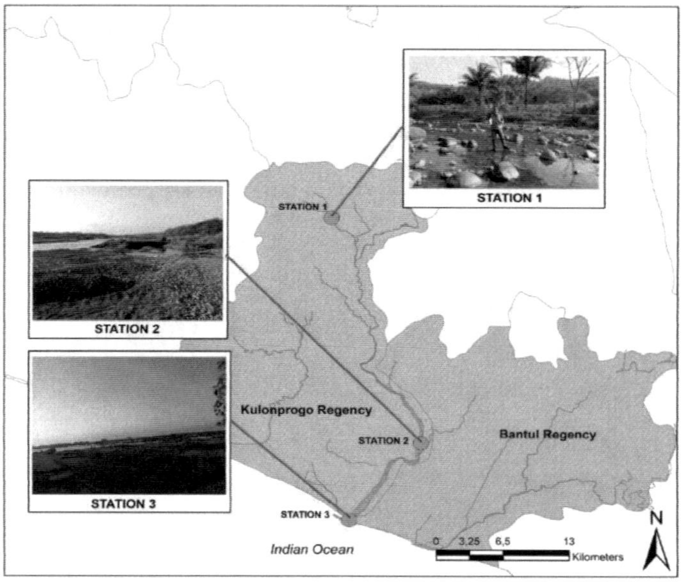

Figure 2. The water sampling locations

3. Result and Discussion

The physicochemical analysis results in the study area are shown in Table 1.

Table 1. The results of the physicochemical analysis in the study area

Parameter	Station 1	Station 2	Station 3	Threshold by the regulation of the Indonesia Minister of Health Number: 492/Menkes/Per/IV/2010 [20]
pH	8.12	8.05	7.77	6.5 - 8.5
EC (μs/cm)	672	661	41800	-
TDS (mg/l)	336	330	20900	500
DO (mg/l)	5.00	4.27	5.10	-
Sulphate (mg/l)	6.18	17.4	1836	250
Phosphate (mg/l)	<0.09	0.24	0.13	-
Nitrate (mg/l)	0.44	3.49	0.26	50

▭ : the value is above the threshold

3.1. Physicochemical parameters

According to Table 1., the pH from the highest to the lowest is Station 1, Station 2, and Station 3. The values were estimated because of sand mining which causes a decrease in the pH value. The acidity or alkalinity of water samples is determined by the pH. In water, pH strikes a balance between hydrogen ions (H+) and hydroxide ions (OH). The pH is also a useful measure of the water's chemical equilibrium [10]. The sand deposited in the Progo River was derived from sand material that is an andesitic Merapi volcano through a river [25]. The decrease in pH at Station 2 and Station 3 is thought to be due to the opening of sand deposits around the river so that minerals such as feldspar, hornblende, and quartz dissolve and mix in water and cause a decrease in the pH value. The pH in the three stations which were within the permissible limit base on Indonesia Minister of Health regulation [20].

The EC value from the highest to the lowest is Station 3, Station 1, and Station 2. Even the EC value is very high on Station 3. EC depends on the dissolved constituents' concentration, form, and degree of ionization. It can be used for the calculation of the estimated dissolved solids concentration. The high EC value indicates an apparently high volume of dissolved ionized dissolved organic compounds in the water and salt enrichment [10, 26]. The EC value was highest at Station 3 because its position is near the ocean. [27] suggested that the seawater EC was approximately 45,000 - 60,000 µs/cm. In addition, EC is the liquid capacity measure for conducting an electric charge. Its ability depends on ionic strength, dissolved ion concentrations, and temperature measurements. The concentration of the dissolved ions is commonly measured as TDS.

The TDS value from the highest to the lowest is the same as the EC value: Station 3, Station 1, and Station 2. According to the result, the TDS value is directly proportional to the EC value. The concentration of TDS explains the presence in water of inorganic salts and small quantities of organic matter. Geological conditions, human activities, seawater, industrial waste, domestic waste, and agriculture can be material sources in the TDS [27]. The high TDS value at Station 3 was estimated due to the erosion increasing at the riverbed and bank associated with sand mining. Those erosion process raises suspended solids in the water at the mining and downstream sites, was responsible for high mean turbidity at the mining and downstream sites. Mining operations also produce fine sand and small silt particles that are present in the stream. Due to their small settling velocity, silt particles (<63 µm) can be carried by the river over vast distances. Turbidity values downstream are usually twice what is obtained upstream [9]. In addition, the highest TDS value at Station 3 can also be derived from ocean currents that carry suspended material toward the river. The TDS value at Station 3 is above the permissible limit of the regulation of the Indonesia Minister of Health [20]. Whereas at Stations 1 and 2, the TDS values are relatively low and within the permissible limit of the regulation of the Indonesia Minister of Health [20]. Station 1 has a low TDS value because there are no mining activities that cause water turbidity, while Station 2 also has a low TDS value even though there are mining activities; this is thought to be because river flows carry material downstream creating more turbidity in the downstream area.

The DO value from highest to lowest is Station 3, Station 1, and Station 2. For aquatic organisms that use it for respiration, DO is an essential water quality parameter. For their existence, most aquatic organisms rely upon dissolved oxygen. The relatively high level of oxygen recorded in the water can be partially attributed to the turbulence caused by sand mining activities in the water. Owing to the trapping of ambient oxygen by exposed water molecules during turbulence, turbulence positively influences dissolved oxygen in water [10]. In addition, geological characteristics such as rock type and weathering rate also affect the chemical composition of water [28]. Owing to increased mining operations, DO was higher. Additionally, flowing water is aerated by bubbles in fast-moving water as it churns over rocks and falls hundreds of tiny waterfalls. These streams are typically saturated with oxygen if they are unpolluted. Oxygen only reaches the top layer of water in sluggish, stagnant water, and deeper water is always poor in DO concentration due to the decomposition of organic matter by bacteria living on or close the bottom of the reservoir [29]. The highest DO value at Station 3 was estimated not only because of mining activities but also because it was located downstream, so the current speed increases and also gets a strong current from the sea. The sulphate value from highest to lowest is Station 3, Station 2, and Station 1. In natural waters, sulphate is one of the main anions and is added by household and industrial discharges. In the form of sulphate-lowering bacteria, these discharges reduce the pH of river water and raise the bacterial load. The concentration of sulphate in natural waters is usually between 2 and 80 mg/l [1]. The values of sulphate recorded in Station 1 and Station 2 were very low and did not reflect any anthropogenic impact, including sand mining activities. Station 1 and 2 values of sulphate are relatively low and within the permissible limit of the regulation of the Indonesia Minister of Health [20]. In addition, sand mining in a river does not lead to an increase of sulphates. Sulphate is measured as milligrams per liter (mg/l) and is a natural constituent of mineral sulphate containing soils and rocks formed in processed industrial water. Sulphates are soluble in water, and concentrations appear to accumulate and increase. Sulphates are usually released from waste from acid mines and from factories that use sulfuric acid or sulphates in their operation. The exchange of ions can eliminate or add sulphate to water, and sulphate can be interconverted by microbiological reduction or oxidation [30]. It means that the high content of sulphate value in Station 3 is not because of sand mining. It assumed that the high value of sulphate content in Station 3 because the location is close to the Indian Ocean. Sulphate ion concentrations contribute to the water composition. Sulphate ions could possibly be derived from the seawater mixing [31]. The sulphate value at Station 3 is above the permissible limit of the regulation of the Indonesia Minister of Health [20].

Phosphate value from highest to lowest is Station 2, Station 3, and Station 1. Due to agricultural activities carried out in the vicinity, high phosphate levels can be attributed to the drainage of fertilizer into the river; and from soaps and detergents used in washing motorcycles, bathing, and laundry activities. Phosphates will harm humans and animals if the amounts are too high. Digestive issues may result from a highly high phosphate level in drinking water, so the high concentration of phosphate would affect it [2]. The enrichment of aquatic plant bodies with phosphate derived from agriculture, industry, and waste has been considered a key driver of aquatic plant loss because it promotes the transition from clear water to highly adverse conditions dominated by phytoplankton and filamentous algae in which macrophytes thrive [32]. All three stations have relatively low phosphate values. The source of phosphate in the three stations was estimated to come from soil erosion. Soil erosion is also associated with high rates of phosphate transfer from land to watercourses in the form of transported suspended sediment in the overland flood [33]. Phosphate will be harmful if the value is above 100 mg/l, so the coagulation processes in drinking water treatment plants should be conducted [34]. So the phosphate value at the three stations is relatively safe for drinking water. Nitrate at the three stations which were within the permissible limit of the regulation of the Indonesia Minister of Health [20]. The nitrate value from the highest to the lowest is Station 2, Station 1, and Station 3. Nitrate levels can have nitrate concentrations up to 5 mg/l but mostly less than 1 mg/l in surface waters when caused by human activities, and concentrations above 5 mg/l typically imply contamination by human or animal waste or fertilizer run-off [1]. In addition, nitrate is also derived from nature, namely soil nitrogen and atmospheric deposition [35]. Nitrate value at the three stations was estimated not to be influenced by sand mining activities and other anthropogenic factors; it was estimated that it derived from nature.

3.2. Sand mining impacts the water quality

According to the results of the physicochemical parameters, not all changes in the Progo River's physicochemical parameters are brought on by sand mining activities. The physicochemical parameters pH, TDS, and DO change due to sand mining activities. The fall in pH value was believed to be the result of sand deposits' opening by mining activities near the river, which allowed minerals including feldspar, hornblende, and quartz to dissolve and mix with water. Due to the high mean turbidity at the mining and downstream locations, which is caused by enhanced riverbed and bank erosion brought on by sand mining and also boosts suspended particles in the water, the change in TDS value was assessed. The change in DO was also estimated because it was positioned downstream, which causes the current speed to increase and also receives a strong current from the sea, in addition to mining activities. Meanwhile, natural processes like soil erosion and seawater mixing impact other metrics, including EC, sulphate, phosphate, and nitrate.

4. Conclusion

Sand mining activities impacted pH, TDS, and DO. The opening of sand deposits along the river is assumed to be the source of the pH drop since this allows feldspar, hornblende, and quartz to dissolve and mix with the water. The increased riverbed and bank erosion caused by sand mining, which raises suspended particles in the water at the mining and downstream sites, was the source of the high mean turbidity at the mining and downstream sites; hence the TDS increase was estimated. Additionally, sand mining operations impact higher DO because bubbles aerate fast-moving water as it churns over rocks and plummets into countless little waterfalls. Almost all metrics, except TDS and Sulphate levels in Station 3, are acceptable for drinking purposes. The Progo River needs to be protected from pollution; hence the stakeholder needs to control sand mining operations nearby.

5. References

1. Donald, A.E., and Blessing, U.A., 2019. Index approach to water quality assessment of a South eastern Nigerian river. International Journal of Fisheries and Aquatic Studies 7(1): 153-159.
2. Oboh, I. P., and Agbala, C.S., 2017. Water quality assessment of the Siluko River, southern Nigeria. African journal of aquatic science 42(3): 279-286.
3. Bayram, A., and Önsoy, H., 2015. Sand and gravel mining impact on the surface water quality: a case study from the city of Tirebolu (Giresun Province, NE Turkey). Environmental earth sciences 73(5): 1997-2011.
4. Ashraf, M. A., Maah, M. J., Yusoff, I., Wajid, A., and Mahmood, K., 2011. Sand mining effects, causes and concerns: A case study from Bestari Jaya, Selangor, Peninsular Malaysia. Scientific Research and Essays 6(6): 1216-1231.
5. Lusiagustin, V., and Kusratmoko, E., 2017, July. Impact of sand mining activities on the environmental condition of the Komering river, South Sumatera. In AIP Conference Proceedings (Vol. 1862, No. 1, p. 030198): AIP Publishing LLC.
6. Rachmawati, Y., & Zamroni, A. (2020). How Indonesian Governments Care for Local People's Education in the Mining Area: Experiences from other Countries. PSYCHOLOGY AND EDUCATION, 57(9), 5924-5934.
7. Barman, B., Kumar, B., and Sarma, A.K., 2019. Impact of sand mining on alluvial channel flow characteristics. Ecological Engineering 135: 36-44.
8. Sreebha, S., and Padmalal, D., 2011. Environmental impact assessment of sand mining from the small catchment rivers in the southwestern coast of India: a case study. Environmental management 47(1): 130-140.
9. Nabegu, A.B., 2013. Effect of Sand mining on Ground Water in Kano river catchment. J. Environ. Earth Sci 3: 81-87.
10. Akankali, J.A., Idongesit, A.S., and Akpan, P.E., 2017. Effects of sand mining activities on water quality of OkoroNsit stream, NsitAtai local government area, AkwaIbom state, Nigeria. International Journal of Development and Sustainability 6(7): 451-462.
11. Padmalal, D., and Maya, K., 2014. Impacts of river sand mining. In Sand Mining (pp. 31-56): Springer, Dordrecht
12. Bhattacharya, R.K., Chatterjee, N.D., and Dolui, G., 2019. Consequences of sand mining on water quality and instream biota in alluvial stream: a case-specific study in South Bengal River, India. Sustainable Water Resources Management 5(4): 1815-1832.
13. Asabonga, M., Cecilia, B., Mpundu, M.C., and Vincent, N.M.D., 2017. The physical and environmental impacts of sand mining. Transactions of the Royal society of South Africa 72(1): 1-5.
14. Asih, A. S., Zamroni, A., Alwi, W., Sagala, S. T., & Putra, A. S. 2022. Assessment of Heavy Metal Concentrations in Seawater in the Coastal Areas around Daerah Istimewa Yogyakarta Province, Indonesia. The Iraqi Geological Journal, 14-22.
15. Zamroni, A., Trisnaning, P. T., Prasetya, H. N. E., Sagala, S. T., & Putra, A. S. 2022. Geochemical Characteristics and Evaluation of the Groundwater and Surface Water in Limestone Mining Area around Gunungkidul Regency, Indonesia. The Iraqi Geological Journal, 189-198.
16. Yen, T.P., and Rohasliney, H., 2013. Status of water quality subject to sand mining in the Kelantan River, Kelantan. Tropical life sciences research 24(1): 19.
17. Bhattacharya, R.K., 2018. Instream sand mining impact on water quality and benthos community in an alluvial reach: A case study on river Kangsabati, West Bengal. International Journal of Current Research in Life Sciences 7(08): 2613-2617.
18. Okorie, P.U., and Acholonu, A.D., 2019. Effect of Commercial Sand Mining on Water Quality Parameters of Nworie River In Owerri, Nigeria. Proceedings of the Nigerian Academy of Science

11(1).
19. Doloksaribu, D.C.N., Barus, T.A., and Sebayang, K., 2020. The impact of marine sand mining on sea water quality in Pantai Labu, Deli Serdang Regency, Indonesia. EandES 454(1): 012086.
20. Indonesia Ministry of Health Regulation No 492/MENKES/PER/IV/2010 about the Requirements of Drinking Water Quality.
21. Ikhsan, J., Legono, D., Rahardjo, A.P., Harsanto, P., and Fujita, M., 2020, February. Dynamics of lahar-affected river tributaries of the Progo river after the 2010 Mt. Merapi eruption. In IOP Conference Series: Earth and Environmental Science (Vol. 437, No. 1, p. 012009): IOP Publishing.
22. Zamroni, A., Putra, B.P., and Prasetya, H.N.E., 2020a. Anthropogenic influences on morphological changes in the Progo River, Daerah Istimewa Yogyakarta Province, Indonesia. Sustinere: Journal of Environment and Sustainability 4(3): 205-22.
23. Rezagama, A., Sarminingsih, A., Zaman, B., and Handayani, D.S., 2019, May. Analysis of land use changes effect on erosion and sedimentation potential in Progo watershed. In Journal of Physics: Conference Series (Vol. 1217, No. 1, p. 012159): IOP Publishing.
24. Brown, A.V., Lyttle, M.M., and Brown, K.B., 1998. Impacts of gravel mining on gravel bed streams. Transactions of the American fisheries society 127(6): 979-994.
25. Nurcholis, M., and Mulyanto, D., 2017. Study on the Residue Resulted from the Metallic Minerals separations to the Coastal Iron-sand of Yogyakarta Indonesia. International Journal of Applied Engineering Research, 12(15): 5218-5225.
26. Zamroni, A., Sugarbo, O., Trisnaning, P. T., Sagala, S. T., & Putra, A. S., 2021. Geochemical Approach for Seawater Intrusion Assessment in the Area around Yogyakarta International Airport, Indonesia. The Iraqi Geological Journal, 1-11.
27. Rusydi, A.F., 2018, February. Correlation between conductivity and total dissolved solid in various type of water: A review. In IOP conference series: earth and environmental science (Vol. 118, No. 1, p. 012019): IOP Publishing.
28. Zamroni, A., Sugarbo, O., Prastowo, R., Widiatmoko, F. R., Safii, Y., & Wijaya, R. A. E. (2020b, July). The relationship between Indonesian coal qualities and their geologic histories. In AIP Conference Proceedings (Vol. 2245, No. 1, p. 070005). AIP Publishing LLC.
29. Obot, O. I., Ekpo, I.F., and David, G.S., 2019. The Effect of Sand Mining on the Physico-Chemical Parameters of Ikot Ekpan River, Akwa Ibom State, Nigeria. J. Aquat. Sci. Mar. Biol 2: 21-24.
30. Maeko, M.P., 2020. An evaluation of the ecological impacts of sand mining on the Mokolo River in Lephalale, South Africa (Doctoral dissertation).
31. Mohanty, A.K., and Rao, V.G., 2019. Hydrogeochemical, seawater intrusion and oxygen isotope studies on a coastal region in the Puri District of Odisha, India. Catena 172: 558-571.
32. Lambert, S.J., and Davy, A.J., 2011. Water quality as a threat to aquatic plants: discriminating between the effects of nitrate, phosphate, boron and heavy metals on charophytes. New Phytologist 189(4): 1051-1059.
33. Fraser, A.I., Harrod, T.R., and Haygarth, P.M., 1999. The effect of rainfall intensity on soil erosion and particulate phosphorus transfer from arable soils. Water science and technology 39(12): 41-45.
34. Pedersen, T.L., 1997. Phosphates. UCD EXTOXNET FAX TEAM. Accessed at http://extoxnet.orst.edu/faqs/safedrink/phos.htm on 5th February 2021.
35. Fadhullah, W., Yaccob, N.S., Syakir, M.I., Muhammad, S.A., Yue, F.J., and Li, S.L., 2020. Nitrate sources and processes in the surface water of a tropical reservoir by stable isotopes and mixing model. Science of the total Environment 700: 134517.

Acknowledgments
We thank to Institut Teknologi Nasional Yogyakarta and Direktorat Riset dan Pengabdian Masyarakat, Deputi Bidang Penguatan Riset dan Pengembangan, Kementerian Riset, Teknologi/Badan Riset dan Inovasi Nasional Republik Indonesia for the support research funding.

Producing Fe and Cu ions and oxides in water with electrolysis as artificial liquid waste

Y Wahyono[1], R Irviandi[2], N K Lo[2], M I A Rahman[2], F Herdiansyah[2], B T Haliza[2], A H Nurauliyaa[3], R A Baihaqi[2], K A Prabahandari[4], Y Hariyono[2] and N I Pratiwi[2]

[1] Doctoral Program of Environmental Science, School of Postgraduate Studies, Diponegoro University, Imam Bardjo SH Street, Pleburan, Semarang, Central Java 50241, Indonesia
[2] Department of Physics, Faculty of Science and Mathematics, Diponegoro University, Prof. Soedarto, SH Street, Tembalang, Semarang, Central Java 50275, Indonesia
[3] Taruna Bangsa School, Babakan Madang Street, Sentul, Bogor, West Java 16810, Indonesia
[4] Department of Biology, Faculty of Science and Mathematics, Diponegoro University, Prof. Soedarto, SH Street, Tembalang, Semarang, Central Java 50275, Indonesia

yoyonwahyono@student.undip.ac.id

Abstract. Water – in the context of an inland water source – is complex when used as an object of research. Often when using river water samples, researchers struggle to find the desired composition. Therefore, a simple and controlled method is needed to produce test samples with specific substance compositions. This study aims to use electrolysis to produce artificial heavy metal waste. Iron (Fe) and copper (Cu) provided the electrodes and water the electrolytes. Electrolysis of water with Fe electrodes produced Fe^{3+} ions and $Fe(OH)_3$ precipitation. Electrolysis of water with Cu electrodes produced Cu^{2+} ions and $Cu(OH)_2$ precipitation. Electrolyte samples were collected at intervals of 30 min for 180 min and were tested with atomic absorption spectroscopy. Fe and Cu concentrations increased during electrolysis. Electrolysis can therefore be used to produce artificial heavy metal waste cheaply and on a small scale.

1. Introduction

A common problem in waters, and especially rivers, is pollution by heavy metals. Heavy metals are toxic when consumed by most organisms and bioaccumulation of heavy metals in the human body will cause health problems. The presence of high concentrations of heavy metals can also damage the balance of the ecosystem [1].

Considering that the presence of heavy metals in waters is often linked to industrial activities, the preferred method for dealing with heavy metals pollution leans towards prevention limiting the potential environmental impact of waste discharge. Common methods for overcoming heavy metal pollution include reverse osmosis, electrodialysis, precipitation, nanofiltration, coagulation, adsorption, distillation, photocatalyst and electrolysis [2,3,4,5].

For the development of a specific method for the treatment of heavy metal concentrations in water, acquiring river water samples presents a challenge. Water pollution has complex parameters, not only heavy metal concentrations but also chemical oxygen demand (COD), biological oxygen demand (BOD), pH, total suspended solids (TSS), the solubility of certain substances and the presence of specific microorganisms. Therefore, for heavy metal pollution research, it could be more effective to use metals concentrated in water as the test material. However, specific metal concentrations in water are not easy to obtain naturally. Creating artificial liquid waste by making a solution using chemical reactions often needed various additives and the by-product contains impurities [6]. Instead, water samples with the desired metal concentrations could be artificially created by utilizing electrolysis. The metal-containing water can then be used as a sample substitute for heavy metal waste. Electrolysis research mainly intends to study hydrogen production, optimization of the electrolysis, or production of chemicals needed in products such as electronics, pharmaceuticals, and food [7]. Prior research on electrolysis has not uncovered the production of artificial liquid waste containing heavy metals in the form of metal ions and metal oxides. Electrolysis delivers ease, economic value, and a high success rate for this work. Therefore, this study aims to produce water artificially polluted by heavy metals by utilizing electrolysis. The water was tested for pH, total dissolved solids (TDS), temperature and heavy metal concentrations.

2. Methodology

2.1. *Tools and materials*

Tools and materials used in this study were: Adapter 12 volts, 1.2 amperes, alligator clip, TDS meter, pH meter, atomic absorption spectroscopy (AAS) instrument, digital scale, 250 mL glass beaker, pipette, 10 mL vial, iron, copper and aquades. The adaptor acts as a rectifier from AC to DC and drops the voltage from 220 volts (source) to 12 volts.

2.2. *Treatment*

The experiment started with the cutting of electrodes to 12 cm in length. The cut electrodes were sanded to remove impurities (corrosion) and then weighed using a digital scale to obtain the electrode's initial mass. After that, the electrode was attached to the alligator clip already connected to the adapter and a sample of distilled water was placed into the vial. Electrolysis was initiated by dipping the electrode into a beaker glass containing 250 mL aquades functioning as the electrolyte. Next, the ppm (parts per million) and temperature were measured with a TDS meter and the pH of the electrolyte with a pH meter. After the initial data was obtained, the electric current were flowed through the adapter. Measurements of ppm, temperature, pH and electrolyte sampling (into vials) were carried out at intervals of 30 min for 180 min. The electrode mass was determined after the electrolysis process had been completed. The samples were tested with the AAS method.

3. Results and discussion

3.1. *Cell potential*

The voltage required for electrolysis to occur in an electrolysis cell is called the cell potential ($E°_{cell}$), which consists of the sum of the reduction potential ($E°_{red}$) and oxidation potential ($E°_{ox}$). The various reduction and oxidation potentials for electrolytic reactions in the voltaic series have been determined and must be met as a condition for electrolysis to occur as a non-spontaneous reaction [8]. Other required voltage factors include ohmic voltage (the voltage that causes energy loss due to resistance of cell components), activation voltage (the voltage required for transfer between chemical specs and electrodes), and concentration voltage (voltages induced by mass transfer such as diffusion and convection). These other voltage factors can be summed algebraically so that the voltage value is obtained for an electrolysis cell [9]. A potential difference between the two electrodes causes an electric current to occur in the electrolysis cell.

However, in this study, the voltage applied to the cell did not take into account the types of voltage mentioned in the previous paragraph. In addition, because the main purpose of this study is to produce the desired substance and the voltage given was sufficient to make the reaction occur, then the comprehensive calculation of the voltage is unnecessary.

3.2. Production of Fe ions and oxides in water electrolyzed with Fe electrodes
Electrolyzing water with iron electrodes results in a brown solution, as shown in figure 1.

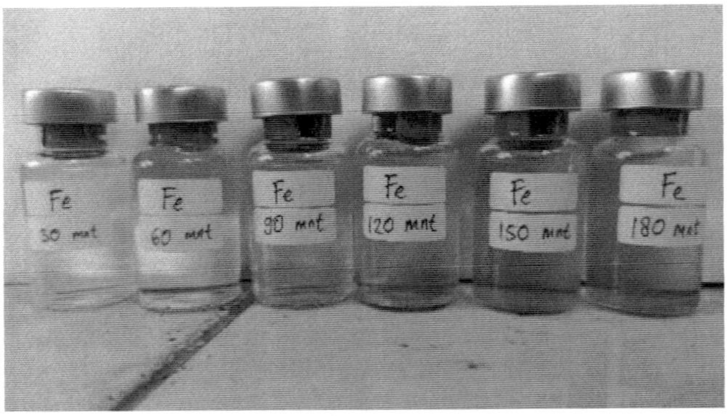

Figure 1. Samples from water electrolyzed with Fe electrodes.

Electrolysis leads to a water reduction at the cathode and iron oxidation at the anode. Water reduction is evidenced by the appearance of hydrogen gas bubbles around the cathode, according to the equation

$$2H_2O_{(l)} + 2e^- \rightarrow H_{2(g)} + 2OH^-{}_{(aq)}, [E°_{red} = -0.83 \text{ volt}] \quad (1)$$

Iron oxidation is evidenced by a reduction in the mass of the iron anode. Considering that iron is a non-inert electrode, the reaction occurs at the anode [10]. When an electric current is applied to the electrolysis cell, the iron anode oxidizes to form Fe^{2+} ions, according to equation (2), considering the Fe electrode is reactive [11].

$$Fe_{(s)} \rightarrow Fe^{2+}{}_{(aq)} + 2e^-, [E°_{ox} = +0.44 \text{ volt}] \quad (2)$$

Fe^{2+} ions can combine with OH^- produced by the reduction of water to form the greenish iron(II) hydroxide ($Fe(OH)_2$). However, in this experiment, no greenish precipitate formed.

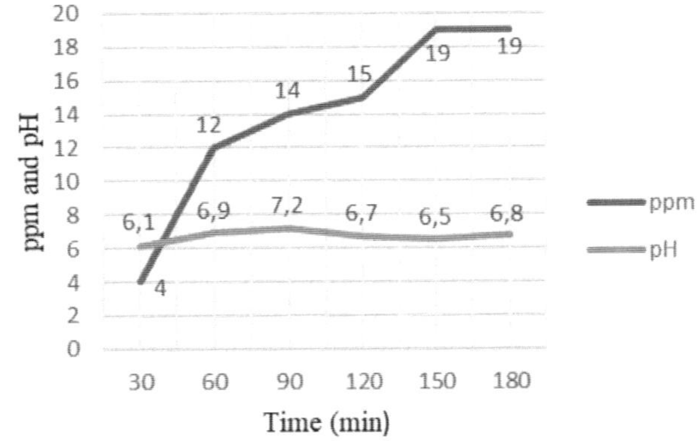

Figure 2. Graph of ppm and pH of water electrolyzed with Fe electrodes against time.

The Fe^{2+} ions produced can undergo further oxidation to form Fe^{3+}. Figure 2 demonstrates that the pH of the water increased over time. According to equation (3),

$$Fe^{2+}{}_{(aq)} \rightarrow Fe^{3+}{}_{(aq)} + e^-, [E°_{ox} = -0.77 \text{ volt}] \quad (3)$$

the increase in pH may indicate the formation of Fe^{3+} ions and the reaction causes the pH and dissolved oxygen in the water to increase [12]. However, as figure 2 demonstrates, the increase in pH fluctuated. This happened because OH^- ions bind with Fe^{3+} ions, thereby reducing the concentration of ions OH^- in the water and decreasing the pH. When Fe^{3+} combines with OH^-, the slightly soluble and brown in colour iron(III) hydroxide is formed. This precipitate was obtained after the experiment [13].

The reduction of water at the cathode aided in increasing the pH. According to equation (1), the reduction of water produces hydrogen gas and OH^-. The longer the electrolysis, the greater the presence of hydroxide ions in the water. As the concentration of hydroxide ions increases, water becomes more alkaline and, therefore, electrolysis increases the pH.

3.3. Production of Cu ions and oxides in water electrolyzed with Cu electrodes

In water electrolyzed with a copper metal electrode, the copper electrode is non-inert, meaning that it can participate in the electrolysis reaction. The copper anode will oxidize to Cu^{2+} while at the cathode there is a reduction of molecules H_2O at neutral pH or a reduction of ions H^+ at acidic pH. The redox equations for water electrolysis with copper metal electrodes in a neutral electrolyte are as follows, with equation (4) for the cathode (reduction), and equation (5) for the anode (oxidation).

$$2H_2O_{(l)} + 2e^- \rightarrow H_{2(g)} + 2OH^-{}_{(aq)}, [E°_{red} = -0.83 \text{ volt}] \quad (4)$$

$$Cu_{(s)} \rightarrow Cu^{2+}{}_{(aq)} + 2e^-, [E°_{ox} = -0.34 \text{ volt}] \quad (5)$$

The redox equations for water electrolysis with copper metal electrodes in acidic electrolytes are as follows, with equation (6) for the cathode (reduction), and equation (7) for the anode (oxidation) [14,15].

$$2H^+{}_{(aq)} + 2e^- \rightarrow H_{2(g)}, [E°_{red} = 0 \text{ volt}] \quad (6)$$

$$Cu_{(s)} \rightarrow Cu^{2+}{}_{(aq)} + 2e^-, [E°_{ox} = -0.34 \text{ volt}] \quad (7)$$

The 180 min electrolysis of water with copper electrodes resulted in a solution with a bluish-coloured precipitate, shown in figure 3.

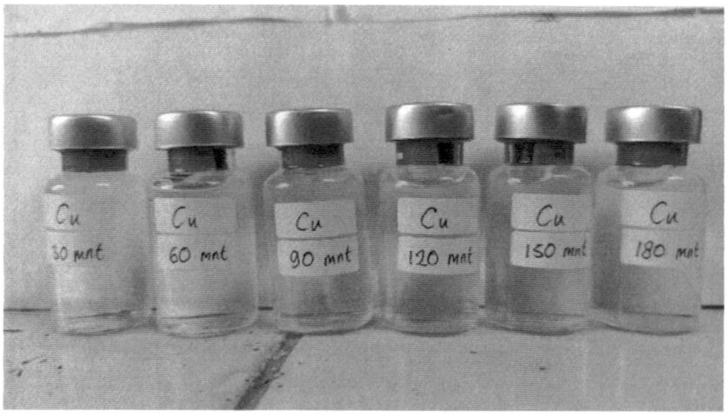

Figure 3. Samples from water electrolyzed with Cu electrodes.

When electrolysis reached 90 min, a light-blue colour appeared in the solution. The light-blue colour and the appearance of a light blue precipitate indicate the production of a copper(II) hydroxide or $Cu(OH)_2$ compound. Additionally, hydrogen gas bubbles formed at the cathode, confirming equation (4) on the reduction of water and equation (5) on the oxidation of copper electrodes.

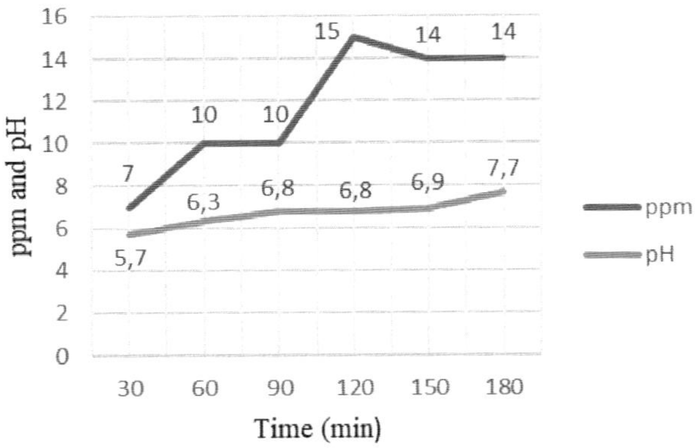

Figure 4. Graph of ppm and pH of water electrolyzed with Cu electrodes against time.

There was a reduction of water at the cathode when the electrode became electrified. The water-electrolyte was initially acidic, as shown in figure 4. Electrolysis involves the simultaneous reactions of the breakdown of water molecules (equation (4)) and the reduction of H^+ (equation (6)), evidenced by the appearance of gas bubbles in the cathode throughout the experiment.

Oxidation of copper to copper(II) ions takes place at the anode. The hydroxide ions produced at the cathode combine with the copper(II) ions to produce copper(II) hydroxide, or $Cu(OH)_2$, which gives the electrolytes a light-blue appearance. $Cu(OH)_2$ is a weak base and its formation causes the pH of the water to increase, reducing its acidity.

3.4. Concentrations of iron and copper and the temperatures during electrolysis

Of the six samples obtained for this experiment, three were tested using AAS to determine the concentrations of dissolved iron and copper in the water.

Table 1. Results of AAS analysis – concentrations of Fe and Cu.

No.	Sample	Concentration (mg/L)	
		Fe	Cu
1	30 min	0.170	1.314
2	90 min	1.421	1.906
3	180 min	1.687	6.466

As table 1 demonstrates, the Fe and Cu concentrations increased as electrolysis went on. The water's Fe concentration exceeded the maximum of 1 mg/L set by Indonesia's water quality standards for sanitation needs, as stipulated in *Peraturan Menteri Kesehatan Nomor 8 Tahun 2014* (The Ministry of Health Regulation No. 8, 2014). The water's Cu concentration exceeded the guideline value of 2 mg/L set by the WHO in Guidelines for Drinking-water Quality, 2017.

This study analysed the concentrations of Fe and Cu in their dissolved form as the AAS method is not practical for analysing Fe and Cu concentrations in suspended or solid form. Analysing the concentrations of $Fe(OH)_3$ and $Cu(OH)_2$ would likely result in higher iron and copper content outcomes.

The electrolyte temperatures were measured in 30 min intervals for 180 min. The results are summarized in table 2.

Table 2. Electrolyte temperatures.

Time (min)	Temperature (°C)	
	Fe	Cu
30	30	31
60	30	31
90	30	31
120	30	31
150	29	30
180	29	31

When electricity was first applied, the water's conductivity was still relatively because there were no ions in the water compared to when both electrodes had produced ions. This means that the electrolyte temperatures should be higher at the start of the electrolysis process. Instead, the temperatures remained at a steady room temperature because the electrodes are conductors and their electrical resistances are relatively small (particularly Cu). The heat dissipated by the electrode to the electrolyte was therefore insignificant and negligible. Moreover, the distance between electrodes was relatively small thus minimizing the energy required for ions in the electrolyte to move and, in other words, improving the electrolysis efficiency [16].

3.5. Comparing synthesis by electrolysis and making solution

Artificial liquid waste containing $Fe(OH)_3$ and $Cu(OH)_2$ can be produced by reacting NaOH with $FeCl_3$ and NaOH with $CuSO_4$ according to the following reactions

$$FeCl_3 + 3NaOH \rightarrow Fe(OH)_3 + 3NaCl \quad (8)$$

$$CuSO_4 + 2NaOH \rightarrow Cu(OH)_2 + Na_2SO_4 \quad (9)$$

With this method, NaCl and Na_2SO_4 are obtained as an unnecessary products. These pollutants can be removed by adding other substances such as NH_4Cl [6]. Materials such as NaOH, $FeCl_3$, $CuSO_4$, and NH_4Cl can not be purchased in small quantities (for the production of liquid waste of 50-250 mL), this implies a higher price. Through electrolysis, $Fe(OH)_3$ and $Cu(OH)_2$ can be synthesized directly by electrolyzing iron and copper electrodes, or other metal electrodes for other liquid waste products, which can be purchased in small quantities at chemical stores. Not only that, the metal electrodes can be used repeatedly considering the metals will be consumed little by little, hence, saving the cost. Regarding the experiment conducted by the authors, the overall materials for the electrolysis (the electrodes, adaptor, and alligator clip) could be obtained at Rp52.000,00, cheaper than the total price required for obtaining the three main chemicals including NaOH, $FeCl_3$, and $CuSO_4$ (Rp176.000,00).

4. Conclusion

The production of heavy-metal polluted water with controlled metal compound contents through electrolysis is possible provided that low concentrations are used. It is also possible to create artificial heavy-metal waste using metal from the voltaic series. Hopefully, these results can help further research on waste management and treatment technologies by providing alternative methods for making small-scale waste with controlled contents and with components that are easily available and inexpensive.

References

[1] Huang Z, Liu C, Zhao X, Dong J and Zheng B 2020 *Environ. Sci. Eur.* **32** 1-9
[2] Lakherwal D 2014 *Int. J. Environ. Res. Dev.* **4** 41-8
[3] Nugroho R A, Ma'Rufah H, Khusna L M, Wahyono Y and Sutanto H 2019 *IOP Conf. Ser. Earth Environ. Sci.* vol 243 (IOP Publishing) p 012120

[4] Rosyad M S, Ramadhani S, Riski I, Sutanto H, Wahyono Y and Indrawan W 2020 *IOP Conf. Ser. Mater. Sci. Eng.* vol 850 (IOP Publishing) p 012008
[5] Wahyono Y, Nofanda T T, Budiyanto I, Nugroho H, Rosyad M S, Halim M A R, Amilia A, Rahmawati D F, Nurauliyaa A H, Baihaqi R A, Syafira F and Lestari P 2021 *J. Phys: Conf. Ser.* vol 1943 (IOP Publishing) p 012161
[6] Cudennec Y and Lecerf A 2003 *Solid State Sci.* **5(11-12)** 1471-4
[7] Grigoriev S A, Fateev V N, Bessarabov D G and Millet P 2020 Int. J. Hydrog. **45(49)** 26036-58
[8] Brown T E, LeMay H E, Bursten B E, Murphy C, Woodward P and Stoltzfus M E 2015 *Chemistry: The Central Science 13th Edition* (New Jersey: Pearson Education, Inc) p 865-9
[9] Ursua A, Gandia L M and Sanchis P 2011 *Proc. IEEE* vol 100 (IOP Publishing) p 410-26
[10] Yuvaraj A L and Santhanaraj D 2014 *Mater. Res.* **17** 83-7
[11] Sugiharta I and Putri A E 2019 *J. Phys. Conf. Ser.* vol 1155 (IOP Publishing) p 012054
[12] Lakshmanan D, Clifford D A and Samanta G 2009 *Environ. Sci. Technol.* **43** 3853-59
[13] Devamani R H P and Alagar M 2013 *Elixir Nanotechnology* **61** 16845-48
[14] Chang R and Kenneth A G 2016 *Chemistry 12th Edition* (New York: McGraw-Hill Education) p 821
[15] Wahyono Y, Sutanto H and Hidayanto E 2017 *Youngster Phys. J.* **6** 353-9
[16] Cahyono H B and Ariani N M 2014 *J. Ind. Res.* **8** 113-22

Analysis of Diversity Level and Vegetation Structural Composition Post Restoration at Resort Cangkringan Mount Merapi National Park

Lia Kusumaningrum[1*], **Prabang Setyono**[1], **Muhammad Amin Sunarhadi**[1], **Muchammad Sholiqin**[1], **Bagus Hermawan**[1], **and Gavriel Enos Berlin**[1]

[1] Environmental Science Program, Faculty of Mathematics and Natural Science, Sebelas Maret University, Surakarta-Indonesia

*Liakusumaningrum@staff.uns.ac.id

Abstract. Mount Merapi eruption in 2010 caused a total of 1,128 ha of damage to Mount Merapi National Park (MMNP). The MMNP Center carried out planting activities as a restoration activity in an area that had a fairly severe impact on Cangkringan Resort. This research is located at Cangkringan Resort in the restoration area of Klangon and Kali Kuning. This study aims to determine the level of diversity and vegetation structure after 10 years of post restoration. Analysis of the composition and structure is important to assess biodiversity and determine the success rate of restoration of the MMNP ecosystem damaged by the eruption. This research uses the line and line transect method with quantitative data analysis by Shanon Weiner and Evenness. This study found 45 plant species in Klangon and 34 plant species in Kali Kuning. The results show that the total level of diversity in Cangkringan Resort MMNP is moderate (1.465), the level of evenness in Klangon is moderate (0.444) and in Kali Kuning moderate (0.346). Therefore, the level of diversity and uniformity of the vegetation at Cangkringan Resort MMNP is at the medium classification level and shows a better change in the improvement of the vegetation.

1. Introduction

Indonesia's geographical condition as an archipelagic country, located between 2 continents and 2 oceans, makes Indonesia a high biodiversity, even being called a mega-biodiversity country [1]. This means that although it is only 1.3% of the total area of the earth, Indonesia has a high level of genetic diversity, species, and ecosystems [2]. Biodiversity includes genes, species of flora / fauna, and ecosystems. However, in this paper, we will focus only on the analysis of the vegetation. Where this vegetation can be interpreted as a collection of several types of plants that live and interact together in one place [3].Vegetation can be analyzed by studying or finding out about the composition of a plant [4].This vegetation analysis will later obtain quantitative data/information regarding the structure of plants. The data that can be analyzed include density, frequency, dominance, and important value indexes. Where data need to be taken such as plant types, number of plants, plant height, plant diameter, number of plots, and area of plots taken. One of the areas that can be analyzed for vegetation is the mountain area because it has unique vegetation that is only found at a certain height [5].

Mount Merapi National Park (MMNP) is a state forest area of strategic and important value as a function of nature conservation (maintaining hydrological, botanical, and aesthetic functions, as well as managing the area specifically as a disaster prone area). The MMNP area is a protected forest area that covers an area of ± 6,410 Ha (± 4,000 Ha of vegetated area) located in the Sleman Regency, DI Yogyakarta Province, and Magelang Regency, Klaten, Boyolali, Central Java Province [6]. Mount Merapi is one of 129 active volcanoes in Indonesia. On October 26, 2010 Mount Merapi experienced a very powerful eruption. The eruption of Mount Merapi in 2010 resulted in the MMNP destruction of the area covering an area of ± 3000 hectares due to hot clouds and being covered by volcanic ash [7]. The eruption of Mount Merapi in 2010 was considered the largest eruption in the last 100 years. The vegetated area affected by the hot clouds that were destroyed and burned in the Sleman Regency was recorded as at least ± 1,128 Ha (Cangkringan Resort and Pakem-Turi Resort).So in the MMNP area there is damage accompanied by the extinction of the ecosystem in the area. According to records from the MMNP Center, the worst damage was in the MMNP area at Cangkringan Resort [6].

MMNP Center carried out planting activities as a restoration activity from 2011-2012 in areas affected by the severe eruption of Mount Merapi at Cangkringan Resort MMNP. During the 10-year post-restoration period in the Cangkringan MMNP Resort area, it is necessary to analyze the composition and structure of the vegetation. This is the reason for researchers to assess biodiversity and determine the success rate of restoration of the MMNP ecosystem that was damaged by the eruption. Therefore, this study aims to determine the physical condition between the time of the eruption in 2010 and the current condition in 2022 at Cangkringan Resort MMNP.

2. Methodology

Data collection was carried out at Cangkringan Resort, namely in Klangon and Kali Kuning with coordinates (-7.57423914, 110.45553924) and (-7.59558296, 110.43889169), data collection was carried out in April 2022. This location is an area that has been restored due to the worst impact of the eruption of Mount Merapi.

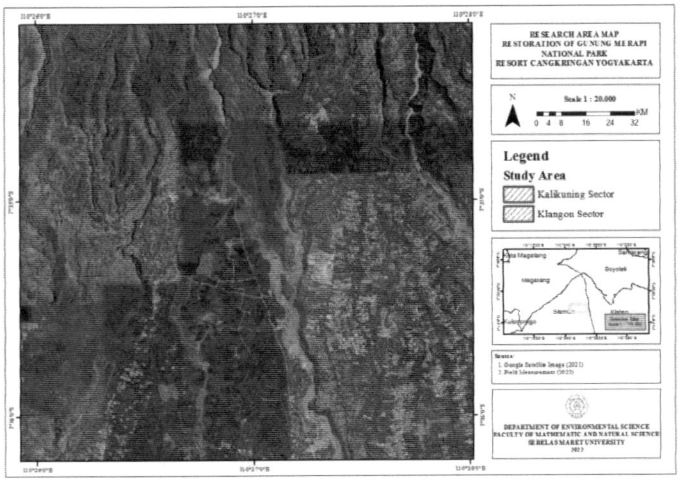

Figure 1. Location of data collection

Data collection using line and line transect method with 8 (eight) plots in each location. The main plot measures 25 x 25 meters, so it can represent the entire post-restoration area with a total area of 1 ha. The main plot measuring 25 x 25 meters was used to record trees (diameter > 20 cm), 10 x 10 meters to record poles (10-20 cm diameter), 5 x 5 meters to record poles, and 2 x 2 meters to record seedlings (height < 1.5 meters). Paths were made in the center of each post-restoration site block to collect all types of vegetation. Vegetation recording was carried out using a randomized complete block design (RCBD) with 8 replications at 2 observation locations. Data analysis using quantitative methods with calculations Shanon Weiner and Evenness / Eveness [8]. These calculations are used to determine the diversity and evenness of species present in the post-restoration area.

- Shanon Weiner

$$H' = \sum_{i=1}^{s} \quad (pi)(in\ pi) \tag{1}$$

Information:
H' = Shanon Weiner
Pi = ni/N
ni = Number of species i
N = Total number of individuals
According [9] the criteria for the diversity index can be understood as follows:
H' < 1 = Low Specific Diversity
1< H'< 3 = Medium species diversity
H' > 3 = High specificity

- Evness

$$\text{Evness} = \frac{D}{D_{max}} \tag{2}$$

$$\text{Evness} = \frac{D-D_{min}}{D_{max}-D_{min}} \tag{3}$$

Information:
D = Observational diversity index value
D_{min} = Minimum value of diversity index
D_{max} = Maximum index value

3. Results and Discussion

MMNP has natural potential as a source of resilience and protection of life. MMNP has volcanic soil types on the slopes of Mount Merapi and its surroundings in the form of andosol, alluvial, regosol, and litosol. With this potential, the level of biodiversity in MMNP varies greatly [10]. However, due to volcanic activity in 2010 the Cangkringan Resort area was affected by the loss of vegetation in the MMNP area. This is one of the reasons for the restoration in the MMNP area. The restoration area is an area that is the center of ecosystem recovery that has decreased due to global damage and destruction. Rehabilitation and remediation efforts are needed to maximize restoration performed so that ecology can return to normal [12]. In this study, vegetation becomes the main thing in restoration efforts. Based on data from the MMNP Center, which carried out reforestation efforts to improve the natural conditions of the area, the species *Acacia decurrens, Schima wallichii, Dodonaea viscose, Inocarpus fagifer, Tarenna Incerta , Eurycoma longifolia* and other local species have been planted. Therefore, during the 10 years, fundamental changes were made in the vegetation found in the restoration area. Vegetation analysis is the main key to knowing and controlling the calculation of the presence of species in the restoration area [13].

In this study reap the results of the calculation of the diversity index based on the Shannon Wiener formula. The restoration area used for data collection is at the Cangkringan Resort MMNP. Then the distribution of data collection points is spread over Klangon and Kali Kuning. Both areas are restoration areas due to volcanic activity from Mount Merapi. From the results of this study, 45 species of plants were found in the Klangon area and 34 species in the Kali Kuning area. Species found in both places include as listed in Table 1 and 2.

Table 1 . Species in the Klangon Restoration Area

No	Species	Amount	No	Species	Amount
1	*Acacia decurrens*	128	24	*Adiantum*	15
2	*Schima wallichii*	21	25	*Pogostemon cablin*	30
3	*Cinnamomum burmanni*	2	26	*Orthosiphon aristatus*	55
4	*Eucalyptus jacksonii*	63	27	*Borassus flabellifer*	6
5	*Dodonaea viscose*	30	28	*Piper aduncum*	7
6	*Quercus sundaica*	1	29	*Inocarpus fagifer*	1
7	*Melastoma malabathricum*	1	30	*Mesua ferrea*	2
8	*Cyperus rotundus*	1379	31	*Dilleniaobovata*	1
9	*Pennisetum purpureum*	991	32	*Ficus Sp.*	5
10	*Ageratum conyzoides*	82	33	*Castanopsis argentea*	15
11	*Erythrina abyssinica*	36	34	*Altingia excelsa*	1
12	*Malabar spinach*	1	35	*Polypodiophyta*	7
13	*Gleichenia linearis*	145	36	*Elymus repens*	85
14	*Centella Asiatica(L)*	76	37	*Cyperus rotundus*	23
15	*Manilkara zapota*	2	38	*Leersia Hexandra*	105
16	*Tarenna incerta*	7	39	*Persea americana*	1
17	*Apium graveolens*	42	40	*Tarenna incerta*	7
18	*Chrysopogon zizanioides*	89	41	*Artocarpus heterophylus*	1
19	*Centella Asiatica L*	94	42	*Syzygium polyanthum*	16
20	*Synedrella nodiflora*	58	43	*Vitex lucens*	1
21	*Elymus repens*	783	44	*Hernandia*	47
22	*Polygala paniculata*	167	45	*Eurycoma longifolia*	9
23	*Charming chinensis*	117		**Total**	**4755**

Table 2. Species in the Kali Kuning Restoration Area

No	Latin Name	Amount	No	Latin Name	Amount
1	*Trema Orientalist B*	4	19	*Erythrina variegata*	1
2	*Peltophorum pterocarpum*	5	20	*Maniltoa grandiflora*	1
3	*Pipturus argenteu*	2	21	*Syzygium polyanthum*	48
4	*Calliandra calothyrsus*	26	22	*Polypodiophyta*	53
5	*Psidium guajava*	1	23	*Digitaria microbachne*	57
6	*Schima wallichii*	148	24	*Themeda arguens*	51
7	*Ficus septica*	1	25	*Ficus heterophylla*	1
8	*Inocarpus fagifer*	39	26	*Syzygium aqueum*	1
9	*Mallotus mollissimus*	12	27	*Calophyllum inophyllum*	1
10	*Pterocarpus indicus*	1	28	*Orthosiphon aristatus*	352
11	*Mimosa pudica*	31	29	*centrosema virginiaum*	82
12	*Cyperus rotundus*	1328	30	*Species A*	7
13	*Chromolaena odorata*	107	31	*Psidium guajava*	3
14	*Elephantopus tomentosus*	414	32	*Peltophorum pterocarpum*	6
15	*Imperata cylindrica*	773	33	*Terminalia catappa*	1
16	*malathima malathirium*	23	34	*Ficus variegata*	4
17	*Altingia excelsa*	8		**Total**	**3595**
18	*Mallotus mollissimus*	3			

Based on Tables 1 and 2, the species found in the Klangon restoration area were 4755 individual species from 45 species in total. Meanwhile, in the Kali Kuning area there were 3595 individuals from

34 species found. From these two areas, the plants found were grouped according to tree species, poles, saplings, and seedlings. This is based on the division of measuring plots defined in the line transect. The total area used as the data collection plot is 1 ha. Then, regarding the species in Klangon, it is dominated by the grass family. This is because grass is indeed one of the fastest growing successions in a state of restoration. However, there are also species of *Acacia Decurenns* , *Dodonaea viscose* , *Schima wallichii* and *Eurycoma longifolia* that are, in fact pioneers in the Klangon area. Of the several species found in the area, it takes a long time to grow to its full potential. This is triggered by several factors, both genetic and environmental [14]. In the Klangon restoration area, it is quite close to the settlements of the surrounding residents who make a living as cattle breeders. Therefore, local residents in this area become grazing land as animal feed. Therefore, grass species such as *Brachiaria mutica* , *Elymus repens, Pennisetum purpureum, Orthosiphon aristatus* and *Cyperus rotundus are included.* Results like this are the same as previous research conducted by [15] in the MMNP which stated that the faster growth of vegetation in the MMNP conservation area included a succession of shrubs and grasses. This is also in terms of the ability of the grass family to adapt to a fairly extreme environment after the eruption under normal conditions [16].

In the Kali Kuning restoration area 34 species of plants were also found, which were still dominated by grasses, both reeds and other grasses. However, the trees that can be found in Kali Kuning include species of *Calliandra calothyrsus, Schima wallichii, Inocarpus fagifer,* and *Syzygium polyanthum.* The existence of vegetation in Kali Kuning is different from that of the Klangon restoration area. Points of difference include humidity, slope, soil conditions, and water sources. The humidity level in the Kali Kuning area is very different because it is under a steep cliff and allows the water to flow. Thus, the soil found is also relatively semi-wet. Then the slope is also very extreme when it comes to the base zone. Meanwhile, the water sources in this area are also relatively easy. This is because the sampling location is close to the river. So, the vegetation is also more fertile because the water supply is easier. However, in this space there are also new plants planted in the area, such as *Peltophorum pterocarpum,* by farmers in the far east. Therefore, the succession in the Kali Kuning area, although still dominated by grass families, the trees in this area are also still balanced and grow at an even height.

This study also discusses the diversity index formulated with the provisions of Shanon Weiner. The diversity index is a calculation of the level of species diversity in an area of an ecological unit as a whole. The following are data on the calculation of diversity in Klangon and Kali Kuning:

Table 3. Calculation of Shanon Weiner in Klangon and Kali Kuning

(H') In 1 and 2 Lines (8 Plots)		
Klangon	2.235279	*Plot 1*
	2,000293	*Plot 2*
	1.719825	*Plot 3*
	1.774637	*Plot 4*
	1.203284	*Plot 5*
	1.483383	*Plot 6*
	1.80381	*Plot 7*
	1.395527	*Plot 8*
H' total	13.61604	*All Plot*
H' total 8 Plot	**1.702005**	**(Currently)**

(H') In 1 and 2 Lines (8 Plots)		
Kali Kuning	1.836893	*Plot 1*
	1.836893	*Plot 2*
	1.215008	*Plot 3*
	1.252316	*Plot 4*
	0.603175	*Plot 5*
	0.87392	*Plot 6*
	0.709967	*Plot 7*
	1.511508	*Plot 8*
H' total	9.839679	*All Plot*
H' total 8 Plot	**1.22996**	**(Currently)**

Table 4. Total Diversity Index of Klangon and Kali Kuning

H' Klangon and Kali Kuning		
	1.70200462	Klangon
	1.22995987	Kali Kuning
H' Total	2.93196449	All
(H') All Twice Place	1.46598225	**(Currently)**

Based on Table 3 which describes the results of the calculation of the diversity index in Klangon and Kali Kuning, the level of diversity in the area is still in moderate stability. This is indicated by the H' in Klangon of 1.702005 and at the Kali Kuning location of 1.22996. This calculation is carried out in two places at once within the same area. Then the number of plots used included eight plots in each area. This study also reviews the differences between the two places and semi-comparisons between the restoration of the Klangon and Kali Kuning areas. These two areas have different locations and natural appearances. Thus, the potential for the diversity owned will highlight its own special characteristics. Similarly to the research carried out by [7] that stated that in the post-eruption area in the Mount Merapi National Park area, after a 3-year gap in reforestation, exotic plant species began to be found that have certain special characteristics according to the potential of the affected area. volcanic eruption. The research also stated that there are foreign invasive species that can grow well in the restoration area of the Cangkringan Resort, namely the *Acacia Decurenss plant species*. Then for the level of diversity, which is currently still considered moderate, it shows that a period in which the development and growth of vegetation species in the restoration area of the Cangkringan Resort is in the adaptation and selection stage [16]. Furthermore, this study is also similar to the research conducted by [4], which explains that the species and also the most successful and rapid adaptation to succession are shrubs, mosses, and herbs in the MMNP area.

Volcanic areas such as the MMNP have fertile soil so that local and foreign species that have successfully adapted to it will grow faster [17]. Therefore, different soil conditions will also affect the diversity of existing vegetation. Regarding the diversity that exists in Klangon, it indirectly brings new species such as *Acacia Decurrents* , *schima wallichii* , *dodonaea viscose* , *Altingia excels,* and *Inocarpus fagifer* [18]. The following is a display of the new tree species planted in the Klangon area.

(a) (b) (c) (d)

Figure 2. *Acacia Decurrents* (a), *Schima wallichii* (b), *Dodonaea viscos* (c), *Altingia excels* (d)

In the Kali Kuning restoration area, many species of shrubs, mosses, ferns, and trees were also found. The place in Kali Kuning is indeed on the bank of the river. So when the research was conducted, many PAM water channels were found for the needs of local residents in the surrounding area. Also, we must walk through steep cliffs to get to this area. The level of diversity in this area is slightly different from that of Klangon in both species and individuals. However, the tree species found in this area include *Schima wallichii* , *Calliandra calothyrsus* , *Syzygium polyanthum,* and *Altingia excels.* These trees are planted to prevent landslides and erosion both from rainwater from the top of the hill and abrasion by Kali Kuning river water. This is because this type of tree has a high absorption capacity and strong roots to hold the soil, as well as erosion and abrasion on the banks of the river [19]. However, the Kali Kuning area has great potential for vegetation diversity, utilization, and tourism zones [20]. However, the level of vegetation and diversity in Kali Kuning is also considered balanced and still safe. This is evidenced by the discovery of ape species as well as other types of hummingbirds and others around the research location. This opinion is also the same as in the research conducted by [21] with the study of diversity and macrofungi in Kali Kuning. In addition, this area is also a place for residents and farmers to find grass or irrigation. It should also be noted that at this location there are tree species that are considered the most common and easy to grow well there, namely *Inocarpus fagifer* and *Syzygium polyanthum*.

The species found in both Klangon and Kali Kuning are still classified in the medium level of diversity classification. However, the distribution or evenness of the species found has also been recorded at once in this study. The evenness index is the level of evenness of a species within the scope of a particular community [22]. In the study of this type of evenness index, the causality of the diversity index, or H',. In this evenness index, it can be found by dividing between H'/Ln(S) and then if the value of this result is less than 1 then the rating strata are declared still moderate. The following is the data calculated in this study:

Table 5. Evenness Index

	Klangon	
H'		**1.702005**
Number of Species found (S)		**46**
E	**(Currently)**	**0.444545**
	Kali Kuning	
H'		**1.22996**
Number of Species found (S)		**35**
E	**(Currently)**	**0.345946**

In the study of this evenness index, it represents the distribution of species in developing and growing evenly in their habitat. Then, for a period of 10 years after this restoration, it requires the possibility that it is still in a new stage and requires a very long time duration so that the diversity index and evenness level can be said to be high. Table 5 shows that the equity in Klangon is only around 0.44454 and in Kali Kuning it is still at 0.345946. According to [23] the results, which are still moderate, may be influenced by several reasons, including natural and non-natural factors. This natural factor is viewed from the basic substance for the instinct of natural development from the internalization of each species and also the supportive environment. Meanwhile, this non-natural factor is the impetus for human behavior to strive for an even distribution of species in each region. The form of this nonnatural factor can be shown in reforestation activities which are carried out in a scattered and separate manner between areas. This is related to the form of local wisdom of the people of the slopes of Mount Merapi as a mitigation and adaptation effort to preserve and maintain the existence of their forests, one of which is the planting of trees [24]. Because the potential shown in this location is dry/volcanic land, it requires a relatively slow spread time [25]. Unlike water intermediaries such as mangroves on the coast with potential water substances, the distribution can be faster if it is not disturbed by anthropogenic activities in coastal areas [26].

Therefore, in the study of the analysis of the composition and structure of the vegetation based on the calculation of the diversity index and also the evenness that exists in the Cangkringan Resort restoration area, which is classified as still in the post-eruption period, it is necessary to have a continuous follow-up effort from both the management and all levels of society to grow well.

4. Conclusion

After restoration at TNMM Resort Cangkringan, 45 species of plants were found in Klangon and 34 species of plants in Kali Kuning, which included the categories of trees, saplings, poles, and seedlings. The level of total diversity in the Cangkringan TNMM Resort is moderate (1,465), with moderate evenness in Klangon (0.444) and moderate in Kali Kuning moderate (0.346). The diversity, which is currently still classified as moderate, indicates that the period in which the development and growth of vegetation types in the Cangkringan Resort restoration area is in the adaptation and selection stage. Therefore, in an effort to increase changes to improve vegetation improvement, continuous follow-up is needed from both management and all levels of society in order to grow well.

References

[1] Hadi UP and Takaoka H 2018 The Biodiversity of Black ies (Diptera: Simuliidae) in Indonesia. *Acta Tropica* **185** 133-137
[2] Fan A 2020 The Study of Biodiversity as a Means of Ecotourism Education in the Baho Majo Waterfall Area, Bawodobara Village *J. Educ. Dev.* **8**(1) 289-293
[3] Fadhillah N, Suwarniati, Zahara M, Aini Q, Muslim M, and Rizki C N 2021 Plant Vegetation Analysis Using the Line Transect Method in Glee Reuleung Area, Leupung District, Aceh Besar. *Biosci. Jo* **1**(2) 75-82
[4] Alima N, Nugroho EC, Rizki EW, Intan A, and Ifani EF 2021 Analysis of Vegetation Around the Bunker Area of Mount Merapi National Park *Biome: Biology Periodic* **22**(2) 110-114
[5] Yamin M, Setiadi D, and Khairuddin 2021 Diet and Behavior of *Macaca fascicularis* for Ecotourism Contributing on Pengsong Area *J. Trop. Biol.* **21**(1) 179-190
[6] BNPB 2011 Action Plan for Post-Disaster Rehabilitation and Reconstruction in the Post-Disaster Eruption of Mount Merapi in Yogyakarta and Central Java Provinces 2011 – 2013. Bappenas – BNPB.
[7] Gunawan H, Heriyanto N M, Subiandono E, Mas A, and Krisnawati H 2015 Invasion of exotic species on degraded land after eruption in Mount Merapi National Park. In *Proc. Natl. Seminar Indonesian Biodiversity Soc.* **1**(5) 1027-1033
[8] Krebs C J and Singleton G R 1993 Indexes of condition for small mammals. *Aust. J. Zool.* **41**(4) 317-323
[9] Odum E P and Barrett G W 1971 *Fundamentals of ecology* (Vol. **3**, p. 5) (Philadelphia: Saunders)
[10] Natalia D and Handayani T 2013 Analysis of bush strata vegetation in Plawangan National Park of Mount Merapi after the 2010 eruption of Merapi *J. Bioedukatika* **1**(1) 62-71
[11] Ianati U A 2013 Community Perception of Puspa, Salam and Gayam as Core Plants for Restoration of Mount Merapi National Park
[12] Conscience S and Christiani L 2019 Curative Preservation of Textual Static Archives After the 2010 Mount Merapi Eruption Natural Disaster in an Effort to Save Archives at the Library and Archives Service of Sleman Regency *J. Lib. Sci.* **6**(4) 361-370
[13] Afrianto W F, Hikmat A, and Widyatmoko D 2017 The floristic community and vegetation succession after the 2010 eruption at Mount Merapi, Central Java *Indonesian J. Biology* **12**(2)
[14] Hidayat R and Rifanjani S 2017 Study of diurnal bird species diversity in the sebadal forest of Gunung Palung National Park, North Kayong Regency *J. Sustain. For.* **5**(3)
[15] Parwati A F, Aptari Z, Saputri R D, Akbarudin AM, Kirana A G and Wahyuni S T 2019 Vegetation analysis in Mount Merapi National Park *J. Dipterocarp Ecosystem Res.* **5**(2) 107-112
[16] Aprilianti, ANM, Ba'it, AAZ, Azizah, RAN, Miranda, D., Rachmayani, M., & Ummah, AAN 2021. Analysis of Vegetation on the Southern Slope of Mount Merapi National Park. *Biomes: Journal of Biology and Learning Biology*, **6**(2) 144-159
[17] Anggana A F, Cahyono S A, and Lastiantoro C Y 2019 Biodiversity in the rehabilitation area of Meru Betiri National Park and its policy implications: the case of Wonoasri Village *J. Environ. Sci.* **17**(2) 283-290

[18] Heriyanto N M, Samsoedin I, and Bismark M 2019 Biodiversity Flora and Fauna in the Bukit Datuk Dumai Forest Region, Riau Province *Sylva Lestari J.* **7**(1) 82-94
[19] Ramadhan R, Mursyid H, Adriyanti D T, Triwanto J, and Triwaskitho N 2020 Growth of Invasive Type Acacia decurrens Willd. and the Effect of Shade on Restoration Plants *Biotropics: J. Trop. Biol.* **8**(2) 71-78
[20] Dunggio I, and Ichsan A C 2022 The Effectiveness of Making Vegetative Plants in Overcoming Erosion and Sedimentation (Case Study in the Limboto River Basin, Gorontalo Province) *Journal of the Wilds* **5**(1) 45-58
[21] Muhammad M 2017 Natural Tourism Ecological Site in the Utilization Zone of Mount Merapi National Park (MMNP) (The Thematic Natural Tourism Development Concept for the Kalikuning-Kaliadem Area as a Volcanic Cultural Area) *Kawistara J.* **7**(3) 238-248
[22] Prasetyaningsih A and Rahardjo D 2015 Diversity and Macrofungi Potential of Mount Merapi National Park. In *Proceedings of National & International Seminars*
[23] Sulistyadi E 2017 The ability of non-conserved areas to protect the conservation of endemic lowland birds on the island of Java, a case study in Kebumen Regency *Indonesian Journal of Biology* **6**(2)
[24] Rachmawati S, Matin H H A, Suhardono S, Setyono P, Kusumaningrum L, Rinawati S, and Sholiqin M 2021 Biodiversity of Rambut Cave Sale Central Java. In *IOP Conference Series: Earth and Environmental Science* **896**(1) 012006
[25] Kusumaingrum L, Rachmawati S, Hashifah D G, Muryanto BS, and Dharmawan W 2021 Local Wisdom of Merapi Slope Village in Ensuring the Availability and Sustainability of Springs in Glagaharjo Village Cangkringan District of Sleman Yogyakarta *J. Glob. Environ. Dyn.* **2**(3) 21-37
[26] Fitria A and Banowati E 2018 Community Participation in the Preservation of the Merapi Slope Forest through the Coffee Agroforestry Program in Tlogolele Village, Selo District. *Edu Geography* **6**(3) 162-169
[27] Sholiqin M, Pramadaningtyas P S, Solikah I, Febriyanti S, Pambudi M D, Mahartika S B and Setyawan A D 2021 Analysis of the diversity and evenness of mangrove ecosystems in the Pacitan Coast, East Java, Indonesia *International Journal of Bonorowo Wetlands* **11**(2) 84-94

Application of anaerobic and aerobic bioreactors in detergent wastewater treatment: A review

Rifqi Ahmad Baihaqi[1]*, Kencana Ayudya Prabahandari[2], Yogi Hariyono[1], Novita Indah Pratiwi[1], Heri Sutanto[1], Yoyon Wahyono[3]

[1]Department of Physics, Faculty of Science and Mathematics, Diponegoro University, Jl. Prof. Soedarto, SH, Tembalang, Semarang, Central Java 50275, Indonesia.
[2]Department of Biology, Faculty of Science and Mathematics, Diponegoro University, Jl. Prof. Soedarto, SH, Tembalang, Semarang, Central Java 50275, Indonesia.
[3]Doctoral program of Environmental Science, School of Postgraduate Studies, Diponegoro University, Jl. Imam Bardjo SH, Pleburan, Semarang, Central Java 50241, Indonesia.

*rifqiab@students.undip.ac.id

Abstract. The laundry business sector is expanding quickly nowadays. However, the laundry industry is still dealing with the issue of processing detergent wastewater. Anaerobic and aerobic bioreactors may efficiently solve this problem. This narrative review aims to assess the feasibility of using anaerobic and aerobic bioreactors for detergent wastewater treatment. Its advantages and disadvantages and the idea of combining multi-media filtration and UV light in detergent wastewater treatment using an aerobic and aerobic bioreactor. The anaerobic bioreactor can reduce chemical and biological oxygen demand to 89.8 % and 94.0 %, respectively. At the same time, aerobic bioreactors can reduce chemical and biological oxygen demand to 99.1% and 71%, respectively. However, some challenges still need to be addressed to make anaerobic ad aerobic bioreactors can be implemented. Suspended solid production, dissolved methane, and temperature-dependent effectiveness are challenges that must be solved. Multi-media filtration can reduce suspended solids and provide ion exchange, while UV light kills excess microorganisms from the bioreactor.

1. Introduction

As people's lives get busier and more mobile, industrial laundry services get increasingly more consumers. However, the laundry industry produces detergent wastewater. Many laundry businesses in Indonesia still dispose of their untreated detergent wastewater into the environment. It can harm the environment because detergents contain phosphate compounds (50%), surfactants (15%), bleach (7%), enzymes (2%), and other ingredients (26%) [1]. Of these compounds, phosphates and surfactants are the most harmful to the environment. They can cause algae bloom [2], [3], which can result in eutrophication and reduced dissolved oxygen in waterways [1]. Untreated detergent wastewater not only causes problems in the water but can also pollute soil and inhibit plant growth [4]. Detergent wastewater can also cause eye and skin irritation in children and the elderly [5], [6]. Several studies have been conducted to solve this problem, such as the chemical coagulation method [7], electrocoagulation [8] [10], and biodegradation methods [11]. Although the chemical coagulation method effectively reduces harmful chemical compounds in detergents, it requires coagulant materials that must be continuously supplied, making it economically unfeasible [12]. Compared to other ways of dealing with detergent wastewater, the biodegradation method is more environmentally friendly and economical [13].

Biodegradation use microorganisms in a bioreactor to degrade detergent's active surface. It is economically feasible and can be implemented on a large scale. There are two types of bioreactors, anaerobic and aerobic bioreactors. Anaerobic bioreactors do not use air in the degradation process, while aerobic bioreactors use air. Anaerobic bioreactor can reduce chemical oxygen demand (COD) and biological oxygen demand (BOD) to 89.8 % and 94.0 %, respectively [14], [15]. Aerobic bioreactors can balance pH and reduce the COD and BOD to 99.1% and 71%, respectively [11], [16]. However, some challenges still need to be addressed to make anaerobic ad aerobic bioreactors can be implemented.

This narrative review discusses the feasibility, advantages, and disadvantages of applying anaerobic and aerobic bioreactors for detergent wastewater treatment in the past ten years to support its implementation for detergent wastewater treatment in Indonesia. Furthermore, this review discusses multi-media filtration and ultraviolet (UV) light as bioreactor support systems. Filtration helps reduce suspended solids and provides ion exchange [17]. UV light is useful for killing excess bacteria from the bioreactor [18]. Combining anaerobic and aerobic bioreactors with multi-media filtration and UV light as supporting features is expected to increase the optimation bioreactor for detergent wastewater treatment. Furthermore, the quality of the processed wastewater can meet the quality of detergent wastewater.

2. Composition of detergent wastewater

The composition of detergent wastewater is 50 % phosphate compounds, 15 % surfactants, 7 % bleach, 2 % enzymes, and 26 % other ingredients [1]. The most commonly used surfactant type is linear alkyl sulfonate (LAS), which has a straight alkyl chain [19]. LAS is a relatively environmentally friendly surfactant when compared to alkyl benzene sulfonate (ABS) because it is biodegradable by *Staphylococcus epidermis*, *Enterobacter gergoviae*, *Staphylococcus aureus*, *Pseudomonas facili*, *Pseudomonas fluoroscens, Pseudomonas putida*, and *Kurthiazopfii* [19] [21]. Table 1 shows the negative impact of laundry detergent components.

Table 1. Negative effects of laundry detergent components.

Ingredients	Negative effects	Source
Surfactants	Causes skin irritation, eye pain, and poisoning of aquatic biotas such as fish and shrimp.	[4]
Phosphate	It can stimulate the overgrowth of freshwater algae resulting in eutrophication and reduced dissolved oxygen in the water channel.	[1]
Foam	Foam caused by detergents can reduce the penetration of oxygen from the air into water resulting in less dissolved oxygen in the water.	[22]
Linear alkyl sulfonate	Adversely impacts fish larvae. In a high enough dose, the mortality rate increases.	[23]

Because of the various adverse effects of detergent wastewater, it is crucial to process and meet the wastewater quality standards before being discharged into waterways. Table 2 below shows the wastewater quality standards for businesses and textile industries expressed in BOD, COD, total suspended solids (TSS), and pH parameters per the Minister of Environment and Forestry of the Republic of Indonesia regarding wastewater quality standards of 2019 [24].

Table 2. Wastewater quality standards.

Debit (m^3/day)	BOD (mg/L)	COD (mg/L)	TSS (mg/L)	pH (mg/L)
≤ 100	60	150	50	6-9
< 100 × < 1,000	45	125	40	6-9
≥ 1,000	35	115	30	6-9

3. Anaerobic bioreactors

In anaerobic bioreactors, wastewater treatment utilizes anaerobic bacterial metabolism [15]. Anaerobic bioreactors can treat detergent wastewater with COD, BOD, and LAS content up to 630 ± 166 mg/L, 172.1 ± 60.4 mg/L, and 20.1 ± 6.0 mg/L, respectively [25,26]. An anaerobic bioreactor can perform the LAS degradation process [27]. During the biodegradation process, microorganisms perform a structural change on the surfactant, resulting in the latter losing its active surface [28]. A study by Fernández et al. [15] shows 60 mg/mL LAS reduction from 85 ng/mL to 25 ng/mL.

Moreover, anaerobic bioreactors can reduce COD and BOD content in detergent wastewater. Hamid et al. [14] used a suspended growth anaerobic baffled reactor method to degrade shopping centre wastewater and showed decreases in COD and BOD of 89.67 % and 88.3 %, respectively. Sawajneh et al. [29] used an anaerobic biofilter and an up-flow anaerobic sludge blanket method to degrade domestic waste in Jordan and showed a decrease in COD and BOD of 81 % and 58 %, respectively. Ji et al. [26], using two anaerobic membrane bioreactors to treat municipal wastewater, showed reductions in COD and BOD of 89.8 ± 1.8% and 94.0 ± 1.3 %, respectively.

Before being allowed to operate in Indonesia, anaerobic bioreactors must meet the Indonesian National Standards (SNI). Table 3 below shows SNI 8455:2017 [30] regarding the planning of domestic wastewater treatment with an insulated anaerobic reactor system (SRAB). The design criteria for anaerobic reactor systems based on the SNI above are shown in table 3.

Table 3. Design criteria for anaerobic reactor systems.

No.	Parameter	Unit	Value
1	Residence time, T_d	h	6–20
2	Organic loading rate (OLR)	kg COD/m^3/day	0.1–8
3	V_{up}, flow rate up	m/h	<2.0
4	Flow area ratio, down flow area to up flow area		1:2
5	The concentration of volatile solids	g/L	4–20
6	% Removal of BOD	%	70–95
7	% Removal of COD	%	65–90

During the degradation process, the pH of detergent wastewater is decreased to 5.5. It is because the degradation process produces dissolved acidic CO_2 gases (H_2CO_3) [31], [32]. This decrease in pH puts the wastewater below the quality standard. Adding an aerobic bioreactor to the system can help overcome the pH problem.

4. Aerobic bioreactors

The degradation process in aerobic bioreactors is conducted in aerobic conditions [33]. The aeration process in aerobic bioreactors increases oxygen levels after anaerobic processes in the anaerobic bioreactors, which lack oxygen. The longer the duration of the aeration process in an aerobic bioreactor, the more oxygen dissolves in the fermented waste. The pH of fermented waste in anaerobic bioreactors gets closer to neutral (pH=7) [16]. Aerobic bioreactors can treat detergent wastewater with COD and BOD content up to 1,000 mg/L and 4,000 mg/L, respectively [34].

Yang et al. [35] used a submerged membrane bioreactor to degrade artificial restaurant waste and showed a reduction in COD of up to 98.3 % and 99.1 %, respectively. Dilafroza et al. [36] used fluidized aerobic bioreactor-based sewage treatment plants and showed a 71 % reduction in BOD. Chan et al. [37] used a thermophilic integrated anaerobic-aerobic bioreactor (IAAB) and showed a decrease in COD of up to 80 %.

5. Advantages and disadvantages of anaerobic and aerobic bioreactors

In its application for detergent wastewater treatment, aerobic and anaerobic bioreactors have advantages and disadvantages. Table 4 shows the advantages and disadvantages of bioreactors for detergent wastewater treatment.

Table 4. The advantages and disadvantages of bioreactors for detergent wastewater treatment.

Advantages	Disadvantages
Affordable cost [13]	Producing biomass and dissolve methane [41]
High COD and COD reduction [14,26,29,35,36]	Producing suspended solid [42]
Producing methane as a renewable biogas source [38,39]	Efficiency depends on temperature [42,43]
It can be utilized on a large scale [40]	Limitation on high COD content treatment [25,26]

Unlike chemical coagulation methods, bioreactors do not need coagulants that must be constantly supplied. So in long-term use, bioreactors in detergent wastewater treatment are more affordable [13]. The fermentation process in anaerobic bioreactors produces biomass and methane. Poorly managed biomass can result in solids accumulation. In the bioreactor that uses membranes, solids accumulation can cause membrane fouling. Solid accumulation can be overcome by optimizing the bioreactor design to provide high shear conditions on the membrane and using a filter [44]. Methane produced by anaerobic bioreactors can be used as a renewable source of biogas. However, not all of the methane produced completely recovered, and dissolved methane in the effluent can be released into the atmosphere and contribute to greenhouse gases [45]. This problem can be solved using degassing membranes, aeration, and air stripping [45]. Bioreactors can be applied on a large scale up to 10 million m3/s in china in 2017 [40]. Temperature fluctuations are a challenge for bioreactor applications. Ferrari et al. [41] reported that COD removal at 23-34°C reached 87%, but if the temperature was reduced to 15-17°C, COD removal was only 70%. Indonesia has a tropical climate with relatively consistent annual temperatures, so temperature fluctuations are not a significant obstacle to applying bioreactors in Indonesia.

6. Multi-media filtration

We recommend adding a multi-media filter to bioreactor systems to optimize the anaerobic and aerobic processes for detergent wastewater treatment. Depending on the material used, there are additional benefits to having a multi-media filter. Table 5 shows the effects and benefits of various filter media.

Table 5. Effects and benefits of various filter media.

Filter media	Effect	Reference
Zeolite	Ion exchange. It can remove ammonia and nitrogen.	[46], [47]
Silica sand	Porous medium to remove phosphorus.	[17]
Carbon Active	Advanced phosphorus removal.	[48]
Gravel	Removal of sediment and heavy metals.	[49]
Powdered *Moringa oleifera* seeds	Natural coagulant reduces COD, BOD, and TSS.	[50]

A study byAssiddieq et al. [51] used silica sand, activated carbon, and zeolite as multi-media filters to filtrate detergent wastewater shows decreases in COD, BOD, and TSS of 87.80 %, 74.49 %, and 88.75 %, respectively.

7. UV light

UV light is electromagnetic waves with wavelengths ranging from 100 to 400 nm. With this wavelength, UV light has enough energy to kill bacteria. UV light with wavelengths between 220 and 300 nm is considered the most effective against microorganisms such as bacteria, viruses, protozoa, and algae [18]. Based on UV light's ability to kill bacteria, it can be beneficial to bioreactor systems, as it kills the bacteria on the bioreactor before they can escape into the environment. UV light is applied at the end of the treatment process to not interfere with the metabolism inside the bioreactor.

8. Prospects and the potential of anaerobic and aerobic bioreactors

Detergent wastewater is increasing in quantity in line with population growth and increasing people's quality of life, meaning that as people's incomes rise, the consumption of detergents also increases. Detergent wastewater containing excessive pollutes and disturbs the ecosystems of aquatic biota. Aerobic and anaerobic bioreactor technology has the potential and prospective for detergent wastewater treatment because it is effective and economical. Although this technology has recently experienced rapid development, some obstacles and challenges limit its wide practical application. Therefore, further research is needed in the following areas.

1. Optimization control of temperature, pH, aeration, nutrients, and osmotic pressure in bioreactors to support the metabolism of microorganisms using the IoT (Internet of Things).
2. Utilization of the metabolism of microorganisms so as not to waste it.

9. Conclusion

Detergent wastewater from the laundry industry causes various environmental issues and must be resolved. The biodegradation treatment method is effectively and economically feasible to solve this problem. It is a highly recommended method, due is Indonesia a developing country with a limited budget. The anaerobic bioreactor can reduce LAS, COD, and BOD up to 70 %, 89.8 % and 94.0 %, respectively. Aerobic bioreactors can reduce chemical and biological oxygen demand to 99.1% and 71%, respectively while balancing pH detergent wastewater from anaerobic bioreactors which lack oxygen. Although, some challenges still need to be addressed to make anaerobic and aerobic bioreactors can be implemented. These challenges can be resolved with good management. Multi-media filters can help suspended solid production problems. Dissolved methane can be solved using degassing membranes, aeration, and air stripping. Temperature fluctuations are not a significant obstacle to applying bioreactors in Indonesia because Indonesia has a tropical climate with relatively consistent annual temperatures. UV light kills excess bacteria, limiting environmental pollution. This method can be further developed in the future, as the addition of automation systems and the utilization of metabolic products can make this method more efficient and further reduce waste.

References

[1] Irawan C, Ratmasari A, Rizaldi F, Nata I. F. and Putra M D 2020 *IOP Conf. Ser. Earth Environ. Sci* **524** 1
[2] Baroi A R, Chowdhury R B, Roy B B, and Sujauddin M 2020 *J. Clean. Prod.* **273** 122865
[3] Schindler D W, Carpenter S R, Chapra S C, Hecky R E, and Orihel D M 2016 *Environ. Sci. Technol.* **50** 8923–8929
[4] Tri Wahyuni E 2016 *Am. J. Appl. Chem.* **4** 174
[5] Settimi L, Giordano F, Lauria L, Celentano A, Sesana F, and Davanzo F 2018 *Inj. Prev.*, **24** 5–11
[6] Nguyen V 2017 *Case Stud. Med. Toxicol.*, Springer, Cham.
[7] Joko T, Agustin D, Putri C, and Lanang H 2016 *Int. J. Sci. Basic Appl. Res.* **4531** 104–111
[8] Wahyono Y, Sutanto H, and Hidayanto E 2017 *Youngster Phys. J.* **6** 4 353–359
[9] Wahyono Y *et al.* 2021 *J. Phys. Conf. Ser.* **1943** 1
[10] Suprihatin and Aselfa F. S. 2020 *IOP Conf. Ser. Earth Environ. Sci.* **472** 1
[11] Mousavi S. A. and Khodadoost F 2019 *Environ. Sci. Pollut. Res.* **26** 26439–26448
[12] Teh C Y, Budiman P M, Shak K P Y and Wu T Y 2016 *Ind. Eng. Chem. Res.* **55** 4363–4389
[13] Chen Y *et al.* 2019 *Bioresour. Technol.* **287** May, p. 121467
[14] Hamid A, Hamid A., and Razif M 2014 *J. Tek. ITS* **3** 2 D85–D88, 2014,
[15] Corada-Fernández C, González-Mazo E and Lara-Martín P A 2018 *J. Hazard. Mater.* **360** 24–31
[16] Sato, P. Utomo, and H. S. B. Abineri 2015 *Semin. Nas. Sains dan Teknol. Terap. III* 185–192
[17] Yang K, Li Z, Zhang H, Qian J. and Chen G 2010 *Environ. Technol.* **31** 601–609
[18] Bintsis T, Litopoulou-Tzanetaki E and Robinson R K 2011 *LWT - Food Sci. Technol.* **44** 2173–2179
[19] Maslahat M, RTM S and Rakhmanika F 2010 *J. Ecolab* **4** 89–96
[20] Asok A K and Jisha M S 2012 *Water. Air. Soil Pollut.* **223** 5039–5048
[21] Costa . F, de Oliveira A M, and de Oliveira Junior E N 2020 *Bioresour. Technol. Reports* **9** 100363
[22] Rajan D S 2015 *Int. J. Fish. Aquat. Stud.* **2** 6 46–48
[23] Rejeki S, Desrina D, and Mulyana A R 2008 *J. Coastal Dev..*
[24] KLHK, 2019 *J. Chem. Inf. Model.* **53** 9 1689–1699
[25] Delforno T P *et al.* 2017 *Sci. Total Environ.* **587**–588 389–398
[26] Ji J *et al* 2020 *Sci. Total Environ.* **745** 140903
[27] de Oliveira L L *et al* 2010 *Bioresour. Technol.* **101** 14 5112–5122
[28] Merrettig-Bruns U and Jelen E 2009 *Materials (Basel).* **2** 1 181–206
[29] Sawajneh Z, Al-Omari A, and Halalsheh M 2010 *Water Sci. Technol.* **61** 9 2399–2406
[30] Badan Standar Nasional Indonesia, 2017 1–25
[31] Bouallagui H *et al* 2004 *Biochem. Eng. J.* 21 2 193–197
[32] Lira-Parada P A, Pettersen E, Pérez-García F, and Bar N 2019 *IFAC-PapersOnLine* **52** 26 231–237
[33] Nugro Rahardjo P 2006 *Jai* **2** 1 66–71
[34] Chan Y J, Chong M F, Law C L, and Hassell D G 2009 *Chem. Eng. J.* **155** 1–18
[35] Yang B, Chen G, and Chen G 2012 *Sep. Purif. Technol.* **88** 184–190
[36] Dilafroza J, Ashok K P, and Azra N K 2013 *African J. Biotechnol.* **12** 17 2224–2233
[37] Chan Y J, Chong M F, and Law C L 2012 *Bioresour. Technol.* **125** 145–157
[38] Lei Z, Yang S, Li Y, Wen W, Wang X C, Chen R. 2018. *Bioresour Technol* **267**:756–68
[39] Liao B Q, Kraemer J T, Bagley D M. 2006. *Crit Rev Environ Sci Technol.* **36**:489–530
[40] Xiao K, Liang S, Wang X, Chen C, and Huang X 2018 *Bioresource Technology* **9** 061
[41] Ferrari F, Balcazar J L, Rodriguez-Roda I, Pijuan M 2019 *Sci. Total Environ.* **670**:993–1000
[42] Smith A L, Skerlos S J, Raskin L 2015 *Environ Sci Water Res Technol* **1** 56–64
[43] Pena ~ M, do Nascimento T, Gouveia J, Escudero J, Gomez A, Letona A, *et al* 2019 *Bioresour*

Technol **285** 93
[44] Gouveia J, Plaza F, Garralon G, Fdz-Polanco F, Pena ~ M. 2015 *Bioresour Technol* **185** 225–233
[45] Velasco P, Jegatheesan V, and Othman M 2018 *Front. Environ. Sci.* **6** 151
[46] Delatolla R, Berk D, and Tufenkji N 2008 *Water Res.* **42** 3082–3088
[47] Stefanakis A I and Tsihrintzis V A 2012 *Microporous Mesoporous Mater.* **155** 106–116
[48] Altmann J, Rehfeld D, Träder K, Sperlich A, and Jekel M 2016 *Water Res.* **92** 131–139
[49] Hatt B E, Fletcher T D, and Deletic A 2007 *Water Res.* **41** 12 2513–2524
[50] Suhartini S, Hidayat N, and Rosaliana E 2013 *Int. J. Recycl. Org. Waste Agric.* **2** 1–11
[51] Assiddieq M, Darmayani S, and Kudonowarso W 2017 *J. Pendidik. Biol. Indones.* **3** 202

An environmental impact assessment of geothermal tourism: A case study of Awit Sinar Alam Darajat, Garut-Indonesia

I Munfarida[1], W Nilandita[1] and S W Auvaria[1]

[1]Environmental Department, Faculty of Science and Technology, UIN Sunan Ampel Surabaya. Jl. Ahmad Yani No. 117, Surabaya, Indonesia.

*munfarida@uinsby.ac.id

Abstract. Geothermal as one of the potential environmental services in conservation areas is increasingly being taken into account. According to the Government Regulation No. 25 of 2021 on the Implementation of the Energy and Mineral Resources Sector, geothermal can be used directly or indirectly. The geothermal tourism of Awit Sinar Alam Darajat is an example of the direct utilization of geothermal energy for tourism. This study aims to assess the impact of geothermal tourism on the environment and develop environmental management and monitoring. The analysis was carried out utilizing a qualitative and quantitative methods approach. Data were collected from secondary data and analyzed to develop environmental management and monitoring. The result showed that the environmental impact of geothermal tourism includes the change of land use, the decrease in geothermal quantity and the generation of quality, and solid waste. We recommend environmental management, such as setting the minimum standard for the Green Base Coefficient of 70%, geothermal conservation, providing the geothermal treatment plant, and solid waste management. Environmental monitoring should be applied when calculating the Green Base Coefficient and the quantity and quality of geothermal and solid waste generation to achieve the standard.

1. Introduction

Geographically, Indonesia is located between three world plates; there are the Indies-Australian Plate, the Eurasian Continental Plate, and the West Pacific Plate, which causes a large number of volcanoes in Indonesia. Indonesia has 129 volcanoes; therefore, Indonesia has a geothermal potential of 40% of the world's geothermal reserves [1]. The geothermal potential can be used as a source of energy and for other uses. The geothermal potential for energy in Indonesia is estimated at around 29,544 MW. However, only about 7.2% are used as domestic electricity [2]. Geothermal as one of the potential environmental services in conservation areas is increasingly being taken into account. According to the Government Regulation No. 25 of 2021 on the Implementation of the Energy and Mineral Resources Sector, geothermal can be used directly or indirectly. Utilization directly utilizes the heat contained in geothermal fluids for various purposes such as space heating, agriculture/agro-industry, fisheries, and tourism. Meanwhile, indirect use, for example, is utilization for power generation. Awit Sinar Alam Darajat geothermal tourism located in Garut Regency, West Java Province, is an example of direct use of geothermal energy for tourism.

In a few decades, the tourism sector has been a key driver of global economic growth [3]. Tourism depends on environmental conditions. The quality of the environment, both natural and man-made, is essential for the development of tourism. Therefore, it is important to maintain and improve the environmental quality of tourist areas. However, tourism development has positive and negative impacts. Positive impacts are creating jobs and improving the economic level, while the negative impacts are the impact on the biophysical environment (such as water, air pollution and ecosystem degradation). Several studies have shown that tourism has an impact on the biophysical environment, including water, air, soil, flora and fauna and noise [4]. Cole's 2012 research stated that the development of tourism in Bali created a conflict on water utilization between the tourism industry and local communities, especially for agriculture [5]. Increasing tourism or visitors leads to the generation of solid waste, pollution, sanitation, and aesthetic problems [6].

The Awit Sinar Alam Darajat tourist attraction is located in the Garut Regency, West Java. Awit Sinar Alam is a natural hot spring tourist spot that uses geothermal resources. There was not only a natural hot spring, there were also many attractions such as archery, horse and outbound vehicle riding. The facility offers restaurants and lounges. The increasing number of tourists in Awit Sinar Alam Drajat will have an impact on the surrounding environment. Therefore, sustainable geothermal tourism is needed. Sustainable tourism is one of the concepts considered by all countries in the world to achieve the 12th SDG, that is creating green tourism [7]. The destruction of nature and fragile ecology has put the state in danger of losing its biodiversity. Although there is a fact that Environmental Impact Assessment (EIA) for Tourism Industry and Development, among others, has been made controlled by the decree of many countries. The increasing call for the adoption of EIA to legislate environmental impacts and mitigation will be conducted and planned [8]. An Environment Impact Assessment is needed to determine the environmental impact of tourism activities in Awit Sinar Drajat Garut. The main factors that must be considered in the environmental impact assessment of the tourism sector are social factors,, economic factors and ecological environmental factors. Ecological environmental impacts are divided into three sub-criteria, namely tourism environmental impacts, ecological environmental impacts, and ecological characteristics [8]. This study aims to assess the impact of geothermal tourism on the environment and develop environmental management and monitoring.

2. Methods

This research is observational and quantitate research based on secondary data. The analysis was carried out utilizing a qualitative and quantitative methods approach. This study uses a qualitative approach, which is used to understand the phenomenon of a sample research in depth and holistically to reveal the tourism development. Data were collected from secondary data and analyzed to develop environmental management and monitoring. Data were collected from tourist attractions managers, BPS, and related agencies. The research results are presented descriptively.

Secondary data including Awit Sinar Alam Darajat Tourist Object map, number of employee, land use, and laboratory result.

The environmental data was collected by conducting library research, observation, and secondary data analysis. Observations were made at the Awit Sinar Alam Darajat Tourism Object, which includes several facilities, including: Pool area, Resorts, Bungalows, Villas, Hotels, ATV Arena, Outbound Area, Fruit Picking Tour, and Open Green Space.

3. Results and Discussions

3.1. General Description of the Tourist Object Awit Sinar Alam Darajat
Tourism Object Activities "Awit Sinar Alam Darajat" by PT. HAGEI DARAJAT GARUT is motivated by the increase in tourism businesses and activities based on the utilization of geothermal potential in the Darajat area, Garut-Indonesia. In the tourism object area "Awit Sinar Alam Darajat", there are several facilities, including: Pool area, Resorts, Bungalows, Villas, Hotels, ATV Arena, Outbound Area, Fruit Picking Tour, and Open Green Space.

• *Pool Area*

The pool area consists of 7 pools for children and adults. The types of pools are ordinary swimming pools and stream pools. This pool contains hot water from geothermal sources in the tourist attraction area. Here is a map of the pool distribution:

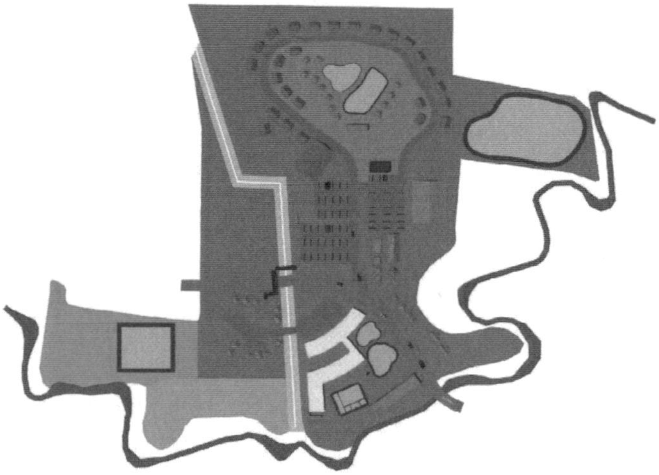

Figure 1. Awit Sinar Alam Darajat Map

• *Resorts, Bungalows, Villas, Hotels*

There are various types of accommodation at Awit Sinar Alam Darajat, such as accommodation which is a type of room for small families or in the form of bungalows / villas; the concept of lodging itself starts from traditional to semi-modern concepts. Resorts, Bungalows, Villas, and Hotels are intended for visitors from the Garut Regency and from outside the Garut Regency. Resorts, Bungalows, Villas and Hotels are built semi- and non-permanently to maintain environmental balance.

Figure 2. Bungalows / villas in Awit Sinar Alam Darajat

• *ATV Arena*

ATV Arena is an out-bound game using an ATV motor. The road used by the ATV motor is a hard road that is made around the "Awit Sinar Alam Darajat" Tourism Object with views of the mountains and natural land around the "Awit Sinar Alam Darajat" Tourism Object.

• *Outbound Area*

The out-bond area in the "Awit Sinar Alam Darajat" Tourist Attraction area includes various types of games, such as ATV Motors, Flying Fox, and children's playgrounds.

• *Fruit picking tour*

In addition to water tourism and games, in the "Awit Sinar Alam Darajat" Tourist Object area there is also a fruit picking tour, where visitors can pick fruit and bring it as souvenirs. The type of fruit grown in the tourist destination "Awit Sinar Alam Darajat" is strawberry.

• *Green Open Space*

The green open space in the "Awit Sinar Alam Darajat" tourist object area has the highest percentage in the tourist attraction land area. This is intended to maintain environmental balance, where the "Awit Sinar Alam Darajat" Tourism Object area in Pasirwangi District is a Provincial Strategic Area as a Geothermal area.

3.2. Environmental Impacts

3.2.1. Land-Use Change. The area of the "Awit Sinar Alam Darajat" Tourism Object is ± 39.558 m^2. The following table is the land use of "Awit Sinar Alam Darajat" Tourism Object.

Table 1. Land Use of the "Awit Sinar Alam Darajat"

				Building Coverage Ratio			
No	**Land Use**	**m2**	**%**	**%**	**Close Space**	**Open Space**	**%**
A	**Close Space**						
1	Bamboo space	98.00	0.25	100	78	-	-
2	Gazebo Type 1	4.80	0.01	100	5	-	-
3	Gazebo Type 2	4.78	0.01	100	5	-	-
4	Gazebo Type 3	6.25	0.02	100	6	-	-
5	Gazebo Type 4	21.28	0.05	100	21	-	-
6	Restaurant	40.00	0.10	100	40	-	-
7	Receptionist room	53.00	0.13	100	53	-	-
8	Hot spring A	249.77	0.63	100	250	-	-
9	Hot spring B	290.81	0.74	100	291	-	-
10	Hot spring C	168.88	0.43	100	169	-	-
11	Information services	5.88	0.01	100	6	-	-
12	Staff room	40.00	0.10	100	40	-	-
13	Mushola Type 1	16.81	0.04	100	17	-	-
14	Mushola Type 2	18.25	0.05	100	18	-	-
15	Attractions area 1	25.00	0.06	100	25	-	-
16	Attractions area 1	25.00	0.06	100	25	-	-
17	Villa 1st floor	300.00	0.76	100	300	-	-
18	Villa 2nd Floor	300.00	0.76	100	300	-	-
19	Helmet room	8.00	0.02	100	8	-	-

				Building Coverage Ratio			
No	Land Use	m2	%	%	Close Space	Open Space	%
20	Bus driver room	15.00	0.04	100	15	-	-
21	Reservation room	73.71	0.19	100	74	-	-
22	Bamboo room semi-permanent	60.48	0.15	100	60	-	-
23	Bamboo restaurant	224.00	0.57	100	224	-	-
24	Rest room 1	33.50	0.08	100	34	-	-
25	Rest room 2	17.60	0.04	100	18	-	-
26	Outbond Rest room	60.00	0.15	100	60	-	-
27	Storage building	27.00	0.07	100	27	-	-
28	Security room	3.00	0.01	100	3	-	-
29	Pump room	5.00	0.01	100	5	-	-
30	Strawberry storage	12.00	0.03	100	12	-	-
31	Main road	3,548.00	8.97	100	3,548	-	-
32	Drainage	1,320.00	3.34	100	1,320	-	-
33	TPSS	50.00	0.13	100	50	-	-
34	WWTP	50.00	0.13	100	50	-	-
	Sub Total A	7,175.80	18.14	100	7,176	-	-
B	**Open Space**						
1	Parking area	5,014.00	12.68	-	-	5,014	100
2	Outbound area	4,838.52	12.23	-	-	4,839	100
3	Green Open Space	22,529.68	56.95	-	-	22,530	100
	Sub Total B	32,382.20	81.86	100	32,382	32,382	100
	Total	39,558.00	100.00	18	7,176	32,382	82

Based on Garut Regency Regulation Number 29 of 2011 on Regional Spatial Planning (RTRW) of the Garut Regency of 2011-2031 stated that the Darajat area is a Provincial Strategic Area (KSP) for Kamojang-Darajat-Papandayan Geothermal with the main function as a geothermal area and additional functions can be in the form of a tourism area and is part of the Central Garut Regency Tourism Strategic Area (KSPK) with the main direction of its business activities including nature and water tourism.

Based on the regulation of the spatial plan as mentioned above, in principle the Awit Sinar Alam Darajat Tourism Object does not break the spatial plan as long as it fulfils the following provisions.

a. Complying with and implementing Garut Regulation No. 19 of 2016 on the Darajat Area Arrangement including its amendments.
b. The maximum height of the building is 8 meters.
c. The number of floors/layers of the building is a maximum of 2 layers.
d. Green Basic Coefficient (KDH) is minimum of 70%
e. The basic building coefficient (KDB) is a maximum of 30%
f. Providing infrastructure, facilities and utilities such as parking areas, drainage, waste management, lighting, hydrants, and fire extinguishers, green open spaces, making infiltration holes (biopori), planting trees, and so on.
g. Create and ratify a site plan as a reference in development, including its changes.
h. Everything that causes significant losses/impacts from development activities is entirely the responsibility of the manager.

Based on the land use of "Awit Sinar Alam Darajat" in Table 1, Green open space is 22,529.68 m^2 or it is about 56.95 %. Therefore, the green open space did not meet the criteria as described on Garut

Regulation, which is stated that the The Green Basic Coefficient (KDH) is minimum of 70%. Green Basic Coefficient (KDH) is a percentage of Green Open Space (RTH) in the area. Green open space is an elongated area or path and clusters whose use is more open, where plants grow, both plants that grow naturally or those that are intentionally planted to support the direct or indirect benefits generated by green open space in an area, namely comfort, welfare, and the beauty of the area. Based on land ownership, Green Open Space (GOS) consists of public green open space and private green open space [9]. Previous studies have shown that green space impacts climate and air quality in cities [10], especially with the reduction of particulate matter (PM) [11] and the acceleration of human well-being [12]. Therefore, inadequate green open space in the area of the "Awit Sinar Alam Darajat" tourism object may result in increasing heat, decreased air quality, and human well-being.

To achieve the green open space standard based on the Garut Regency regulation, it is important to calculate and provide the minimum standard for the Green Base Coefficient (GBC) of 70%.

$$GBC = \frac{70}{100} x\ 39.558 = 27.690{,}6\ m^2 \quad (1)$$

3.2.2. Decreasing in Geothermal Quantity and Quality. We received data on laboratory result of geothermal quality that are used as pool sources in "Awit Sinar Alam Darajat". Laboratory results are described in Table 2.

Table 2. Laboratory Result on Geothermal Quality

Paremeter	Unit	Results	Max. Quality Standards	Method
Physical				
Colour	Pt Co scale	<2.68	50	2120 C
Odour	-	Odourless	Odourless	Organoleptic
Taste	-	Tasteless	Tasteless	Organoleptic
Turbidity	NTU	0	25	2130 B
Dissolved Solid	mg/L	465	1500	2540 C
Chemical				
pH	-	3.03	6.5-9.0	4500-H-B
Iron	mg/L	4.65	1	3111 B
Manganese	mg/L	0.16	0.5	3111 B
Zinc	mg/L	0.05	15	3111 B
Chloride	mg/L	39.59	600	4500-Cl-D
Flouride	mg/L	1.02	1.5	4500-F-D
Nitrate	mg/L	2.53	10	4500-NO_3-B
Nitrite	mg/L	<0.01	1	4500-NO_2-B
Sulphate	mg/L	31.02	400	4500-SO_4-B
Arsenic	mg/L	<0.002	0.05	3114 C
Cadmium	mg/L	<0.003	0.005	3111 B
Cyanide	mg/L	<0.01	0.1	4500-CN-E
Chromium Hexavalent	mg/L	<0.002	0.05	3500-Cr-B
Lead	mg/L	<0.02	0.05	3111 B
Mercury	mg/L	<0.0008	0.001	3112 B
Selenium	mg/L	<0.001	0.01	3114 C
Total Hardness as $CaCO_3$	mg/L	68.6	500	2340 B
Surfactants Anionic as MBAS	mg/L	<0.05	0.5	5540 C
Organic matter by $KMnO_4$	mg/L	0.32	10	SNI 01-3554-2015
Microbiology				
Coliform	Coloni/100 mL	0	50	9222 B

IOP Conf. Series: Earth and Environmental Science 1098 (2022) 012035 doi:10.1088/1755-1315/1098/1/012035

The results of laboratory testing showed that the quality of the geothermal source in pool water, in general, the parameters do not exceed the quality standards based on the Regulation of the Minister of Health No 907/Menkes/SK/VII/2002 concerning the Terms and Conditions for the Supervision of Drinking Water Quality, except for the pH and iron parameter. It indicates that a lower pH and high iron concentration in the pool can create an environmental and health problem. A pH level below 7, acidic water, may cause tooth erosion, eye irritation, and pool destruction [13]. In addition, high iron concentration may result in browning in freshwater [14], and its presence at elevated levels can cause aesthetic problems in the environment.

As described in the Garut Regency Regulation, the "Awit Sinar Alam Darajat" tourism object is located in a geothermal area. The utilization of geothermal resources without conservation can result in a decrease in the amount of geothermal, therefore, it is important that the Manager of the Tourism Object 'Awit Sinar Alam Darajat" implement geothermal conservation. Moreover, the manager should apply chemical agent in order to neutralize the pool water. In addition to reducing iron concentration, the manager should provide a geothermal treatment plant until iron concentration meets the standard. Monitoring can be done by measuring the quantity and quality of geothermal resources in the area.

3.2.3. Solid Waste Generation. The waste generated from the activities of the "Awit Sinar Alam Darajat" tourism object is in the form of organic and inorganic waste, including the activities of employees and visitors. Organic waste is produced from the remains of food and plants. Meanwhile, inorganic waste comes from the remains of food wrappers, drinks, and others. The following table is a concentration of solid waste generation from the tourism object "Awit Sinar Alam Darajat".

Table 3. Concentration of Solid Waste Generation

NO	Usage Type	Usage		Waste Generation Standard per Day		Waste Generation per Day (m^3/day)
1	Staff	29	People	3.25	L/people/day	0.09
2	Visitors	104	People	3.25	L/people/day	0.34
3	Area	39,558	m^2	0.15	L/m2/day	5.93
		TOTAL				**6.37**

The volume of solid waste generation is 6.37 m^3/day. The accumulation of solid waste in the area can cause an environmental problem. The lack of solid waste management and disposal is causing significant environmental problems. This includes soil, air, and aesthetic pollution [14].

There are several strategies for solid waste management in a tourist object, such as a collection system: source collection system and segregation of organic (wet) and inorganic (dry) waste at the point source and solid waste minimization through increased recycling. The other method or strategies are solid waste processing (biodegradable waste composting and vermicompost), solid waste (by avoiding open dumping of solid waste and creating a sanitary landfill site), and waste management. The "Awit Sinar Alam Darajat" tourism object may implement a municipal or industrial solid waste management. Industrial waste management can be carried out by monitoring hazardous waste management in accordance with legislation and the adoption of separation treatment and hazardous waste disposal facilities with increased financial provision for the solid waste management system [15]. Solid waste monitoring in the "Awit Sinar Alam Darajat" Tourism Object can be done by solid waste generation to achieve the standard.

4. Conclusions

This study aims to assess the impact of geothermal tourism on the environment and develop environmental management and monitoring. The analysis was carried out utilizing a qualitative and quantitative methods approach. Data were collected from secondary data and analyzed to develop environmental management and monitoring. The result showed that the environmental impact of geothermal tourism includes the change of land use, the decrease in geothermal quantity and the generation of quality, and solid waste. We recommend environmental management, such as setting the minimum standard for the Green Base Coefficient of 70%, geothermal conservation, providing the geothermal treatment plant, and solid waste management. Environmental monitoring should be applied when calculating the Green Base Coefficient and the quantity and quality of geothermal and solid waste generation to achieve the standard.

5. References

[1] Tarmidzi F and Setyawan A 2014 Energy Procedia 47 101–7
[2] Iwan G, Windarta J and Harmoko 2021 U JEBT **2** 60–73
[3] Greco G, Cenciarelli V G and Allegrini M 2018 J, Clean Prod. **177** 62–8
[4] Zhong L, Deng J, Song Z and Ding P 2011 J. Environ Manage **92** 2972–83.
[5] Cole S 2012 Ann Tour Res. **39** 1221–41
[6] Nayan N I M, Din S A M and Abdullah A 2015 Malaysia. Env Sci. **9**
[7] Arbain A and Lenggogeni S 2019 Dampak: J Teknik Lingkungan Universitas Andalas **16** 86-94
[8] Tian C, Peng J juan, Zhang W yu, Zhang S and Wang J qiang 2019 Technol Econ Dev Econ **26** 355–78.
[9] Rabbani A.A, Tistogondo J and Nasihien R.D 2021 IJIEEB **4** 128-135
[10] Szaraz L 2014 JOUR **2** 326-254
[11] Diener A and Mudu P 2021 Sci. Total Environ. **796** 1-18
[12] Jabbar M, Yusoff M.M and Shafie A 2021 GeoJournal **1** 1-19
[13] Lumbantobing D.A., Hutagalung M.H.P and Erawati S 2022 Prima J. of Oral and Dental Sci. **5**, 1-2
[14] Kritzberg E.S, Hasselquist E.M, Škerlep M, Lo¨fgren S, Olsson O, Stadmark J, Valinia S, Hansson L-A and Laudon H 2020 Ambio **49**, 375–390
[15] Hussein I, Shafy A and Mansour M.S.M 2018 Egypt. J. Pet. **27** 1275-1290
[16] Bano N, Shamim S.K and Ahmad A 2018 The Geografer **65** 1-8

Waste management policy: a study of Malang waste bank in implementing the green economy concept

S Muljaningsih[1], **N K Indrawati**[1], **and D A Nur Asrofi**[1*]

[1]Economics Department, Economics and Business Faculty, Brawijaya University. MT.Haryono 165 Street, Malang 65145. Indonesia

*dienamalinanur@student.ub.ac.id

Abstract. The problem of waste has become a concern almost all cities in Indonesia. Therefore, comprehensive waste management that is beneficial for the environment, economy, social, becomes an important issue. This is inseparable from the fact that population growth in Indonesia is correlated with the volume of waste. If this is allowed, it will threaten aspects of environmental sustainability. This study aims to evaluate the role of the Malang Waste Bank in implementing Law Number 18 of 2008 through Malang City Regional Regulation Number 10 of 2010 concerning waste management. The approach used in this research is a quantitative approach to the Malang Waste Bank (BSM) which is analyzed using descriptive and evaluative analysis based on the concept of the green economy. The results showed that the Malang Waste Bank has an important role in changing the mindset and behavior of people in Malang City regarding waste management. The implication of this research is the need to improve and optimize the role of the Malang Waste Bank in waste management activities to improve the welfare of the community in Malang.

1. Introduction

The development of waste increases along with population growth. The more the population, the higher the amount of waste generated. It is known that the development of the urban population in Indonesia per year is 4%. Based on the Indonesian Waste Statistics (2008), the waste generated per person per day in Indonesia is 0.45 kg. Thus, the existing waste generation in Indonesia is multiplied by the total population of Indonesia. For example, if Indonesia's population is 231.8 million, the amount of waste is 104.31 million kg/day. So if in one year as many as 365 days. Therefore, a lot of waste is generated by the people of Indonesia. Whereas landfill capacity is limited, it can only accommodate about 30% of the total waste.

On the island of Java, the population is 137.2 million. The total waste generated is 21.2 million tons/year. It means that the waste generated per person is 0.42 kg/day. The actual ability to transport waste is 12.49 million tons/year, with a total population that can be served as many as 80.8 million people. Thus the waste produced is not transported as much as 8.71 million tons/year[1]. Therefore, it is necessary to think about ways to overcome the problem of untransported waste.

Malang City is one of the big cities located in the province of East Java, with a population of 807,136 people and a population growth rate of 3.9% per year[2]. This, of course, impacts the amount of waste generated in Malang City. Based on data from the Sanitation Department from Malang in 2001, the volume of waste is 1,790.5m^3. Considering that the volume of waste is positively correlated with population growth[3]. Therefore, if the population increases, the volume of waste also increases. The types of waste include organic and non-organic waste such as paper, plastic, glass, metal, and others. Therefore, the volume of waste that is ranked first is organic waste, and the second is paper waste.

This waste needs to be managed. Organic waste can be composted, and non-organic waste can be recycled. Thus, the city of Malang has implemented the 3R concept (*Re-Duce, Re-Use, Re-Cycle*) in waste management. The implementation of this concept requires the help and participation of the community. In order to attract the public, the concept of a Waste Bank was created as an effort to empower the community. The Malang Waste Bank (BSM) was first initiated by those who have the concept of caring for the environment and how to convert this waste into something of economic value. This is in accordance with the concept of *green economics*, namely *zero waste*[4].

Malang Waste Bank (BSM), which was initiated in 2011. The application of Malang Waste Bank is to launch unique programs such as buying groceries and paying for them with garbage, borrowing money to pay with trash, and saving money by depositing trash. This makes the attractiveness of the people of the city of Malang extraordinarily high, as evidenced by the establishment of various Malang Waste Bank units in several urban villages. Malang Waste Bank already has thousands of customers equipped with account books, where they can deposit valuable waste such as plastic, iron, cans, bottles, and so on at any time to the Malang Waste Bank unit in their immediate environment.

Malang Waste Management has a professional business system and has complete facilities such as weighing equipment, plastic bottle crushing machines, cleaning machines, small and large-scale trucks/transportation equipment, offices and branch units, warehouses, and a distribution network for landfills. Now the members of Malang Waste Management are no longer littering because it turns out that the waste has economic value and generates good cash. Used papers, mineral water bottles, used iron, used bottle caps, used cans, and others can be collected, and if there are enough, they can be weighed directly in a vehicle around Malang Waste Management which comes from time to time according to demand/needs.

Based on this background, to find out the success of Malang Waste Management in influencing people's behavior, there is a purpose in this study, namely to evaluate how the role of the Malang Waste Bank in implementing Law Number 18 of 2008 through Malang City Regional Regulation Number 10 of 2010 regarding management waste based on the concept of a green economy.

2. Literature Review

2.1. Implementation of Malang City Regulation Number 10 of 2010 concerning Waste Management

The implementation of the Malang City regional regulation number 10 of 2010 concerning waste management is the establishment of the Malang Waste Bank institution. One of them in Hamlet 3 Urban village Sukun Malang City shows a positive impact, namely waste reduction and economic improvement[5]. With a cooperative legal entity, the Malang Waste Bank is intended as a forum for fostering, collecting, and managing household waste aims are as follows:

2.1.1. Environmental Aspect. The aims are to assist the city government in reducing the volume of waste in Malang City, especially landfills and to change people's perspectives and behavior towards waste, where waste was previously shunned or hated, is now approached by processing and utilizing it and turning it into "rupiah" by saving it in Waste Bank Management. In this way, it is hoped that people will not litter anymore.

2.1.2. Social Aspect. The community's sense of caring and mutual cooperation was born by forming a Malang Waste Bank unit in each Neighborhood/Hamlet and sub-district in order to make the environment clean and cool.

2.1.3. Educational Aspect. There is environmental education for the community and school students who are members of the Malang Waste Bank unit so that they will know the dangers of untreated waste and the benefits of waste from managing waste directly from the source (households).

2.1.4. Empowerment Aspect. There is empowerment in all elements at the family level (parents, children) to the Neighborhood/Hamlet level by joining the Malang Waste Bank unit in waste management.

2.1.5. People's Economic Aspects. There is a waste saving system valued in rupiah by the Malang Waste Bank in all circles of society who are members of the Malang Waste Bank unit, and there is a system of borrowing money in installments using the saved waste. In addition, the system will add new job opportunities thanks to the waste management, especially for housewives and youth groups.

In addition, implementing the Malang City regional regulation number 10 of 2010 concerning waste management is the establishment of the Malang Waste Bank (BSM) institution. One of them is Hamlet 3, the Sukun sub-district of Malang City, which shows a positive impact, namely waste reduction and economic improvement[6]. To find out the flow of waste management system in Malang City is shown in Figure 1 below:

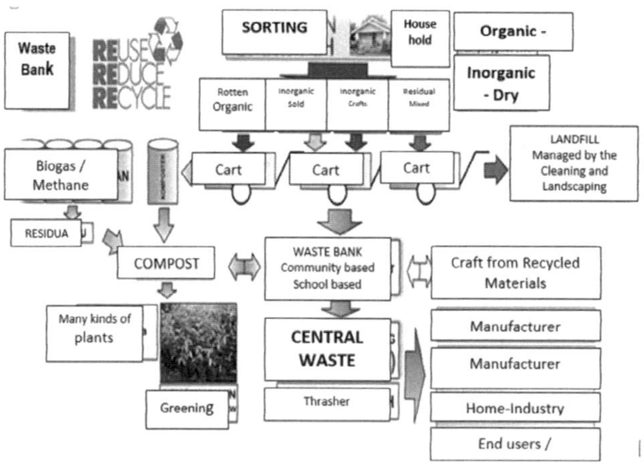

Figure 1. Malang Waste Management Flow through Malang Waste Bank (Source : BSM,2012 in Naditya et al.,2014)

The establishment of Malang Waste Bank has objectives related to several aspects, namely environmental, social, empowerment, and economic aspects. Thus, the objectives are: 1) Assisting the city government in reducing the volume of waste in Malang City, especially garbage dumps and landfills; and 2) Changing communities's perspective and behavior toward waste.

2.2. Successful Model of Waste Bank Implementation

The story of the Waste Bank, the case in DKI Jakarta. One of the successful Waste Banks is the Karya Peduli Waste Bank. This Bank has made changes in its citizens' lives, namely overcoming economic problems, especially for residents who cannot afford the cost of education. Likewise, for people who cannot afford treatment. This Waste Bank specifically aims to help residents who cannot afford school fees and medical treatment. The Bank's success led Semper Barat to receive provincial and national awards in 2010[7].

2.3. Social Engineering Efforts with Inter-Organizational Participation in the Waste Bank Program in Batam City

This study aims to discuss the problem of waste management in urban areas, especially in Batam City, which borders neighboring countries. The launch of the waste bank program in Batam City is one of the government's efforts to reduce waste. This is a social engineering effort through the waste bank program, which aims to build public awareness and independence in the hope of providing social change[8].

2.4. A Study on Determining the Success of the Waste Bank Implementation with the Analytical Hierarchy Process Method

According to Aldilla et al, [9] Waste Banks can provide solutions for the community to the impact of waste generation in Banjarbaru City. In order to determine the success of the implementation of the Waste Bank, there are several factors. The influencing factors are understanding of waste management, management techniques, economics, the existence of buildings, community leaders and the role of local government, and competition (prestige). As a form of problem-solving using the Analytical Hierarchy Process method).

2.5. Theoretical Review

2.5.1. Application of the theory of the three pillars of sustainability (green economy) in the Malang Waste Bank. The theory of the three pillars of sustainability, which includes economic, social and environmental studies. The three studies are interrelated which form a wedge, as shown in Figure 2.

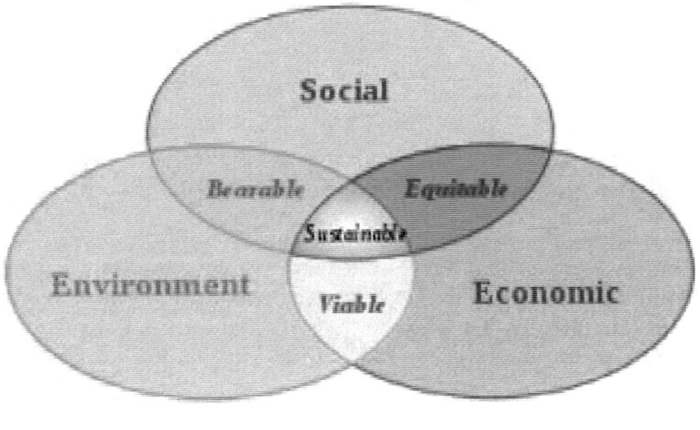

Figure 2. Three pillars of sustainability
Source: adopted from UNEP (2009)

Sustainable development is based on the thinking of these three pillars. This is the starting point for green economy thinking, which suggests efficient use of resources with efforts to minimize waste. This effort is applied to industrial activities/products with the concept of "zero-emission," namely the utilization of all raw materials. To implement the principles of the green economy, one's creativity is needed, which is based on science [10] . This is following the idea that the Malang Waste Bank was formed, which is expected to be able to minimize waste generation and even go "zero." However, achieving this requires the efforts of humans as environmental managers. Thus, it is necessary to examine the relationship between economic activities that are part of human activities with nature.

2.5.2. Relationship between Economic and Nature Activity. Economic activity is the relationship between consumer and producer households (firm). Of course, these activities cannot be separated from nature. Thus, there is a reciprocal relationship between nature and economic activity. The hope is that there will be a synergy that provides prosperity for human life. This can be shown in Figure 3. below.

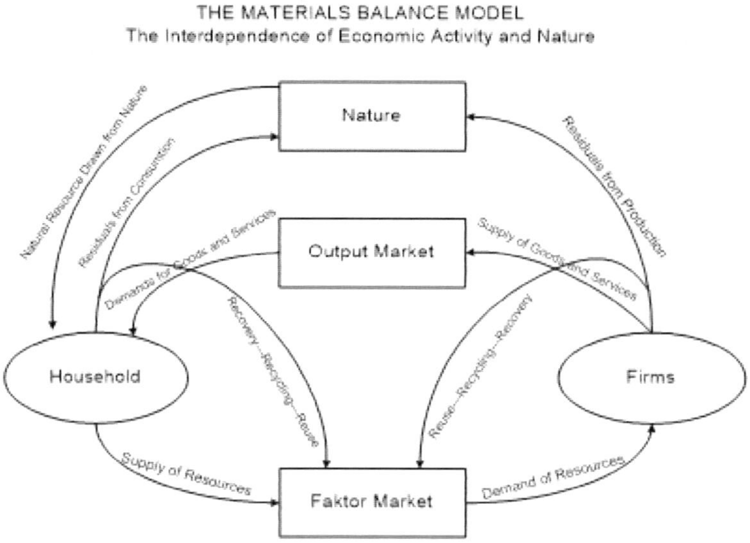

Figure 3. Model of Relationship between Economic and Nature Activity
Source: Adopted from Tientenberg (2003) [11]

From the material balance model, humans and nature cannot be separated. This means that humans should not ignore natural conditions, but in reality, humans as economic actors forget the importance of maintaining natural conditions. This is shown in various economic activities that trigger water, air, and soil pollution. Some time ago, that pollution was ignored because it was considered not to provide economic value, which was called external diseconomies. Currently, the market is familiar with the 3R concept, so factories have produced many recycled products.

3. Methods

This study aims to evaluate the role of the Malang Waste Bank on community behavior in waste management. This study uses a descriptive statistical approach. A descriptive statistical approach is used to systematically describe the data collection related to the dynamics of the waste obtained, which is then given analysis and comparison following applicable normative standards. The change in community behavior in waste management is assessed by comparing the waste generation before and after the Malang Waste Bank. Malang Waste Bank is considered effective in improving community waste management if waste generation in the city of Malang has decreased.

In its development, the complete data regarding waste lead in a city is very lacking. Therefore, in this study, waste generation is calculated assuming that the volume of urban waste will be directly proportional to the total population. This argument is based on the fact that everyone will produce daily waste to fulfill their needs. Therefore, waste generation will be calculated using a certain formula as follows:

$$VTS = WG \times Pop \tag{1}$$

where, VTS = Volume of waste generation, WG = Estimated Waste/individual/day, Pop= Population

Table 1. Parameters for calculating waste generation in Malang City.

No	Parameter	Score
1	Residents of Malang City 2021	843.810
2	Waste Generation Coefficient	0,291
3	Waste Service Level of Malang City	74%

In addition, the composition of waste in Malang City in 2019 was dominated by Wood-Twig waste 61% (Figure 4). This is because the city of Malang is a Kota Kembang with a city atmosphere that is still beautiful and cool and lots of trees. Making wood twigs is a majority of waste in Malang. However, wood-twig waste is organic waste that is relatively easy to recycle. Followed by paper-cardboard waste and food waste by 18% and 11.80%. The high percentage of food waste, further supports the government in optimizing the role of the Malang Waste Bank.

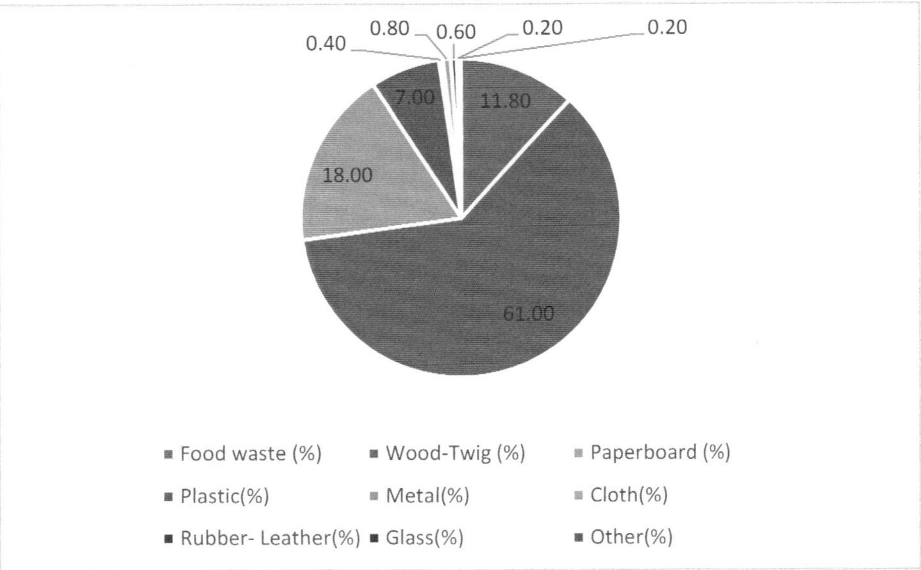

Figure 4. Composition of Waste by Type of Waste in Malang City in 2019
Source: Processed by Author

4. Results and Discussion

The establishment of waste banks in Indonesia is based on the need for changes in the waste management concept. Law No. 18 of 2008 concerning Waste Management and Government Regulation No. 81 of 2012 is the legal basis that indicates an overhaul of the waste management model. So far, waste management has focused on waste management at the final place oriented toward landfills. Government Regulation Number 81 of 2012 plays a role in encouraging waste producers to process waste through a 3R approach. At the household level, this policy encourages people to sort their waste early before it is disposed of in landfills. In addition, as an implementation of the Malang City regional regulation number 10 of 2010 concerning waste management, which was formed, the Malang Waste Bank (BSM) institution was established in 2011.

In its development, the contribution of the Malang Waste Bank (BSM) in reducing the volume of waste in Malang City is still relatively small. The growth of waste volume is positively correlated with population growth, so the volume of waste in Malang City since 2001 has continued to increase.

However, since the establishment of the Malang Waste Bank, the volume of waste generated has decreased, although by a small amount. This means that the presence of the Malang Waste Bank can at least change the community's mindset toward waste management. The decline in the volume growth of waste generation between 2008 and 2020 is estimated to be due to the economic crisis that occurred at that time. The economic crisis will reduce people's purchasing power, and the low level of consumption will result in a relatively lower volume of waste generation.

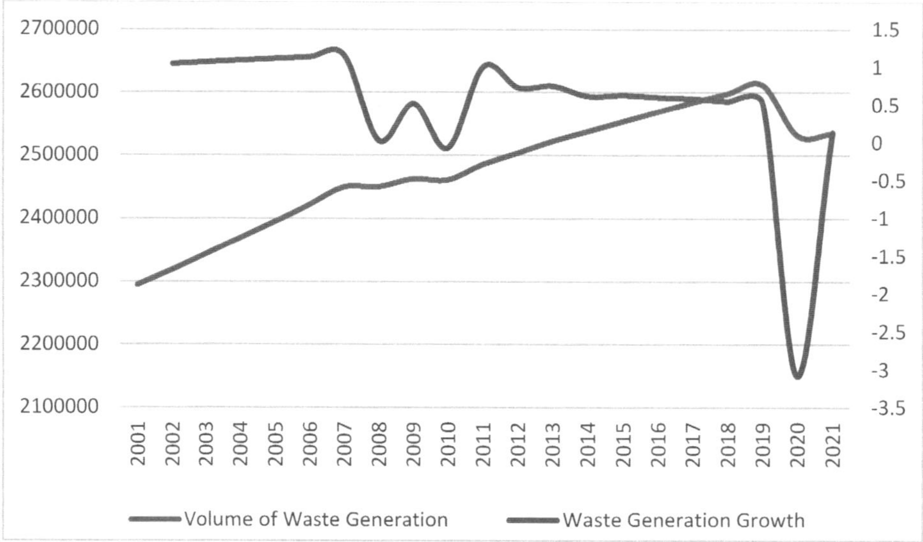

Figure 5. Development of Waste Volume in Malang City
Source : Statistics Indonesia, processed

Several factors influence the effectiveness of the Malang Waste Bank in reducing waste. Among them are to overcome weaknesses and maximize opportunities, namely to increase capacity and train managers by utilizing CSR funds, recruit waste bank managers, and conduct periodic research to develop campus waste banks. In addition, managers communicate with the private sector and local governments to support the procurement of facilities and waste bank infrastructure[12]. In line with this research, Hasnam et al. (2017)[13] stated that the strategy for developing waste banks in the West Nusa Tenggara region includes increasing human resource capabilities with training programs, expanding networks, increasing the use of IT, and maintaining collaboration with outside parties.

In addition, internal and external factors affect the success of the Malang Waste Bank. Internal factors include legal identity, organizational structure, human resources, Malang Waste Bank activities, waste management mechanisms, facilities and infrastructure, and Malang Waste Bank finances. In contrast, the external factors analyzed include the role of the central and local governments and the role of CSR PT. PLN (Persero) East Java distribution, market opportunities, and community participation to become Malang Waste Bank customers[14].

Thus, efforts to optimize the role of the Malang Waste Bank (BSM) are carried out through 3 main parties who play the most role. Figure 6 shows the three main parties in optimizing the Malang Waste Bank. Where the Malang Waste Bank is not only based on individuals but also needs synergy between the community, the Malang Waste Bank, and the government through the triple helixmodel.

Figure 6. Three Main Parties in Optimizing the Malang Waste Bank (BSM).
Source: Processed by Author

Intercept (A) is a collaborative and cooperative community with Government and society. This section is a form of the active role of the community to support the Government Program in community economic empowerment and environmental programs. On the other hand, the Government's support for the community shows itself in supporting programs to improve human resources through supervision, assistance, training, and dissemination.

Intercept (B) is a collaboration between waste banks and the Government. This is manifested in togetherness in realizing the local economy, community empowerment, and alternative models to overcome waste and social and economic problems. The Government must also take part in efforts to manage waste banks. For example, providing waste processing equipment.

Intercept (C) is a collaboration between the community and the waste bank where the community must play a role and participate actively in the management of the waste bank. So that the goal of the waste bank can be realized optimally, namely reducing the volume of waste generation and re-managing waste that has the potential to be reused. It will benefit both the economic side of the community and the environment. So this is the most prioritized point with the collaboration between the community and the Malang Waste Bank in particular. When the waste in Malang city can be partially managed by the community and the community gets additional benefits from the sale of waste and can create jobs in the community.

Intercept (D) is a collaboration of third parties (community, waste bank, and Government) in realizing regional economic development. Through the synergy of these elements, the economic empowerment program makes the community more independent and improves the economy, society, and environment. In accordance with Birkholzer (2009)[15] describes that local communities need common goals, not only individual goals (for the common good). The local economy starts based on the community's commitment or the surrounding community. In addition, an integrated, holistic approach is important in defining economic problems. This means that it does not only produce the production of goods and services but also pays attention to environmental, social, and cultural aspects to create benefits for the community. This is important because increasing social capital can increase the capacity, knowledge, and abilities of local communities [16].

5. Conclusions

The results of this study indicate that the Malang Waste Bank (BSM) has an important role in changing the mindset and behavior of community in Malang regarding waste management. Although at a relatively small level, the growth in the volume of waste due to population growth can be reduced. Even so, the increase in waste management capacity by the Malang Waste Bank needs to be improved both in terms of management capacity, synergy with policymakers, and the community to improve facilities and infrastructure. The implication of this research emphasizes the importance of optimizing the role of the Malang Waste Bank in waste management activities to improve welfare. the people of Malang City, as well as the quality of the environment.

References

[1] KLH 2008. Indonesian Waste Statistics. SWM Scoping Study. Jakarta.
[2] BPS 2006 Malang City Center in Figures 2006. BPS Catologue: BPS – Statistics of Malang City.
[3] Rahmawati, Tri. R. 2006 Strategi Pengelolaan Sampah Di Kota Malang (Studi Di Dinas Kebersihan Kota Malang. UMM. https://eprints.umm.ac.id/12072/1/STRATEGI_PENGELOLAAN_SAMPAH_DI_KOTA_MALANG.pdf [in Indonesian Language]
[4] UNEP 2009 United National Environment Program: The Green Economy Initiative is one of the nine UN- wide Joint Crisis Initiatives, the UN System's Chief Executives Board.
[5] Naditya R 2013 Implementasi Peraturan Daerah Kota Malang Nomor 10 Tahun 2010 Tentang Pengelolaan Sampah (Doctoral dissertation, Brawijaya University) [in Indonesian Language]
[6] Naditya R 2013 Suatu Studi Di Dinas Kebersihan Dan Pertamanan (DKP) Dalam Pelaksanaan Program Bank Sampah Malang (BSM) Di Kelurahan Sukun Kota Malang (Doctoral dissertation, Brawijaya University) [in Indonesian Language]
[7] Bergita E 2013 Dulu dibuang, kini jadi uang: Kisah Dua Bank Sampah. Jurnal Prakarsa Infrastruktur Indonesia 23-25 [in Indonesian Language]
[8] Kurnianingsih F, Iskandar Z, and Okparizan O 2021 Social Engineering Attempt with Inter-Organization Participation of Waste Bank Program in Batam City *Intl. Res. J. Adv. Eng. Sci.* 6(3) 64-69
[9] Aldila Rr M A, Chairul A, and Firmasyah M 2015 Kajian Faktor Penentu Keberhasilan Bank Sampah Dengan Metode AHP (*Analytical Hierarchy Process*) & SWOT di Kota Banjarbaru. *Jukung Jurnal Teknik Lingkungan* **1**(1) 22-32 [in Indonesian Language]
[10] Djajadiningrat S T *et al* 2014 *Ekonomi Hijau* (Bandung: Rekayasa Sains) [in Indonesian Language]
[11] Tientenberg T 2003 *Environmental and Natural Resource Economics* (New York: Pearson Education)
[12] Halid A, Yulianto K, Saleh M 2022 Strategi Pengelolaan Bank Sampah Di Ntb (Studi Kasus Bank Sampah Bintang Sejahtera) *J. Innov. Res. Knowl.* **1**(8) 763-770 [in Indonesian Language]
[13] Hasnam L F, Syarief R, and Yusuf A M 2017 Strategi Pengembangan Bank Sampah di Wilayah Depok. *Jurnal Aplikasi Bisnis Dan Manajemen* **3**(3) 407 [in Indonesian Language]
[14] Hidayat R, Yanuwiadi B, and Prasetyo B 2017 A Study on The Independence of Malang Waste Bank (BSM) Based on Swot Analysis *Indonesian J. Environ. Sustain. Dev.* **8**(1) 53-60
[15] Birkhölzer K 2009 *The role of social enterprise in local economic development*. EMES CONFE.
[16] Wulandari D, Utomo S H, and Narmaditya B S 2017 Waste bank: Waste management model in improving local economy *Int. J. Energy Econ. Policy* **7**(3) 36-41

Does water accounting support sustainable water management? A review

A Mahmud[1*], **N Susilowati**[1], **A Susanti**[1], **P N Sari**[2]

[1]Faculty of Economics, Universitas Negeri Semarang, Sekaran, Gunungpati, Semarang, Indonesia
[2]Faculty of Teacher Training and Education, Universitas Sebelas Maret Surakarta, Ir. Sutami 36A, Surakarta, Indonesia

amirmahmud@mail.unnes.ac.id

Abstract. The business risks experienced by clean water companies include losses and decreased loyalty of water users. This risk occurs due to the absence of accountability, reliability, and transparency in sustainable clean water management. It can be overcome by implementing water accounting. This narrative review explains a sustainable water management model using a water accounting approach. The narrative review method consists of compilation, tabulation, comparison of research results, and summarizing. The literature review study collected articles from accredited journal sites (*Sinta, Emerald, Science Direct*). After searching for research keywords criteria, 29 articles were taken as samples. The results show that water accounting, in addition to reducing business risk for clean water service providers, also supports sustainable water management. Water accounting components such as gross inflow, net inflow, available water, water depletion, and outflow provide transparent and accountable information. The components of water accounting can analyze water use, scarcity, and productivity to predict the need for clean water because the amount of water discharge produced and used in one day can be known with certainty. Management and community commitment are needed to realize sustainable groundwater use conservation to ensure that the community enjoys the long-term supply of clean water.

1. Introduction

Water is an essential need and cannot be separated from the life of living things on the surface of the Earth [1]. On the surface of the Earth, 70% of the composition is water, which means that clean water should not depend on clean water management companies [2]. From year to year, Indonesia experiences an increase in population and development, so the need for clean water has increased [3]. The increasing availability of clean water does not offset the rise in population and growth. A platform for collaboration between ministries was established as mandated by the Presidential decree of the Republic of Indonesia Number 185 of 2014 regarding provisions to accelerate the provision of clean water supply to achieve the availability of clean water for all. Therefore, provinces and districts reaffirm their commitment to achieving targets by strengthening and enabling environmental policies through the implementation of regional policies, budgets, and objectives [4].

A poor clean water management model will harm the chain of water availability in the future [5]. Then, water accounting is needed for clean water management to reduce negative impacts [6]. Water accounting is the identification, recording, and communication of information related to water resources

IOP Conf. Series: Earth and Environmental Science 1098 (2022) 012037 doi:10.1088/1755-1315/1098/1/012037

to allow companies to manage water responsibly and openly [7]. The components of water accounting are gross inflow, net inflow, available water, water depletion, and outflow. The management of water flowing and distribution to the community can be seen from these components [8]. It means that it can integrate various aspects ranging from inputs, outputs, and impacts on the environment so that there is no scarcity of clean water availability [9]. Thus, clean water management can be effective and efficient and create water management continuity [10].

The study results also explained that water accounting provides information for managers to assess business risks in the management of clean water [6], thus facilitating decision making [11]. Clean water management companies can identify, reduce business risks, and predict water availability [10]. For example, 12 states in the western United States, such as California, Arizona, Colorado, Idaho, Kansas, Nebraska, Nevada, New Mexico, Oregon, Texas, Utah, and Washington, implemented water accounting to achieve clean water management that is transparent, reliable, accountable, quality and provides guarantees of water availability and community loyalty [11].

The problem of water availability also penetrates rural areas near water sources. It means that the community has not been able to utilize water sources and has not managed them properly. When the water source is used optimally, there will be no water shortage, even in the dry season. Even in some places in large cities, clean water sources managed by Local Water Companies (*PDAMs/ Perusahaan Daerah Air Minum)* have been polluted by industrial waste and domestic waste [12], so the burden of clean water management is increasing [13]. The water will be treated through complex stages to become clean water ready to be consumed for daily needs. The management of clean water is a priority to be completed [5]. The results showed that in 2019 *PDAM Tirta Pakuan Bogor* experienced an increase in demand for clean water needs reaching 2066 liters /second [1], similar to the clean water management in Konawe Regency, southeast Sulawesi Province also responded to the same [14], [15].

The demand for clean water is continuously increasing, but it is not offset by the increasing availability of water sources [16]. The water demand in Indonesia in 2016 increased by 0.3937 liters / second and is predicted in 2036 to be 0.5654 liters / second [17], so good water management is needed to achieve sustainable water management. Therefore, this article explains in detail the support of water accounting in creating sustainable water management around us.

2. Methodology

The narrative review method consists of compilation, tabulation, comparing research results, and summarizing. The literature review study collected the articles from accredited journal sites (*Sinta, Emerald, Science Direct*). The journal articles used are open access, full text journal, publication date in the past ten years, May 2012-May 2022, relevant to the research topic, and indexed journal. Since August 2021 until June 2022, the authors searched for reputable international journal articles using the keyword 'water accounting' and obtained up to 17 titles (17 titles), 'water management' obtained two titles (2 titles), and 'sustainable environment' obtained one title (1 title). Meanwhile, articles in national journals using the keyword "water management" obtained six titles (6 titles), and "sustainable environment" obtained three titles (3 titles). In addition to the relevant articles, the search proceeds by reviewing relevant literature and books. This study examines water management and a sustainable environment. However, to the best of the researcher's knowledge, the water accounting approach in the management of sustainable water resources has not been widely used. Usually, they did not use a water accounting approach; on the other hand, in water accounting research, they are concerned with financial reporting and examination or auditing [18].

3. Result and Discussion

3.1. Business Risks of Clean Water Management

Risk is defined as the chance of loss, and the risk is the possibility of loss, uncertainty, the dispersion of actual from expected, and risk is the probability of any outcome different from the one expected [10]. Research [9] explained that risk is classified as speculative and pure risk, which can come from within or outside of the company. The high growth of the population, either naturally or implicitly migratory, adds to the increasing pace of the water crisis and ineffective management of clean water [19]. In research [11] in Australia, clean water management has not been implemented effectively, with the finding that there is no explicit calculation of water availability in the present and future. Sustainable clean water management by applying water accounting will anticipate water shortages during the dry season [19]. In the future, the water crisis will make clean water management stop distributing clean water and become a significant challenge in dry areas [10].

The accounting of water for clean water management will minimize business risks [10], as in the western United States, which often experiences drought even though they have applied the accounting of water to clean water management [20]. There are several models of clean water management to overcome the water deficit; According to the combined model, where the result is that the availability of groundwater in 2050 is predicted to be 0.89 x 106 m3 and does not experience a groundwater deficit, and the moratorium model on groundwater use, the result is that since 2020 the availability of water is predicted to increase. In 2025 it has reached stability [11].

3.2. Components of Water Accounting
Water accounting can be used to analyze clean water management [19]. Water accounting is identifying, recording, and communicating information related to water resources, enabling companies to carry out water management more responsibly [21]. This identification allows for analysis of water use, scarcity, and productivity [18]. Here are the components of water accounting commonly used in research [4].

- Gross inflow is the total discharge of water entering the system consisting of precipitation, surface water, and subsurface water.
- Net inflow is the gross inflow plus storage changes. If there is a reduction in storage, the net inflow is more significant than the gross inflow. If there is an increase in storage, the net inflow is smaller than the gross inflow. If there is no addition to storage, then net inflow = gross inflow.
- Available water is available water that can be utilized, is the difference between net inflow and committed water
- Water depletion is the use of water that will eventually reduce the availability of water (available) for other uses. Water depletion can be categorized as process depletion and nonprocess depletion. Process depletion uses water to produce specific products, for example, water for evapotranspiration, drinking water needs, and industry. Nonprocess depletion is the use of water other than process depletion, for example, evaporation from the free water level and or from the surface of the soil that is not cultivated and deep percolation.
- Outflow is a certain amount of water that comes out of the system. Outflows can be grouped into committed water and uncommitted water. Committed water is part of the outflow agreed to be used or given to users at the bottom of the system. It is to be utilized by irrigation areas downstream, used for ecosystem purposes, for example, as a minimum discharge that must be left to the digger of the river flow and or for the needs of downstream fisheries, or as part of the debit to prevent seawater intrusion. Uncommitted water is a million water that comes out of the system as excess from the user.

3.3. Application of Water Accounting
Water accounting is the systematic quantitative assessment of the status and trends in water supply, demand, distribution, accessibility, and use in specified domains, producing information that informs water science, management, and governance to support the results of sustainable development for society and the environment [22], [23]. The benefits of water accounting itself (1) can improve understanding of costs in terms of water for sustainable development, (2) can help identify water problems across sectors, and (3) increase transparency over water allocation. Water accounting is a system companies use to measure and calculate their water use and cover issues related to water quantity and storage, water quality, and impacts of use [4].

Although water accounting has many benefits when implemented, the application has several obstacles because most developing countries still have difficulty managing water resources in clean water management companies [6]. In terms of water sources, water accounting has an undetermined amount and quality from time to time. Furthermore, the absence of hydrometric infrastructure (e.g., measuring devices) in the management of clean water will result in the absence of a calculation of the quantity of water in the water sources in the present and future [24]. It makes the clean water management carried out by the company very vital.

3.4. Sustainable Water Management Model

Water management models can positively impact clean water management in the present and future, especially at water sources [25]. The following is a model of sustainable water management [26].

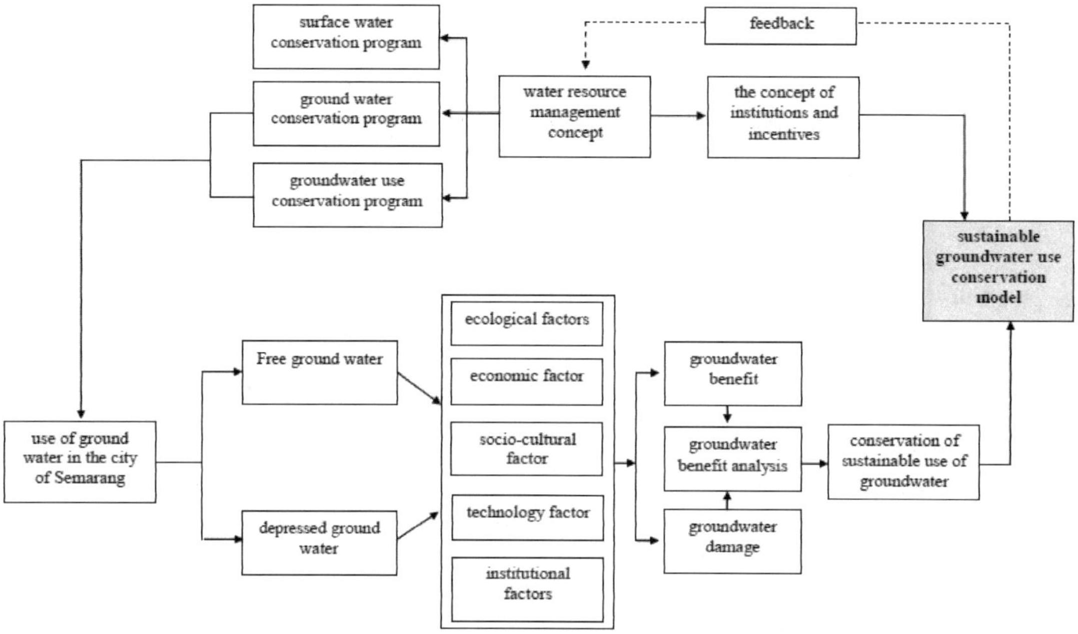

Figure 1. Sustainable Water Management

The model explains that groundwater use in Semarang City can be done by conducting surface water conservation programs, groundwater conservation, and groundwater utilization conservation. In its implementation, it is influenced by ecological factors, economic factors, sociocultural factors, technological factors, and institutional factors. These five factors greatly determine the conservation of sustainable groundwater utilization by analyzing water use and groundwater damage. The concept of water resource management will provide feedback from the user community. Therefore, it requires the commitment of managers and communities to strengthen institutions and appropriate incentives [27]. All these actions will achieve the conservation of sustainable groundwater use so that it can overcome the scarcity of water availability in the future [28]. Clean water management by implementing water accounting must pay attention to several aspects [29] to support sustainable clean water management. The aspects in question are as follows.

- Aspects of community participation include the need to increase the provision of clean water, perceptions of the relationship between benefits and increased provision of clean water, a sense of responsibility and ownership, culture, habits, and beliefs related to clean water.

- Technical aspects include current and future water needs, clean water treatment, technical standards, organizational procedures, and water quality management.
- Environmental aspects include the quality and quantity of water sources and the protection of water sources.
- Financial aspects include cost–revenue analysis, ability and willingness to pay, and tariff structure.
- Institutional aspects, namely strategies at the national level and policies/legal foundations

4. Conclusion

Each year, population growth and development have increased, which will impact the demand for clean water. In the dry season, many regions in Indonesia experience scarcity of clean water availability due to poor clean water management. Clean water management companies play a crucial role in maintaining the availability of clean water needed by the community. Applying water accounting to clean water management companies can solve this problem. Water accounting can identify business risks and calculate future water use and availability. Companies can manage water in a transparent and accountable manner. Business risks that are often experienced are losses and loss of customer trust. Implementing water accounting can minimize these risks. In water accounting, the amount of water discharge produced and used in one day can be known with certainty, so it will bring many benefits. Therefore, community involvement in clean water management is needed.

Many communities lack water, especially during the dry season. It is necessary to increase community involvement to maintain water sources so that natural sustainability is maintained. Community involvement begins with the implementation of surface water conservation programs, groundwater conservation programs, and groundwater utilization programs. Water resource management focuses on ecological, economic, sociocultural, technological, and institutional factors. If all these factors are considered, there will be a sustainable management of water resources and the environment. Finally, the community can enjoy the water supply in the long term.

References

[1] Armadi D, Hidayat A, and Simanjuntak SM 2019 Analisis Pengelolaan Air Bersih Berkelanjutan Di Kota Bogor (Studi Kasus: Pdam Tirta Pakuan) *Journal of Agriculture, Resource and Environmental Economics*, **2** pp 1–12

[2] Hazar M 2015 Suplai Air Bersih Yang Murah Dan Berkelanjutan (Eco-Efficient).

[3] Hendrik P 2018 *Pengelolaan Sumber Daya Air Yang Berkelanjutan Di Kota Sorong II* p. 25-31.

[4] Escriva-Bou A, McCann H, Hanak E, Lund J, and Gray B. 2016 Accounting for California Water *California Journal of Politics and Policy*, **8**

[5] Christ K, and Burritt, R 2017 Supply chain-oriented corporate water accounting: a research agenda Sustainability Accounting, Management and Policy Journal 8

[6] Holyoak J 2019 Water accounting knowledge pathways *Pacific Accounting Review* **31** pp 258–274

[7] Christ KL and Burritt RL 2018 The role for transdisciplinarity in water accounting by business: reflections and opportunities *Australasian Journal of Environmental Management* **25** pp 302–320

[8] Cummings E 2015 Potential Users' Perceptions Of General Purpose Water Accounting Reports. *Journal of Islamic Marketing* **6** pp 1–18.

[9] Pawsey N, Ananda J, and Hoque Z 2018 Rationality, accounting and benchmarking water businesses: An analysis of measurement challenges *International Journal of Public Sector Management* **31** pp 290–315

[10] Gibassier D 2018 Corporate water accounting, where do we stand? The international water accounting field and french organizations *Advances in Environmental Accounting and Management* 7 pp 31–65

[11] Escriva-Bou A, McCann H, Hanak E, Lund J, Gray B, Blanco E, Jezdimirovic J, Magnuson-Skeels B, and Tweet A 2020 Water Accounting in Western US, Australia, and Spain: Comparative Analysis *Journal of Water Resources Planning and Management* **146** pp 1–15
[12] Susanto A, Rusdianto E, and Sawir I 2014 Model Konservasi Pemanfaatan Air Tanah yang Berkelanjutan di Kota Semarang Jurnal Matematika, Saint, Dan Teknologi 15 pp 29–41
[13] Tortajada C and Biswas AK 2018 Achieving universal access to clean water and sanitation in an era of water scarcity: strengthening contributions from academia. *Current Opinion in Environmental Sustainability* **34** pp 21–25
[14] Surya RA, Purwanto MYJ, Sapei A and Widiatmaka 2018 Analisis Kelembagaan Pengelolaan Air Baku Berkelanjutan dengan Metode Interpretative Structural Modelling (ISM) di Kabupaten Konawe Sulawesi Tenggara Jurnal Ecogreen 4 pp 99–109
[15] Irawan AWP 2019 Eksplorasi Air Tanah Di Kampus Universitas Siliwangi Dalam Rangka Pengelolaan Sumber Daya Air Berkelanjutan *Jurnal Siliwangi* **5**
[16] Sri S, Eny A and Ummi C 2019 Konservasi Badan Air Berkelanjutan Sebagai Upaya Ketersediaan Sumber Daya Air (Kabupaten Temanggung) Sri 5 p 1–19.
[17] Cindy I, Supit J and Hendratta L 2017 Perencanaan Sistem Penyediaan Air Bersih di Desa Soyowan Kecamatan Ratatotok Kabupaten Minahasa Tenggara Jurnal Sipil Statik 5 pp 31–40
[18] Charles Batchelor (FAO Consultant) 2017 Water Accounting and Auditing: A Sourcebook FAO Land and Water Division. ISSN 1020-1203
[19] Elmahdi A 2019 Road Maps for Water Accounting Designing and Institutionalizing for Sustainable Water Management in MENA Region *International Journal of Environmental Sciences & Natural Resources* **22** pp 8–11.
[20] Christ KL and Burritt RL 2018 The role for transdisciplinarity in water accounting by business: reflections and opportunities *Australasian Journal of Environmental Management* **25** pp 302–320
[21] Tortajada C 2020 Contributions of recycled wastewater to clean water and sanitation Sustainable Development Goals. *Npj Clean Water* **3**
[22] Hazelton J 2015 Developments in corporate water accounting and accountability. Developments in Corporate Governance and Responsibility 8 pp 27–55.
[23] Momblanch A, Pedro-Monzonís M, Solera A and Andreu J 2018 Water Accounting for Integrated Water Resources Management In *Advances in Chemical Pollution, Environmental Management and Protection* **3**
[24] Elmahdi A, Hafeez M, Smith A, Frost A, Vaze J, and Dutta D. 2015 Australian Water Resources Assessment Modelling System (AWRAMS)- informing water resources assessment and national water accounting *The Art and Science of Water - 36th Hydrology and Water Resources Symposium HWRS 2015 December* pp 979–986
[25] Sallata MK 2015 Konservasi dan Pengelolaan Sumber Daya Air Berdasarkan Keberadaannya sebagai Sumber Daya Alam *Info Teknis EBONI* **12** p 75–86
[26] Susanto A, Rusdianto E and Sawir I 2014 Model Konservasi Pemanfaatan Air Tanah yang Berkelanjutan di Kota Semarang *Jurnal Matematika, Saint, Dan Teknologi* **15** pp 29–41
[27] Rahmi YK, Paulus W, Haryoto K, Setyo SM 2017 Citizen engagement to sustaining community-based rural water supply in Indonesia *International Journal of Development Issues* **16** pp 276-288
[28] Putra D 2017 Evaluasi Keberlanjutan Sistem Pengelolaan Sumberdaya Air "HIPPAM" pada Masyarakat Desa Pandanrejo Kecamatan Bumiaji, Batu Jawa Timur *Jurnal Pendidikan Geografi* **22** pp 22–31
[29] Hendrik P 2018 *Pengelolaan Sumber Daya Air Yang Berkelanjutan Di Kota Sorong* **II** p 25–31

Improving quality of COD analysis using $AgNO_3$ for wastewater samples with high chloride content

T S Pertiwi[1*], T Purwanti [1], N Widagdo[1]

[1]Technical Department, Badak LNG, Bontang, Kalimantan Timur 75324, Indonesia

tiarasari@badaklng.com, nugrahantow@badaklng.com

Abstract. Chemical oxygen demand, or COD, is one of the important environmental parameters that provides the extent of pollution in the water bodies. Valid analysis of COD is needed as a basis to evaluate the condition of water bodies or a wastewater plant. Therefore, any potential deviations suspected during analysis are required to be thoroughly investigated. The problems generally found for COD analyzes came from interferences and contaminations. One challenge that the Badak LNG's Laboratory had, was the wastewater interferences from inorganic substance. It was reflected from the results of COD analyses which prevented to deliver accurate and reliable results for high chloride wastewater samples. It is also mentioned on the Standard Method that high chloride ions (> 2000 mg/L) on wastewater can cause problems on COD determination. Further evaluations are conducted to ensure whether pre-treatment with the addition of specific compounds can compensate the interferences. On this opportunity, mercuric sulfate and silver nitrate are evaluated to determine whether they deliver effective measures to remove the chloride interference and provide valid and accurate results. Later, it was silver nitrate that had more capability to remove chloride interference with additional benefits, less cost and less hazardous wastes generated.

1. Introduction

The chemical oxygen demand (COD) is one of parameters commonly used for determining the degree of organic pollutant in the wastewater. Particularly in oil and gas industries, typically wastewater with anaerobic system is installed to reduce the COD identified in the water. The anaerobic system uses microorganism that can degrade the waste biologically without the presence of oxygen [1]. The depletion of oxygen in the wastewater is commonly indicated with higher value of COD. When the anaerobic microorganism can perform satisfactorily, the COD value in the water system will be lowered. Therefore, COD can also be stated that as the extent of biodegradability of the waste.

Depend on the wastewater itself, the types of pollutant may be varied. These variances of pollutant might require another system, the aerobic system. In a combined system of anaerobic-aerobic treatment, there are several advantages of this system arrangement are the improving quality of the wastewater effluent, reduction of waste quantity and requiring smaller area for the system [2]. COD in particular has the benefit over this combined system arrangement of wastewater, by recognizing both oxygen from aerobic and anaerobic system [3]. Therefore, typically COD measurement are adequate for identifying the wastewater plant performance.

The above explanations obviously describe the importance of COD as the key parameter or indicator for the wastewater condition. Reliable analysis of COD is required to observe the organic matters in the wastewater plant [4]. Then, any potential problems that may cause doubts on accuracy and validity of analysis must be treated carefully and thoroughly solved. Common problems related with COD analysis are the interferences of any inorganic substance. Widely known problem for these kinds of disturbances is the chloride ions [5]. As also found in Badak LNG, the sea water outlet shares the similar stream with blowdown drums sludge, which made the result of COD analysis inaccurate. Other typical resources for wastewater in similar type of industries include chemical industry and oil produced wastewater.

The chloride interference usually can be suppressed up to concentration of 2,000 mg/L [4,6]. Chemical used for this purpose is commonly mercuric sulfate ($HgSO_4$). However, for industries which have higher chloride content than 2,000 mg/L, such as typical industrial coffee factories that ranges 4,000-4,600 mg/L, and wastewater in oilfield that exceeds 120,000 mg/L, the capability of mercuric sulfate needs to be evaluated further [7,8]. Similarly at our wastewater sample, the chloride ion may reach up to 18,000 mg/L. Therefore, the purpose of this study was to test the capability of mercuric sulfate to inhibit the higher chloride concentration in our wastewater sample. Furthermore, this paper also investigates the alternatives of mercuric sulfate, when the substance reached the limit in removing the chloride ions.

2. Efforts to Overcome High Chloride Interference

The Standard Methods which describe the COD Analysis has 3 optional procedures to measure the COD values, namely 5220B, 5220C and 5220D [9]. Currently our laboratory preference is using the 5220D. According to the methods, 5220C and 5220D is cost-effective in using the chemicals and produce less amount of hazardous waste. Besides these benefits, the supporting vials and standard chemicals can be found relatively easy. Another factor of selection is the safety aspect. From safety point of view, 5220D is considered safer to analysts since it has no potential blow out of sulfuric acid during mixing in 5220B and heat produced during sulphuric acid adding in 5220C. In addition, 5220D can handle higher COD value than others, which at this moment is considered fit to be used to measure COD for our wastewater samples.

Based on the Standard Methods, chloride seems to be the highlight of interference and particularly for higher content, it is considered difficult to be overcame [9]. Mercuric sulfate which is also used in the method can reduce the chloride content before the method is applied. Several others have been proposed to replace mercuric sulfate such as silver nitrate ($AgNO_3$) and Cr(III), not to mention the more environmentally friendly attempt that uses calcium hydroxide [10]. However, all of them have their own limitation of usage.

Mercuric sulfate has proven to treat the higher chloride content during COD determination [9,11]. The pre-treatment using this substance is considered successful, when wastewater has approximately 1,000-2,000 mg/L Cl- [8,9]. Although the chemical is considered toxic, but since it is still recommended by many governmental regulations and standards, not to mention much more practical, the procedure remains usable and accepted [10]. Our further research will then depart from this point of view to find the limitation of mercuric sulfate in handling chloride interference.

Other alternatives of mercuric sulfate mostly cannot handle wastewater samples with varied salt concentration [10]. The experiment of Cr (III) addition to overcome chloride ions has been found to be promising [5], although the procedure prevents the sample from being determined colorimetrically. Our previous research also found that cationic resins are normally used for water treatment to remove dissolved solids and ions. However, the research encountered problems such as the need for the loaded resin to regenerate, and the chemicals used in the regenerations along with the removed ions generate more wastes. Calcium hydroxide, as one of the greener approaches for treating this chloride problem, requires a further bismuth-based bed to capture the ions [10], which may not be available commercially. Lastly, silver nitrate, which is considered less toxic than mercuric sulfate, has also been mentioned to work effectively to precipitate chloride ions [9,12]. With the current research result reported on chloride interference that can be removed of up to 6,000 mg / L, it is an opportunity to examine whether silver nitrate can be used to exceed the salt removal limit [13].

By considering all options, the research proceeded with mercuric sulfate and silver nitrate, although Axan and Morrison (2006) stated that mercuric sulfate and silver nitrate may work to reduce the chloride content in the sample, but they will not be completely removed [10,14]. Nevertheless, an opportunity has emerged in some papers to investigate the capability of these substances in reducing chloride ions.

3. Materials and Method

3.1. Materials

Mercuric sulfate and silver nitrate, which were examined, took place prior to COD determination, or it could be stated that these chemicals were used for pre-treatment of samples. At first, synthetic wastewater samples were prepared. Before the samples were added with the chloride ions, the COD of the wastewater was determined. These samples had chloride content more than 2,000 mg/L, or in this case approximately 18,000 mg/L. Initial mercuric sulfate test with 0.25 grams/2,000 mg/L chloride ions were added to the sample. The sample was centrifuged with the addition of this 2.25 grams of mercuric sulfate to ensure homogeneity. Three (3) variances of rotation speed were applied. Later, the amounts of mercuric sulfate added to the samples were also varied. The alteration of the reagent was started from 0.1 grams/2,000 mg/L with 0.2 grams increments until 1 gram/2,000mg/L reached, except for 0.25 grams used earlier which excluded 0.3 grams/2,000 mg/L. Then, CODs were measured after this pre-treatment.

The next substance, silver nitrate, used several concentrations to be varied in complexing the solution. The concentration ranged from 0.1N, 0.5N, and 1N. Similarly, to ensure the homogeneous solution, the sample was centrifuged with several options of rotation speed. Once the solution was ready, COD could be determined.

3.2. Method

Following the sequence of pre-treatment steps with mercuric sulfate and silver nitrate added to the samples, observations were conducted. One of the observations was to identify whether white salt sediment was formed during the pre-treatment phase. The method of this study to determine chloride removal was continued with determination of COD after salt was formed. COD analysis was compared with the previous measurement before the chloride ions were added. With this comparison, the difference in COD values could be observed. If possible, further optimization could be conducted to refine the performance of the removal of chloride ions, such as less reagent utilization.

The following steps were to conduct verification and validation of the pre-treatment procedures. Verification of a procedure was an examination aimed at how the pre-treatment procedure could be fit for its purpose under certain actual conditions through environment, personnel, and equipment. The verification step obtained reliable data of the chloride ions removal procedure for the first time.

On the other hand, validation of this chloride ions removal procedure would assess the performance characteristics of this procedure through various analytical testing and conform the suitability for implementation. Certain targets had to be set as the criteria of acceptance from the validation. The variables used in verification and validation includes precision and accuracy. Once verification and validation analysis were performed, the best method to be further implemented could be determined. The overall steps of the method are described in the diagram in Figure 1.

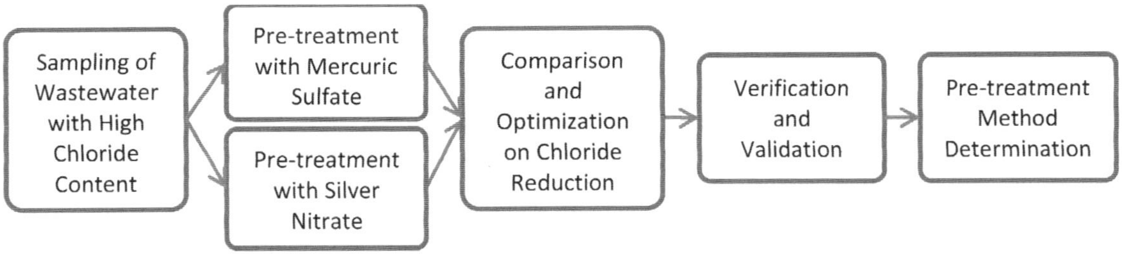

Figure 1. Method for Determining Procedure for Chloride Ions Removal on COD Samples.

Subsequent actions after determining the appropriate procedure included formalizing the procedures, developing the management of change documentation, and disseminating the pre-treatment procedure.

4. Results and Analysis

The pre-treatment procedure as mentioned previously was conducted through two alternatives. The first is to use mercury sulfate and the second is silver nitrate. Both reagents will reduce the number of chloride ions available in the synthetic wastewater samples. The COD value set on the synthetic wastewater sample is 23 mg/L. With the additional interference from chloride 18,000 mg/L, the COD value determined after the pre-treatment can assess the effectiveness removal of chloride content.

4.1. Chloride Removal Results

The first opportunity to remove the chloride ions for mercuric sulfate. The substance, which was also used during COD determination, was added as a pre-treatment of COD samples. The addition of mercuric sulfate to minimize the chloride content was followed by COD analysis to confirm the preset value. Unfortunately, the results were not promising, as the COD value measured from the sample is 409 mg/L, which is far beyond the target value. Therefore, another proposed procedure of pre-treatment for mercuric sulfate addition was discontinued.

Next, in the second pre-treatment procedure, silver nitrate was added to reduce the chloride content. Various variations of the silver nitrate concentration were performed to determine the effectiveness of the removal. The concentrations were varied to 0.1, 0.5 and 1N. After the addition of silver nitrate, the solutions were centrifuged with varying speed. The variation of speed ranged from 500, 1,000 and 1,500 rpm. It was expected that the reaction will accelerate with the rotations. The duration of the centrifuge was limited to 1 hour with each 15 minutes the data were taken. The result of this pre-treatment procedure is shown in Table 1 below.

Table 1. Chloride Content on the Liquid Samples

Time (min.)	Chloride Content at 500 rpm (ppm)	Chloride Content at 1,000 rpm (ppm)	Chloride Content at 1,500 rpm (ppm)	Silver Nitrate Concentration
15	2,140	2,102	1,941	0.1 N
30	2,128	2,068	1,931	
45	2,098	2,026	1,921	
60	2,085	1,990	1,909	
15	2,089	2,003	1,945	0.5 N
30	2,003	1,958	1,895	
45	1,934	1,912	1,826	
60	1,889	1,856	1,760	
15	1,971	1,925	1,833	1.0 N
30	1,873	1,773	1,716	
45	1,738	1,626	1,574	
60	1,644	1,483	1,401	

Table 1 shows that the silver nitrate can precipitate the chloride content in the sample into approximately 2,000 mg/L. In certain cases, it was lower than 2,000 mg/L. The table also shown be shown that at higher speed of centrifuge, more chloride can be removed. The higher the silver nitrate concentration also removed more chloride content from the sample. It can be seen from the table that at 1N silver nitrate, the chloride content was lower than expected. A higher concentration of silver nitrate can remove more chloride, while a higher rotation of the centrifuge can speed up the reaction.

The next discussion considered how the amount of silver nitrate used can be reduced to optimize the usage of chemicals. Therefore, further efforts were taken to determine whether a smaller sample quantity or less than 25 ml can still provide a promising result. Then, considering the practicality of the measurement, the sample was reduced to 10 ml. The silver nitrate 1N was used for this purpose.

With similar parameters for concentration and higher rotation for centrifugation, the target would be a higher chloride removal within a certain time. Then the variance used for this step is the time that will record the chloride removal from the first 30 minutes and is followed by 15 minutes increments, up to 2 hours. The result for this procedure is shown in Table 2.

Table 2. Chloride Content Reduction

No.	Time (min.)	Chloride Content that Can be Reduced (ppm)
1	30	4,605.06
2	45	4,705.17
3	60	5,038.87
4	90	4,905.39
5	120	4,505.95

It can be seen in Table 2 that generally the longer time required for the chloride to precipitate can remove more chloride content. However, research showed that removal peaked at 1 hour. Commonly, the time required for removing a higher content of chloride will be used for the routine analysis. However, considering the higher removal of chloride with the optimum time required for analysis, the required time was determined from the average chloride content that can be reduced. The average chloride content that can be reduced from samples is 4,751.89 mg/L, then the closest number to the average came from 45 minutes duration.

Once the optimum time has been determined, the next action would be to perform a linearity test on the reduction of chloride content in the sample. This effort was made to identify how much silver nitrate was required to reduce every 1,000 mg/L of chloride in the sample. The linearity test will plot the volume of silver nitrate with the reduced chloride. From the calculation, the linear equation was found to be y = 34,752x + 66.868 with $R^2 = 0.9965$, where y is the value of the reduced chloride and x is the required volume of silver nitrate. Figure 2 shows the linear correlation for silver nitrate with chloride reduction.

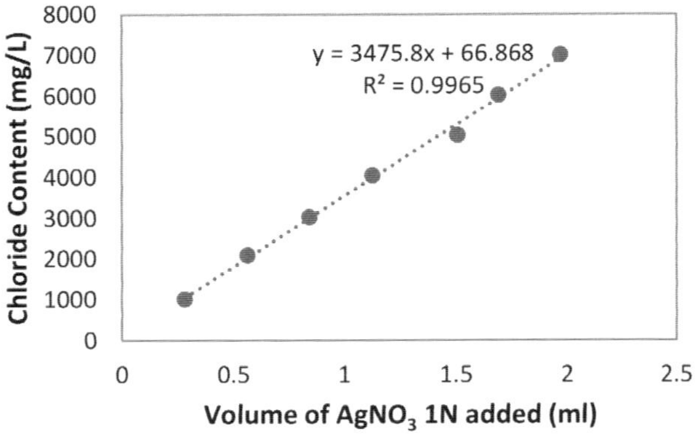

Figure 2. Linearity Test of Chloride Reduction

Linear correlation is generally accepted if the R value is approaching 1. Based on the criteria of R > 0.997 [15], then the graph showed that the linear equation can be accepted with the result of R = 0.998. Furthermore, with this linear equation, the required amount of silver nitrate volume can be estimated to remove certain chloride ions content in wastewater samples.

4.2. Verification and Validation

Verification and validation analysis was performed as the sequence of the trial procedures. The pre-treatment method needs to be verified and validated to see whether the procedure can be suitable for the actual condition and accepted for further implementation. Both analyses were conducted with variables of assessment, which includes linearity, precision, and accuracy. As the basis to determine the sample concentration, the linearity graph between the standard COD concentration measurement and the absorbance value obtained from the spectrophotometer must be prepared. The standard concentrations that were prepared ranged from 0 to 250 mg/L. All the variances in COD concentration were then measured on a spectrophotometer to obtain the absorbance values. The linearity graph that describes the relation of COD concentration and absorbance is shown in Figure 3.

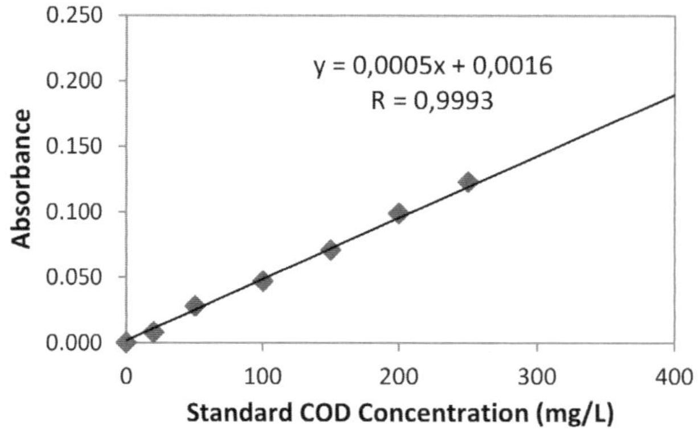

Figure 3. Linear Correlation of COD and Absorbance

In Figure 3 it is shown that the correlation between COD and absorbance has the equation of $y = 0.0005x + 0.0016$ with $R = 0.9993$. The linear correlation between these 2 variables has met the criteria ($R > 0.997$). Then, any changes to variables will be correlated linearly. From this range of concentration, it was previously mentioned that a standard concentration that was further used in one of the verification and validation analyses is 23 mg/L.

The next verification step will be the precision test. Precision test describes the closeness of the analyses results through series of repetition for similar samples. The test used the repeatability and reproducibility variables with the Horwitz function as the following [16,17]:

$$CV\ Horwitz\ (\%) = 2^{(1-0.5\log C)} \tag{1}$$

Where C is the COD concentration in percentage. The repeatability and reproducibility have requirements to be considered acceptable. The repeatability uses the %RSD value to be compared with Horwitz equation. The %RSD value must be lower than 0.5 of Horwitz value to be determined acceptable [18].

Table 3. Repeatability Data of the Proposed Method

Repetition	COD Analysis 1 (mg/L)	COD Analysis 2 (mg/L)
1	26	27
2	27	26
3	26	26
4	26	27
5	27	27
6	26	26
7	27	26
Average	26	27
Standard Deviation	0.55	0.43
%RSD	2.10	1.62
Horwitz Value	9.77	9.77
0.5 * Horwitz	4.89	4.89

Table 3 shows that two data analyses were performed in 7 repetitions. Both data analysis has a similar Horwitz value, 9.77. By using the criteria mentioned above, the% RSD values are 2.10 and 1.62, and both are less than the Horwitz value of 0.5, which is 4.89. Then these results indicate that the proposed method for pre-treatment is repeatable. Reproducibility, on the other hand, has criteria of 0.67 Horwitz value to be categorized as acceptable [18]. From further calculation, we can obtain that% RSD = 1.80 and the Horwitz value = 9.77. With the limit of 0.67 Horwitz value, which is 6.55, the %RSD value is also lower than the criteria. It means that the reproducibility of the method is acceptable.

Further validation with the Certified Reference Material (CRM) also provided similar results. Both repeatability and reproducibility on CRM of COD are accepted since all the values of calculated %RSD are still lower than requirement. With a series of verification and validation, all the results showed favourable results for the proposed pre-treatment procedure. Hence, this procedure is justified to be implemented.

This pre-treatment procedure using silver nitrate to minimize chloride interference can be applied with several notable implications. The chloride content that can be reduced can be implemented in any laboratory that shares similar problems. The benefit of using silver nitrate in the procedure is also supported by other similar research [13,19], so the utilization of the substance is proven. The replication of this pre-treatment procedure is expected to stimulate further research and development of chloride ions removal in wastewater samples.

Additional advantages which were observed during the implementation of this new method include the delivery of analysis, cost reduction, and minimization of hazardous waste generation. The proposed procedure evidently can shorten the time required to complete the full analysis of COD determination. Previously, more time was required to remove the higher chloride content and often the COD determination was repeated, due to unsatisfactory results. Fortunately, this new pre-treatment procedure could also reduce the cost and generate minimum hazardous waste due to a smaller sample quantity.

As previously noted by Freire and Sant'Anna (1998), the maximum chloride interference that can be eliminated is up to 6,000 mg/L [13]. By implementing this current method, the proposed procedure can extend the ability to remove more chloride ions. It is obvious that wastewater samples vary among industries; however, the intrusion of sea water into the wastewater stream may often occur. Therefore, the range of chloride ion interference that silver nitrate can handle up to 18,000 mg / L is expected to accommodate most chloride interference issues on wastewater samples. However, it is also an opportunity for further research in investigating higher chloride ion removal possibilities.

The use of silver nitrate also has shortcomings. Silver nitrate, as well as mercuric sulfate, is a toxic chemical [10,14], although they are considered practical, easier to obtain commercially, and supported by standard methods to be utilized. Eventually, greener alternatives are desired, therefore, future opportunities for research in finding more environmentally friendly substances for these pre-treatment efforts are available. Once the alternatives are found, the challenges would be the support of authorities and governments to promote the implementation.

5. Conclusions

Wastewater samples with a high chloride content produce interferences in COD determination. Previous research has made various efforts to address this problem. This paper proposed reduction attempts through the utilization of mercuric sulfate and silver nitrate. However, only silver nitrate is shown to demonstrate the effectiveness of chloride content removal in the COD samples. Additional efforts to reduce the volume of COD samples were also made without sacrificing the effectiveness of the removal and the validity of COD determination were also performed. These activities provided additional benefit in less analysis time, reduced analysis cost, and less generated hazardous waste. Further verification and validation activities has provided the acceptance of the new pre-treatment procedure. This will justify the procedure to be implemented in the routine COD determination.

References

[1] Cazaudehore G, Schraauwers B, Peyrelasse C, Lagnet C and Monlau F 2019 Determination of chemical oxygen demand of agricultural wastes by combining acid hydrolysis and commercial COD kit analysis *Journal of Environmental Management* **250** 109464. https://doi.org/10.1016/j.jenvman.2019.109464

[2] Barik D ed 2019 *Energy from Toxic Organic Waste for Heat and Power Generation* (United Kingdom: Woodhead Publishing). https://doi.org/10.1016/C2017-0-01876-1
[3] Zaher K and Hammam G 2014 Correlation between biochemical oxygen demand and chemical oxygen demand for various wastewater treatment plants in Egypt to obtain the biodegradability indices *International Journal of Sciences: Basic and Applied Research* (IJSBAR) **13** 1. [cited 2022 May 19] Available from: https://www.gssrr.org/index.php/JournalOfBasicAndApplied/article/view/1382
[4] Saral A and Goncaloğlu B I 2008 Determination of real COD in highly chlorinated wastewaters *Clean Soil Air Water* **36** pp 996-1000. https://doi.org/10.1002/clen.200800073
[5] Kayaalp N, Ersahin M E, Ozgun H, Koyuncu I and Kinaci C 2010 A new approach for chemical oxygen demand (COD) measurement at high salinity and low organic matter samples *Environmental Science and Pollution Research* **17** pp 1547–1552. https://doi.org/10.1007/s11356-010-0341-z
[6] Vyrides I and Stuckey D C 2009 A modified method for the determination of chemical oxygen demand (COD) for samples with high salinity and low organics *Bioresource Technology* **100** 2 pp 979-982. https://doi.org/10.1016/j.biortech.2008.06.038
[7] Zayas Péerez T, Geissler G and Hernandez F 2007 Chemical oxygen demand reduction in coffee wastewater through chemical flocculation and advanced oxidation processes *Journal of Environmental Sciences* **19** 3 pp 300-305. https://doi.org/10.1016/S1001-0742(07)60049-7
[8] Zhang S, Chen W, Liu Y, Luo P and Gu H 2018 A modified method for the accurate determination of chemical oxygen demand (COD) in high chloride oilfield wastewater *Open Journal of Yangtze Oil and Gas* **3** pp 263-277. https://doi.org/10.4236/ojogas.2018.34023
[9] Baird R B, Eaton A D and Rice E W ed 2017 *Standard Methods for the Examination of Water and Wastewater 23rd Edition* (Washington, D.C.: American Public Health Association, American Water Works Association, Water Environment Federation)
[10] Geerdink R B, van den Hurk R S and Epema O J 2017 Chemical oxygen demand: historical perspectives and future challenges *Analytica Chimica Acta* **961** pp 1-11. https://doi.org/10.1016/j.aca.2017.01.009
[11] Dobbs R A and Williams R T 1963 Elimination of chloride interference in the chemical oxygen Demand Test *Analytical Chemistry* **35** 8 pp 1064-1067. https://doi.org/10.1021/ac60201a043
[12] Jeffery G H, Bassett J, Mendham J and Denney R C 1989 *Vogel's Textbook of Quantitative Chemical Analysis 5th Edition* (New York: John Wiley & Sons Inc.)
[13] Freire D D C and Sant'Anna G L 1998 A proposed method modification for the determination of COD in saline waters. *Environmental Technology*, **19** 12 pp 1243-1247. https://doi.org/10.1080/09593331908616784
[14] Axán E and Morrison G M 1995 A mercury-free microwave method for the chemical oxygen demand analysis of sewage *International Journal of Environmental Analytical Chemistry*, **59** 1 pp 69-78. https://doi.org/10.1080/03067319508027637
[15] Moosavi S M and Ghassabian S 2018 Linearity of calibration curves for analytical methods: a review of criteria for assessment of method reliability ed Stauffer M T *Calibration and Validation of Analytical Methods - A Sampling of Current Approaches* (London: IntechOpen) [cited 2022 May 19] Available from: https://www.intechopen.com/chapters/58596
[16] Zuas O, Budiman H, Mansur D and Mulyana M R 2018 Measurement of HDO products using GC-TCD: towards obtaining reliable analytical data *Catalysis for Sustainable Energy* **5** 1 pp 1-11. http://dx.doi.org/10.1515/cse-2018-0001
[17] Linsinger T P J and Josephs, R D 2006 Limitations of the application of the Horwitz equation *TrAC Trends in Analytical Chemistry* **25** 11 pp 1125-1130. https://doi.org/10.1016/j.trac.2006.11.002
[18] Hadi A and Asiah 2018 *Statistika Pengendalian Mutu Internal: Mendukung Penerapan ISO/IEC 17025: 2017* (Bogor: IPB Press Printing)
[19] Cardona I, Park H I and Lin L -S 2016 Improved COD measurements for organic content in flowback water with high chloride concentrations *Water Environment Research* **88** pp 210-216. https://doi.org/10.2175/106143016X14504669767571

Analyzing the environmental sustainability of rural water management in community self-help group (CSHG) – a case study of Gunungpati district

N Susilowati1[1*], A Mahmud[1], L Latifah[1], A Susanti[1], P N Sari[2]

[1]Faculty of Economics, Universitas Negeri Semarang, Sekaran, Gunungpati, Semarang, Indonesia
[2]Faculty of Teacher Training and Education, Universitas Sebelas Maret Surakarta, Ir. Sutami 36A, Surakarta, Indonesia

nurdiansusilowati@mail.unnes.ac.id

Abstract. Economic developments and the increasing population have recently led to a rapid increase in the consumption of clean water. The objectives of this study are to analyze the management of available water and develop a model to support the distribution of rural water services. It was a qualitative study with informants from the manager and users of the Community Self-Help Group (CSHG) in *Gunungpati* District, Semarang. The data were collected through in-depth interviews and observations, and then processed using the Huberman data analysis technique. Physically, the water supply managed by CSHG was sufficient to fulfill the community needs. Every morning in a month produces more than 1,000 cubic meters of water to distribute to 150 households. Management could meet the water daily needs of the community in the dry and rainy seasons. They also perform intensive maintenance on water pumping machines and checking water capacity to be distributed to the public. The sustainability of the water supply in the CSHG depends on the participation of the community in protecting the environment. Community participation was elaborated from the sociological perspective that the basic elements of social life consisting of structure, culture, and processes in the realm of environmental communities.

1. Introduction

Economic developments and the increasing population in recent decades have led to a rapid increase in the consumption of clean water. About one third of the world's population is threatened by a lack of fresh water to meet their daily needs [1]. In Indonesia, the commitment to universal access to clean water and sanitation has been mandated by Presidential Decree of the Republic of Indonesia No. 185 in 2014 with respect to provisions and acceleration of water supply and sanitation. Thus, provinces and districts are committed to achieving targets through the adoption of regional policies, budgets, and targets. An area that experiences problems with the lack of clean water sources and the uneven distribution of clean water services is in *Semarang* City. This is due to changes in land use and environmental damage. *Gunungpati* district receives fresh water from the *Garang* and *Babon* rivers and shallow groundwater. Some areas do not receive water from the Local Water Company (*PDAM*) because the water is not sufficient and strong enough to flow to the upper regions. It causes several regions to experience a water crisis.

A form of community empowerment in the *Gunungpati* district is called the *Gunungpati* Community Board (GCB). GCB has a program to provide and manage clean water through CSHG (Community Self-Help Group) *Mandiri Berkah,* which has been operating since 2004. From 2004 until 2017, almost all Gunungpati houses received services from CSHG *Berkah Mandiri,* providing clean water for daily needs. [2] found that transparent, high-quality, and reliable water disclosure was required. Water is a public good and needs transparent and accountable management and protection in water use. Therefore, stakeholders such as local governments, villages, communities, and environmental activists should use their responsibility to ensure that the existence of water in the future is maintained and also exists for future generations. Next, [3] found that environmental management accounting was needed to manage environmental costs to create sustainable development. Therefore, community empowerment was needed to handle it [4].

Community participation and empowerment positively contribute to the achievement of sustainable development goals. This approach has increased precision, transparency, and accountability, reduced poverty, service quality, community capacity to manage the environment, maintain infrastructure, and create public spaces [5–7]. Community involvement has proven to be successful in infrastructure, but it was not effective in addressing the needs of poor and rural communities. The factors that determined its success were facilitator, elite support, local government, and village government [8].

Community participation and community empowerment are the responsibility of the government [9]. Citizens give the government a mandate to regulate people's lives simultaneously and require accountability and certainty that the government works to fulfill community needs [2,10]. Citizens and governments have the same position, binding commitment, and shared responsibility. There is no blueprint for implementing the concept. All depend on the social-political context of its users [11]. Therefore, understanding citizen participation requires understanding of the basic elements of social life. Sustainability is explored through the cycle of rural water service provision employing strategic plans, budget and budget allocations, construction/expansion, operations and maintenance, and support systems to maintain services.

The objectives of this study are to analyze the available water management and develop a model to support the distribution of rural water services. The participation of community members is elaborated through a sociological perspective of the basic elements of social life consisting of structure, culture, and processes in the realm of environmental communities. This research contributes to managing the sustainability of rural water, calculates daily water needs, and the role of trust in the water business activity. Customer satisfaction and excellent service are the main consideration.

2. Methods

It is a qualitative study with a case study. The location of this research was CSHG *Berkah Mandiri* in *Gunungpati* District, Semarang City. The main data sources of this qualitative study were words and actions[12], the rest are additional data such as documents and others. Words mean that the qualitative researcher deals with data that are in the form of words rather than numbers and statistics. Actions mean data sources obtained from informants through the dept-interview. Data in this study were primary and secondary data. Primary data are collected through in-depth interviews. This interview technique is an unstandardized interview; it means that the interview guidelines used are not absolute from loose, so it is possible to be carried out more personally to enable the extraction of the deepest information. Secondary data is obtained from books, papers, and documentation studies. Then, the study of documents is used to collect nonhuman data or any project that does not involve research, a human subject, or a clinical investigation, such as the water data available at CSHG *Berkah Mandiri*; this technique can be used to obtain supporting data efficiently.

This research used Miles and Huberman's qualitative data analysis techniques. There were three paths of qualitative data analysis, i.e. data reduction, data display, and concluding or verification. Validation of the research results needed to be done to obtain the credibility of the research results. They could be an extension of the observation time that the researcher made by checking whether the data obtained from the informant were the correct data or not. The researchers returned to the field and conducted in-depth interviews with informants who had provided information, as well as with new informants who had never been met/interviewed. Furthermore, researchers also made a schedule of observation extension so that the available time could be used as much as possible and organized. The triangulation process that the researchers conducted was the triangulation of data collection sources. The way to do the triangulation was for the researchers to conduct in-depth interviews to obtain data on the impacts obtained from social, economic and environmental aspects.

A member check was performed after the recorded data were transcribed or after the draft research report was completed. Researchers came to informants who had been interviewed by providing interview transcripts derived from interviews. The researcher asked the informant to read it; if the transcript did not match what the informant, the researcher would fix it again. After being repaired, it was returned to the informant. The audit trail is an examination of the suitability of research findings with field data, by tracking field records, data collection methods, and analysis techniques. Additionally, the expert's opinion was validated by requesting responses from experts, academics, and practitioners. In this context, the researcher placed the thesis advisers as experts. For validation purposes, the researchers would conduct in-depth interviews and discussions with the manager and employees of CSHG *Berkah Mandiri*.

3. Result and Discussion

3.1. Result

Physically, the rural water supply managed by CSHG *Berkah Mandiri* is sufficient to meet the community needs. During the rainy season, water remains abundant and in the dry season even water is less, but management can provide clean water as the daily needs of the community. The water managed by CSHG *Berkah Mandiri* is sufficient for the needs of the community. They can provide water in the rainy and dry seasons. The water is also clean. Unlike the water in other CSHGs, some of them are dirty. The parameters used in clean water are physical, biological, and chemical parameters [13]. The water in CSHG *Berkah Mandiri* is clean because it has no taste, no color, no odor, and has a normal temperature (physical parameters); CSHG water does not contain E. Coli bacteria (biological parameters); and the water is below the maximum limit of quality standards set by the Regulation of the Minister of Health of the Republic of Indonesia Number 32 of 2017 on Quality Standards of Environmental Health and Water Health Requirements (chemical parameters) [14].

The household has been using this water since 2009. They paid IDR 45,000 to 60,000 for water every month. It is cheaper than PDAM water. The available water meets daily needs and has never been problems. The water is clean and fresh, not turbid. The turbidity concentration of water at CSHG *Berkah Mandiri* on June 2022 reached 0.811 NTU. If it is rain or the long drought, they still have water for bathing, cooking, and washing. Also, you do not need to buy a gallon of water. Although the amount of water is not as much as during the rainy season, it can meet daily needs. The water is clean and clear, and there is no dirt.

The CSHG *Berkah Mandiri* manager is committed to providing excellent service to the community. However, people need water for their daily needs. It cannot be imagined that there is no water. This commitment is evidenced by the level of professional management; starting from performing intensive maintenance on water pumping machines, checking water discharge, and water capacity to be distributed to the public. The manager said that their job is to serve the community on water channels. Community satisfaction is shown by the absence of complaints related to the water they consume. They also always take care of the engine pumps so that the water distributed to the community is well distributed and there is no shortage.

The manager also provides excellent service to ensure that the community does not feel disadvantaged. They have paid every month on time to pay for water. The standard operating procedures that they have well implemented. The district head also has the commitment to provide excellent service. There is a water reservoir in front of the village office that the community uses to meet daily needs. It is well managed and benefits all parties. It should not only benefit managers. Therefore, existing water is used for community welfare. Every month, the community pays for the amount of cubic water they use. The officer records and calculates the water meter and the cubic of water used daily. There is no cheating or any other harmful act.

The cost of cubic meters is IDR 1,500. The household uses approximately 30,67 cubic meters of water every month. The officer was greatly facilitated by the presence of clean water and the supply well. Management is professional and conscientious. They never wrong to calculate the nominal payment. The water management delivery by CSHG *Berkah Mandiri* is very good. The district head gives highly supervision, has a strategic plan, and targets. Planning is related to all efforts and efforts made to achieve the goal of meeting water needs. In addition, the objective is to provide water and provide satisfaction to the community. Communities provide opinions and suggestions for managers to develop business activity. For example, if you get a tariff increase, the community can provide suggestions and discussions.

The CSHG *Berkah Mandiri* budget financing priority is for the maintenance of pumping machines, water reservoirs, and water hoses. The CSHG budget comes from the district budget. All assets and income managed by a professional person. It has returned the capital to the district. The income from water management is included in the district income. Using a revenue sharing system, the the income creates community welfare. The water can be used by 150 households. The pumping machine is maintained, checked every month, and cleaned so that no dust enters. Maintenance costs come from clean water revenues. The CSHG can cover all the costs of engine maintenance.

3.2. Discussion

3.2.1. Available water and sustainability environment. The rate of population growth and the increasing community needs lead to a worse environment to meet the needs. The community and its environment have a reciprocal relationship. The environment affects the life of the community and vice versa; the community is influenced by the environment of his life. Humans exist in their environment and cannot be separated from it. The environment becomes an important part of human life to meet its needs [15]. In the development of science and technology, the environment and the social environment should always be considered to bring various types of disasters. For this reason, it is necessary to assume responsibility from all elements of the community to maintain the environmental and social environment, so it is hoped that a better perspective will be created in viewing the environment itself. An awareness of the importance of protecting the environment that includes the social environment is to maintain sustainability. In achieving a sustainable environment primarily on ecological emphasis, where every ecological component, starting from the smallest should not go unnoticed. The sustainability on ecological emphasis will be an important component of environmental sustainability, not only due to its potential to reduce the impact of environmental damage, but also due to its potential to introduce new concepts about human relations with nature [5,16]. A sustainable environment has some benefits or contributions to emphasize sustainability below [17].

- Protecting the life support system
- Protecting and increasing the biotic diversity
- Maintaining or enhancing ecosystem integrity and developing and applying rehabilitation measures for severely damaged ecosystems
- Developing and implementing preventive and adaptive strategies to respond to the threat of a changing global environment

In the ecological sphere, ecological sustainability proposed that for renewable resources, the harvest rate should not exceed the regeneration rate (sustainable yield). The level of waste generation from the projects should not exceed the environmental assimilation capacity (sustainable waste disposal). For non-renewable resources, depletion of non-renewable resources should require the development of comparable renewable substitutes for these resources [18,19].

Ecological sustainability is the responsibility of all people in the environment; it is similar to the case with the presence of water resources. Physically, the water supply managed by the Community Self-Help Group (CSHG) is sufficient to meet the community needs. During the rainy season, water remains abundant and the dry season can still meet the daily needs of the community. The manager is committed to providing the best service to the community. This commitment is evidenced by the level of professional management. Management performs many tasks; starting from performing intensive maintenance on water pumping machines, checking water discharge and water capacity to be distributed to the public [20,21].

The water management carried out is good and works professionally, as evidenced by the satisfied response from users or households on water management since 2009. In terms of good management, [21] found that the characteristic of the internal community consisting of collective initiatives, strong leadership, and institutional transparency should exist in the management of the community water of rural water. Collective initiative is evidenced by the existence of ideas based on mutual cooperation to achieve certain goals, responsibility, and community participation in making decisions. Strong leadership comes from an extraordinary individual or group of individuals from the community who have been able to provide supervision, monitoring, and evaluation of systems and workers, as well as taking on day-to-day and strategic decision-making roles. Institutional transparency is the existence of an accountability mechanism for water supply, including democratic procedures and disclosure of financial and other performance data.

The characteristics of the internal community can be found in the water management at CSHG *Berkah Mandiri*. Under the supervision of the district director, the objective is to provide water services to the community. Therefore, they have strategic plans and goals. The plans and objectives are implemented by involving the community [22]. Approximately 150 households can use water. Furthermore, the aspect of institutional transparency is proven by the existence of reports on the use of water for each household that follow the actual water cost of users.

Table 1. Water Available at CSHG *Berkah Mandiri*

Time	Hour	Amount of Water	Usage Capacity	Notes
Morning	4 – 8 am	1,500 cubic meters. The water pump is always running well, so the water is always filled.	1,000 cubic meters	Morning is the highest intensity of water use that is used to meet needs such as washing clothes, dishes, cooking, and bathing
Noon	11am-2 pm	985 cubic meters The water pump is always running well, so the water is always filled.	550 cubic meters	During the day, the use of water is relatively small, half of the use in the morning. This happened because most of the people were in the office. There are only housewives and children at home who need water.
Afternoon	3 – 7 pm	910 cubic meters The water pump is always on so water is always filled.	950 cubic meters	The intensity of water use in the afternoon is almost the same as in the morning. There is only a difference of 50 cubic meters. Because in the afternoon many people go home from the office to bathe, wash, and use the toilet. In the afternoon, they also cook; it is a habit of people at Gunungpati to serve at night.

Night	7.30 pm – 3 am	The volume of water at night can reach 1,440 cubic meters.	350 cubic meters	At night; there are a few people who use water to wash clothes, so only a little water is used.

3.2.2. Rural Water Service Distribution. Community participation greatly determines the sustainability of rural water. The sustainability of rural water management is the ability to recover from technical interruptions in the scheme, minimize external support, provide village-level financing, and ensure the continuation of useful services over time [23,24]. The criteria or factors for sustainability of rural water management are public trust in the management, public awareness of the environment, public awareness of the sustainability of water in the policies of the future, support from the Semarang City government policies, and transparent and accountable water management. Public trust in management is the main factor determining the sustainability of water management [1]. Complaints to management regarding water issues rarely occur. In addition, the maintenance of machines and water pipes is also guaranteed by management, so that engine damage rarely occurs. If there is water damage, the community does not have to pay. Humans are always aware of their existence in the environment and nature, since environmental damage makes water supply worse [15]. Environmental awareness is balanced with awareness of using water wisely and economically. Then there is no shortage of water in the dry season. The way to evaluate whether rural water management is sustainable or not is to campare the criteria of the sustainability of rural actual water management at CSHG *Berkah Mandiri*.

Furthermore, the Semarang City Government also provides regulations that independently regulate water management. It means that not just any party can drill wells for fresh water, which will ultimately benefit certain individuals and groups. Finally, there is transparency and accountability in financial management [25]. In every subdistrict meeting, the cash flow income and expenditure are always presented to the community. The community knows exactly how to use it. Also, every three months there is financial accountability to the public in the form of financial reports disseminated through a decree signed by the head and manager. Good water management will sustainably provide a good water supply that will be used long term by all members of the community [26,27].

CSHG *Berkah Mandiri* has a strategic plan. They also have budget and financial preparation so that the organization runs well and abundantly. CSHG *Berkah Mandiri* carries out maintenance of pumping machines, reservoirs and hoses; then, water can flow well to every house, does the market expansion, not only for the community at *Gunungpati*, and gives a full support for the management.

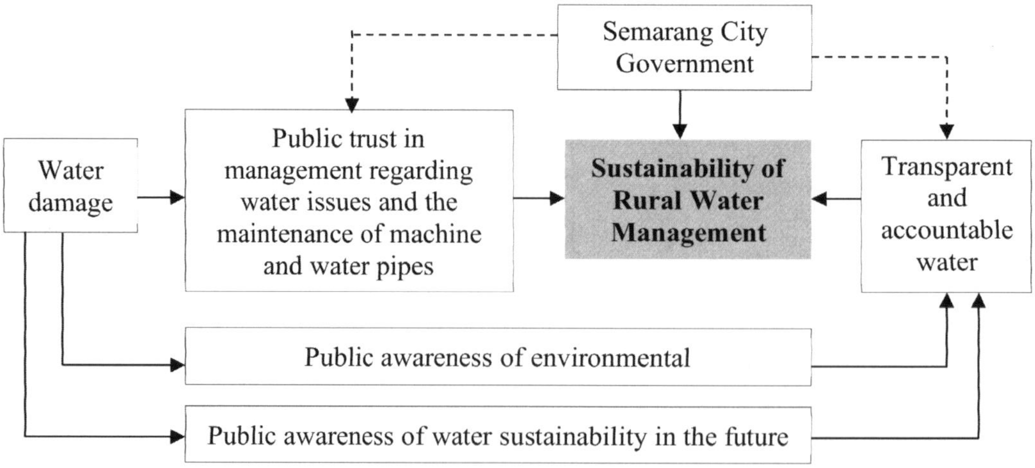

Figure 1. Model of Sustainability of Rural Water Management

IOP Conf. Series: Earth and Environmental Science 1098 (2022) 012039 doi:10.1088/1755-1315/1098/1/012039

4. Conclusion

The Berkah Mandiri Community Self-Help Agency (CSHG) has excellent rural water management. Water can be distributed well to meet community needs. During the rainy season, water is abundant and in the dry season, management can still meet the daily water needs of the community. The level of management of CSHG *Berkah Mandiri* can be seen in their commitment to satisfy the users of water. The characteristic of the internal community, including collective initiative, strong leadership, and institutional transparency, plays a role in terms of management success. The CSHG *Berkah* Mandiri manager plays a role in providing the best service to the community to manage water supply, so that it can gain a trust from public; starting from performing intensive maintenance on *water* pumping machines, checking water discharge, to providing the fresh water to be distributed to the public. The sustainability of water resources can be maintained with optimal community participation in protecting the environment and nature of the community. The community can also be accessed by the community, so that the public can know the current financial situation at CSHG *Berkah Mandiri.*

This research was taken in a qualitative approach and cannot be generalized. The recommendation for further research is to use a quantitative method to explore factors that support environmental sustainability, water resources, excellent service, and the role of the community.

Reference

[1] Kasri R Y and Moersidik S S 2018 Citizen engagement: An approach to sustaining Indonesian rural water supply and sanitation? *Compet. Coop. Soc. Polit. Sci.* 297–305

[2] Signori S and Bodino G A 2013 *Water management and accounting: Remarks and new insights from an accountability perspective* vol 26 (Emerald Group Publishing Limited)

[3] Chang H-C 2013 Environmental management accounting in the Taiwanese higher education sector: Issues and opportunities *Int. J. Sustain. High. Educ.* **131** 33–145

[4] Kasri R Y, Wirutomo P, Kusnoputranto H and Moersidik S S 2017 Citizen engagement to sustaining community-based rural water supply in Indonesia *Int. J. Dev. Issues* **16** 276–88

[5] Tyndale-Biscoe P, Crawford P and Bailey B 2020 Engaging with the wash enabling environment *J. Water Sanit. Hyg. Dev.* **10** 124–35

[6] Tingey-Holyoak J and Pisaniello J D 2019 Water accounting knowledge pathways *Pacific Account. Rev.* **31** 258–74

[7] Shoko C, Dube T and Clark D 2017 Total evaporation estimation for accurate water accounting purposes: an appraisal of various available estimation methods *Geocarto Int.* **32** 1333–51

[8] Galdeano-Gómez E, Pérez-Mesa J C and Godoy-Durán Á 2016 The social dimension as a driver of sustainable development: the case of family farms in southeast Spain *Sustain. Sci.* **11** 349–62

[9] McNeil M A game change of development? Citizen engagement: history and definition, the world bank

[10] Batchelor C, Hoogeveen J, Faurès J M and Peiser L 2016 *Water accounting and Auditing: A sourcebook*

[11] Grandvoinnet H, Aslam G and Raha S 2015 *Opening the Black Box: The Contextual Drivers of Social Accountability*

[12] Seawright J 2016 *Strategies for Social Inquiry Multi-Method Social Science: Combining Qualitative and Quantitative Tools* (Cambridge University Press)

[13] World Health Organization 1993 *Guidelines for drinking-water quality: volume 1 Recommendations*

[14] Menteri Kesehatan Republik Indonesia 2017 *Peraturan Menteri Kesehatan Republik Indonesia Nomor 32 Tahun 2017 Tentang Standar Baku Mutu Kesehatan Lingkungan Dan Persyaratan Kesehatan Air Untuk Keperluan Higiene Sanitasi, Kolam Renang, Solus Per Aqua dan Pemandian Umum*

[15] Basin D, Affairs F and Affairs F 2018 *Case Study: Water Management and Addressing Water Scarcity in Australia– Achieving SDG6 ?*

[16] Christ K L and Burritt R L 2017 *Supply chain-oriented corporate water accounting: a research agenda* vol 8
[17] Gann G D, McDonald T, Walder B, Aronson J, Nelson C R, Jonson J, Hallett J G, Eisenberg C, Guariguata M R, Liu J, Hua F, Echeverría C, Gonzales E, Shaw N, Decleer K and Dixon K W 2019 International principles and standards for the practice of ecological restoration. Second edition *Restor. Ecol.* **27** S1–46
[18] Steffen W and Stafford Smith M 2013 Planetary boundaries, equity and global sustainability: Why wealthy countries could benefit from more equity *Curr. Opin. Environ. Sustain.* **5** 403–8
[19] Callicott J B and Mumford K 1998 Ecological Sustainability as a Conservation Concept 31–45
[20] Ulubeyli S and Kazanci O 2018 Holistic sustainability assessment of green building industry in Turkey *J. Clean. Prod.* **202** 197–212
[21] Zhang D and Guo P 2016 Integrated agriculture water management optimization model for water saving potential analysis *Agric. Water Manag.* **170** 5–19
[22] Elmahdi A 2019 Road Maps for Water Accounting Designing and Institutionalizing for Sustainable Water Management in MENA Region *Int. J. Environ. Sci. Nat. Resour.* **22** 1–3
[23] Parry-Jones S, Reed R and Skinner B H 2001 Sustainable Hanpump Projects in Africa: A Literature Review *Sustain. Handpump Proj. Africa* 1–57
[24] Haysom A 2006 a Study of the Factors Affecting Sustainability of Rural Water Supplies in Tanzania Msc Water Management , Community Water Supply *Water Supply* **2** 31–6
[25] Stojkovic Zlatanovic S, Stojkovic M and Mitkovic M 2018 Current state and perspective of water management policy in terms of climate change: Case study of the Velika Morava River *Int. J. Clim. Chang. Strateg. Manag.* **10** 796–811
[26] Misiurski P 2016 Potensial users' perceptions of general purpose water accounting report *Accounting, Audit. Account. J.* **29** 471–84
[27] Gibassier D 2018 Corporate water accounting, where do we stand? The international water accounting field and french organizations *Adv. Environ. Account. Manag.* **7** 31–65

Learning from Contextual Material Practices in Architecture: Exploring Nature-Based Materials in Indonesia and Thailand

Kristanti Dewi Paramita[1], Paramita Atmodiwirjo[1]*, Soranart Sinuraibhan[2]

[1]Department of Architecture, Faculty of Engineering, Universitas Indonesia, Indonesia
[2]Faculty of Architecture, Kasetsart University, Thailand

paramita@eng.ui.ac.id

Abstract. This paper outlines pedagogical strategies in architectural design studio to construct awareness of local material practices towards architecture which rooted in its environmental context. Current architectural discourse perceives the notion of material as something separate from its environmental context. Through a joint design workshop conducted between Universitas Indonesia and Kasetsart University, this paper explores existing material practices happening in the local context of Indonesia and Thailand and how it informs material-driven architectural programming. In a group, students were tasked to reconstruct their understanding of materiality, and develop collections of material practices from both countries which revealed various local knowledge, spatiality, and environmental understanding. The workshop then demonstrates important lessons in the ways material triggers programmatic relation between architecture and its surrounding environment. Among these relations are shown in the way nature-driven community events triggers specific material craftsmanship, the connection between material utilisation in the community's spaces and natural cycles of seasons, and the interdependence between material practices and different geographical positions. Strategies in reconstructing the material knowledge towards material-based programming cross-culturally enables students to reflect on the unique material possibilities, as well as build awareness of local community strategies toward sustainable and material-driven architecture.

1. Introduction

This paper's objective is to outline the learning process of contextual material practices in an architectural studio, and address the possibilities of developing such practices as part of architectural programming. The discussion regarding material is undervalued over form and conceptual matter in architectural discourses (1) creating a gap in material-driven design learning. The existence of material is tightly connected to its sources as cultural artefact constructed in a particular context (2). Nevertheless, it is argued that such contextual relation of material is rarely discussed in architectural practice and pedagogy. The process of understanding and encountering the materials in education has been largely reduced in the latter part of the design process, despite its potential in form development and performance (3), as well as significance in shaping the sensorial encounter (4). This paper takes particular interest in exploring pedagogical strategies to construct the connection of material with its context. In addition, it follows with further exploration of the possibilities of utilising such connections as part of architectural design programming within the architectural design studio.

This study begins by exploring current discourses on architectural education, materiality, and context. It outlines the theoretical discourses which elaborate on current pedagogical perspectives on materiality and context within the learning processes in design studios. These studies then reflect on how such processes take place in the joint studio workshop conducted between the graduate students of Universitas Indonesia, Indonesia and Kasetsart University, Thailand under the theme Materiality, Nature and Locality. This paper explores how students identify traditional material practices distinct from their context, and how it is tightly related to the community. In addition, this paper discusses how the students develop propositions of architecture based on analysis resulting from the comparison of material practices found in Indonesia and Thailand. This paper then reflects on the lesson learned from such pedagogical strategies and discusses how it may expand methods of architectural education that are driven by materiality and context.

2. Architecture education, materiality, and context

2.1. Deconstructing the idea of materiality, nature, and locality in an architectural studio

The current architectural design studio has commonly started with a form of research to enable students to acquire knowledge and apply it throughout the design process (5). Nevertheless, despite the wide understanding of architecture as a material thing (6), addressing material as the trigger of design thinking requires some process of dismantling and reconstructing the idea of the material itself (1) within the studio. In this sense, material needs to be perceived beyond its technical specificities and its visual presence (6), but also as an important basis of creative thinking and making, a shift from object to process (7).

Tracing the existence of a material as a contextual artefact, be it social or environmental artefact benefit such a process of reconstruction, as a material is no longer perceived as something that is merely applicable or ready at hand (1) but instead as something that is produced through extensive processes of which intersect with conditions of society as well as its environment (8). For example, research on the traditional brick-making process in Central Java demonstrates society's local knowledge in choosing types of soil resources and timing of making the brick production to ensure optimum results (8,9). Such local knowledge reflects ecological values that exist from the dialogue between the articulation of the cultural and environmental surroundings (6,8), which then embody architecture material that is deeply connected with its context.

The introduction of such contextuality and ecological values in architectural pedagogies is no longer sufficient to simply introduce instrumental knowledge, but also to critically inquire about the meaning and appropriateness of such values in a particular design context (10). Some architectural pedagogy also builds such values by referring to the existence of material in the design context as 'material custodies', that 'reposition our buildings from unchanging, static objects, to organic systems in an ever-changing state of transition..' (11). Shifting and deconstructing the idea of materiality in connection with nature and local communities in design studios therefore can be done by interrogation of material and its role, both in the built environment and as part of the design process.

2.2. Assembling the material collection as a process of contextual discovery

The process of sorting through different kinds of material and cataloguing a thorough collection as part of learning strategies creates the organisation of information that is important to generate or support the design intent (12). The collection of information generates a record that contains operational relationships (13). For example, material collection can be sorted based on the different qualities of the material, annotating qualities such as texture, transparency, fluidity, fragility, and so on (14). Apart from the visual and performative qualities of the material, the processes which the material undergoes in relation to how it is sourced and produced in relation to community and context becomes significant (1). This information may provide different depth for the collection of material practices of a community, where relevant information includes how a material artefact is informed by community values, ideas, attitudes, and assumptions (15). In such a sense, the collection not only exists as a compilation of information but also as a means of thinking and discovery.

The materials collection exists as teaching tools produced by students, where different materials are subject to discovery of their physical and cultural performance, as well as possibilities of design operations (14). some pedagogical strategy aims to divide the collection based on hierarchy and layers, be it from the material itself or from the context it sits in, such as the use of the material in individual space to the neighbourhood (16). These layers of data construct the different material stories, which outline the practices of material relevant to the event in society (17). The material collection also annotates the local ritual of which such material is used by society, which then includes the different habits or daily activities performed as part of cultural traditions (18). The flow of material throughout these activities potentially creates territorialisation and other forms of spatial qualities (19). The structure of material collection therefore should not only discuss the physicality and performance of the material, but also the contextual stories discovered through the process of making the material collection. Processes of assembling a catalogue start by collecting and filtering relevant information, followed by analysing patterns in the array of available information, which may exist on different scales and perspectives (12). In subsequent, the study highlights possibilities in how such patterns may inform design statements towards alternative architectural programming (20), as will be discussed in the following paragraph.

2.3. *Materiality in context as the basis of architectural programming*

There are various possibilities of how materiality may inform architectural programming. One of the importance of interrogation and re-interpretation of the idea of material and contextuality in architectural studios is also related to the changing climate and the rise of natural disasters globally (21). The cycles of seasons and certain flows of matter in the environment have contributed to the availability of natural resources for nature-based material production (8) nevertheless such cycles and flows may change or become uncertain in the future. The future programmatic response must recognise such uncertainty, creating values towards the interdependence between material systems that requires critical reflection towards context, or even adopting different thinking paradigms such as apocalyptic and climate-adaptive thinking (21,22).

Programmatic discussion driven by materiality may also reflect and be driven by particular temporalities, particularly when the material underwent specific biological growth and generating process (23). The programmatic architecture is driven by such material and temporal processes that organise the ecological system that then transforms particular spatialities. Another programmatic discussion of materiality also highlights the sensorial quality of material, where the reaction between materials composes a form of the atmosphere (24). In this discussion, the materiality serves as an interweave between tectonic and tactile qualities which manifest together towards a particular architectural proposition, valuing the experience of bodies in doing daily activities (7,24). The programmatic exploration of architecture projects must link inquisitive and applicable knowledge, with sensitivity to context and society (5).

3. Methods

This study is based on qualitative observation and analysis from the works of a joint online studio conducted between Universitas Indonesia and Kasetsart University. The joint studio takes place for around three consecutive weeks in October 2021. The workshop has three main objectives, which are (1) to investigate distinctly, yet hidden material practices from local contexts in Indonesia and Thailand, (2) to speculate on material-driven approaches to architecture, and (3) to develop cross-cultural learning collaboration. The online workshop was followed by 17 students from Indonesia and Thailand, which were divided into three different groups that consist of a mix of both Indonesia and Thailand students.

The workshop started through mapping studies and online research on hidden and distinct material practices in Thailand and Indonesia. This research is then compiled in a form of a catalogue that compares materials practices based on local knowledge in Indonesia and Thailand. The study discusses how this catalogue enables students to learn material practices cross-culturally and analyse any community's similarities and differences in response to different contexts. The catalogue is then used to propose architectural programming based on their understanding of the contextual material practices that exist in both countries. In parallel with the joint studio, a lecture series on materiality and context was performed to provide a useful theoretical basis for the study.

The study explores processes of such programming development, annotating ideas arising from the mix of students and reflecting the learning process elaborated from such a design process. The study then addresses the contribution of such programmatic proposition toward the learning process of design based on materiality and context.

4. Learning Contextual Material Practices in Indonesia and Thailand

4.1. Mapping the idea of material, nature, and locality

At the beginning of the workshop, the Indonesian and Thailand students are invited to discuss and interrogate the notion of material, nature, and locality, informed by the theoretical perspectives from the lecture series. The students were asked to produce mind maps that utilise text and images about material, culture and context that become relevant in each country and compare both altogether. Each group tends to demonstrate a different approach to starting and producing their mind maps. One of the groups tries to map the similarity and differences available between both countries, annotating different traditional events, community types, and knowledge systems. Another group produces mind maps that try to weave between materiality and everyday life, paying attention to the process and tools of the materials, animals and relations with material and human daily activities as well as rituals concerning material(s). The last group discusses the connection between materiality and context in terms of its physical characteristics, such as contour topographies and water bodies, and then explores how such characteristics shape communities' livelihood and subsequently their materiality.

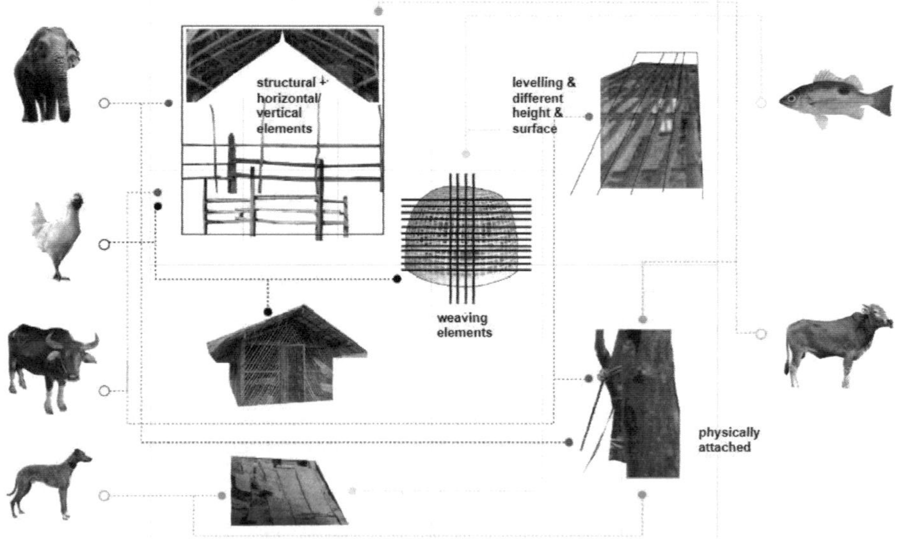

Figure 1. Example of a mind map exploring group's detailed perspectives on materiality and everyday life in a traditional context of animal rearing

4.2. Cataloguing material practices of Indonesia and Thailand

Based on the structured focus and raw materials explored from the mind mapping stages, the workshop followed through by assembling a catalogue that recorded material practices from Indonesia and Thailand. The catalogue must contain at least 10 (ten) different kinds of materials, and then format each of their catalogues based on their designated focus (See Figures 2 and 3). Each group then produces different strategies for compiling information, analysing them, and structuring them in a form of a material collection catalogue. For example, one of the groups compares similar material based on its relation with animals in Indonesia and Thailand, highlighting related equipment and objects of both countries in parallel. Another group explores cycles and processes materials with regards to the community tradition, dismantling parts and modules of materials that utilise local techniques, paying attention to the local craftsmanship of each country in dealing with similar materials.

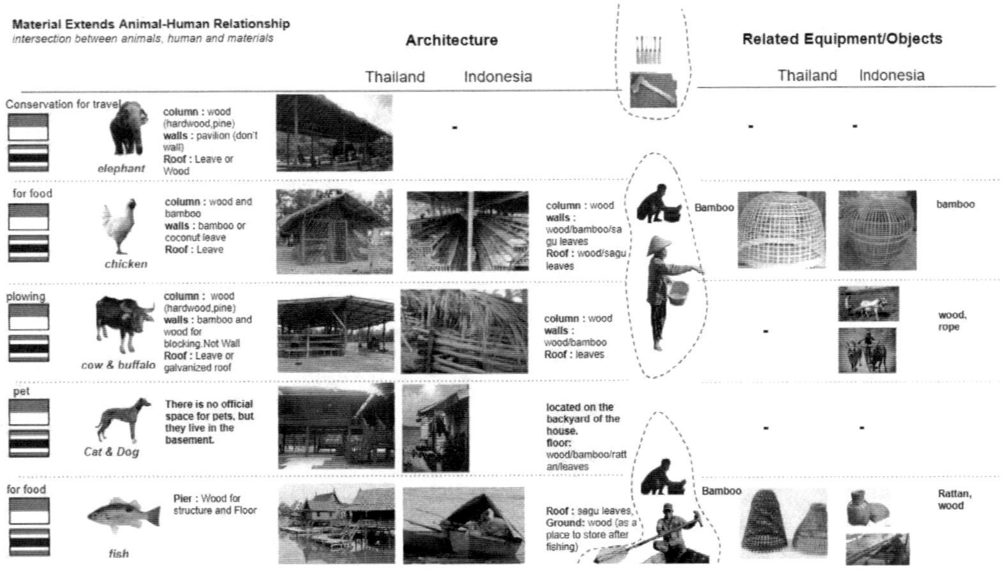

Figure 2. Example of the material catalogue that investigates relation between animal, material, and their living setting and objects

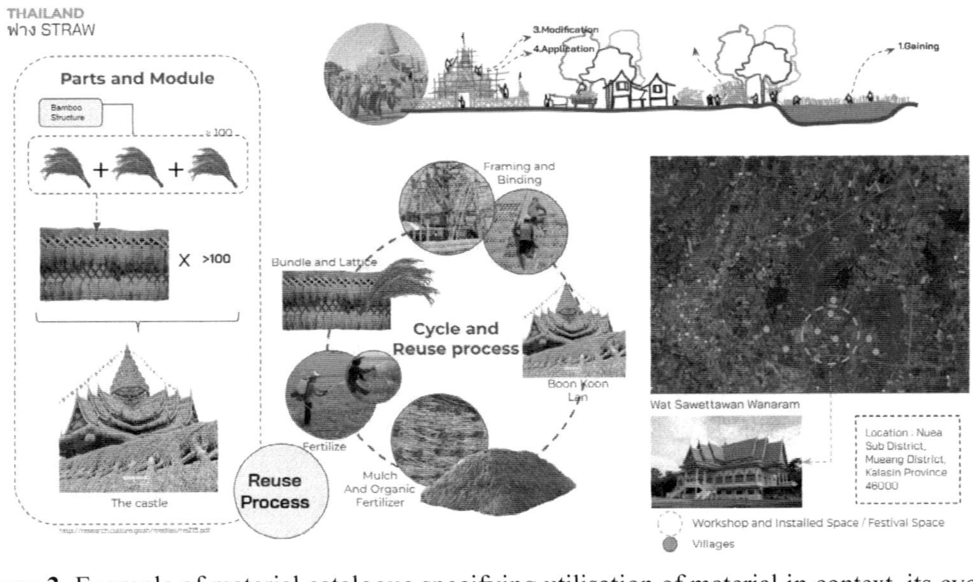

Figure 3. Example of material catalogue specifying utilisation of material in context, its cyclical process as well as the parts and modules of the materials as crafted by local communities

4.3. *Exploring programmatic possibilities: a reflective thinking*

The workshop followed through by exploring architectural programmatic possibilities informed by the material catalogue produced by each group. This process tends to be non-linear, where the group was involved in a collective and iterative discussion to analyse and respond to the catalogue and generate architectural programming propositions. The students were not expected to produce a finished design output due to time constraint and instead were tasked to create three programmatic propositions at the end of the workshop. The programmatic propositions take place in the form of creative diagrams. The first group proposes an architecture for fishing activities by the community that allows engagement with different kinds of fish and seawater encounters. Materially, the fishing structure is aimed to appreciate decay and destruction through time, existing as a medium between humans and fish in a challenging environment (See Figure 4). The second group explores architecture as a material celebration within the community, creating water structures by utilising seasonally harvested materials for multiple traditional events that also reduce environmental waste in the long run. The third group proposes an architecture that conforms to different topographical needs, consisting of material modules that can be assembled and disassembled depending on location and transportation needs.

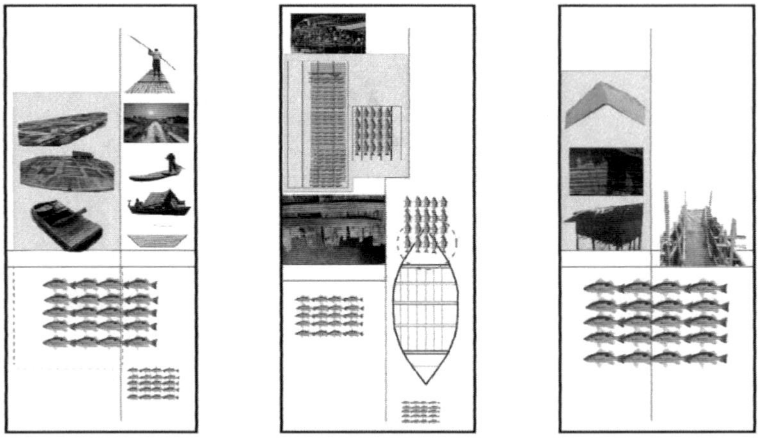

Figure 4. Example of a programming diagram of architecture which accepts material decay

4.4. *Discussion: Pedagogical Strategies of Material-Driven Architectural Studio*

The process of mind mapping challenges the groups to choose the focus of which they will construct their interpretation of materiality, nature, and locality, based on their interest, prior knowledge, and the result of their discussion itself. Each member of the group, be it from Indonesia or Thailand can exchange important aspects that are similar or particularly vital for each country, particularly about its traditional context. For example, some students may reflect that some traditions take place around the harvesting season, as both countries have a significant proportion of agricultural communities. Other groups take interest in how the communities demonstrate a tight connection between humans and animals due to the diversity of animals in both countries. This interest then is developed further in a map of the interconnection of such topics within the traditional setting of the community (Figure 1). This focus demonstrates critical thinking and decision-making abilities to deconstruct and reinterpret the idea of materiality, not only as a ready-made object but as a product of the interrelation of community and environmental systems based on the obtained cross-cultural contextual knowledge.

In the subsequent process of mapping, important keywords discussed in the mind maps were discussed further to be used as the structure for the material practices catalogue that will be the next stage of the workshop. Students were then asked to perform preliminary searches of raw material regarding such material practices, following such structure. In these stages the students will practice their abilities to filter through the relevant information, as well as sharpen their focus further, moving their design thinking forward in response to the issue of materiality and context. The structure of the material catalogue being produced demonstrates a deep investigation of materiality in context further. Each group creates their hierarchies and layers of information, using collaged images, maps, sections, and texts to elaborate their collection of material reading from both countries. Some hierarchies were based on the part and whole structure, exploring the origin and wholeness of the material which is then dismantled into parts and techniques. Another strategy was to create a comparative study where information is placed in parallel with one another, enabling discussion on variations and diversity found in the objects or the setting in which the material is applied, originated, and used in everyday life. In this sense, the situated contextuality of material is deconstructed with different means of thinking, from highlighting bodily encounters to highlighting making techniques within a particular temporality.

The contextuality of material then lead to different programmatic propositions, which demonstrate the reflective capabilities of students concerning the existence of databases constructed by their focused modes of inquiry following material and context. The first group appreciates cycles of material as an acceptable part of architectural programming, while the second group promotes community-based material events as a source of craftsmanship skills. The third group, on the other hand, explores processes of material flows and construction in response to the environmental context, thus requiring flexible structures for different geographical terrain. The notion of material cycles, material events, material flows, and material flexibility becomes the main programmatic ideas that become a form of response from contextual discovery in the material catalogue. Such a response will then become the catalyst for certain spatial components and spatial qualities, creating architecture driven by materiality with a deep connection to its context.

5. Conclusion: Towards Architecture Driven by Materiality in Context

The study outlines pedagogical strategies to construct awareness of local material practices that can be the basis of material-driven architectural design methodologies. Exploration in a cross-cultural joint design studio utilises creative mind-mapping, cataloguing, and reflective activities to deconstruct ideas of materiality in architecture, collect information and assemble contextual knowledge that may be initially hidden. In addition, the reflective and responsive pedagogical activities also enable alternative possibilities of architectural programming that are rooted in an understanding of nature and local knowledge.

The cross-culture joint workshop reflects how knowledge richness regarding local material practice can be celebrated. The process of interaction between students from two countries also benefits their exploration as students become exposed to a new perspective of reading the material in other contextual traditions and different kinds of communities. The students' reflection of the workshop acquired at the end of the workshop also demonstrates their appreciation of the rich variation material practices that exist interconnectedly between two countries, while also having the awareness of the different challenges perceived by local communities in acquiring, producing, or utilising materials. Lessons of understanding and utilising the notion of materiality in architectural design studios, while demonstrating criticality and reflexivity, can be expanded further with additional timeframe and with direct learning experiences in the field. Such discussion potentially strengthened students' embodied knowledge of materiality in context, contributing toward material-driven architectural design methods.

6. Acknowledgements

This research is supported by the Ministry of Education, Culture, Research, and Technology under the research grant titled Penelitian Dasar Kompetitif Nasional Tahun Anggaran 2022, with Grant Number: NKB-993/UN2.RST/HKP.05.00/2022. The authors would like to thank students and lecturers from the Department of Architecture Universitas Indonesia and Kasetsart University who have participated in the 2021 Joint Architectural Design Workshop under the theme Materiality, Nature, and Locality.

7. References

[1] Thomas KL, editor. Material matters: architecture and material practice. London: Routledge; 2007. 262 p.
[2] Thomas KL. Building Materials: Material Theory and the Architectural Specification. 1st edition. Bloomsbury Visual Arts; 2021. 249 p.

[3] Chandler A. A Philosophy of Engagement: Developing A StrategicAbility Through Direct Engagement With Material, Process, And Collective Action. In: Thomas KL, editor. Material matters: architecture and material practice. London: Routledge; 2007.
[4] Butterworth C, Vardy S. Site-Seeing: Constructing the 'Creative Survey'. Field Journal. 2008;2(1).
[5] Saghafi MR. Teaching strategies for linking knowledge acquisition and application in the architectural design studio. Archnet-IJAR: International Journal of Architectural Research. 2020 Jan 1;15(2):401–15.
[6] Dernie D. Material imagination in architecture / David Dernie and Jacopo Gaspari. Routledge; 2016.
[7] Couchez E, Heynickx R, editors. Architectural Education Through Materiality: Pedagogies of 20th Century Design. 1st edition. Abingdon, Oxon ; New York: Routledge; 2021. 256 p.
[8] Atmodiwirjo P, Johanes M, Saginatari DP, Yatmo YA. Ecological aspects of the traditional brick making process in pedurungan kidul, central Java: 3rd International Tropical Renewable Energy Conference 'Sustainable Development of Tropical Renewable Energy', i-TREC 2018. E3S Web of Conferences [Internet]. 2018.
[9] Harahap MMY, Suryantini R, Paramita KD, Yatmo YA. Sun-Drying in Traditional Brickmaking: Strategies for Achieving Efficiency. IJTech. 2020 Dec 16;11(7):1414.
[10] Khan AZ, Vandevyvere H, Allacker K. Design for the Ecological Age: Rethinking the Role of Sustainability in Architectural Education. Journal of Architectural Education. 2013 Jul 3;67(2):175–85.
[11] Satterfield B, Swackhamer M. Material Custodies: Embracing Loss, Failure, and Death as Opportunities in Wood Construction. Journal of Architectural Education. 2021 Jan 2;75(1):121–8.
[12] Karimah A, Atmodiwirjo P. Catalogue drawing: A framing device for design thinking. ARSNET [Internet]. 2021 Apr 30 [cited 2022 Mar 13];1(1). Available from: https://architecture.ui.ac.id/arsnet/index.php/ojs/article/view/6
[13] Vidler A. Diagrams of Diagrams: Architectural Abstraction and Modern Representation. Representations. 2000;(72):1–20.
[14] Schroepfer T, Margolis L. Integrating Material Culture. Journal of Architectural Education. 2006;60(2):43–8.
[15] Prown JD. Mind in Matter: An Introduction to Material Culture Theory and Method. Winterthur Portfolio. 1982;17(1):1–19.
[16] Saunders A. Material Manifestations. Journal of Architectural Education. 2013 Mar 7;67(1):86–95.
[17] Alonso CP. Deep-Time Architecture: Building as Material-Event. Journal of Architectural Education. 2021 Jan 2;75(1):142–4.
[18] Atmodiwirjo P, Yatmo YA. Interiority in Everyday Space: A Dialogue between Materiality and Occupation. Intry. 2019 Jan 30;2(1):1–4.
[19] Suryantini R, Paramita K, Yatmo Y. Investigating the Food-Based Domestic Materiality of Nuaulu People, Seram Island: The Multiple Roles of Sago. J Phys: Conf Ser. 2019 Nov 1;1351(1):012115.
[20] Yatmo YA, Paramita KD, Atmodiwirjo P. Future Landscape of Architecture: Data, Methods, and Programming. Depok: Department of Architecture; 2020.
[21] Keller E, Main KL. After the Flood. Journal of Architectural Education. 2020 Jan 2;74(1):126–8.
[22] Clouse C. The Himalayan Ice Stupa: Ladakh's Climate-adaptive Water Cache. Journal of Architectural Education. 2017 Jul 3;71(2):247–51.
[23] Oxman N, Burckhardt A, Steiner HA. Neri Oxman: Material Ecology. Antonelli P, editor. New York: The Museum of Modern Art, New York; 2020. 181 p.
[24] Atmodiwirjo P, Yatmo YA. 'Tanahku Indonesia': On Materialscape as the Materiality of a Nation. Architecture and culture. 2020;8(2):328–49.

IOP Conf. Series: Earth and Environmental Science 1098 (2022) 012041 doi:10.1088/1755-1315/1098/1/012041

Noise Intensity Analysis and Mitigation of Workers Health in home industry Steamed bun Arum Jaya Nusukan

Siti Rachmawati[1], Sisca Indriyanti[1], Muhammad Kukuh Apriyanto[1], Zahra Hanun[1], Thalita Aldila Pramitasari[1], Fathoni Firmansyah[2], Iwan Suryadi[3]

[1]Department of Environmental Science, Faculty of Mathematics and Natural Science, Universitas Sebelas Maret, Surakarta, Indonesia
[2]Occupational Safety and Health D4 Study Program, STIKes Mitra Husada Karanganyar, Indonesia
[3]Department of Environmental Health, Politeknik Kesehatan Makassar, Makassar, Indonesia

siti.rachmawati@staff.uns.ac.id

Abstract. Home Industry has a positive impact on absorbing labor, but is still lacking in supervision. Bakpao Arum Jaya's home industry is a small-scale industry that uses noise-producing equipment. The purpose of this study is to determine the description of noise intensity when the machine is working during the production process. This research uses the quantitative descriptive method. The data used are primary data from field measurements and secondary data. The measurement points were carried out in the grinding room and steam area. The results showed that the Leq of the milling room was 89.36 dB(A) which exceeded the TLV and the Leq of the steam area was 77.34 dB(A) which did not exceed the TLV. Playing music through speakers can also increase the noise in the work area. In the steam area, if measurements are taken on three steam equipment that is turned on simultaneously, the noise value will be greater than the measurement results and may exceed the allowable TLV. Exposure to noise every day will have an impact on health problems for workers. Forms of mitigation that can be done are administrative mitigation, worker health checks, use of protective equipment, and reducing noise from the source.

1. Introduction

The Industry is part of the development of the sociocultural component. Short-term economic movements alone by humans can greatly alter the sociocultural component [1]. Activity of the economy that creates goods or services is the core of the industry. Born from the development of human activities that only meet all their own needs and develop into a specific industry producing one type of product or service. The next industry is one of the causes of socioculture becoming very influential on an environment. The industry that was carried out at first was based on human needs that increased rapidly and even tended to be exponential. This need that must be met can no longer be met only by utilizing human labor. As a result, the industry is experiencing developments using tools that are getting higher in productivity. The increasing complexity of technology and the economy is creating people's need for raw materials and energy to soar [2].

Indonesia is a developing country, and the acceleration of growth in the number of industries is also high. Especially a large number of industries are carried out by direct people on a smaller scale. The Home Industry as one of the informal economic sectors is a type of industry that many people do, with ease of doing it anywhere and with a smaller amount of capital. Economies in developing countries have a more rapidly developing informal economy [3]. So that many also give birth to home industries for residents in residential locations, on the other hand, this is a good thing because it can empower the surrounding community, especially-fact Indonesia is currently preparing for the arrival of a wave of the labor force which is very much called a demographic bonus. Home Industry can absorb the labor force and be productive for the advancement of the country's economy. The quality of human resources also determines the opportunities for upcoming investments [4].

One of the many home industries in Indonesia is the steamed food industry. The bun industry is considered to be one of the profitable businesses that can be chosen as a snack business that promises profits for its actors [5]. Home Industry steamed bun Arum Jaya Nusukan not only brings a positive impact, but losses are also present in it. Because of the small-scale industry, attention and supervision are seriously lacking. Especially the impact of the industry that we all know is in the form of pollution that can come from the production process, resulting in different types of pollution. Existing pollution can be in the form of pollution of water, air, sound, and much more. Many pollutions that are not widely realized turn out to have a bad impact [6]. This can be a serious threat because of the location of the home industry, the majority of which is in settlements, and the pollution is directly exposed to humans.

Noise generated from industrial activities is a major factor in the high increase in ambient noise levels around residential areas. Home Industry developed and switched to using machines to increase the volume of production, resulting in pollution produced. The use of machines is closely related to the level of sound pressure produced and becomes noise pollution for the community around the industry, especially workers in the industry who are exposed to high noise throughout working hours. Manufacturing companies have indeed become one of the workplaces that are at risk of noise exposure and have become a concern in various countries especially, developing countries. In some studies, it has been shown that many developing countries have an average noise value exceeding the threshold and it is quite possible to continue to increase because noise control and noise protection are not always implemented as industrialization develops [7]. Noise pollution is an unwanted sound with a high degree of intensity [8]. The noise produced by the Home Industry not only has an impact on the surrounding community, but workers have a higher risk because they are closer and longer to the noise. Even by working in noisy environments, workers are at a higher risk of hypertension [9]. This happens depending on the noise level received and the hours worked.

Based on interviews that have been conducted with workers at the home industry of Arum Jaya Nusukan buns, workers complain of noise from production equipment. According to workers in the home industry of Arum Jaya Nusukan buns, they are used to hearing noise from the production equipment. The habit of being exposed to daily high levels of noise will result in significantly not being felt noise damage [10]. When working, they also do not wear hearing protective equipment. Thus, the impact felt such as decreased hearing power has not been felt by the workers because the workers are used to being exposed to the noise every day. In addition, the inability of workers to identify early indications of hearing damage and ignorance to recognize the impact of hearing loss are also causes of hearing loss by workers. Mild hearing loss is experienced by workers, but in the future, it may be very dangerous for hearing, for example causing hearing loss or other related diseases. Workers' hearing loss can take years to have a noticeable impact on the body [11]. Therefore, personal protective equipment is needed when performing work related to production equipment that has noise.

The research purpose is to examine the intensity of noise caused by the bun home industry located in Nusukan Village, Banjarsari District, Surakarta City, Central Java Province as well as noise mitigation for the safety and health of its workers.

2. Research Methods

2.1. Research Location

The research was conducted in April which is working hours. The research location was carried out at the Arum Jaya Steamed bun Home Industry, which is located in West Nayu RT 05 RW 08, Nusukan, Banjarsari District, Surakarta City, Central Java Province with location coordinates (-7.551507, 110.825640). At the location of the steamed bun home industry, it has an area of 60 m^2 which consists of 2 rooms, namely the milling room and steam room. The grinding room has an area of 24 m^2 while the steam room has a room area of 36 m^2.

2.2. Data Collection and Research Method

The research method uses a quantitative descriptive method. The method of descriptive analysis is a survey method that clearly explains the state of affairs in the absence of a connection with the object under study. The descriptive method serves as a factual explainer of the nature and relationship between the facts under study.

The data collection method uses primary data and secondary data. Primary data were obtained from the results of direct noise measurements obtained in the field and interviews with workers. The location point is used as a measurement of the noise level, namely in the milling room and steam area. Measuring noise is carried out at a height of about 1 to 1.5 meters above ground level. The study was conducted at 2 points with a measuring time of about 10 minutes and every 5 seconds recorded measurement results at each sampling point, and 120 data will be generated. There are several tools needed to measure noise, namely the Sound Level Meter tool, the stopwatch serves as a logger for the length of time when measuring, and GPS functions as a record for coordinate locations. Secondary data, includes various literature sources such as articles, journals, and books. The literature studies used are pre-existing literature and similar to the research theme.

2.3. Data Analys

Procedure the noise measurement procedure refers to SNI 7231:2009. Data processing is carried out by Leq noise values. This refers to KEP-48/MENLH/11/1996 [12], then 120 noise intensity values obtained are processed into Leq. Leq calculation is carried out at each measurement point because the sound level meter tool used is not equipped with a Leq value reading. Then the value of the equivalent noise level can be calculated using the following formula:

$$L_{eq} = 10 \text{ Log } \left(\frac{1}{120} \sum_{1}^{120} 10^{\frac{L1}{10}}\right) \text{ dB (A)}$$

Description

Leq : Equivalen noise level

L : Noise measurement results

3. Discussion

Noise is one of the common physical hazards in industrial activities [13], especially in industries that utilize large production equipment during the manufacturing process. According to [14], the industrial sector that uses tools that produce higher noise results in a higher prevalence of hearing loss in its workers. Such production equipment has the potential to cause noise caused by engine noise, old machinery, engine vibration, and exhaust and lines in the engine [15]. Noise exposure can hinder communication between workers and have an impact in terms of physical, physiological, and psychological aspects on workers [16]. High noise levels in the work environment adversely affect the health and safety of workers [17], such as the inconvenience of the work environment, stress, fatigue, and difficulty in communicating between workers which can trigger the occurrence of work accidents [7].

The potential for these risks prompted governments in various countries to make regulations that limited exposure to the level of sound produced for workers [15]. As for the regulation regarding the threshold value of workplace noise levels in Indonesia based on the Decree of the Minister of Manpower Number PER.13 / MEN / X / 2011 [18], concerning the threshold value of physical factors and chemical factors in the workplace, there is a noise limit of 85 dB with exposure to 8 hours a day and 40 hours a week. The threshold is the highest intensity and the average acceptable value of labor with continuous exposure within 8 hours a day or 40 hours a week without causing permanent deafness [19]. Home Industry Steamed bun Arum Jaya is one of the industries that produce noise intensity. The source of noise comes from steam tools and milling tools used during the production process. Measurements were made at 2 points close to the milling device and the steam device which were suspected to produce noise intensity. The map of the Arum Jaya Steamed bun Home Industry along with the location of the noise measurement point in Figure 1.

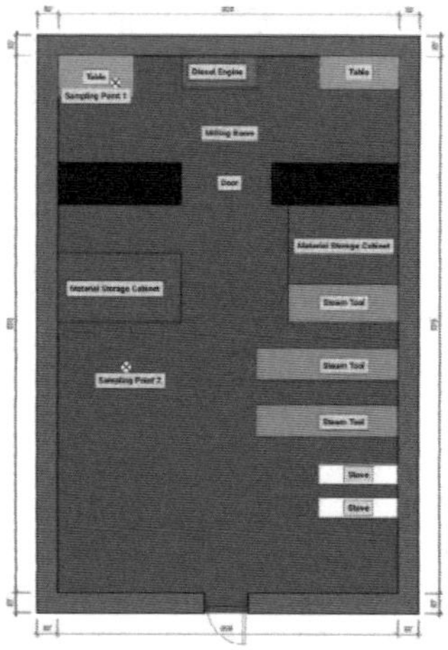

Figure 1. Measurement Point Location Plan

The noise intensity measurement of this study was carried out on the weighting scale A with a measurement range of 30-130 dB(A) using a Sound Level Meter. Scale A is a scale that describes the limits of hearing and the human ear's response to noise, including noise that can cause deafness [20], and is commonly used in data collection related to noise-induced hearing loss [21]. Noise measurement is carried out in the middle of working hours, namely at 07.00 - 08.00 WIB and 09.00 - 10.00 WIB when the production machine is operating. Noise measurement at 07.00 – 08.00 WIB is carried out to obtain a noise value when the diesel engine in the milling room is operating. Meanwhile, noise measurements at 09.00 – 10.00 WIB are carried out to get noise values when the stove in the steam room is operating. Noise measurement at 2 points, each carried out for 10 minutes with noise data retrieval every 5 seconds, so that at each point 120 noise intensity values are generated.

The equivalent noise level in the milling chamber is 89.36 dB(A), so it has exceeded the noise threshold value. The equivalent noise level in the steam room is 77.34 dB(A), so it does not exceed the noise threshold value.

Table 1. Arum Jaya Steamed Bun Noise Measurement Results

Measurement Location	Coordinate Point		Noise Intensity (dB(A))	Length of Exposure (hour)	TLV (dB(A))	Description
	X	Y				
Milling Room	110.825789	7.551804	89,36	4	88	Exceeding the threshold limit value
Steam Area	110.825759	7.551774	77,34	4	88	Meet threshold limit values

Based on the results of measuring noise intensity in the home industry of Arum Jaya buns it can be seen that the equivalent noise value of the milling room is 89.36 dB (A) was obtained with noise exposure for 4 hours, while in the steam room an equivalent noise level of 77.34 dB (A) was produced with noise exposure for 4 hours. Based on the Regulation of the Minister of Manpower of the Republic of Indonesia Number 5 of 2018 [22], the threshold value of noise in industrial areas for exposure for 4 hours is 88 dB (A). In industrial work environments, the noise level produced is quite high, so there must be a time limit on noise exposure.

Based on the table 1, it is known that the equivalent noise value in the milling chamber has exceeded the allowable threshold value. This is due to the presence of a diesel engine used to grind dough. The diesel engine has been operated for a long enough period so that the sound of the diesel engine is very noisy because there is a combustion system problem that causes the diesel engine not to work optimally. The accumulation of dirt in the machine can also be the cause of the problem. According to, [23], one of the disadvantages of using machines and work tools to facilitate work is the potential to cause noise. Based on the Regulation of the Minister of Manpower and Transmigration Number PER. 13/MEN/X/2011 [24], noise is an unwanted sound that comes from a production machine that can cause health problems, especially hearing. Where the noise due to the means of production and machines will have an impact physically, psychically, on communication, and balance [25]. In addition to diesel engines, music playback through speakers also contributes to the noise in the milling room. The music playback is done so that workers are not too tense in doing work, but this causes hearing loss in workers. Diesel engines with lower noise will provide better comfort for workers when doing their jobs [26]. The relationship between noise levels and the possibility of hearing loss in workers is influenced by the intensity of noise and the length of time workers are in places where noise is experienced. In this bun industry home, workers experience noise exposure every day while working, so the possibility of worker health problems is very high, for example, deafness, errors in receiving information, cardiovascular problems, mental health disorders, performance disorders, emotional ease, and stress [27]. The effect of noise exposure on workers in the industry is divided into two, namely the effect on hearing and the effect on non-hearing. Hearing effects include decreased hearing power and permanent hearing loss due to excessive noise exposure. Non-auditory effects include stress and physiological effects [28].

Mitigation steps that can be taken at Home Industry Bakpao Arumjaya, the first are mechanical noise can be mitigated by adding sound-absorbing materials or dampening noise directly on the device [29]. Noise originating from diesel engines includes mechanical noise, namely the noise generated from mechanical sources peaks around a certain frequency and is very loud for the human ear. Creating a special room for the device that can absorb sound noise or in that room a noise barrier is installed [30] so that the noise is only in that room without spreading to other production rooms. The second is to control noise with administrative methods. The form of administrative noise control is by measuring and monitoring noise in the workplace periodically every 6 months including routine maintenance of the machines used. Machine maintenance is carried out a maximum of once a year by PER.04/MEN/1985 by replacing the generator silent box with better quality or installing an enclosure on a generator engine that has a damping capability of 84.66 dB (A) [31]. The third is periodic health checks on workers who are exposed to noise. Based on the Government Regulation concerning the health examination of workers in the implementation of occupational safety, it is stated that there are three kinds of health checks for workers, namely pre-employment health checks, periodic health checks, and special examinations. Health checks that can be carried out are by conducting audiometric examinations on workers to determine their hearing function of workers so that workers' hearing loss can be detected early [32]. Based on the Minister of Manpower and Transmigration No. PER.02/MEN/1980, periodic inspections are carried out at least once a year. The fourth is to rotate workers to reduce noise exposure, the rotating work system can protect workers from declining health threats due to exposure to high noise intensity in the long term [33]. Working hours rotation or also known as Shift Working Hours is administrative mitigation. Constant noise exposure to workers causes fatigue in light, moderate, and severe stages [34]. The rotation of workers' working hours is needed because they work in non-ideal conditions such as high noise, workers need rest time and work differently from human working hours in general. And lastly, equip workers with adequate personal protective equipment in the form of hearing protection devices in the form of ear muffs and ear plugs. Ear muffs are more effective than ear plugs. The use of ear plugs has been shown to significantly reduce a hearing loss [35] [36]. The use of earplugs and earmuffs is adjusted to the worker's ear so that the size of the earplug or earmuff for each worker will be different. This is done to maximize the function of the ear protector in blocking too loud sound and pay attention to the comfort of the worker. Apart from the size of the earplug and earmuff, the selection of this ear protector also needs to pay attention to the NRR of the earplug or earmuff. Noise Reduction Rating (NRR) is a measure of the ability of an ear protector to reduce noise levels expressed in dB units proposed by the United States Environmental Protection Agency [37]. The higher the NRR value on the label of ear protection equipment, the greater the ability of the device to reduce exposure to noise. Remembering Home Industry Bakpao Arum Jaya buns is a medium-scale industry, so a foam-type ear plug (NRR 33 dB (A)) can be used which can reduce the noise level received by workers by 76.8 dB (A). The 1000 earmuff model made of dielectric plastic which has an NRR of 20 dB (A) can also reduce the noise level received by workers by 73.65 dB (A) [32].

4. Conclusion

Based on the explanation above, it can be seen that the equivalent noise value in the milling room is 89.36 dB(A) and the equivalent noise value in the steam room is 77.34 dB(A). In the milling room, the noise value has exceeded the threshold value allowed in the Regulation of the Minister of Manpower of the Republic of Indonesia Number 5 of 2018, which is 88 dB(A) for a 4-hour exposure period. The factors that affect the noise are diesel engines that have been operated for a long period and music playback through speakers by workers can also increase the noise in the room. If workers experience noise exposure every day while working, it will allow workers' health problems to arise. Therefore, mitigation is required. Some forms of mitigation that can be done include administrative mitigation, routine health monitoring of workers, and the use of personal protective equipment in the form of hearing protection like ear muff and ear plugs. The tool aims to provide direct mitigation to the hearing of workers exposed to noise from the machine. Another alternative is to mitigate the equipment by adding materials that can reduce noise directly on the equipment. The next mitigation is to create a special room for the tool that can absorb the noise.

References

[1] Dumasari, B. Dharmawan, I. Santosa, W. Darmawan, P. Utami, and D. D. Aisyah. 2019. Sociocultural and Economic Conditions as Causes of the Weakened Social Cohesion in Landless Peasants. Journal of Arts & Humanities. 8 (6) : 11-18.

[2] Brand-Correa, L. I. and J. K. Steinberger. 2017. A Framework for Decoupling Human Need Satisfaction from Energy Use. *Ecological Economics*. 141: 43-52.

[3] Horvath, J. 2018. Business Cycles, Informal Economy, and Interest Rates in Emerging Countries. *Journal of Macroeconomics*. 55: 96-116.

[4] Kheng, V., S. Sun, and S. Anwar. 2017. Foreign Direct Investment and Human Capital in Developing Countries: A Panel Data Approach. *Economic Change and Restructuring*. 50(4): 341-365.

[5] Hidayat. R. R., 1. M. Sugitha, dan A. A. I. S. Wiadnyani. 2019. Pengaruh Perbandingan Tepung Beras Hitam (*Oryza Sativa* L.Indica) Dengan Terigu Terhadap Karakteristik Steamed bun. Jurnal Ilmu dan Teknologi Pangan. 8 (2) : 207-215.

[6] Ukaogo, P. O., E. Ewuzie, and C.V. Onwuka. 2020. Environmental Pollution: Causes, Effects, and the Remedies. Microorganisms *for Sustainable Environment and Health*. 21: 419-429.

[7] Wang, X., Orelaja, O. A., Ibrahim, D. S., & Ogbonna, S. M. 2020. Evaluation of Noise Risk Level and Its Consequences On Technical Operators of Tobacco Processing Equipment in A Cigarette Producing Company in Nigeria. *Scientific African, 8*, 1-14.

[8] Jariawala, H. J., H. S. Syed, M. J. Pandya, and Y. M. Gajera. 2017. Noise Pollution & Human Health: A Review. *Noise and Air Pollutions: Challenges and Opportunities*. 32 (5) : 122-127.

[9] Zhou, F., A. Shresth, S. Mai, Z. Tao, J. Li, Z. Wang, and X. Meng. 2019. Relationship Between Occupational Noise Exposure and Hypertension: A Cross-sectional Study in Steel Factories. *American Journal of Industrial Medicine*. 62 (11) : 961-968.

[10] Febriyanto, K., S. Sunarti, and F. F. Rahman. 2019. Assessing Noise-Exposure and Daily Habits Can Cause Hearing Loss among Ladies Club at Nightclub. *Indian Journal of Public Health Research & Development*. 10 (3): 827-832.
[11] Masterson, E. A., J. A. Deddens, C. L. Themann, S. Bertke, and G. M. Calvert. 2015. Trends in Worker Hearing Loss by Industry Sector, 1981-2010. *American Journal of Industrial Medicine*. 58 (4): 392–401.
[12] Keputusan Menteri Negara Lingkungan Hidup Nomor: KEP-48/MENLH/11/1996 tentang Baku Tingkat Kebisingan.
[13] Jiang, Z., B. Fa, X. Zhang, J. Wang, Y. Feng, H. Shi, Y. Zhang, D. Sun, H. Wang, and S. Yin. 2021. Identifying Genetic Risk Variants Associated with Noise-Induced Hearing Loss Based on A Novel Strategy for Evaluating Individual Susceptibility. *Hearing Research*. 407: 1-8.
[14] Masterson, E. A., J. A. Deddens, C. L. Themann, S. Bertke, and G. M. Calvert. 2015. Trends in Worker Hearing Loss by Industry Sector, 1981-2010. *American Journal of Industrial Medicine*. 58 (4): 392–401.
[15] Silviana, N. A., Siregar, N., Banjarnahor, M., & Munte, S. (2021). Pengukuran dan Pemetaan Tingkat Kebisingan pada Area Produksi. *JIME (Journal of Industrial and Manufacture Engineering), 5*(2), 160-166.
[16] McCloskey, K. P., K. E. Chapman, L. Chapuis, M. I. McCormick, A. N. Radford, and S. D. Simpson. 2020. Assessing and Mitigating Impacts of Motorboat Noise on Nesting Damselfish. *Environmental Pollution*. 266 : 1-10.
[17] Indrayani, R. dan K. Aryatika. 2021. Keluhan Pendengaran dan Pemetaan Kebisingan pada Industri Penggergajian Kayu UD. Mayora Kabupaten Jember. *Ikesma: Jurnal Ilmu Kesehatan Masyarakat*. 17 (1) : 14-21.
[18] Peraturan Menteri Tenaga Kerja dan Transmigrasi Nomor PER.13/MEN/X/2011 tentang Nilai Ambang Batas Faktor Fisika dan Kimia di Tempat Kerja.
[19] Hendrawan, A. dan A. K. Hendrawan. 2020. Analisa Kebisingan di Bengkel Kerja Akademi Maritim Nusantara. *Jurnal Saintara*. 5 (1) : 1-5.
[20] Aini, A. N., I. F. Anwar, A. M. Sufanir, dan Y. Astor. 2018. Survei dan Pemetaan Zona Kebisingan Arus Lalu Lintas pada Kawasan RSUP Dr Hasan Sadikin Bandung. *Potensi: Jurnal Sipil Politeknik*. 20(1): 1-5.
[21] Neitzel, R. L., and B. J. Fligor. 2019. Risk of Noise-Induced Hearing Loss Due To Recreational Sound: Review and Recommendations. *The Journal of the Acoustical Society of America*. 149 (5):3911–3921
[22] Peraturan Menteri Ketenagakerjaan Republik Indonesia Nomor 5 Tahun 2018 tentang Keselamatan dan Kesehatan Kerja Lingkungan Kerja
[23] Isliko, V., N. Budhiarti, dan E. Adriantantri. 2022. Analisa Kebisingan Peralatan Pabrik dalam Upaya Meningkatkan Kesehatan Keselamatan Kerja dan Meningkatkan Kinerja Karyawan di PT. Wangi Indah Natural. *Jurnal Valtech (Jurnal Mahasiswa Teknik Industri)*. 5(1): 101-106
[24] Peraturan Menteri Tenaga Kerja dan Transmigrasi Nomor PER. 13/MEN/X/2011 tahun 2011 tentang Nilai Ambang Batas Faktor Fisika dan Kimia di Tempat Kerja
[25] Lestari, I, dan Harmon. 2018. Pengaruh Lingkungan Kerja terhadap Kinerja Karyawan. *Jurnal Riset Bisnis & Investasi*. 3(2): 94-103
[26] Panda, K. C. 2016. Dealing with Noise and Vibration in Automotive Industry. *Procedia Engineering*. 144 : 1167-1174.
[27] Pamingotan, S dan T. Mulyanto. Analisis dan Evaluasi Pengaruh Temperatur, Pencahayaan, dan Kebisingan terhadap Jumlah Produksi pada Fabrikasi Ekstruksi Alumunium dengan Menggunakan Metode Sugeno dalam Logika *Fuzzy*. *Jurnal Teknologi Manufaktur*. 12 (1) : 44-54.
[28] Essers, E., L. P. Crespo, M. Foraster, A. Ambros, H. Tiemeier, and M. Guxens. 2022. Environmental Noise Exposure and Emotional, Aggressive, and Attention-Deficit/Hyperactivity Disorder-Related Symptoms in Children from Two European Birth Cohorts. *Environmental International*. 158 : 1-8.

[29] Deshmukh, S., S. Bhattacharya, A. Jain, and A. R. Paul. 2019. Wind Turbine Noise and its Mitigation Techniques: A Review. *Energy Procedia.* 160 : 633-640.

[30] Tumavice, A., A. Laurinavicius, A. Jagniatinskis, and A. Vaitkus. 2016. Environmental Noise Mitigation Measures for Lithuanian Railway Network. *Transportation Research Procedia.* 14 : 2704-2713.

[31] Xaw, A. K., A. M. Myat, M. Thandar, Y. M. Htun, T. H. Aung, K. M. Tun, and Z. M. Han. 2020. Assesment of Noise Exposure and Hearing Loss Among Workers in Textile Mill (Thamine), Myanmar : A Cross-Sectional Study. *Safety and Health at Work.* 11 : 199-206.

[32] Septiana, N. R. and E. Widowati. 2017. Gangguan Pendengaran Akibat Bising. *HIGEIA.* 1(1): 73-82.

[33] Tan. Y., Y. Fang, T. Zhou, V. J. L. Gan, and J. C. P. Cheng. 2019. BIM-Supported 4D Acoustics Simulation Approach to Mitigating Noise Impact on Maintenance Workers on Offshore Oil and Gas Platforms. *Automation in Construction.* 100: 1-10.

[34] Septio, Y. R., Suhardi, B., Astuti, R. D., and Adiasa, I. (2020). Analisis Tingkat Kebisingan, Beban Kerja dan Kelelahan Kerja Bagian Weaving di PT. Wonorejo Makmur Abadi Sebagai Dasar Untuk Perbaikan Proses Produksi. *Performa: Media Ilmiah Teknik Industri.* 19(1) : 19-26.

[35] Gong, W., Zhao, L., Li, L., Morata, T. C., Qiu, W., Feng, H. A., & Zhu, B. 2021. Evaluating the effectiveness of earplugs in preventing noise-induced hearing loss in an auto parts factory in China. International Journal of Environmental Research and Public Health, 18 (13) : 7190.

[36] Akbari, J., Safari, S., Kazemi, M., Nourmoradi, H., Mahaki, B., & Dehghan, H. 2018. The Impact of Hearing Protection Devices (HPDs) on Blood Pressure in Workers Exposed to Noise: A Cross-sectional Study in a Textile Industry. Iranian Journal of Health, Safety and Environment, 5 (3) : 1066-1071.

[37] Biabani, A., M. Aliabadi, R. Golmohammadi, and M. Farhadian. 2017. Individual Fit Testing of Hearing Protection Devices Based on Microphone in Real Ear. *Safety and Health at Work.* 8: 364-370.

The Bias in SPARCLE Measurements

Moch Syarif Romadhon[1*]

[1]Pusat Riset Iklim dan Atmosfer, Badan Riset dan Inovasi Nasional, Jl. Dr. Djundjunan 133 Bandung 40173, Indonesia.

moch.syarif.romadhon@brin.go.id

Abstract. SPARCLE is an optical particle counter (OPC) designed to measure particle size and refractive index in the size range between 500 nm and 2500 nm. The knowledge of the parameters is essential for understanding the aerosol impact on the Earth's radiative forcing. There are other instruments which try to measure the parameters simultaneously, however, those instruments need either ad hoc scaling factors or further manipulation to their outputs to retrieve the parameters. In contrast to those instruments, no manipulation is needed in retrieving the parameters from SPARCLE outputs. SPARCLE measurements suffer a bias due to various factors. In this paper, those factors are evaluated and the impact of the bias on simulated SPARCLE measurements is demonstrated. The results indicate that the bias introduces additional structures to measured size distribution.

1. Introduction

A report by the Intergovernmental Panel on Climate Change indicates a large uncertainty in determining the impact of atmospheric aerosols on the Earth's climate[1,2,3]. One contribution to the uncertainty is the lack of aerosol particle size and refractive index *in situ* measurements essential for understanding the impact of atmospheric aerosol on the Earth's radiative forcing. Most *in situ* aerosol measurements use OPCs that assume particle refractive index to estimate particle size[4]. There are at least two OPC designs that attempt to measure the two parameters simultaneously however, they require ad hoc scaling factors or further data manipulation to retrieve the size and refractive index of particles[4][5]. SPARCLE is an OPC that measures the parameters with no ad hoc scaling factors or data manipulation needed[6]. However, the bias in SPARCLE measurement has not been studied yet. The knowledge of the bias is essential to recover the actual distributions of the parameters using parameter distributions measured by SPARCLE.

2. Sparcle Design

The basic principles of SPARCLE design are to sample particles, illuminate them and measure the light they scatter. Particles are sampled into SPARCLE sensing volume and illuminated by a compressed laser beam. Two detectors are used in the design: one detector to detect the presence of particles in the sensing volume and one detector to measure the pattern of light scattered by the particles. A Photomultiplier Tube (PMT) was selected as the first detector due to its extreme sensitivity and fast response time. A Charge-coupled Device (CCD) camera was selected as the second detector due to its high sensitivity and large pixel number. Particle size and refractive index are retrieved from the outputs of the detectors. The design of SPARCLE is shown in Figure 1. As shown in the figure, the laser beam is designed to be as thin as possible with the width covering all possible paths of particles in the sensing volume. A thin beam is preferred to reduce the possibility of two particles in the sensing volume at the same time. To achieve this objective, the beam is manipulated using a slit and a cylinder lens. A laser block is installed to avoid the reflection of the beam back to particles.

The intensity of light elastically scattered by spherical particles illuminated by linearly polarized light can be calculated using Mie Theory[7]. The PMT measures light scattered over a 4.8 sr solid angle defined as the angle subtending the PMT window area from particle position. The PMT circular window is on the x-z plane with its midpoint on the x-axis. The CCD camera measures angularly resolved light over scattering angles between 40° and 144° with scattering angles defined as the angle between scattering light and the propagation direction of the laser beam. The laser beam is propagated along the z-axis.

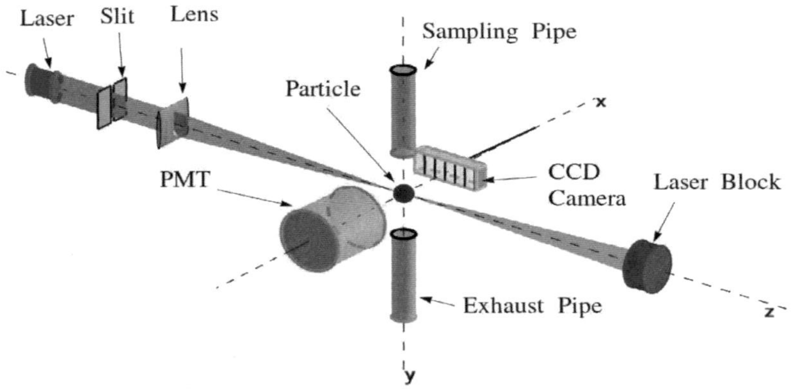

Figure 1. SPARCLE design with each component indicated.

The two detectors in SPARCLE are coupled: once the PMT detects a particle in the sensing volume then the CCD camera will be triggered to measured scattering light. The detection of particles is based on the depth of PMT pulse: if the pulse is lower than 0.4 V then it is assumed that there is a particle in the sensing volume. The examples of SPARCLE outputs are shown in Figure 2. The plot on the left is the outputs of the PMT while that on the right is the outputs of the CCD camera. In the figure, the red lines are measured outputs while the blue lines are calculated outputs for the light scattered by particles with the size of 1809 nm and the refractive index of 1.59 illuminated at the position x=-0.22 mm and z=-0.43 mm. The blue lines fit with the red lines indicating that the measured outputs are the light scattered by a particle whose parameters are the same as that used for the calculated outputs.

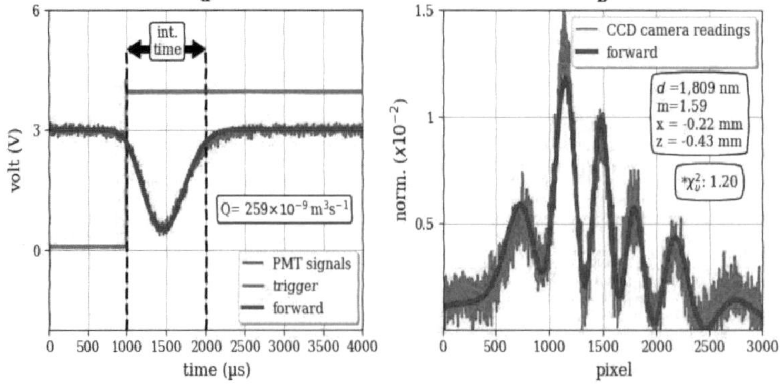

Figure 2. The outputs of SPARCLE measurements. In the left is the ouputs of the PMT while in the right is the outputs of the CCD Camera.

3. Method And Results

The bias in measurements using SPARCLE was calculated numerically based on physics involved in the measurements. In general, there are two types of bias in SPARCLE measurements: the bias due to the efficiency of SPARCLE to sample and detect particles and the bias due to the selection of SPARCLE outputs qualified for particle size and refractive index retrievals. The efficiency of SPARCLE is determined by two factors: particle loss when sampling and transporting particles to SPARCLE sensing volume and the ability to detect them based on the intensity of light they scatter. In this study, the particle loss was calculated using Particle Loss Calculator[8]. In detecting particles, the intensity of scattering light is calculated using Mie Theory. The theory describes that the intensity is determined by particle size and refractive index, and the intensity of light illuminating particles. Since the beam is spatially distributed, particle position determines the intensity of light illuminating the particle. The probability of particles being at a particular position when they are illuminated in the sensing volume is a function of SPARCLE sampling rates. The probability is equal to the ratio between the flow rate at the position and the sampling rate. Furthermore, a background was detected in the sensing volume hence, a threshold was applied to distinguish scattering signals from the background. The threshold can make SPARCLE omits light scattered by small particles.

Some qualifications are applied to select SPARCLE outputs from which particle size and refractive index are retrieved. The outputs should have Signal-to-Noise-Ratio (SNR) bigger than 4.2 so that the outputs contain sufficient scattering information. Also, saturated PMT outputs are omitted. The saturation is due to the limited range of the digital oscilloscope used in the design. The range is between -5 V to 5 V.

The probability of SPARCLE to detect particles and to have corresponding outputs qualified for the retrieval was calculated. To illustrate, the probability is shown in Figure 3 as two-dimensional map of the probability as a function of particle size and refractive index. As shown in the figure, the probability is a complex function of particle size and refractive index. The effect of the threshold can be seen in the low sensitivity of SPARCLE to detect particle sizes smaller than 600 nm with various refractive indices. Meanwhile, SPARCLE has a high probabilities detecting a particle size of 2100 nm with refractive indices from 1.30 to around 1.50. For detecting particles bigger than 6000 nm, the effect of the saturated PMT outputs is observed as indicated by the low probability to detect the particles with refractive indices higher than 1.45 as indicated by the blue colour on the right side of the figure.

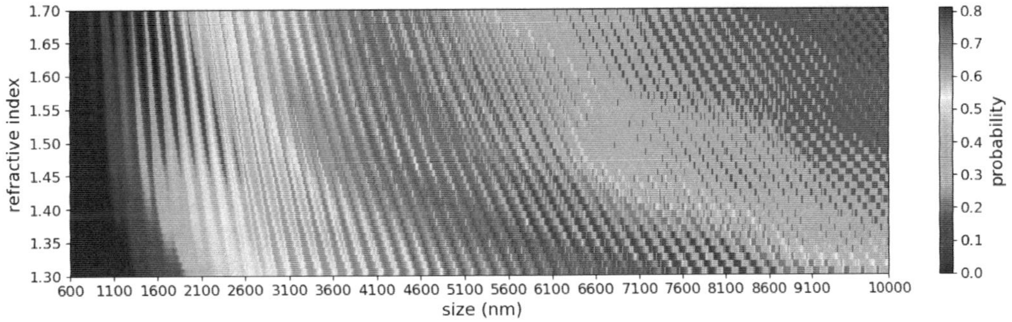

Figure 3. The probability of SPARCLE to detect particle and to have the corresponding outputs qualified for particle size and refractive index retrieval.

The effect of the bias is illustrated by the simulation of SPARCLE measurements of an aerosol with its particle size and refractive index normally distributed. The normally distributed particle size and refractive index are shown in Figure 4. The number of particles is 10,000 particles. The particle size distribution peaks at 2000 nm while the refractive index peaks at 1.50. Due to the bias in SPARCLE measurements, different distributions are expected when measuring the aerosol. The expected distributions are shown in Figure 5. As shown in the figure, the expected particle size distribution has additional high-frequency structures in addition to the low frequency of normal distribution. Meanwhile, the expected distribution of the refractive index is slightly different from the normal distribution. The number of particles to be detected by SPARCLE with the corresponding outputs qualified for the retrievals is 3,253 particles. So, the efficiency of SPARCLE in this particular simulation is 32.53%.

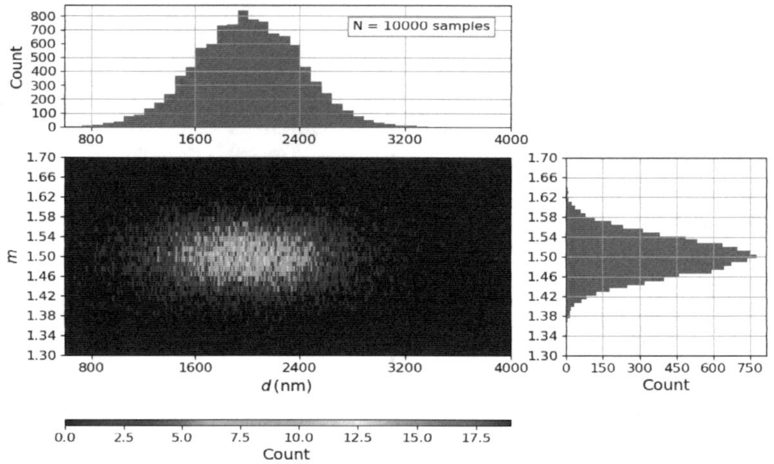

Figure 4. An illustration of an aerosol with its partice size and refractive index normally distributed

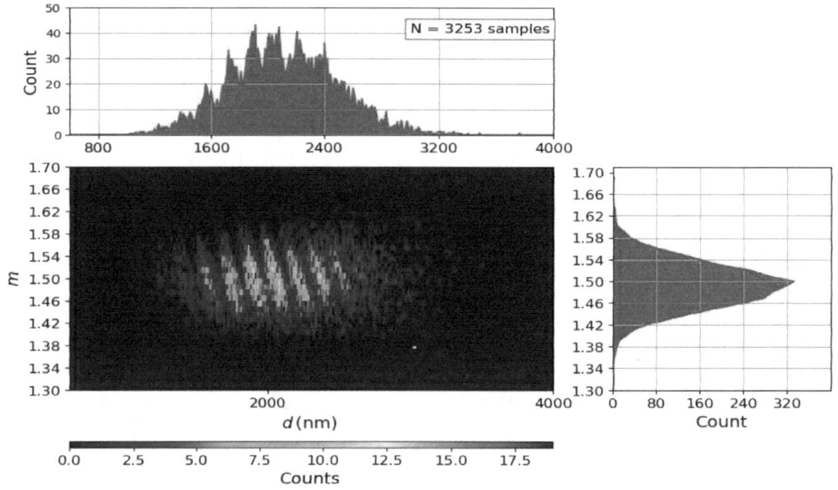

Figure 5: The distribution of particle size and refractive index measured by SPARCLE for an aerosol with its particle size and refractive index normally distributed as shown in Figure 4.

4. Conclusion

The bias in SPARCLE measurements was studied and its effect on SPARCLE measurements was simulated. The probability of SPARCLE to detect particles with the corresponding outputs qualified for particle size and refractive index retrieval was calculated. The results indicate a complex relationship between the probability and particle size and refractive index. The smallest particle SPARCLE can detect for various particle refractive indices is 600 nm. This limit is due to the threshold applied to differentiate between scattering signals and background. The probability of SPARCLE in detecting particles as big as 10,000 with qualified output is low due to the limited range of the digital oscilloscope to measure PMT outputs. The simulation of SPARCLE measuring an aerosol with its particle size and refractive index normally distributed was done. The results indicate that the bias introduces additional structures in the expected size distribution. While the expected refractive index distribution is expected slightly different from a normal distribution.

Acknowledgments
I like to thank my supervisor Don Grainger from University of Oxford and Dan Peter from Rutherford-Appleton Laboratory for enriching discussion and support. I like to thank as well to Lembaga Pengelola Dana Pendidikan, Kementerian Keuangan, Indonesia for the financial support during my study. I like to thank Badan Riset dan Inovasi Nasional as my work institution. Lastly, I would like to thanks my dear brave friends: Bidhari, Eni, Rika, Rusdi, Rusti, Dewa and Heru for a remarkable support.

References
[1] IPCC, CLIMATE CHANGE 2014: Synthesis Report. Technical report, Geneva, 2014. URL http://ipcc.ch.
[2] Valenzuela, A., Reid, J. P., Bzdek, B. R., and Orr-Ewing, A.J. 2018: Accuracy Required in Measurements of Refractive index and Hygroscopic Response to Reduce Uncertainties in Estimates of Aerosol Radiative Forcing Efficiency, J. Geophys. Res. Atmos., 123, 6469-6486.
[3] Myhre, G., et. al. 2013: Radiative forcing of the direct aerosol effect from AeroCom Phase II simulations, Atmos. Chem. Phys., 13, 1853-1877.
[4] A.Nagy, W.W. Szymanski, P. Gal, A. Golczewski, and A. Czitrovszky. Numerical and experimental study of the performance of the dual wavelength optical particle spectrometer (DWOPS). *J. Aerosol Sci.,* 38(4):467-478, apr 2007.
[5] K.B. Aptowicz, R. G Pinnick, S. C. Hill, Y. L. Pan, and R. K. Chang. Optical scattering patterns from single urban aerosol particles at Adelphi, Maryland, USA: A classification relating to particle morphologies. *J. Geophys. Res. Atmos*., 111(12), 2006.
[6] Moch Syarif Romadhon, SPARCLE: an instrument for aerosol size and refractive index measurements. PhD Thesis. *Oxford University Research Archive* (ORA) 2021: https://ora.ox.ac.uk/objects/uuid:3f586dca-7d46-42e9-a468-d8d26f894ce2.
[7] C. F. Bohren and D. R. Huffman. *Absorption and scattering of light by small particles*, volume 1. John Wiley & Sons Inc., 1983. ISBN 9783527618156
[8] S. L Von Der Weiden, F. Drewnick, and S. Borrmann. *Atmospheric Measurement Techniques*, Particle Loss Calculator-a new software tool for the assessment of the performance of aerosol inlet systems. 2009. www.atmos-meas-tech.net/2/479/2009/.

Adsorption of Oil and Grease in Wastewater using Activated Carbon Derived from Sewage Sludge

Eka Masrifatus Anifah[1*], Ismi Khairunnissa Ariani[1], Rina Noor Hayati[1], Satria Arya Nugraha[1]

[1]Environmental Engineering, Institut Teknologi Kalimantan, Balikpapan 76127, East Kalimantan, Indonesia

ekamasrifatus@lecturer.itk.ac.id

Abstract. Oil and grease are organic contaminants that are detrimental to the environment. The high concentration of oil and grease can cause clogging in the sewer. Clogged sewers can deteriorate sanitary conditions. Adsorption is one technique to reduce oil and grease concentration in wastewater. Adsorption using natural resources can overcome the high cost of commercial activated carbon. Sewage sludge has potential as raw material for activated carbon because of the high carbon content. The studies aim to investigate the capability of activated carbon derived from sewage sludge to remove oil and grease from restaurant wastewater. The studies were carried out in the batch experiment. Adsorbent dosage ranged from 2-6 g/L, and a contact time range of 5-45 minutes was performed in the experiment. Sewage sludge was chemically activated using KOH 4M and physically activated at a temperature of 700°C for 60 minutes. The result shows that the optimum adsorbent dosage and contact time are 4 g/L and 25 minutes, respectively. The highest adsorption capacity of sewage sludge to remove oil and grease was 63.75 mg/g. Under observed data, the adsorption fit the second-pseudo-order, and Langmuir isotherm indicated that the adsorption process was chemisorption.

1. Introduction

Oil and grease are organic pollutants that contain long fatty acids. The primary sources of oil and grease contamination are the hotel, restaurant, kitchen, food processing, and slaughtering industry. Many industries discharge high oil and grease wastewater into the sewer without proper treatment. Oil and grease contamination is a severe problem for sewer and water treatment. Deposit of oil and grease can cause clogging and sanitary sewer overflow [1]. A high concentration of oil and grease causes deterioration of the performance of the sewage treatment system. Oil and grease disrupt not only the process of aeration [2], clarification [3], and dewatering [4] but also increase aeration demand [5]. Oil and grease produce layers on the water surface that lowering dissolved oxygen in water [6] and obstruct light penetration for photosynthesis [7]. The low dissolved oxygen level destroys the aquatic ecosystem [8].

Oil and grease wastewater treatment has been accomplished using various physical, chemical, and biological processes. The most common and low-cost technique is the grease trap. However, the grease trap only removes free oil, not emulsified oil, so the quality of effluent still exceeds quality standards [9]. The past studies investigated hybrid treatments that combine chemical and physical processes. The results show that combining dissolved air flotation with chemical treatments such as coagulation [10] or electrocoagulation [11] could remove oil and grease. Although chemical treatments are highly effective in decreasing oil and grease concentration, the treatments require a high cost and generate chemical sludge. Biological treatments are found that can degrade oil and grease in wastewater. Aerobic biodegradation can remove oil and grease with an efficiency of 54.6 to 80%.[12]. Lipase-producing bacteria can decompose oil and grease with a removal of 59% [13]. Biological methods are environmental-friendly, but the treatment needs constant nutrient requirements and a longer time to process.

Adsorption is the physicochemical process using adsorbents to remove contaminants in oily wastewater. Adsorption offers the benefits of a rapid kinetic process and high removal efficiency. [14]. Activated carbon is produced from materials that undergo thermal carbonization. Biomass wastes have been developed as adsorbent material because it is affordable, sustainable, and renewable. Natural adsorbents had high removal efficiency for oil and grease, such as sugarcane bagasse of 63,4% [15], the banana pith of 97%[16], corn cobs of 60.38% [17], and palm kernel shells of 99.89% [18]. Moringa oleifera reached the oil and grease adsorption capacity of 294.41 mg/g [19]. The result indicated that natural resources are promising alternative adsorbents.

The wastewater treatment plants produce sewage sludge as the by-product of the process. The pyrolyzed sludge contains highly porous and carbonaceous material that can be used as an adsorbent. Numerous studies reported that chemical activation using salts, acid, and base enhanced the adsorption capacity. Potassium hydroxide (KOH) improves the surface area of sludge significantly ten times higher than without activation [20]. The activated carbon from sewage sludge demonstrated a high capacity in removing color [21], heavy metal [22], ammonium, and phosphate [23]. The literature has not reported the study of sewage sludge for oil and grease adsorption. This research aims to investigate the adsorption capability of activated carbon from sewage sludge to remove oil and grease in wastewater.

2. Method

2.1. Preparation of activated carbon
Sludge obtained from a domestic wastewater treatment plant in Balikpapan was used as the raw material for activated carbon. Initially, the sludge was dried at a temperature of 110°C for 24 hours. The oven-dried sludge was crushed and sieved into 0.4-0.6 mm particle size. The dried sludge was soaked in KOH 4M with a ratio of 1:1.5 (w/w) for 16 hours for chemical activation. The impregnated sludge was dried at a temperature of 110°C for 24 hours until a constant mass was obtained. Finally, the sludge was heated at a temperature of 700°C for 60 minutes in a muffle furnace.

2.2. Wastewater characterization
A wastewater sample was collected from the kitchen and restaurant of one hotel in Balikpapan. The wastewater was characterized by several parameters, including oil and grease concentration, pH, and temperature. The gravimetric analysis using n-hexane as solvent extractor was used to determine the oil and grease concentration [24].

2.3. Optimization of pH
The optimum pH for oil and grease adsorption was determined to optimize the oil and grease removal process. The influence of pH on oil and grease adsorption was conducted in the pH range of 3-8. Sodium hydroxide 0.1 M and hydrochloride acid 0.1 M were utilized to adjust the pH of the solution. Dosage of 4 g/L adsorbents was added to 100 mL of wastewater, pH of the solution was adjusted and agitated at 90 rpm using a stirrer a room temperature for 25 minutes. The supernatant was filtered using filter papers and was measured for oil and grease concentration.

2.4. Adsorption Analyses
The adsorption studies were performed in a batch experiment. The batch adsorption was carried out on 100 mL with a varied dosage of adsorbent and contact time. The batch experiment was conducted at

room temperature and a constant mixing speed of 110 rpm. The filtrate was filtered using Whatman filter paper 42 and was measured for oil and grease concentration. The removal percentage and capacity of the adsorbent were calculated using the equation as follows [19],

$$q_t = \frac{(C_0 - C_t)V_L}{m_{ads}} \quad (1)$$

$$Oil\ Removal = \frac{(C_0 - C_t)}{C_0} \times 100\% \quad (2)$$

C_0 and C_t represent the concentration of oil and grease at the beginning and the time (mg/L), V_L indicates the quantity of solution (L), and m_{ads} describes the mass of adsorbent.

The adsorption process at the equilibrium was evaluated using the Langmuir and Freundlich isotherms. The non-linear Langmuir and Freundlich were determined by equations (3) and (4), respectively [19]. The adsorbate-adsorbent interaction constant in the Langmuir isotherm was calculated by equation (5).

$$q_e = K_F C_e^{1/n} \quad (3)$$

$$q_e = \frac{q_{max} K_L C_e}{1 + K_L C_e} \quad (4)$$

$$R_L = \frac{1}{1 + K_L C_0} \quad (5)$$

where q_e and q_{max} (mg/g) represent the adsorption capacity, oil and grease at equilibrium, and the maximum oil and grease adsorption capacity. C_e (mg/L) corresponds to the equilibrium residual oil and grease concentration, and n implies the degree of nonlinearity among solution concentration and adsorption. K_L (L/mg) and K_F (L/g) indicate the rate constants corresponding to the isotherm model.

The pseudo-first-order and pseudo-second-order were used to explain the kinetic adsorption. The non-linear kinetic models were defined by the equation of (6), (7), and (8) [25].

$$\frac{dq_t}{d_t} = k_1 (q_e - q_t) \quad (6)$$

$$\frac{dq_t}{d_t} = k_2 (q_e - q_t)^2 \quad (7)$$

where q_t is the amount of adsorbed adsorbate at t time (mg/g), the adsorption capacity at equilibrium (mg/g), k_1 and k_2 are constant kinetic for pseudo-first-order and pseudo-second-order, respectively. The initial adsorption rate (h) of the pseudo-second-order model (mg/g/h) at t = 0 can be described as:

$$h = k_2 q_e^2 \quad (8)$$

2.5. SEM Analysis

The surface morphology observation of the adsorbent was analyzed by Scanning electron microscopy (Thermofisher Scientific, Phenom ProX). The magnification of 4000 times and voltage of 15kV were used in the analysis. The SEM analysis was applied to the adsorbent before and after the adsorption process.

3. Result and Discussion

3.1. Wastewater Characteristics

Table 1 shows the kitchen and restaurant wastewater characterization in this study. The oil and grease concentration in wastewater was 300 mg/L, relatively high and exceeding the Indonesia quality standard of restaurant wastewater. The wastewater had a slightly acidic pH of 5.75 because of contained the long chain fatty acid [26].

Table 1 Wastewater characteristics

Parameter	Concentration	Unit
Oil and Grease	330	mg/L
pH	5.29	-

3.2. Effect of pH on oil and grease adsorption

The pH is an essential aspect that affects the adsorption process. The initial pH significantly influenced the degree of ionization, adsorbent surface charge, and adsorbate solubility [27]. After equilibrium, pH could neither rise nor drop. Figure 1 presents the influence of pH on oil and grease adsorption. The graph shows that increasing pH could increase oil and grease adsorption capacity. A similar result also reported that increased solution pH promoted the oil adsorption on sludge biochar [28]. The lower oil concentration after adsorption at higher pH was related to the side reaction of excessive NaOH used for pH adjustment. Sodium hydroxides react with oil to produce water-soluble soap [29]. The initial pH of 6 resulted in an adsorption capacity of more than 47.5 mg/g. The initial pH of 7 obtained the maximum adsorption capacity, but the final pH increased significantly to 8.5. Thus, the adsorption experiment was conducted at the initial pH of 6.

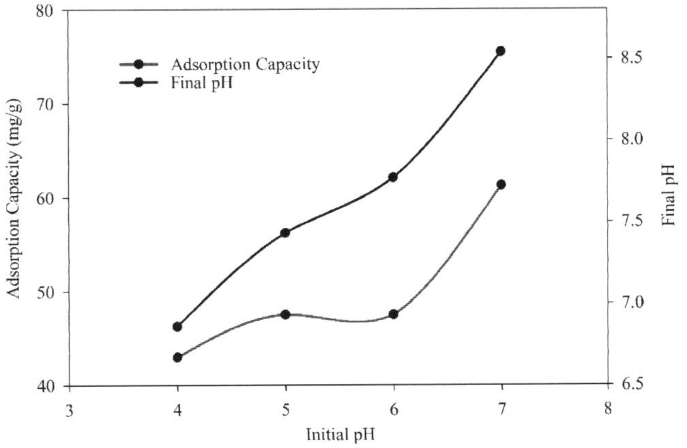

Figure 1. Effect of pH on oil and grease adsorption

3.3. Effect of adsorbent dosage on oil and grease adsorption

Figure 2 illustrates the influence of adsorbent dosage on oil and grease adsorption. The adsorbent dosage of 2-4 g/L had a removal efficiency of 20-80%. The optimum dosage for oil and grease removal was 4 g/L. The higher dosage resulted in higher removal efficiency. The initial increase at a higher dosage provides sufficient surface area and an active binding site [30–32]. The adequate surface area allows adsorbate to penetrate the adsorbent easily [33]. However, the higher dosage decreased the oil and grease adsorption after equilibrium. Adsorption capacity decreased because of the increased saturated oil active site and the saturation effect that occurs when oil removal efficiency reaches maximum capacity [34,35].

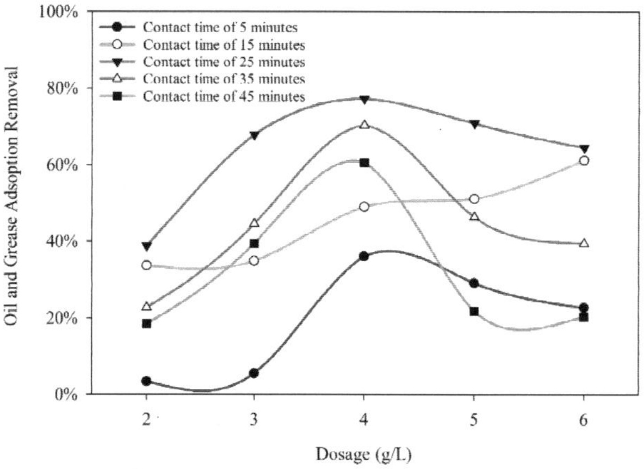

Figure 2. Effect of adsorbent on oil and grease adsorption

3.4. Effect of Contact Time on Oil and Grease Adsorption

Figure 3 shows the influence of contact time on oil and grease adsorption removal. In the beginning, oil and grease removal increased due to a longer contact time. The fast adsorption in the initial phase because of the enormous vacant active site results in easy access to the adsorbent site[36,37]. The adsorption process reached equilibrium at a contact time of 25 minutes for all adsorbent dosages. After equilibrium, the oil and grease removal experienced a decline. Previous research has achieved a similar outcome: desorption occurred after the rapid adsorption phase[38]. The change of pH in the solution reverses the process of desorption [39]. This phenomenon occurred because of saturated adsorbent, which adsorption occurred on the inner surface instead of the outer adsorbent surface [40].

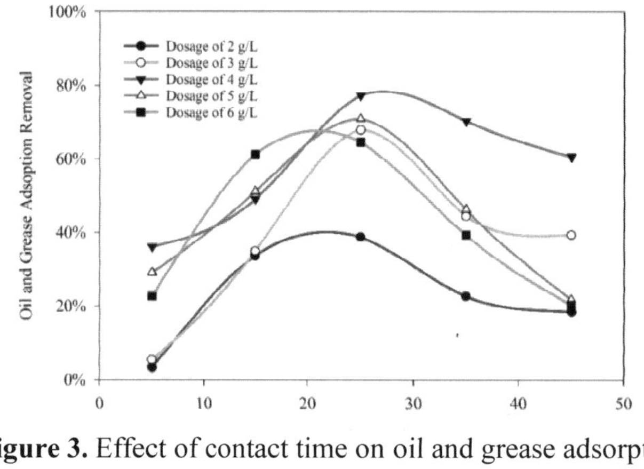

Figure 3. Effect of contact time on oil and grease adsorption

3.5. Adsorption Isotherm and Kinetic

The adsorption isotherm interprets the correlation between adsorbates that were adsorbed by an adsorbent at equilibrium. Isotherm was modeled based on oil and grease adsorption data. Table 2 describes the constant obtained from the Freundlich and Langmuir model. The Langmuir model has a higher value of R^2 than Freundlich, so the Langmuir isotherm was more fit for explaining oil and adsorption on the activated carbon derived from sewage sludge. Langmuir isotherm indicates that the adsorption process forms a monolayer on a similar activation energy site [41]. Langmuir model also implies adsorption that occurs was chemisorption. The Langmuir isotherm had RL value of 0.1, indicating that adsorption is favorable for adsorbate and adsorbent. The result of adsorption kinetic was shown in table 3. The adsorption kinetics shows that the second-pseudo-order was best fitted for oil and grease adsorption. Furthermore, the fitted of second-pseudo kinetic supported adsorption was chemisorption [42].

Table 2 Adsorption isotherm

Freundlich	
R^2	0.0023
K_F (mg$^{1-(1/n)}$ g^{-1} L$^{1/n}$	46.2381
n	25.5754
Langmuir	
R^2	0.5104
q_{max} (mg/g)	70.4225
K_L (mg/L)	0.0271
R_L	0.1005

Table 3 Adsorption Kinetics

First-pseudo-Order	
R^2	0.1322
q_e	20.2862
k_1	-0.0322
Second-pseudo-Order	
R^2	0.9334
h (mg/(g.minute)	15.9744
q_e (mg/g)	58.8235
k_2 (10^{-3} g/mg.minute)	4.6166

3.6. SEM Analysis

Figure 1 (a) illustrates the SEM figure on the surface of sewage sludge-activated carbon before and after adsorption. The SEM image showed that the surface of sewage sludge has waved and irregular structures. The previous studies also reported that the morphology of activated carbon from sewage sludge had these structures [43]. Figure 1 (b) displays the activated carbon of sewage sludge after oil and grease were adsorbed. Figure 1(b) shows that the surface of the sludge was covered by oil after adsorption. The circle mark indicates that the pore structures were connected after the adsorption.

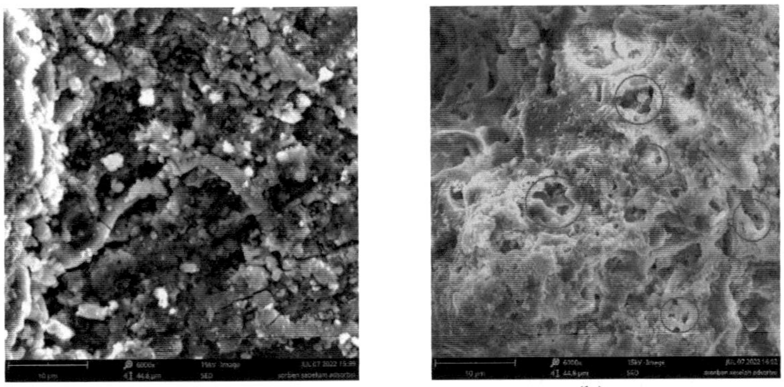

(a) (b)

Figure 4 Scanning Electron Microscopy image of activated carbon obtained from sewage sludge (a) before adsorption (b) after adsorption, at a magnification of 4000 times

4. Conclusion

The study found that the sewage sludge-activated carbon can remove oil and grease in kitchen or restaurant wastewater. The activated carbon derived from sewage sludge has optimum removal efficiency of 77% at a dosage of 4 g/L and a contact time of 25 minutes.

Acknowledgment

The author expresses gratitude to Research Center and Community Development (LPPM), Institut Teknologi Kalimantan, for funding this research, and Lusi Ernawati for assisting the SEM results.

References

[1] He X, de los Reyes F L, Leming M L, Dean L O, Lappi S E and Ducoste J J 2013 Mechanisms of Fat, Oil and Grease (FOG) deposit formation in sewer lines *Water Research* **47**

[2] Brooksbank A M, Latchford J W and Mudge S M 2007 Degradation and modification of fats, oils and grease by commercial microbial supplements *World Journal of Microbiology and Biotechnology* **23**

[3] Wallace T, Gibbons D, O'Dwyer M and Curran T P 2017 International evolution of fat, oil and grease (FOG) waste management – A review *Journal of Environmental Management* **187**

[4] Wang Y, Feng S, Bai X, Zhao J and Xia S 2016 Scum sludge as a potential feedstock for biodiesel production from wastewater treatment plants *Waste Management* **47**

[5] Collin T D, Cunningham R, Asghar M Q, Villa R, MacAdam J and Jefferson B 2020 Assessing the potential of enhanced primary clarification to manage fats, oils and grease (FOG) at wastewater treatment works *Science of the Total Environment* **728**

[6] Abd El-Gawad H S 2014 Oil and Grease Removal from Industrial Wastewater Using New Utility Approach *Advances in Environmental Chemistry* **2014**

[7] Sanghamitra P, Mazumder D and Mukherjee S 2021 Treatment of wastewater containing oil and grease by biological method- a review *Journal of Environmental Science and Health - Part A Toxic/Hazardous Substances and Environmental Engineering* **56**

[8] Khwakaram A I 2016 Effects of Fat, Oil and Grease (FOG) Discharge Pollutants on Water Quality of Qalyasan Stream, Tanjero River and Impact of Fat, Oil and Grease on Darbandikhan Reservoir in Sulaimani City-Kurdistan Region of Iraq-Iraq *International journal of environment, ecology, family urban studies* **6** 1–7

[9] Yau Y H, Rudolph V, Lo C C man and Wu K C 2021 Restaurant oil and grease management in Hong Kong *Environmental Science and Pollution Research* **28**

[10] Zheng T, Wang Q, Shi Z, Huang P, Li J, Zhang J and Wang J 2015 Separation of Pollutants from Oil-Containing Restaurant Wastewater by Novel Microbubble Air Flotation and Traditional Dissolved Air Flotation *Separation Science and Technology (Philadelphia)* **50**

[11] Ji M, Jiang X and Wang F 2015 A mechanistic approach and response surface optimization of the removal of oil and grease from restaurant wastewater by electrocoagulation and electroflotation *Desalination and Water Treatment* **55**

[12] Sanghamitra P, Mazumder D and Mukherjee S 2020 A study on aerobic biodegradation of oil and grease containing wastewater *Journal of the Indian Chemical Society* **97**

[13] Haselroth K J, Wilke P, Dalla Costa I M, Lustoza Rotta V M, Rosado A F and Hermes E 2021 Effectiveness of Aeromonas hydrophila for the removal of oil and grease from cattle slaughterhouse effluent *Journal of Cleaner Production* **287**

[14] Dotto G L and McKay G 2020 Current scenario and challenges in adsorption for water treatment *Journal of Environmental Chemical Engineering* **8**

[15] Boni H T, de Oliveira D, Ulson de Souza A A and Ulson de Souza S M A G 2016 Bioadsorption by sugarcane bagasse for the reduction in oil and grease content in aqueous effluent *International Journal of Environmental Science and Technology* **13**

[16] Abdul Hamid N S, Che Malek N A, Mokhtar H, Mazlan W S and Mohd Tajuddin R 2016 Removal of oil and grease from wastewater using natural adsorbents *Jurnal Teknologi* **78**

[17] Igwegbe C A, Umembamalu C J, Osuagwu E U, Oba S N and Emembolu L N 2020 Studies on Adsorption Characteristics of Corn Cobs Activated Carbon for the Removal of Oil and Grease from Oil Refinery Desalter Effluent in a Downflow Fixed Bed Adsorption Equipment *European Journal of Sustainable Development Research* **5**

[18] Ramli A N and Ghazi R M 2020 Removal of Oil and Grease in Wastewater using Palm Kernel Shell Activated Carbon *IOP Conference Series: Earth and Environmental Science* vol 549

[19] dos Santos Bispo M, dos Santos J P L, dos Santos L C L, dos Santos Freitas L, Bispo D F and da Silva G F 2021 Synthesis and characterization of activated carbon of Moringa oleifera Lam pod and evaluation of its performance in the removal of oils and greases *Journal of Environmental Chemical Engineering* **9**

[20] Zhang J, Shao J, Jin Q, Li Z, Zhang X, Chen Y, Zhang S and Chen H 2019 Sludge-based biochar activation to enhance Pb(II) adsorption *Fuel* **252**

[21] Kutty S R M, Almahbashi N M Y, Nazrin A A M, Malek M A, Noor A, Baloo L and Ghaleb A A S 2019 Adsorption kinetics of colour removal from palm oil mill effluent using wastewater sludge carbon in column studies *Heliyon* **5**
[22] Xue Y, Wang C, Hu Z, Zhou Y, Xiao Y and Wang T 2019 Pyrolysis of sewage sludge by electromagnetic induction: Biochar properties and application in adsorption removal of Pb(II), Cd(II) from aqueous solution *Waste Management* **89**
[23] Kończak M and Huber M 2022 Application of the engineered sewage sludge-derived biochar to minimize water eutrophication by removal of ammonium and phosphate ions from water *Journal of Cleaner Production* **331**
[24] Clesceri L S, Greenberg A E, Eaton A D, Franson M A H, Association A P H, Association A W W and Federation W E 1998 *Standard Methods for the Examination of Water and Wastewater* (American Public Health Association)
[25] Parus A 2018 Copper(II) ions' removal from aqueous solution using green horse-chestnut shell as a low-cost adsorbent *Chemistry and Ecology* **34**
[26] Klaucans E and Sams K 2018 Problems with fat, oil, and grease (FOG) in food industry wastewaters and recovered FOG recycling methods using anaerobic co-digestion: A short review *Key Engineering Materials* vol 762
[27] Akpomie K G and Conradie J 2020 Efficient synthesis of magnetic nanoparticle-Musa acuminata peel composite for the adsorption of anionic dye *Arabian Journal of Chemistry* **13**
[28] Sohaimi K S A, Ngadi N, Mat H, Inuwa I M and Wong S 2017 Synthesis, characterization and application of textile sludge biochars for oil removal *Journal of Environmental Chemical Engineering* **5**
[29] Ahmad A L, Sumathi S and Hameed B H 2005 Adsorption of residue oil from palm oil mill effluent using powder and flake chitosan: Equilibrium and kinetic studies *Water Research* **39**
[30] Wahi R, Chuah Abdullah L, Nourouzi Mobarekeh M, Ngaini Z and Choong Shean Yaw T 2017 Utilization of esterified sago bark fibre waste for removal of oil from palm oil mill effluent *Journal of Environmental Chemical Engineering* **5**
[31] Al-Zuhairi F, Azeez R, Mahdi S, Kadhim W and Al-Naamee M 2019 Removal oil from produced water by using adsorption method with adsorbent a Papyrus reeds *Engineering and Technology Journal* **37**
[32] Chukwuemeka-Okorie H O, Ekemezie P N, Akpomie K G and Olikagu C S 2018 Calcined corncob-kaolinite Combo as new sorbent for sequestration of toxic metal ions from polluted aqua media and desorption *Frontiers in Chemistry* **6**
[33] Mohammed R R and Chong M F 2014 Treatment and decolorization of biologically treated Palm Oil Mill Effluent (POME) using banana peel as novel biosorbent *Journal of Environmental Management* **132**
[34] Ibrahim S, Wang S and Ang H M 2010 Removal of emulsified oil from oily wastewater using agricultural waste barley straw *Biochemical Engineering Journal* **49**
[35] Rajaković-Ognjanović V, Aleksić G and Rajaković L 2008 Governing factors for motor oil removal from water with different sorption materials *Journal of Hazardous Materials* **154**
[36] Mastral A M, García T, Murillo R, Callén M S, López J M and Navarro M v. 2003 Measurements of polycyclic aromatic hydrocarbon adsorption on activated carbons at very low concentrations *Industrial and Engineering Chemistry Research* **42**
[37] Okoya A A, Ochor N O, Akinyele A B and Olaiya O O 2020 The Use of Chicken Feather Waste as an Adsorbent for Crude Oil Clean Up from Polluted Water *Journal of Agriculture and Ecology Research International*
[38] Nasaruddin N F N, Halim H N A, Rozi S K M, Mokhtar Z, Tan L S and Jusoh N W C 2022 Potential of Pretreated Spent Coffee Ground as Adsorbent for Oil Adsorption *Lecture Notes in Civil Engineering* vol 214

[39] Darama S E, Gürkan E H, Terzi Ö and Çoruh S 2021 Leaching Performance and Zinc Ions Removal from Industrial Slag Leachate Using Natural and Biochar Walnut Shell *Environmental Management* **67**

[40] Yu B, Zhang Y, Shukla A, Shukla S S and Dorris K L 2000 The removal of heavy metal from aqueous solutions by sawdust adsorption - Removal of copper *Journal of Hazardous Materials* **80**

[41] Ghosal P S and Gupta A K 2017 Determination of thermodynamic parameters from Langmuir isotherm constant-revisited *Journal of Molecular Liquids* **225**

[42] Bae M, Lee H, Yoo K and Kim S 2021 Copper(I) selective chemisorption on magnetite (Fe3O4) over gold(I) ions in chloride solution with cyanide *Hydrometallurgy* **201**

[43] Almahbashi N M Y, Kutty S R M, Ayoub M, Noor A, Salihi I U, Al-Nini A, Jagaba A H, Aldhawi B N S and Ghaleb A A S 2021 Optimization of Preparation Conditions of Sewage sludge based Activated Carbon *Ain Shams Engineering Journal* **12**

The Efficiency of Reducing COD and Turbidity of Tofu Wastewater using A Combination of Electrocoagulation and Ozone

Wiharyanto Oktiawan[1], Budi Prasetyo Samadikun[1], Afan Sulton Ashari[1], Purwono Purwono[2]

[1]Department of Environmental Engineering, Faculty of Engineering, Diponegoro University, Semarang – Indonesia 50275
[2]Center for Science and Technology, UIN Raden Mas Said Surakarta, Jl. Pandawa, Pucangan, Kartasura 57168, Indonesia

Corresponding author: wiharyantooktiawan@lecturer.undip.ac.id

Abstract. We present an investigation of tofu wastewater that contains organic substances that can cause rapid microbial growth in the water. It can lead to a decrease in oxygen levels. In addition, tofu wastewater also contains suspended substances that cause turbidity. This study aims to evaluate changes in pH, temperature, and efficiency of reducing turbidity & chemical oxygen demand (COD) of tofu wastewater using a combination of electrocoagulation and ozone. This research method was carried out on a laboratory scale using tofu wastewater. The ozone generator is used for ozonising tofu wastewater and an acrylic batch type electrochemical reactor. The electrode material used for the anode and cathode is made of aluminum and iron sheets. There were also differences in the highest turbidity removal of 78.3% at the 60th min when the wastewater was treated using a combination of electrocoagulation followed by ozone. The highest COD removal of 51.9% resulted from the treatment of tofu wastewater using ozone for 60 min. The initial temperature was 27.6°C then increased by 28.1°C (1.8%). Combination of electrocoagulation followed by ozone was more effective in reducing pollutants in tofu wastewater.

1. Introduction

The majority of tofu industries in Indonesia lack waste treatment facilities and do not practice proper waste management.. Tofu wastewater contains organic substances that can cause the rapid growth of microbes in the water. Wastewater COD and BOD concentrations from tofu treatment facilities range from 5,000 to 8,500 and 3,500 to 4500 mg/L, respectively [1]. Such a high concentration of organic compounds can cause unpleasant odors, pollution, groundwater, and rivers [2]. This can lead to a decrease in oxygen levels. In addition, tofu wastewater also contains suspended substances that cause turbidity. Therefore, tofu wastewater treatment becomes important. Several attempts have been made to lower the levels of organic matter in the wastewater of the tofu industry using biological aerobic methods [1,3]. However, the application of such methods is relatively difficult. This is because the biological processing process requires a large area and a relatively long processing time. Physics-chemical methods for treating tofu waste are a solution to overcome biological weaknesses [4]. Electrocoagulation and filtration technology processes have been used to treat domestic wastewater, which can remove 95% COD in the Al-Fe electrode configuration with a current of 2 A and sampling at 27 min [5]. However, researchers still rarely find tofu wastewater treatment using a combination of electrocoagulation and ozone.

In this study, the combination of electrocoagulation-ozone to remove organic pollutants from tofu waste. The efficiency of removal using this method is investigated in various operating parameters, namely the time of electrolysis and the sequence of processing processes using ozone, electrocoagulation + ozone, and electrocoagulation followed by ozone. The benefits of this study will aid readers in acquiring sufficient knowledge regarding the treatment of waste with a high organic matter content.

2. Method

This research was conducted at the water laboratory, Department of Environmental Engineering, Faculty of Engineering, Diponegoro University. Tofu wastewater is obtained from the tofu industry effluent in Ungaran Regency, Semarang, Central Java, Indonesia. The electrode material used for electrocoagulation experiments is made of aluminum and ferrous plates. The electrodes are cleaned with sandpaper No. 120 to maintain their efficiency. Analytical reagent quality chemicals (p.a) obtained from Merck (Darmstadt, Germany) were used for all experiments.

Ozone generator air purifier formaldehyde 36G 220V - SO160 (China) is used for the process of ozonation of tofu wastewater with airflow rate of 0.5 L/min. The rate of ozone formation usually ranges from 0 to 3 mgs^{-1}. Ozone is supplied to the reactor through a diffuser, resulting in small bubbles. The volume of tofu wastewater taken 500 mL for each analysis, and the capacity of the acrylic reactor is 1080 mL. The ozonation process follows the procedure carried out by Das et al., [7] for 40 min. All analyzes are carried out at ambient temperature. The sample was analyzed at intervals of 15 minutes, and changes in pollutant were measured using standard methods for water and wastewater examination [8].

In this study, a batch-type electrochemical reactor made of acrylic with dimensions of 6 cm x 10 cm x 18 cm with an overflow on the side with a width of 5 cm and a height of 5 cm was used. The electrode material used for the anode and cathode is made of aluminium and ferrous plates, with a length of 15 cm and 9 cm (effective surface area 0.0135 m^{-2}). The number of electrodes is one sheet. Variations in the treatment sequence are carried out to treat tofu wastewater to effectively achieve the desired effluent target. Tofu wastewater is treat to achieve the desired effluent target effectively (TE + OZ) and electrocoagulation followed by ozone (LTETOOZ). Different electrolysis times of 0 to 60 minutes are performed to determine the best percentage of pollutant removal. The electrocoagulation process is carried out using a fixed voltage (20 V)

Sample analysis Laboratory tests are used to calculate the efficiency of removal of organic pollutants from tofu wastewater. Physical parameters including turbidity (NTU) measured using a turbidimeter (Hanna, USA), temperature using a thermometer (SNI 06-6989.23-2005). pH was measured with pH meter (Trans Instrument, Singapore) (SNI 06-6989.11-2004). COD is measured using the method contained in the standard method for water and wastewater [8].

3. Results and Discussion

3. 1 Turbidity

The turbidity of the tofu wastewater occurs due to suspended particles. Turbidity of wastewater causes stratification or coating of water and protects microorganisms from the effects of disinfection and stimulates the growth of bacteria [9]. **Fig. 1** shows a turbidity graph of tofu wastewater removal during the treatment process using ozone (OZ), a combination of electrocoagulation + ozone (LTE+OZ), and electrocoagulation followed by ozone (LTETOOZ). A variance analysis (ANOVA) was performed to verify the presence of significant differences in turbidity reductions as a result of the treatment.

The most interesting aspect of the results of this study is that the combination of electrocoagulation followed by ozone is able to remove turbidity up to 78.3% at the 60th minute. Partially, the timing of tofu wastewater treatment using a combination of electrocoagulation followed by ozone had a significant effect α 0.049 < 0.05. Overall, the percentage of turbidity allowance using ozone is lower than that of electrocoagulation + ozone or a combination of electrocoagulation followed by ozone.

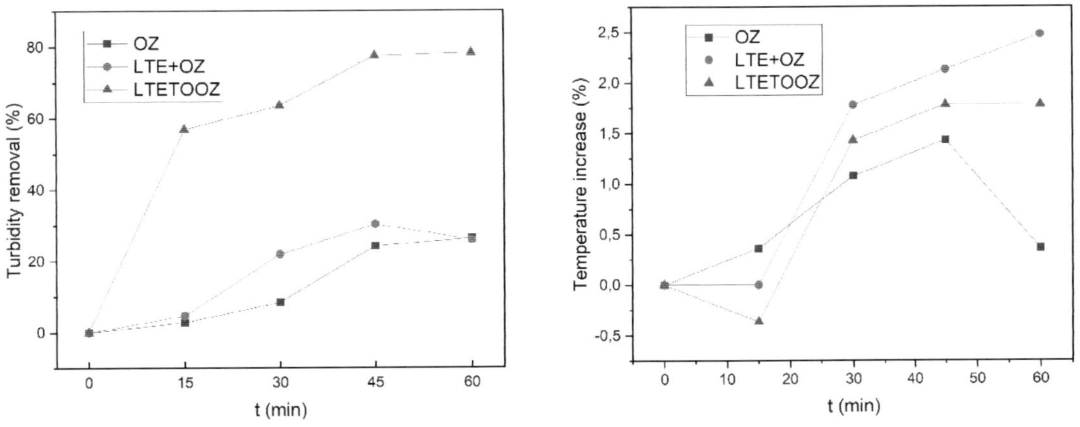

Figure. 1 Turbidity removal chart, temperature increase of tofu wastewater during the treatment process using ozone (OZ), a combination of electrocoagulation + ozone (LTE + OZ), and electrocoagulation followed by ozone (LTETOOZ)

Electrocoagulation in the process generates coagulants that function to form floc, and the electrocoagulation process runs more and more total suspended solid over time so that it can remove more pollutants [10]. This is also the case in a study conducted by Hajiali et al. in their research obtained the results that ozone can reduce the turbidity rate which is quite good with an initial turbidity of 20 NTU successfully remove up to the number 9 NTU [11]. Then the research conducted by Daghrir et al. [12] who conducted the study by comparing the waste treatment process between ozonation and electrocoagulation continued by ozonation in his study reported that ozonation can reduce the turbidity rate by about 16% this is still relatively low because by comparing to electrocoagulation followed by ozonation can reduce the turbidity rate by up to 97%. Likewise, what Hernandez-Ortega et al., reported in his study the combination of electrocoagulation and ozonation can reduce turbidity by up to 98% [13]. The conclusion of the study was that the highest turbidity allowance was 78.3% at the 60th minute when wastewater was treated using a combination of electrocoagulation followed by ozone.

3. 2 Temperature

The increase in the temperature of electrocoagulation leads to a decrease in the efficiency of removal of pollutants because the increase in temperature causes the volume of the colloidal solution $M(OH)_3$ to decrease and the pore production on the metal anode to become closed [14]. Fig. 1 shows a graph of the temperature increase (%) of tofu wastewater during the treatment process using ozone (OZ), a combination of electrocoagulation + ozone (LTE + OZ), and electrocoagulation followed by ozone (LTETOOZ).

The most interesting aspect of the results of this study is that the combination of electrocoagulation + ozone significantly increases the temperature of tofu wastewater. The initial temperature of 27.6°C then increased by 28.1°C (1.8%). Statistically, the increase in substrate temperature in the electrocoagulation process has no significant effect (α > 0.05). The temperature value of the processing results using this technology meets the required quality standards. According to the Regional Regulation of Central Java Province Number 5 of 2012 concerning Amendments to the Regional Regulation of Central Java Province Number 10 of 2004 concerning Wastewater Quality Standards, the maximum temperature of tofu wastewater is 38°C.

In the electrocoagulation process, there is a release of energy in the form of thermal energy that can increase the temperature during the electrocoagulation process [15]. The release of such energy is caused by several factors such as an increase in electric current. The increase in electric current at the electrodes directly proportional also increases the temperature of the waste during the electrocoagulation process running [16]. Ribordy et al. [17] argue that temperature has little significant effect on the electrocoagulation process. It is reported that the temperature rises from 20ºC to 60ºC did not affect anything in the degradation process using the electrocoagulation method using platinum electrodes.

3. 3 pH

Tofu wastewater treatment using a combination of electrocoagulation + ozone and electrocoagulation followed by ozone can increase the pH value. The initial pH of tofu wastewater was 3.74 then increased to 4.57 resulting from electrocoagulation followed by ozone for 60 minutes **(Fig. 2).** The pH value of the processing results using this technology meets the required quality standards. According to the Regional Regulation of Central Java Province Number 5 of 2012 concerning Amendments to the Regional Regulation of Central Java Province Number 10 of 2004 concerning Wastewater Quality Standards, the maximum temperature of tofu wastewater is 6-9.

The occurrence of pH increase can be found in the combination of electrocoagulation and ozonation with the production of hydroxyl ions at the cathode which occurs due to water reduction which eventually causes an increase in pH. Keep in mind that the initial pH state greatly affects the production of hydroxyl radicals produced [12]. The initial pH state also greatly affects the efficiency of pollutant removal [18]. Therefore, many make pH adjustments before processing either using ozonation or a combination of electrocoagulation with ozonation [12]. This is in line with research conducted by Öztürk et al. [19] who performed phenol processing using a combination method of electrooxidation and electrocoagulation reported that the phenol removal efficiency was directly proportional to the increase in pH to pH 7 the removal efficiency decreased. Same as done by Muryanto et al [20], it was reported in the study that there was an increase in pH in tofu waste treated by electrocoagualation for 8 hours. This is due to the large number of hydroxide ions during the electrocoagulation process.

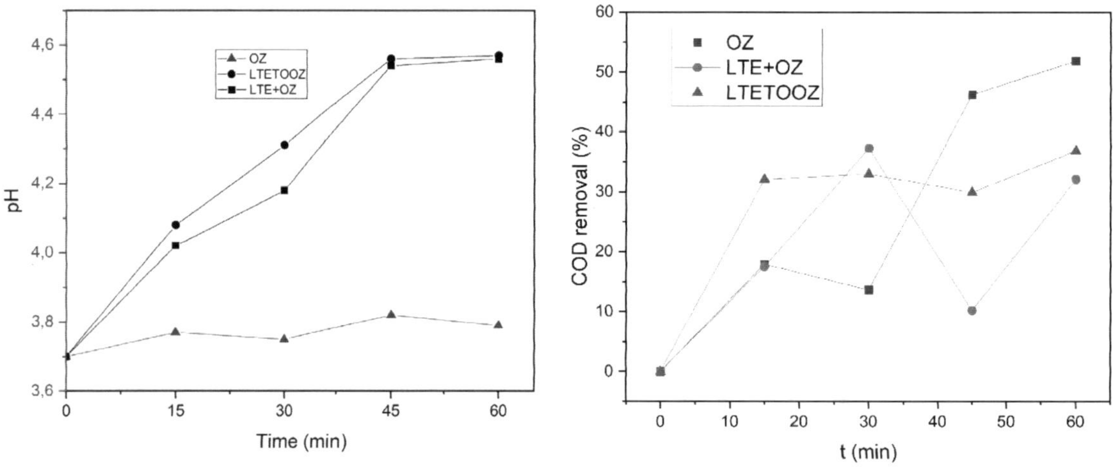

Figure. 2 Graph of pH and COD removal of tofu wastewater during the treatment process using ozone (OZ), a combination of electrocoagulation followed by ozone (LTETOOZ), and electrocoagulation + ozone (LTE + OZ)

4. COD

The highest COD removal of 51.9% resulted from the treatment of tofu wastewater using ozone for 60 minutes **(Fig.2).** The time of ozonization had a significant effect on reducing COD with a value of α= 0.017, smaller than 0.05. The combination of electrocoagulation + ozone and electrocoagulation followed by ozonization had an insignificant effect with α values of 0.299 and 0.138, respectively. According to the Regional Regulation of Central Java Province Number 5 of 2012 concerning

Amendments to the Regional Regulation of Central Java Province Number 10 of 2004 concerning Wastewater Quality Standards, the maximum COD level of tofu wastewater is 275 mg / L.

This is in line with the research conducted by Daghrir et al. who conducted the study by comparing the ozonation method with a combination of electrocoagulation and ozonation in the processing of grey wastewater [12]. The study found that the combination of electrocoagulation and ozonation can reduce COD_T and CODs, respectively, from 645 mg/L and 519 mg/L before the treatment to 51 mg/L and 50 mg/L. Before ozonation, the concentrations of COD_T and COD_S were 651 mg/L and 514 mg/L, respectively. After ozonation, these concentrations decreased to 350 mg/L and 246 mg/L. Similar to hernandez-Ortega et al., who processed industrial waste, 651 mg/L and 514 mg/L were reduced to 350 mg/L and 246 mg/L, respectively. Similar to what Hernández-Ortega et al., who process industrial waste, reported, the COD level of industrial waste was successfully reduced by 60% using the combination of electrocoagulation and continued ozonation [13]. The above proves that the combination of electrocoagulation and ozonation is more effective compared to ozonation itself (Li et al. 2013). In the electrocoagulation process, Aluminum hydroxide coagulants are formed from Al^{3+} and OH^- and also Iron hydroxide derived from Fe^{2+} and OH^- to destabilize pollutants in wastewater [21]. Likewise with the combination of electrocoagulation and ozonation, coagulants formed from electrocoagulation and reactive oxygen species derived from ozonation become intermediaries to remove pollutants both organic and inorganic [12].

5. Conclusion

This study aims to evaluate efficiency of reducing turbidity, COD, changes in pH, temperature, of tofu wastewater using a combination of electrocoagulation and ozone. The highest turbidity removal of 78.3% and COD removal of 51.9% at the 60th minute when the wastewater was treated using a combination of electrocoagulation followed by ozone. pH increases to 4.5 from the initial pH of 3.75. The combination of electrocoagulation+ozone significantly increased the temperature of the tofu wastewater even though it was tiny. The initial temperature was 27.6°C then increased by 28.1°C (1.8%). Combination of electrocoagulation followed by ozone was more effective in reducing pollutants in tofu wastewater.

6. Acknowledgments

The authors would like to thank you to the Department of Environmental Engineering, Faculty of Engineering, Diponegoro University for funding this Research through strategies grant for the financing year 2022

References

[1] Seroja R, Effendi H and Hariyadi S 2018 Tofu wastewater treatment using vetiver grass (Vetiveria zizanioides) and zeliac *Appl. Water Sci.* **8** 1–6
[2] Faisal M, Mulana F, Gani A and Daimon H 2015 Physical and chemical properties of wastewater discharged from Tofu industries in Banda Aceh city, Indonesia *Res. J. Pharm. Biol. Chem. Sci.* **6** 1053–8
[3] Tay J-H 1990 Biological Treatment of Soya Bean Waste *Water Sci. Technol.* **22** 141–7
[4] Maulana M R and Marsono B D 2021 Penerapan Teknologi Membran untuk Mengolah Limbah Cair Industri Tahu (Studi Kasus: UKM Sari Bumi, Kabupaten Sumedang) *J. Tek. ITS* **10** F54–60
[5] Oktiawan W, Priyambada I B, Aji S and Budi F S 2021 Effect of current strength on electrocoagulation using Al-Fe electrodes in COD and TSS removal of domestic wastewater *IOP Conf. Ser. Earth Environ. Sci.* **623**
[6] Das P P, Anweshan A, Mondal P, Sinha A, Biswas P, Sarkar S and Purkait M K 2021 Integrated ozonation assisted electrocoagulation process for the removal of cyanide from steel industry wastewater *Chemosphere* **263** 128370
[7] Mallick N, Bagchi S K, Koley S and Singh A K 2016 Progress and challenges in microalgal

biodiesel production *Front. Microbiol.* **7**
[8] APHA 2017 *Standard Methods for the Examination of Water and Wastewater* (Washington)
[9] Tomperi J, Isokangas A, Tuuttila T and Paavola M 2022 Functionality of turbidity measurement under changing water quality and environmental conditions *Environ. Technol.* **43** 1093–101
[10] Deshpande A M, Ramakant and Satyanarayan S 2012 Treatment of Pharmaceutical Wastewater by Electrochemical Method: Optimization of Operating Parameters by Response Surface Methodology *J. Hazardous, Toxic, Radioact. Waste* **16** 316–26
[11] Hajiali A and Pirumyan G P 2014 Evaluation of Turbidity and Color Removal in Treatment of Wastewater Containing Resistant Pollutants with Ozonation *IERI Procedia* **9** 8–12
[12] Daghrir R, Gherrou A, Noel I and Seyhi B 2016 Hybrid Process Combining Electrocoagulation, Electroreduction, and Ozonation Processes for the Treatment of Grey Wastewater in Batch Mode *J. Environ. Eng.* **142** 1–13
[13] Hernández-Ortega M, Ponziak T, Barrera-Díaz C, Rodrigo M A, Roa-Morales G and Bilyeu B 2010 Use of a combined electrocoagulation-ozone process as a pre-treatment for industrial wastewater *Desalination* **250** 144–9
[14] Al-Rubaiey N A, Albrazanjy M G and Kadhim W A 2022 Combined electrocoagulation and photocatalytic for oily wastewater treatment using TiO2 nano-catalysts *Egypt. J. Chem.* **65** 55–64
[15] Hernaningsih T 2016 Tinjauan Teknologi Pengolahan Air Limbah Industri Dengan Reviews of Electrocoagulation Process on Waste Water Treatment *J. Rekayasa Lingkung.* **9** 31–46
[16] Green D W and Southard M Z 2019 *Perry's chemical engineers' handbook* (McGraw-Hill Education)
[17] Ribordy P, Pulgarin C, Kiwi J and Péringer P 1997 Electrochemical versus photochemical pretreatment of industrial wastewaters *Water Sci. Technol.* **35** 293–302
[18] Sharma D, Chaudhari P K, Dubey S and Prajapati A K 2020 Electrocoagulation Treatment of Electroplating Wastewater: A Review *J. Environ. Eng.* **146**
[19] Öztürk H, Barışçı S, Turkay O and Veli S 2019 Electrocatalytic Degradation of Phenol by the Electrooxidation–Electrocoagulation Hybrid Process: Kinetics and Identification of Degradation Intermediates *J. Environ. Eng.* **145** 1–9
[20] Muryanto M, Sari A A, Pertiwi S, Prasetyo D A and Sudarno S 2021 Solar-Powered Electrocoagulation System for Tofu Wastewater Treatment and its Characteristic *J. Presipitasi Media Komun. dan Pengemb. Tek. Lingkung.* **18** 338–48
[21] Bote M E 2021 Studies on electrode combination for COD removal from domestic wastewater using electrocoagulation *Heliyon* **7** e08614

Study of Aquifer Distribution Based on Rock Resistivity Data in Warureja Village and Surroundings, Warureja District, Tegal, Central Java

N Santi[1], I Barala[1] and T T Putranto[1]

[1]Geological Engineering Department, Universitas Diponegoro

Diponegoro University, Prof. Soedarto street, Tembalang, Semarang City 50275, Indonesia.

narulita.santi@live.undip.ac.id

Abstract..The need for clean water in the dry season is often higher because many people would experience drought. The drought occurred in a Central Java area, Tegal, precisely at Warureja Village and its surroundings. The purposes are to determine the aquifer distribution and to provide recommendations for well drilling points. The study used the geo-electricity method of the Schlumberger configuration and field surface observation. Moreover, it used rock resistivity analysis to see how the subsurface conditions are. The relief in this area is wavy with denudational landforms, and the lithology is the alluvium unit, which was resulting from sedimentary deposits. The groundwater flows from south to north, following the morphology. The subsurface conditions consisted of two layers, the clay as the aquiclude and the sand as the aquifer. There are three recommended well drilling points. They are P-1 in Banjar Agung Village with a confined aquifer at a depth of 60 m, P-2 in Warureja Village at a depth of 20 m, and P-3 in Kreman Village with a depth of 18 m. In conclusion, all reserve points, static and dynamic, are sufficient.

1. Introduction

Water is a very important compound in supporting human life on earth. People have to consider its needs because the population on earth is getting bigger [1]. The most widely used water is groundwater, where more than 90% of clean water comes from there [2]. It provides many benefits for daily activities, such as cooking, bathing, washing, and industrial processing [3]

Water resources are essential to support everyone in the world, and one of its sources is groundwater. Many areas in Indonesia rely on groundwater to meet their daily needs due to the lack of surface water resources. However, not all-districts have detailed information about groundwater resources such as in Tegal, Central Java, Indonesia. During the dry season, the need for clean water is indispensable for the daily needs of the community [4]. This problem occurred in Tegal, precisely in Warureja District (Figure 1).

Water wells are experiencing drought, and the distribution of water sources is uneven, so groundwater exploration needs to be carried out to increase the clean water resources.

There are several ways to explore groundwater, one of which is using the geo-electrical resistivity method. It is a method of the electric current's flow to investigate subsurface conditions based on the resistivity data [5]. The interpretation of the aquifer is based on the material properties of the ground [6]. This method has advantages over others because of its low cost and rapid processing [7]. Geoelectric is a geophysical method often used for the exploration of groundwater. It has been widely used to find porous aquifers [8].

The objects were to determine the subsurface condition and the resistivity value. In addition, it would lead to the distribution of aquifers in Warureja and its surroundings.

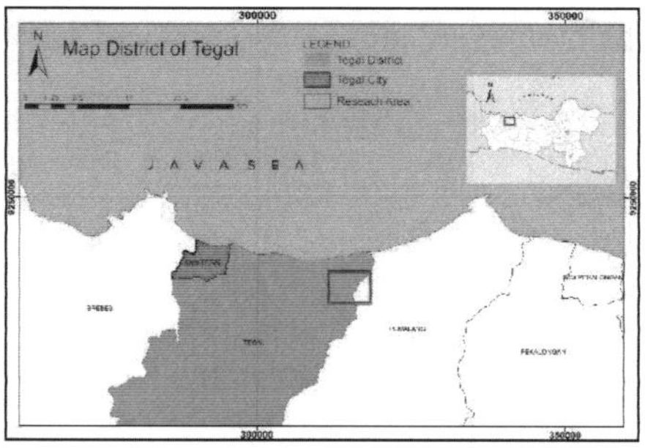

Figure 1. Location.

2. Methods

The research used two methods, namely descriptive and analytical. The first method was used to analyze data, both primary and secondary. The descriptive methods included describing lithology, hydrology conditions, and investigating geo-electricity. Then, the analytical method consisted of matching curve analysis, thickness and depth layer, and subsurface correlation. Meanwhile, the analysis method was for the further step. A geoelectrical method was conducted using the Schlumberger Method to estimate the type and vertical thickness of layers. The vertical electrical sounding (VES) was used to identify the resistivity in the subsurface. Furthermore, the measurement of the groundwater level was conducted in the nearest dug well to support the geoelectrical interpretation. The survey used two electrodes to deliver the electric current and two other potentials for reading the voltage and resistivity after the current passed to the earth materials [12]. They are installed in the ground with a certain and regular distance between electrodes and potentials. The two electrodes have a linear relationship with their depth of penetration towards the layers of material. The apparent resistivity value is controlled by measured potential (V), the electrical current (I), and geometric factor (K). It can be calculated using the following equation:

$$T = K \times b \quad (1)$$

where T is transmissivity (m2/day), K is permeability(m/day), and b is the thickness of the saturated aquifer(m).

$$p = K \Delta V I \quad (2)$$

3. Results and Discussion

Geomorphological conditions in the study area could be determined from slope and height differences parameters. The result from the mapping showed that the altitude morphology is relatively low because it is near the coast [9]. The slope percentage is low, 3%, with a height difference of about 5 to 10 m. Based on classification of van Zuidam [10] the relief in this area is gently undulating with denudational landforms (Figure 2). Statistics Indonesia of Tegal in 2019 stated that 71,54% of Warureja is used as a rice field [11]. The rest are for plantation and residential.

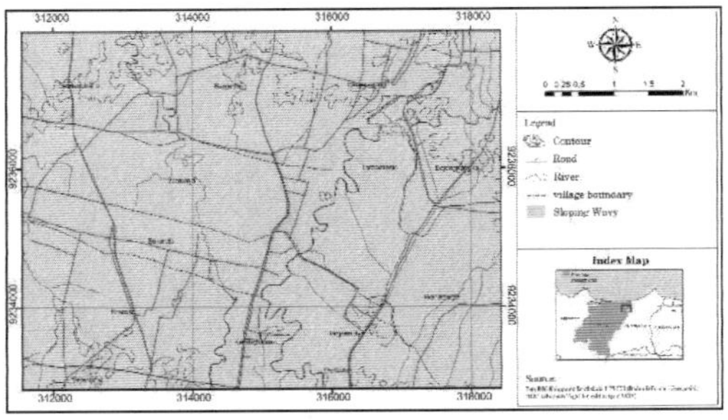

Figure 2. Morphology Map of the Area

Furthermore, based on Brahmantyo and Bandono (2006), it could be interpreted as river and lake plains. Then, based on the mapping at several stop sites, it is a form of a floodplain; characterized by the presence of floodplains on the banks of the river (Figure 3).

Figure 3. A floodplain. Location: Warureja Village, Warureja District, Tegal, Central Java. UTM WGS 84 Zone 49S: 311621, 9235685

The location consists of the Alluvial unit, which is part of the North Java Alluvial Plain zone, referring to regional geological conditions. Based on field observation at several stop sites, the lithologies are alluvium units (figure 4).

Figure 4. An alluvium unit. Location: Warureja Village, Warureja District, Tegal, Central Java. UTM WGS 84 Zone 49S: 311621, 9235685

This alluvium unit was formed due to the weathering of previous rocks. Then the rock is transported and deposited as alluvium. The Alluvium unit in addition, its formation can also be indicated due to the presence of sediment deposits from the coastal area, considering that the research location is close to the coast and is part of the alluvial plain zone of North Java. The lithology distribution is on the lithological map (Figure 5).

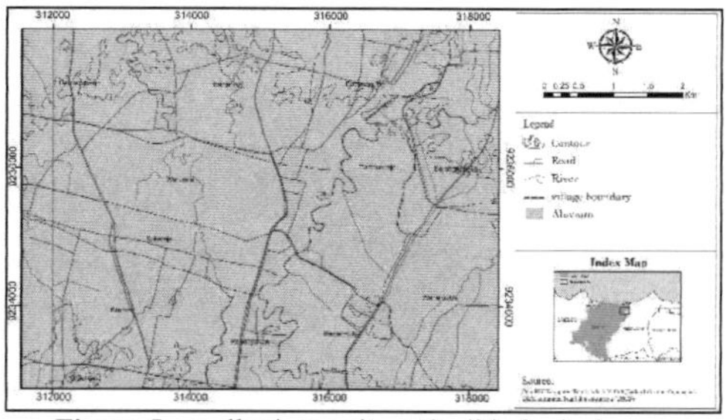

Figure 5. an alluvium unit on the lithological map.

The determination of lithostratigraphy and hydrostratigraphy was derived from analyzing resistivity logs of geo-electrical measurement points. There were seven points in three villages: Banjar Agung, Kreman, and Warureja. After that, to see the subsurface conditions related to lithology and hydrogeology, all points were correlated. There are four correlations below:

Line A - A' passes through KR-01 and KR-02 points. The location is in Kreman Village, which is the southwest part of the study area. The result showed that the subsurface is majorly layers of clay. Its thickness ranges from 17 to 42 m. The layers are in the first and last, which are aquicludes. Next, there is a layer of sand between the clay layers, precisely at the KR-01, depth of 5.3 to 9.5 m. It is not only that but also at a depth of 12.5 m to 26 m, which was getting thicker towards point KR-02 (figure 6). The layer is interpreted as a confined aquifer because its position is between two aquicludes layers that could not drain water, so the water pressure in the aquifer is restrained.

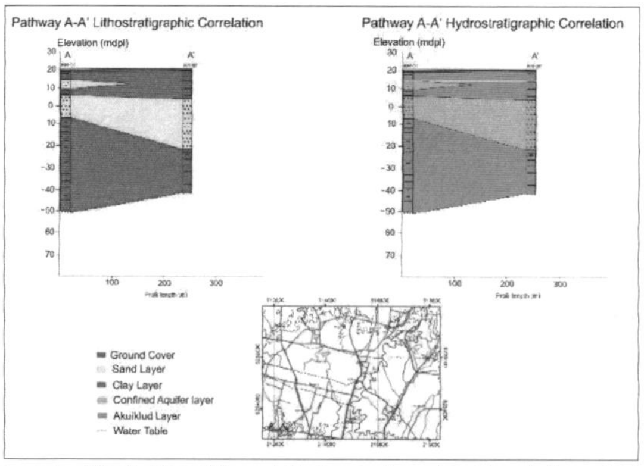

Figure 6. Litostratigraphic and Hydrostratigraphic Correlation of Line A-A'

Line B - B' passes through three points, WR-01, WR-02, and WR-03. As a result of lithostratigraphic correlation, the first layer is a fairly-large clay thickness, ranging from 7 to 37 m (Figure 7). Like the previous explanation, it is an aquiclude. Next, the second layer is sand. From the correlation results, it is located at the WR-01 point at a depth of 18 m to 51 m. Furthermore, it is an aquifer layer, particularly a confined aquifer.

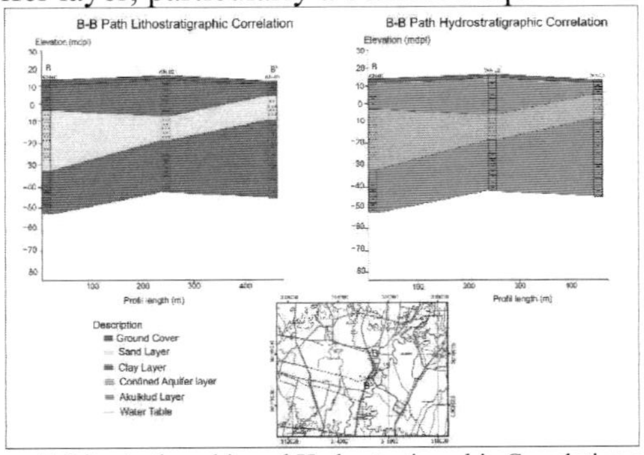

Figure 7. Litostratigraphic and Hydrostratigraphic Correlation of Line B-B'

Line C - C' passes through two points, which are BA-01 and BA-02. It is dominated by thick clay, ranging from 5.5-36 m. The first clay layer is at a depth of 1 to 9 m, where it is getting thinner to a thickness of 5.5 m at BA-02 (Figure 8). Meanwhile, the second layer is getting thicker from BA-01 to BA-02. These layers are aquicludes. Then, there are two layers of sand at different depths. The first one is at BA-01, depth of 11.5 to 15.5 m. It is getting thicker towards BA-02 to a thickness of 14 m. The second layer is at BA-01, depth of 41.5 to 60 m. As in lines A - A' and B - B', this sand layer has the properties of a confined aquifer.

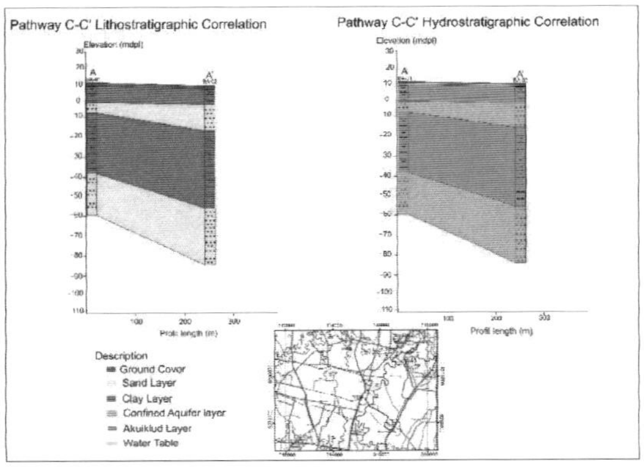

Figure 8. Litostratigraphic and Hydrostratigraphic Correlation of Line C-C'

Line D - D' passes through seven geoelectric measurement points. Correlation at those seven points has indicated that the presence of a sand layer at point KR-02 is not thick enough. Then, closer to WR-03 and WR-02, it gets thinner but thickens again at the WR-01(Figure 9). Based on the character of groundwater rocks, the sand layer has a property of an aquifer. In conclusion, only a type of aquifer was found at all geoelectric measurement points in this area. It is a confined aquifer.

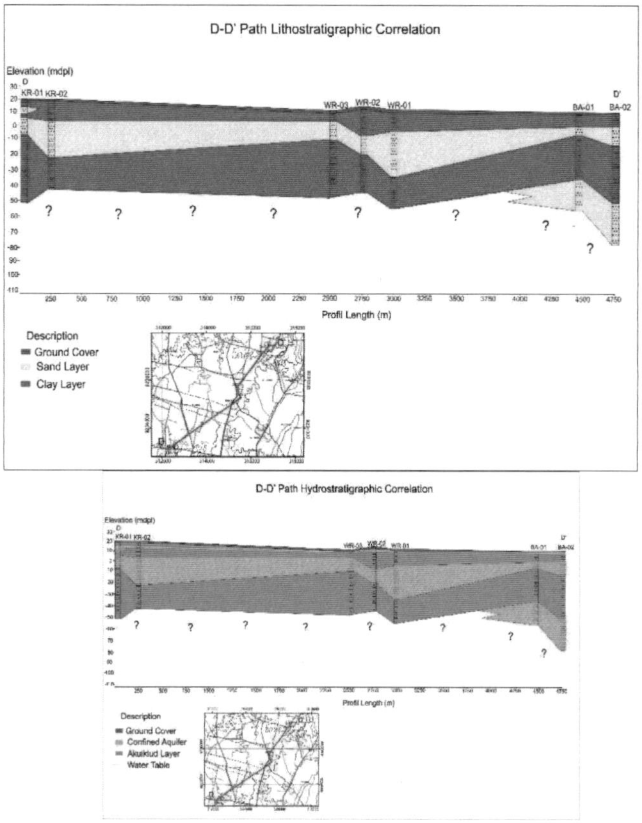

Figure 9. Litostratigraphic and Hydrostratigraphic Correlation of Line D-D'

Groundwater conditions in the area were determined from the hydrogeological field observation. It was related to the groundwater level (MAT) at dug and drilled wells (Figure 10). Groundwater level (MAT) could determine the direction of the groundwater flow. There are six dug wells and also 17 drilled wells with different depths.

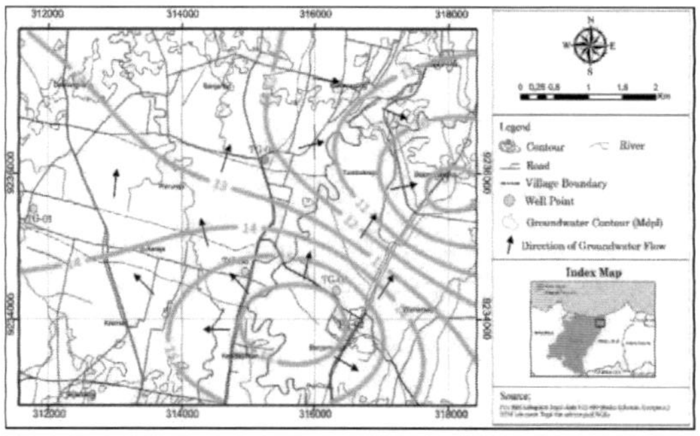

Figure 10. Dug well Groundwater flow map of the Area

MAT in dug wells has a depth of 0.5 m to 1.2 m. The difference in depth is not too big because the elevation of the study area tends to be flat, and the altitude interval is not too high. The direction of groundwater flow on the interpolated dug well map is from south to north. Next is the result of interpolation on 17 drilled wells. The depth is from 35 to 67 m. The direction varies from the southwest to the middle and from the northeast to the middle.

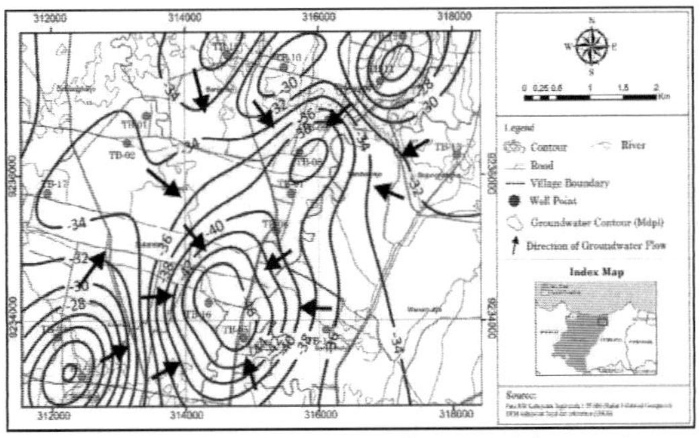

Figure 11. Bore well groundwater flow map ini research area

Hydrogeological mapping could be for determining the chemical conditions of groundwater too. It came from the TDS, DHL, and pH analysis of samples (Table 1). There were 14 samples, six from dug wells and eight from drilled wells. The following are the results of TDS, DHL, and pH values:

Table 1 TDS, DHL and pH Results.

Wells	District (Kecamatan)	Lithology	Type	DHL (μS/cm)	TDS (mg/L)	pH
TG-01	Warureja	Alluvium	dug	1.495	957	7,17
TG-02	Pemalang	Alluvium	dug	1.490	954	7,2
TG-03	Pemalang	Alluvium	dug	1.500	960	7,12
TG-04	Warureja	Alluvium	dug	1.150	736	7,27
TG-05	Warureja	Alluvium	dug	590	378	7,96
TG-06	Warureja	Alluvium	dug	662	424	6,78
TB-02	Warureja	Alluvium	drilled	862	552	7,7
TB-04	Warureja	Alluvium	drilled	2.800	1.984	7,18
TB-05	Warureja	Alluvium	drilled	1.303	834	7,3
TB-08	Warureja	Alluvium	drilled	1.187	760	7,89
TB-09	Warureja	Alluvium	drilled	645	413	8,07
TB-10	Warureja	Alluvium	drilled	2.100	1.344	7,33
TB-12	Warureja	Alluvium	drilled	663	424	8,1
TB-17	Warureja	Alluvium	drilled	985	630	7,6

From the chemical properties of groundwater at the DHL and TDS values in Table 1, the dug wells are classified as fresh water. Unlike drilled wells, there are two wells with high DHL and TDS, namely TB-04 and TB-10. Possibly, this is due to seawater intrusion. The value of water reserves, it is necessary to get how large the area is and its population. Then, the calculation of domestic water needs volume was carried out on a village scale, and the amount was 60 liters/person/day.

Several parameters on previous explanations, such as aquifer type, groundwater flow, chemical properties, and reserves, help to decide recommendations for groundwater availability points Table 2. Finally, there are three recommended points for making water sources: Banjar Agung Village, Warureja Village, and Kreman Village.

Table 2 Groundwater Source Recommendation Points

Point	Location	Type	Depth	DHL	TDS	pH	Static Reserves (m^3)	Dynamic Reserves (m^3/day)
P-1	Banjar Agung	Confined Aquifer	30 m	Good	Fresh Water	Fit Consuming	35.604.000	10.476
P-2	Warureja	Confined Aquifer	50 m	Good	Fresh Water	Fit Consuming	45.194.000	5.310
P-3	Kreman	Confined Aquifer	25 m	Poorly good	Brackish Water	Fit Consuming	27.936.000	6.880

4. Conclusion

Geomorphologically, it is a denudational landform with a gently undulating morphology. Lithologically, it consists of alluvium unit lithology, which is part of the Alluvium Unit formation. The rock resistivity varies from 0.43 Ωm to 86.4 Ωm. Based on the results of lithostratigraphic analysis, the subsurface consists of clay and sand layers. Then, it consists of aquifers and aquicludes. The type of aquifers is confined aquifers. In conclusion, there are three recommendations for drilling points for making water sources based on the type of aquifer, groundwater flow conditions, chemical properties, and water reserves in the research area.

References

[1] Ismawan, M. F 2016 *Kajian intrusi air laut dan dampaknya terhadap masyarakat di pesisir Kota Tegal* Geo-Image, 5(1)

[2] González, J. A. M., Comte, J. C., Legchenko, A., Ofterdinger, U., & Healy, D 2021 *Quantification of groundwater storage heterogeneity in weathered/fractured basement rock aquifers using electrical resistivity tomography: Sensitivity and uncertainty associated with petrophysical modeling* Journal of Hydrology, 593, 125637

[3] Harjito 2013 Metode Vertical Electrical Sounding (VES) untuk menduga potensi sumberdaya air Jurnal Sains dan Teknologi Lingkungan Vol. 5 No.2 p 127-140

[4] Wilopo W, Rachman M. H and Putra D. P. E 2020 Assessment of groundwater resources potential using geoelectrical method and slug test in Tegal District, Java Province, Indonesia. Central E3S Web of Conferences 200, 02003, 1-5

[5] Usman, B., Manrulu, R. H., Nurfalaq, A., & Rohayu, E 2017 *Identifikasi Akuifer Air Tanah Kota Palopo Menggunakan Metode Geolistrik Tahanan Jenis Konfigurasi Schlumberger* Jurnal Fisika Flux: Jurnal Ilmiah Fisika FMIPA Universitas Lambung Mangkurat, 14(2), 65-72.

[6] Araffa S.A.S., Mohamadin M.I., Sabet H.S. and Takey M.S 2019 Geophysical interpretation for groundwater exploration around Hurghada area, Egypt NRIAG Journal of Astronomy and Geophysics 8, 1

[7] Wilopo W, Putra D.P.E., Susatio R. 2020 Aquifer distribution and groundwater geochemistry in Bojonegoro Sub-district, Bojonegoro District, East Java Province, Indonesia Journal of Degraded and Mining Lands Management 7, 4

[8] Hasan m, Shang Y.J., Jin W.J., and Akhter G., G 2019 Investigation of fractured rock aquifer in South China using electrical resistivity tomography and self-potential methodsJournal of Mountain Science 16, 4

[9] Brahmantyo B and Bandono 2006 *Klasifikasi Bentuk Muka Bumi (Landform) untuk Pemetaan Geomorfologi pada Skala 1:25.000 dan Aplikasinya untuk Penataan Ruang* Jurnal Geoaplika Vol. 1 p 71-78

[10] Rahanjani, Y. E., Setianto, A., and Srijono, S 2015 *Pemanfaatan Citra Digital Elevation Model (DEM) Untuk Studi Evolusi Geomorfologi Gunung Api Merapi Sebelum dan Setelah Erupsi Gunung Api Merapi 2010* In Seminar Nasional Informatika (SEMNASIF) (Vol. 1, No. 5)

[11] Badan Pusat Statistik Kabupaten Tegal 2019 Kabupaten Tegal Dalam Angka 2019, (Badan Pusat Statistik Kabupaten Tegal, Kabupaten Tegal

[12] Goldman M and Neubauer F.M 1994 *Surveys in Geophysics*, 15

Acknowledgments

This research was funded by The Faculty of Engineering, Diponegoro University, Indonesia Through Strategic Research Grant 2022. Also, the author thanks CV. Geodipa that supports do research the author thanks all parties who have been willing to provide corrections, criticisms, suggestions, and inputs so the study could be resolved.

A comparison of rap-tourism method and multi attribute aggregation in sustainability assessment of tourist destination

A A A Putri[1], R Purwaningsih[2] and S Hartini[3]

[1,2,3] Department of Industrial Engineering, Diponegoro University, Semarang, 50275, Indonesia

[1]aisyaharifna@gmail.com, [2]ratna.purwaningsih@ft.undip.ac.id, [3]srihartini@lecturer.undip.ac.id

Abstract. The development of the tourism industry causes emergence of positive and negative impacts on tourist destinations. The development of the tourism industry has led the government to implement the concept of sustainable tourism. To assist the implementation of sustainable tourism, it is important to know the sustainability status of tourist destinations. The measuring tools used for calculated sustainability index values and identification of sensitive indicators. The method used must be easy to use and provide accurate results. This study compares between the rap-tourism and the multi attribute aggregation method. The aim is to choose which method better in measuring sustainability status when the objects of measurement are a lot of tourist destinations, better means easy to use and represent indicators condition briefly. The results of the assessment using both methods obtain the same sustainability status but there are differences in the sustainability index value. These two methods also use different approaches in determining sensitive indicators that should be improved to increase the sustainability index of tourist destinations.

1. Introduction

The tourism industry plays an important role in the economic growth of a country [1], [2], [3]. 6,1% of Indonesian GDP contributed by tourism sector [4]. The development of the tourism industry leads to positive and negative impacts on the local livelihoods and tourist destinations [5]. Economic growth [6], preservation of the local culture [7], and improvement in quality of life [8] are some positive impacts of tourism industry. However, with the increase in tourism activities, environmental pollution [9], [10] and social conflict also increases [8].

Countries whose economies are heavily influenced by the tourism industry are becoming more concerned about the environmental and socio-cultural impacts caused by the tourism industry [11]. In Indonesia, sustainable tourism has been implemented by the national and local governments involving all tourism stakeholders and tourists [12].

Sustainable development leads not only to prosperity and profit but also to achieve environmental sustainability and social welfare [13]. Therefore, a sustainability assessment is needed to measure the sustainability performance [14]. Sustainability assessment can be performed using a set of indicators which function as a measuring tool for evaluating the impact of tourism on sustainability [15].

Result of measuring the sustainability status of tourist destination is an index produce from multidimensional indicators that reflect the condition of the tourism destination. When the value of indicators was combined or aggregated, the result of the measurement is an index [16]. The aggregation method that easy to use and represent indicators condition briefly are needed.

This research aims to compare between the two aggregation methods, rap-tourism (rapid appraisal-tourism) and multi attribute aggregation. Rap-tourism is an adaptation of rap-fish, a statistical technique that uses a multidimensional scaling approach [17], while multi attribute aggregation used a composite index technique [18].

Indicators used in this study and calculation of data using rap-tourism are done in previous research conducted by Purwaningsih, et. al., (2020). Then, assessment result in the previous research will be compared with the assessment result using multi attribute aggregation to know the differences in value and the identification of weak indicator. Weak indicator should be improved by tourism management to increase the sustainability level. The object of this assessment is the Baturaden tourism site, the assessment is performed by assessing four dimensions of sustainability. The four dimensions of sustainability consist of economic dimension, environmental dimension, social and culture dimension, and institutional dimension [19]; [20].

2. Literature Review

2.1. Sustainable Tourism

Sustainable tourism is tourism that accommodate the needs of tourist, the tourism industry, the environment, and local communities while considering the impact of tourism activities on the state of economic, social, and environment in the present and the future [21]. Traditionally, three dimensions are included in the sustainable development paradigm which includes economic, socio-cultural, and environmental [22]. However, without an institutional perspective to manage and facilitate it is difficult to establish a balance between these three dimensions [23]. Hence, four dimension is used to assessing the sustainability status of tourist destination, each dimension is consisting of element, and each element consist of indicators. There are 12 element and 34 indicators used in assessing sustainability status of tourism, shown in Table 1.

Table 1. Dimensions, elements, and indicators of sustainable tourism

No	Element	Indicator	Reference
		Environmental Dimension	
1	Environmental Protection	1. Environmental risk management system	[12]; [24]; [25]; [26]; [27]
		2. Environmental management and protection system	
		3. Waste treatment program	[12]; [24]; [25]; [26]; [27]; [28]; [29];
		4. Transport planning and programs to improve green transportation	[12]; [26]
2	Flora Protection	1. Flora data collection	[12]; [24]; [27]
		2. Prohibition of flora destruction	
3	Energy Management	1. Energy consumption report	[12]; [24]; [25]; [28]; [29]; [30]; [31]
		2. Water quality and safety management system	[12]; [24]; [32]
		Economical Dimension	
1	Participation and access for the community	1. Communication to accommodate the aspirations of local-residents	[12]; [24]; [27] [28]; [29]; [33]
		2. Communication regarding tourism issues	[12]; [24]; [27]
2	Employment opportunities for local communities and supporting local businesses	1. Provide employment for local residents	[12]; [24]; [25]; [27]; [29]; [32]; [34];
		2. Engage & support local businesses	
3.	Economic continuity	1. Regular reporting and monitoring of expenses, income and investments	[12]; [25]

No	Element	Indicator	Reference
		2. The existence of attractions and tourism destination programs	[24]; [27]
		Socio-Cultural Dimension	
1.	Visitor management and public relations	1. Visitor management system	[12]; [24]; [27]
		2. Tourist and cultural guide	
		3. Volume and type of tourist trips to destinations	[25]; [27]
		4. Relations with the local community	
2.	Interpretation service	1. Accurate interpretive information	[12]; [24]; [35]
		2. Availability of information in various media forms	[12]; [24]; [30]; [35]
		3. Management system to protect natural and cultural sites	[12]; [24]; [35]
		Institutional Dimension	
1	Destination management organization	1. Sustainable tourism destination development strategy	[12]; [25]
		2. The involvement of the community, government, private sector, and entrepreneurs as stakeholders	[12]; [25]; [33]
		3. Responsible organization and clear coordination	[12]; [25]
		4. Tourist destination management system	
		5. Tourism promotion strategy	
2	Safety and security	1. Emergency handling and response systems for safety and security	[12]; [24]; [25]; [26]; [27]; [29];
		2. Inspection and preventive systems against hazards	[12]; [24]; [25]; [26]
		3. Policies for mitigating seasonal changes	
3	Visitor satisfaction	1. Responsive management for visitor satisfaction	[12]; [24]; [25]; [29]; [36]
		2. Visitor satisfaction level	[12]; [24]; [25]; [29]; [33]
4	Planning arrangement	1. Periodic asset inventory	[12]
		2. Destination regulations and policies which include analysis of resources, infrastructure, zoning, as well as the facilities and services provided	
		3. Accessibility that supports tourist accommodation including persons with disabilities and special needs	[12]; [27]

The environmental dimension focused on the availability of natural resources. The economic dimension refers to infrastructure, buildings and structures, human well-being, availability of jobs or livelihoods. The social dimension refers to education, human ability, human behavior, experience, and human rights. The institutional dimension refers to social institutions, government organizations, interpersonal relationships, partnerships, or collaborations [22]; [37].

2.2. Rap-tourism

Rap-tourism is an adaptation of rap-fish, a rapid appraisal technique developed to evaluate the sustainability status of fisheries [38]. Rap-tourism is a rap-fish application which used to assess the sustainability status of tourist destination [19]; [20]. It uses multidimensional scaling (MDS) approach to visualize the result of sustainability assessment, quantify similarity judgement among a group of items [39].

Rap-fish transforms multidimensional concept to a lower dimension, in this case two dimension. The horizontal x dimension is to plot the fisheries from bad to good, while the y axis represents other factors. Rap-fish uses ALSCAL scaling algorithm, it generates an R2 measure which represent the proportion of variance of the scaled data [39].

Monte Carlo is used in rap-tourism to measure error from the ordination. Sensitives attribute or indicators can be identified by perform the leveraging calculation. Leverage values range between 2% to 6% in a multivariate condition. The lower the leverage value of an attribute, the lower its contribution to the sustainability status of the tourist destination. The final results of sustainability status are expressed on a scale of 0 to 100% and then visualized using a kite diagram [20].

2.3. Multi Attribute Aggregation

To aggregate the value of indicators into an index, there are several methods such as additive aggregation and geometric aggregation, and multi attribute aggregation [40]. Multi attribute aggregation uses the concept of multi criteria utility theory (MAUT). MAUT was developed by Keeney & Raiffa in 1976, MAUT generates an ordinal ranking of alternatives using a linear or multiplicative model [41]. MAUT uses multicriteria approach to generate an ordinal ranking of alternatives, it ranks the alternatives from best to worst. MAUT is a utility-based model which calculate the value of each criterion and then the values are aggregated into a final utility value [42].

In this research multi attribute aggregation is used to aggregate indicators from four sustainability dimension to calculate a sustainability index on tourism destination. The multi-attribute utility function in this study is linear and in the form of an additive model. The sustainability index is obtained by sum up multiplication value of the indicator's utility times the weight of the indicator.

3. Research Methodology

Comparison of the two methods used in measuring sustainability status will be seen from the aspect of the method or steps of implementing the measurement, the measurement results are in the form of a sustainability index and identification of indicators that are weak and need improvement. Then the steps for implementing the research are:

1. Identify the sequence of steps for carrying out measurements, especially in the data processing section to obtain a sustainability index,
2. Comparing the resulting sustainability index values, also the value of each dimension.
3. Comparing indicators that are considered weak from the two methods.

The primary data is obtained from a questionnaire filled out by stakeholders from the tourist destination. Whereas the secondary data is given by the tourist destination management. The index value found from data processing are categorized in four class of sustainability status, the scale for determining the status of sustainability is shown in table 2 [34]; [43].

Table 2. Sustainability category status

No	Sustainability Index	Status
1	0,00 – 25,00	Bad (unsustainable)
2	25,01 – 50,00	Less (less sustainable)
3	50,01 – 75,00	Enough (quite sustainable)
4	75,01-100,00	Good (very sustainable)

The literature review and the previous research found that the stages of data processing using rap-tourism are [19]; [20]:

1. Determining the measurement scale to quantify each indicator
2. Calculating the index of sustainability for each dimension using multidimensional scaling technique
3. Performing conformity test by calculating stress value and coefficient of determination
4. Calculating leverage value to identifying sensitive indicator
5. Performing Monte Carlo simulation at 95% confidence interval to validate the model
6. Calculating the final sustainability index
7. Categorizing the index of sustainability into sustainability status scales

The second method is multi attribute aggregation, the stages of data processing using multi attribute aggregation are:

1. Collecting utility data of each indicator
2. Performing data normalization to standardize the data into 0 to 1 scale using min-max method [44]
3. Weighting each dimension based on expert opinion
4. Aggregating the indicators using multi attribute utility function
5. Categorizing the index of sustainability into sustainability status scales
6. Analyzing sensitive indicators based on the utility value

The weighting method is based on expert opinion can increase the legitimacy of the result and can lead to a consensus for policy action [40]. The scale for determine the status of the sustainability is based on the scale developed by Allen (1997) and was used in research on tourism sustainability assessment conducted by Weng (2019).

4. Results and Discussion

4.1 Compare the step

There are different stages of data processing between the two methods, table 3 will show the differences and similarities between the two methods.

Table 3. Comparison of data processing stage

Data Processing Stage	Rap-tourism	Multi Attribute Aggregation
Data Transformation	Ordination using the ALSCAL Algorithm	Normalization using min-max procedure, feature scaling, z-score
Conformity Test (Goodness of Fit)	Using the stress value and the coefficient of determination	-
Sensitivity Analysis	Using leverage analysis which will generate the root mean square	-
Model Validation	Using the monte Carlo simulation	-
Weighting Method	The weighting is done by ranking method, the only weight required is the weight of each dimension	The weighting is carried out by the ranking method, generally in the composite index method the weight value is sought for each attribute or indicator
Aggregation Method	Determining the weighted index by multiplying the value of the sustainability index by the dimensional weight	Determining the value of the sustainability composite index with a multi-attribute utility function, which is multiplying the utility value by the weight.

4.2 Compare the index

The following are the results using rapid appraisal technique and multi attribute approach. The comparison of index is shown in Table 4.

There is a deviation in the value of the sustainability index between rap-tourism method and multi attribute aggregation. With the value of 85,84 and 81,92, Baturaden tourism site can be categorized as sustainable. The deviation in the sustainability index value is caused by using different data transformation methods. In the rap-tourism method the data transformation is carried out by ordination using the ALSCAL algorithm, whereas the multi-attribute approach using normalization with the min-max procedure for the data transformation. Using the same measurement tool, even though the data processing method is different still resulting the same sustainability status. Increasing the sustainability index can be done by improving the performance of its indicators.

Table 4. Index value comparison

Data Processing Method	Rap-tourism	Multi Attribute Aggregation	Deviation
	Baturaden Tourism Site		
Environmental Dimension	39.72	41.36	1.64
Economic Dimension	14.11	14.6	0.49
Socio-Cultural Dimension	9.17	9.85	0.68
Institutional Dimension	18.92	20.03	1.11
Average of Deviation			0.98
Global Index Value	81.92	85.84	
Sustainability Status	Sustainable	Sustainable	

4.3 Compare the weak indicators

Rap-tourism and multi attribute aggregation have different way to determine which indicators should be improved. The rap-tourism method uses leverage analysis to identify sensitive indicators. Increasing the sustainability index can be done by improving the sensitive indicators Sensitive indicator is an indicator whose has a major contribution on the value of sustainability index.

Leverage analysis calculate at the root mean square, the bigger the root mean square, the greater the influence of the indicators on the value of sustainability index [38]. In the multi-attribute aggregation, the indicator that must be improved is the indicator that has the lowest utility value. The lower the utility value of the indicator, the worse the performance of the indicator [45]. To increase the sustainability index of the tourist destination site, indicators that have low utility values must be improved. Table 5. shows the indicators that must be improved on each dimension.

Table 5. Indicators that need improvement

Dimension	Indicators that need improvement	
	Multi Attribute Aggregation	**Rap-tourism**
Environmental	Transport planning and programs to improve green transportation	Transport planning and programs to improve green transportation
Economy	Engage & support local businesses	Engage & support local businesses
Socio-Cultural	Availability of information in various media forms	Visitor management system
Institutional	Sustainable tourism destination development strategy	Inspection and preventive systems against hazards

Table 5 shows that the two methods result on different finding on the weak indicators. These differences caused by different approach in identify the weak indicators. The rap-tourism used leverage value and the multi aggregations used utility value.

5. Conclusion

Based on the comparison done in this study, the stages of data processing using rap-tourism are more complex compared to multi attribute aggregation. Rap-tourism has a conformity test and validation test. While rap-tourism result more reliable, multi attribute aggregation is more common and easy to conduct, it can combine qualitative judgement and quantitative data to effectively rank indicator and aggregate indicator value in multidimensional concept.

Although there is deviation of 0.98 in the sustainability index value between the rap-tourism method and multi attribute aggregation. The result of the two methods still falls into the same sustainability category with the final value in the range of 75.01-100. To help stakeholders increase the value of the sustainability index, identification of indicators that need improvement is performed. The basis for selecting indicators that need to be improved is different for the two methods, rap-tourism using leverage factor value while multi attribute aggregation using utility value.

References

[1] WTTC 2014 *WTTC report* (Cusco: World Travel & Tourism Council)
[2] Christie I T 2002 *Tour. Rev.* **57** 35-41
[3] Dogru T and Bulut U 2018 *Tour. Manag.* **67** 425-34
[4] WTTC 2022 *Economic impact reports* (London: World Travel & Tourism Council)
[5] Archer B, Cooper C and Ruhanen L 2004 *Global tourism 3rd edition*, ed William F Theobald (London: Routledge) pp. 79-102
[6] Min C K, Roh T S and Bak S 2016 *Appl. Econ.* **48** 7-16
[7] Shahzalal M 2016 *J. Tour. Hosp. Sports* **20** 30-34
[8] Ibanescu B C, Stoleriu O M, Munteanu A and Iatu C 2018 *Sustainability* **10**
[9] Burak S, Dogan E and Gazioglu C 2004 *Ocean Coast. Manag.* **45** 515-27
[10] Holden A 2003 *Ann. Tour. Res.* **30** 94-108
[11] Neto F 2003 *Nat. Resour. Forum* **27** 212-22
[12] Kementrian Pariwisata 2016 *Peraturan menteri pariwisata RI no 14 tahun 2016 tentang pedoman destinasi pariwisata berkelanjutan* (Jakarta: Kementerian Pariwisata RI) [in Indonesian Language]
[13] Purvis B, Mao Y and Robinson D 2019 *Sustain. Sci.* **14** 681-95
[14] Ko T G 2005 *Tour. Manag.* **26** 431-45
[15] UN 1992 *Agenda 21* (Rio de Janerio: United Nations)
[16] Ness B, Urbel-Piirsalu E, Anderberg S and Olsson L 2007 *Ecol. Econ.* **60** 498-508
[17] Kavanagh P and Pitcher T J 2004 *Implementing microsoft excel software for rapfish: a technique for the rapid appraisal of fisheries status* (Vancouver: University of British Columbia)
[18] Ferretti V, Bottero M and Mondini G 2014 *Adv. Electron. Forum.* **11** 536-41
[19] Purwaningsih R, Santoso H and Khasanah U Rap 2020 *Rap-Tourism Method to Assess Tourism Objects Sustainability Proc. Int. Conf. on Engineering and Technology for Sustainable Development* (Bristol: IOP Publishing)
[20] Purwaningsih R, Annisa H N, Susanty A and Puspitaningrum D D 2020 *Assessment of Tourism Destination Sustainability Status using Rap-Tourism Case of Natural Based Tourism Proc. The 5th Int. Conf. on Industrial, Mechanical, Electrical, and Chemical Engineering* (Surakarta) vol 2217 (Maryland: AIP Publishing)
[21] UNWTO 2015 *Annual report* (Madrid: World Tourism Organization)
[22] Spangenberg J H 2002 *Ecol. Indic.* **2** 295-309
[23] Eden M, Falkheden L and Malbert B 2000 *Plan. Theory Pract.* **1** 260-72
[24] UNEP & WTO 2005 *Making tourism more sustainable: a guide for policy makers* (Madrid: United Nations Environment Progamme and World Tourism Organization)
[25] Swarbrooke J 1999 *Sustainable tourism management* (Sheffield: CABI Publishing)
[26] Kozic I and Mikulic J 2011 *University of Zagreb Working Paper Series*
[27] WTO & Ministry of Tourism and Environment of Montenegro 2007 *Sustainable tourism indicators and destination management* (Montenegro: World Tourism Organization & Ministry of Tourism and Environment of Montenegro)
[28] Asadpourian Z, Rahimian M and Gholamrezai 2020 *Soc. Indic. Res.* **152** 289-315
[29] Asmelash A G and Kumar S 2019 *Heliyon* **5** e01335

[30] Lozano-Oyola M, Blancas F J, Gonzalez M and Caballero 2019 *J. Environ. Manage.* **250**
[31] Huang Y and Coelho V R 2017 *Tour. Manag.* **59** 510-527
[32] Agyeiwaah E, McKercher B and Suntikul W 2017 *Tour. Manag. Perspect.* **24** 26-33
[33] Ashok S, Behera M D and Tewari H R 2019 *Environ. Monit. Assess.* **191** 801
[34] Weng L, He B J, Liu L, Li C and Zhang X 2019 *Sustainability* **11**
[35] Gkoumas A 2019 *Heliyon* **5** e02707
[36] Gallardo J A F, Ocerin J M C and de la Torre M G M V 2019 *Sustainability* **11**
[37] Valentin A and Spangenberg H 2000 *Environ. Impact Assess. Rev.* **20** 381-92
[38] Pitcher T J and Preikshot D 2001 *Fish. Res.* **49** 255-70
[39] Hout M C, Papesh M H and Goldinger S D 2013 *WIREs Cogn. Sci.* **4** 93-103
[40] OECD 2008 *Handbook on constructing composite indicators* (Paris: Organisation for Economic Co-Operation and Development)
[41] Bard J F 1992 *IIE Transactions* **24** 111-21
[42] Guarini M R, Battisti F and Chiovitti A 2018 *Sustainability* **10**
[43] Prescott-Allen R 1997 *Barometer of sustainability: measuring and communicating wellbeing and sustainable development* (Gland: The World Conservation Union)
[44] Mateo J R S C 2012 *Multi criteria analysis in the renewable energy industry* (New York: Springer) pp 63-72
[45] Tzeng G H and Huang J J 2011 *Multiple attribute decision making: methods and applications* (Boca Raton: CRC Press)

IOP Conf. Series: Earth and Environmental Science 1098 (2022) 012047 doi:10.1088/1755-1315/1098/1/012047

Investigation of time impact on electrocoagulation process to treat ablution wastewater

Mudofir Mudofir[1], Moh. Taufik[2], Ilzamha Hadijah Rusdan[2], Wahyu Dian Silviani[3], Purwono Purwono[3*]

[1]Sharia Faculty, UIN Raden Mas Said Surakarta, Jl. Pandawa, Pucangan, Kartasura 57168, Indonesia
[2]Center for Halal Research, UIN Raden Mas Said Surakarta, Jl. Pandawa, Pucangan, Kartasura 57168, Indonesia
[3]Center for Science and Technology, UIN Raden Mas Said Surakarta, Jl. Pandawa, Pucangan, Kartasura 57168, Indonesia

*purwono.ga@gmail.com

Abstract. The ablution ritual performed by Muslims before praying requires large amounts of clean water (3.9 L – 4.42 L/person). This is a challenge that needs to be addressed, as in some places, water is a scarce commodity. This study aims to treat ablution wastewater at Masjid Al-Bukhari UIN Raden Mas Said Surakarta using electrocoagulation. The treated water is expected to be reused for ablution rituals in the mosque. Laboratory-scale research methods were performed to treat ablution wastewater. The effect of electrolysis time was evaluated to determine efficiency of the method on the parameters of pH, temperature, total dissolved solid (TDS), conductivity, and ammonia level. Interestingly, ablution wastewater treated using electrocoagulation meets clean water quality standards. This means that the water is safe to reuse for ablution activities in mosques. The electrolysis time has a significant effect on the increase of pH value and the decrease of ammonia level. TDS levels decreased to 65.3% when treated for 40 min. Meanwhile, the conductivity value decreased to 526 μS/cm from the initial value of 1108 μS/cm when the water was treated for 60 min. Statistically, electrolysis time has a significant effect on the decrease of TDS, conductivity, and ammonia levels ($\alpha<0.05$).

1. Introduction

The ablution ritual performed by Muslims before performing prayers five times a day requires large amounts of clean water. The amount of clean water needed will increase tremendously in mosques, including the Al-Bukhari Mosque of UIN Raden Mas Said Surakarta. Based on studies conducted by Mafra [1] and [2] in Makassar and Palembang, Muslims estimate that 3.9 L – 4.42 L of water/ person are used by Muslims to perform ablution every day. The Al-Bukhari Mosque can accommodate around 400 people. Provided that pilgrims perform prayers five times a day, the mosque will need 7,800 – 8,840 L of clean water every day. This means that the mosque needs a continuous supply of clean water every day. However, the source of water in the mosque is well water that is affected by the seasons. As a result, when the dry season arrives, water discharge drops, and in turn disrupting the ablution ritual.

The wastewater of the ablution ritual can be treated using several methods, one of which is electrocoagulation methods. The electrocoagulation method is an environmentally friendly, effective, efficient, and inexpensive method that can be applied to the treatment of wastewater so that the wastewater can be reused for other purposes, including the ablution ritual. The basic principle of the electrocoagulation (EC) method is an oxidation and reduction reactions that will destabilize small colloidal particles so that metal hydroxides are formed [3]. There are several parameters that affect the electrocoagulation process, namely pH, electrode type, electrical current, duration, etc. [3]. Among these parameters, the duration of processing is an important parameter that affects the efficiency and cost of processing. However, studies on the ablution wastewater treatment using the electrocoagulation method are still limited. Therefore, the purpose of this study is to determine the effect of the electrocoagulation time of the ablution wastewater on the removal of physical and chemical parameters. The treated water is expected to be reused for ablution activities in mosques.

2. Method

This study was carried out using a tubular electrocoagulation reactor (EC) with a total volume of 600 mL. The reactor was equipped with an inlet and an outlet. An adapter (Kayama, Japan) of 15.1 V was used to match the electric current with the voltage. The adapter was connected to stainless steel (diameter of 5 cm x 10 cm) and iron (Fe) (cylindrical shape, diameter 2 cm × 10 cm). The entire electrode bodies were submerged in the wastewater. All research procedures were carried out at room temperature. The electrodes were washed with ethanol and water to remove impurities on their surface before starting each process. The wastewater was streamed into the EC reactor using a 12 V DC pump. The treated water was collected in a 5 L capacity water reservoir. At this stage is aimed to determine the effect of the electrolysis time on physical and chemical parameters. The analysis was carried out on the ablution wastewater before and after processing using EC. Some of the parameters analyzed were conductivity (μS/cm) using a conductivity meter, pH using a pH meter (Trans Instrument, Singapore) and ammonia-nitrogen (NH3-N) using the Nessler method [4]. The effect of the electrolysis time of the ablution wastewater were analyzed using Origin® software. The results of the analysis were tabulated using Microsoft Excel® software.

3. Results and Discussion

3.1. Effect of electrolysis time on the pH of wudu wastewater

In this study, the initial pH of the ablution wastewater from the Al-Bukhari Mosque of UIN Raden Mas Said Surakarta was 8.4. Treatment results showed that the electrolysis time had a significant effect on the pH value of the wastewater. The longer the time used to treat the wastewater, the more the pH of the wastewater increased **(Fig. 1).** The pH reached a pH of 10 when electrolysis was carried out for 50 minutes. This increase corresponds to a statistical test that shows that the value of the calculated significance is lower than the set value (0.033< 0.05). Therefore, the treatment of the ablution wastewater with the electrocoagulation method should not be carried out for more than 20 minutes to prevent the pH of the treated wastewater from exceeding the standard determined by the Minister of Health of Indonesia and WHO. According to the regulation of the Minister of Health No. 32/2017, the pH of water is between 6.5 and 8.5. This value is not much different from the WHO standard for fresh water, which is 6.5 – 9.

The change in the pH of the ablution wastewater is closely related to the type of electrodes used, which in this case is iron and stainless steel. The use of iron and stainless steel as the anode will cause the formation of Fe^{2+} ions due to oxidation reactions in the anode [5]. The reaction is as follows:

$$Fe \rightarrow Fe^{2+} + 2\,e^- \quad (1)$$

Meanwhile, at the cathode there is a reduction of water that produces hydroxide ions and H_2 gas.

$$2\,H_2O + 2\,e^- \rightarrow 2\,OH^- + H_2 \quad (2)$$

The hydroxide ions formed in Equation 2 are the cause of pH increase in the solution. The longer the electrolysis process is carried out, the more OH^- ions formed at the cathode, thus the higher the pH of the solution. The OH^- ions will then bind to the Fe^{2+} ions formed due to oxidation at the anode producing $Fe(OH)_2$. $Fe(OH)_2$ has a high affinity for pollutants in wastewater [6].

Another reaction that is also responsible for the process of separating pollutants is the formation of Fe^{3+} ions that have a higher charge density [5], which means they can better support the coagulation-precipitation mechanism. Fe^{3+} ions can be formed directly at the flow anode when the voltage is appropriate (equation 3) or due to the oxidation of Fe^{2+} ions by oxygen (equation 4) [7].

$$Fe \rightarrow Fe^{3+} + 3\ e^- \quad (3)$$

$$4Fe^{2+} + 10\ H_2O + O_2 \rightarrow 4Fe(OH)_3 + 8H^+ \quad (4)$$

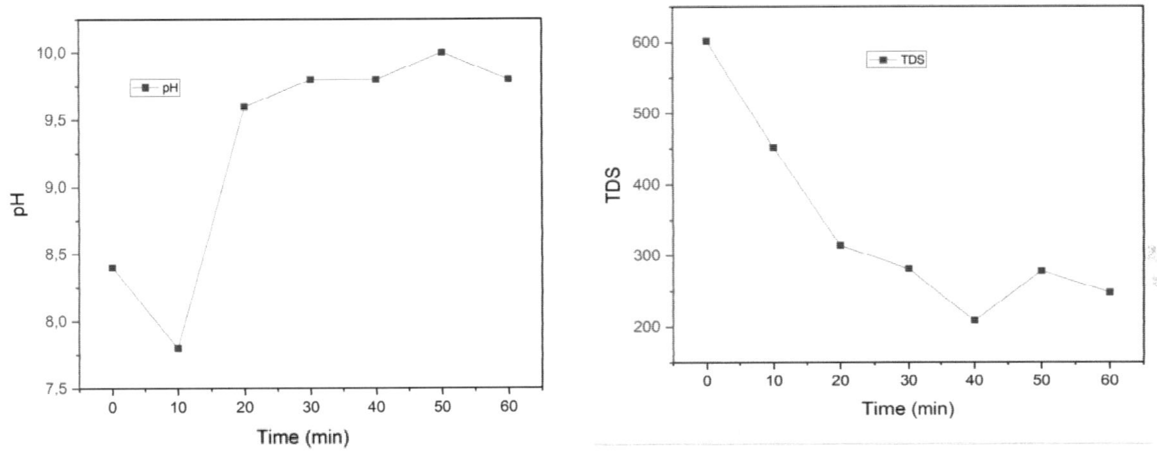

Figure 1. Graph of changes in pH and TDS values of the ablution wastewater when treated using electrocoagulation with a power of 15.1 V, a current of 1.2 A for 60 minutes

3.2. Effect of electrolysis time on Total Dissolved Solid (TDS)

TDS is one of the parameters of water quality [8,9] that describes the total concentration of solutes in the water. The solute is in the form of inorganic salts and a small amount of organic substances [10]. High TDS will lead to a decrease in acceptance of a water [8]. In this study, electrocoagulation of the ablution wastewater was carried out to reduce the TDS value. **Figure 1** shows a graph of the changes in the TDS value of the ablution wastewater when treated by electrocoagulation with a power of 15.1 V and an electric current of 1.2 A at a time of 0 to 60 minutes. Based on statistical tests, it was found that the electrolysis time had a significant effect on the TDS value with a significance level of 0.01669 < 0.05. The initial TDS value was 602 mg/L and the lowest value was 209 mg/L resulting from an electrolysis time of 40 minutes. The TDS value of the ablution wastewater before electrocoagulation was 602 mg/L. This value is still in accordance with the standard established by the Ministry of Health of Indonesia through Regulation No. 32/2017 on water for sanitary hygiene purposes, which is a maximum of 1000 mg/ L. However, the TDS value is still higher than the drinking water standard of the US Environmental Protection Agency (USEPA) which is a maximum of 500 mg/ L. The results show that the electrocoagulation time has a significant effect on the TDS value with a signivicance level of 0.01669 < 0.05. The TDS value generally decreases with the length of the electrocoagulation time. This result is in accordance with the results of Osman's et al. [11] and El-Ezaby et al. [12] which successively performed electrocoagulation on tanning wastewater and wastewater from an orange juice industry. The lowest TDS value was found when the wastewater was treated for 40 minutes, which was 209 mg/L. The percentage decrease in the TDS value was 65.3%. The decrease in the TDS value was higher than the value obtained by Bagastyo et al. [13] (49%) but lower than those obtained by Bakry et al [11,14,15] (78%, 70.40%, and 91.50% respectively). According to Galvo et al. [16] and

Guo et al. [17] there are several factors that affect the result of electrocoagulation, which are the material and shape of the anode, pH, current density, electrode distance, conductivity, current strength, electrocoagulation time and power.

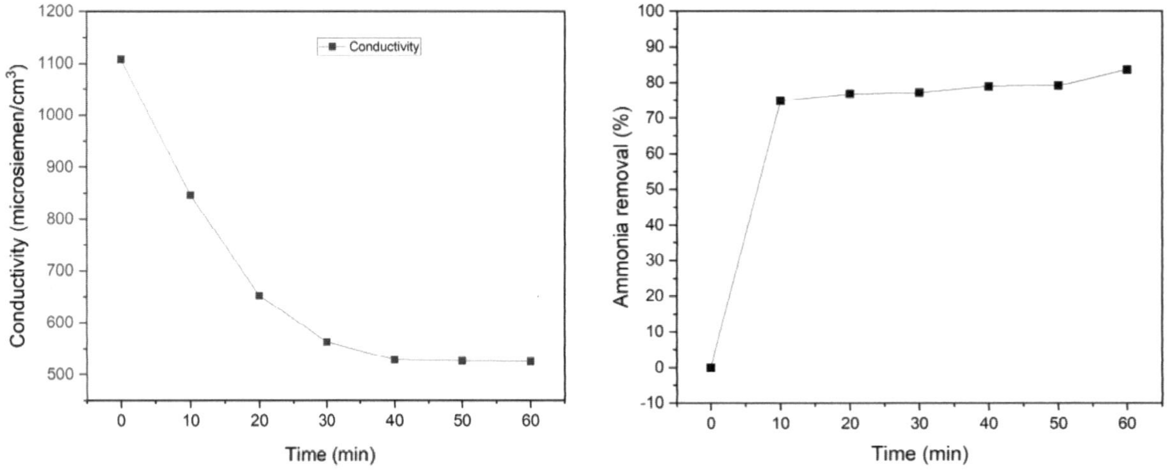

Figure 2 Graph of changes in conductivity and ammonia values of wudu wastewater when treated using electrocoagulation with a power of 15.1 V, current 1.2 A for 60 minutes

3.3. Effect of electrolysis time on conductivity

Conductivity is a measure of the ability of water to be an electrical conductor, this ability is influenced by salt or the number of ions contained in water, the higher the salt content, the higher the conductivity value, causing the purity of the water to be lower [18]. Electrocoagulation is a flocs formation that happens because positive ions bond with impurities and/ or pollutants on the surface of the water and on the bottom of the water [19]. In this study, the flocs were formed as a result of ions released by the iron and stainless-steel electrodes during the electrocoagulation process. **Figure 2** shows a graph in changes of the conductivity value of the ablution wastewater when treated with electrocoagulation with a power of 15.1 V and a current of 1.2 A. Based on statistical tests, the duration of electrocoagulation was obtained to significantly affected the decrease of conductivity value with significance level of 0.0105< 0.05. The initial conductivity value was 1108 μS/cm and after treatment with electrocoagulation the lowest value obtained was 526 μS/cm resulting of the electrolysis run for 60 min. The Indonesian Ministry of Health Regulations No. 32/2017 which regulates environmental health quality standards and water health requirements for sanitary hygiene purposes, does not directly mention conductivity as one of the parameters in determining the quality of water quality standards used for sanitary hygiene purposes. However, the provision of a good conductivity value is mentioned by WHO (2022) in the guidelines for drinking water quality, that the maximum conductivity limit in drinking water is 1500 μS/cm. Meanwhile, according to Mokhtar et al.. [22] and Nishanthiny et al [23] water is classified into 4 parts according to the danger of salinity or its salt content, which is low salinity (< 250 μS/cm), medium salinity (250-750 μS/cm), high salinity (750¬-2250 μS/cm), and very high salinity (>2250 μS/cm). These shows that the treated ablution wastewater in this study still meets the conductivity requirements as clean water and has a degree of salinity classified as moderate.

3.4. Effect of electrolysis time on ammonia level

The contamination of nitrogen (ammonia, nitrates, nitrites) in water can cause serious problems, such as deterioration in water quality, eutrophication of waters, and potential harm to animal and human health. Nitrogen compounds are nutrients that can cause algae bloom, which reduces the amount of oxygen in the water, while ammonia nitrogen is toxic to aquatic life [24]. **Figure 2** is a graph of the the efficiency of removal (%) ammonia content of the ablution wastewater when treated with electrocoagulation with a power of 15.1 V and a current of 1.2 A for 60 minutes. Based on statistical tests, it was obtained that the electrolysis time had a significant effect on ammonia levels with a signivicance level of 0.03< 0.05. The initial ammonia level was 4.72 mg/ L and the lowest ammonia level after being treated was 0.48 mg/ L which was obtained from the electrocoagulation treatment for 60 minutes. Ammonia levels are not listed as environmental health quality standards and water health requirements for sanitary hygiene purposes in the regulation of the Ministry of Health of Indonesia No.32/2017. However, wastewater produced from ablution rituals will enter water bodies such as rivers, lakes, and seas. Therefore, ammonia levels must be properly managed to pollute the environment. According to Indonesia's Government Regulation No. 22/2021 concerning the Implementation of environmental protection and Management, ammonia levels in river water are 0.5 mg/ L. Ammonia removal is possible due to the formation of Fe^{3+} ions from the electrodes and the adsorption and precipitation/ floatation of contaminants through iron hydroxide flocs produced through an electrocoagulation process [25]. The duration of electrolysis of the ablution wastewater significantly reduces the ammonia levels of the wastewater. As a result, the treated wastewater meets the required quality standards.

4. Conclusion

This study aims to treat wastewater ablution at the Al-Bukhari Mosque of the UIN Raden Mas Said Surakarta using electrocoagulation. The results showed that the ablution wastewater that was treated using electrocoagulation meets water quality standards for sanitary hygiene purposes based on the parameters pH, TDS, conductivity, and ammonia. This water is safe when reused for ablution activities in mosques. Statistically, the electrolysis time of 60 minutes has a significant effect on reducing the levels of TDS, conductivity, and ammonia levels ($\alpha<0.05$). The treated water is expected to be reused for ablution activities in mosques. The challenge of implementing the ablution wastewater electrolysis method is a fiqh study of treated water to determine whether treated water with electrocoagulation is in accordance with the rules of law in Islam, and the study of consumer acceptance of wastewater that has been reprocessed to be used as wastewater.

Acknowledgement
Thank you to Litapdimas Ministry of Religion and Institute for Research and Community Service (LP2M) UIN Raden Mas Said Surakarta who have fully funded this research for the 2022 budget year.

References

[1] Mafra R 2018 Pengukuran Durasi Waktu Berwudhu dan Volume Penggunaan Air Pada Masjid Masjid di Kota Palembang Measurement of Duration of Ablution Time and Volume of Water Use In mosques in the city of Palembang **2** 71–9 [in Indonesian Language]
[2] Natsir M, Agus M, Rachmadani A, Mushbir A, Fahsa A and Fachry A 2020 Analisis Kuantitas Air Bekas Wudhu Pada Masjid Kota Makassar 2020 *Nas. Ilmu Kesehat.* **3** 44–60 [in Indonesian Language]
[3] Bener S, Bulca Ö, Palas B, Tekin G, Atalay S and Ersöz G 2019 Electrocoagulation process for the treatment of real textile wastewater: Effect of operative conditions on the organic carbon removal and kinetic study *Process Saf. Environ. Prot.* **129** 47–54
[4] APHA 2017 *Standard Methods for the Examination of Water and Wastewater* (Washington)
[5] Garcia-Segura S, Eiband M M S G, de Melo J V and Martínez-Huitle C A 2017 Electrocoagulation and advanced electrocoagulation processes: A general review about the fundamentals, emerging applications and its association with other technologies *J. Electroanal. Chem.* **801** 267–99
[6] Tahreen A, Jami M S and Ali F 2020 Journal of Water Process Engineering Role of electrocoagulation in wastewater treatment : A developmental review *J. Water Process Eng.* **37** 101440
[7] Dura A and Breslin C B 2019 Electrocoagulation using stainless steel anodes : Simultaneous removal of phosphates , Orange II and zinc ions *J. Hazard. Mater.* **374** 152–8
[8] Maliki A A A, Chabuk A, Sultan M A, Hashim B M, Hussain H M and Al-Ansari N 2020

Estimation of Total Dissolved Solids in Water Bodies by Spectral Indices Case Study: Shatt al-Arab River *Water. Air. Soil Pollut.* **231**

[9] Rusydi A F 2018 Correlation between conductivity and total dissolved solid in various type of water: A review *IOP Conference Series: Earth and Environmental Science* vol 118 (Institute of Physics Publishing)

[10] Hussain I, Shakeel M, Faisal M, Soomro Z A, Hussain M and Hussain T 2014 Distribution of Total Dissolved Solids in Drinking Water by Means of Bayesian Kriging and Gaussian Spatial Predictive Process *Water Qual. Expo. Heal.* **6** 177–85

[11] Osman A T, Elamin M R and Almalki M H 2017 Treatment of Tannery Wastewater with Nano-Electrocoagulation Process *J. Environ. Anal. Toxicol.* **07**

[12] El-Ezaby K H, El-Gammal M I and Shaaban Y A 2021 Using electro- and alum coagulation technologies for treatment of wastewater from fruit juice industry in New Damietta City, Egypt *Environ. Monit. Assess.* **193**

[13] Bagastyo A Y, Sidik F, Anggrainy A D, Lin J L and Nurhayati E 2022 The Performance of Electrocoagulation Process in Removing Organic and Nitrogenous Compounds from Landfill Leachate in a Three-Compartment Reactor *J. Ecol. Eng.* **23** 235–45

[14] Bakry S A, Matta M E and Zaher K 2018 Electrocoagulation process performance in removal of TOC, TDS, and turbidity from surface water *Desalin. Water Treat.* **129** 127–38

[15] Padmaja K, Cherukuri J and Anji Reddy M 2020 A comparative study of the efficiency of chemical coagulation and electrocoagulation methods in the treatment of pharmaceutical effluent *J. Water Process Eng.* **34**

[16] Galvão N, de Souza J B and de Sousa Vidal C M 2020 Landfill leachate treatment by electrocoagulation: Effects of current density and electrolysis time *J. Environ. Chem. Eng.* **8**

[17] Guo Z, Zhang Y, Jia H, Guo J, Meng X and Wang J 2022 Electrochemical methods for landfill leachate treatment: A review on electrocoagulation and electrooxidation *Sci. Total Environ.* **806** 150529

[18] Astuti A D 2014 Kualitas Air Irigasi Ditinjau dari Parameter DHL, TDS, pH pada Lahan Sawah Desa Bulumanis Kidul Kecamatan Margoyoso *J. Litbang Media Inf. Penelitian, Pengemb. dan IPTEK* **10** [in Indonesian Language]

[19] Novita S 2017 Pengaruh Variasi Kuat Arus Listrik Dan Waktu Pengadukan Pada Proses Elektrokoagulasi Untuk Penjernihan Air Bakupdam Tirtanadi Ipa Sunggal *Semin. Nas. Pendidik. Dasar Univ. Negeri Medan* [in Indonesian Language]

[20] MENKES R 2017 *Peraturan Menteri Kesehatan Republik Indonesia Nomor 32 Tahun 2017 Tentang Standar Baku Mutu Kesehatan Lingkungan dan Persyaratan Kesehatan Air Untuk Keperluan Higiene Sanitasi, Kolam Renang, Solus Per Aqua, dan Pemandian Umum* (Indonesia) [in Indonesian Language]

[21] WHO 2022 Guidelines for drinking-water quality: Fourth edition incorporating the first and second addenda *World Heal. Organ.*

[22] Mokhtar M Bin, Aris A Z, Abdullah M H, Yusoff M K, Abdullah M P, Idris A R and Raja Uzir R I 2009 A pristine environment and water quality in perspective: Maliau Basin, Borneo's mysterious world *Water Environ. J.* **23**

[23] Nishanthiny S C, Thushyanthy M, Barathithasan T and Saravanan S 2010 Irrigation water quality based on hydro chemical analysis, Jaffna, Sri Lanka *Am. J. Agric. Environ. Sci.* **7**

[24] Han B, Butterly C, Zhang W, He J and Chen D 2021 Adsorbent materials for ammonium and ammonia removal: A review *J. Clean. Prod.* **283** 124611

[25] Choudhury M R, Rajagopal R, Meertens W and Rahaman M S 2022 Nitrogen and organic load removal from anaerobically digested leachate using a hybrid electro-oxidation and electro-coagulation process *J. Environ. Manage.* **311** 114761

IOP Conf. Series: Earth and Environmental Science 1098 (2022) 012048 doi:10.1088/1755-1315/1098/1/012048

Tracing flow directions of the Pentung Allogenic River in Gunungsewu Karst Area, Gunungkidul, using artificial tracer test

A Cahyadi[1,2,3], T N Adji[2,3], E Haryono[2,3], M Widyastuti[2,3], R F Agniy[4]

[1]Geography Study Program, Faculty of Geography, Universitas Gadjah Mada, Yogyakarta, Indonesia
[2]Karst Research Group, Faculty of Geography, Universitas Gadjah Mada, Yogyakarta, Indonesia
[3]Department of Environmental Geography, Faculty of Geography, Universitas Gadjah Mada, Yogyakarta, Indonesia
[4]Geographic Information Science, Universitas Mahakarya Asia, Yogyakarta, Indonesia

ahmad.cahyadi@ugm.ac.id

Abstract. Previous studies in the Gunungsewu Karst Area categorized the Pentung Allogenic River catchment that drains to the Sawah Ombo Pond as the upper part of the Bribin Karst Drainage Basin (KDB). However, more detailed results from natural isotope studies after 2010 have provided reasons to postulate that the groundwater does not flow from the catchment to Bribin KDB. Therefore, the research aimed to trace the direction of Pentung's groundwater flows in Gunungsewu's karstic hydrogeological system. To achieve this, an artificial tracer test was used. There were two fluorescent dye injection namely uranine and tinopal. The results confirmed that the allogenic river's catchment supplied groundwater to Beton KDB instead of Bribin KDB. While the artificial tracers were not detected in the water at Gilap Cave and Bribin Underground River, they appeared in Jomblangan Cave and Sriti Spring, parts of Beton KDB.

1. Introduction

Although hydrogeological studies in the Gunungsewu Karst Area started in 1979, many questions and debates on a thorough and detailed understanding of the area's hydrogeology remain unresolved. MacDonalds and Partners' research [1] in 1979–1984 is the first comprehensive study on this subject (part of the results was published in Waltham et al., 1983[2]), but it is still far from sufficiently describing the area's specific conditions. After mapping hundreds of caves, springs, and karst windows and assessing the connectivity between underground rivers using an artificial tracer test at ten locations, it produced a prediction map of the underground river flows known as the Bribin-Baron Karst Drainage Basin (KDB). However, the Pentung Allogenic River is still missing from the study.

In 1993, the Faculty of Forestry, Universitas Gadjah Mada (1993)[3], tried to delineate Bribin KDB. The research is the first to state that the Pentung Allogenic River recharges the Bribin Underground River. It added another point of observation to the analysis, i.e., Jomblangan Cave/ Karst Window (west of Sawah Ombo Pond to which the allogenic river drains), to link Sawah Ombo Pond with Bribin Underground River. However, even though an artificial tracer test was conducted, the results are not discussed in the report; thus, it cannot be decided if the KDB boundaries were based on estimates or what the test had found. Adji and Sudarmadji (2008)[4] also share some doubt on the delineation results for the same reason.

Despite the shortcomings, several previous studies have referred to the 1993 boundaries to define Bribin KDB, including Adji [5], Widyastuti [6], Pratama [7], and Endarto [8]. Later, Gunawan (2013)[9] revised the 1993 boundaries using remote sensing image interpretation and digital elevation models to delineate Bribin-Baron KDB. This study inserted the catchment of the Pentung Allogenic River into the Bribin-Baron KDB system.

Since an incident during an artificial tracer test in 2012, there have been reservations about excluding the Pentung Allogenic River from the recharge system of Bribin-Baron KDB. A tracer (possibly, Sulforhodamine B) injected in Sawah Ombo Pond was expected to emerge in the Bribin Underground River; instead, it appeared in Beton Spring and caused the water in Beton Dam to turn red. Subsequent hydrogeochemical studies using natural isotope and injection of an artificial tracer (uranine), e.g., Sidauruk et al. [10, 11], found that Sawah Ombo Pond was connected to Sriti Spring. Sidauruk et al. [11] mentioned that the spring is a leak of the underground river system from Sawah Ombo Pond to Beton Spring. However, it ignored the existence of Jomblangan Cave, which the Faculty of Forestry UGM [2] claimed was connected to Sawah Ombo Pond, and Seropan Ponor, which is the headwater of Sriti Spring (unknown lower course). Also, based on the natural isotope analysis, Sidauruk et al. [10] explained that Sawah Ombo Pond does not issue water to the Bribin Underground River even though no artificial tracer test was performed in this section. Later, Setiawan et al. [12] corroborate the findings of Sidauruk et al. [10] through natural isotope analysis.

An artificial tracer test by Cahyadi [13, 14] discovered that the western part of the allogenic river in the Ponjong Hydrogeological System supplies groundwater to Gremeng Spring instead of to Bribin KDB and that Gremeng KDB is separate from Beton KDB. According to another artificial tracer test by Cahyadi [15] at Seropan Ponor, it turns out that the river draining to the ponor (including from Sriti Spring) flows to Beton Spring through a single conduit of reasonably large size. These findings then raised a question as to whether the water from Sawah Ombo Pond flows directly to Beton Spring and partially leaks at Sriti Spring before reaching Seropan Ponor or entirely flows through Sriti Spring to Seropan Ponor and then Beton Spring.

Based on the problems described above, this study sought to trace the movement of groundwater flow from the catchment of the Pentung Allogenic River emptying into Sawah Ombo Pond using an artificial tracer test. In detail, it tried to answer several questions related to (1) connectivity and characteristics of the tunnels or passages along the Sawah Ombo Pond-Beton Spring system and connectivity between these feature pairs: (2) Jomblangan Cave-Sawah Ombo Pond, (3) Jomblangan Cave-Sriti Spring, (4) Sawah Ombo Pond-Gilap Cave (part of Bribin KDB), and (5) Jomblangan Cave-Gilap Cave.

2. Methodology

The research's interest focused on the Pentung Allogenic River catchment area and Gunungsewu Karst Area, especially in the Ponjong Hydrogeological Subsystem (Figure 1). The catchment area is located in Panggung Masif Hills, which are old (tertiary) volcanic rocks that are part of the Semilir Formation and are primarily composed of sandstone and volcanic ash deposits from past explosive eruptions. It drains to the Sawah Ombo Pond in the transition zone between Semilir and Wonosari Formations consisting mainly of reef limestone and layered limestone. Administratively, the study area is in Sawahan, Umbulrejo, and Tambakromo Villages, Ponjong District, Gunungkidul Regency, Province of Daerah Istimewa Yogyakarta. It is in the northeastern part of the Gunungkidul Regency.

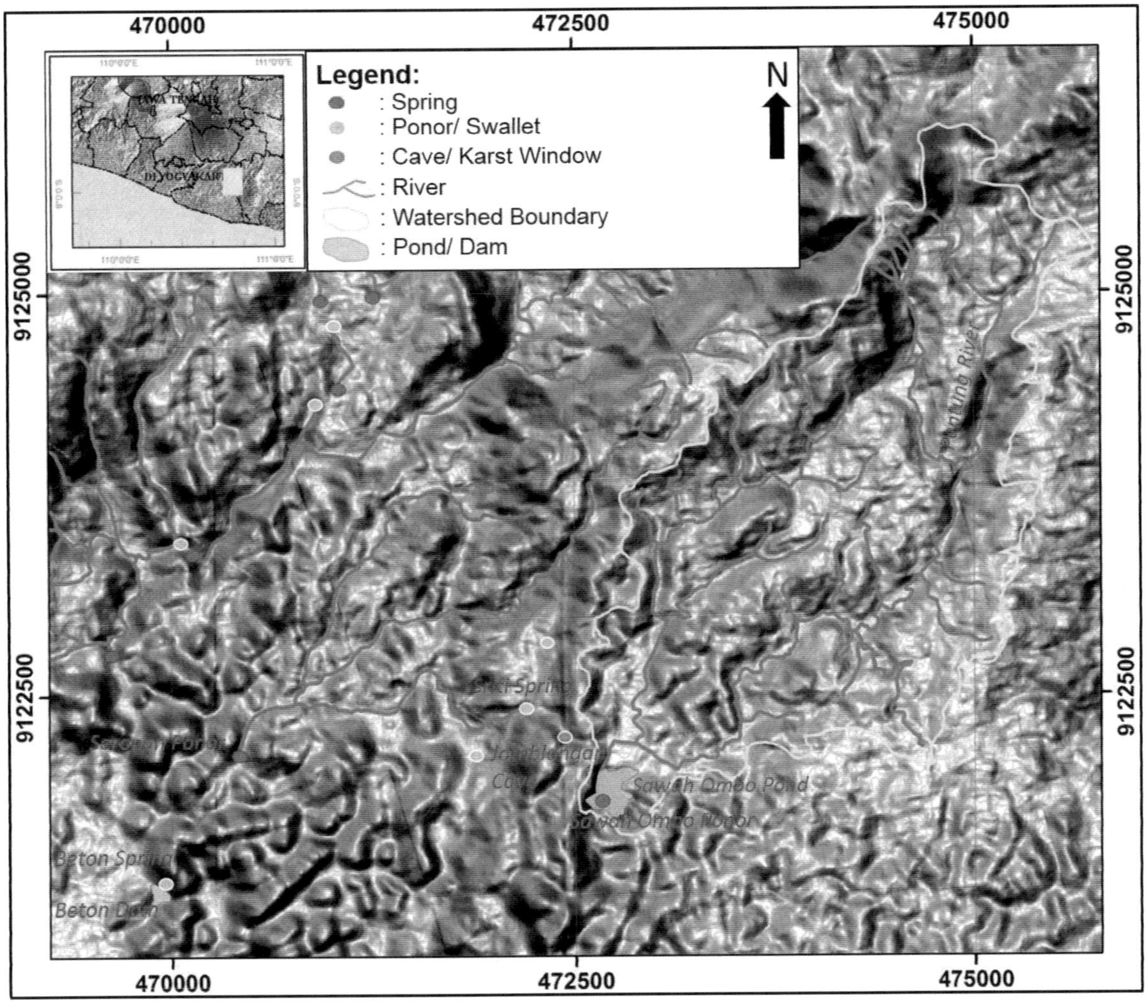

Figure 1. Map of the Study Area.

The research used an artificial tracer test with two dyes namely uranine and tinopal. Both are fluorescent dyes that are safe to use in aquatic environments and can be detected with a fluorometer even in minute concentrations. The fluorometers used were FL-24 and FL-30, with a minimum detection limit of 2×10^{-11} gram/ml and the ability to capture and record data for quite a long time at a predefined interval. During the study, the data collection interval was set at 5 minutes.

There were two fluorescent dye injection sites namely Sawah Ombo Pond (using tinopal) and Jomblangan Cave (uranine). The fluorometer was installed at Jomblangan Cave, Beton Spring, and Gilap Cave. In addition, changes in the water color were observed visually, and charcoal was placed in several locations suspected of being connected with Sawah Ombo Pond and Jomblangan Cave. Some of these locations are Selonjono, Sriti, and Gedong Springs. The results of the hydrochemograph data recording by fluorometer were then analyzed using Fluo software.

3. Results and discussion

The artificial tracer test analysis discovered connectivity between Sawah Ombo Pond and Jomblangan Cave (Figure 2). Jomblangan Cave is a karst window formed after the cave's roof collapsed. It is a hole through which an underground river can be seen from the surface. The tinopal solution injected in Sawah Ombo Pond was detected in Jomblangan Cave about 35 minutes after pouring. The relatively fast travel time with a one-peaked breakthrough curve (BTC) indicated an underground tunnel of a fairly large size in the form of a single conduit that connected Sawah Ombo and Jomblangan Cave. This finding corresponds to Palmer [16, 17], which explained that the underground tunnels influenced by allogenic rivers are generally single or in the form of caves with numerous branched passages (maze caves). The current research confirmed the artificial tracer test results conducted in the Seropan Ponor-Spring Beton system [15] and the cave passages in the upper course of the Gremeng KDB fed by the allogenic flows from the Panggung Masif Hills.

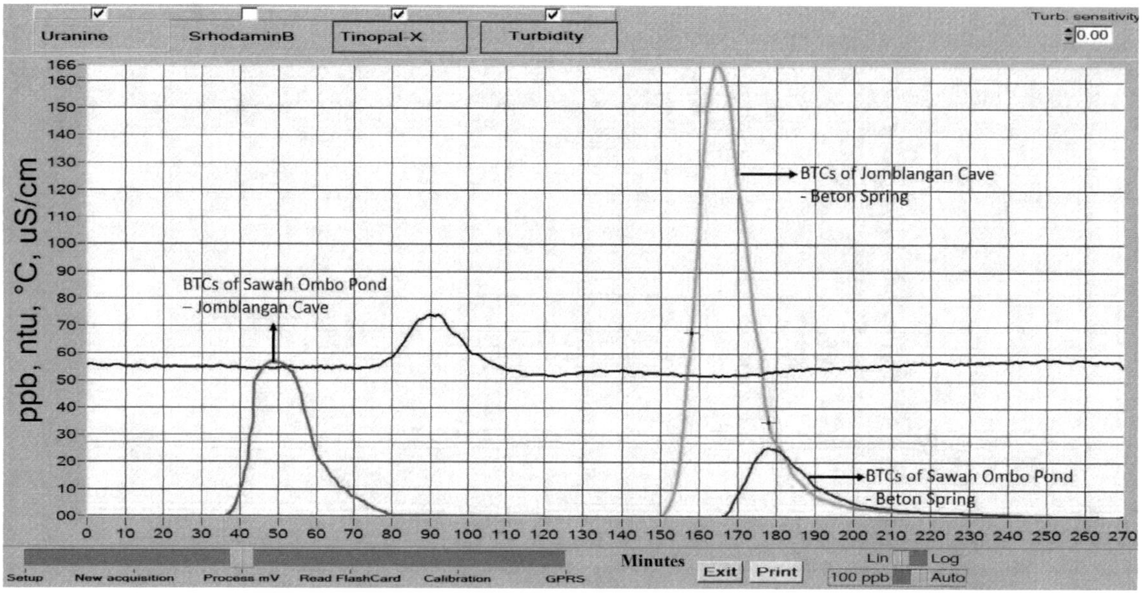

Figure 2. Breakthrough Curves of the Tracer/Dye Concentrations Recorded by Fluorometers at the Jomblangan Cave and Beton Spring.

The BTC presented in Figure 2 also showed connectivity between Jomblangan Cave and Beton Spring. During the visual observation, the water in Sriti Spring turned green after the dye injection in Jomblangan Cave (Figure 3), meaning that the cave's water flows to it before reaching Beton Spring, which corresponds to Sidauruk et al. [11]. However, Figure 2 refutes the conclusion of Sidauruk et al. [11], stating that Sriti Spring is a leak of an underground river flowing from Sawah Ombo Pond to Beton Spring. The reason is that the BTCs generated from the artificial tracer test results in Jomblangan Cave (uranine) and Sawah Ombo Pond (tinopal) had one peak, suggesting a single conduit. Therefore, it can be inferred that the groundwater flows from Sawah Ombo to Jomblangan Cave through a developed single conduit before it appears in Sriti Spring and flows as a surface river. The surface river then enters Seropan Ponor, which is connected to Beton Spring through a single large conduit [15]. The prediction of the flow from Pentung River to Beton Spring is presented in detail in Figure 4.

Figure 3. Green Color Indicating the Presence of Uranine in the Water at Sriti Spring after the Dye Injection in Jomblangan Cave.

It is most likely that the emergence of an underground river at Sriti Spring is due to the Sumbergiri fault. This fault displaced the terrain in its east upward and west downward, cutting the groundwater level—and the underground river—to appear as springs. Two other springs were found along the fault line namely Selonjono Spring (north of Sriti Spring) and Gedong Spring (south of Sriti Spring). Unlike Sriti Spring, which is a resurgence of an underground river, both are epikarst springs. It is evident from their after-rain conditions: murky water in Sriti Spring, while year-round clear water in Selonjono and Gedong Springs. Moreover, the charcoal placed in both epikarst springs showed no uranine or tinopal content, which further confirms their hydrogeological characteristics and, at the same time, indicates that they are not directly connected with Sawah Ombo Pond through a conduit flow. Topographically, Sawah Ombo Pond is located upstream (east) of the springs.

Fluorescent dyes content recording by the fluorometer suggested no connectivity between Sawah Ombo Pond and Gilap Cave and between Jomblangan Cave and Gilap Cave. These results confirmed previous findings that the Bribin Underground River is not fed by Sawah Ombo Pond [10, 12]. More specifically, it does not receive groundwater from the Pentung Allogenic River, especially through Gilap Cave. Therefore, the Bribin KDB boundaries proposed by the Faculty of Forestry, Universitas Gadjah Mada [3], in 1993 should be redefined.

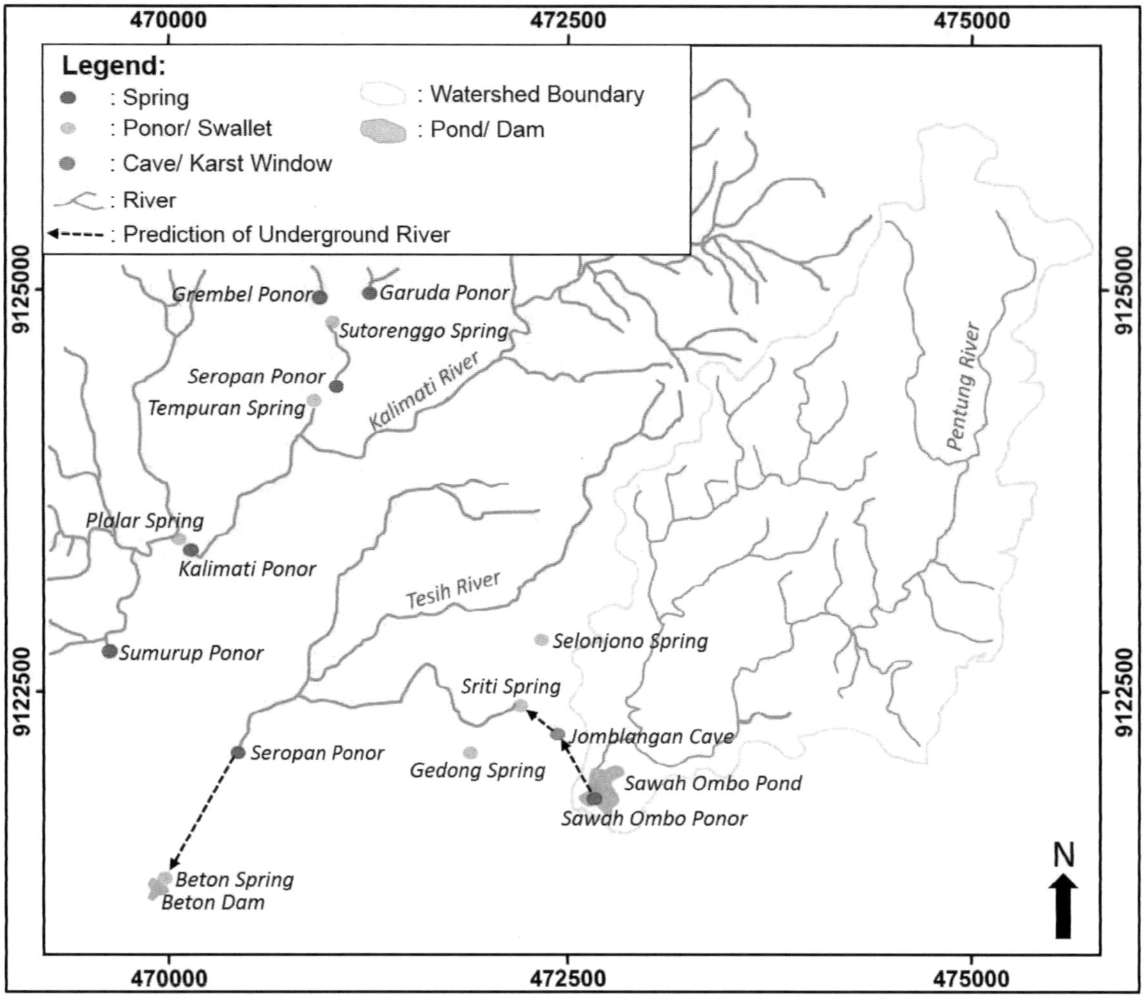

Figure 4. Flow Prediction Map of the Underground River Based on the Breakthrough Curve (BTC) Interpretation.

4. Conclusion

The artificial tracer test revealed a connection between several karstic features in the underground river system of Gunungsewu Karst Area: Sawah Ombo Pond-Jomblangan Cave-Sriti Spring-Seropan Ponor-Beton Spring. The passages linking Sawah Ombo Pond to Jomblangan Cave and Seropan Ponor to Beton Spring are a single conduit with large size. Also, based on the breakthrough curves(BTCs) analysis, Sriti Spring supplies groundwater to Beton Spring through a surface river that re-enters the underground system through Seropan Ponor, which contradicts the previous research finding that Sriti Spring is a leak of the underground river system between Sawah Ombo Pond and Beton Spring. It is evident in the one-peaked BTCs. The current study also found that the Pentung Allogenic River does not issue groundwater to the Bribin Underground River, primarily through Gilap Cave, because the fluorometer installed at the cave during the artificial tracer test did not detect any fluorescent dyes contents (uranine or tinopal) in the water.

Acknowledgment

This research is a small part of the results of the first author's dissertation in Faculty of Geography, Universitas Gadjah Mada. This research was funded by the 2022 Final Project Recognition (*Rekognisi*

Tugas Akhir, RTA) Grant, Universitas Gadjah Mada, with the assignment letter No. 3550/UN1.P.III/Dit-Lit/PT.01.05/2022, as part of the research entitled Characterization of Hydrodynamic and Hydrogeochemical Properties of Karst Drainage Basins in the Gunungsewu Karst Area for a Karst Critical Zone Management Model in the Tropical Karst Regions of Indonesia (*Karakterisasi Sifat Hidrodinamika dan Hidrogeokimia pada* Karst Drainage Basin *di Kawasan Karst Gunungsewu sebagai Model Pengelolaan* Karst Critical Zone *di Kawasan Karst Tropis Indonesia*).

References

[1] MacDonald and Partners. 1984. *Research Report.* Directorate General of Water Resources Development of the Republic of Indonesia.
[2] Waltham A C, Smart P L, Friederich H, Eavis A J and Atkinson T.C. 1983. *Cave Science, 10(2)*: 55–96.
[3] Faculty of Forestry, Universitas Gadjah Mada. 1993. *Research Report.* Forestry Service of Yogyakarta Special Province, Indonesia.
[4] Adji T N and Sudarmadji. 2008. *Proceedings of Integrated Water Resources Management Seminar.* Faculty of Geography, Universitas Gadjah Mada.
[5] Adji T N. 2009. *Dissertation.* Faculty of Geography, Universitas Gadjah Mada.
[6] Widyastuti M. 2014. *Dissertation.* Faculty of Geography, Universitas Gadjah Mada.
[7] Pratama W H. 2014. *BSc. Thesis.* Faculty of Geography, Universitas Gadjah Mada.
[8] Endarto, R. 2015. *MSc. Thesis.* The Graduate Program of Universitas Gadjah Mada.
[9] Gunawan, T. 2013. *Research Report.* Faculty of Geography, Universitas Gadjah Mada.
[10] Sidauruk P, Pujiinditanti E R and Satrio. 2015. *Jurnal Ilmiah Aplikasi Isotop dan Radiasi,* **11(1)**, 65-72.
[11] Sidauruk P, Satrio, Pujiindiyati E R and Aliyanta B. 2017. *Eksplorium,* **38(2)**,81-88.
[12] Setiawan T, Isnaini S, Asghaf N M A, and Effendi I. 2018. *Jurnal Lingkungan dan Bencana Geologi*, **9(3)**, 143-155.
[13] Cahyadi A, Haryono E, Adji T N, Widyastuti M, Riyanto I A, Naufal M, and Ramadhan F. 2020. *IOP Conference Series: Earth and Environmental Science*, **448.**
[14] Cahyadi A, Adji T N, Haryono E and Widyastuti M. 2021. *E3S Web of Conferences*, **325**.
[15] Cahyadi A, Haryono E, Adji T N, Widyastuti M, Naufal M, Ramadhan F, Agniy R F, and Riyanto I A. 2020. *Jurnal Geografi*, **12(2)**, 105 – 114.
[16] Palmer A N. 2001. *Acta Carsologica,* **30(2)**, 13-32.
[17] Palmer A N. 2007. Dayton, Ohio: Caves Books.

The Effect of *Allogenic Recharge* on Multi-Temporal Water Quality Variations in the Pindul Cave Underground River, Gunungkidul

N F Tastian[1], T N Adji[2] and A Cahyadi[2]

[1] Karst Research Group, Faculty of Geography, Universitas Gadjah Mada, Yogyakarta, Indonesia
[2] Department of Environmental Geography, Faculty of Geography, Universitas Gadjah Mada, Yogyakarta, Indonesia

naufal.tastian@gmail.com

Abstract. The Pindul Cave Underground River is a source of water that is used for irrigation, fulfilling domestic needs, and also tourism. The source of water in the Pindul Cave Underground River comes from the surface river flow that enters the ponor (Allogenic Recharge) so that it affects the water quality in the Pindul Cave Underground River. The research was conducted with the aim of temporally analyzing the water quality in the Pindul Cave Underground River. River water samples were taken periodically for one year. Water quality was analyzed by comparing samples from laboratory and field tests with quality standards and hydrogeochemical analysis using triangle piper diagrams and rectangular piper diagrams. The results show that allogeneic affixes have an effect in the form of a relatively fast dilution process in the rainy season, thereby reducing the calcium value. This condition causes the water quality in the rainy season to be better than the dry season because the calcium level decreases and does not exceed the water quality standard. The results are corroborated by triangular and rectangular piper diagram analysis which shows that the water in the Pindul Cave Underground River is classified as calcium enrichment, unpolluted water.

1. Introduction

Karst is a unique landform with a more developed subsurface hydrology system than surface hydrology [1]. The subsurface hydrological system stores and drains water in karst aquifers with soluble rock formations [2]. This rock formation is characterized by physical, morphological, and structural characteristics of carbonate rocks [3]. Karst aquifers are different from aquifers in other rock formations due to their very complex hydrological conditions [4]. The complexity of karst hydrological systems is caused by infiltration dualism (seepage and concentrated), porosity/flow dualism (conduit and fracture), and recharge mechanism (allogenic and autogenic) [5][6]. Karst tunnels will also be formed along with the development of the karst area [2].

The Gunungsewu Karst area is one of the karst areas in Indonesia that has an important role in providing water resources for the community in Gunungkidul Regency, Special Region of Yogyakarta. The importance of this water resource is influenced by the condition of the karst area which has developed porosity so that water that falls on the surface will flow through gaps or tunnels and collect in karst aquifers or underground rivers [7]. This causes dry conditions on the soil surface [8]. Therefore, subsurface water resources in karst areas are quite abundant and can be used for community water needs.

Underground rivers are karst landforms resulting from the dissolution process [9]. The Pindul Cave Underground River, located in Karangmojo District, Gunungkidul Regency, is one of the geosites of the Gunungsewu Geopark (Figure 1). Physiographically, Pindul Cave is located in the Wonosari Basin [10][11]. Pindul Cave has two types of karst aquifer systems, namely a diffuse system as evidenced by a pumping test and flownet construction [9] and a conduit system with an interconnected cave system [12].

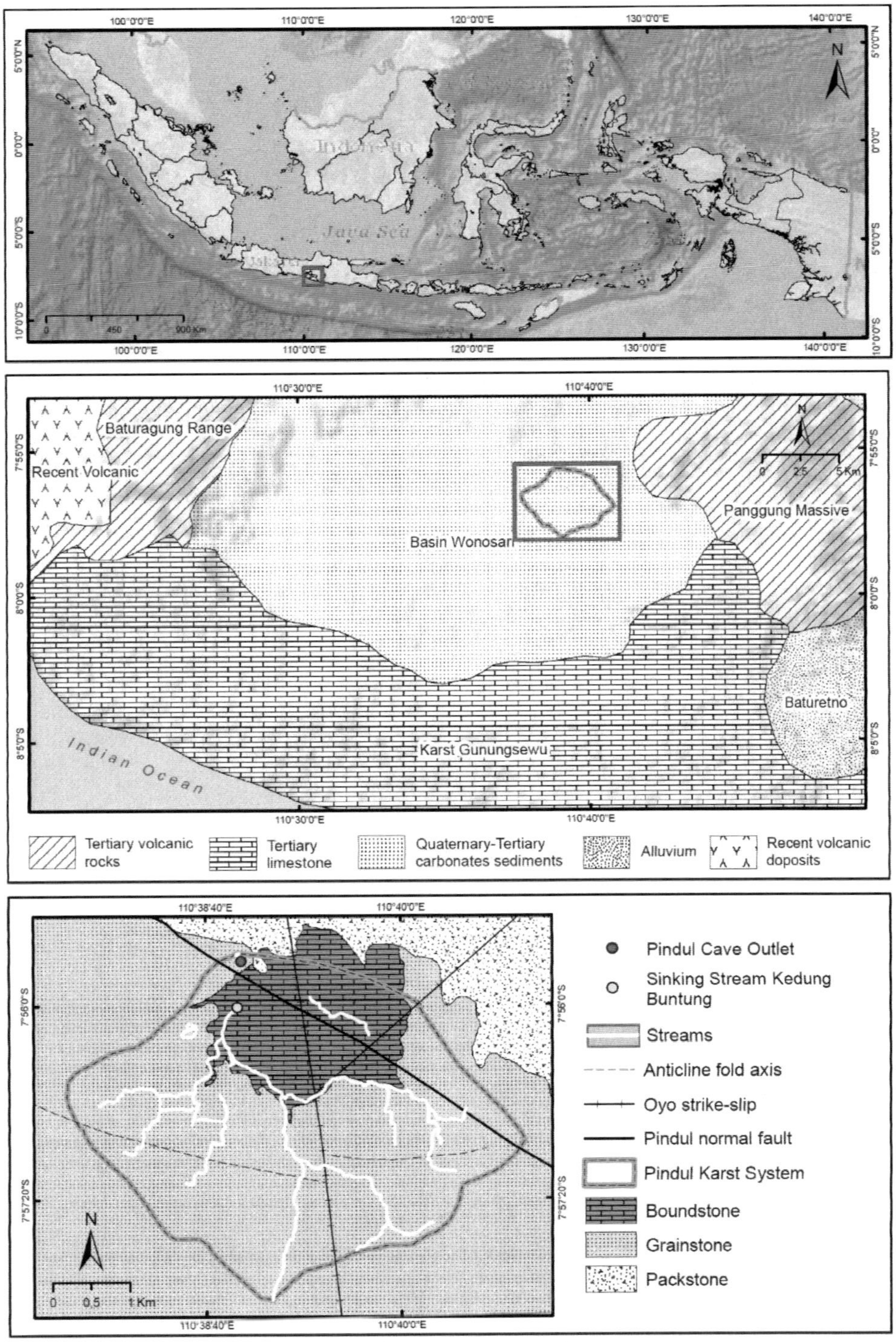

Figure 1. Map of Study Area.

Water resources in Pindul Cave are supplied by allogenic recharge from the Sinking Stream Kedung Buntung [13] (Figure 2). The Pindul Cave Underground River with water sources from surface rivers and developed porosity causes contaminants to easily reach groundwater so that the karst aquifer is vulnerable to pollution [14]. This needs to be a concern because the water resources in Pindul Cave are used for domestic purposes, water tourism, and agricultural irrigation. Therefore, understanding the characteristics of allogeneic recharge is very important because it supplies karst groundwater [15][16], particularly with regard to water quality.

Figure 2. Pindul Cave Underground River (a) and Sinking Stream Kedung Buntung (b)

The conceptual model of groundwater and the evolution of groundwater that affects water quality can be described by hydrogeochemistry [17][18]. This is because hydrogeochemical studies can provide an overview of the conditions of pollution and ion enrichment in water [19], chemical processes that occur in water, and applied studies for water quality analysis [20]. In addition, hydrogeochemical properties are also influenced by geological conditions [21] so that each location will have different water quality characteristics.

This study aims to analyze variations in water quality in the Pindul Cave Underground River which are affected by multi-temporal allogeneic recharge. The study location is in Karangmojo District, Gunungkidul Regency, Special Region of Yogyakarta which is in the Wonosari Formation with rock geology composed of reef limestone [22] (Figure 1).

2. Methodology

This research was conducted temporally within 1 year from November 2020 - October 2021. Water samples were taken once a month. Sampling was carried out at one location in Pindul Cave with coordinates 110°38’54,495” E, 7°55’41,852” S. The tools used in this research are water checker, alkalinity test kit, calcium test kit, and sample bottles. A water checker is used to measure water quality in the field directly including parameters of temperature, pH, EC, and TDS. Alkalinity test kit is used to measure HCO_3^- ions while calcium test kit is used to measure Ca^{2+} ions. Sample bottles are used to take water samples to be tested in the laboratory including Mg^{2+}, Na^+, K^+, Cl^-, and SO_4^{2-}. Data equipment in this research show in Tabel 1.

Table 1. Data Equipement

No	Data	Type	Source	Equipment
1	Ion			
	Ca^{2+}, HCO_3^-	Primary	Direct measurement in the field	Alkalinity test kit and calcium test kit
	Mg^{2+}, Na^+, K^+, Cl^-, SO_4^{2-}	Primary	Sampling and laboratory analysis	sample bottles
2	Physical And Chemical Parameters			
	pH	Primary	Direct measurement in the field	Water Quality Checker
	Electric Conductivity	Primary		
	Total Dissolved Solid	Primary		
	Temperature	Primary		

Water quality treatment is carried out by comparing the quality standard values from the Regulation of the Minister of Health Number 90 of 2002 using Microsoft Excel. Parameters that are not regulated in the previous Minister of Health Regulation use the standards set by experts [23]. Hydrogeochemical data processing was carried out using Easy Quim v5.0 software to obtain triangular piper diagrams and rectangular piper diagrams.

Data analysis in this study was carried out by comparing the measurement data in the field and the results of laboratory data with water quality standards. Analysis was also carried out by looking at the distribution pattern of the sample on the triangular piper diagram and the rectangular piper diagram. The classification of triangular piper diagrams is based on [24], while the classification of rectangular piper diagrams is based on [25]. The use of Stif Diagram to determine the uniformity of the hydrochemical properties of existing groundwater. The data used in the Stiff Diagram analysis is the major ion concentration in meq/l units.

3. Result and Discussion

Multi-temoral data on water quality in the Gua Pindul Underground River which includes major ions and water physical and chemical parameters are shown in Table 2. 02 – 7.79 which belongs to the classification of alkaline water. This is because the study location is in a karst landform which is dominated by limestone. The total dissolved solid (TDS) values ranged from 108 – 274 ppm while the electric conductivity (EC) values ranged from 216 – 559 s/cm. The water temperature calculated during the study period ranged from 24.9 – 29.6 °C.

The results of field measurements and laboratory tests show that major ions have high values in April – October. Calcium ions (Ca^{2+}) have a value range of 46 – 120 mg/L, magnesium ions (Mg^{2+}) 2.95 – 18.47 mg/L, sodium ions (Na^+) 6 – 20 mg/L, and potassium ions (K^+) 1 – 2 mg/L. The anions in the Pindul Cave Underground River are bicarbonate ions (HCO_3^-) which have a value range of 152.54 – 378.30 mg/L, chloride ions (Cl^-) 3.0 – 10.4 mg/L, and sulfate ions (SO_4^{2-}) 7 – 41 mg/L.

Based on the analysis of water quality, the Pindul Cave Underground River complies with the water quality standard, except for calcium ions in samples 1, 6, 7, 9, 10, 11, and 12 which have values above the quality standard with a maximum threshold of 100 mg/L. . The high value of calcium in some samples occurred in the dry season and the beginning of the rainy season. This is due to the interaction process of rocks with water in water aquifers during the dry season so that water mineral enrichment occurs. Meanwhile, the value of calcium ions tends to be smaller in December to March. This condition is due to the dilution process of rainwater and allogenic recharge that enters through the Sinking Stream Kedung Bunting.

Table 2. A slightly more complex table with a narrow caption.

No	Date	Ion (mg/L)							Physical And Chemical Parameters			
		Ca^{2+}	Mg^{2+}	Na^+	K^+	Cl^-	HCO_3^-	SO_4^{2-}	pH	EC	TDS	Temp
1	22/11/2020	**110**	2,95	14	1	10,0	341,69	12	7,71	528	256	28,2
2	20/12/2020	93	12,27	15	1	10,0	298,98	15	7,54	357	189	27,2
3	27/01/2021	46	2,43	6	2	3,0	152,54	7	7,02	259	130	24,9
4	14/02/2021	95	17,41	14	1	8,4	317,28	7	7,43	406	224	26,8
5	20/03/2021	48	4,88	7	2	4,0	146,44	18	7,22	216	108	25,9
6	25/04/2021	**112**	11,12	12	1	11,4	341,69	41	7,66	438	231	28,7
7	23/05/2021	**120**	13,48	16	1	10,0	378,30	14	7,69	495	245	28,7
8	20/06/2021	72	5,83	14	2	8,4	256,27	22	7,45	346	173	27,1
9	18/07/2021	**118**	18,47	17	1	9,9	378,30	11	7,69	505	257	28,6
10	29/08/2021	**119**	12,15	20	1	9,9	366,10	8	7,79	559	274	28,7
11	26/09/2021	**118**	4,86	17	1	10,4	359,99	12	7,79	559	260	29,6
12	24/10/2021	**112**	14,89	15	1	9,0	353,89	10	7,71	548	259	29,3
Standards		100[b]	50[b]	200[a]	50[b]	250[a]	500[b]	250[a]	6.5-8.5[a]	1000[b]	500[a]	

Source: [a]The Regulation of Indonesian Minister of Health No. 90/2002, [b](Todd and Mays, 2005)

Based on the multitemporal water quality concentration pattern (Figure 3), there are similar patterns between the values of calcium ions, bicarbonate ions, electric conductivity, and total dissolved solids. The pattern of these four parameters tends to be high and stable in the dry season but low and fluctuating in the rainy season. Ions Mg^{2+}, K^+, Na^+, Cl^-, SO_4^{2-} have fluctuating values during the study period. However, Na^+ ions tend to be higher in the dry season. The results of this study indicate that there are similarities with research conducted in several locations in the Gunungsewu Karst Area where the values of Ca^{2+} and HCO_3^- during the dry season tend to be high compared to the rainy season such as Beton Spring, Petung Alogenic River, Gilap Cave, Jomblangan Karst Window, and Petoyan Spring. [26][27].

The results of the analysis using a triangle piper diagram show that the Pindul Cave Underground River water is classified as unpolluted groundwater and calcium enrichment (Figure 4). The main triangle with SO_4+Cl and Ca+Mg ions obtained G classification (unpolluted groundwater) for all samples in both dry and rainy seasons. The lower left triangle with Mg, Na+K, Na ions is classified as C (calcium enrichment) while the right triangle with HCO_3, Cl-, and SO_4 ions is classified as D (unpolluted groundwater) for all samples.

Groundwater from the Pindul Cave Underground River which is classified as calcium enrichment indicates that the aquifer in the study area is composed of limestone. The dominant calcium ion comes from the water-rock interaction. This also shows that the role of diffuse flow through the voids between rock grains is still quite dominant. The classification of groundwater as unpolluted water indicates that the source of water flowing into Pindul Cave, especially the Kedung Buntung allogenic river, is included in a good quality water source. Groundwater sources classified as Unpolluted Groundwater and Calcium Enrichment are also found in other karst water sources in the Gunungsewu Karst Area in Gunungkidul Regency and the Jonggrangan Karst Area in Kulonprogo Regency, Special Region of Yogyakarta [26][28].

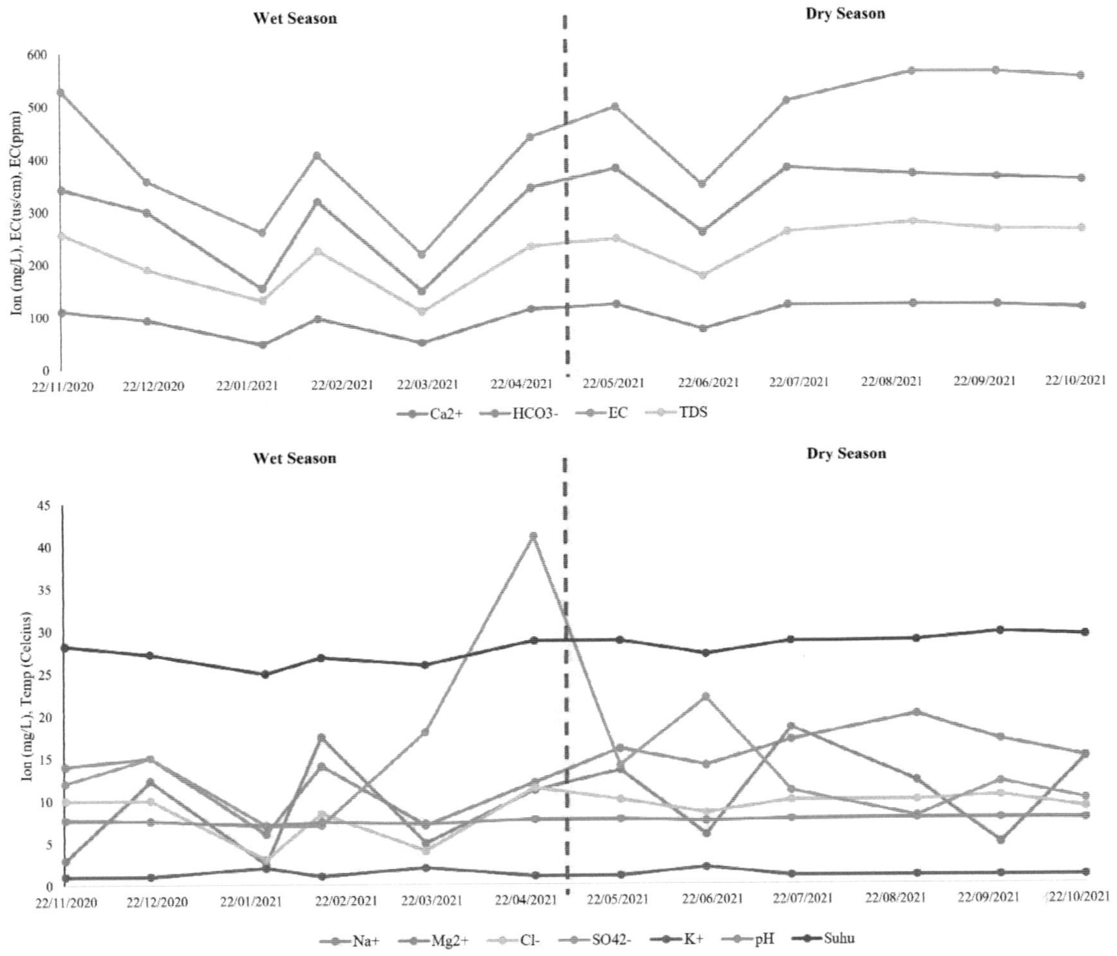

Figure 3. Chemical comparison of the Pindul Cave Underground River in the dry season and the rainy season

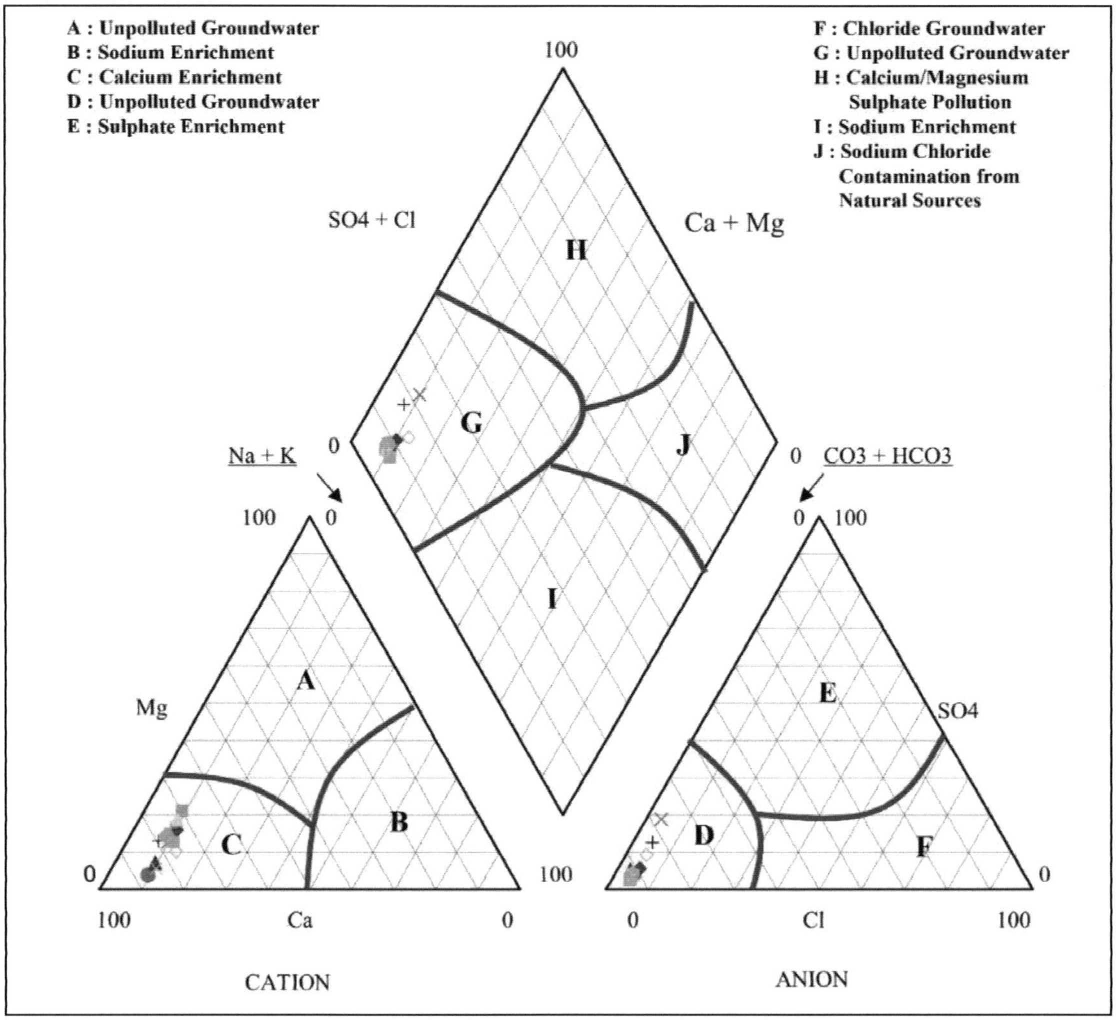

Figure 4. Triangle piper diagram of Pindul Cave Underground River

The rectangular Piper diagram of Pindul Cave Underground River supports the results of hydrochemical characteristics of the previous discussion (Figure 5). All of the water samples that marked with red dots, occupy the section of average bicarbonate water. This is also showed by the red dots on the mixing line. This type of water develops because of the high bicarbonate ion (HCO_3^-). This classification has the characteristics of fresh and good quality water. This is in accordance with the results of the piper triangle diagram which is included in the classification of unpolluted water.

In general, the Underground River of Pindul Cave has good water quality. However, when viewed in terms of seasons, the quality of water in the rainy season is better than the dry season. This is due to the effect of allogenic recharge which brings rainwater into the underground river system of Pindul Cave. This process is called dillution by precipitation by water from the surface river entering the Kedung Buntung Sinking Stream as an allogenic river. On the other hand, during the dry season, the water in the underground river of Pindul Cave undergoes a water-rock interaction process so that the chemical content of the water tends to be high.

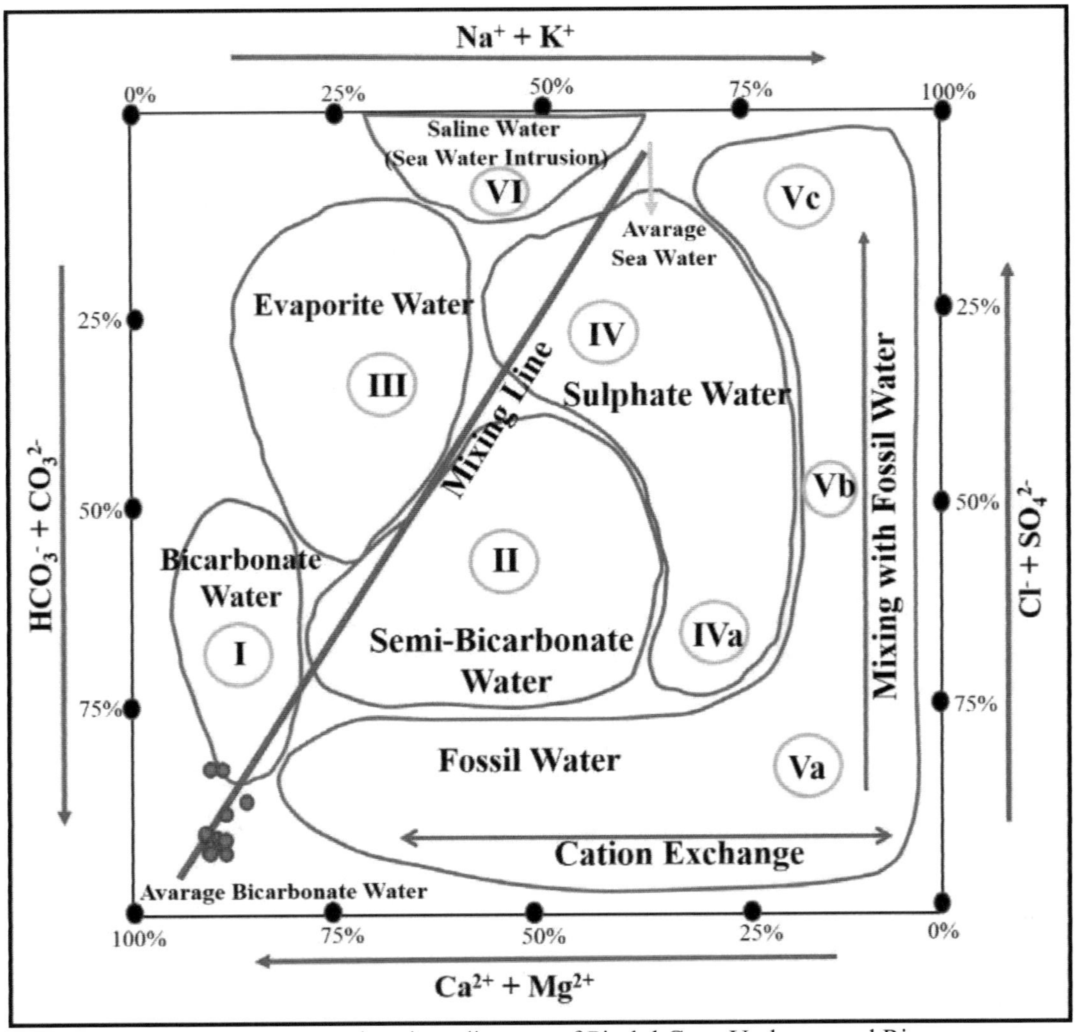

Figure 5. Rectangular piper diagram of Pindul Cave Underground River

4. Conclusion

Based on the analysis results, the Pindul Cave Underground River in multi-temporal shows good water quality reinforced by the classification of unpolluted groundwater. This explains that the water source that feeds the Pindul Cave Underground River has good water quality. However, during the dry season and rainy season there are variations in water quality that are influenced by chemical content. During the dry season, the dominant calcium ion exceeds the water quality standard due to the interaction of rocks with water. This condition is evidenced by a triangular piper diagram showing the classification of calcium enrichment. During the rainy season, the effect of allogenic recharge from the Sinking Stream Kedung Buntung is very dominant which affects water quality that meets clean water quality standards. The water quality in the Pindul Cave Underground River is classified as good for use as a domestic water source, water tourism, and agricultural irrigation.

Acknowledgements

This research is a small part of the results of the first author's B.Sc. Thesis in Faculty of Geography, Universitas Gadjah Mada. This research was funded by the 2022 Final Project Recognition (*Rekognisi Tugas Akhir*, RTA) Grant, Universitas Gadjah Mada, with the assignment letter No. 3550/UN1.P.III/Dit-Lit/PT.01.05/2022, as part of the research entitled Characterization of

Hydrodynamic and Hydrogeochemical Properties of Karst Drainage Basins in the Gunungsewu Karst Area for a Karst Critical Zone Management Model in the Tropical Karst Regions of Indonesia (*Karakterisasi Sifat Hidrodinamika dan Hidrogeokimia pada* Karst Drainage Basin *di Kawasan Karst Gunungsewu sebagai Model Pengelolaan* Karst Critical Zone *di Kawasan Karst Tropis Indonesia*).

References

[1] Ford D and Williams P 2007 *Karst Hydrology and Geomorphology* (John Wiley and Sons, Ltd)
[2] Haryono, E and Adji T N 2004 *Pengantar Geomorfologi dan Hidrologi Karst* (Yogyakarta: Kelompok Studi Karst Fakultas Geografi Universitas Gadjah Mada)
[3] Nohegar A, Norouzi A, Hosseinzadeh M M 2012 The role of karst formations in groundwater resources development. *Int J Geol Earth Environ*
[4] Stevanovic Z 2015 *Karst Aquifers–Characterization and Engineering* (Switzerland: Springer International Publishing)
[5] Goldscheider N, and Drew D 2007 *Methods in Karst Hydrogeology* (London: Taylor and Francis)
[6] Pratama A D 2019 Kajian Spasial-Temporal Hidrodinamika dan Hidrogeokimia Untuk Karakterisasi Akuifer Di Sebagian Wilayah Karst Jonggrangan, Kabupaten Kulonprogo-Purworejo *Skripsi* (Yogyakarta: Fakultas Geografi Universitas Gadjah Mada)
[7] Cahyadi A 2010 *Pengelolaan Kawasan Karst dan Peranannya dalam Siklus Karbon di Indonesia. Makalah dalam Seminar Nasional Perubahan Iklim di Indonesia* (Yogyakarta: Sekolah Pasca Sarjana UGM)
[8] Cahyadi A, Ayuningtyas E A, and Prabawa B A 2013 Urgensi Pengelolaan Sanitasi Dalam Upaya Konservasi Sumberdaya Air Di Kawasan Karst Gunungsewu Kabupaten Gunungkidul. *Indonesian Journal of Conservation* 2 (1)
[9] Agniy R F 2016 Kajian Hidrogeologi Karst Sistem Gua Pindul, Kecamatan Karangmojo, Kabupaten Gunungkidul *Skripsi* (Yogyakarta: Fakultas Geografi UGM)
[10] Bemmelen R W V 1949 *The Geology of Indonesia Vol I A General Geology and Adjacent Archipelagoes* (The Hague: Government Printing Office)
[11] Nurkholis A 2017 Variasi Spasial Dan Temporal Sifat Aliran, Perilaku Banjir, dan Respons Debit Terhadap Hujan Di Sistem Drainase Karst Pindul, Kabupaten Gunungkidul *Skripsi* (Yogyakarta: Fakultas Geografi Universitas Gadjah Mada)
[12] Haryono E 2014 Speleogenesis Gua Pindul dan Sekitarnya. *Laporan Penelitian Hibah Fakultas* Fakultas Geografi UGM
[13] Nurkholis A, Adji T N, Haryono E, Cahyadi A, and Suprayogi S 2019 Time Series Analysis Application for Karst Aquifer Characterisation In Pindul Cave Karst System, Indonesia *Acta Carsologica* **48** (1) 69-84
[14] Widyastuti M 2010 Karakterisasi Daerah Tangkapan Ponor Karst Gunungsewu sebagai Variabel Penentu Kerentanan Airtanah terhadap Pencemaran (Studi Kasus di DAS Bribin). *Laporan Akhir Kegiatan Penelitian Hibah Disertasi Doktor* (Yogyakarta: Universitas Gadjah Mada)
[15] Adji T N 2012 Wet Season Hydrochemistry of Bribin Cave in Gunungsewu Karst, Indonesia *Environ Earth Sciences* **67** 1563–1572
[16] Agniy R F, Adji T N, Cahyadi A, Nurkholis A, and Haryono E 2019 Characterizing the Cavities of Anjani Cave in Jonggrangan Karst Area, Purworejo, Central Java, Indonesia. *IOP Conference Series: Eart and Environmental Science* **256** (1) 012009
[17] Giambastiani B M S, Colombani N, Mastrocicco M, and Fidelibus M D 2013 Characterization of the lowland coastal aquifer of comacchio (ferrara, italy): Hydrology, hydrochemistry and evolution of the system *Journal of Hydrology* **501** 35–44
[18] Yang P, Cheng Q, Xie S, Wang J, Chang L, Yu Q, Zhan Z, and Chen F 2017 Hydrogeochemistry and geothermometry of deep thermal water in the carbonate formation in the main urban area of Chongqing China *Journal of Hydrology* **549** 50–61

[19] Kim Y T, Hyun S G, Cheong J Y, Woo N C, and Lee S 2018 Hydrogeochemistry in the coastal area during construction of geological repository Journal of Hydrology **562** 40–49

[20] Yang P, Luo D, Hong A, Ham B, Xie S, Ming X, Wang Z, and Pang Z 2019 Hydrogeochemistry and geothermometry of the carbonate-evaporite aquifers controlled by deep-seated faults using major ions and environmental isotopes *Journal of Hydrology* **579**

[21] Reyes F A P, Crosta G B, Frattini P, Basiricò S, and Pergola R D 2015 Hydrogeochemical overview and natural arsenic occurrence in groundwater from alpine springs (upper Valtellina, Northern Italy) *Journal of Hydrology* **529** 1530–1549

[22] Rahardjo W, Sukandarrumidi, Rosidi H M D 1995 *Peta geologi lembar Yogyakarta, Jawa.*

[23] Todd D K and Mays L 2005 *Groundwater Hydrology* (John WIley and Sons, Ltd)

[24] Appelo C A J and Postma D 2005 *Geochemistry, Groundwater, and Pollution (2nd ed)* (A.A. Balkema Publisher)

[25] Kloosterman, F. H. (1983). Reconnaissance Study of Groundwater Resources in the Kabupaten Cirebon. Provincial Health Service Irectorate CDC

[26] Adji T N, Haryono E, Fatchurohman H., and Oktama R 2017 Spatial and temporal hydrochemistry variations of karst water in Gunung Sewu, Java, *Indonesia Environmental Earth Sciences* **76** (20), 1–16.

[27] Cahyadi A Haryono E, Adji T N, Widyastuti M, Riyanto I A, Muhammad D T N, and Tastian N F 2021 Rainfall Variability in Gunung Sewu Karst Area, Java Island, Indonesia *Indonesian Journal of Forestry Research* **8** (1) 23–35

[28] Pratama A D, Adji T N, and Dwiputra D S 2020 Hydrochemistry temporal variation of Anjani Underground River, Jonggrangan Karst, Java Island, Indonesia *IOP Conference Series: Earth and Environmental Science* **451** (1)

Developing a Visual Counting Method to Quantify Riverine Plastic Litter: A Case Study of Rivers in Semarang City, Indonesia

A Sarminingsih, P Andarani*, and W D Nugraha

Department of Environmental Engineering, Faculty of Engineering, Diponegoro University, Semarang – 50275, Indonesia

*andarani@ft.undip.ac.id

Abstract. Indonesia has established a national action plan on marine plastic debris to tackle the issue of marine plastic debris. Because most of the marine plastic debris in Indonesia is likely due to mismanaged plastic waste, it is important to address the problem of riverine plastic river problem in Indonesia. The objectives of this study are to visually observe the condition of the river in terms of plastic pollution, to quantify plastic litter in the river, and to develop a visual counting model of plastic litter quantification. The visual counting of plastic debris (> 2.5 cm) method based on the European standard method was applied to a river in Indonesia. The main challenge was the high turbidity of the water that obstructed visual observation. The deployment of the fishing net was also not appropriate to collect debris. Most of the plastic debris was stranded on the riverbanks, indicating that the plastic flow was likely elevated due to runoff during the rainy season. Four types of plastic litter polymers were obtained, i.e., polyethylene terephthalate (PET), low-density polyethylene (LDPE), polypropylene (PP), and polystyrene (PS). The highest percentage of macroplastics is LDPE, which usually used as daily packaging (90%).

1. Introduction

Since its invention in 1907, the use of plastic has increased rapidly. In 2019, annual plastic production reached 368 million tons, which are likely to double in the next few decades [1]. Plastic is favorable because it has a low production cost, is easy, comfortable, and has good durability. However, if plastic waste is not properly managed, its durability can potentially damage the environment, aesthetics, and human health. No exceptions occurred in the city of Semarang, Indonesia. The amount of waste generated in Semarang City reaches 1.18 million tons per day with a percentage of plastic waste of 17.2% [2]. Most plastic waste comes from land [3] with various sizes. Macro-debris is a type of waste with a size that can still be seen visually from 2.5 to 100 cm. The types of waste that are usually found at this size are plastic waste (plastic bags, plastic cups, plastic gloves, and plastic bottles). Marine debris of this size can be found in coastal areas both at the bottom and at the surface of the waters [4]. Plastic debris found in the sea and exposed to sunlight may decompose and change in size. These sizes can be divided into 4 levels, namely macroplastic (>25 mm), mesoplastic (5–25 mm), microplastic (5 mm–1 μm), and nanoplastic (<1 μm) [5].

Plastic litter is transported by river flow with different amounts, types and frequencies in each river in Semarang City. The complex river system requires monitoring and sampling of macroplastic and microplastics in the Semarang City rivers. This study aims to estimate floating macroplastic fluxes in river water, specifically in Kanal Banjir Timur (KBT) and Babon River, Semarang City, Indonesia.

The city of Semarang with a population density of 4,431 people/km^2 [6] has an estimate of waste management service coverage of 79% [7]. Plastic illegally dumped near the riverbank has the potential to enter water bodies and may be transported to the sea. Visual counting is one method to determine the load of macroplastic pollutants in the river. In this study, the applicability of the visual counting method is elaborated.

2. Materials and Method

2.1. *Study Area*

The headwater of Babon Watershed originates from Mount Butak in Ungaran, Semarang Regency. Administratively, the flows are divided into three subwatersheds including Kanal Banjir Timur, Dombo-Sayung River, and Babon River. These divisions were made to prevent floods in the urban area (downstream section). In this study, a point was located in Dombo-Sayung River (P1) and three sampling points were selected in Kanal Banjir Timur (P2–P4), as described in Table 1. All sampling points are located in urban area of Semarang City, specifically in Semarang Selatan Subdistrict. The coverage of the waste management service was expected to be more than 75% in the vicinity of rivers (Table 1).

Table 1. Description of sampling point

Sampling point (st.)	Description
P1	Located behind a residence area, and illegal dumping was found (Bengkung Bridge, Pedurungan)
P2	Located in an urban area, but about 10 m from the residence (Gemah Bridge, Kedungmundu, Tembalang)
P3	Residential area located in the urban of Semarang City. A traditional market is available in the morning (Gotong Royong Bridge, Sendangguwo, Tembalang)
P4	Located in a densely populated area (Talang Londo Bridge, Jl Lamper Tengah IX, Semarang Selatan)

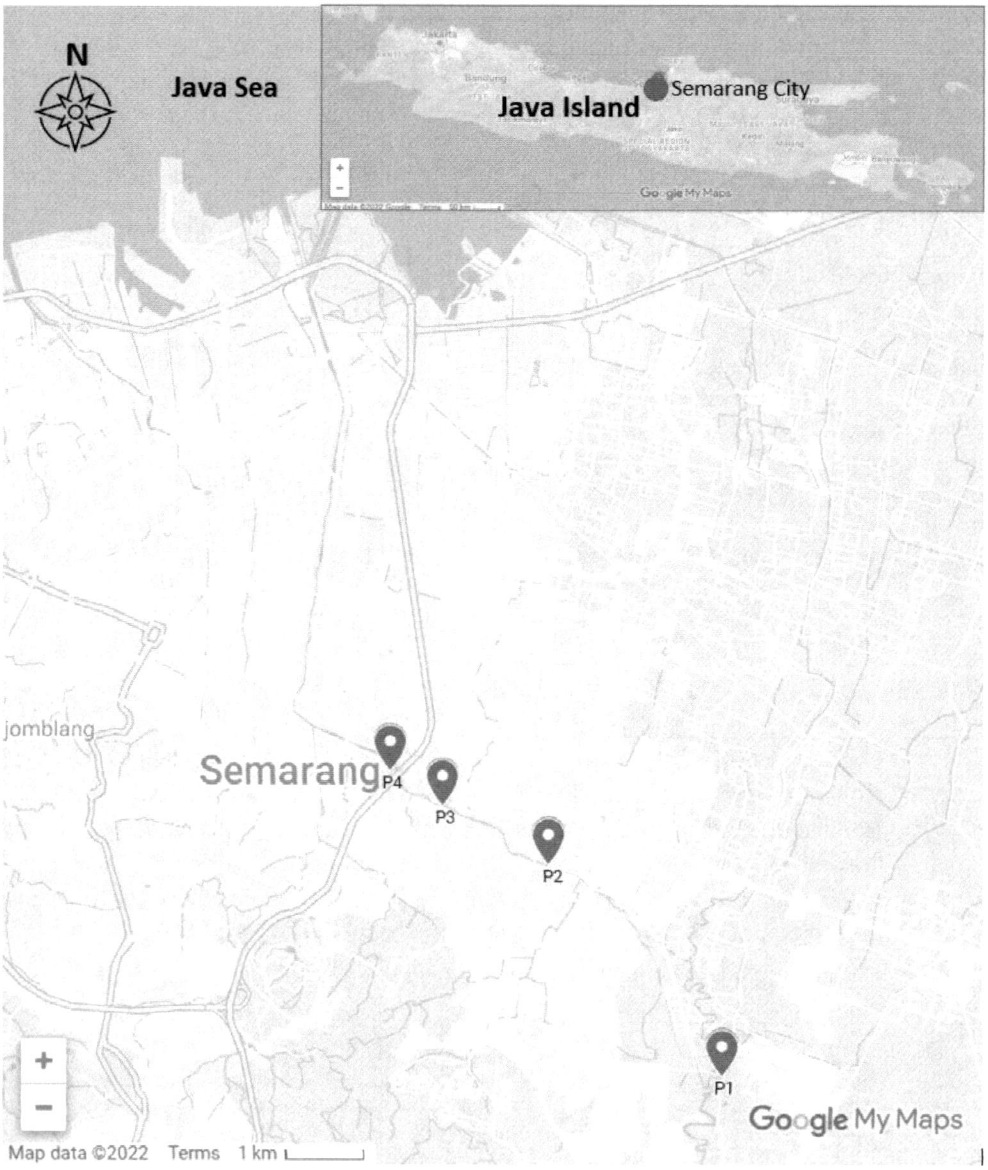

Figure 1. Study area in rivers located in Semarang City. This map was created using Google My Maps.

2.2. *Visual counting method and trawling*

The visual counting method is a standardized method to quickly estimate the amount of macroplastic in rivers [8]. Visual counting can be performed on the bridge facing the upstream part of the river. This method consists of counting the number of macroplastic items visible on the water surface and flowing downstream over a certain period of time from the bridge. For large rivers, several observation points are generally determined along the bridge due to the influence of local hydrodynamic conditions across the river [9]. In this study, the river width was not greater than 8 meters. Therefore, two observers stood on both sides of the river and counted only plastic litter. This study focused on the plastic litter with a size over 2.5 cm. Only plastic floated and easily observed could be counted. The duration of the observation was 30 minutes which is divided into 3 sessions (1 session with a duration of 10 minutes).

To avoid double counting, the observers had to verify the plastic counting with each other. Then the observed plastic flux was extrapolated to the plastic load per day [10].

2.3. Plastic litter trawling
Plastic litter sampling was done using fish net. To make the net static in the water, we equipped it with a load made of selenium. Trawling deployments lasted between one and ten minutes, depending on the velocity of the water flow [8]. In this study, a 4-m long net with a square mesh size of 2.5 cm was held by hand on both side of the river for 10 minutes at the same time as visual counting.

2.4. Classification method
The weight of the waste was calculated after it passed through the drying process. The collected macroplastic waste is plastic waste from rivers, so it is necessary to dry to remove the water content in macroplastics, because it affects the weight of river waste and macroplastics. The volume of waste was done using compaction by dropping the container three times as required by the Indonesian National Standard of the Solid Waste Sampling and quantification method (SNI 19-3964-1994). The classification followed the plastic polymer classification (seven types: Polyethylene terephthalate (PET), Low Density Polyethylene (LDPE), polypropylene (PP), and Polystyrene (PS). Additionally, the plastic litter was classified according to its functional use. Plastic litter was reported in abundance.

3. Results and Discussion

3.1. Spatial variation of plastic debris in KBT River
Table 2 shows the characteristics of each sampling point in this study. It can be seen that the majority of the population (76–100%) has been served by the municipal solid waste management of Semarang City.

The amount of plastic litter collected from the river may depend on population and estimated shortest distance between the sampling point and the residential housing. P2 has the highest plastic abundance of all sampling points, that is, 864 ± 204 plastic pieces/day. Interestingly, it has 100% served by the municipal solid waste management (MSWM) system, according to the official figure. P4, which has the lowest MSWM also presented second highest amount of plastic litter (504 ± 102 pieces/day). Meanwhile, the highest river discharge point (P3) has the lowest plastic abundance (288 ± 204 pieces/day). P1 (360 ± 305 pieces/day) and P4 have a similar river discharge of 0.03 m^3/s, but have a different abundance of plastic. It should be noted that more data are needed to derive conclusive evidence.

Table 2. Characteristics of each sampling point

Sampling point (st.)	River width (m)	Population served by MSWM* [7]	Estimated shortest distance (m)**
P1	9.2	91%	35
P2	6.4	100%	32
P3	7.8	91%	30
P4	7.5	76%	10

* MSWM: Municipal Solid Waste Management. The population rates served by MSWM were referred to COWI [7] using data in 2017.
** Estimated shortest distance between each sampling point and the nearest housing.

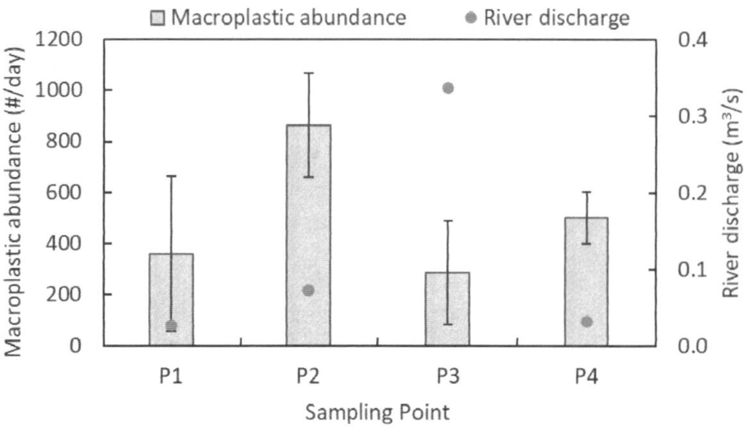

Figure 2. Macroplastic abundances and river discharges at each sampling point. The error bars represent standard deviations (n = 3).

Macroplastics can be distributed via different transport routes, and this distribution in the environment. It generally occurs through deviation, which is defined as passive dispersion by wind and water currents. According to Lechthaler et al. [11] plastic can be distributed in water bodies horizontally and vertically, horizontally influenced by river flow, wind velocity, and direction and navigation. They also take notes different distribution between high and low tides in a coastal area. Vertical distribution was affected by density changes after development biofilm due to water turbulence. [12]. Its distribution is affected by the seasonal changes of the surface water and the flow rate of tributaries. In addition, plastic waste is associated with temperatures changes temperatures associated with freezing and thawing cycles, solar radiation, and mechanical abrasion [13].

Tracing the source of plastic debris should start from identifying the plastic polymers. In this study, the most common type of polymer found in the river was LDPE with 90.14% of all macroplastic pieces, followed by PP (4.22%) and PS (4.22%). The lowest proportion was PET (2.5%). LDPE is often used in food packaging films, agricultural films, plastic bags, and plastic bottles because of its strong chemical stability. Most macroplastics (> 2.5 cm) included everything commonly identified as plastic waste such as bags (90.14%) or bottles, the samples found were quite diverse, but most commonly found in the form of sheets or packaging. As the most common use was packaging, the colors found also vary, such as white, blue, green, and transparent. Packaging might originate from daily household activities.

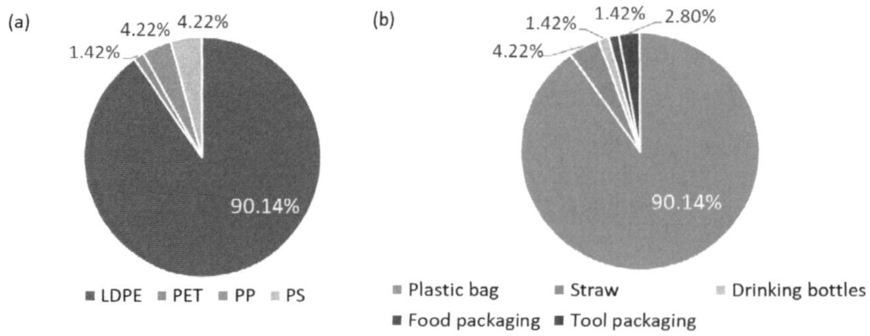

Figure 3. Plastic debris classification: (a) proportion of plastic polymers; (b) use of plastic product

Macroplastics have been found in surface water. Evidence of illegal dumping and open burning of plastic waste was also observed near riverbank. Although a waste reduction strategy has been implemented in Semarang City, plastic bags still dominated the proportion of riverine plastic litter.

Currently, the plastic ban policy was only applied in supermarkets and convenience stores, implying that this mitigation is not enough into the prevent plastic leakage to water body.

The visual counting method applied in this study has proven to be a quick and easy way to estimate plastic fluxes. However, there were drawbacks that one should consider before applying this method, i.e., the turbidity hindered plastic observation; plastic litter stranded in riverbanks and the canal structure (in case of P3). The primary obstacle was the severe turbidity, which impeded visual observation. The deployment of the fishing net to recover the debris was also difficult. A stronger structure to capture riverine litter is preferable despite its expensive cost.

4. Conclusions

The sampling point served by the 100% MSWM system had the highest plastic loading of all sampling locations, with 864 ± 204 plastic fragments per day. The highest proportion of plastic litter was LDPE, which commonly use as daily packaging or plastic bags (90%). The presence of macroplastics could be due to waste input from land such as in East Ungaran District, Banyumanik District, and Mranggen District, Pedurungan District because of anthropogenic activities. More studies are needed to estimate the plastic flux not only by river discharge but also population, existing waste management system, and river basin characteristics.

Acknowledgement

This study was funded by the Strategic Research grant from *RKAT Fakultas Teknik*, Diponegoro University in 2022. The authors sincerely thank all students involved (*Kelompok 6*), particularly Mr. Dhega Bayu Susilo and Ms. Rahma Safitri, and Mr. Arsa Maradinata for their dedication to do field survey and lab works.

References

[1] Geyer R, Jambeck J R and Law K L 2017 *Sci. Adv.* **3** 3–8
[2] Indonesian Ministry of Environment 2022 National Information System of Solid Waste Management (SISPN)
[3] Meijer L J J, van Emmerik T, van der Ent R, Schmidt C and Lebreton L 2021 *Sci. Adv.* **7** 1–14
[4] Lippiatt S, Opfer S and Arthur C 2013 *Marine Debris Monitoring and Assessment : Recommendations for Monitoring Debris Trends in the Marine Environment*
[5] Iwasaki S, Isobe A, Kako S, Uchida K and Tokai T 2017 *Mar. Pollut. Bull.* **121** 85–96
[6] BPS-Statistics of Semarang Municipality 2022 *Semarang Municipality in Figures 2022* (Semarang)
[7] COWI 2018 *Master Plan Revision of Municipal Solid Waste Management in Semarang City* (Semarang City)
[8] van Emmerik T, Kieu-Le T C, Loozen M, Oeveren K van, Strady E, Bui X T, Egger M, Gasperi J, Lebreton L, Nguyen P D, Schwarz A, Slat B and Tassin B 2018 *Front. Mar. Sci.* **5** 1–11
[9] van Emmerik T, Loozen M, Van Oeveren K, Buschman F and Prinsen G 2019 *Environ. Res. Lett.* **14**
[10] Cordova M R and Nurhati I S 2019 *Sci. Rep.* **9** 1–8
[11] Lechthaler S, Waldschläger K, Stauch G and Schüttrumpf H 2020 *Environ. - MDPI* **7** 1–30
[12] Schwarz A E, Ligthart T N, Boukris E and van Harmelen T 2019 *Mar. Pollut. Bull.* **143** 92–100
[13] Battulga B, Kawahigashi M and Oyuntsetseg B 2020 Abundance of microplastics in sediments from the urban river in Mongolia *Geogr. reports Tokyo Metrop. Univ.* **55** 35–48

Community based integrated sustainable waste management in Lerep tourism village

Maya Damayanti[1*], Wido Prananing Tyas[1], Luluh Cahya Pangestu Ningtyas[1]

[1]Department of Urban and Regional Planning, Diponegoro University. Tembalang Campus, Jl. Prof. Soedarto, Semarang - Indonesia 50275

maya.damayanti@pwk.undip.ac.id

Abstract. The waste generated in tourist villages is not only from household activities but also from tourism activities. If not managed wisely, it can effect on health and the surrounding environment and are usually thrown away without a chance to return to the economy. The waste management system in the Lerep Tourism Village has started paying attention to the concept of sustainability and being integrated by empowering the community through BUMdes to achieve zero plastic. This article aims to explore the efforts of the community of this village in managing waste of tourism and domestic activities. This research applied a qualitative study by interviewing the key informants in this village, The villagers reduce the use of plastic in one of the tourist attractions, namely the rural culinary market by replacing plastic containers with leaf containers. Other efforts made such as early segregation of waste by residents from their respective homes, reducing burning waste, provision of 3R facility and waste banks to recycle waste become a more economical product. These efforts not only have an impact on quality environment, but also on the community's economy where the results of waste banks are able to help people pay health insurance.

1. Introduction

Rural tourism is one of the most popular tourist destinations in the world today [1,2]. Tourism activities in rural areas are considered as an alternative economic activity to agriculture and animal husbandry. In addition, rural communities themselves become a tourist attraction by taking advantage of their cultural differences from people living in urban areas [3]. Although tourism activities have a good impact on the local community's economy, they also have negative impacts including on the environment. The presence of tourist visits will cause the generation of waste generated to increase [4, 5]. If this waste is not handled wisely, it can cause various problems such as damaging the environment, natural resources and cultivation [6,7]. Waste processing is carried out so as not to cause a negative impact on the environment [8]. The community is the main actor in the waste management system [9], so it needs to be empowered to be able to work on waste management in their own environment. A waste management system that empowers the community should be integrated with various elements (government, NGOs, schools and others) by making the community the main object and subject to be able to create a sustainable tourism village waste management system and improve the community's economy [10]. The concept of community participation in waste management in a tourism village is in line with the concept of community-based tourism [11-13] which focusing on the empowerment the local people in managing their tourism activities in order to leverage their livelihood.

Waste in various regions in Indonesia is still managed traditionally and does not implement an integrated system, where the process of waste generation, collection and disposal is still carried out separately. Lerep Tourism Village is one of the tourist villages in Semarang Regency, Central Java, starting to pay attention to the concept of sustainable waste management that empowers the community managed by local institution called as BUMdes in order to create a zero plastic tourism village. In 2021, the Lerep Tourism Village received an award from the Ministry of Tourism and Creative Economy as a Sustainable Tourism Village. Interestingly, the Lerep Tourism Village also implements the Climate Village Program (Proklim) which is a certification program initiated by the government to increase community participation in implementing climate change mitigation and adaptation [14]. One of its flagship programs is the reduction of plastic waste, namely replacing plastic containers with leaf containers at tourist attractions in rural culinary market. Based on this, this article aims to explore the efforts of the community in Lerep Tourism Village to reduce waste generation and how they play a role in a sustainable waste management system.

2. Method

This research applied qualitative case study [15, 16] as the research method. The data were collected in June-July 2022 with Lerep Tourism Village as the case study. Lerep is one of the tourist villages in Semarang Regency, precisely in West Ungaran District, Semarang Regency, Central Java. The majority of the people's livelihood is in agriculture with its main commodities, namely fruits, cassava, sweet potatoes and vegetables. Meanwhile, in the livestock sector, the main activity is dairy farming.

In terms of data collection, the researchers interview key informants who were selected based on their capacity and knowledge related to this topic. The key informants are the representative of the local group concerning on tourism (called as pokdarwis / kelompok sadar wisata), manager of the TPS 3R (temporary waste station applying reduce, reduce, and recycle), and the representatives of local seller in the culinary market. Additionally, the researchers also applied naturalistic observation [17] to explore village's facilities related to waste management and community's activities related to this topic.

3. Results and Discussion

3.1. Characteristics of the waste

As a Tourism Village, the waste generated in Lerep Village is not only from household activities, but also from tourism activities. Household waste is generated from daily activities in each house. The characteristics of the waste lifted by the Lerep Tourism Village waste transportation services consist of organic waste, plastic, paper, glass bottles and metal. Organic waste that is transported such as leftover cooking activities, yard sweeping waste in the form of leaves. The daily amount of waste generated from household activities is around 70m3. Meanwhile, the characteristics of waste from tourism activities are mostly organic waste. This is because tourism activities in Lerep Village have reduced the use of plastic waste. In addition, inorganic waste such as drink bottles and plastic bag were also found which were brought and left by tourists during their visit to Lerep Village.

3.2. Community behaviors related to waste management

This research found that the community has already aware on the importance of protecting the environment, including in sustainable waste management. This is illustrated, among other things, by changes in the behavior of the people of Lerep Village in processing waste. Previously, people processed waste by burning it. However, after the socialization through the climate village program, they have realized that this activity can cause air pollution due to the waste burning process. Although the community already understand how to manage sustainable waste, in its application it has not been fully implemented by the people in the Lerep Tourism Village. Some of them still combine the waste and use

IOP Conf. Series: Earth and Environmental Science 1098 (2022) 012051 doi:10.1088/1755-1315/1098/1/012051

the plastic containers/bags. Hence, the willingness of the community to be actively involved in the process of sustainable waste management is needed, especially in the process of reducing plastic waste, sorting waste from an early age and so on.

3.3. Village's facilities related to waste management

Facilities and infrastructure to support sustainable waste management in Lerep Tourism Village include:

3.3.1 Waste Transport Vehicles. The process of transporting waste in the Lerep Tourism Village has been integrated where waste transportation vehicle has been provided. The vehicle is motorbike tubs operated every two days.

Figure 1. Motorbike tubs to collect waste in Lerep Village

3.3.2. TPS 3R /Tempat Pembuangan Sementara (Temporary Waste Station applying 3R/reduce, reuse, recycle). The TPS 3R has been provided to manage waste sustainably. The TPS 3R of Lerep Tourism Village is named as TPS Sukaku Asri. Nowadays, this facility only covers one hamlet, namely Dusun Suko (Suko Hamlet). This TPS 3R implements a membership system in which people who are the members of this facility will be provided with a trash can in each house. In addition, TPS 3R Sukaku Asri has also provided garbage transportation services that will transport waste from each member's house and the operational costs come from members' fees. The Sukaku Asri 3R TPS has also been equipped with various tools to process waste such as chopping machines and so on.

Figure 2. *TPS* 3R */Tempat Pembuangan Sementara* (Temporary Waste Station)

3.3.3. Waste Bank. The efforts to implement sustainable waste management in the Lerep Tourism Village are also supported by a waste bank facility that functions to collect sorted plastic waste from residents' homes to then be sold to collectors and some are also used as more economical products. The collection of plastic waste in the waste bank is only done once a week, which is every Sunday. The results obtained by residents from waste bank savings are up to Rp. 500,000 per month and able to help the local community's economy.

3.4. Mechanism of waste management

3.4.1 Waste Collection. Waste generated from household activities in Lerep Tourism Village is collected in front of each house. As for tourism activities, the waste generated from the culinary market attractions is collected by each trader in the trash cans provided at each trade stall. There is a communal 3R trash can provided at the culinary market for tourist visitors. However, based on the results of observations, there are still many visitors who throw garbage out of place.

3.4.2 Waste Sorting. Waste sorting is carried out by the community only separating plastic bottle waste from other waste. There is no further sorting such as sorting paper waste, organic waste and so on. Further sorting of waste is only carried out at 3R TPS.

3.4.3 Waste Transportation. Waste sorting is carried out by the community, only separating the waste resulting from household activities which has been collected by residents in front of their respective homes and then transported by waste transportation services to be taken to the TPS every two days. Meanwhile, the waste generated from tourism activities is partly transported directly by garbage collectors and partly transported by waste transport services provided by the BUMdes of Lerep Village. between plastic bottle waste and other waste. There is no further sorting such as sorting paper waste, organic waste and so on. Further sorting of waste is only carried out at TPS 3R.

3.4.4 Tourism waste reduction. The waste reduction carried out in the Culinary Market by replacing plastic goods with organic container, particularly from leaf. This strategy is a compulsory for the all sellers in this market. The aim of this program is to achieve zero plastic from tourism activities and also to improve the value and attractiveness of this culinary market.

3.5. Community Participation in the Waste Management
Community participation in waste management is the willingness of the community to help the success of the waste management program that has been determined according to their respective abilities without inflaming their own interests. Waste management in Lerep Tourism Village has empowered the community as the main actor.

3.5.1 Waste Reduction. The community in the Lerep Tourism Village reduces the use of plastic materials, one of which was in the tourist attractions of the rural culinary market. This effort is being made to replace plastic containers with banana leaves and teak leaves by traders who are local people of the Lerep Tourism Village. The existence of this concept itself is a special attraction for tourists because in addition to serving typical village snacks, it also teaches environmental care. Even though they already have an understanding of sustainable waste management, not many people are implicated in reducing plastic waste in their daily activities. In household activities, the villagers still use plastic because it is considered more effective and efficient.

3.5.2 Waste Sorting. The Lerep Village Government has conducted socialization to the community to start sorting waste from an early age from their respective homes. It aims to make people aware of the types of waste. However, in practice, only a small part of the community has started sorting waste from their respective homes. The form of sorting that is carried out is not comprehensive, where people only separate plastic bottle waste to be saved in a waste bank. According to the community, the existence of the 3R TPS makes it no longer necessary to sort waste from the house. The lack of public interest in sorting waste is very unfortunate because if it has been implemented thoroughly it can make it easier for 3R TPS managers at the next processing stage.

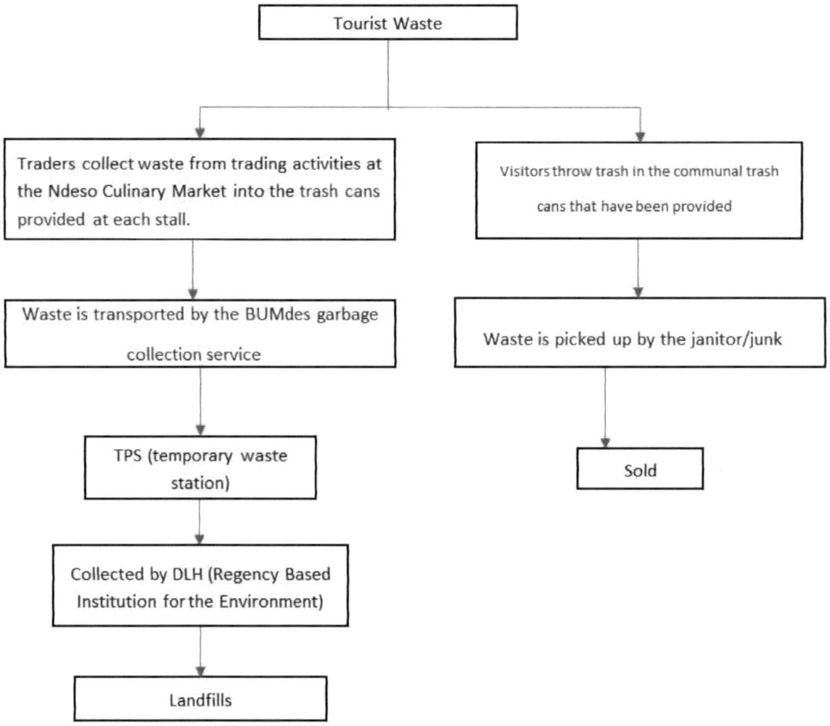

Figure 3. Mechanism of tourism waste management in Lerep village

3.5.3 Waste Recycle. Through the waste bank provided by BUMdes, the people of Lerep Tourism Village, especially mothers and young women, are empowered to recycle plastic waste into more economical goods. Products resulting from recycling activities are fruit baskets, shopping baskets, mats and so on. The product is then marketed through BUMdes where the results are able to help the community's economy, one of which is to cover the cost of health insurance. However, for organic waste both generated from tourism and household activities, it has not been recycled by the community.

3.6. Challenges in implementing solid waste management in Lerep Tourism Village

Based on the interviews with the informants and observation in the field, there are two main challenges in implementing solid waste management in Lerep Village. Those challenges are:

- The people of Lerep Village have not been fully engaged in the implementation of this concept. Initially, the community was expected to routinely sort waste from their households before it was taken to the 3R TPS, but in reality, the community was reluctant to do this with the assumption that the waste would be sorted at the 3R TPS. Based on the interview with the representative of *pokdarwis* and the manager of the TPS 3R, only 10-15% of the community are actively involved in this program. The community's willingness to be active will affect the time and costs that must be spent at 3R polling stations. In addition, not all communities are willing to pay regular fees for waste management. Various things can be the cause of this community's reluctance, including the assumption that waste is leftover goods that must be disposed of and also the assumption that people can process their own waste, either through making compost from organic waste and burning un-organic waste.

- TPS 3R operations are not yet independent as expected. The operational funds referred to waste transportation vehicles, staff salaries, and unexpected budgets. However, based on the results of interviews, the money generated from TPS 3R activities, both from villagers' fees and from sales of TPS 3R products, often still cannot cover routine operational costs that must be incurred, so they must be subsidized from the village funds managed by BUMDes.

In order to overcome this challenge, the government can increase efforts to empower the community through socialization from the lowest level of community, such as groups of women in the neighborhood to the wider level, such as meetings in the village level. Additionally, the manager of waste bank can seek for collaboration with private sector to buy the sorted waste from the community. This strategy can increase the value of sorted waste and make waste bank program more interesting for the community.

4. Conclusion

The waste management system in the Lerep Tourism Village has begun to pay attention to the concept of sustainability and is integrated by empowering local communities. It aims to support the realization of a sustainable tourism village. The community as the main actor has an important role in its implementation [9, 10]. The efforts made by the people of Lerep Tourism Village are to reduce waste burning, sorting plastic bottle waste for further savings in a waste bank, paying waste fees and recycling plastic waste through a waste bank. Although it has not been fully implemented, the efforts that have been made by the community have been able to have a good impact on the environment such as reducing air pollution and also an initial step to implement the concept of circular economy in the tourism activities [18] at the community level. In the future, the local government, particularly BUMDes and *pokdarwis* as the main actors in the development of this village can empower more people in this community-based waste management through outreach activities, improving the performance of waste banks, and scaling up strategies implemented in the culinary market at other tourist attractions.

Acknowledgement
This research is funded by the Strategic Research Grant 2022, Engineering Faculty, Diponegoro University.

References

[1] Kastenholz E, Carneiro M J, Peixeira Marques C and Lima J 2012 Understanding and managing the rural tourism experience — The case of a historical village in Portugal *Tourism Management Perspectives* **4** 207-14
[2] Irshad H 2010 *Rural Tourism – an Overview* (Alberta: Government of Alberta)
[3] Doshita M 2014 Re-evaluating rural environments: rural tourism development in Japan 2014 *J Tour Consum Pract* **6** 28 -51
[4] Abbas I, Purwaningrum P and Indrawati D 2021 Evaluation of waste management in tourism area of Luar Batang Village, outside area of the Historic Old Jakarta Kota, Penjaringan District, North Jakarta. In: *IOP Conference Series: Earth and Environmental Science*: (IOP Publishing) p 012012
[5] Kaseva M E and Moirana J L 2009 Problems of solid waste management on Mount Kilimanjaro: A challenge to tourism *Waste Management & Research* **28** 695-704
[6] Juniartha I P, Antara M and Sudarma I M 2019 Pengelolaan Sampah Berbasis Masyarakat di Desa Pakraman Padangtegal Kecamatan Ubud Kabupaten Gianyar *Ecotrophic* **13** 205-14

[7] Guo Y, Wang P and Zhang D 2021 Challenges and Prospects of Solid Waste Management in Rural Tourism Areas in China *E3S Web Conf.* **251** 02082
[8] Reni V V and Anna P 2016 Sustainable waste management of traditional craft industry in Lopati Tourism Village, Yogyakarta *DIMENSI (Journal of Architecture and Built Environment)* **43** 123-30
[9] Sewak A, Deshpande S, Rundle-Thiele S, Zhao F and Anibaldi R 2021 Community perspectives and engagement in sustainable solid waste management (SWM) in Fiji: A socioecological thematic analysis *Journal of Environmental Management* **298** 113455
[10] K.C S, Rao D U M, Srivastava S, Sharma D, Varghese E and Rashmi I 2018 Impact of community based waste management effort in the socio-economic upliftment of a rural tourism village in Kerala *International Journal of Basic and Applied Agricultural Research* **6** 123 - 127
[11] Đurkin J and Perić M 2017 Organising for community-based tourism: Comparing attitudes of local residents and local tourism entrepreneurs in Ravna Gora, Croatia *Local Economy* **32** 678-91
[12] Tolkach D and King B 2015 Strengthening Community-Based Tourism in a new resource-based island nation: Why and how? *Tourism Management* **48** 386-98
[13] Rocca L H D and Zielinski S 2022 Community-based tourism, social capital, and governance of post-conflict rural tourism destinations: the case of Minca, Sierra Nevada de Santa Marta, Colombia *Tourism Management Perspectives* **43** 100985
[14] Dewi A E, Maryono M and Warsito B 2019 Potential sustainability of "Kampung Iklim" program in Surakarta municipal. In: *E3S Web of Conferences*: EDP Sciences) p 02002
[15] Yin R K 2009 *Case study research: Design and methods* (Thousand Oaks: SAGE Publications)
[16] Stake R E 2005 *The SAGE handbook of qualitative research,* ed N K Denzin and Y S Lincoln (Thousand Oaks: SAGE Publications) p 443-66
[17] Wells B M 2010 Naturalistic observation. In: *Encyclopedia of research design* ed N J Salkind (Thousand Oaks: SAGE Publications) p 886-91
[18] Rodríguez C, Florido C and Jacob M 2020 Circular Economy Contributions to the Tourism Sector: A Critical Literature Review *Sustainability* **12** 4338

Effect of drying duration on the water content of durian peel waste for bio pellet

Yustina M Pusparizkita[1*], Alif F Hidayatullah[1], Norrisal F Anwar[1], Junaidi J[1], Sudarno S[1]

[1]Department of Environmental Engineering, Diponegoro University, Jl. Prof. Soedharto, S.H. Tembalang, Semarang, Indonesia

yustinapusparizkita@gmail.com

Abstract. Bio pellet is popular as an alternative solid fuel for replacing firewood and coal. In this study, the biomass used as raw material for making bio pellets is durian peel. Due to the high-water content, durian peel must be dried before being processed into a bio pellet. Therefore, this study aimed to analyze three variable time effects of drying duration at a temperature of 100°C on the water content of durian peel. The results showed that the water content of durian skin after drying for 6, 12, and 24 hours was 11,04%; 8,09%; and 3,75%. The most extended drying duration will produce the driest raw materials and the lowest moisture content. Nevertheless, the drying time that has the appropriate moisture content for the raw material for making pellets was 6 and 12 hours.

1. Introduction

Pellet fuel is a biofuel made from compressed biomass or organic matter. The pellets are usually made from five basic categories: food waste, by-products, industrial waste, agricultural residues, wood, and energy crops. Moreover, the pellets can be categorized based on their calorific value, ash content, moisture content, and dimensions. Usually, the pellets formed can be used for various purposes such as fuel for power plants, residential and commercial heating, cooking, and much more. Currently, pellet fuels are widely used as large-scale boilers in retail operations and home heating applications [1]. Even some industries consider using pellet fuel as a substitute for coal in terms of renewable behaviour. The selection of pellets as fuel is always based on their quality. The quality of the pellet fuel depends on its chemical composition, such as water content, ash weight content, and mineral composition [2].

Different uses of biomass have been considered for the production of pellet fuel, such as oil palm and rubber tree residues [3], mixed bark (*eucalyptus*), mangosteen rind, and papaya rind [4], residues of milling olives [5], a mixture of pine sawdust, wheat straw and rapeseed straw [6], Norwegian spruce (*Picea abies*), European beech (*Fagus sylvatica*), and wheat straw (*Triticum aestivum*), bamboo sawdust, eucalyptus sawdust, corn cobs, rubber tree twigs, palm fibre and lippia grass, oil palm empty fruit bunches, oil palm fronds, oil palm shells and oil palm mesocarp [7], waste from oil palm plantations and the palm oil industry [8], biomass waste from canola meal, rice straw, and sunflower husk residue, corn stover, barley straw [9], waste from microalgae *Scenedesmus*, coconut, canola [10], wheat and peat [11].

Besides various types of biomasses that have been studied, durian peel waste has the potential to be used as raw material for making bio pellets [12]. Durian (*Durio zibethinus*) is one of the leading fruit crops cultivated in Indonesia. The Indonesia Central Statistics Agency (BPS) noted that national durian fruit production reached 1.13 million tons in 2020. The high level of durian production also causes high waste generation, where most fruit except the flesh is treated as waste. Only 20.52% of the fruit is used, and about 79.48% is the part not utilized for consumption, such as durian skin and seeds. Durian skin that is not utilized and is difficult to degrade causes waste problems and becomes a collision that ultimately pollutes the environment. Alternatively, the high calorific value of durian skin, around 15.91 MJ/Kg, is considered an excellent alternative fuel source [13]. However, the utilization of the durian shell as a significant fuel source is challenged by its seasonal production, technological feasibility, and social acceptability [14]. Also, if the durian skin will be used as raw material for bio pellets, the drying process as a pretreatment needs to be done because the durian skin has a high-water content. In other experimental studies, it has been found that the optimum moisture content for palletization of cellulosic materials is 8-12 % [15]. The moisture content of raw material is one of the most important factors that negatively influence the properties of pellets, such as bulk density or mechanical durability during storage and transportation. Studies found a positive correlation between pellets' durability and optimal moisture (10 %) because water, together with raw material's temperature, pressure, and chemical composition, acts like a binding agent that increases pellet quality. They recommended that 5-12 % moisture content is required to produce a good quality product (in terms of optimal density and properties on long-term deposit). Conversely, moisture higher than 15 % damages pellets during storage [15].

In previous studies, durian skin had to be processed by pyrolysis before being used as raw material for pellet fuel [12]. In this study, three variations of drying time were used as a pretreatment of durian skin to convert durian skin into bio pellets to increase the economic value of the material. Furthermore, by varying the drying duration, this research is expected to describe the optimum time to produce durian skin suitable for making bio pellets.

2. Material and methods

The material used for these experiments was durian peel waste obtained from durian seller centres in Semarang, Indonesia. Samples of durian peel (a dimensional size of approximately 9 cm^2) were placed on trays to be dried in the oven. The oven capacity used is 53 L (Memmert UN 55, Germany). For the experiment, three variations of drying time were set, namely 6 hours, 12 hours, and 24 hours, with a drying temperature of 100^oC. After drying, all samples were analyzed for moisture content using a moisture analyzer (Shimadzu MOC63u, Japan). Water content was measured with different samples (minimum of 5 repetitions). Statistical analysis was performed with ANOVA using Microsoft Excel 2019, and significance level (α) alpha was set at 0.05.

3. Results and discussion

The moisture content is one of the most crucial factors affecting the efficiency, stability and completeness of combustion of solid biofuel. The higher calorific value seemed to be provided by pellets with low ash and moisture content and were mechanically durable [16]. Moreover, the high moisture content of wood pellets may negatively affect the performance of pellets by reducing the net energy output during combustion and generating high emissions of air pollutants [17]. Generally, the optimum moisture content of feedstock during palletization seems to be 8–13% [18].

3.1. The durian skin moisture before the drying process

The water content measurement of the durian skin waste before drying was divided into the outer and inner skin. The results can be seen in Table 1. Based on the measurement results on ten samples of the inner skin, the highest moisture content value reached 83.67%, with an average moisture content value of 82.05%. While the results of measurements on ten samples of outer skin obtained the highest moisture content value, reaching 82.98%, with an average moisture content value of 82.26%. Based on

ANOVA calculations, the p-value was 0.66. Since $p \geq 0.05$, these results indicated that the water content value before drying on the inner durian skin is not much different from that of the outer durian skin.

Table 1. The moisture content of durian inner and outer skin

Inner Skin		Outer Skin	
Sample	Moisture Content (%)	Sample	Moisture Content (%)
1	82.45	1	82.01
2	80.19	2	81.22
3	80.93	3	82.69
4	81.65	4	82.71
5	83.54	5	82.34
6	82.5	6	82.49
7	80.12	7	81,87
8	81.27	8	81.68
9	84.15	9	82.64
10	83.67	10	82.98

3.2. The durian skin moisture after the drying process

There were three levels for placing the tray in the oven. In this study, three levels were used. Tray 1 was in the topmost position. Tray 2 was in the middle, and tray number 3 was at the bottom. The water content results after 24 hours of drying can be seen in Table 2.

Table 2. The moisture content of durian skin was placed in three different trays after drying

Sample	Inner Skin Moisture Content (%)			Outer Skin Moisture Content (%)		
	Tray 1	**Tray 2**	**Tray 3**	**Tray 1**	**Tray 2**	**Tray 3**
1	2.07	1.95	2.02	2.91	2.76	3.85
2	2	2.38	1.83	3.37	3.67	3.86
3	2.36	2.1	3.14	4.53	4.14	4.21
4	2.43	2.44	2.68	4.3	4.42	4.48
5	2.27	1.55	2.94	5.17	4.97	4.86

According to Table 2, which shows the comparison of the water content, the difference in the tray position during the drying process did not affect the water content of the final product. The average water content in trays 1, 2 and 3 for the durian inner skin was 2.22%, 2.1% and 2.52%. At the same time, the average water content in trays 1, 2 and 3 for the durian outer skin was 5.06%, 4.71%, and 4.63%. As the p-value for part of durian skin after drying was less than our significance level (5.04E-8), this factor was statistically significant. However, the p-value for tray positions was greater than the significance level (0,458), so this factor was not statistically significant, indicating that the oven's hot air was perfectly distributed. Besides that, the interaction effect between the part of durian skin and tray level was insignificant because the p-value was 0.879 ($p \geq 0.05$).

Due to the same water content between the outer and inner skin of the durian, the samples used in the drying process were mixed. Overall, the various drying duration affected the water content of durian skin (p-value ≤ 0.05). Table 3 and Fig. 1 compared the durian skin water content after drying for 6, 12, and 24 hours. According to Table 3, the average moisture content of durian skin samples used for the drying process was 81.94%. After the drying process, the average water content for 24 hours of drying time was 3.76%, while for 12 hours was 8.1%.

On the other hand, the water content of durian skin with the shortest drying time of 6 hours was 11.04%. Therefore, the faster the drying time, the higher the water content remaining in the durian skin. The long drying time allowed more water trapped in the durian skin to evaporate.

Table 3. The moisture content of durian skin before and after drying

Sample	Moisture Content (%)			
	Before drying	24 hours	12 hours	6 hours
1	81.33	4.1	10.54	10.42
2	80.89	5.81	7.14	8.94
3	81.77	4.56	9.93	8.81
4	82.33	3.62	5.23	13.2
5	81.45	3.39	6.43	12.81
6	81.92	2.92	8.25	11.41
7	81.44	3.4	8.86	11.6
8	82.37	1.74	9.03	11.49
9	83.12	4.09	6.3	10
10	82.79	3.93	9.2	11.7

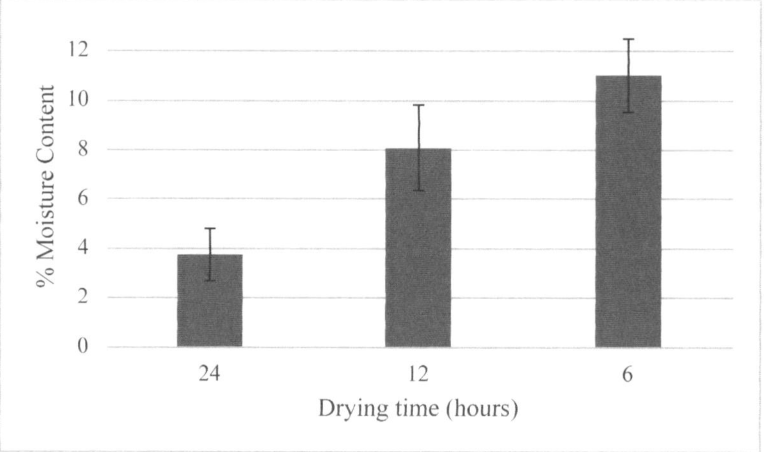

Figure 1. The average durian skin moisture content after drying three different times

Based on several previous studies, the durian skin that has been dried for 6 and 12 hours was suitable for raw material for good quality pellets. In contrast, durian peel that has dried for 24 hours is not ideal for making pellets, even though it has the lowest water content. The pellets became fragile just a few days after production if the moisture content used was less than 4 % due to environmental moisture absorption. Also, pellets with 5 % moisture had low strength, became brittle, and large amounts of dust were produced during storage and transportation [19].

4. Conclusion

As the potential raw material for solid pellet fuel, the three variations of drying time for durian skins will produce products with average water content below 12%. Drying for 6 and 12 hours makes durian skin suitable for pellet raw materials because it contains 11.04% and 8.1% water. Although the drying process is considered adequate as a pretreatment process in the context of utilizing durian skin as raw material for pellet fuel, an economic review of the process is necessary, considering that the drying process at a temperature of 100°C takes a relatively long time. Alternatively, to shorten the drying time is to increase the drying temperature. Nonetheless, further studies need to be done.

5. Acknowledgement

The authors would like to express their special thanks to the Strategic Research Faculty of Engineering Diponegoro University, RKAT Grants Funds 2022 (grant number 356/UN7.5.3.2/HK/2022), which has funded this research.

References

[1] Anwar A H S, Saimon N N, Ali M W, Kamarizan K, Jusoh Y M M, Mazura J and Zakaria Z Y 2018 Effect of particle size on the explosive characteristics of grain (wheat) starch in a closed cylindrical vessel *Chemical engineering transactions* **63** 571–6

[2] Chandrasekaran S R, Hopke P K, Rector L, Allen G and Lin L 2012 Chemical Composition of Wood Chips and Wood Pellets *Energy Fuels* **26** 4932–7

[3] García R, Pizarro C, Lavín A G and Bueno J L 2017 Biomass sources for thermal conversion. Techno-economical overview *Fuel* **195** 182–9

[4] Shen D K, Gu S, Luo K H, Bridgwater A V and Fang M X 2009 Kinetic study on thermal decomposition of woods in oxidative environment *Fuel* **88** 1024–30

[5] Tauro R, García C A, Skutsch M and Masera O 2018 The potential for sustainable biomass pellets in Mexico: An analysis of energy potential, logistic costs and market demand *Renewable and Sustainable Energy Reviews* **82** 380–9

[6] Bilal M, Asgher M, Iqbal H M N, Hu H and Zhang X 2017 Biotransformation of lignocellulosic materials into value-added products—A review *International Journal of Biological Macromolecules* **98** 447–58

[7] Serrano C, Monedero E, Lapuerta M and Portero H 2011 Effect of moisture content, particle size and pine addition on quality parameters of barley straw pellets *Fuel Processing Technology* **92** 699–706

[8] Ishii K and Furuichi T 2014 Influence of moisture content, particle size and forming temperature on productivity and quality of rice straw pellets *Waste Management* **34** 2621–6

[9] Gil M V, Oulego P, Casal M D, Pevida C, Pis J J and Rubiera F 2010 Mechanical durability and combustion characteristics of pellets from biomass blends *Bioresource Technology* **101** 8859–67

[10] Senneca O 2007 Kinetics of pyrolysis, combustion and gasification of three biomass fuels *Fuel Processing Technology* **88** 87–97

[11] Demirbas A 2009 Political, economic and environmental impacts of biofuels: A review *Applied Energy* **86** S108–17

[12] Selvarajoo A, Lee C W, Oochit D and Almashjary K H O 2021 Bio-pellets from empty fruit bunch and durian rinds with cornstarch adhesive for potential renewable energy *Materials Science for Energy Technologies* **4** 242–8

[13] Wilaipon P 2011 Durian Husk Properties and its Heating Value Equation *American Journal of Applied Sciences* **8** 893–6

[14] Foo K Y and Hameed B H 2011 Transformation of durian biomass into a highly valuable end commodity: Trends and opportunities *Biomass and Bioenergy* **35** 2470–8

[15] Liu C and Wyman C E 2005 Partial flow of compressed-hot water through corn stover to enhance hemicellulose sugar recovery and enzymatic digestibility of cellulose *Bioresource Technology* **96** 1978–85

[16] Proskurina S, Junginger M, Heinimö J, Tekinel B and Vakkilainen E 2019 Global biomass trade for energy— Part 2: Production and trade streams of wood pellets, liquid biofuels, charcoal,

industrial roundwood and emerging energy biomass *Biofuels, Bioproducts and Biorefining* **13** 371–87

[17] Liu Z, Quek A and Balasubramanian R 2014 Preparation and characterization of fuel pellets from woody biomass, agro-residues and their corresponding hydrochars *Applied Energy* **113** 1315–22

[18] Sgarbossa A, Costa C, Menesatti P, Antonucci F, Pallottino F, Zanetti M, Grigolato S and Cavalli R 2014 Colorimetric patterns of wood pellets and their relations with quality and energy parameters *Fuel* **137** 70–6

[19] Junfeng L and Runqing H 2003 Sustainable biomass production for energy in China *Biomass and Bioenergy* **25** 483–99

IOP Conf. Series: Earth and Environmental Science 1098 (2022) 012053 doi:10.1088/1755-1315/1098/1/012053

Effect of Glass Cover Thickness and Inclination Angle on Distillate Efficiency of Single-Stage Solar Still

Felicia[1], Riana Ayu Kusumadewi[1*], Winarni[1], and Rositayanti Hadisoebroto[1]

[1] Environmental Engineering Department, Faculty of Architecture Landscape and Environmental Technology, Universitas Trisakti, Jakarta, Indonesia

*rianaayu.kusumadewi@trisakti.ac.id

Abstract. Solar desalination is one of many methods to separate excess salt from seawater. In this work, solar energy is applied because of its economical and environmental value. Additionally, solar energy is practical to use in the coastal region. To increase the distillate produced by the solar still, four variations of cover thickness and three variations of inclination angle are applied to the still. The cover thickness and inclination angle used in this work are 3 mm, 5 mm, 8 mm, 10 mm and 25°, 30°, 45°. The type of still that is used in this work is a rectangular-shaped single-stage solar still with an area of 0.8 m^2. The desalination processes in this work were done from 09.00 until 16.00 and the data was collected every hour in the processes. The result of this work shows that the distillate efficiency produced by the solar still is 1.03% - 6.91%. The highest solar intensity is 1231.97 W/m^2 and the lowest solar intensity is 101.03 W/m^2. The most effective efficiency was obtained with the variation of 3 mm cover thickness and 25° inclination angle, while the lowest efficiency was obtained with the variation of 10 mm cover thickness and 30° inclination angle.

1. Introduction

Water is one of the most important components for the human being. According to the *United States Geological Survey* (USGS), ninety-seventh percent of available water resources on earth are saline water and the other three percent is fresh water. Of the three percent of fresh water, only around one percent can be accessed by humans to fulfil their daily needs [1]. Out of one percent water that can be accessed, not all water can be used directly. The water needs treatment as it still contains pollutant and other harmful microorganisms. Because of the limited availability of fresh water, one of the alternatives that can be done is to utilize the abundantly available seawater.

To utilize the abundantly available seawater, the excess salt must be separated from the water. One of the methods to separate excess salt from seawater is solar desalination. Although the efficiency dan the distillate output of the solar desalination method is fairly low compared to the other desalination method, solar desalination is still being used because of its economical and environmental value. Also, solar desalination is more practical to use in the coastal region because it uses simple materials and simple operating procedures. Solar desalination can be classified into two types, direct solar desalination and indirect solar desalination [2]. In this work the types of desalination used is passive direct solar desalination. The passive direct solar desalination only used solar as it main source of energy, and the heat is directly transferred to the still [2].

The efficiency of the solar desalination is affected by several climate factors such as solar irradiation, weather, ambient temperature, wind speed, and humidity [3]. Solar desalination efficiency is also affected by brine depth, vapor leakage, the slope of the transparent cover, the thickness of the transparent cover and the material used as the transparent cover [4]. The operation of solar desalination is done by using an evaporation basin that was covered using a transparent cover. The sunlight was used to increase the basin and water temperature so that the water temperature rise and the water will be evaporated until it touches the transparent cover and turned into a water droplet that slides down the cover [4].

Previously, several research on desalination had been conducted. Wijaya (2021) has done the water desalination using the same desalination set used in this work using flannel color variation and different artificial saline water concentration. The maximum distillate efficiency obtained by Wijaya was 4,94% and the volume of the distillate produced was 570 ml with 645 W/m^2 solar irradiance using black colored flannel and the research was done from 09.00-15.00 [5]. Mulyanef (2015) has also done similar work to Wijaya (2021) using single stage solar still. The volume of the distillate produced on Mulyanef's research was 2012 ml/day with 451 W/m^2 solar irradiance and the research was done at 1 day time [6]. To improve the distillate efficiency in this work, glass cover thickness and inclination angle variation will be applied to the still to know the best cover thickness and inclination angle that can improve the efficiency of the distillate that is produced by the still.

2. Methodology

In this work, desalination processes were done on the 10th floor of Building I at Trisakti University, Jakarta and it was done from April until June 2022. Also, the desalination was done on 7 hours of observation from 09.00 until 16.00. The type of water used in this work was artificial saline water that was made by using fresh water and sea salt until the salinity reach 35‰.

The solar still used in this work is rectangular-shaped solar still with an area of 0.8 m^2. The solar still was insulated using styrofoam, while the still was made of aluminium so that the heat will be transferred equally inside the still. Also, black flannel is used as an absorber because the water flowed continuously into the still. The use of black flannel was done because the black color is easier to receive heat than the brighter colored flannel. The top of the still is covered with a transparent glass cover. The desalination set used in this work can be seen in Figure 1.

Figure 1. Solar Desalination Set

There are also several variations of glass cover thickness and inclination angle that are applied to the still. The glass cover thickness used in this work are 3 mm, 5 mm, 8 mm, and 10 mm. The inclination angle of the solar still used in this work are 25°, 30°, and 45°. In this work, the data was collected on every hour of operation starting from 09.00 to 16.00 and for every variation, the data were collected on two days of operation. The solar still and reservoir temperature that were measured during the operation can be seen as in Figure 2.

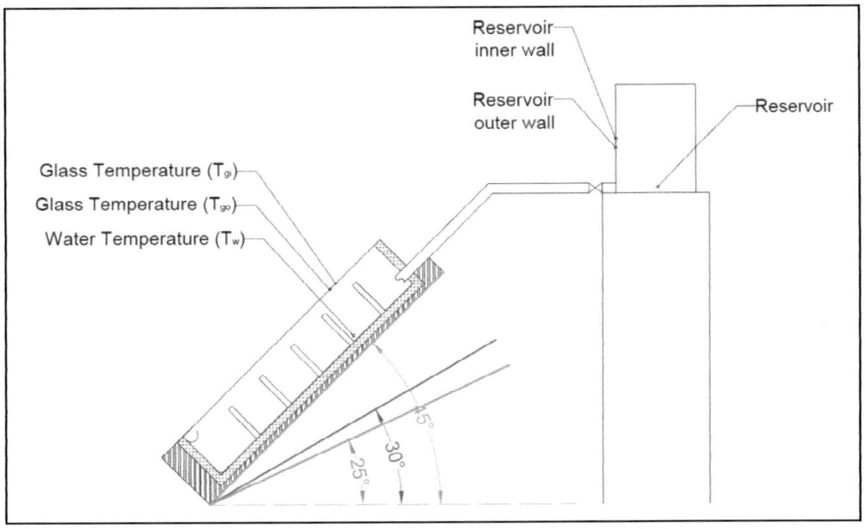

Figure 2. Temperature Measurement Point

The actual efficiency of the solar still is measured by using Equation 1 [7].

$$\text{Actual efficieny} = \frac{\text{SS distillate volume}}{\text{Volume of artificial saline water entered SS}} \text{ x } 100\% \quad (1)$$

The theoretical efficiency in this work is analyzed by using Equation 2 [8].

$$\text{Theoretical efficiency} = \frac{Q_u}{Q_{in}} \text{ x } 100\% \quad (2)$$

$$Q_u = M_d \text{ x } C \text{ x } \Delta T + M_d \text{ x } h_{fg} \quad (3)$$

$$Q_{in} = \propto \text{ x } I(t)_s \text{ x } A_c \text{ x } t \quad (4)$$

Where C is the specific heat of water (J/kg°C), ΔT is temperature differences (°C), M_d is distillate water (kg), h_{fg} is latent heat (J/kg), $\propto$ is Absorptivity, $I(t)_s$ is total solar irradiance (W/m^2), A_c is collector area (m^2), and t is the time of operation (s).

Water quality testing were also done on the distillate and brine produced by the solar still. The checked in-situ parameter was salinity, electric conductivity (EC), total dissolved solids (TDS) and pH. While the checked ex-situ parameter was turbidity, Cl^- ion, total hardness, magnesium hardness, calcium hardness, iron, and Escherichia coli. The ex-situ measurement in this work was done at Trisakti University Environmental Laboratory, Jakarta, Indonesia.

3. Results and discussion

The data of the distillate amount on this work was measured every hour from 09.00 until 16.00 for every variation and can be seen in Figure 3.

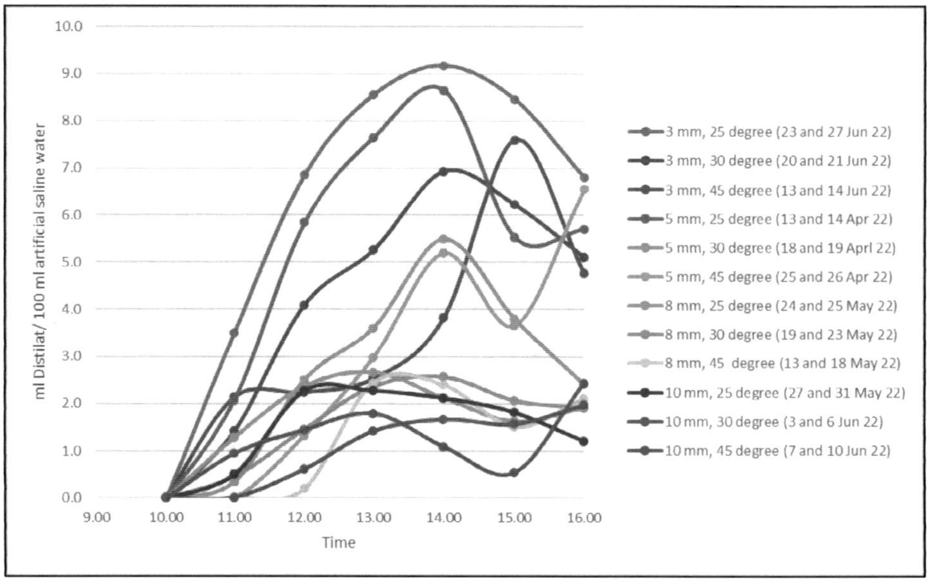

Figure 3. Distillate produced per hour

Figure 3 shows distillate that was produced per 100 milliliter artificial saline water which the value varies for every cover thickness and inclination angle. The highest distillate output was obtained using variation of 3 mm cover thickness and 25° inclination angle. While the lowest distillate output was obtained using variation of 10 mm cover thickness and 30° inclination angle. It could also be seen that the distillate only starts to produce at 11.00. The delay in the production of distillate was affected by the low water flow rate that is used in this work. It could also be seen that the distillate volume produced by the kept increasing until around 14.00 and then it would significantly gone lower affected by the sun movement that is also reducing the solar irradiation received by the still.

3.1. Distillate efficiency

Actual distillate efficiency that was produced by the solar still is calculated by using Equation 1 and the theoretical efficiency is calculated by using Equation 2. The actual and the theoretical efficiency can be seen in Table 1.

Table 1. Actual and Theoretical Efficiency

Cover Thickness	Angle	Actual Efficiency (%)	Theoretical Efficiency (%)	Actual Efficiency (%) Per Liter Artificial saline water
3 mm	25	6.19	7.13	0.91
	30	4.15	4.63	0.60
	45	3.30	6.12	0.47
5 mm	25	5.06	5.87	0.63
	30	2.59	2.71	0.36
	45	2.81	3.01	0.34
8 mm	25	1.71	2.40	0.24
	30	1.55	1.85	0.22
	45	1.23	1.43	0.17
10 mm	25	1.46	1.71	0.21
	30	1.03	1.43	0.15
	45	1.18	1.52	0.17

It can be seen from Table 1 That the highest actual efficiency and theoretical efficiency were obtained using variation of 3 mm cover thickness and 25° inclination angle amounting to 6.19% and 7.13% While the theoretical efficiency was obtained using variation of 10 mm cover thickness and 30° inclination angle amount to 1.03% and 1.43%. If compared to the previous work done by Wijaya (2021), it could be known that adjusting the cover thickness and inclination angle will affect the distillate produced by the still [5]. In Wijaya's work the highest actual distillate efficiency was 4,94% using 5 mm glass cover thickness and 25° inclination angle, while in this work the highest actual distillate efficiency was 6,19% using 3 mm cover thickness and 25° inclination angle.

From Table 1, it could also be seen that the actual efficiency was lower than the theoretical efficiency. The 3 mm cover thickness and 25° inclination angle has higher efficiency than the 10 mm cover thickness and 30° inclination angle was affected by the energy loss because the thicker glass cover and the higher inclination angle will result in more energy lost [9]. The actual efficiency was calculated based on the distillate produced by the solar still, while the theoretical efficiency was calculated based on the thermal processes inside the still. Percent actual efficiency achieved to theoretical efficiency can be seen in Figure 4.

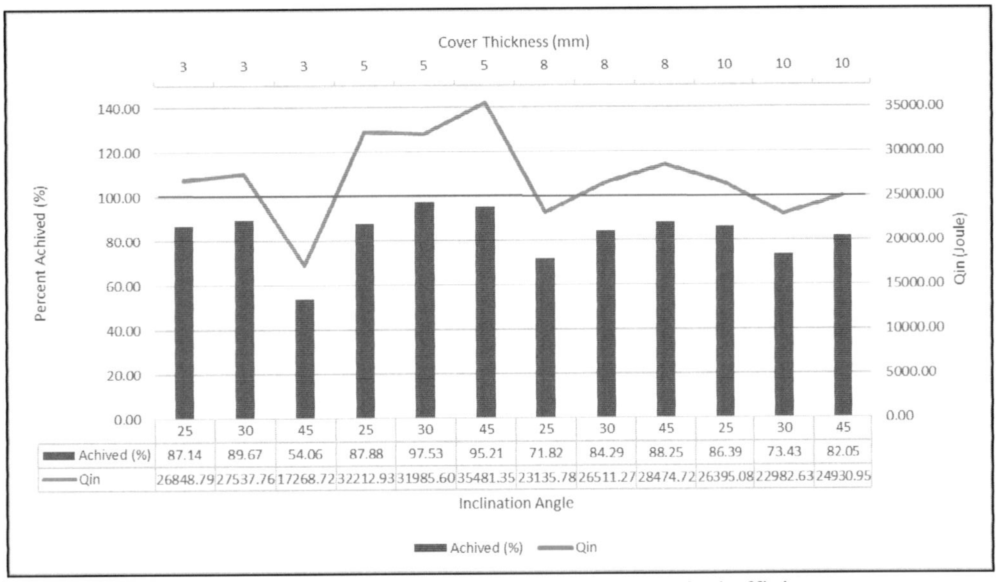

Figure 4. Achieved actual efficiency to theoretical efficiency

From Figure 4, the percentage efficiency achieved follows the fluctuation of incoming energy to the still. The incoming energy was calculated based on the solar irradiation as shown in Equation 4. Therefore, it could be known that the percentage of actual efficiency to theoretical efficiency achieved by the still was affected by the fluctuation of solar irradiation. It could also be seen from Figure 5 that it is true that the distillate efficiency and the incoming energy to the evaporator has a linear relationship [10].

3.2. Cover Thickness

The optimum cover thickness in this work was analyzed by calculating the distillate volume per 10.000-joule incoming energy and can be seen in Figure 5. The incoming energy was calculated based on the solar irradiation as shown in Equation 4.

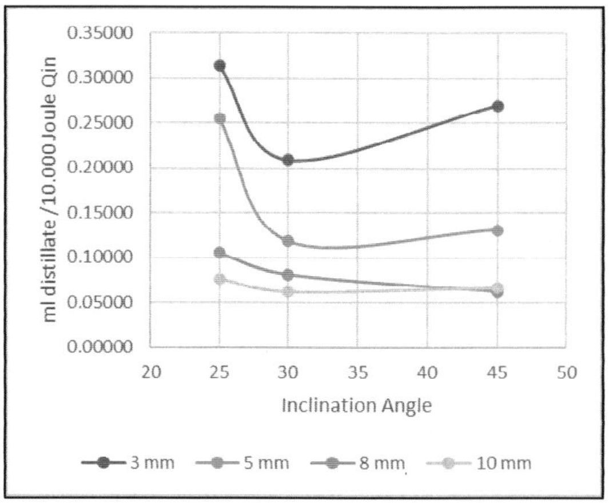

Figure 5. Distillate volume per 10.000 joule Q_{in}

Figure 5 shows that the most distillate volume produced by the still was obtained by using 3 mm, 5 mm, 8 mm, and 10 mm cover thickness respectively. The difference in the produced distillate for every cover thickness was affected by the still temperature. The thicker glass cover has lower water temperature than the thinner glass cover [11]. So that the distillate water produced by using 10 mm glass cover was significantly lower than the 3 mm, 5 mm. and 8 mm glass cover. Although the more thinner glass cover has higher distillate efficiency, it is also important to note that the thinner glass cover is more prone to breaking, so that the thinnest glass cover used in this work is 3 mm.

3.3. Inclination Angle

The optimum inclination angle in this work was analyzed by calculating the distillate volume per 10.000-joule incoming energy and can be seen in Figure 6.

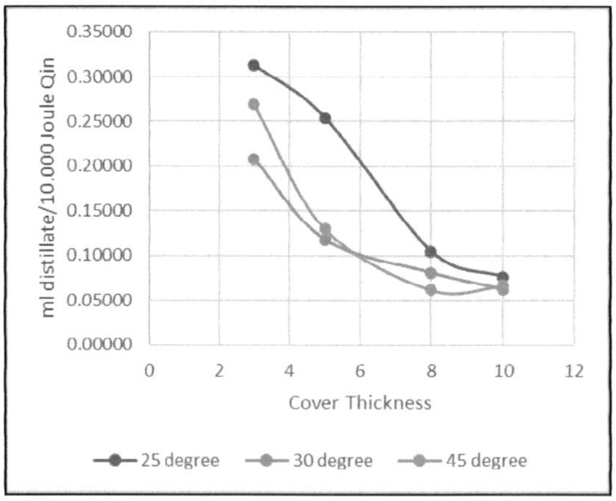

Figure 6. Distillate volume per 10.000 joule Q_{in}

Figure 6 shows that the most distillate volume produced by the still was obtained by using 25° cover angle. The difference in the produced distillate for every inclination angle was affected by energy loss The higher inclination angle will reflect more heat energy to the environment so that the energy that enters the still would be lower than the 25° inclination angle [9]. If the inclination angle used was lower than a certain degree, it could also be possible that the still would produce fewer distillate because the distillate would fall to the still before it reaches the effluent channel [12].

3.4. Water Quality

Water quality is also tested in this work. The result of the water quality tested in this can be seen in Table 2. The tested parameters are pH, turbidity, Cl^- ion, total hardness, calcium hardness, magnesium hardness, iron, electric conductivity, total dissolved solids, salinity, and Escherichia coli.

Table 2. Water Quality

Parameter	Artificial Saline Water	Brine	Distillate	Standard*
pH	8.16 – 8.73	8.4 – 8.86	7.4 – 8.66	-
Turbidity (NTU)	8.86 – 35.6	0 – 7.69	0 – 3.55	4
Cl^- ion (mg/L)	20,593.61 – 27,591.44	23,392.75 – 35,389.07	4 – 53.98	200-300
Total hardness (mg/L)	1,452 – 2,120	1,464 – 2,608	0	-
Ca hardness (mg/L)	40.4 – 486.4	81.6 – 707.2	0	100-300
Mg hardness (mg/L)	116.16 – 379.2	111.36 – 464.64	0	-
Fe (mg/L)	0 – 4.17	0 – 2.18	0 – 0.68	0.3
EC (μS)	56,000 – 58,100	58,500 – 100,500	14-214	<600
TDS (mg/L)	28 – 29.2	20.6 – 52.5	7 - 83	-
Salinity (‰)	34.8 – 35.7	35.6 – 69.2	0	-
E-coli	0	0	0	0

*WHO drinking-water guideline

Based on the WHO drinking-water guideline [13], the water produced by this solar desalination cannot be consumed directly because the water still contained an excess iron parameter and further processing is required for the water to lower the iron parameter.

4. Conclusion

The highest actual efficiency was 6.19%, obtained using variation 3 mm cover thickness and 25° inclination angle. While the lowest actual efficiency was 1.03%, obtained using variation 10 mm cover thickness and 30° inclination angle. The highest theoretical efficiency was 7.13%, obtained using variation 3 mm cover thickness and 25° inclination angle. While the lowest theoretical efficiency was 1.43%, obtained using variation 10 mm cover thickness and 30° inclination angle

Distillate water produced by the desalination processes met the quality standard for turbidity, Cl^- ion, calcium hardness, TDS, and Escherichia coli parameters. While the iron parameter is still above the quality standard, so the water has to undergo more processes to be consumed directly as drinking-water

References

[1] Manju S and Sagar N 2017 Renewable Energy Integrated Desalination: A Sustainable Solution to Overcome Future Fresh-Water Scarcity in India *Renew. Sustain. Energy Rev.* **73** 594–609
[2] Tiwari G N and Sahota L 2017 Review on the energy and economic efficiencies of passive and active solar distillation systems *Desalination* **401** 151–179
[3] Kusumadewi R A, Notodarmodjo S, and Helmy Q 2018 Heat Transfer Analysis and Water Quality in Saline Water Desalination Using Solar Energy in Vacuum Condition *Indones. J. Urban Environ. Technol.* **2**(1) 66–87
[4] Kucera J 2014 *Desalination* (Massachusetts: Scrivener Publishing)
[5] Wijaya D, Kusumadewi R A, Wijayanti A, and Hadisoebroto R 2021 Saline water desalination with multilevel solar distillation *IOP Conf. Ser.: Earth Environ. Sci.* **802** 012041
[6] Mulyanef, Saputra R A, Kaidir, and Duskiardi 2015 Kaji Eksperimental Alat Pengolahan Air Laut Menggunakan Energi Surya untuk Memproduksi Garam dan Air Tawar, In: Seminar Nasional Tahunan Teknik Mesin Indonesia XIV [in Indonesian Language]
[7] Hasanah F 2016 *Desalinasi Berbasis Tenaga Surya Untuk Menghasilkan Air Tawar* (Jakarta

Universitas Indonesia) [in Indonesian Language]
[8] Iqbal S, Sukmawaty, Putra G M D, and Setiawati D A 2019 Analisis Kinerja Alat Desalinasi Air Laut Penghasil Air Tawar dan Garam dengan Menggunakan Tenaga Surya *J. Agrotek.* **6**(1) 29-34 [in Indonesian Language]
[9] Syahrul M 2021 Pengaruh Ketebalan Kaca Terhadap Efisiensi Kolektor Surya Pada Proses Desalinasi Air Laut," (Medan: Universitas Muhammadiyah Sumatera Utara) [in Indonesian Language]
[10] Kusumadewi R A, Wijayanti Y, and Fittkow M 2021 Saline water desalination using solar energy: Performance analysis and implementation barriers *IOP Conf. Ser. Earth Environ. Sci.* **894**(1) 012020
[11] Saputro A E, Tarigan B, and Jafri M 2016 Pengaruh Tebal Kaca Penutup terhadap Efisiensi Kolektor Surya Pelat Gelombang Tipe V pada Proses Destilasi Air Laut *Lontar J. Tek. Mesin Undana* **3**(1) 1–10 [in Indonesian Language]
[12] Aljubouri A A 2017 Design and Manufacturing of Single Sloped Solar Still: Study the Effect of Inclination Angle and Water Depth on Still Performance *J. Al-Nahrain Univ.* **20**(2) 60–70
[13] WHO 2022 *Guidelines for Drinking-water Quality Fourth Edition Incorporating the First and Second Addenda*, vol. 33, no. 33 (Geneva: World Health Organization)

Selection of design criteria for the coagulation and flocculation unit in the cimanggis drinking water treatment plant

Anggi Tasya Megawati Putri[1], Riana Ayu Kusumadewi[1*], Winarni[1]

[1]Department of Environmental Engineering, Faculty of Landscape Architecture and Environmental Technology, Universitas Trisakti, Jakarta, Indonesia

*rianaayu.kusumadewi@trisakti.ac.id

Abstract. Coagulation and flocculation are essential components of drinking water treatment, because they provide a reliable process for treating water turbidity. This study aims to use design criteria that can be used to design Cimanggis Water Treatment Plant (WTP), so it can produce water that fulfills the Drinking Water Quality Standard No. 492 Year 2010 by the Health Minister of Republic of Indonesia. The Cimanggis WTP development design refers to the existing condition of Legong Konvensional 2 WTP, considering that the raw water quality of the Legong Konvensional 2 WTP is almost the same as the Cimanggis WTP. Determination of design criteria for Cimanggis WTP was carried out by literature study and evaluating the existing conditions of the Legong Konvensional 2 WTP. The evaluation of the Legong Konvensional 2 WTP is done by direct measurements of the plant dimensions. From the results of the data analysis, the design criteria that will be used for Cimanggis WTP, the G value is 393.78/sec for the coagulation unit; 40-10/sec for the flocculation unit, detention time (t_d) value is 112.5 seconds for the coagulation unit; 18.8 minutes for the flocculation unit, $G.t_d$ value is 44,311 for the coagulation unit; 27,158 for the flocculation unit.

1. Introduction

As the population in Depok City increases, the water demand also increases. It will require a new treatment plant to meet the demand for water supply demand. Based on the Depok City Drinking Water Supply Master Plan (Rencana Induk Penyediaan Air Minum Kota Depok), a new water treatment plant with a design capacity of 150 L/s is to be constructed to meet the water demand projection until the end of the planning year 2040 [1]. The development of drinking water services is closely related to the development of drinking water facilities in the form of enhancement capacity from the main facilities of the Drinking Water Supply System (SPAM) piping. The enhancement capacity program of drinking water facility is based on efforts to fulfill community water needs in achieving 100% access to safe drinking water according to SDG's [2]. The treatment plant was set to be built in the Cimanggis Subdistrict, Depok City, using surface water from the Ciliwung River [1]. The water that can be consumed by the Indonesian population meets the Drinking Water Quality Standard No. 492 Year 2010 by the Health Minister of the Republic of Indonesia [3].

The reason for choosing Legong Konvensional 2 WTP as a reference for design criteria in designing WTP is because it comes from the same raw water source, with a distance of 5.9 kilometers from the Cimanggis WTP. In addition, along the distance between Legong Konvensional 2 WTP and Cimanggis

WTP, it is dominated by the household. The existing plant performed well and the treated water meets drinking water standards according to Drinking Water Quality Standards Number 492 Year 2010.

The water treatment plant consists of several water treatment units that are commonly used: intake, coagulation, flocculation, sedimentation, filtration, and reservoir [4]. Coagulation and flocculation are essential components of drinking water treatment, because they provide a reliable process for treating water turbidity, which is a key to water quality. Coagulation and flocculation are processes that occur sequentially to destabilize suspended particles, causing particle collisions and growing into flocs. Coagulation is the process of mixing certain chemical compounds (coagulants) quickly and evenly through rapid stirring to destabilize colloids and suspended particles in raw water [5]. Generally, hydraulic coagulation uses the gravitational force of the difference in tank height that causes mixing power in the floc formation process in the coagulation unit. In addition, there is also a type of rapid mixing in the pipe [6]. The detention time of coagulation is 10-300/sec [7]. Flocculation is a slow and continuous stirring process, in which particles suspended in water coalesce into a larger mass so that they can be removed by further processing, particularly by sedimentation [6]. There are two types of flocculation, hydraulic and mechanical. Hydraulic flocculator is preferred in developing countries due to ease of operation and maintenance [8]. Tapered flocculation is usually accomplished by providing a high G value during the first stages of the flocculation period, then a lower G value, and a much lower G value during the last stage [5].

2. Methodology

The design of the coagulation and flocculation unit for Cimanggis WTP was carried out in several planning stages, mainly; literature study, data collection, and data analysis. The research flow diagram can be seen in Figure 1.

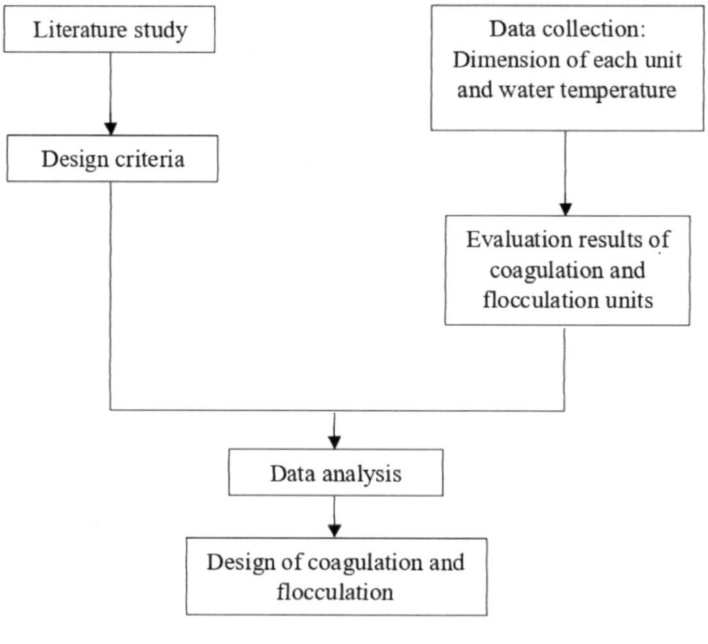

Figure 1. Research flow diagram.

2.1. Data collection

Data collection was done by collecting data in the form of the dimension of each unit and the water temperature. This study was carried out on Legong Konvensional 2 WTP, which is located in Jl. Legong Raya No. 1, Depok 2 Tengah, Mekar Jaya, Sukmajaya Subdistrict, Depok City, throughout August 2021.

In this part of the study, both the literature study and the data collection were analyzed to provide useful information in the design of the coagulation and flocculation unit of Cimanggis WTP.

2.2. Data analysis

The dimensions of each unit and the water temperature were used to evaluate the coagulation and flocculation, which is then used to design the coagulation and flocculation unit of Cimanggis WTP. There are six design parameters calculated with the following equations [7]:

Pipe velocity $$v = \frac{Q}{A\,pipe} \quad (1)$$

Gradien velocity $$G = \sqrt{\frac{P}{\mu \times Vol}} \quad (2)$$

Detention time $$t_d = \frac{Vol}{Q} \quad (3)$$

$G.t_d$ value $$G.td = G \times t_d \quad (4)$$

Headloss in channel $$h_L = n \times \frac{v1^2}{2\,g} \quad (5)$$

Headloss in channel turn $$h_L = k \times n \times \frac{v1^2}{2\,g} \quad (6)$$

3. **Results and discussion**

Evaluation of the existing coagulation and flocculation unit was carried out at Legong Konvensional 2 WTP of PDAM Tirta Asasta, Depok City, with a production capacity of 160 L/s. The source of treated raw water comes from the Ciliwung River, which is pumped to Legong Konvensional 2 WTP. The existing plant performed well and the treated water meets drinking water standards according to Drinking Water Quality Standards Number 492 Year 2010.

3.1. Existing coagulation and flocculation at Legong Konvensional 2 WTP

In designing a WTP, it is necessary to have the reference existing WTP, so that it can estimate the water treatment unit that can be used and the water treatment process runs optimally and can produce water that meets the Drinking Water Quality Standard No. 492 Year 2010 of the Health Minister of the Republic of Indonesia. The type of coagulation used by Legong Konvensional 2 WTP is hydraulic coagulation, where the system does not use a machine, but rather uses the gravitational force from the difference in the height of the tank that causes mixing power. Coagulant that used by Legong Konvensional 2 WTP is poly aluminium chloride (PAC). PAC is a type of coagulant that is not too affected by pH, [13] PAC can work well in the pH range of 4.5-9. Before entering the coagulation tank, the water is collected first in the receiving tank, and then drops to the coagulation tank. Coagulation mixing occurs in three stages. The first stage is in the raw water transmission pipe before it goes to the receiving tank, through the injection pipe. The second stage comes from the difference in the height of the water falling on the raw water transmission pipe before it enters the receiving tank, and with the water level of the receiving tank itself. The third stage is in the coagulation tank, where the water drop comes from the rectangular weir. The design coagulation unit can be seen in Figure 2. The results of the coagulation process evaluation are shown in Table 1.

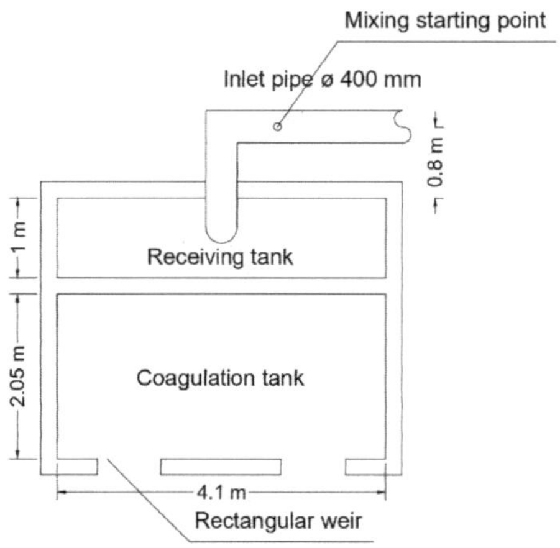

Figure 2. Plan view hydraulic coagulation

Table 1. Evaluation results of coagulation

Parameter	Design Criteria	Source	Result	Remarks
a. Injection Pipe				
a.1 Detention time	-	-	5 s	-
a.2 Gradient velocity	-	-	10.40/s	-
a.3 $G.t_d$ value	-	-	49.01	-
b. Receiving Tank				
b.1 Detention time	-	-	47.41 s	-
b.2 Gradient velocity	-	-	480/s	-
b.3 $G.t_d$ value	-	-	22,762.91	-
c. Coagulation Tank				
c.1 Detention time	-	-	60.41 s	-
c.2 Gradient velocity	-	-	356/s	-
c.3 $G.t_d$ value	-	-	21,498.93	-
Detention time total	10-300	Qasim, 2000 [7]	112.5	appropriate
$G.t_d$ total	10,000-100,000	Qasim, 2000 [7]	44,311	appropriate
G average	100-1000	Darmasetiawan, 2001 [12]	393.78	appropriate

The type of flocculation used by Legong Konvensional 2 WTP is hydraulic flocculation horizontal baffle channel, and the turbulence comes from the use of 180° rotation. In the flocculation unit, the coagulation water will flow horizontally zigzag from one canal to another until it flows into the sedimentation basin. There are 15 chambers in the flocculation unit. The chamber is then divided into 6 sections to obtain a range of G values to plan the flocculation unit. Tapered flocculation is usually accomplished by providing a high G value during the first stages of the flocculation period, then a lower G value, and a much lower G value during the last stage [5]. The design flocculation unit can be seen in Figure 3. The evaluation results of the flocculation process are shown in Table 2.

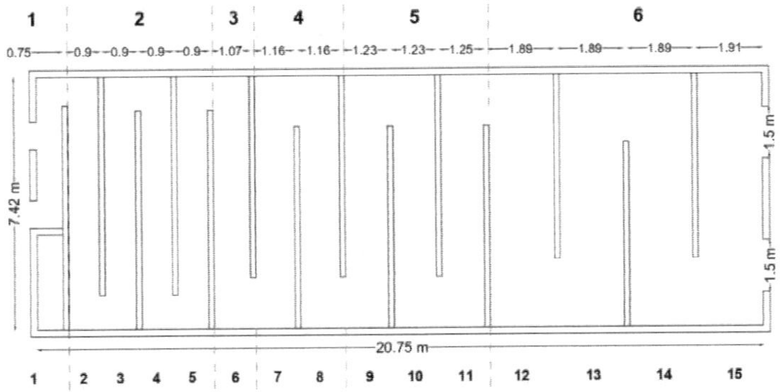

Figure 3. Plan view hydraulic flocculation horizontal baffle channel

Table 2. Evaluation results of flocculation

Parameter	Design Criteria	Source	Result	Remarks
Detention time	15-30	Schulz & Okun, 1984 [6]	22.9 min	appropriat e
Gradient velocity	60-5	Schulz & Okun, 1984 [6]	38.43 – 15.84/s	appropriat e
G.t_d value	10,000 – 150,000	Qasim, 2000 [7]	27,158	appropriat e

3.2. *Design of coagulation and flocculation at Cimanggis WTP*

In designing the Legong Konvensional 2 WTP the type of coagulation that will be used is hydraulic coagulation that utilizes the gravitational force from the difference in the height of the tank that causes mixing power. Hydraulic coagulation is a type of coagulation that is very effective in the rapid mixing process, and is easy to operate and maintain [8]. The parameter of design criteria that will used in designing coagulation is detention time total and the average of G value. The design parameters must cover the entire process that occurred in the coagulation unit, which is in the raw water transmission pipe, receiving tank, and coagulation tank. The proposed design and the water level in each tank can be seen in Figure 4. Table 3 shows the summary design of the coagulation.

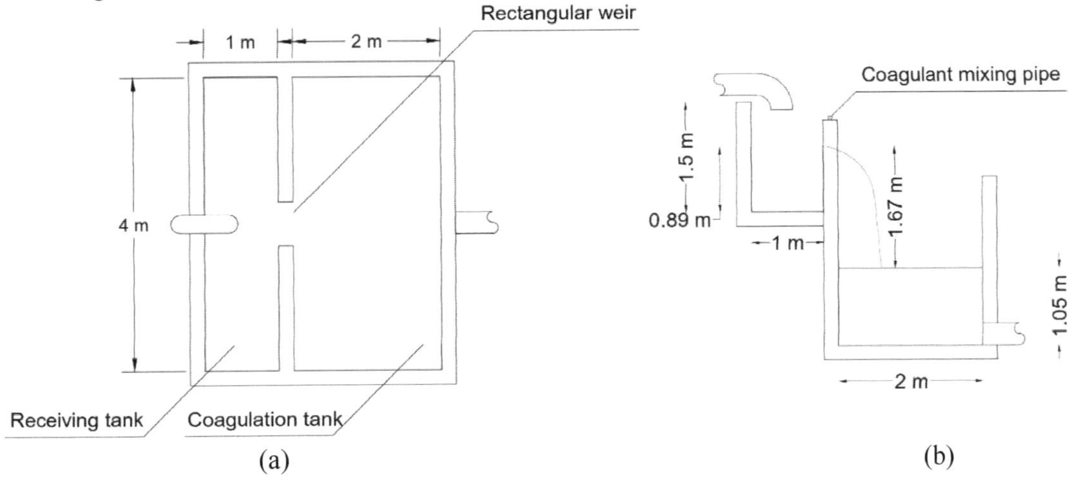

Figure 4. Hydraulic coagulation (a) plan view (b) longitudional cross-section

Table 3. Hydraulic coagulation design summary

Parameter	Unit	Design Result
Coagulation tank length	m	2
Receiving tank length	m	1
Tank width	m	4
Water height in receiving tank	m	0.89
Water height in coagulation tank	m	1.05
Tank height	m	2.3
Headloss	m	1.67
Detention time	s	112.5
Gradien velocity	/s	393.78
$G.t_d$ value		44,311

Flocculation is a water treatment unit that uses slow stirring that considers speed to prevent floc rupture due to excessive pressure, so it must consider the difference in the water levels in each compartment [9]. The type of flocculation that will be used by Cimanggis WTP is a vertical hydraulic flocculation baffle channel; this type of flocculation was chosen because it does not require a large area. The water moved from the bottom of the basin to the first compartment or flocculation chamber and crossed the first baffle between to reach the second compartment. The water then flowed downward to arrive at the under baffle towards the third compartment. The example stream was rehashed over and under [10] until it reached the last chamber where the water exits from the base side of the chamber to the sedimentation bowl. The parameter of design criteria that will be used in designing flocculation is the range of the G value and the total G.td, while the detention time will be set. The $G.t_d$ value is an important factor in the planning of flocculation [11] because these numbers represent the mixing intensity. In Legong Konvensional 2 WTP, the G value used is the range from 38.43-15.84/s, so in designing the flocculation, the G range has been set to the range 40–10/s. $G.t_d$ is an important design parameter to get the desirable floc, if the G is too large and the detention time is too small, floc will be small and dense, and if the G is too small and the detention time is too high, floc will be large and light. Factors that affect floc form are [14] turbidity in raw water, type of suspended solids, pH, alkalinity, coagulant used, and duration of mixing. The proposed design and the water level in each chamber can be seen in Figure 5. Table 4 shows the summary design of the flocculation.

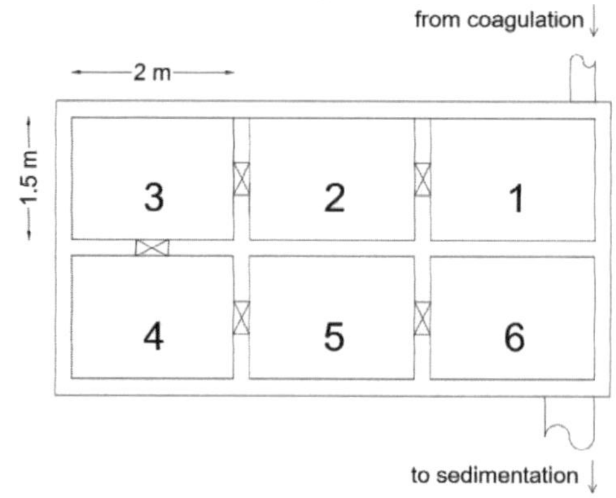

(a)

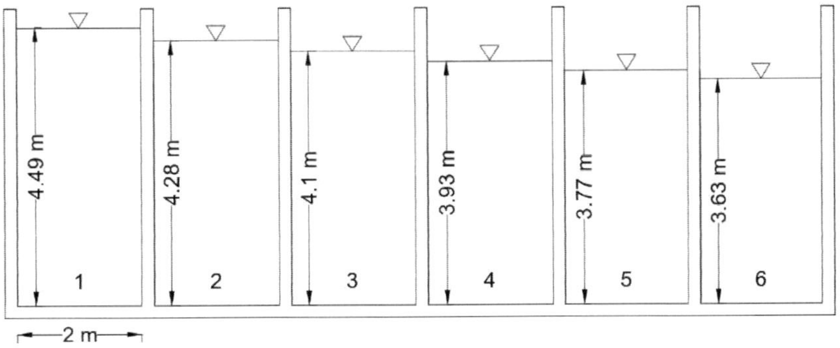

(b)

Figure 5. Hydraulic flocculation vertical baffle channel (a) plan view (b) longitudional cross-section

Table 4. Hydraulic flocculation design summary

Parameter	Unit	Chambers					
		1	2	3	4	5	6
Gradient velocity	/s	40	30	25	20	15	10
Detention time	s	209.50	199.57	191.39	183.50	176.07	169.50
Water height	m	4.49	4.28	4.10	3.93	3.77	3.63
Opening height	m	0.05	0.06	0.06	0.06	0.06	
Opening width	m	0.4					
Detention time total	min	18.80					
$G.t_d$		27,158					
Chamber length	m	2					
Chamber width	m	2					
Chamber height	m	4.80					

4. Conclusion

The design criteria to be used at Cimanggis WTP is as follow: in the coagulation unit, the detention time (t_d) used is 112.5 seconds, the G value used is 393.78/sec, and for the $G.t_d$ value used is 44,311. In the flocculation unit, the detention time (t_d) used is 18.8 minutes, for the G value used it is the range from 40 to 10/sec, and for the $G.t_d$ value used it is 27,158.

References

[1] Drinking Water Supply Master Plan of Depok City of 2020

[2] Ramadhan F, Siami L and Winarni 2019 *Prosiding Seminar Nasional Pembangunan Wilayah dan Kota Berkelanjutan*

[3] Drinking Water Quality Standard No. 492 Year 2010 by the Health Minister of Republic of Indonesia

[4] Kawamura S 2000 *Integrated Design of Water Treatment Facilities* (New York: Jorg/harin Willey and Sons, Inc)

[5] Reynolds T D and Richards P A 1996 *Unit Operations and Processes in Environmental Engineering* (2nd ed.) (PWS Publishing Company)

[6] Schulz C A and Okun D A 1984 *Surface Water Treatment for Communities in Developing*

Countries (WASH Technical Report No. 29, Water and Sanitation for Health (WASH) Project of the United States Agency for International Development)
[7] Qasim S R and Motley G Z 2000. *Water Works Engineering: Planning, Design, and Operation*. (Prentice-Hall, Inc.)
[8] Ghawi A H 2018 *Scientific Review – Engineering and Environmental Science* 27(4) 438-451
[9] Jannah F H S, Kusumadewi R A and Ratnaningsih R 2020 *International Journal of Scientific & Technology Research* 9(01) 3500-05
[10] Solihin, W S, Kusumadewi R A, and Winarni Solid State Technology **63**(06)
[11] Gabrielle H, Kusumadewi R A, and Winarni *IOP Conference Series: Earth and Environmental Science* **894**
[12] Darmasetiawan M 2001 *Teori dan Perencanaan Instalasi Pengolahan Air* (Jakarta: Ekamitra Engineering)
[13] Davis M L 2010 *Water and Wastewater Engineering* (McGraw-Hill, Inc)
[14] Sutrisno 2002 *Teknologi Penyediaan Air Bersih* (Jakarta: PT Rineka Cipta)
[15] Statistics Center of Depok City of 2022
[16] Development Planning Agency of Depok City of 2020
[17] SNI 6774:2008 *Tata Cara Perencanaan Paket Instalasi Pengolahan Air* (Jakarta: Badan Standarisasi Nasional)
[18] WHO 1997 Guidelines for Drinking-Water Quality 2nd ed **3** (Malta: World Health Organization)

Potential Carbon Storage In The Forest Area Of Mount Merbabu National Park (MMbNP) Resort Selo Central Java

Lia Kusumaningrum[1*] **and Rinoa Salsabila Izdihar**[1]

[1] Environmental Science Program, Faculty of Mathematics *and Natural Science,* Sebelas Maret University, Surakarta-Indonesia

*Liakusumaningrum@staff.uns.ac.id

Abstract. Forests play a very large function in the solution of climate problems. Trees are capable of absorbing carbon dioxide (CO2) in the photosynthesis process and store it in the form of biomass, thereby reducing the concentration of carbon in the atmosphere. The forest area of Mount Merbabu National Park (MMbNP) has great potential to absorb carbon emissions and mitigate climate. The purpose of this study was to determine the potential for carbon storage in MMbNP. The research method used is quantitative descriptive and analyzed by allometric method adapted to tree species at low, medium, and high densities. The estimated carbon storage refers to the Indonesian National Standard (SNI) 7724 of 2011. The potential for total carbon storage in the forest area is 129.357,71 tons. In detail, the total carbon storage in low density land cover types is 10.201,87 tons, medium density is 23.729.66 tons, and high density is 95.426,18 tons. This type of land cover with high density has the highest carbon absorption capacity; this is relevant to the presence of the number of trees that are able to absorb more carbon and have a larger area than other types of land cover.

1. Introduction

Forests are the dominant terrestrial ecosystem on earth and are spread throughout the world [1]. Forests play a very massive function in solving climate problems. Forest conservation, stopping deforestation and replanting forests are the most cost-effective ways, especially in terms of reducing carbon emissions on the earth. This is because trees are one of the carbon storage places. Forests help to increase carbon absorption, where with the help of sunlight, water from the soil is absorbed through the roots, and vegetation that has chlorophyll is able to absorb carbon dioxide (CO_2) from the atmosphere through the process of photosynthesis [2]. After carbon dioxide is absorbed in the process of photosynthesis, the carbon is stored in the form of biomass, thereby helping to reduce the concentration of carbon in the atmosphere. Forest management by halting deforestation, reforestation, and improving forest practices is estimated to reduce 7 billion metric tons of carbon dioxide per year, which is equivalent to emissions from 1.5 billion cars [3]. For this reason, natural forest management is expected to reduce carbon emissions and increase carbon storage.

Indonesia is one of the countries that has the largest forest on earth. One of the forest areas with fairly extensive forest in Indonesia is the Forest Area of Mount Merbabu National Park (MMbNP). MMbNP is a conservation area with an area of approximately 5.725 ha, which was previously a protected forest area and a forest for nature tourism [4]. The MMbNP forest area has great potential to absorb carbon emissions as a form of mitigating carbon emission reduction in the national park area or in the surrounding area. Therefore, "Potential Carbon Storage in the Forest Area of Mount Merbabu National

Park (MMbNP) Resort Selo Central Java" was carried out to determine the ability of forest areas to deal with carbon emissions problems.

2. Methodology

The location of data collection for research on carbon storage potential is focused on the Selo Resort forest area, which is located in Tarubatang Village, Selo District, Boyolali Regency, Central Java. Data collection was carried out in April 2022. Data collection was carried out on land cover types with high, medium, and low density in the Mount Merbabu National Park Resort Selo area.

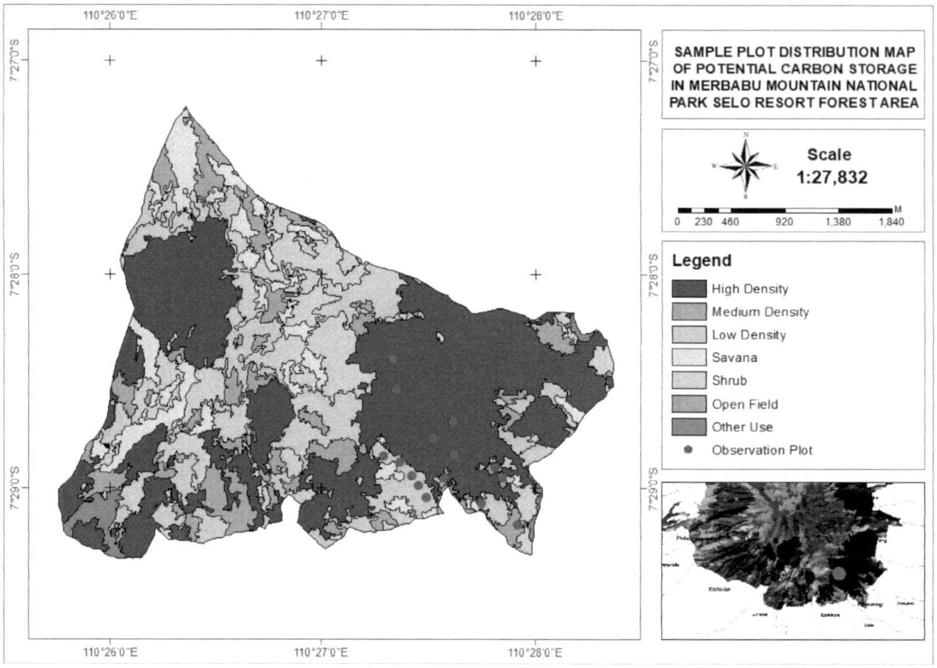

Figure 1. Location of data collection

This research uses a quantitative description method. This research uses a non-destructive approach. The sample used is vegetation from three types of land cover, namely high density, and medium density, low density forest. For each type of cover, five data collection plots were made, with the area of each plot being 20 x 20 m with a distance between the plots of 100 meters. Analysis of biomass data for various types of land cover using allometric methods for each tree species. The allometric equations of these tree species can be seen in Table 1 below:

Table 1. Allometric Equations of *Pinus merkusii, Schima wallichii*, and *Casuarina junghuniana*

Species	Allometric Equations	References
Pinus merkusii	$B = 0{,}0936 \times D^{2{,}4323}$	[5]
Schima wallichii	$B = 0{,}4594 \times D^{1{,}9978}$	[6]
Casuarina junghuniana	$B = 0{,}016 \times D^{2{,}663}$	[7]

Allometric equations according to Brown [8]:

$$B = 42{,}69 - 12{,}8\,(D) + 1{,}242\,(D^2) \quad (1)$$

Definitions :
B = Tree biomass (kg)
D = Diameter at chest height (cm)
Estimation of carbon storage refers to the Standarisasi Nasional Indonesia (SNI) 7724 [9]:

$$C = B \times 0{,}47 \quad (2)$$

Definitions :
C = Carbon (ton/ha)
B = Biomasss (kg/tree)
0,47 = Carbon estimation conversion factor

3. Results and Discussion

3.1 Vegetation in the Forest Area of Mount Merbabu National Park (MMbNP) Resort Selo

From the results of research conducted in the forest area of Mount Merbabu National Park Resort Selo, with a total of 15 observation plots on 3 types of land cover, there are a total of 158 trees with 8 different species, namely *Pinus merkusii, Acacia decurrens, Engelhardia serrata, Cupressus sp., Glochidion rubrum, Schima wallichii, Casuarina junghuniana*, and *Nauclea lanceolata* as presented in Table 2.

Table 2. Composition of Tree Vegetation in the Forest Area of Mount Merbabu National Park

No	Species	Family	Amount of Vegetation		
			Low Density	Medium Density	High Density
1.	*Pinus merkusii*	Pinaceae	31	2	0
2.	*Acacia decurrens*	Fabaceae	5	19	1
3.	*Engelhardia serrata*	Juglandaceae	1	2	16
4.	*Cupressus sp.*	Cupressaceae	0	19	13
5.	*Glochidion rubrum*	Euphorbiaceae	0	1	0
6.	*Schima wallichii*	Theaceae	0	8	34
7.	*Casuarina junghuniana*	Casuarinaceae	0	1	0
8.	*Nauclea lanceolata*	Rubiaceae	0	1	4
	Total vegetation		37	53	68

The level of diversity of vegetation types can be known from the number of individuals in each type. In each type of land, low, medium, and high density have different types and amounts. In the type of land cover with low density there were 37 individuals, the dominant tree vegetation type was *Pinus merkusii* with a total of 31 trees. While in the type of land cover with medium density there were 53 individuals, the dominant vegetation type was *Cupressus sp*. and *Acacia decurrens* with 19 trees each. In the type of land cover with high density there were 68 individuals, the dominant tree vegetation type was *Schima wallichii* with 34 trees. At high density *Schima wallichii* shows a clumped pattern. This indicates the availability of sufficient nutrients and nutrients for a species. According to Kusumaningrum [10] if the number of species on a land is large, this causes the diversity in the type of land to be high, so the higher the value of the diversity of vegetation indicates that the ecosystem is improving. In addition, species that can dominate an area are species that can control the place to grow according to environmental conditions.

(a) (b) (c)

Figure 2. Types of Low Density Land (a), Types of Medium Density Land (b), and Types of High Density Land (c)

3.2 Biomass Estimation of the Mount Merbabu National Park Resort Selo Forest Area

From the results of research conducted in the forest area of Mount Merbabu National Park Resort Selo, the amount of plant biomass above ground level in each plot of land cover types with low density can be seen in Table 3 below:

Table 3. Aboveground Plant Biomass from Different Types of Land Cover

No. Plot	Land Area (Ha)	Biomass (Kg/plot)		
		Low Density	**Medium Density**	**High Density**
1	0,04	4.470,98	43.848,88	9.368,28
2	0,04	4.783,63	6.415,22	18.082,70
3	0,04	5.059,67	19.519,47	13.587,70
4	0,04	5.754,09	84.27,97	22.850,83
5	0,04	5.890,92	30.081,31	25.971,89
Average		5.191,86	21.658,57	17.972,28

Forest biomass plays an important role in the biogeochemical cycle, especially in the carbon cycle [11]. The carbon cycle absorbs carbon in the atmosphere/air into living organisms such as plants through the processes of photosynthesis and respiration. The absorption of carbon in the atmosphere by plants helps to overcome climate problems, such as the greenhouse effect. Based on Table 3, it can be seen that the average biomass in each plot of low, medium and high density land cover was 5.191,86 kg 2.1658,57 kg, and 17.972,28 kg. From the results of these calculations, the average amount of biomass in the entire land cover area is calculated by multiplying the average biomass/plot by 25 because each plot has an area of 0,04 ha. Results of the calculation of the average biomass in the forest area of the Mount Merbabu National Park Resort Selo on land cover types with a low density of 129,8 tons/ha, a medium density of 541,46 tons / ha and a high density of 449,31 tons/ha.

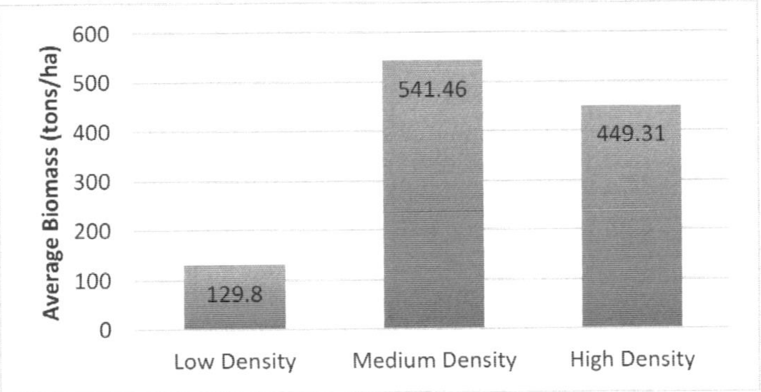

Figure 3. Graph of average biomass on land cover types MMbNP Resort Selo

3.3 Estimation of Carbon Storage in the Mount Merbabu National Park Resort Selo Forest Area

The calculation of carbon storage can be obtained by multiplying the biomass and the conversion factor of the estimated carbon, which is 0,47. The results of the calculation of potential carbon storage in low-density land cover types are as shown in Table 4 below:

Table 4. Potential Above Ground Carbon Storage in Different Types of Land Cover

No. Plot	Land Area (Ha)	Carbon Storage (Kg/plot)		
		Low Density	Medium Density	High Density
1	0,04	2.101,36	20.608,97	4.403,09
2	0,04	2.248,30	3.015,15	8.498,87
3	0,04	2.378,04	9.174,15	6.386,22
4	0,04	2.704,42	3.961,14	10.739,89
5	0,04	2.768,73	14.138,21	12.206,79
Average		**2.440,17**	**10.179,53**	**8.446,97**

Based on Table 4, it can be seen that the average carbon storage in each plot of land cover with low density, medium density, and high density is 2.440,18 kg, 10.179,53 kg, and 84.446,97 kg. From the results of these calculations, the average carbon storage in the entire area of land cover density is calculated by means of the average biomass/plot multiplied by 25 because each plot has an area of 0,04 ha. The results of the calculation show that the average amount of carbon storage in the forest area of Mount Merbabu National Park Resort Selo for low-density land cover types is 61 tons/ha, medium-density land cover types are 254,49 tons/ha and high-density land cover types are 211,17 tons/ha.

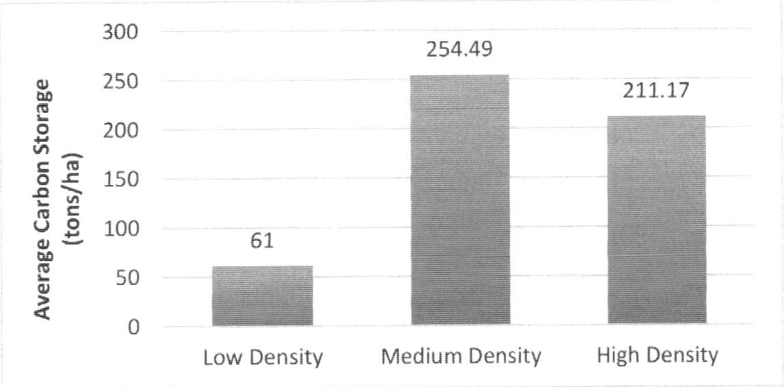

Figure 4. Graph of average carbon storage on land cover types MMbNP Resort Selo

3.4 Potential Total Carbon Storage of Mount Merbabu National Park Resort Selo Forest Area
Mount Merbabu National Park Resort Selo has a land area of about 1.024,731 ha. Based on the analysis, the potential total carbon storage of the Mount Merbabu National Park Resort Selo forest area was obtained at low, medium, and high density land cover types, as shown in Table 5 below:

Table 5. Biomass and Total Carbon Storage in Each Type of Land Cover

Types of Land Cover	Land Area (Ha)	Average Biomass (Tons/ha)	Total Biomass (Tons)	Average Carbon Storage (Tons/ha)	Total Carbon Storage (Tons)
Low Density	167,23	129,8	21.706,31	61	10.201,87
Medium Density	93,25	541,46	50.488,64	254,49	23.729,66
High Density	451,88	449,31	203.034,71	211,17	95.426,18

From Table 5 it is known that different types of land cover have different land areas. In the low density type of land cover, it is known that the area is 167,23 ha, so that the estimated total carbon storage is 10.201,87 tons. The type of land cover with medium density is known to have an area of 93,25 ha, so the estimated total carbon storage is 23.729,66 tons. The high-density land cover type is known to have an area of 451,88 ha, which is much wider than the land area in other land cover types. So, the estimated total carbon storage is 95.426,18 tons. From the results of the total carbon storage in the forest area of Mount Merbabu National Park (MMbNP) as a whole on various land covers, the total carbon storage area of the area is 129.357,71 tons.

Based on the results of this study, it is known that the type of high density land cover has the highest estimated total carbon storage compared to other types of land cover. Meanwhile, for low-density land cover types, the estimated total carbon storage has the least amount. In fact, if taken from the results of the calculation of average carbon storage, land cover with medium density has a higher value than land cover with high density. This is because the type of land cover with high density has a much larger area, which is 451,88 ha compared to the area of land cover with medium density which is only 93,25 ha. Therefore, the type of high density land cover becomes the area that has the most potential to store carbon.

Less average carbon storage in high-density land cover types compared to average carbon storage in medium-density land cover types is known because species diversity in high density land cover types is less, where there are only five species in 5 observation plots. In fact, the number of individuals in high density land cover types was found to be more than in other types of land cover. Furthermore, the type of high density land cover is dominated by the species puspa or *Schima wallichii,* which has a smaller potential for carbon storage compared to other species. According to Millang and Yuniati [12] the various values of biomass and carbon stored in an area are influenced by several factors such as the number of trees, the characteristics of the tree trunk and tree age. This is in line with the statement of Indrajaya and Mulyana [13] that the average carbon stock is not only influenced by one parameter, but is also influenced by the dimensions of the trees, the diversity of plant species, and individual density, all of which contribute to the high value of the carbon stock of a stand.

4. Conclusion

The total carbon storage in the entire forest area of Mount Merbabu National Park (MMbNP) on various land covers is 129.357,71 tons. The potential for biomass and total carbon storage in various types of land cover has various differences, the type of land cover with high density has the highest potential for biomass and carbon storage. This is because the estimated total carbon storage in this area is higher than that of low- and medium-density land cover types. The estimated amount of carbon storage in high density land cover types is because it has a large number of trees and a larger land area compared to other land cover types.

References

[1] Pan Y, Birdsey RA, Phillips OL, and Jackson RB 2013 The Structure, Distribution, and Biomass of the World's Forests *Annu. Rev. Eco.l Evol. Syst.* **44** 593-622

[2] Tuah N, Sulaeman R, and Yoza D 2017 Calculation of Biomass and Carbon Above Ground in the Rumbio Prohibition Forest, Kampar District *JOM Faperta UR*. **4**(1) 1-10

[3] Forest Conservation Minnemeyer S, Harris N, and Payne O 2018 *Could Reduce Carbon Emissions Equivalent To All Cars On Earth*. Accessed on 28 May 2022, from WRI Indonesia Website: https://wri-indonesia.org/id/blog/conservation-forest-can-combat-carbon-emissions-equivalent-to-emissions-all-cars-on-earth

[4] Ministry of Forestry 2004 *Decree of the Minister of Forestry Number 135/Menhut-II/2004 dated May 4, 2004 concerning Changes in the Functions of Protected Forest Areas and World Tourism Parks in the Merbabu Forest Group of 5,725 Hectares* (Jakarta: Ministry of Environment and Forestry of the Republic of Indonesia)

[5] Siregar CA 2007 Biomass Estimation in Pine Plantation Forests (Pinus merkusii Jungh et de Vriese) and Cianten Soil Carbon Conservation, West Java *J. For. Res. Nat. Conserv.* **4**(3) 251-256

[6] Suhendang E 2002 *Introduction to Forestry* (Bogor: Yayasan Penerbit Fakultas Kehutanan)

[7] Yuniaty D and Kurniawan H 2013 Allometric Equation of Biomass and Carbon for Estimating Carbon Stocks in Supporting Corypha Savanna Conservation Efforts *J. Soc. Econ. For. Res.* **10**(2) 75-84

[8] Brown S 1997 *Estimating biomass and biomass change of tropical forests: a Primer* (Rome: FAO Forestry Paper 134)

[9] Indonesian National Standardization 2011 *Measurement and Calculation of Carbon Stock – Field Measurement for the Assessment of Forest Carbon Stocks (Ground Based Forest Carbon Accounting)* (Jakarta: National Standardization Agency (BSN) Indonesia)

[10] Kusumaningrum L, Suminar L, Karim FF, and Sulton MN 2022 Vegetation Inventory, Biomass and Carbon Stock in The Government Office Complex of Sleman Regency *J. Glob. Environ. Dyn.* **3**(2) 6-10

[11] Windusari Y, Sari NA, Yustian I, Zulkifli H 2012 Alleged Carbon Stocks of Understorey Biomass and Litter in Natural Succession Areas in Tailings Deposition Areas in PT Freeport Indonesia *Biospecies* **5**(1) 22-28

[12] Millang S and E Yuniati 2010 Carbon Sequestration Potential of Several Types of Plants in Green Open Space Hasanuddin University Makassar *Biocelebes* **4**(2) 113-122

[13] Indrajaya Y and S Mulyana 2017 Carbon Storage in Tree Biomass in City Forest Bandung Zoo *Proceedings of the National Seminar on Geography of UMS*. **VIII** 550-560.

Planning for the 3R-based waste processing site in Aimas District, Sorong Regency

Hermin Poedjiastoeti[1*] **and Benny Syahputra**[1]

[1] Departement of Civil Engineering, Sultan Agung Islamic University, Jl. Raya Kaligawe Km 4 Semarang 50112, Indonesia

*hermin@unissula.ac.id

Abstract. A design prototype will be created as part of this research project for the 3R-based Waste Processing Site (TPS 3R) in the Aimas District. The design carried out refers to the Minister of Public Works No. 03 of 2013 concerning the Implementation of Waste Facilities and Infrastructure and Technical Guidelines for 3R-based Waste Management by the Director General of Human Settlements in 2021. According to the study's findings, the average quantity of waste produced daily in the Aimas District was 1.69 liters per person, or 40.87 m^3 per day, with a composition of 38.13 % organic waste, 30.13 % recyclable inorganic waste, and 31.75 % inorganic waste that cannot be recycled. The 3R-based Waste Processing Site building is expected to require an area of 237.75 m^2, made up of a room for managing organic waste of 129.75 m^2 with composting techniques using seven bamboo aerators, a room for managing plastic waste of 29.50 m^2, a space for managing inorganic waste of 6.25 m^2, and supporting buildings covering an area of 72.25 m^2. At the 3R-based Waste Processing Site, processing is divided into five series: waste collecting, waste sorting, organic waste recycling, inorganic waste recycling, and residue recycling.

1. Introduction

Garbage is all solid waste consisting of organic and inorganic materials which are considered useless and must be managed so as not to endanger the environment and protect development investment [1]. Waste production will rise as a result of population expansion and growth, which are both accompanied by increased activity [2]. The imbalance between waste production and waste management processes, as well as the dwindling capacity of nature as a disposal site, can lead to waste concerns. The capacity for processing waste is still insufficient, despite the fact that the amount of waste is increasing quickly [3].

According to Law no. 18 of 2008 concerning Waste Management, waste handling by throwing waste directly to the landfill is no longer allowed. Waste must first be processed before being returned to environmental media in a safe manner for humans and the environment. The law encourages waste management from the source by developing processing in the form of changing the characteristics, composition, and amount of waste. One of them is by making a waste processing facility, namely a waste processing site with the principle of reduce, reuse, recycle (TPS 3R). The 3R concept (Reduce, Reuse and Recycle) is oriented to waste prevention through reduction, reuse, and recycling. TPS 3R work involves community participation to make waste reduction sustainable and environmentally friendly [4].

In general, the management system for household waste and similar household waste in Aimas District, Sorong Regency is still limited to waste handling efforts and has not made any efforts to reduce waste. The waste management procedures used are "collect-transport-disposal" activities and has not yet carried out sorting, processing and final processing of waste. The proportion of waste transported to the landfill only ranges from 20 – 30%. Environmental sanitation conditions in Aimas District can be impacted by the lifestyle of the locals, who frequently accumulate trash on unused land, as well as by the absence of waste management equipment and facilities. This needs to be taken into account because one of the factors that define slum settlements is the lack of suitable waste infrastructure and facilities that do not adhere to technical requirements [5]. On the other hand, there are no district-level profiles or plans for reducing and managing trash because there is no waste management master plan.

Therefore, it is intended to implement TPS 3R in Aimas District, Sorong Regency, in an effort to address the issue of waste generation that has arisen at this time. TPS 3R is expected to play a role in ensuring the increasingly critical need for land for the provision of waste landfills in urban areas. With this plan, it is hoped that it will reduce the volume of waste generated by the community, reduce the operational costs of transporting waste and also extend the life of the landfill. Based on these issues, a waste processing facility using the TPS 3R (Reduce, Reuse, Recycle) approach will be built in the Aimas District of the Sorong Regency, due to the land's availability for the TPS 3R.

2. Methodology

2.1. Study area description

In designing the 3R TPS Prototype, the location used as a basic pilot design is Aimas District, Sorong Regency, West Papua, Indonesia (Figure 1). Sorong Regency consists of 33 districts with Aimas district as the capital with a population of 24,148 people in 2021 [6]. As the new City of Aimas grows and developed into a center of government, economy and trade as well as education, where almost 30% of the total population of Sorong Regency is domiciled in Aimas. Of course, this will have an effect on future sanitary issues, one of which is the issue of solid waste. Based on the data contained in the Sorong Regency Sanitation Strategy Book 2020, Aimas District is one of the districts included in the risk area for waste sanitation in the Risk category 4 [7].

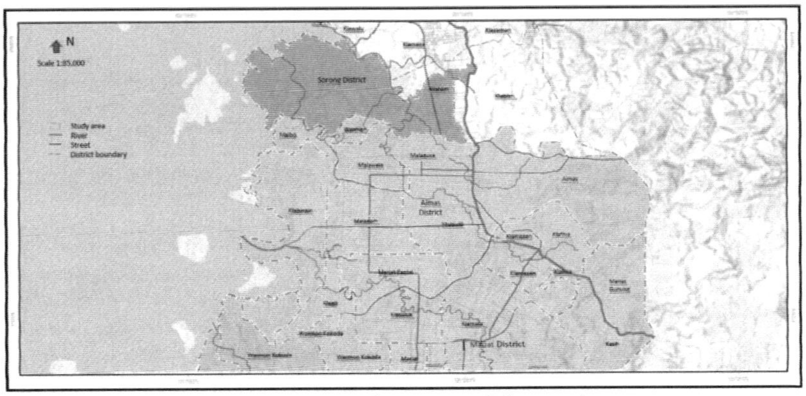

Figure 1. Location map of the study area

2.2. Data Collection and Analysis

Both primary and secondary data were gathered for this study using methods like measurement, observation, and literature reviews. The amount of waste generated and the volume of waste produced were measured over the course of one week by counting the number of waste hauling vehicles, the capacity of those vehicles, and the number of rituals performed each day. A literature review was done to gather waste management reference information. To calculate the capacity and dimensions required in planning TPS 3R, an analysis of predicted population data and waste generation is conducted [8].

2.3. TPS 3R planning

The TPS 3R building floor plan is designed during planning, and the dimensions of each processing unit are determined. According to Module E.3 on Integrated Waste Management Sites 2017's, planning process for TPS 3R is as follows:

- Material balance analysis to knowing the amount of waste that enters the location of the waste processing site. Mass balance calculated using equations 1 and 2 [9].
  Waste recovery weight = Waste generation per component x % Recovery (1)
  Weight of waste to TPA = Waste generation per component - Weight recovery (2)
- Identify all possible uses of the material
  Knowing the characteristics of waste and its utilization to make a material balance flow chart.
- Waste accumulation calculation
  Determine and calculate the amount of accumulated waste, how much waste will be handled by TPS 3R and the rate of accumulation by determining the operating time of TPS 3R
- Calculation of material loading rate.
  This calculation is used to determine the number of workers and tools needed as well as working hours and equipment operation at TPS 3R.

$$Loading\ rate = \frac{waste\ volume\ (\frac{m^3}{day})}{processing\ time\ (\frac{hour}{day})} \quad (3)$$

- Layout and design
  The amount of waste generation to be managed is estimated to be dependent on the land area and TPS 3R design. The amount of land needed for TPS 3R will also depend on how many units are planned, including compost houses, inorganic waste management, storage, various support buildings such office space, an equipment warehouse, restrooms, and parking lots. AutoCAD software will be used to create the design. The research mechanism can be seen in Figure 2.

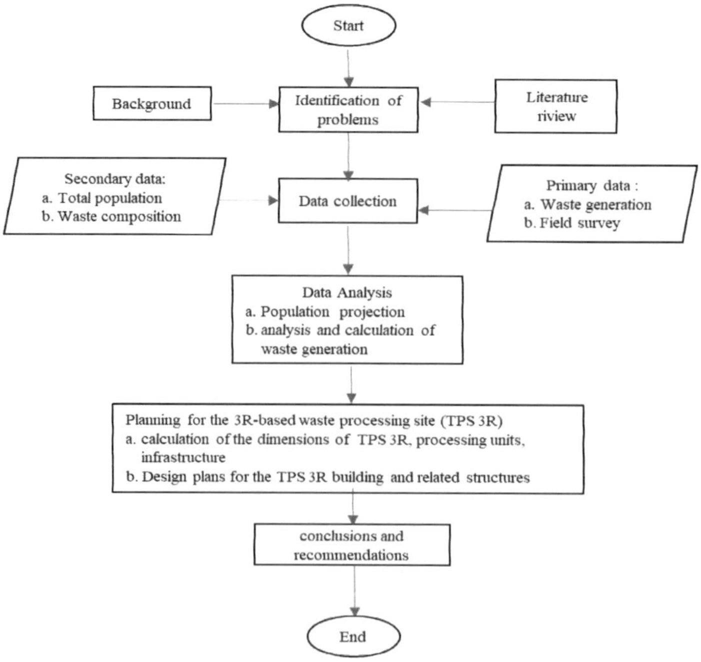

Figure 2. Research Flowchart

3. Results and Discussion

3.1. Existing Condition of Waste Management
The amount of garbage brought to the landfill represents just 35.16 percent of the total amount of waste generated in the Aimas region, which totals 22,678 kg/day (22.8 tons/day) (according to an early assessment by the Department of Housing and Settlement Areas of Sorong Regency, 2017). This is due to the inadequate resources, fleet, equipment, personnel, and community involvement in the Aimas District, which has minimal garbage services that are only provided in a few locations in the city center. In terms of final waste disposal, Sorong Regency as a whole has one Final Processing Site / landfill, specifically TPA 36, which is situated in Mariyat Gunung Village about 17 kilometers from the center of Aimas. TPA 36 has been operated since 2018 with a plan to implement the sanitary landfill method, although until now it is still using the open dumping system.

According to the findings of an interview with the Head of the Environmental Service of Sorong Regency, there are presently 4 motorcycles, 2 dump trucks, and 2 arm roll trucks under the administration of the Environment Agency at the Aimas District waste management facility (previously managed by the Housing and Settlement Areas Office). Motorized rickshaws are built for short distance travel and are used to transfer waste from residents' homes to Temporary Shelters (TPS).

Domestic and non-domestic waste, specifically domestic waste and similar domestic waste, enters the TPS. Depending on the condition of the existing waste, garbage transportation in the Aimas District is typically done twice or once per day. One truck can typically transport up to 7 tons of garbage.

3.2. Population Projection
Planning for waste processing site with the principle of reduce, reuse, recycle (TPS 3R) in Aimas District requires data on the amount of household waste generation that can be identified with population growth projections for the next 5 years. The results of the projected population of Aimas District, Sorong Regency tend to increase and in 2026 the total population is 26,427 people as shown in Figure 3.

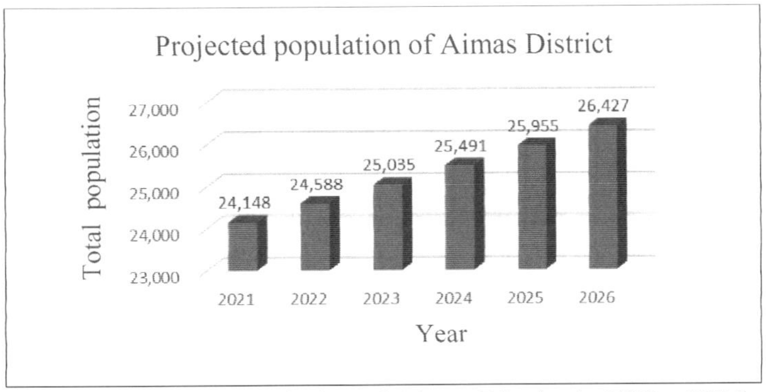

Figure 3. Projected population of Aimas District in 2022 - 2026

3.3. Waste generation
The amount of waste generated in Aimas District based on data on the amount of waste that came to TPA 36 for 7 consecutive days and interviews, the amount of waste per day was 40.87 m^3. According to Nur Lailis (2018) in [10], the density of waste is 0.284 kg/l. The total weight of waste produced in Aimas District per day is calculated by the equation:

$$Weight\ of\ waste = amount\ of\ waste\ x\ density \quad (4)$$

The weight of waste per day in Aimas District using this equation is 1.69 l/person/day. The projected waste generation in Aimas District is presented in Table 1.

Table 1. Waste generation forecast for the Aimas District from 2022 to 2026

	2022	2023	2024	2025	2026
Waste generation (l/person/day)	1.69	1.69	1.69	1.69	1.69
Total population	24,588	25,035	25,491	25,955	26,427
Waste generation (m^3/day)	41.55	42.31	43.08	43.86	44.66

3.4. Waste composition and Projected Waste Generation
The rate of waste generation during the predicted period is exactly proportional to the growth in population [11]. The waste generation projection is made in order to foresee TPS 3R's capacity so that it can operate for the anticipated operational time, which is five to ten years in the future [12]. Based on the results of the JAKSTRADA study of Sorong Regency, Sorong City and Raja Ampat Regency 2019, the composition of waste in Aimas District in general is as shown in Figure 4, while the projected composition of waste for the next 5 years can be seen in Table 2 [13].

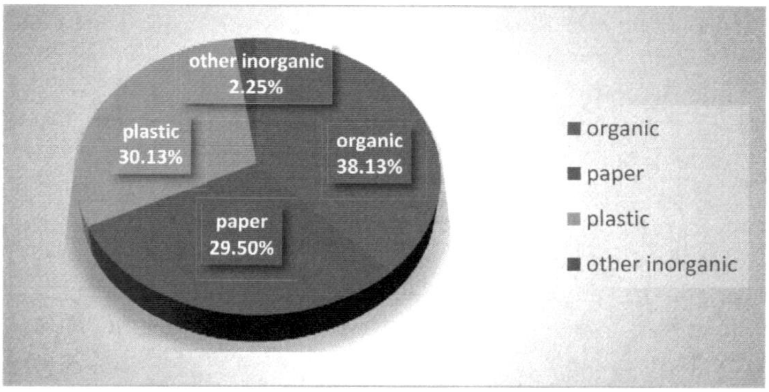

Figure 4. The typical waste composition in the Sorong Regency

The highest percentage of trash in Aimas District is organic waste (38.13%), plastic waste (30.13%), paper waste (29.50%) and other inorganic waste (2.25%). The proportion of organic waste is the biggest in comparison to other types of garbage because it weighs more than the others and because each individual produces more organic waste every day, such as food scraps and cooking residue [11]. The findings of the expected waste generation over the following five years demonstrate that the composition of organic waste still accounts for the biggest percentage (Table 2).

Table 2. Projections on the composition of total waste generation

Year	Waste generation projection (m^3/day)	Waste generation projection (per waste type)			
		Organic	Paper	Plastic	Other inorganic
2022	41.55	15.84	12.26	12.52	0.93
2023	42.31	16.13	12.48	12.75	0.95
2024	43.08	16.42	12.71	12.98	0.97
2025	43.86	16.72	12.94	13.21	0.99
2026	44.66	17.03	13.17	13.45	1.00

3.5. Loading Rate
According to Law No. 13 of 2003, the TPS 3R will be operating for 7 hours during the hours of 8 a.m. to 12 p.m. and 13 p.m. to 16 p.m., processing 0.86 m^3 of trash each hour [14].

3.6. Waste Material balance

The quantity of waste in Aimas District that can be recycled and the amount of waste that becomes residuals are calculated using the waste material balance. Calculating the waste weight from the recovery factor for each waste component yields the material balance. The amount of waste that may be utilised or converted into residue is represented by the recovery factor [9]. In table 3, the total computation is displayed. From the waste material balance data, it can be seen the percentage of waste that can be composted is 28.60% and waste that can be recycled is 31.62%. Meanwhile, the total waste from composting and the sorting that will be transported to the landfill is 39.79%.

Table 3. Waste Material Balance in TPS 3R

Treatment	Component	Waste composition (%)	Factor of treatment	Waste composition (%)
Composting	organic	38.13	75	28.60
	total	**38.13**		**28.60**
Residue	organic	38.13	25	9.53
	total	**38.13**		**9.53**
Sorted	Paper	29.50	50	14.75
	Plastic	30.13	50	15.07
	Other inorganic	2.25	80	1.8
	total	**61.88**		**31.62**
Residue	Paper	29.50	50	14.75
	Plastic	30.13	50	15.07
	Other inorganic	2.25	20	0.45
	total	**61.88**		**30.26**

3.7. Requirements for the area of the TPS 3R

The following equation can be used to determine the amount of land needed [15]; [16] and [17].

- Organic waste

  Quantity of organic waste entering each hour

  $$V_{organic\ waste} = \%_{organic\ waste}\ x\ total\ garbage\ in \quad (5)$$

  Weight of organic waste entering each hour

  $$B_{organic\ waste} = \%_{organic\ waste}\ x\ \sum resident\ served\ x\ generated\ waste\ weight\ /person/day \quad (6)$$

  $$V_{composting} = \frac{time \times Bweight\ of\ trash\ chopped}{waste\ dentsity\ chopped\ site} \quad (7)$$

- Plastic waste

  $$Plastic\ waste\ volume = \%\ plastic\ waste\ x\ Incoming\ trash\ volume \quad (8)$$

  $$Plastic\ waste\ weight = \%\ plastic\ waste\ x\ number\ of\ users\ x\ trash\ weight \quad (9)$$

- Inorganic waste

  $$Paper\ waste\ volume = 50\%\ x\ \%paper\ composition\ x\ Incoming\ trash\ volume \quad (10)$$

- Total Land Requirement

  Based on the calculation results of waste characteristics, the land area required in this design is 237.75 m^2. The land requirement for TPS 3R is intended for several units such as processing organic waste 129.75 m^2, plastic waste management 29.5 m^2, inorganic waste management 6.25 m^2 and supporting facilities 72.25 m^2.

Table 4. Total land requirement

No.	Room	Land Requirement
1	Organic Waste Management	
	a. Organic waste storage room	5 m^2
	b. Sorting room	7.5 m^2
	c. Organic waste chopping room	7.5 m^2
	d. Composting room	85.75 m^2
	e. Sieving room	24 m^2
	Total	**129.75 m^2**
2	Plastic Waste Management	
	a. Plastic waste storage room	5 m^2
	b. Plastic waste sorting room	3.25 m^2
	c. Plastic waste washing room	2.5 m^2
	d. Plastic waste drying room	15 m^2
	e. Plastic waste grinding room	3.75 m^2
	Total	**29.5 m^2**
3	Inorganic Waste Management	
	a. Inorganic waste storage room	2.25 m^2
	b. Inorganic waste sorting room	4 m^2
	Total	**6.25 m^2**
4	Supporting facilities	
	a. Warehouse	5.25 m^2
	b. Toilet	9 m^2
	c. Office	6 m^2
	d. Motorbike cart parking	30 m^2
	e. Guardhouse	2 m^2
	f. Maggot nursery, garden and fish pond	20 m^2
	Total	**72.25 m^2**
	TOTAL	**237.75 m^2**

3.8. Capacity and processing flow for TPS 3R

The waste management system is being used in the Aimas District in accordance with the analysis that was done. The waste processing system that can be used at TPS 3R Aimas District includes turning organic waste into compost, processing inorganic waste so that it can be used as plastic seeds, selling metal or goods that can still be used to collectors or third parties, and discarding the remaining residual waste in landfills.

According to the 2021 TPS 3R Technical Instructions, the TPS 3R must be able to serve or process 4-6 m^3/day of trash. The 3R TPS is intended to serve up to 3,900 people (using the population of Malagusa Village as an example), or can process 6 m^3 of garbage each day, based on the quantity of waste produced in the Aimas District. Figure 5 can be used to depict the processing flow at TPS 3R.

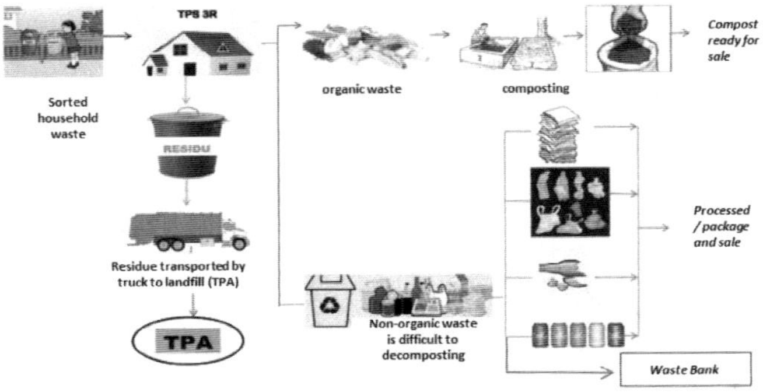

Figure 5. Waste Process Flow at TPS 3R

3.9. TPS 3R building design

The planning of a TPS 3R prototype is as follows: (a) In the form of: walled courtyard; (b) the transport system is HCS / Hauled Container System ; (c) There is a waste selection process; (d) TPS type 2, according to Indonesian Nasional Standard SNI 19-2454-2002 (Badan Standardisasi Nasional, 2002); and (e) Collecting waste generation > 6m^3. The TPS 3R building prototype is shown in figure 6.

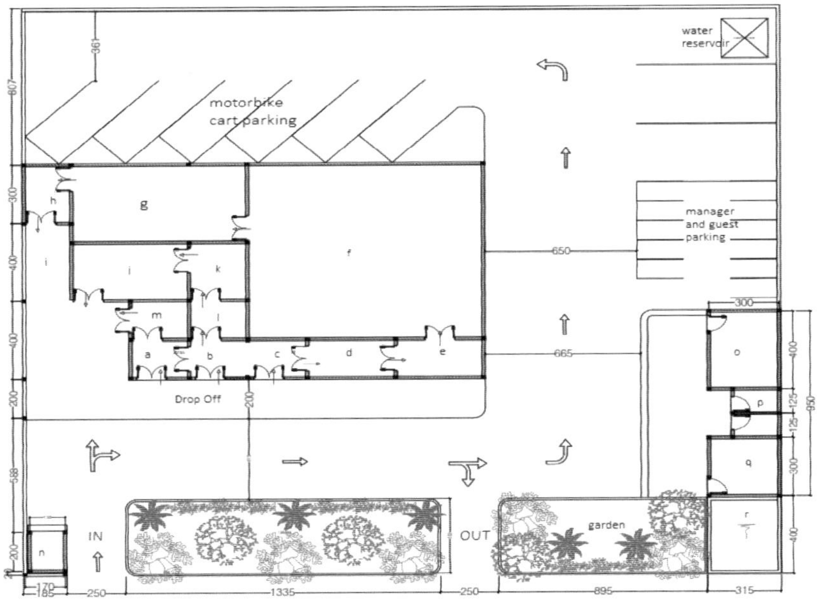

Figure 6. Planning for The TPS 3R Building Prototype (Siteplan)

Explanation:

a : inorganic waste storage room
b : plastic waste storage room
c : organic waste storage room
d : sorting room
e : organic waste chopping room
f : composting room
g : sieving room
h : warehouse
i : loading dock
j : plastic waste drying room
k : plastic waste washing room
l : plastic waste sorting room
m: inorganic waste sorting room
n : guardhouse
o : office
p : toilet
q : maggot nursery
r : fish pond

4. Conclusion

From the waste material balance data in Aimas District, Sorong Regency, it can be seen the percentage of waste that can be composted is 28.60% and waste that can be recycled is 31.62% and the total waste from composting and the sorting that will be transported to the landfill is 39.79%. In order to improve waste management efforts in Aimas District, a land disposal site (TPS) 3R Prototype has been planned which requires an area of 237.75 m^2, with a maximum handled waste generation of 8 m^3. There are several areas in TPS 3R, namely:1) an organic waste management area consisting of an organic waste storage room, sorting room, organic waste chopping room, composting and sieving room; 2) inorganic waste management area, consisting of storage and sorting rooms, 3) plastic waste management area consisting of plastic waste storage room, sorting, washing, drying and grinding rooms; and 4) supporting buildings consisting of warehouse, toilets, office, motorbike cart garages, guard posts and magot nursery areas, gardens and fish ponds. Because there is currently no available land ready for use in Aimas District, the calculations are based on planning, and field measurements have not been followed up. In order to make the planning results more accurate, further surveys are needed on the condition of the village area.

References

[1] Badan Standardisasi Nasional, "SNI 19-2454-2002 :Tata Cara Teknik Operasional Pengelolaan Sampah Perkotaan," *Standar Nas. Indones.*, no. ICS 27.180, pp. 1–31, 2002 [in Indonesian Language].

[2] Firmansyah M and Noor R 2016 Perencanaan Pengelolaan Sampah Terpadu Perumahan Kota Citra Graha Provinsi Kalimantan Selatan Design of Integrated Solid Waste Management in Kota Citra Graha Residence South Kalimantan Province *Jukung (Jurnal Tek. Lingkungan)*, 2 (2) 73–82 [in Indonesian Language].

[3] Kazerooni Sadi M A K, Abdullah A, Sajoudi M N, Kamal M F, Torshizi F, and Taherkhani R 2018 Reduce, reuse, recycle and recovery in sustainable construction waste management," *Adv. Mater. Res.* **446–449** 937–944

[4] Rachmawaty A, Nasution S M, and Juliawati P 2019 Penerapan Teknologi Pengomposan Pada Pengelolaan Sampah Rumah Tangga Di Kabupaten Garut *Tematik* **6**(1) 54–63 [in Indonesian Language].

[5] Menteri PUPR, "Peraturan Menteri PUPR RI Nomor 02/PRT/M/2016 tentang Peningkatan Kualitas terhadap Perumahan dan Permukiman Kumuh." JDIH Kementerian PUPR, Jakarta, p. 36, 2016 [in Indonesian Language].

[6] BPS, *Kabupaten Sorong Dalam Angka*. Kabupaten Sorong: Badan Pusat Statistik Kabupaten Sorong, 2021 [in Indonesian Language].

[7] PokjaSanitasi, *Strategi Sanitasi Kabupaten/Kota Kabupaten Sorong Tahun 2020-2024*. Kabupaten Sorong: Pemerintah Daerah Kabupaten Sorong, 2019 [in Indonesian Language].

[8] Wilson D C, Velis C, and Cheeseman C 2006 Role of informal sector recycling in waste management in developing countries *Habitat Int.* **30**(4) 797–808

[9] Tchobanoglous G, Theisen H, and Vigil S 1993 *Integrated Solid Waste Management: Engineering Principles and Management* (Singapore: Mc. Graw Hill)

[10] Lawa J I J, Mangangka I R, and Riogilang H 2021 Perencanaan Tempat Pengolahan Sampah (TPS) 3R Di Kecamatan Mapanget Kota Manado *Tekno.* **19** 77–89 [in Indonesian Language].

[11] Basri N A, Basri H, and Stentifor E 2005 An Expert System to Design Composting Facilities for Municipal Solid Waste *J. Kejuruter.* **17** 85–99

[12] Damanhuri E and Padmi T 2016 *Integrated waste management* (Bandung: Publisher ITB)

[13] DLH 2019 *Jakstrada Pengelolaan Sampah Rumah Tangga dan Sampah Sejenis Rumah Tangga (Daerah Studi: Kota Sorong, Kab. Sorong dan Kab. Raja Ampat)* Dinas Lingkungan Hidup dan Pertanahan provinsi Papua Barat, 2019 [in Indonesian Language].

[14] Muhashiby M I N, Hasibuan H S, and Wahyono S 2021 Waste Management in Jakarta Recycle Centre: Case Study of Pesanggrahan, Jakarta *IOP Conf. Ser. Earth Environ. Sci.* **940**(1)

012034.
[15] Memon M A 2010 Integrated solid waste management based on the 3R approach *J. Mater. Cycles Waste Manag.* **12** 30–40
[16] Maryanti D F 2017 Performance of Community-Based Solid Waste Management for Integrated and Sustainable Solid Waste Management. The Case of Bogor City, Indonesia," *Unesco-Ihe*, no. 49947, 95
[17] Bovea M D and Powell J C 2006 Alternative scenarios to meet the demands of sustainable waste management *J. Environ. Manage.* **79**(2) 115-312

Addition of Local Microorganisms (MOL) Organic Waste as Compost Bioactivator

S Sumiyati[1*], I B Priyambada[1], S A F Zahra[1*], D R Pradhana[1], R T Haritsa[1], T Rahman[1], M F Q Haq[1], A W P Harjanti[1]

[1]Departement of Environmental Engineering, Faculty of Engineering, Diponegoro University, Semarang, Central Java, Indonesia 50275

noesri07@gmail.com
anisaafitria13@gmail.com

Abstract. Waste is a crucial problem because it causes serious impacts in terms of social, economic, cultural, and environmental aspects. With the dominance of organic waste in the composition of waste in Indonesia, there is a potential to recycle organic waste into useful products, namely compost. Composting occurs naturally so it takes a long time and is slow, therefore bio activators such as local microorganisms (MOL) are needed that can accelerate the decomposition of organic waste. The purpose of this study was to determine the effect of the use of MOL on the quality of the compost. This research method is experimental with two treatments, the first treatment with the addition of MOL and the second treatment without MOL. The results showed that the quality of compost with the addition of MOL were temperature 30°C, pH 7.4, moisture content 47.01%, C/N ratio 9.55%, phosphate 0.208%, potassium 0.027%. Meanwhile, the quality of compost without the addition of MOL were temperature 31°C, pH 7, water content 63.49%, C/N ratio 10.54%, phosphate 0.080%, potassium 0.015%. Compost with the addition of MOL was better because it met the criteria set by SNI No. 19-7030-2004 compared to compost without MOL.

1. Introduction

Waste production that continues to increase along with population growth, inefficient processing technology, and limited landfill are the main problems in big cities in Indonesia. Waste can be sourced from households, offices, commerce, markets, and public facilities. According to SIPSN data in 2021, the composition of the types of waste mostly came from food waste at 30.01% (45.843,121 tons), followed by plastic at 15.73% (24,031,572 tons), wood and twigs at 12.5% (19,096). 400 tons), paper/cardboard 11.93% (18,226,625 tons), metal 6.33% (9,671,948 tons), fabric 6.04% (9,227,635 tons), glass 5.9% (9,015. 705 tons), rubber/leather 3.29% (5,025,214 tons), and others 8.27% (12,636,054 tons) [1]. From these data, it can be seen that organic waste dominates the composition of waste types.

Processing of solid waste in the form of vegetables needs to be done, one way to treat solid waste is by making compost. Compost is organic fertilizer, the use of organic fertilizer is very widely used because it has 3 advantages, namely: benefits for the environment, soil, and plants. Compost is very helpful in solving environmental problems, especially waste. The raw material for composting is garbage, so the problem of house waste and municipal waste could be overcome [2].

Composting is the process of weathering organic materials biologically by microbes. The success of making compost can be judged by the length of time of composting, the faster the compost is produced the higher the success rate. Compost can form naturally, but it will take quite a long time, which is 2-3 months. Some reach 6-12 months, depending on the organic material used. Therefore, various treatments need to be applied to speed up the composting time [3]. Several things must be considered to accelerate the composting process, including the selection of bio activators.

One of the bio activators in composting is Local Microorganisms (MOL). MOL is a liquid made from natural organic ingredients. MOL contains a collection of microorganisms that help speed up the process of decomposition of organic matter to shorten the composting time. This type of microbe can be cultured using various sources of organic matter. Vegetable waste can be a source of organic matter and a good medium for the proliferation of decomposing microorganisms and can be used as a bio activator [4]. Making a MOL is done by adding brown sugar as a source of glucose/bacterial energy and rice washing water as a source of carbohydrates. The materials are stirred until homogeneous and placed in a closed bath. Then fermented for 14 days in an aerobic process [5].

Many studies regarding the use of MOL as a bio activator in composting. Among them is a study conducted by Suwatanti et al (2017), with the title “Utilization of Vegetable Waste MOL in the Compost Making Process”. The results of the study showed that the physical quality of the compost in both treatments (MOL vegetable waste and EM4) had a blackish color category, a soil-like odor, and a fine texture according to the criteria of SNI 19- 7030-2004. However, the C/N ratio of compost using MOL vegetable waste is better because it meets the criteria set by SNI compared to compost using EM4 [4]. There is another study conducted by Sipayung (2019), entitled "Composting Using Garden Waste and Goat Manure with Vegetable Waste Mole". The results of this study indicate that the addition of MOL to vegetable waste has a significant effect on the performance and yield of composting. This was indicated by the compost that was ripe in the MOL addiction treatment with a C/N ratio that met SNI 19 – 7030 – 2004, while in the treatment without the addition of MOL (control) the compost was immature and did not meet SNI 19 – 7030 – 2004 [6]

From the description above, it is necessary to research the addition of fruit waste-based MOL bio activator in composting organic waste. The purpose of this study was to compare the parameters of the compost quality with the addition of MOL with the quality standards of SNI 19-7030-2004 concerning Compost Specifications from Organic Waste.

2. Methodology

2.1. Tools and Materials

The tools needed are used cardboard, plastic, shovels, mineral water bottles, knives, cutting boards, basins, paint buckets, blenders, measuring cups, scales, and wooden storage. While the materials used are 8 liters of rice washing water, 2 liters of coconut water, 2 ounces of brown sugar, and 2 kg of organic waste as the basic ingredients for making local microorganisms (MOL), as well as husks and other organic waste as the main composting material.

2.2. Preparation Stage

The stages in the manufacture of a local microorganism (MOL) bio activator are as follows:

2.2.1. Preparation stage

1) Prepare the tools and materials needed.
2) Chopping 2 kg of organic waste, then smoothing the chopped organic waste with a blender.
3) Chop 2 ounces of brown sugar with a knife and cutting board.
4) Measure 2 liters of coconut water and 2 liters of rice water using a measuring cup.

2.2.2. Stage of manufacture

1) Put the crushed organic waste into the paint bucket.

2) Put the chopped brown sugar into the paint bucket.
3) Put 2 liters of coconut water and 8 liters of rice water into the paint bucket.
4) Stir all ingredients by hand to mix evenly.
5) Close the paint bucket tightly so that no air enters.
6) Open the paint bucket every day for 2 weeks to release the fermented gas from getting trapped in the bucket.

2.3. Organic Waste Treatment

After the manufacture of local microorganisms (MOL) is completed, local microorganisms (MOL) will later be mixed with husks and other organic waste as the main ingredients for composting in wooden storage. The composting material is mixed in layers, the first layer is compost and the second layer is filled with chopped waste. The collected waste is then stirred until smooth. Then, a local microorganism (MOL) was added as a bio activator to speed up the compost decomposition process and stir until smooth. After thoroughly mixed, then the garbage is covered with a layer of husk so that the garbage does not cause a strong odor.

The analysis and observations carried out in this study were the physical quality of the compost which included color, odor, texture, determination of pH, temperature, and percentage of water content, as well as the chemical quality of the compost which included fluctuations in the content of C/N, phosphorus, and potassium.

3. Result and Discussion

3.1. Compost Physical Quality

Observation of the physical quality of compost refers to the criteria of SNI: 19-7030-2004, namely compost produces an odor like the smell of earth because it already contains soil nutrients, black color due to the impact of stable organic matter, and smooth compost texture due to the decomposition of microorganisms.

Table 1. Compost Physical Quality

Treatment	Color	Smell	Texture
Control	Black	Earthy Smell	Smooth
Local Microorganism (MOL)	Black	Earthy Smell	Smooth

At the beginning of composting, the compost is blackish green and over time it will resemble the color of black soil. The color in the composting process of vegetable waste without and with the addition of MOL is in accordance with the established standard, namely SNI 19-7030-2004, which is black. During observations, reversal and watering were carried out every 3 days to keep the decomposing bacteria alive. Entering the 2nd week, the color of the compost with the addition of MOL is dark brown, the texture is loose, loose, and smells of earth. Meanwhile, organic fertilizer without the addition of MOL bio activator is brown, has an earthy smell, and has a crumb texture but is coarser.

3.2. Compost Chemical Quality

3.2.1. Determination of pH, Temperature, and Percentage of Moisture Content

Table 2. Temperature, pH, and Percentage of Moisture Content in Compost

Date	Variable	pH	Temperature (°C)	Moisture Content
May 31st, 2022	Control	8.6	29	92.04
	MOL	6.2	30	80.78
June 2nd, 2022	Control	6.8	29	67.27
	MOL	6.9	29	77.4
June 6th, 2022	Control	7	29	71.85
	MOL	7.3	29	66.41
June 8th, 2022	Control	7.4	31	65.05
	MOL	7.3	30	42.06
June 10th, 2022	Control	7	30	61.17
	MOL	7.3	30	60.94
June 13th, 2022	Control	7	31	63.49
	MOL	7.4	30	47.01

In the composting process without the addition of MOL, the first measurement has an alkaline pH, while the composting with the addition of MOL is acidic. The pH during composting increases from acid/base to neutral. The increase and decrease in pH is a marker of the activity of microorganisms in the decomposition of organic matter. At the beginning of composting, acid-forming bacteria will lower the pH so that the compost is more acidic. Furthermore, microorganisms begin to convert inorganic nitrogen into ammonium so that the pH increases rapidly to become alkaline. Some of the ammonia is released or converted into nitrate and nitrate is denitrified by bacteria so that the pH of the compost becomes neutral [7]. The pH measurement results on the 14th day reached the pH value of the compost with the addition of MOL of 7.4, while the pH value of the compost without the addition of MOL was 7. It could be concluded that both of them had met the SNI standard. 14-7030-2004, namely the pH value of 6.8-7.49 [8].

The temperature in the composting process of vegetable waste without the addition of a MOL bio activator and with the addition of a MOL bio activator always changes throughout the measurement. This situation indicates that the decomposition process has begun because several bacteria convert organic waste into simpler materials that are easily absorbed by plants, where the higher the temperature, the more oxygen consumption and the faster the decomposition process of waste. Furthermore, in the next measurement, the temperature decreased because the organic matter described in the compost had begun to decrease and began to shrink. During the aerobic composting process there are 2 phases consisting of a mesophilic phase that occurs at a temperature of 23°C-45°C and a thermophilic phase at 45°C-65°C. The ideal temperature for a compost heap is 55°C-65°C because at that temperature the optimal proliferation of microorganisms so that the population is good and produces the most effective enzymes in the decomposition of organic matter. Meanwhile, according to the criteria of SNI 14-7030-2004, the ideal temperature for the composting process has a maximum value of 50°C. Based on the above criteria, it can be seen that the composting process in this study has not yet reached the ideal stage in the formation of compost because the temperature that occurs is only between 29°C-31°C. In this composting process, the two treatments did not reach the temperature for the thermophilic phase which could be caused by the amount of organic matter that was not high enough and affected the temperature of the compost pile.

The highest humidity occurs in composting with MOL reaching 80,78%. Humidity greater than 60% will decrease oxygen supply and nutrients are leached, consequently, microbial activity will also decrease. Decreased microbial activity will cause the composting process to run slower. However, this can be overcome by reversing the compost pile, so that air can supply oxygen for the composting process and reduce the moisture of the material [9]. In the following days, from 80,78% to 66,41% on the 7th day, and likewise on the following days, the humidity drops until the compost has matured. In the second week, compost with MOL reached a moisture content of 47.01% while compost without MOL reached a moisture content of 63.49%. It can be concluded that only compost with MOL meets the standards of SNI 14-7030-2004, namely the maximum moisture content of 50%.

3.2.2. Fluctuations of C/N Content

The following are the results of laboratory tests for fluctuations in the C/N content of the compost samples taken on June 6 and 13, 2022. The C/N ratio describes the maturity level of the compost. The higher the C/N value, the higher the amount of ammonia and nitrogen trapped in the pores of the compost pile so that the compost is declared to have not decomposed completely.

Table 3. C/N Fluctuations in Compost

	C/N Fluctuations	
Date	Control (%)	Local Microorganism (MOL) (%)
June 6th, 2022	12.900	10.54
June 13th, 2022	9.76	9.55

The higher the value of the C/N ratio in the compost indicates the compost has not decomposed completely or has not yet matured, due to a large amount of ammonia and nitrogen trapped in the pores of the compost pile. Based on the laboratory data and observations above, all samples of pure compost and those using local microorganism activator (MOL) have met SNI number 19-7030-2004 with a percentage below 20%, but compost using local microorganism activator (MOL) has the lowest percentage occurred on the 14th day, which was 9.55%, which means that it experienced a more complete decomposition.

3.2.3. Fluctuations of Phosphorus Content

The following are the results of laboratory tests for fluctuations in phosphorus content of compost samples taken on June 6 and 13, 2022. Based on the SNI 19-7030-2004 document regarding compost specifications from domestic waste, the quality standard for phosphorus content in compost is that it has a higher percentage value of 0,1%. This is because the weathering or decomposition process that occurs in the compost material causes the phosphorus content to be high. At the compost maturation stage, the microbes will die and the phosphorus content in the microbes will be mixed with the compost material, thereby increasing the phosphorus content in the compost.

Table 4. Phosphorus Fluctuations in Compost

	Phosphorus Fluctuations	
Date	Control (%)	Local Microorganism (MOL) (%)
June 6th, 2022	0.080	0.080
June 13th, 2022	0.584	0.208

On day 14, the laboratory results stated that the two compost samples were in accordance with the quality standard of phosphorus in the compost which indicated that decomposition had occurred with higher concentrations. The highest phosphorus content was found in control samples or in composting without an additional activator. Meanwhile, the compost sample with a local microorganism activator (MOL) has phosphorus content with the lowest percentage concentration of 0.208%.

3.2.4. Fluctuations of Potassium Content

The following are the results of laboratory tests for fluctuations in the phosphorus content of compost samples taken on June 6 and 13, 2022. Based on the SNI 19-7030-2004 document regarding compost specifications from domestic waste, the quality standard for potassium content in compost is that it has a higher percentage value of 0,2%. The microbes in the starter use potassium for their activities and the decomposition process of organic matter is simpler during the composting process and will produce elemental potassium.

Table 5. Potassium Fluctuations in Compost

	Potassium Fluctuations	
Date	Control (%)	Local Microorganism (MOL) (%)
June 6th, 2022	0.007	0.015
June 13th, 2022	0.035	0.027

It can be observed that from the laboratory data on day 7 and day 14 there is no calculation of the potassium concentration in the compost sample that reaches the quality standard. With the final result, the smallest percentage of potassium concentration is in the compost sample using a local microorganism activator (MOL) which is 0.027% and the highest percentage of potassium concentration is in the pure compost sample or without an activator, namely 0.035%.

Compost with the addition of MOL made for 14 days showed a change in color, texture, and smell of the compost that had started to cook according to SNI 19-7030-2004 which was blackish brown, textured, and smelled like dirt, the ratio of C/N was 9,55% which is greater than compost without MOL. This is in accordance with previous research by Suwatanti et al (2017) which had the same research results with a C/N ratio of 14.13%. This shows that the organic waste compost product with the MOL of fruit waste in this study is still in the quality range which is not much different from the results of other studies.

Judging from the results of the study, composting with fruit waste-based MOL is effective in accelerating the composting process, which in this study. Compost forms naturally, but it will take quite a long time, which is 2-3 months. However, using this fruit waste-based MOL, composting can be shortened, and in this study, the fastest time obtained was 14. However, even so, the quality of compost produced from composting with MOL is good and in accordance with SNI 19- 7030-2004.

4. Conclusion

In this study, local microorganisms (MOL) were used as activators in making compost. A local microorganism bio activator (MOL) is useful for accelerating the decomposition process, improving compost quality, and providing additional nutrients for plants. The materials used in the manufacture of local microorganisms (MOL) are a mixture of 8 liters of rice washing water, 2 liters of coconut water, 2 ounces of brown sugar, and 2 kg of organic waste.

The addition of local microorganisms (MOL) to compost according to laboratory tests proved that local microorganisms (MOL) were proven to be effective in compost decomposition as evidenced by an odor like the smell of earth because it already contained soil nutrients, blackish color due to the impact of stable organic matter, and a good texture of compost. smooth due to the decomposition of microorganisms.

The addition of local microorganisms (MOL) also makes the compost sample contain phosphorus in accordance with quality standards which illustrates that decomposition has occurred with higher concentrations. In addition, local microorganisms (MOL) also caused a more complete decomposition of ammonia and nitrogen in the compost, which was indicated by the compost sample added with local microorganisms (MOL) reaching the lowest percentage of C/N content at day 14 of 9.55%. Meanwhile, in the potassium content test, neither the control sample nor the sample with the addition of local microorganisms (MOL) matched the quality standard, which had a percentage value higher than 0.2%. The compost sample using a local microorganism activator (MOL) was 0.027% and the highest percentage of potassium concentration was found in pure compost samples or without a bio activator, namely 0.035%.

Acknowledgments
The author would like to acknowledge the Faculty of Engineering Diponegoro University for the support of this study with the acceptance of the research proposal Grants Competing Fund Faculty of Engineering in budget 2022. We also thankfull to Laura Haqq Syuura, Anis Sulistianingsih, Anisa Ikhtiar Febrianti, Baktida Revi, Inez Zahra Talitha, Jema Nawa Agatha Putri, Kurniawan Dwi Novanto, Muhammad Iqbal Julianggara, Muhammad Nafizaldi Saputra, Olivia Stefany Sihombing, Tarisa, Farrelia Febriana, Yulia Cahyani Putri, Muhammad Imaduddin, and other team members for the support and helpful feedback.

References
[1] SIPSN 2021 *Sistem Informasi Pengelolaan Sampah Nasional* Menlhk
[2] Febria F A, Rahayu S 2021 *Development of Local Microorganism from Organic Waste as an Alternative Product for EM4* IJPSAT 30(1) 98-105
[3] Suwahyono U 2018 *Cara Cepat Buat Kompos Dari Limbah (3rd Ed)* Penebar Swadaya: Jakarta Timur
[4] Suwatanti E P S, Widiyaningrum P 2017 *Pemanfaatan MOL Limbah Sayur pada Proses Pembuatan Kompos* J MIPA 40(1) 1-6
[5] Said M I et al 2020 *Quality of Compost Produced from Different Types of Decomposer Substrate and Composition of Straw* IOP Conf. Ser.: Earth Environ. Sci. 492 012088
[6] Sipayung H D 2019 *Pembuatan Kompos Menggunakan Sampah Taman Dan Kotoran Kambing Dengan Mol Limbah Sayur* USU Repository Skripsi Sarjana: Sumatera Utara
[7] Astari L P, Salundik, Siagian P H 2011 Kualitas pupuk kompos bedding kuda dengan menggunakan activator mikroba yang berbeda IPB Repository Undergraduate Thesis: Bogor
[8] BSN 2004 *Spesifikasi Kompos dari Sampah Organik Domestik* SNI 19-7030-2004
[9] Yulianto A B, Ariesta A, Anggoro D P, Heryadi H, Bahrudin M, and Santoso G 2009 *Buku Pedoman Pengolahan Sampah Terpadu : Konversi Sampah Pasar Menjadi Kompos Berkualitas Tinggi* Yayasan Danamon Peduli: Jakarta
[10] Surtinah 2013 *Pengujian Kandungan Unsur Hara Dalam Kompos yang Berasal Dari Serasah Tanaman Jagung Manis (Zea mays saccharata)* J Ilm Pertan 11(1): 16-26

Utilization of magnetic silica as tofu wastewater treatment absorber (influence of mass and adsorption time)

Badrus Zaman[1], Nurandani Hardyanti[1*], Purwono Purwono[2], and Jane Ivana[1]

[1] Department of Environmental Engineering, Faculty of Engineering, Diponegoro University, Semarang - Indonesia 50275
[2] Center for Science and Technology, UIN Raden Mas Said Surakarta, Jl. Pandawa, Pucangan, Kartasura, Indonesia 57168

nurandanihundip@gmail.com

Abstract. Wastewater from tofu production generally has many pollutant parameters, such as nitrogen, phosphorus, turbidity, and total suspended solids (TSS). These pollutants can cause a decrease in oxygen levels and pollute water bodies. This study aims to determine changes in pH and removal efficiency of turbidity and TSS in tofu wastewater using the size of 80 mesh magnetic silica. The research used a laboratory-scale batch system with variations in the absorbent mass of 5 g, 7.5 g, and 10 g per 1 liter of tofu wastewater and adsorption times of 15, 30, 45, and 60 minutes. The stirring speed used is 100 rpm, and the operating temperature is adjusted to the laboratory temperature. As a result, the TSS removal has the most effective results of 80%, with 7.5 g of magnetic silica added per liter of wastewater and 45 minutes of adsorption time with a TSS value of 90 mg/L. Meanwhile, the optimum turbidity removal was 82.52% with 7.5 g adsorbent and adsorption time of 15 minutes resulting in a turbidity value of 82.1 NTU. The resulting pH varies in the range of 3.28-4.27, depending on the mass of the adsorbent and the adsorption time.

1. Introduction

Tofu is a widely found and popular food in Southeast Asia [1]. Tofu is a traditional food favored by various people in Indonesian society because of its taste and nutrition. In addition, tofu production is also cheap, so the purchase price is low and can be bought by various groups [2]. In its production, every kg of soybeans needed to make tofu requires 17–54 kg of water [1]. Making tofu is simple and can be done by home producers because the ingredients needed to make it are pretty simple, consisting of soybeans, vinegar (CH_3COOH), coagulant, as well as tofu brick ($CaSO_4nH_2O$), and tofu seed solution. Tofu seed solution is a solution made from a previously produced tofu that has been left overnight [3].

The production of tofu produces wastewater, which could pollute the environment. Parameters of tofu wastewater in Central Java, Indonesia, can be seen in the Central Java Provincial Regulation No. 5 in 2012 concerning Industrial Wastewater Quality Standards. Tofu wastewater is produced from processing tofu seeds in the form of soaking, washing, boiling, filtering, froth taking, and disposing of seed water. The wastewater contains high organic compounds, including total suspended solids (TSS). High TSS blocks the entry of sunlight into the water so that it will interfere with the photosynthesis process, which causes a decrease in dissolved oxygen (DO) in the water [4]. The usage of acidic compounds to coagulate soybean solution during the tofu-making process causes the wastewater to have a pH in the range of 3–5 [5]. The organic material in tofu wastewater includes carbohydrates, protein, fat, potassium, and others. The high organic matter in tofu wastewater, generated from its production activities, is one of the reasons that tofu wastewater must be treated [6].

Various negative impacts on the environment can arise if there is no treatment for tofu wastewater that is discharged into the environment. These impacts include water pollution, disease, odor, increased microorganisms, insects, and reduced environmental aesthetics [7]. One of wastewater treatment method with low operating costs, efficient, easy to operate, and safe is adsorption [8]. The availability of various materials raises the need to find new adsorbents. Silica, one of considered a good adsorbent material, has a large surface area, adjustable pore size, good mechanical stability, and can reduce toxic materials [9]. The silica-based absorbent is also cheap. Another treat is that silica can be an excellent absorbent when combined with other materials [10]. Si-OH bonds on the silica surface make silica can be used as organic and inorganic adsorbents [9].

In this study, silica was modified with ferric chloride ($FeCl_3$), glucose ($C_6H_{12}O_6$), and sodium hydroxide (NaOH) to form a synthesis of magnetic silica. Magnetic silica has high adsorption properties and can easily and quickly be separated from the solution by an external magnetic field [9]. Therefore, adsorption with magnetic silica can be an alternative treatment in an integrated water treatment process. However, there is little literacy regarding using magnetic silica adsorbents to remove organic pollutants. Therefore, this study appears to bridge the lack of literacy related to magnetic silica adsorption, especially in tofu wastewater. This study aims to determine changes in pH and the efficiency of turbidity and TSS removal from tofu wastewater due to variations in magnetic silica's adsorption time and mass with a fixed size of the adsorbent.

2. Method

This research was conducted using an experimental method to find the effects of adsorption time and magnetic silica mass on several pollutant levels reduction under controlled conditions. The method used in this study was a laboratory scale conducted at the Environmental Engineering Laboratory, Department of Environmental Engineering, Faculty of Engineering, Diponegoro University. The tofu wastewater came from one of the tofu industries in Semarang: ABAH ABU Silk tofu. Jl. Gayamsari IV No.16, Jl. Majapahit RT.1/RW.12, Gemah, Kecamatan Pedurungan, Semarang City, Central Java 50191. This study determines the efficiency of removing turbidity and TSS from tofu wastewater while looking at pH changes.

2.1. Synthesis of magnetic silica

Synthesized magnetic silica was made by modifying the green coprecipitation method developed by Marimón-Bolívar & González (2018). The synthesis of magnetic silica was modified from silica, ferric chloride ($FeCl_3$), glucose ($C_6H_{12}O_6$), and sodium hydroxide (NaOH). Briefly, a solution of Ferric Chloride ($FeCl_3$) 0.36 M; 5 mL was prepared, heated to 70°C, and then 5 mL of 0.05 M glucose solution was added. The solution was heated at 85 ±5 C °C for 15 minutes, then silica was added and kept at 85 ±5 C °C for 30 minutes with continuous mechanical stirring at 100 rpm. The pH adjustment was made by adding 6N NaOH to reach a pH value > 8.5, where the solution turned black. Finally, the magnetic silica was washed three times with deionized water and alcohol for further drying at 60 °C for 12 hour.

2.2. Implementation of absorption research

The pollutant absorption process of tofu wastewater was brought out using magnetic silica. Magnetic silica was ground and sieved to obtain sieve sizes 80 mesh, washed with demineralized water, heated at 100°C for 30 minutes, and then filtered to remove inorganic impurities. Magnetic silica was then dried at 200°C for 4 hours. Surface area and pore volume before and after activation can be evaluated using Autosorb Brunauer-Emmett-Teller (BET). The absorption method uses various masses of magnetic silica (5 grams, 7.5 grams, and 10 grams) to achieve the best pollutant removal results in 1 liter of tofu wastewater. Then 10 mL samples were taken at 15, 30, 45, and 60 minutes to determine pH, turbidity, and TSS removal levels.

2.3. Data analysis stage

Initial testing was carried out at the preparation stage to determine the most effective removal of turbidity and TSS from several processing techniques. The data obtained were pH, turbidity, and TSS. From the tables and graphs, it could be seen that the efficiency was based on the decrease in parameters in tofu wastewater.

At the tabulation stage, the collected data was transferred to the data processing application, ensuring that all data had been filled in correctly until the data presentation was in the form of numeric or numbers, tables, and graphs/pictures. The data obtained were pH, turbidity, and TSS.

At the stage of applying the data according to the approach, a quantitative approach was used. This approach was chosen due to the results of measurements. Generated data was in the form of numbers. In this study, the data analyzed was the value of the removal efficiency of turbidity and TSS and changes in pH in each reactor.

3. Result and Discussion

3.1. Effect of Adsorption Time on pH

Figure 1 shows a graph of changes in the pH of tofu wastewater during the adsorption process using 80 mesh of magnetic silica with a stirring speed of 100 rpm for 60 minutes. The result of the initial pH measurement of tofu wastewater was 3.4. After the adsorption process, the pH values were in the range of 4.14 to 4.27. The final expected pH value is under the Central Java Provincial Regulation No. 5 in 2012 concerning industrial wastewater quality standards of 6.0–9.0.

As seen in Table 1, for the adsorbent mass of 5 grams, there was an increase in pH up to 4.12 in minute-15. Then the pH dropped to 4.03 at minute-60. For the 7.5 grams adsorbent added in wastewater, there was an increase in the pH value up to 4.27 from minute-0 to minute-15. However, the pH value decreased to 4.14 at minute-60. For the adsorbent mass of 10 grams, the pH value fluctuated. At the minute-15, the pH value rose to 3.68 but continued to decrease to 3.28 at minute-45. The pH value rose at minute-60, although the value was not much different from the initial, with a value of 3.49. These results indicated an increase in the pH value in the first 15 minutes. In this study, the adsorption time of tofu wastewater using magnetic silica for 15 minutes was the optimum for increasing tofu wastewater.

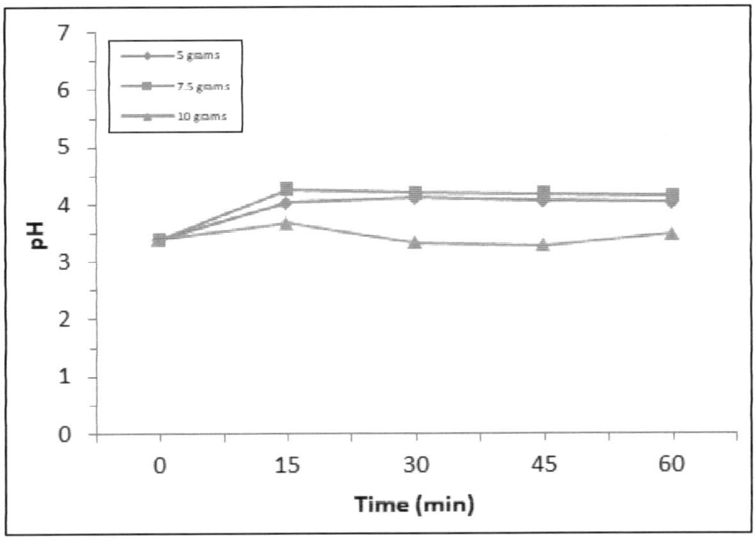

Figure 1. Graph of the pH concentration of tofu wastewater during the adsorption process using 80 mesh of magnetic silica with a stirring speed of 100 rpm for 60 minutes.

Table 1. TSS results, turbidity, and pH due to absorbance influenced by the time and mass of the adsorbent

Adsorban Type	Mass	Stirring Time	pH	Turbidity (NTU)	TSS (mg/L)
Magnetic Silica 80 mesh	0 gram	0 minute	3.4	567	450
	5 grams	15 minutes	4.03	456	350
		30 minutes	4.12	258	225
		45 minutes	4.07	252	219
		60 minutes	4.03	242	204
	7.5 grams	15 minutes	4.27	82.1	105
		30 minutes	4.2	148.9	99
		45 minutes	4.19	159	90
		60 minutes	4.14	274	260
	10 grams	15 minutes	3.68	114	225
		30 minutes	3.33	341	230
		45 minutes	3.28	351	280
		60 minutes	3.49	362	365

3.2. Effect of adsorption time on turbidity

Figure 2 shows a graph of the turbidity removal rate of tofu wastewater during the adsorption process using 80 mesh of magnetic silica with a stirring speed of 100 rpm for 60 minutes. Turbidity relates to the presence of suspended solids in the water, which causes water to have a muddy or cloudy appearance and makes the water less aesthetic [11]. The Regional Regulation of Central Java Province No. 5 in 2012 concerning the quality standard of industrial wastewater does not regulate the maximum level for turbidity. Although the regulation does not include turbidity as a wastewater parameter, turbidity affects various problems in wastewater treatment [11]. Turbidity can affect the light's entry into the water [11], which aquatic plants need [12]. Moreover, turbidity can also raise the surface water temperature because of the suspended particles that can absorb heat from sunlight [12].

The turbidity of the initial tofu wastewater was 567 NTU. After treatment, the turbidity value in wastewater was in the range of 82.1 NTU to 456 NTU. The TSS value of wastewater added with 5 grams adsorbent tend to decrease over time with the increasing removal percentage. Initially, the removal rate was at 19.58% from minute-0 to minute-15. The turbidity removal rate continued to rise to the highest value for the mass added, 57.32% at minute-60. This shows an increase in the percentage of turbidity removal with increasing time. For the 7.5 grams of adsorbent added to wastewater, a maximum increase of turbidity removal value happened in the first 15 minutes, reaching 85.52% on 82,1 NTU. The removal value began to decrease, reaching only 51.68%, with an increasing number in turbidity on 274 NTU. As for the adsorbent mass of 10 grams added to wastewater, there was also a drastically decreased turbidity value in the first 15 minutes, with removal efficiency reaching 79.89%. The percentage of turbidity adsorption decreased by 40.05% at 15 to 30 minutes, causing the turbidity value in wastewater greater than at 15 minutes. The turbidity value continued to increase, reaching 362 NTU at minute-60. Although the final result was higher than the previous minutes, it was still smaller than the initial turbidity value. For a mass of 7.5 grams and 10 grams of magnetic silica added to water, the absorbance showed an increase in the percentage of turbidity removal both at the first 15 minutes. However, the adsorption capacity would decrease gradually as the adsorption time increases. This result can occur due to a vacant site on the adsorbent. This causes an increase in the adsorption up to a particular time (equilibrium point) [13].

In general, there is a decrease in turbidity in wastewater, so the turbidity value at each test time is smaller than the initial turbidity value. In this study, it was seen that with magnetic silica 80 mesh adsorbent, the most optimum results were obtained at a stirring time of 15 minutes with a turbidity value of 82.1 NTU.

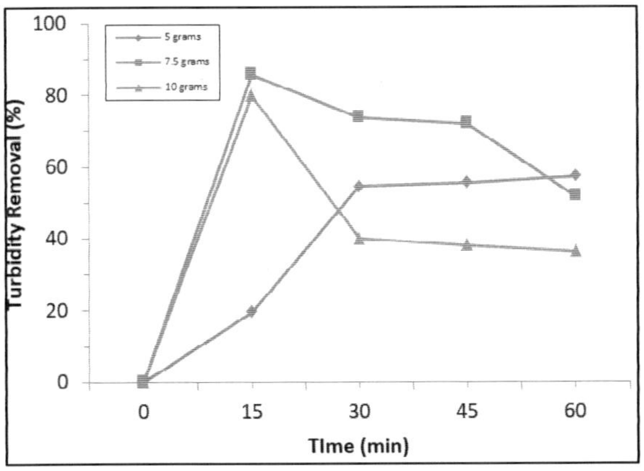

Figure 2. Graph of turbidity removal (%) of tofu wastewater during the adsorption process using 80 mesh magnetic silica with a stirring speed of 100 rpm for 60 minutes.

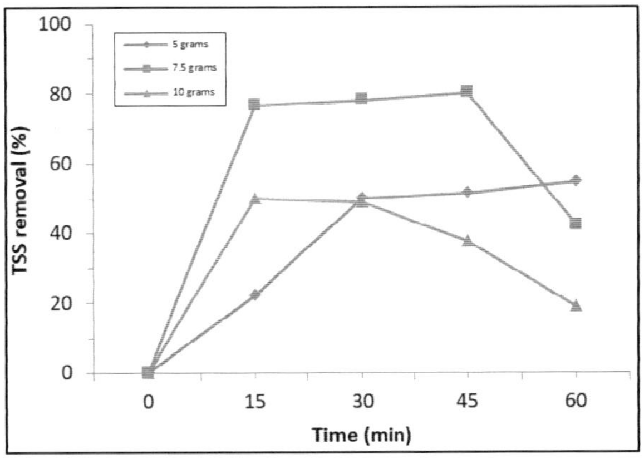

Figure 3. Graph of TSS removal (%) of tofu wastewater during the adsorption process using 80 mesh magnetic silica with a stirring speed of 100 rpm for 60 minutes.

3.3. *Effect of adsorption time on TSS*

Figure 3 shows a graph of the TSS removal rate of tofu wastewater during the adsorption process using 80 mesh of magnetic silica with a stirring speed of 100 rpm for 60 minutes. The TSS value of the initial tofu wastewater was 450 mg/L. The desired final TSS value is 100 mg/L under the quality standard of Central Java Provincial Regulation No. 5 in 2012 concerning the quality standard of industrial wastewater.

For 5 grams of adsorbent added to wastewater, there was a rapid increase in TSS removal levels from 22.22% to 54.67% in minute-15 to minute-60. This also indicated a decrease in TSS value from 350 mg/L in minute-15 to 204 mg/L in minute-60. This value shows that the TSS level also decreases as the adsorption time increases. For the 7.5 grams of adsorbent added, the TSS adsorption percentage increased from minute-15 until minute-45 from 76.67% to the highest value of 80%. The value of TSS then decreased from 105 mg/L in minute-15 to 90 mg/L in minute-45. However, the TSS value in minute-60 began to rise almost triple the value from the previous 30 minutes. The final value was 260 mg/L, and the adsorption percentage decreased to 42.22%. As for the 10 grams of magnetic silica added to wastewater, a gradual decrease in TSS removal percentage happened in the range of 50% in minute-15 to 18.89% in minute-60.

This fluctuation indicates that the removal rates rise drastically in the first 15 minutes and then decrease gradually until the end of the research, resulting in the TSS value not changing much from the initial. An increase in absorbency can occur due to a vacant site on the adsorbent. This causes an increase in the adsorption up to a particular time (equilibrium point). And if the space starts to fill up, no adsorption will occur, causing a stagnant value [13]. For the overall mass, it is generally known that until a certain time range, adsorption will result in decreased TSS values in wastewater. However, if the adsorption time is prolonged, it can also increase the TSS value in wastewater even though the ending results are smaller than the initial TSS value. In this study, the time for the adsorption of tofu wastewater using magnetic silica for 45 minutes was the optimum for removing TSS from tofu wastewater.

3.4. *Effect of magnetic silica mass on pH*

Variations in the mass of magnetic silica were 5 grams, 7.5 grams, and 10 grams. The desired final pH is under the quality standard of Central Java Provincial Regulation No. 5 in 2012 concerning Industrial Wastewater Quality Standards in the range of 6.0 – 9.0. The pH value of the initial tofu wastewater was 3.4. Adding 5 grams of adsorbent increased the pH value, with the highest value being 4.12 and the lowest value being 4.03. The addition of 7.5 grams of adsorbent also caused an increase in the pH value with the highest value was 4.27, and the lowest value was 4.14. Meanwhile, adding an adsorbent mass of 10 grams caused an increase and decrease in the pH value. The highest value was 3.68 and the lowest value was 3.28, which was lower than the initial pH. In this study, the mass of magnetic silica, as much as 7.5 grams, was the most optimum to increase pH in tofu wastewater.

3.5. *Effect of magnetic silica mass on turbidity*

Figure 4 shows a graph of turbidity removal (%) of tofu wastewater due to variations in the mass of magnetic silica with a stirring speed of 100 rpm for 60 minutes. The turbidity of the initial tofu wastewater was 567 NTU.

The value of turbidity in wastewater, when viewed from figure 4 has an optimum tendency at a certain mass. For a mass of 5 grams added to wastewater, 15 minutes had the lowest turbidity removal value of 19.58%. Meanwhile, the adsorption values were not too different at 30 minutes, 45 minutes, and 60 minutes, in the range of 54.5% to 57.32**%.** If 7.5 grams of adsorbent mass was added, the TSS removal efficiency increases to 85.52%, 73.74%, and 71.96% in minute-15, minute-30, and minute-45 respectively, which were the highest TSS removal values in each period. Meanwhile, for 60 minutes, the mass decreased to 51.68%. This result is in line with previous research that the increase in adsorbent capacity was caused by the addition of the adsorbent dose due to the addition of surface area and active surface number for adsorbing pollutants [13]. When 10 grams of adsorbent added, the efficiency in the entire time span decreased to 79.89%, 39.86%, 38.10%, and 36.16% at minute-15, minute-30, minute-45, and minute-60 respectively. These values were the lowest removal efficiency values in the 30, 45, and 60 minutes time range.

This result indicates that adding adsorbent mass to wastewater does not increase the efficiency of TSS removal. In this study, the mass of magnetic silica, as much as 7.5 grams, was the optimum for removing turbidity in tofu wastewater.

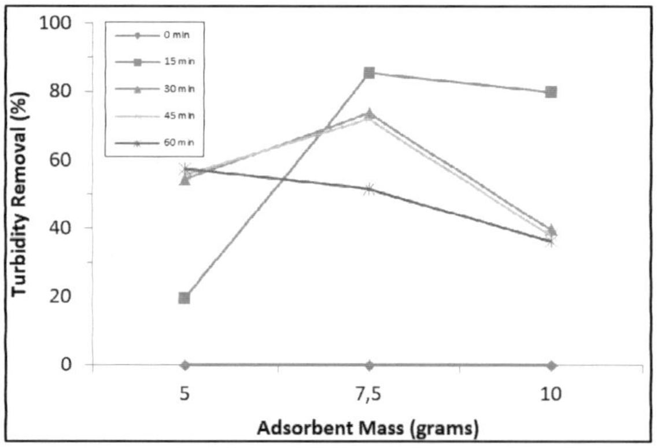

Figure 4. Graph of turbidity removal (%) of tofu wastewater due to variations in the mass during the adsorption process using 80 mesh magnetic silica with a stirring speed of 100 rpm for 60 minutes.

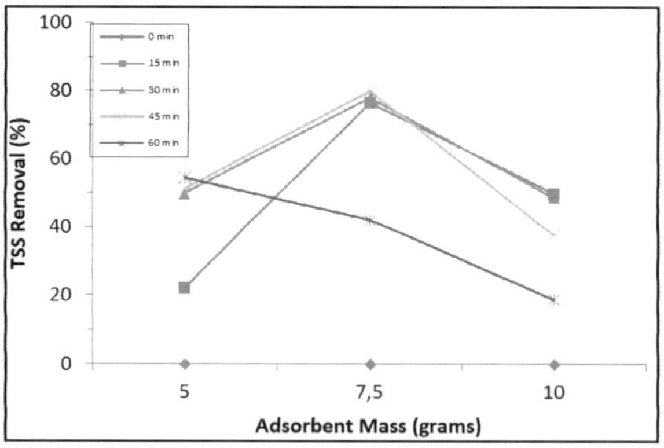

Figure 5. Graph of TSS removal (%) of tofu wastewater due to variations in the mass during the adsorption process using 80 mesh magnetic silica with a stirring speed of 100 rpm for 60 minutes.

3.6. Effect of magnetic silica mass on TSS

Figure 5 shows the graph of TSS removal (%) of tofu wastewater due to variations in the mass of magnetic silica with a stirring speed of 100 rpm for 60 minutes. The TSS value of the initial tofu wastewater was 450 mg/L. The desired final TSS value is under the quality standard of Central Java Provincial Regulation No. 5 in 2012 concerning the quality standard of industrial wastewater of 100 mg/L.

Wastewater added with 5 grams of magnetic silica had the highest percentage of TSS removal of 54.67%, which had a period of 60 minutes with a TSS value of 204 mg/L, and the lowest percentage of 22.22% with a TSS value of 350 mg/L in 15 minutes. If 7.5 grams of adsorbent was added, the percentage of TSS removal was at its peak for 15, 30, and 45 minutes with values of 76.67%, 78%, and 80%, respectively. The lowest TSS value was 90 mg/L, which had met the applicable TSS quality standards. This result is in line with previous research that the increase in adsorbent capacity was caused by the addition of the adsorbent dose due to the addition of surface area and active surface number for adsorbing pollutants [13]. When 10 grams of adsorbent was added, the percentage of TSS removal in the four time periods of 15, 30, 45, and 60 minutes decreased to 50%, 48.89%, 37.78%, and 18.89%, respectively. From these results, it is known that when the adsorbent mass of 7.5 grams was added, the entire period experienced the most optimal percentage of TSS removal.

4. Conclusion

Magnetic silica as an absorbent in tofu wastewater treatment was used and tested using experimental methods to determine the effect of changes in absorbent mass and adsorption time under controlled conditions. With 80 mesh magnetic silica size, the results show that the optimum amount of adsorbent added to 1 liter of tofu wastewater in a batch system is 7.5 grams. The optimum time required to decrease the turbidity value in the tofu wastewater was 15 minutes, reaching 82.1 NTU at 85.52% removal. In comparison, the optimum time required to reduce the TSS level in the tofu wastewater was 45 minutes, with the final result reaching 90 mg/L at 80% removal. The pH value during the test was in the range of 3.28-4.27.

References

[1] Widyarani *et al* 2018 IOP *Conf. Ser.: Earth Environ. Sci.* **160** 012014
[2] N I Said, Heru D W 2009 *Teknologi Pengolahan Air Limbah Tahu – Tempe dengan Proses Biofilter Anaerob dan Aerob* (Jakarta : BPPT)
[3] Sunandar Priya Karnanta *et al* 2021 *J. Phys.: Conf. Ser.* **1858** 012003
[4] T Ruhmawati, Denny S, Mimin K and Tatang R S *Jurnal Pemukiman* **12** 25–32
[5] Widyarani *et al* 2018 *IOP Conf. Ser.: Earth Environ. Sci.* **160** 012014
[6] Arnita Chaorlina *et al* 2021 *IOP Conf. Ser.: Earth Environ. Sci.* **775** 012049
[7] L Sriwuryandari *et al* 2019 *IOP Conf. Ser.: Earth Environ. Sci.* **277** 012010
[8] M M El Bendary, Emad K R, M F El-Shahat 2021 *Environ. Nanotechnol. Monit. Manag.* **15** 100455
[9] Roxana N, S G Muntean, M A Nistor, A M Putz, Lázló A, Liviu S 2020 *Chemosphere* **261** 127737
[10] Roxana N *et al* 2020 *Appl. Sci.* **10** 2726
[11] Harashit K M 2014 *IJERD* **4** 105–114
[12] Vijay S K 2016 *IARJSET* **3** 186–190
[13] C Irawan *et al* 2020 *IOP Conf. Ser.: Earth Environ. Sci.* **499** 012009

Solidification/Stabilization of waste containing lead using flyash and bentonite

T L Simangunsong[1,*], Y Fransiscus[1], M I Prayitno[1], S W H Pratama[1]

[1]Chemical Engineering Department, [2]Center for Environmental and Renewable Energy Studies, University of Surabaya

tuani@staff.ubaya.ac.id

Abstract. In this study, the Stabilization/Solidification was used to treat Pb as heavy metal waste using flyash and bentonite, as binder and mixing agent. Flyash and bentonite was mixed with the waste containing Pb and formed to cube called "mortar" with a size of 5x5x5 cm. The variations of flyash and bentonite were 100:0, 90:10, 80:20, 70:30, 60:40, 50:50, 40:60, 30:70, 20:80, 10:90, and 0:100. Then the mortars had a curing time process for 7 days, 14 days, and 28 days. After curing time process, the products tested using a compressive strength test and TCLP test. Result of compressive strength test indicated that the optimum compressive strength value occurred in 100% fly ash composition with 28 days curing time. The estimation of compressive strength test was 3220 tons/m^2. The result of the TCLP test showed the optimum TCLP test results occurred at 100% fly ash composition and the rate was 0,0338 mg/L Pb. The composition of fly ash and bentonite showed a significant effect on the compressive strength and TCLP test.

1. Introduction

Small and medium electroplating industries are one of Pb containing wastewater sources. The process in the electroplating are cleaning the material with acid (HCl or H_2SO_4), rinsing with water, neutralizing with lime (CaO), glossing and polishing which produces dust, second rinsing with water that produces $Ca(OH)_2$ and HCl containing wastewater. After that selected material is coated with copper, then the coating material is rinsed with water, and finally the coating material is dried. In the electroplating process, mostly wastewater is generated from rinsing with water and washing tubs. The process makes the composition of electroplating wastewater heterogeneous. One of the compositions of electroplating wastewater is lead (Pb). Pb is found in the electroplating process at the washing stage because at the washing stage the container is coated with Pb and filled with HCl or H_2SO_4. By the time being the Pb will be dilute and react with the acid. In the certain time there will be sludge containing Pb at the bottom of the tub and the acid has to be disposed and replaced with the new acid. Most small and medium-sized industries have not treated the wastewater and disposed Pb containing sludge without treatment.

The lead containing wastewater can be treated using several methods which are physically, chemically and biologically. One of the physically treatments is adsorption. It is used to remove parameters of plate printing washing wastewater [1]. The result shows that the concentration of Pb and Zn can be reduced at the levels of < 0.01 mg/L. Another treatment is activated carbon application to eliminate heavy metals containing wastewater. The process is able to deflate 95% of Pb(II) and

98% for Cd(II) [2]. Physically-chemically treatment research conducted by [3] employed cement, lime and bentonite for the S/S of waste containing heavy metals Cd, Cr, Cu, Ni, Pb and Zn. The solidification was carried out using 100 ml of sediment and mixed with 10 g of cement, bentonite and lime. It stirred for 10 minutes and curing time was 7 days. The result showed that according to TCLP test the use of cement could reduce the concentration on all metals up to 80% but not for Pb. Another research conducted by [4], the aim of the study was to gain S/S met cement-fly ash, a small amount of bentonite, and a mixture concrete as reinforcement. The tests carried out are the compressive strength test and TCLP test. The compressive strength test was carried out with a cube mold (100 mm) as an object test. The sequential extraction scheme developed by Tessier was used to analyze variations in heavy metal chemistry in the form of different substrates compacted. Scanning electron microscopy and X-ray diffraction were used to tested the development of the solid phase and the microstructure stabilization/solidification (S/S). Samples were tested in the laboratory for 7,14,28, and 60 day. The result of this research was the S/S method with cement-fly ash obtained, a small amount of bentonite, and the concrete mix can effectively treat high concentrations of heavy metals, such as Cu, Cd, and Pb.

Solidification/stabilization(S/S) is a method that usually utilized to removed heavy metals from waste containing metals. S/S refers to the process of encapsulating waste into mass monolith with a solid structure. S/S provides many advantages such as cost-effective, easy to employ, time-saving, and high compatibility with mixed contaminants [5]. According to [6], there are 2 encapsulation mechanisms in the process stabilization/solidification, namely macroencapsulation and microencapsulation. Macroencapsulation is a mechanism by which hazardous waste is bound in the large structure of matrix, while microencapsulation is a mechanism by which hazardous waste is trapped in microscopic crystal structures. Basically harmful/toxic substances are solidified in cementitious materials or pozzolan materials. Cement is the standard binder used on the S/S system. Cement solidification is ideal for inorganic waste, especially heavy metal containing waste. Cement-based solidification lessened all the heavy metals of the solidified matrices of the waste. Alternative material beside cement that can be applied for heavy metals S/S is flyash. Flyash is the main solid waste released from coal-fired power plants. In many countries, coal that produces a lot of flyash is taken as the important foundation energy. The accumulation of large amount of the flyash will cause serious environmental pollution every year [7] so the utilization of flyash as stabilizing agent will reduce its amount and the effect to environment. Sometimes, the main stabilizing subtance such as cement, lime, flyash, etc may not be effective for immobilization of contaminants. In order to increase the effect of S/S in such conditions, a combination of main elements (e.g., flyash or cement) with another binder can be considered. Stabilization/solidification (S/S) requires additional binder to physically encapsulate contaminants or chemically to form a stable product. Bentonite is a common clay mineral which was also used in the experiment as an auxiliary additive for the solidification.

In this study, stabilization/solidification (S/S) will be used to treat sludge containing Pb. S/S product is a stable product so the toxicity of hazardous waste can be reduced. Mostly the concentration of the waste can meet the government standard so the products will be safe to dispose to the environment. The binder is a mixture of bentonite and flyash. The utilization of bentonite-flyash mixture can be applied in the process stabilization/solidification (S/S). Flyash is by product that need to utilize to reduce its discarding to environment. Bentonite can be applied as a partial material replacement of cement which is relatively cheap and easy to obtain. The aim of the research is to determine the optimum composition of flyash-bentonite to meet the government standard no. 22/2021 according to compressive strength test and TCLP test. Then the products will be safe to dispose or turn into another product such as paving block, brick mixture, etc.

2. Materials and Methods

2.1. Materials

Flyash is from Tjiwi Kimia paper industry, Mojokerto. The type of bentonite is Ca-bentonite. The hazardous waste used is artificial waste ($PbSO_4$)

2.2. Methods

- The variations of flyash and bentonite were 100:0, 90:10, 80:20, 70:30, 60:40, 50:50, 40:60, 30:70, 20:80, 10:90, and 0:100.
- The compositions of the S/S product were mixed, and the mixture was poured into the mold which size of 5x5x5cm .
- Then the S/S product must be settled in the mold for 24 hours.
- After 24 hours, the S/S products (mortars) were released from the mould.
- After that the mortars had a curing time process by soaking it in a tub filled with water for 7 days, 14 days, and 28 days.
- To verify the effectiveness of the flyash-bentonite based of solidification/stabilization (S/S) treatments to contaminated samples investigated in this study, the compressive strength and the TCLP of Pb tests were carried out.

3. Results and Discussion

3.1. Compressive Strength Result

The results show that the 28 days of curing time has the highest compressive strength value in all variations of flyash-bentonite. Compressive strength test is carried out on a variation of 40 - 100 % fly ash because at 0% - 30% compositions of flyash, the products cannot dry in 24 hours. Table 1. represent the result of compressive strength of S/S product at 28 days curing time.

Table 1. The Result of Compressive strength

Compositon (% flyash)	Compressive strength (ton/m^2)
100	3220
90	1092
80	416
70	304
60	196
50	202
40	154

The table represent that 100 % flyash give the highest compressive strength rate. Flyash and bentonite can yield hydration reactions between pozzolans to form $CaSiO_2H_2O$. A material can be classified as pozzolan if it contains SiO_2, Al_2O_3, Fe_2O_3 and a low concenration of CaO. Pozzolanic material itself is not cement-like, but can act like cement when the pozzolanic material reacts with calcium hydrate and water. The following is a hydration reaction that occurs in pozzolans:

$$CaO + H_2O \rightarrow Ca(OH)_2$$
$$Ca(OH)_2 \rightarrow Ca^{+2} + (OH)^{2-}$$
$$Ca^{+2} + (OH)^{2-} + SiO_2 \rightarrow CaSiO_2H_2O \text{ (C-S-H)}$$

The formation of $CaSiO_2H_2O$ or commonly called C-S-H, in the reaction between pozzolans so the generation of C-S-H become slower than the formation of C-S-H in the reaction of cement and pozzolan. The factor that affect the reaction rate of the formation of C-S-H is % weight of CaO. The increasing of flyash composition by % weight of CaO will be escalated as the same as by % weight of CaO from pure cement (58%). The formation of C-S-H will be enlarged by the addition of flyash if it is combine with cement. It is discovered that the % weight of CaO in flyash (6.72%) is greater than that of bentonite (3.68%). The addition of fly ash will be rised the rate of hydration reaction and more C-S-H will be produced so that the compressive strength value of the mortars will be expanded [8].

In this study, curing time was carried out by immersing the S/S products in a tub filled with water. The purpose of curing is to maintain the humidity of the products, environment temperature, and decreasing of the products hydration temperature. The longer curing time will create longer hydration reaction time so the production of C-S-H compound will be enhancing . The impact of the fenomena is the value of the compressive strength test will escalate. The non-curing treatment causes the hydration reaction not optimal because the heat from the hydration reaction can cause the difference of the product's and the environment's temperature. It will induce the water content of the products evaporated which will disrupt the hydration reaction. The composition of fly ash 100% and 90% has a significant difference value of compressive strength among curing times. However, there are no significant difference of compressive strength value at 80%-40% flyash composition in the same curing time duration. This is due to the decreasing composition of fly ash will effect to the rate of hydration reaction. The slower of hydration reaction will lead to the slower of C-S-H formation. The decelerate formation of C-S-H evolve no notable rate of compressive strength at the composition of fly ash 80-40%. Research by [9] was found that the significantly effect of solidification depended upon the composition of the binder material, type of leaching extractant, characteristic and concentration of waste (especially in the context of heavy metals). Vul and Gowripalan [10] stated that efficient immobilisation of toxic metals, the synthesis and curing conditions of geopolymers (a feasible alternative to portland cement, such as fly ash) and the long term stability of the products are important considerations. These parameters are rules useable for applying an effective stabilisation/solidification of harmful metals.

3.2.. TCLP test Result

The TCLP is designed to discover the movability of both organic and inorganic subtances that containing in liquid and solid wastes. TCLP testing is applied to confirm the safety of the environment when dumping possibly hazardous waste. Table 2. represent the result of TCLP Test of S/S product at 28 days curing time. 28 days of curing time derive low concentration of Pb leachate.

Table 2. The Result of TCLP

Compositon (% flyash)	TCLP (mg/L)
100	0,0338
90	0,0405
80	0,0502
70	0,0583
60	0,0652
50	0.0806
40	0,1042

Table 2 shows the results of the TCLP test. Pb waste concentration that leaching is decreased as the composition of fly ash increased. The fenomena relate to the formation of C-S-H gel. C-S-H gel along with the adsorption of heavy metals will be limited the migration of ions and ultimately reduced the concentration of leachate [8]. During the process of S/S, hazardous heavy metals in the waste enhance to the various stable morphologies. The convestion of heavy metals is affected by extracted liquation, stabilizers and leaching characteristics, etc. The morphological transformation and migration characteristics of heavy metals have an effect on the long term leaching behavior of hazardous compounds [11]. The solidification process indicates that the waste compounds are trapped in the porous gel structure's pores, and heavy metal securing generates insoluble metal hydroxides. The binders or cement (or flyash) increase the resistance and the compressibility and permeability of the waste. Stabilisation is the other S/S component that uses a physiochemical process to turn contaminants into less leaching and inactive forms to create a chemically stable condition [12].

All the compositions of the objects have met the TCLP quality standard according to regulations of Indonesia Government Regulation No. 22/2021 Appendix XII concerning TCLP Test Quality Standards for Pb waste is 0.5 mg/L Pb.

4. Conclusions

Result of compressive strength test indicated that the optimum compressive strength value occurred in 100% flyash composition with 28 days curing time which was 3220 tons/m^2. The result of the TCLP test showed that the optimum TCLP test results occurred at 100% fly ash composition which was 0,0338 mg/L Pb. The composition of flyash and bentonite showed a significant effect on the compressive strength and TCLP test.

References

[1] Syauqiah I, Kusuma F I, Mardiana 2020 *Konversi* **9(1)** 28-34
[2] Singanan M 2011 *Science Asia* **37** 115-119
[3] Bao J, Wang L, Xiao M 2016 *Environmental Science Pollution Research* **23(9)** 1-8
[4] Yan M, Zeng G, Li X, He J, Chen G, Huang D, Tang L, Lai C, Zhang C, Li X, Wang L, Guo Z, Tao W 2017 *Environmental Science Pollution Research* **24** 892-901
[5] Hou D, Gu Q, Ma F, O Connell S 2016 *Journal of Cleaner Production* **139** 949–956.
[6] Nindyapuspa A, Alifah T, Virgo I T 2019 *MATEC* **276** 1-6
[7] Wang, XY, Park, K B 2015 *Journal Construction and Building Materials* **98** 810–819.
[8] Tang Q, Liu Y, Gu F, Zhou T 2016 *Advance in Materials Science and Engineering* 1-10.
[9] Mao Y, Muhammad F, Yu L, Xia M, Huang X, Jiao B, Shiau Y C, Li D 2019 *Int. J. Environ. Res. Public Health* **16** 1121 doi:10.3390/ijerph16071121
[10] Vu1 T JH, Gowripalan N 2018 *Journal of Advanced Concrete Technology* **16** 124-135. doi:10.3151/jact.16.124
[11] Wang Y, Tang J, Li Z 2021 *Environmental Research* **197** https://doi.org/10.1016/ j.envres. 2021. 111053
[12] Qian G R, Shi J, Cao Y L, Xu Y F, Chui P C 2008 *J Hazard Mater* **152(1)** 196–203https://doi.org /10.1016/j.jhazmat.2007.06.118

Environmental disclosure and its determinants

I F S Wahyuningrum[1*], M I Amal [1], S Oktavilia[2], A Setyadharma[2], M Khafid[1], M Lina[1]

[1]Department of Accounting, Faculty of Economics, Universitas Negeri Semarang
[2]Department of Economics, Faculty of Economics, Universitas Negeri Semarang

*i.fajarini@mail.unnes.ac.id

Abstract. Nowadays, companies are required to report not only about their economic performance but also about the social and environmental aspects of their operations. This is because social and environmental disclosures are seen as increasingly significant in terms of their influence on the survival of the company. The purpose of this study is to see how firm size, profitability, leverage, and managerial ownership affect environmental disclosure. The population of this study comprises non-financial sector companies listed on the Indonesia Stock Exchange (IDX) between 2017 and 2020. Purposive sampling was used, yielding 160 units to be analysed. A documentation technique was used to collect the research data that were used. The data came from the IDX website as well as from the official websites of each company. Descriptive statistical analysis and regression testing using SPSS were used. The results show that profitability and leverage have a negative effect on environmental disclosure. A suggestion stemming from this research is that companies should increase their environmental disclosure in order to ensure the transparency and accountability of their information in the eyes of the public. Future research is expected to be able to use all financial ratios to assess the impact of financial performance on company environmental disclosures.

1. Introduction

Climate change has emerged as one of the most pressing problems facing the world since the late twentieth century. Global warming refers to conditions in which there is climate change that is directly or indirectly related to human activities, and where these activities are able to change the composition of the atmosphere over a predetermined period of observation [1]. Another issue that is related is sustainability of the environment and natural resources. These problems will have an impact on how companies carry out their operating activities. At the same time, the public is becoming increasingly aware of how operational activities of companies can pollute the surrounding environment.

This growing public awareness has had an impact in terms of increased pressure being placed on companies to take responsibility for the environment. Other stakeholders, including governments, international organizations and other similar organizations, have been applying strong pressure to companies to pay attention to their environmental performance through the issuing of regulations and rules. As a consequence of pressure from stakeholders, companies must be accountable to the community for the impacts resulting from their operating activities.

Through Law No.32 of 2009 concerning Environmental Protection and Management, Law No.40 of 2007 concerning Limited Liability Companies (Article 66 paragraph 2C and Article 74 paragraph 1), Government Regulation No.47 of 2012 concerning Social Responsibility for the Environment of Limited Liability Companies and the Indonesian Capital Market, and the Financial Institutions Supervisory Agency, the government requires companies to be responsible for the environment [2].

Qiu et al. [3] state that companies with large profits have the ability to finance expenditures related to environmental disclosure. A company's large resources will provide an opportunity for it to demonstrate its commitment to the environment. Doing this has the aim of reducing the pressure that comes from the community and it can create a good impression in the eyes of stakeholders. Previous research has found that profitability has a positive influence on environmental disclosure [4,5,6].

Leverage is a ratio that can reflect a company's capacity to meet its obligations. Choi et al. [7] state that a company with high leverage will choose to focus on paying all debts rather than making voluntary disclosures. On the other hand, management has the responsibility to prepare the annual report addressed to the company's owner and its stakeholders. Therefore, management is expected to disclose all relevant information in the annual report for the basis of efficient decision making. A high level of managerial ownership will influence managers to pay more attention to their performance. The result of that study found managerial ownership had positive influence on environmental disclosure [8].

Company size is one of the observable characteristics of a company. People will pay more attention to large companies. As a result, large companies will get pressure from society that will grow increasingly stronger. In addition, large companies have the resources available to finance all corporate expenses related to environmental disclosure. Some past research found that firm size positively affects environmental disclosure [9,10,11].

The following hypotheses are formulated based on the evidence presented above:
H1 = Environmental disclosure is significantly positively influenced by profitability.
H2 = Environmental disclosure is significantly negatively affected by leverage.
H3 = Environmental disclosure is significantly positively influenced by managerial ownership.
H4 = Environmental disclosure is significantly positively influenced by firm size.

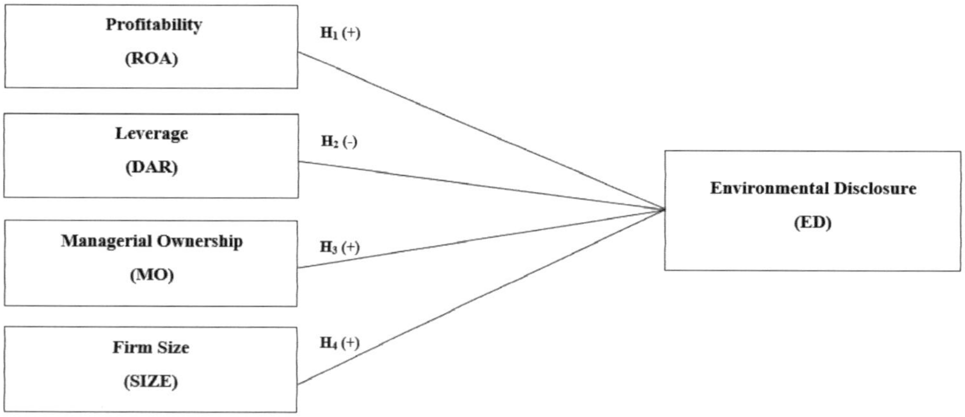

Figure 1. Theoretical framework.

2. Methodology

The population for this study comprises non-financial sector companies in Indonesia. These companies must be listed on the Indonesia Stock Exchange (IDX) and provide information related to the environment in their annual reports and on their official websites for the period 2017 to 2020. Purposive sampling was used in this study. The research sample consisted of 160 units of analysis. SPSS v24 was used to perform multiple linear regression analysis. The multiple linear regression equation is shown in equation (1).

$$CED = \alpha + \beta 1\ ROA - \beta 2\ DAR + \beta 3\ MO + \beta 4\ SIZE + e \qquad (1)$$

Table 1. Measurement variables.

Variables	Codes	Measurements
Environmental Disclosure	ED	Analyze the content
Profitabilitas	ROA	Net income/total assets
Leverage	DAR	Liabilities/ total assets
Managerial Ownership	MO	Number of shares owned by management/ number of shares outstanding
Firm Size	SIZE	Log of total assets

3. Results and Discussion

3.1. Descriptive and regression result

Descriptive statistical analysis yielded a mean of 0.318 and a standard deviation of 0.235. The mean value, being greater than the standard deviation, implies that the data are normally distributed. However, the minimum value of environmental disclosure is 0 which means that there are several companies that do not disclose environmental information.

Table 2. Results of descriptive statistics.

Variables	N	Minimum	Maximum	Mean	Std. Dev
ROA	160	-0.451	0.526	0.052	0.102
DAR	160	0.126	0.961	0.519	0.212
MO	160	0	0.029	0.002	0.005
SIZE	160	19.547	26.587	23.870	1.169
ED	160	0	0.933	0.318	0.235

The significance level of the normality test using the Kolmogorov Smirnov One-Sample is 0.281 > 0.05, indicating that the data are normally distributed. The tolerance value is greater than 0.10 and the VIF value is greater than 10, which indicates that the regression model does not have symptoms of multicollinearity. Heteroscedasticity was tested using the Glejser Test. Each variable has a significance value greater than 0.05, which indicates that the data are free from heteroscedasticity. The Cochrane Orcutt approach was used in the autocorrelation test. The DW value is 1.907 where dU < DW < 4-dU (1.7925 < 1.907 < 2.2075), which indicates that there is no autocorrelation in the data. A simultaneous test using ANOVA and produces sig value. of 0.046. This means that ROA, DAR, MO, and SIZE have a simultaneous effect on the ED variable.

Table 3. Results of a statistical test.

Model	Unstandardized Coefficients		Standardized Coefficients	t	Sig
	β	Std. Error	Beta		
(Constant)	.214	.392		.546	.586
ROA	-.397	.184	-.173	-2.164	.032
DAR	-.184	.089	-.166	-2.081	.039
KM	-2.954	3.675	-.065	-.804	.423
SIZE	.010	.016	.047	.582	.561

3.2. Hypothesis Discussion

3.2.1 The effect of profitability on the environmental disclosure. Profitability has a significance level of 0.032 with coefficients indicating the negative direction of the dependent variable, which, in turn, indicates that environmental disclosure is negatively and significantly affected by the profitability variable. The difference in direction between hypothesis and results means the **H1 is rejected**. Companies believe that environmental disclosure is a burden that can reduce the nominal income that they earn. Meanwhile, the low value of the company's profitability indicates that its financial performance is poor. Thus, the company will make decisions to disclose environmental information in order to gain legitimacy in the eyes of stakeholders. This is supported by research conducted by Wahyuningrum et al. and Udo [12,13].

3.2.2. The effect of leverage on the environmental disclosure. Leverage has a significance value of 0.039, which indicates that the second **hypothesis (H2),** which explains the negative effect of leverage on the disclosure of carbon emissions, **is accepted**. Companies with high leverage need to be more conservative when creating some of their policies, especially regarding company spending; one of these is by not making environmental disclosure. The company choses not to disclose its environmental performance due to the lack of financial resources, although, on the other hand, the company feels pressure from various parties. Additionally, the company is under pressure to pay off debts to creditors immediately. The company will choose to reduce the intensity of environmental disclosure aimed at not to attracting the attention of debtholders. This study supports previous research [14,15], which has found results that environmental disclosure negatively affecte the leverage variable.

3.2.3. The effect of managerial ownership on the environmental disclosure. The testing of the third hypothesis (H3), indicating a significance value of 0.423, demonstrates that managerial ownership is not able to have an effect on environmental disclosure which means that H3 is rejected. Managerial ownership of non-financial companies in Indonesia is not able to influence company policy regarding conducting environmental disclosure. This is because the number of managerial ownerships in Indonesia is relatively small so there is no comformity between the interests of managerial and corporate owners. According to previous research, managerial ownership cannot affect environmental disclosure [16,17].

3.2.4. The effect of firm size on the environmental disclosure. The findings show that company size has no effect on environmental disclosure. SIZE indicates significance value of 0.561 which is > 0.05 and so proves that the **H4 is rejected**. Companies that have a lot of assets do not necessarily make extensive environmental disclosure; this is because many companies in Indonesia still do not understand the importance of contributing to the protection of the environment. The impact of large companies on the environment is greater than that of small companies, therefore the government needs to impose stricter regulations related to the obligations of large companies to carry out environmental disclosure. This is supported by other studies [18,19].

4. Conclusions

This study has examined the factors that influence environmental disclosure in non-financial sector companies listed on the Indonesia Stock Exchange between 2017 and 2020. The results show that only one of the four hypotheses has been proven. Based on the test results, it is found that profitability and leverage have a negative effect on environmental disclosure, while managerial ownership and firm size have no effect. The study's conclusions may be valuable to companies, investors, and regulators in developing rules to make environmental disclosure decisions. A suggestion stemming from this research is that companies should increase environmental disclosure in order to ensure the transparency and accountability of their information in the eyes of the public. Future research is expected to be able to use all financial ratios to measure the influence of financial performance on company environmental disclosures.

References

[1] Declarations of the United Nations Conferences on the Human Env 1992 Retrieved from www.unep.org/documents.multilingual/default.asp?documentid=97&articleid=1503
[2] Deswanto R B and Siregar S V 2018 *Soc.l Resp. J.* **14**(1) 180-193
[3] Qiu Y, Shaukat A, and Tharyan R 2014 *The British Acc. Rev.* 1-15
[4] Kansal M, Joshi M, and Batra G S 2014 *Adv. Account.* **30**(1) 217-229
[5] Lu Y and Abeysekera I 2014 *J. Clean. Prod.* **64**(1) 426-436
[6] Muttakin M B and Khan A 2014 *Adv. Account.* **30**(1) 168-175
[7] Choi B B, Lee D, Psaros J 2013 *Pac. Account. Rev.* **25**(1) 58-79
[8] Sari G A C N, Yuniarta G A, and Wahyuni M A 2018 *J. Ilmiah Akuntansi* **9**(3) 145-155
[9] Welbeck E E, Owusu G M Y, Bekoe R A, and Kusi J A 2017 *Int. J. Corp. Soc. Resp.* **2**(1) 11
[10] Khalid T B, Kouhy R, and Hassan A 2017 *J. of Acc. and Auditing: Res. Pract.* 369352
[11] Ismail A H, Rahman A A, and Hezabr A A 2018 *Int. J. Ethnics Syst.* **34**(4) 527-563
[12] Wahyuningrum et al 2021 *IOP Conf. Ser.: Earth Environ. Sci.* **623** 012065
[13] Udo E J 2019 *Int. J. Account. Finance* **8**(1) 1-20
[14] Cormier D and Magnan M 2003 *J. Account. Public Policy* **22** 43-62
[15] Paramitha B W and Rohman A 2014 *Diponegoro J. Account.* **3**(3) 1-11
[16] Junita N L and Yulianto A 2018 *Account. Analysis J.* **7**(3) 144-150
[17] Nilam P and Wahyuningrum I F S 2021 *Jurnal Akuntansi Bisnis* **14**(1) 58-72
[18] Elshabasy Y N 2018 *J.Bus.Retail Manag. Res.* **12**(2)
[19] Gatimbu K K and Wabwire J M 2016 *Int. J. Sustain. Manag Inform. Tech.* **2**(1) 1-6

Rapid cultivation of aerobic granular sludge in laboratory scale sequencing batch reactor

Nur Novilina Arifianingsih[1,2]*, Andrianto Ansari[3,4], Pei-Hsun Wu[1]

[1]Graduate Institute of Environmental Engineering, National Taiwan University, Taipei 10617, Taiwan
[2]Faculty of Civil and Environmental Engineering, Institut Teknologi Bandung, Bandung 40116, Indonesia
[3]Department of Bioenvironmental Systems Engineering, National Taiwan University, Taipei 10617, Taiwan
[4]Department of Agronomy, Faculty of Agriculture, Universitas Gadjah Mada, Yogyakarta 55281, Indonesia

nurnovilina@gmail.com

Abstract. Aerobic granular sludge (AGS) is known as excellent biological treatment for wastewater industry, high organic load wastewater and adsorption of heavy metal. AGS have some advantages such as higher density, higher biomass retention, simultaneous nutrient removal and higher tolerance to toxicity. The aim of this research is to use sequencing batch reactor (SBR) to cultivate AGS in the short time (28 days) in laboratory scale. Experimental condition that we use to cultivate AGS are based on this parameter: 1.5 hours of total cycle time, 50 % of exchange ratio, 3 hours of HRT and 3 l/min air flow rate. Even the AGS cultivation successfully achieve in 28 days with pH around 7-9, DO 3.6 mg/l, SVI 232.6 mg/l, and MLSS 500 mg/l, it is still need more time to generate more mature AGS to breakdown organic molecule in the wastewater.

1. Introduction

Microorganisms can exist as biofilm and biogranules in biological wastewater treatment systems. A biogranule is an entirely immobilized microbe, whereas a biofilm is made up of bacteria adhered to an inert substrate [1]. Due to the repellent electrostatic forces and interactions caused by their mutual hydration, bacteria are unlikely to assemble naturally. Biogranulation must be induced in reactors by strong selective pressure [2]. In the sludge of anaerobic sewage treatment plants, observations of this phenomena were made for the first time in 1980 [3]. In the Sequencing Batch Reactor (SBR), the first granular sludge to be found in an aerobic system, was discovered in 1997 [4]. Aerobic granular sludge (AGS) is an unusual microbial community that enables for the simultaneous removal of carbon, nitrogen, phosphorus, and other contaminants in a single sludge system [5].

There are four phases in the process through which aerobic particles are formed [5]. First, intercellular interactions with microbes. Second, in response to an external disturbance, microbes combine with one another to create first aggregates in a process similar to quorum sensing. Third, microorganisms collaborate and produce EPS to improve adherence. One of the crucial physiological skills of microbes, which may break down environmental chemicals into smaller molecules in a test tube before ingesting them, is the production of extracellular polymers (EPS). And the last, by using aeration bubbles to create hydrodynamic shear force, particles will gradually develop. The single reactor tank will then alternate between an aerobic and anaerobic cycle. A visible layering develops gradually because the differential of oxygen concentration decreases with particle depth.

AGS has a number of benefits over typical activated sludge systems, including high biomass retention, strong settling ability, and toxin tolerance [6–8]. AGS also has a number of benefits over anaerobic granules, including good tolerability and increased synthesis of extracellular polymer substances (EPS) [9]. Proteins, polysaccharides, humic acids, and lipids found in bacteria's secretions are known as EPS, which start the granulation process [10]. About 20 years have passed since the discovery of AGS, and in that time it has been studied for a wide variety of purposes, including the treatment of domestic and high organic load wastewaters, the bioremediation/biotransformation of toxic aromatic pollutants like phenol, toluene, and pyridine, and the treatment of industrial effluents like those from the textile, dairy, and brewery industries, the adsorption of heavy metals, and the recovery of high value-added products [9].

Most studies use hydraulic retention time (HRT) in the range of 5- 52 hours [8,11–13]. However, very little is currently known about the impact of HRT under 5 hours. The study experiment used a speedy HRT time of 3 hours. Therefore, the objective of this study was to cultivate AGS SBR on a laboratory scale in a relatively short time.

2. Experimental design

2.1 Reactor setup and operation

Experimental design based on prior study [1]. Aerobic granules were cultivated in 2.0 l column. The diameter of reactor is 5 cm with height 120 cm. The reactor made by acrylic, with supplying fine air bubbles at the rate of 3 l min^{-1} through a ceramic diffuser at the bottom reactor. Overall, SBR systems was performed through four stages namely, feed, aeration, settle and decant. This study used stable time for feeding and decanting which is 5 minutes since the first week until fifth week. Gradual changes were setting for aeration time and settle time. Aeration time increase from 15 minutes in first and second week, 72 minutes in third and fourth week and 75 minutes in fifth week. The total cycle time for whole stage is 1.5 hours with exchange ratio 50% and HRT was maintained 3 hours in the reactor.

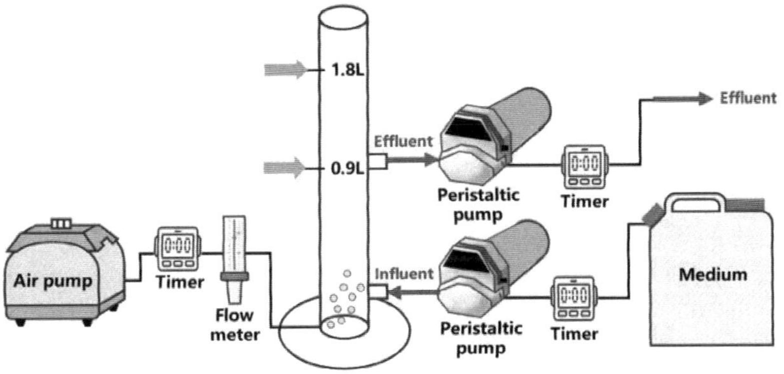

Figure 1. Experimental condition-configuration of reactor

2.2 Medium

Based on prior study [1], the following compounds such as Sodium acetate (CH_3COONa) as COD 1000 mg l^{-1}, ammonium chloride (NH_4Cl) 100 mg l^{-1}, K_2HPO_4 22.5 mg l^{-1}, $CaCl_2.2H_2O$ 15 mg l^{-1}, $MgSO_4.7H_2O$ 12.5 mg l^{-1}, $FeSO_4.7H_2O$ 10 mg l^{-1} were used as extracellular medium for substrate utilization phase (feast phase). In addition, the following chemical compounds such as $(NH_4)_6Mo_7O_{24}$

0.05 mg l^{-1}, $MnSO_4{\cdot}H_2O$ 0.05 mg l^{-1}, H_3BO_3 0.05 mg l^{-1}, $ZnCl_2$ 0.05 mg l^{-1}, $CuCl_2{\cdot}2H_2O$ 0.03 mg l^{-1}, were used for substrate depleted phase (famine phase).

2.3 Seeding
Activated sludge from a local municipal wastewater treatment plant in Taipei City was taken in this research, which was further cultivated with the same wastewater as used in the reactors for four days in the barrel. Hereafter, the acclimated activated sludge was inoculated into the reactor

2.4. Analytical methods
At the end of each cycle, a predetermined quantity of sludge was removed from the reactors and used to determine the sludge volume index (SVI) and the mixed liquid suspended solid (MLSS) content of the reactors on a weekly basis, respectively. Standard 2710 D was used for the analysis, and SVI was calculated [14]. Every day, samples of the effluent produced by the reactors were taken throughout the day so that pH and DO levels could be determined.

3. Result and Discussion

3.1 Daily measurement (pH-DO)

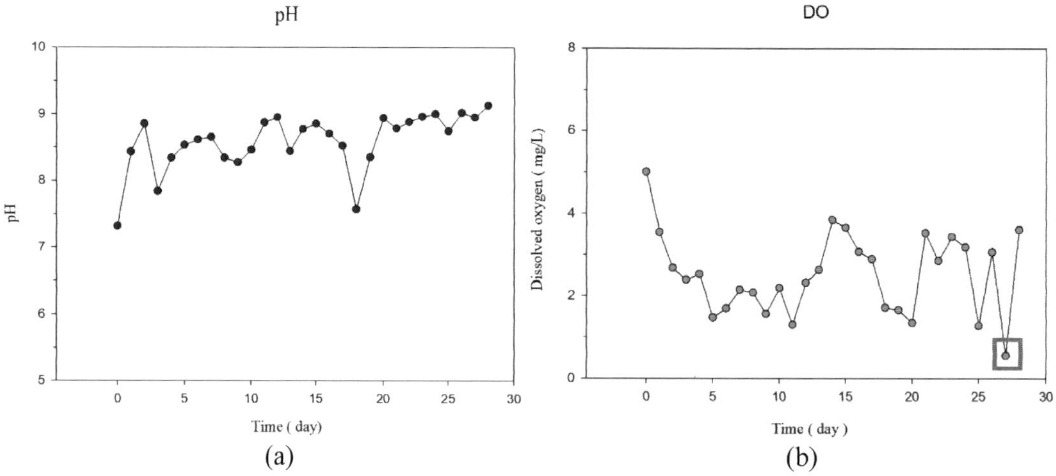

Figure 2. (a) pH value (b) DO concentration of AGS SBR in 28 days.

Figure 2a shows no significant fluctuation during the period of cultivation. The daily pH showed steady in 7~9. The pH value follows a previous study that pH must be 7.0 or more significant in the aerobic zone [15]. It also showed that the trendline went up gradually. Increasing pH in the aerobic system is caused by removing phosphate from the liquid phase, which can remove acidity from the medium [15].

Figure 2b shows that DO peaked on the 1st day (5.0 mg/L), followed by 1.0~4.0 mg/L until the 30th day. Low DO concentration caused the bloom of filamentous fungus [16] . On the 27th day, DO dropped to 0.55 mg/L. It detected a number of filaments in the effluent. On the 28th day, DO reached 3.6mg/L since most filaments got drained.

3.2 Weekly measurement- SVI and MLSS
The settling characteristics of activated sludge are commonly described using the sludge volume index (SVI) [16]. As shown in Figure 3, in the first week, SVI had obvious increase from 149.4 mg/l to 410.9 mg/l, which could represent the growth of filamentous fungus. From the 1st week to the 4th week, SVI

had decrease to 232.6 mg/l in the 4th week. It is shown that filamentous fungus disappears by the time and AGS growth successfully. According to earlier studies, a declining SVI can therefore only be attributed to the impact of operational conditions (i.e., intermittent feeding) and specific loading rate [17].

From the first to the second week, the MLSS declines, and then rises in the third week. Following that, in the following week, MLSS concentration dropped once more. The MLSS value fluctuated during the early stages of the process with short HRT. This is consistent with the earlier study [12]. However, after 20 days of operation, the MLSS value starts to gradually stabilize.

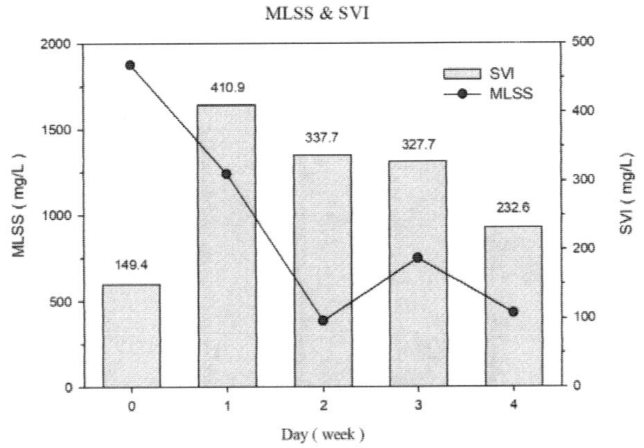

Figure 3. Concentration of MLSS and SVI of AGS SBR in 4 weeks

3.3 Sludge performance in reactor

Based on the Figure 4, the AGS changes over time based on the observation of the SBR reactor in the laboratory. In the first week, the reactor filled with the dark brown sludge from the treatment plant. The color turned darker in the second week. In this stage, sludge is more aggregated, causing to settle in the bottom of the reactor quickly. Compared to the initial state, the sludge color became lighter in the third week, and the structure became fluffy. In the final week, the color of the sludge is still light brown. And the sludge aggregates formed in this week.

The findings demonstrated that even the sludge aggregates formed in the 4th week but the particle is relatively small. Short HRT, on the other hand, causes late growth of microorganisms, resulting in small particles [18]. Reducing HRT gradually helped to build up AGS biomass. However, when the HRT was too short, it led to a rise in the organic load, making it difficult to build AGS biomass [12].

First week

Second week

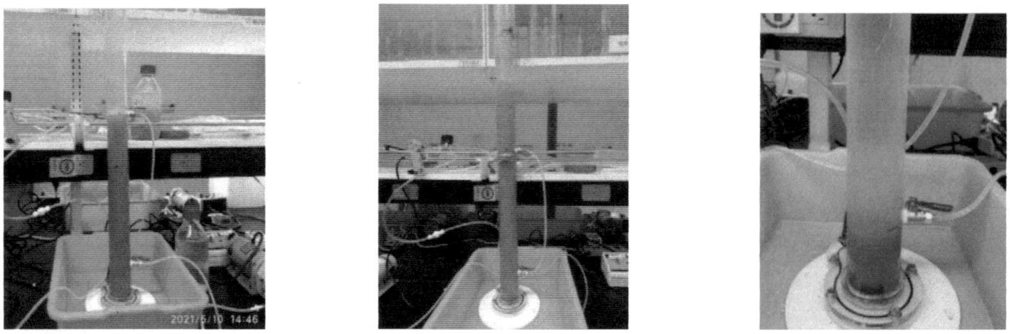

Third week

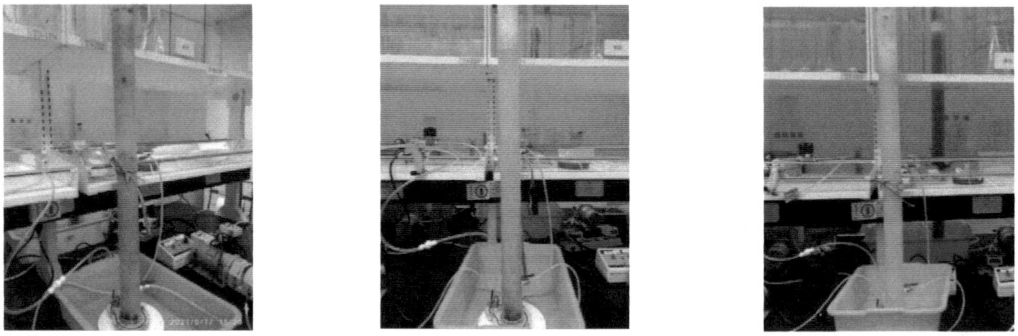

Fourth week

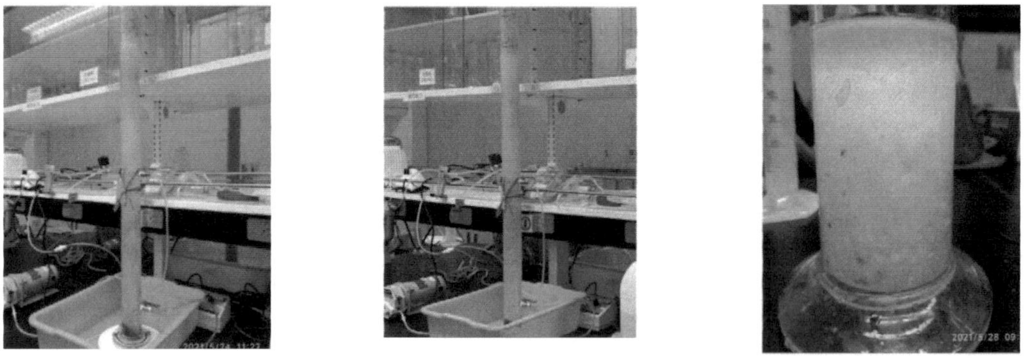

Figure 4. Sludge performance in SBR in 4 weeks

3.4 Sludge performance (under microscope)

According to Figure 5, in the first week, low F/M ratio and DO mainly result in filamentous fugus. Well-distributed around the reactor. Protists grew a lot and acted vivid of small sizes. We could observe protists eating and swimming in sludge. In the second week, we saw more filamentous fugus and clustered. So far, AGS was still undeveloped. Protists grew larger and seemed cultivated to certain groups. In the third week, filaments decreased drastically in the reactor. The majority of protists changed. Sludge started to get aggregated. However, it was still not significantly observed.

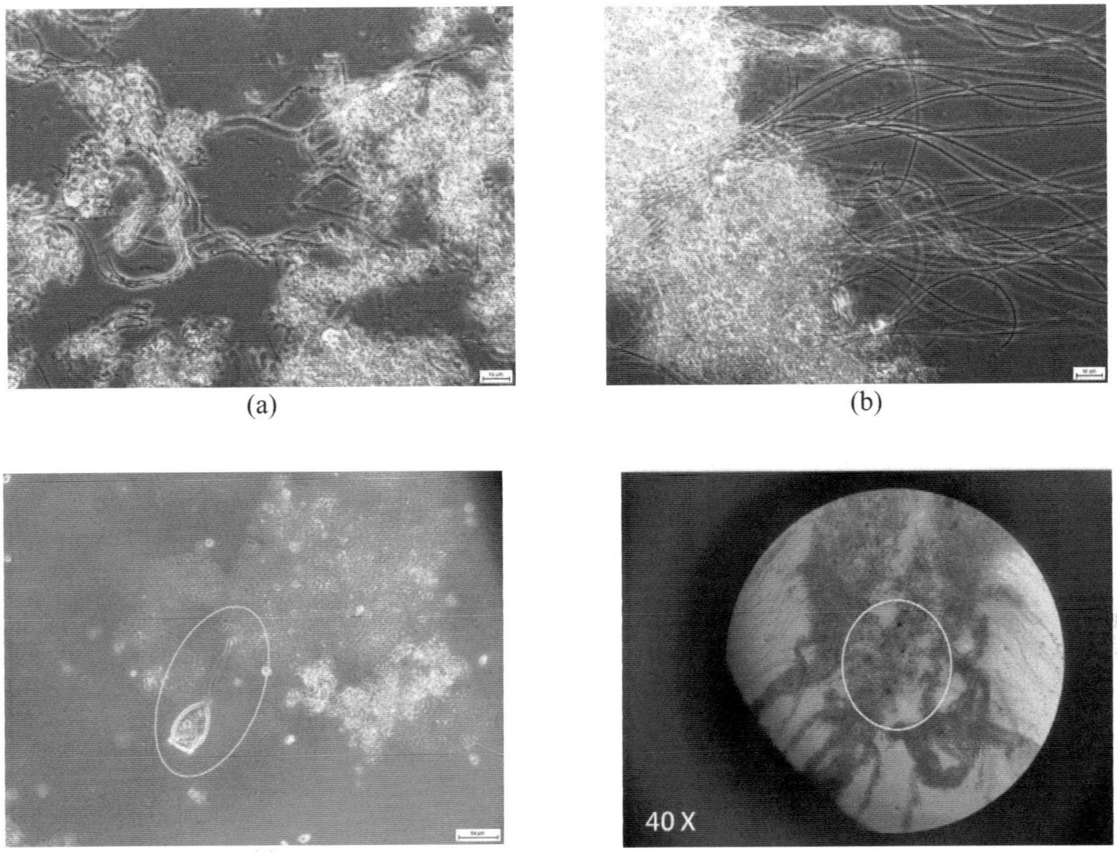

Figure 5. Sludge performance under microscope in (a) first week (b) second week (c) third week and (d) final day

4. Conclusion

Short cycle time improvement effectively enhanced AGS development. The settling time reorganization experienced a fault in the second week. As a result, some additional sludge in the effluent was lost, which could have caused the immature AGS. The sludge reactor needs more time to run in order to generate a well-developed AGS with strong capacity to breakdown organic molecules.

5. References

[1] Liu YQ, Tay JH 2007. Influence of cycle time on kinetic behaviors of steady-state aerobic granules in sequencing batch reactors. *Enzyme Microb Technol*. **41**:516–22.

[2] Liu Y, Tay JH 2004. State of the art of biogranulation technology for wastewater treatment. *Biotechnol Adv*. **22**:533–63.

[3] Lettinga G, F.M. van V, S.W. H, W.J. de Z, A. K 1980. Use of the upflow sludge blanket (USB) reactor concept for biological wastewater treatment. *Biotechnol Bioeng*. **22**:699–734.

[4] Morgenroth E, Sherden T, Van Loosdrecht MCM, Heijnen JJ, Wilderer PA 1997. Aerobic granular sludge in a sequencing batch reactor. *Water Res*. **31**:3191–4.

[5] Nancharaiah Y V., Kiran Kumar Reddy G 2018. Aerobic granular sludge technology: Mechanisms of granulation and biotechnological applications. *Bioresour Technol*. **247**:1128–43.

[6] Adav SS, Lee DJ, Show KY, Tay JH 2008. Aerobic granular sludge: Recent advances. *Biotechnol Adv*. **26**:411–23.

[7] Moy BYP, Tay JH, Toh SK, Liu Y, Tay STL 2002. High organic loading influences the physical characteristics of aerobic sludge granules. *Lett Appl Microbiol.* **34**:407–12.

[8] Tay J, Liu Q, Liu Y 2001. Microscopic observation of aerobic granulation in sequential aerobic sludge blanket reactor. *J. Appl. Microbiol*. **639798**:168–75.

[9] de Sousa Rollemberg SL, Mendes Barros AR, Milen Firmino PI, Bezerra dos Santos A 2018. Aerobic granular sludge: Cultivation parameters and removal mechanisms. *Bioresour Technol*. **270**:678–88.

[10] Schmidt JEE, Ahring BK 1994. Extracellular polymers in granular sludge from different upflow anaerobic sludge blanket (UASB) reactors. *Appl Microbiol Biotechnol*. **42**:457–62.

[11] Qin L, Liu Y, Tay J 2004. Effect of settling time on aerobic granulation in sequencing batch reactor. *Biochem. Eng. J*. **21**:47–52.

[12] Wang X, Li J, Zhang X, Chen Z, Shen J, Kang J 2021. Impact of hydraulic retention time on swine wastewater treatment by aerobic granular sludge sequencing batch reactor. *Environ Sci Pollut Res*. **28**:5927–37.

[13] Lotito AM, Fratino U, Mancini A, Bergna G, Di Iaconi C 2012. Effective aerobic granular sludge treatment of a real dyeing textile wastewater. *Int Biodeterior Biodegrad*. **69**:62–8.

[14] APHA. 2017. Standard Methods for the Examination of Water and Wastewater 23rd Ed. Baird RB, Eaton AD, Rice EW. *American Public Health Association.* Washington, DC, New York.

[15] Filipe CDM, Daigger GT, Grady CPL 2001. pH as a Key Factor in the Competition Between Glycogen-Accumulating Organisms and Phosphorus-Accumulating Organisms. *Water Environ Res*. **73**:223–32.

[16] Liu Y, Liu QS 2006. Causes and control of filamentous growth in aerobic granular sludge sequencing batch reactors. *Biotechnol Adv*. **24**:115–27.

[17] Chiesa SC, Irvine RL, Manning JF 1984. Feast / Famine Growth Environments and Activated Sludge Population Selection. *Biotechnol. Bioeng.* **27**: 562-569.

[18] Lan H, Chen Z, Chen Y 2004. Effect factors on formation and performance of aerobic granular sludge. *Eng Technol*. 28–31.

IOP Conf. Series: Earth and Environmental Science 1098 (2022) 012062 doi:10.1088/1755-1315/1098/1/012062

The Effect of Bioflocculants and Cationic Inducers Concentration on Harvesting *Chlorella vulgaris* Biomass Using Fungal *Aspergillus niger* Bioflocculation Method

L Isaroyati[1], H Hadiyanto[2*]

[1] Department of Chemical Engineering, Faculty of Engineering, Diponegoro University, Semarang, Central Java, 50275, Indonesia

[2] Department of Chemical Engineering, Diponegoro University, Jl. Prof. Soedarto, SH, Semarang, 50275, Indonesia

hadiyanto@live.undip.ac.id

Abstract. Energy consumption from non-renewable fuels is increasing, so there is a need for a new energy source from microalgae. The microalgae *Chlorella vulgaris* has a high lipid content ranging from 2-77% of the dry weight of the biomass, but it small size, ranging from 2-10 micrometers, which makes it uneconomical at the harvesting stage. One of the cheaper, effective, and environmentally-friendly harvesting methods is bioflocculation. The bioflocculation was carried out using the help of fungi by activating *Extracellular Polymeric Substances* (EPS) on positively charged fungi which would stick to negatively charged microalgae so that the charge became neutral and formed flocs. Previous research on the efficiency of harvesting bioflocculation of *Chlorella vulgaris* using *Aspergillus niger* pellets still takes a long time. Harvesting of *Chlorella vulgaris* with *Aspergillus niger* pellets was able to increase the efficiency of bioflocculation with the optimal parameters which were at pH 8, cationic inducer of 60 mM, bioflocculant dose of 3.5% w/v, and stirring speed of 150 rpm by 93% with a biomass of 333 mg within 48 hours. The bioflocculation harvesting method with fungi pellets can reduce time, save energy, increase harvesting efficiency, and not produce side products so harmful to the environment.

1. Introduction

Increased energy consumption from fuels such as fossil fuels, coal, and natural gas can have negative impacts such as climate change, ecological degradation, and global warming [1]; [2]. Therefore, there is a need for renewable energy sources such as biomass from microalgae [3];[4]. Microalgae is a renewable fuel source because it contains carbohydrates (11–56%), lipids (8–70%), protein (40–70%), and pigments (3–5%) higher than other biomass [1];[2]. *Chlorella vulgaris* is a microalga in the *Chlorellaceae* family that is dark green, spherical, cell diameter 2-10 m, unicellular eukaryotic [5];[6]. *Chlorella vulgaris* has cell growth and high lipid productivity as an alternative source for animal feed, dietary supplements, and pharmaceuticals [6]. However, the cell diameter of *Chlorella vulgaris* is very small and the low concentration in the culture media makes it difficult to cultivate at the harvesting stage [2];[7];[8]. The process of harvesting microalgae requires a high cost of about > 30% of the total production cost [3];[8]. The method of harvesting with a simple, cost-effective, and environmentally friendly process is bioflocculation. Bioflocculation is a method of harvesting microalgae using a flocculant of fungal microorganisms by activating Extracellular Polymeric Substances (EPS) [9]. *Aspergillus niger* contains a positively charged extracellular polymer (EPS) in the form of galactosaminogalactan (-COH) which is located in the cell wall of hyphal tissue. The fungal EPS will stick to the negatively charged microalgae so that there will be collisions between cells that can form flocs [9]. The floc will settle due to gravity [10]. EPS fungi have long chains and large molecular weights, so they can increase the sedimentation rate and density of microalgae [6];[11]. *Aspergillus niger* is a filamentous fungus capable of forming pellets due to granulation [12]. Pellet formation occurs through mycelium interaction by fungal hyphae branches [11]. The method of bioflocculation of *Chlorella vulgaris* with the help of *Aspergillus niger* pellets has the following advantages: the process is simple, low cost, non-toxic, environmentally friendly, easy to separate, and increases the mass transfer rate [13];[14];[11].

Several studies bioflocculation of the microalgae *Chlorella vulgaris* of pellets *Aspergillus niger* obtained a harvest efficiency of >80% for 72 hours [15];[16]. Based on this research, it was found that the efficiency of bioflocculation with pellet formation was quite good but it took longer. The purpose of this study was to increase the harvesting efficiency of *Chlorella vulgaris* using *Aspergillus niger* pellets with several parameters is the dose of fungal flocculant, and concentration of cationic inducer ($CaCl_2$).

2. Methodology

2.1. Chlorella vulgaris Cultures

Chlorella vulgaris seeds were obtained from UgoPlankton in Jepara, Central Java, Indonesia. *Chlorella vulgaris* culture was carried out at UPT Laboratory C-BIORE, Diponegoro University, Semarang, Indonesia. Cultivation was carried out in an aquarium with an aerator (BS-410, Amara, Shanghai, China) and an 8 W Philips lamp with a light intensity of 5000 lux (light/dark ratio of 8/16 hours) [17]. To maintain the growth of *Chlorella vulgaris* given a nutritional source, namely F_2 *Guillard* and Artificial Sea Water (ASW) salt 2 mL^{-1} every 1 week. Optical density was measured at a wavelength of 680 nm using a spectrophotometer (Spectroquant® Prove 100, Merck KGaA, Darmstadt, Germany).

2.2. Isolation of Aspergillus niger Fungal Pellets

Aspergillus niger culture was obtained from the Central Laboratory of Food and Nutrition Studies, Gajah Mada University, Yogyakarta, Indonesia. The manufacture of fungal pellets begins with filling 100 mL of bean sprout extract liquid media in an Erlenmeyer. The addition of the source of carbon nutrients ($CaCO_3$) and nitrogen (peptone) each 9 mL^{-1}, stir until homogeneous. *Aspergillus niger* cultures were inoculated in liquid media. The results of the liquid media culture were diluted, namely 1 x 10^4 spores L^{-1} using sterile distilled water. *Aspergillus niger* dilution sample was added to the bean sprout extract agar liquid medium. Stirring using a shaker (Digital Orbital Shaker Nesco Type SK-0330-PRO, China) for 48 hours, fungal pellets will form and can be stored in a 4°C refrigerator as stock.

2.3. Analysis of Bioflocculation Activity

Bioflocculation experiments were carried out on microalgae during the stationary growth phase. *Chlorella vulgaris* was bioflocculated at 100 mL with a concentration of 12 mL^{-1} ($OD_{680} = 2$). Several parameters for bioflocculation include the concentration of bioflocculants, cationic inducer ($CaCl_2$), and stirring speed. The optimal pH setting was 8 using 0.1 N HCl and 0.1 M NaOH. The addition of a cationic inducer, $CaCl_2$ of 40, 50, 60, 70 mM, stirred until homogeneous. Fungal pellets 1.5, 2.5, 3.5, 4.5 % w/v were added and washed with sterile distilled water. The culture was stirred at 150 rpm with an orbital shaker (Digital Orbital Shaker Nesco Type SK-0330-PRO, China) for 48 hours. Optical density at 680 nm (OD_{680}) was measured using a spectrophotometer (Spectroquant® Prove 100, Merck KGaA, Darmstadt, Germany). Microalgae will be trapped in fungal pellets which can be separated by filtering. The concentration of microalgae biomass was determined gravimetrically by drying the sample in a dish for 24 hours at a temperature of 50°C until it was constant. The dry microalgae biomass can be calculated using the final cup weight. The bioflocculation efficiency equation is shown as follows [15];[18] :

$$\% \, Biofloculation \, Efficiency = \left(1 - \frac{OD_t}{OD_0}\right) \times 100\% \qquad (1)$$

FE = bioflocculation efficiency (%), OD_t = final optical density, OD_0 = initial optical density. The yield of biomass can be calculated using the following equation :

$$Biomass = Weight \, of \, microalgae \, pellet \, dish - Weight \, of \, empty \, dish \qquad (2)$$

3. Result and discussion

3.1. Effect of Bioflocculation on Bioflocculant Dosage

Figure 1. shows that the highest bioflocculation activity is the bioflocculant dose of 3.5 %w/v of 96% with a biomass of 332.8 mg. However, the high dose of 4.5 %w/v bioflocculant resulted in a 91% decrease in bioflocculant efficiency with 188.2 mg of biomass. The low bioflocculant dose of 1.5 ; 4.5 %w/v also affected the bioflocculation activity of 89 ; 91%. This is because low or high doses of fungal bioflocculants affect bioflocculation due to electrostatic repulsion, so it is necessary to balance the charge of microalgae and fungal pellets [15]. Electrostatic repulsion can be prevented by the presence of extracellular compounds from *Aspergillus niger* pellets. A previous study investigating various fungal inoculation doses of 30 mg/L found an optimal harvest efficiency of 97% [19]. Research on the bioflocculation efficiency of *Spirulina sp* with fungal pellets 5% and 7.5% w/v can cause a decrease in bioflocculation activity [20].

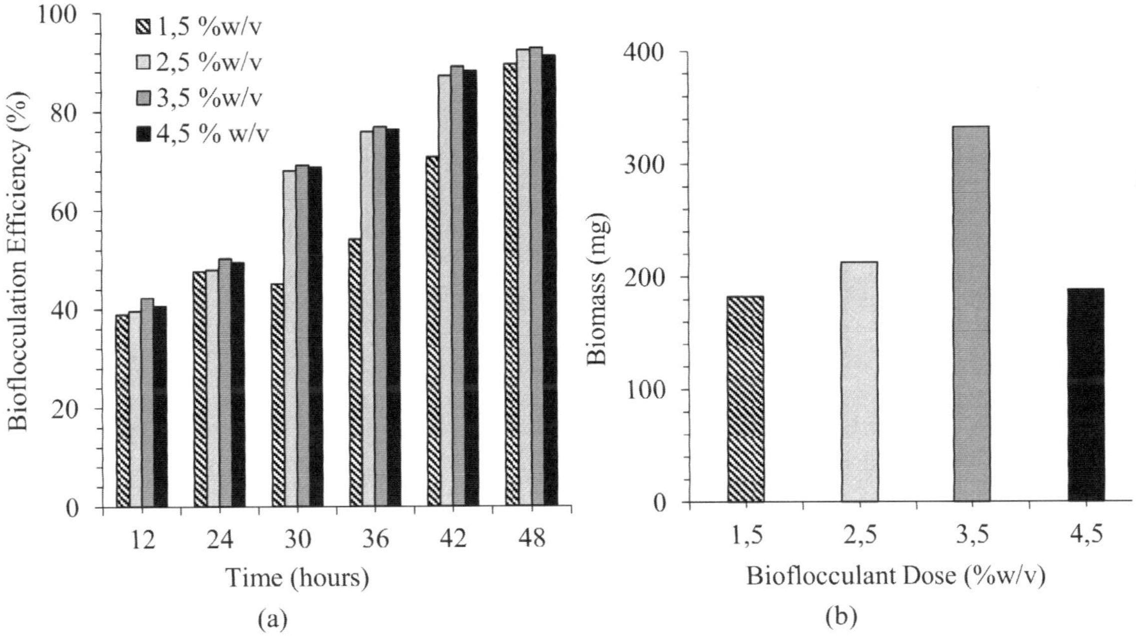

Figure 1. Effect of Bioflocculant Dosage on a) Bioflocculation Efficiency b) *Chlorella vulgaris* Biomass

Aspergillus niger pellet size with a diameter of 1-2 mm and an inoculation concentration of 1 x 10^4 spores/mL affected the absorption of fungal pellets against microalgae. Smaller inoculum concentrations can result in smaller pellet sizes due to the large surface area so that the adsorption of microalgae into the pellets increases [14]. Several studies of microalgae bioflocculation with pellets of *Aspergillus oryzae* UMN F07 1.0×10^4 spores/mL resulted in smaller and more stable pellets that were able to increase bioflocculation activity [14]. The bioflocculation activity of microalgae with pellet concentrations greater than 2.0×10^7 spores/mL prevented the microalgae cells from being completely trapped in the fungal pellet [13]. Based on Figure 2. *Aspergillus niger* pellets were marked with white color, round in shape, and there were no microalgae cells trapped in the pellet. The bioflocculated *Aspergillus niger* pellets changed color from white to green. This indicates that the microalgae have been associated with *Aspergillus niger* pellets. Microalgae from bioflocculation can be harvested by simple filtering because the size of the fungal pellets is relatively large, which is 2-5 mm, and is denser. Filtration of microalgae pellets can use a mesh sieve with a larger size than the US standard sieve, namely No. 18. The simple filtering process does not require large energy requirements, so it is more effective and environmentally friendly.

Aspergillus niger pellets have polysaccharide extracellular substances with long chains and large molecules, so they have strong adsorption bonds [10]. The extracellular component of *Aspergillus niger* functions as an adhesive that has a positive charge in the presence of a polysaccharide hydroxyl group (glucan, glycoprotein, chitin, and melanin) [10];[21]. The hydroxyl cations from the fungus will absorb the negative charge from around the microalgae surface so that it has a bridging effect [11]. The positive charge of the fungal pellet will attract negative microalgae ions by coagulation so that the charge undergoes a neutralization process [14];[11]. The neutralization process will eliminate the electrostatic repulsion between the particles, then they will agglomerate to form flocs [14]. Therefore, the dose of bioflocculant can affect the bioflocculation of microalgae by fungal packets.

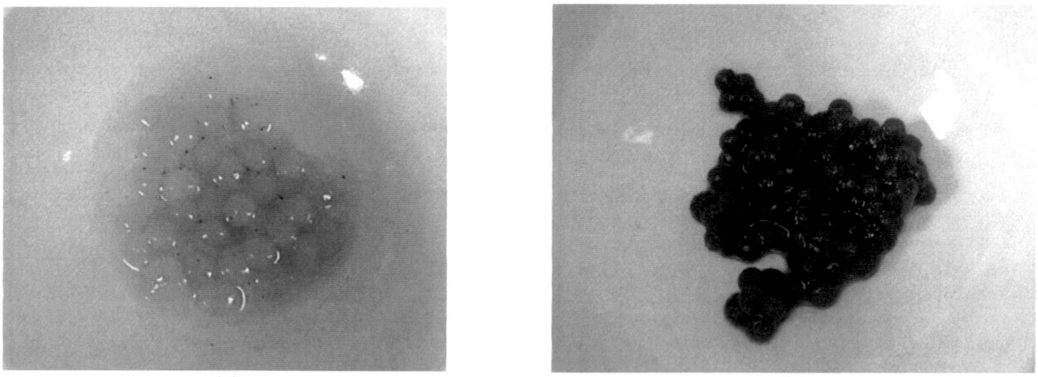

(a) (b)

Figure 2. *Aspergillus niger* Pellets a) Before b) After Bioflocculation

3.2. Effect of Bioflocculation on the Cationic Inducers Concentration
Cationic inducers as coagulants can help increase the efficiency of microalgae bioflocculation [1]. The effect of cationic inducers on the bioflocculation of microalgae with different concentrations added to the microalgae culture along with the fungal pellet is shown in Figure 3.

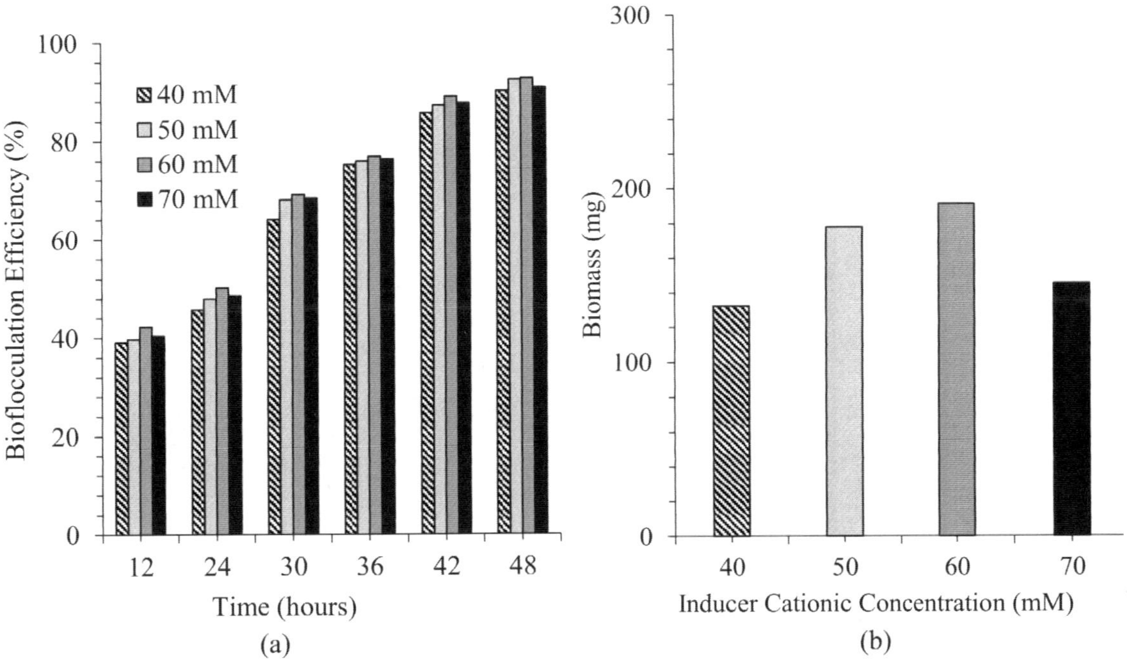

Figure 3. Effect of Cationic Inducers on a) Bioflocculation Efficiency b) *Chlorella vulgaris* Biomass

Based on Figure 3. shows the addition of a cationic inducer of 60 mM resulted in the highest bioflocculation activity of 93% with biomass of 191 mg. The addition of a cationic inducer concentration of 40 mM had a bioflocculation activity that was not optimal, namely 90% with a biomass of 132 mg. This is because the cationic distribution of *Aspergillus niger* has not been balanced to meet the anionic needs of microalgae. Therefore, microalgae were still scattered in the culture and had not been completely trapped in the fungal pellet. Increasing the concentration of cationic inducers of 50 and 60 mM did not significantly increase the bioflocculation activity of 92 : 93% and the yield of microalgae biomass of 178 : 191 mg. The ion neutralization process that occurs has almost fulfilled the cationic needs of microalgae by fungal pellets. However, an increase in the concentration of cationic inducer by 70 mM resulted in a decrease in the bioflocculation efficiency. Excess cationic fungal pellets trapped in microalgae cells resulted in electrostatic repulsion. In addition, the addition of a cationic inducer that is too high can cause turbidity of the culture.

The concentration of cationic inducer added was too high and too low affecting the bioflocculation process. This is because the metal ions that are distributed too high cause the electrostatic repulsion to increase. In addition, Ca^{2+} ions from $CaCl_2$ that are too high can cause turbidity in microalgae cultures [22]. The cationic inducer functions as an ion-binding agent helping the charge neutralization process [23]. Divalent cationic Ca^{2+} and Cu^{2+} have stronger anionic ion absorption from microalgae than monovalent K^+ and Na^+ [23]. Performed bioflocculation of microalgae by *Bacillus sp* RP1137 with the help of calcium-binding assay and showed that the added ions affected the aggregation process in neutralizing the surface charge of microalgae [24]. Bioflocculation of *Chlorella vulgaris* with *Aspergillus niger* using the addition of 5 mM $CaCl_2$ ion was able to increase the bioflocculation activity up to 90% [23].

Bioflocculation occurred neutralization of cationic hydroxyl from fungal pellets distributed to the anionic surface of microalgae [11];[10]. However, the process of neutralization by uneven ions requires a charge patch on the surface of the microalgae. The uneven distribution of ions resulted in an unbalanced electrostatic repulsion between the surface of the microorganism and the fungal pellet, so the bioflocculation activity decreased [10]. The $CaCl_2$ ion helps the process of attracting the charge of negatively charged microalgae which can meet the cationic needs of microalgae [15]. The addition of these ions helps reduce electrostatic repulsion so that the bioflocculation activity increases.

4. Conclusion

Chlorella vulgaris bioflocculation with the help of *Aspergillus niger* fungal pellet bioflocculant was able to increase the efficiency by 93% with a biomass of 333 mg within 48 hours. Optimum bioflocculation efficiency with several parameters, namely the bioflocculant dose of 3.5% w/v, and the cationic concentration of the inducer was 60 mM with a stirring speed of 150 rpm. From the parameters carried out in the study, too low and too high did not affect the bioflocculation activity because it could lead to the stability of the cationic requirements of microalgae cells. The harvesting method of microalgae bioflocculation is feasible to be developed for sustainable cultivation on a large scale because it is efficient and environmentally friendly.

Acknowledgement

The authors would like to thank the staff of the Laboratory of Biomass and Renewable Energy Group (C-BIORE), Diponegoro University, Semarang, Indonesia for the funding assistance to carry out the current research.

References

[1] Vasistha S, Khanra A, Clifford M, and Rai M P 2020 Current advances in microalgae harvesting and lipid extraction processes for improved biodiesel production : A review *Renewable and Sustainable Energy Reviews* 110498

[2] Li S, Hu T, Xu Y, Wang J, Chu R, Yin Z, Mo F, and Zhu L 2020 A review on flocculation as an efficient method to harvest energy microalgae : Mechanisms, performances, influencing factors and perspectives *Renewable and Sustainable Energy Reviews* **131** 110005

[3] Gul M, Alam M A, Mofijur M, Jahirul M, Yongkun, Wenlong X, Hwal C O, and Jingliang X 2020 Modern developmental aspects in the field of economical harvesting and biodiesel production from microalgae biomass *Renewable and Sustainable Energy Reviews* **135** 110209

[4] Abraham F, Moreno-Garcia, Neri-Torres E E, Mena-Cervantes, Violeta Y, Altamirano R H, Pineda-Flores G, Luna-Sanchez R, Garcia-Solares M, Vazquez-arenas J, and Suaste-Rivas J K 2021 Sustainable biorefinery associated with wastewater treatment of Cr (III) using a native microalgae consortium *Fuel* 119040

[5] Wa I, Citra U, and Abdul M B, 2019 The Growth of Chlorella vulgaris Cultured in Liquid Organic Fertilizer of Water Hyacint H (Eichhornia crassipes) at Different Salinities *Aquaculture Indonesiana* **20(2)** 117–126

[6] Dong T, Nguyen P, Thi V A L, Pau L S, Thanh T N, Minh H T, Thi N T, and Sze Y L 2018 Bioflocculation formation of microalgae-bacteria in enhancing microalgae harvesting and nutrient removal from wastewater *Bioresource Technology*

[7] Luo S, Wu, Jiang H, Yu M, Liu Y, Min A, Li W, and Ruan R J 2019 Edible Fungi-Assisted Harvesting System for Efficient Microalgae Bio-flocculation *Bioresource Technology* **282** 325-330

[8] Wang S, Yerkebulan M, Abomohra A E, El-Khoday S, and Wang Q 2019 Bioresource Technology Microalgae harvest influences the energy recovery : A case study on chemical flocculation of Scenedesmus obliquus for biodiesel and crude bio-oil production *Bioresource Technology* **286** 121371

[9] Yi L, Zhiyu Z, Yingxing D, and Hailei W 2019 The effect of recycling culture medium after harvesting of Chlorella vulgaris biomass by fl occulating bacteria on microalgal growth and the functionary mechanism *Bioresource Technology* **280** 188–198

[10] Li S, Hu T, Xu Y, Wang J, Chu R, Yi, Z, Mo F, and Zhu L 2020 A comprehensive review on cultivation and harvesting of microalgae for biodiesel production : Environmental pollution control and future directions *Bioresource Technology* **301** 122804

[11] Chu R, Li S, Yin Z, Hu D, Zhang L, Xiang M, and Zhu L 2021 Science of the Total Environment A fungal immobilization technique for ef fi cient harvesting of oleaginous microalgae : Key parameter optimization , mechanism exploration and spent medium recycling *Science of the Total Environment* **790** 148174

[12] Zhang Z, Gao P, Guo L, Wang Y, She Z, Gao M, Zhao Y, Jin C, and Wang G 2020 Elucidating temperature on mixotrophic cultivation of a Chlorella vulgaris strain: different carbon source application and enzyme activity revelation *Bioresource Technology* 123721

[13] Gultom S O, Zamalloa C, and Hu B 2018 Microalgae Harvest through Fungal Pelletization—Co-Culture of Chlorella vulgaris and Aspergillus niger *Energies Springer Switzerland*

[14] Zhao Y, Guo G, Sun S, Hu C, and Liu J 2019 Co-pelletization of microalgae and fungi for efficient nutrient purification and biogas upgrading *Bioresource Technology* **289**

[15] Mohd N, Yunos F H M, Jusoh H H W, Lam A, and Juson A 2019 Subtopic : Advances in water and wastewater treatment harvesting of Chlorella sp . microalgae using Aspergillus niger as bio-fl occulant for aquaculture wastewater treatment *Journal of Environmental Management* **249** 109373

[16] Bhattacharya A, Mathur M, Kumar P, Kumar S, and Malik A 2017 A rapid method for fungal assisted algal flocculation : Critical parameters & mechanism insights *Algal* **21** 42–51

[17] Irene T, Ru K, Yeong Y K, Malinna J, Mohd E A W, and Thilahgavani N 2020 Chlorella vulgaris : a perspective on its potential for combining high biomass with high value bioproducts *Applied Phycol*ogy 1–10

[18] Chen J, Leng L, Ye C, Lu Q, Addy, Min, Wang J, Liu J, Chen P, Ruan R, and Zhou W 2018 A comparative study between fungal pellet- and spore-assisted microalgae harvesting methods for algae bioflocculation *Bioresource Technology* **259** 181–190

[19] Nasir M N, Bakar N S A, Lananan F, Hamid S H A, Lam S S L, and Jusoh A 2015 Treatment of African catfish, Clarias gariepinus wastewater utilizing phytore-mediation of microalgae Chlorella sp. with Aspergillus niger bio-harvesting *Bioresource Technology*

[20] Mateus T N, Cesar V T R, Alan R, and Luciane M C 2021 Harvesting of Spirulina platensis using an eco-friendly fungal bioflocculant produced from agro-industrial by-products *Bioresource Technology* **322**

[21] Akira Y, Ken M, and Keietsu A 2017 Cell wall structure and biogenesis in Aspergillus species *Bioscience, Biothecnology, and Biochemistry*

[22] Hadiyanto H, Widayat W, Marcelinus C, and Monica E P 2022 Current Research in Green and Sustainable Chemistry The fl occulation process of Chlorella sp . using chitosan as a bio- fl occulant : Optimization of operating conditions by response surface methodology *Current Research in Green and Sustainable Chemistry* **5** 100291

[23] Li Y, Xu Y, Liu L, Li P, Yan Y, Chen T, Zheng T, and Wang H 2017 Flocculation mechanism of Aspergillus niger on harvesting of Chlorella vulgaris biomass **25** 402–412

[24] Ryan J, Powell, and Ruseell T H 2014 Mechanism of Algal Aggregation by Bacillus sp Strain RP1137 *Applied and Environmental Microbiology* **80(13)** 4042-4050

Characteristics of Banana Leaves as Gaseous Biosorbent

Rahmatia Sarah Wahyudi[1], Haryono Setiyo Huboyo[1*], Endro Sutrisno[1], Badrus Zaman[1]

[1]Environmental Engineering Department, Faculty of Engineering, Diponegoro University, Semarang-Indonesia

*E-mail: haryono.setiyo.huboyo@ft.undip.ac.id

Abstract. The quality of air is vital for our existence. Increasing industrial development and household activities will increase the risk of air pollution and environmental damage. Some air pollution in ambient air quality is very toxic and lethal. One method that can remove gaseous pollutants is adsorption. Previous study about banana plant as adsorbent of gaseous pollutant, banana leaves has the highest adsorption efficiency value of 76.52%. Purpose of this study to discuss characteristics and potential of banana leaves as gaseous adsorbent. This research was carried out by studying literature and testing the Scanning Electron Microscope (SEM) of the banana leaf surface morphology. The structural properties of adsorbent were characterized, which exhibited the adsorbent's porous surface or stomata. There are several studies of the SEM analysis results that explain stomata are able to trap air pollution. Banana leaves contain lignin, cellulose, and cuticle which have been found in several studies to be useful as gas adsorbents. Based on these studies, it can be concluded that banana leaves have potential as an adsorbent of gaseous pollutants.

1. Introduction

The quality of air is vital for our existence. Increasing industrial development and household activities will increase the risk of air pollution and environmental damage [1][2][3]. This can result in damage to the basic structure and function of the ecosystem. Industrial development can pollute through technological advances such as motor vehicles [4], electronic devices [5][6], and various other uses. Henceforth, household air pollution from the use of industrial goods and improper waste disposal changes in ambient air quality usually include parameters such as NH_3, O_3, CO, NO_2, SO_2, H_2S, and hydrocarbons [7][8]. If there is an increase in the level of these materials in the ambient air, which exceeds the specified quality standard, it can cause health problems.

Some air pollution in ambient air quality is very toxic and lethal[9], some workers in oil and gas drilling are at great risk from the release of this gas[10]. Then hydrogen sulfide gas effluent is also found in fertilizer production[11] or sewers[12]. H_2S comes from industrial wastes that undergo a bacterial decomposition process under anaerobic conditions. A location such as a pond or pond with mud contaminated with H_2S is characterized by a characteristic odor[13][14].

Removal of gaseous pollutant can be done by making pyrolysis gas[15], desulfurization mechanism[16], and using fenton reagent[17]. One method that also can be used to overcome the dangers of gaseous is adsorption [18][19][20]. Many materials can be used to test the adsorption capacity, such as polymer adsorbents [21], natural adsorbents (zeolite, clay minerals, oxides) [22], and activated carbon [23]. Despite their reasonable performance, the range of applications especially on an industrial scale is limited due to the cost of the adsorbent. Scientific community is conducting research to develop a low cost, environmentally friendly adsorbent material that can be used on an industrial scale [24].

Hence, this research using agricultural by-products for the development of low-cost adsorbent material. Banana plants (Musa sp.) are the easiest plants to find and breed in tropical areas such as Indonesia [25]. Banana plant waste has not received serious attention to convert it into adsorbent. However, large amounts of it are generated for multiple uses. Musa species is one of the well known plants of the musaeae family that have been used as craft materials [26], packaging materials [27], traditional medicine [28], ingredients in food industry [29], and potential biosorbent in adsorption process [30]. Banana plant as adsorbent of hydrogen sulfide adsorption [31], results of adsorption efficiency found 6.44% on banana peel, 51.83% on its stem, and 76.52% on banana leaves. [32]. According to previous study about banana plant as adsorbent of gaseous hydrogen sulfide (H_2S), banana leaves has the highest adsorption efficiency value of 76.52%. Then in this research continue on the previous study on why banana leaves can be an effective gas adsorbent through its physical and chemical characteristics.

2. Research Methodology

The method of writing is literature study. Various literatures on gas adsorption by biosorbent and the characteristics of banana leaves which will be the adsorbent were collected. The information obtained is compiled based on the results of the study obtained. This research was carried out in 2022. Several stages were required in this research, including literature study, data collection as well as identification, and laboratory tests located at one of the campuses in Indonesia, namely Diponegoro University.

The literature study stage in this research is used to determine the variety of banana plant varieties in Indonesia where these plants can be biosorbents of gaseous pollutants, the impact of plants on the environment, how banana leaves are prepared as biosorbent, as well as physical and chemical characteristics of banana leaves as biosorbents, in addition to find out the techniques or steps needed to make a potential gas adsorbent.

The physio-chemical characteristics were determined in this study through literature study and SEM analysis. The Scanning Electron Microscope (SEM) analysis of banana leaves were studied to investigate its morphology characteristics. The banana leaf used in the SEM test in this study came from the banana plant, which scientific name is *Musa acuminata.* Preparation of banana leaves before the SEM test, was cut to a maximum size of 1 cm and dried at 60°C for 1 day. Drying is not carried out at high temperatures to keep the leaf tissue from being damaged [33].

3. Result and Discussion

3.1 Varietes of Banana Plant in Indonesia

Bananas are harvested at the age of 12-15 months after planting or 4-6 months after flowering, depending on the type of banana [34]. There are several types of bananas that have a short harvest life [35]. However, there are also several other types that have a longer harvest life [36]. In contrast to other annual fruit plants, banana plants only bear fruit once and after fruiting they will die. Therefore, banana plants that have been harvested for fruit must be immediately dismantled and replaced with new banana plants or keep the growing saplings [37].

Due to the relatively short life span of banana plants, utilization of those that are no longer able to grow and develop is carried out. The leaves of the banana plant that will be used are from plants that have been harvested so that the waste can be utilized. Some of the banana plants whose leaves possible to be used as because of its existence are taken from Table 3 below.

Table 1. Varietes of Banana Plants in Indonesia

No.	Indonesian Name	Scientific Name
1.	Pisang Raja	*Musa textilia*
2.	Selayar/ Pisang Raja Pulut	*Musa acuminata* Var. Colla
3.	Pisang Burung	*Musa acuminata*
4.	Pisang Barangan Kuning	*Musa acuminata* Var. Linn
5.	Pisang Ambon	*Musa paradisiaca* Var. Sapientum
6.	Pisang Batu	*Musa brachycarpa*
7.	Pisang Timbago/Merah	*Musa velutina*
8.	Pisang Raja Bulu	*Musa textilia* Var. Grindy
9.	Pisang Raja Sereh/ Pisang Susu	*Musa acuminata* Var. Colla
10.	Pisang Tanduk/Pisang Gebyar	*Musa acuminata* Var. Eumusa
11.	Pisang Kepok Putih	*Musa paradisiaca* forma typica
12.	Pisang Kepok	*Musa paradisiaca* Var. Balbisina Colla.
13.	Pisang Raja Nangka	*Musa paradisiaca linn*
14.	Pisang Tonto	*Musa acuminata*

Source: [38][39][40][41]

3.2 Plant Leaves on The Environment

Air pollution can be overcome or combated by plants in different ways [42]. During photosynthetic gas exchange, the stomata on plant leaves can be penetrated by air pollution. In addition, the deposition of these pollutants on the surface of the leaf cuticle can also occur or be taken directly by the roots to the soil [43]. Gaseous pollutants by stomata can be controlled and limited leaf gas exchange, thus forming the first line of defense [44].

Saranya 2016 found potential low-cost adsorbent for hexavalent chromium removal using *Melaleuca diosmifolia* leaves. This natural adsorbent proved to be a useful and valuable means for controlling pollution caused by toxic metal ions. Total 3.02 and 3.95 mg Cr sorbed by 100 mg leaf powder [45]. Adsorption of heavy metals onto plant leaves derived adsorbent occurs on a monolayer with each molecule attaching itself to a specific noncompeting site on the adsorbent either in pores, functional complexes or interstitials in the microbiomass matrix.

3.3 Banana Leaves into Biosorbent

A simple technique is required in the preparation of biosorbents or adsorbents from plant leaves. The biosorbents can be made to undergo modification or used as is. In further modification, the biosorbent may improve its functional properties. The techniques can be summarised into three broad spectrums: [46]

1. To produce unmodified biosorbents, the initial steps are cleaning, drying, and grinding.
2. Production of biosorbents by further chemical and/or physical modification.
3. To produce activated carbon (AC), chemical activation and carbonization of biosorbent.
   A high specific surface area in the range of 798.51-1466.7 m^2/g and an average pore size in the range of 2-3 nm are resulted from the conversion of banana leaf waste into activated carbon through chemical activation [47].

The next preparation of plant leaf biosorbent is the reduction of moisture through drying. Three popular techniques are used by researchers are oven drying, room temperature drying, and sun drying or it can be a combination from those three methods [46]. Sieving and grinding are the third major step in the preparation. Upon grinding, the biosorbents are sieved to particle sizes through a variety of mesh sizes [46].

3.4 *Physical Characteristics of Banana Leaves as Biosorbent*

(a) (b)

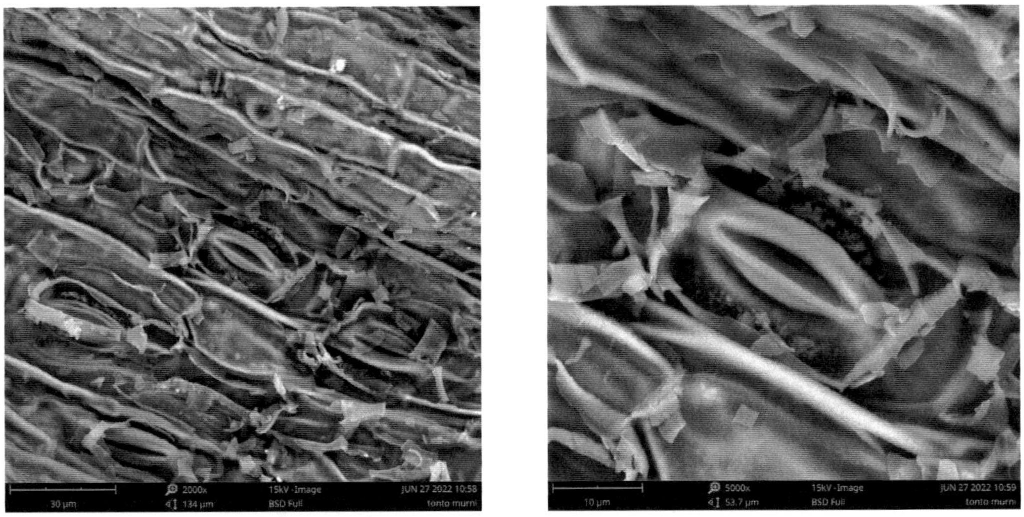

Figure 1. Scanning Electron Microscope (SEM) of banana leaves before adsorption (a) Magnification 2000 times (b) Magnification 5000 times

The results of the characterization by Scanning Electron Microscope (SEM) on Figure1 showed that the morphology of the banana leaf. The structural properties of adsorbent were characterized using SEM micrographs, which exhibited the adsorbent's porous surface or stomata. Gaps, slits or channels can be seen as well as get the useful information about the pore structure by SEM image [48].

To compare this finding with several studies using leaves as adsorbents that have similar surface morphology. There was a study by Wenxia 2022, where The SEM image with 300x of magnification revealed that the leaf surface of *C. lanceolata* was covered by waxy reticulate structure, and most smoke PM was blocked by reticulate structure [49].

There are several studies of the SEM analysis results that explain stomata are able to trap air pollution. For example, The magnification 600x SEM leaf images showed stomata were 10 μm long inside clearly indicating that Pb entered the Chinese cabbage leaves [50]. The surface of tea leaves, study by Nur E Alam 2018, was found smooth and with uniform microporous structure and after adsorption the Cr molecules cover the most of the available pores present in spent leaves adsorbent causing the surface to become saturated [51]. Several large pores in a honeycomb shape were found on the surface of *Moringa oleifera* leaf, these pores provided a good surface for dyes, heavy metals and waste effluents to be trapped and adsorbed into [52]. From several studies that have been passed regarding the results of SEM photograph on the leaves, banana leaves can have the potential to become an adsorbent of pollutant.

3.5. *Chemical Characteristics of Banana Leaves as Biosorbent*

Banana leaf waste has a relatively high content of lignocellulose [53]. Lignin is a natural O-containing aromatic amorphous polymers from lignocellulosic biomass, and has abundant O-functional groups including Ar-OH, –OH, $-O-CH_3$, and even –C– –O, etc. [54]. These features enable them to be used as dispersants[55], adsorbents[56], adhesives[57], coatings[58], and cosmetics after chemical modification [59]. As the potential adsorbents, lignin and its derivatives have lots of advantages such as low-cost, allowable flexible modification, and extensive resources [60]. Various forms of carbon materials such as carbon fiber and concrete can be found in lignin content, however for adsorption application, lignin-derived activated carbons are by far the most common [61].

In recent years, adsorbents derived from cellulose earogels have been studied more extensively. In terms of oil absorption, the researchers through hydrophobizing ultra-lightweight and highly porous cellulose aerogels obtained a highly oil-absorbing material, and a wide range of organic solvents and oils can be collected with absorption capacities up to many times their own weight [62][63]. Recent studies have shown that nanofabrillated cellulose sponges could absorb up to 45 times their own weight in organic solvent or oil, depending on the liquid [64]. Cellulose aerogels modifed by cross-linking with PEI, its composite has a great CO_2 adsorption capacity [65][66]. Cellulose has great potential as a bio-adsorben based on the findings.

Banana plants have various cuticle thicknesses [67]. Although the plants were grown under dark conditions, the plants could still grow and uptake benzene only through cuticular wax because the stomata was closed under dark conditions [68][69]. Benzene concentration in the atmosphere was found to be higher than the local standard guidelines in many places around the world [70].

4. Conclusion

Some of the physico-chemical characteristics of banana leaves as biosorbent for gaseous pollutant potentially related with the adsorption process. The results of the characterization using Scanning Electron Microscope (SEM) showed that the morphology of the banana leaf has porous surface. This morphology through some studies could trap pollution inside its pores or stomata. The chemical characteristics of banana leaves contain lignin, cellulose, and cuticular wax. This content can be used as a material in the manufacture of gas adsorbents according to several studies.

Acknowledgement

The authors acknowledge Faculty of Engineering Diponegoro University for providing research funding for this research work through contract number 356/UN7.5.3.2/HK/022.

References

[1] M. Sarmento, D. Durão, and M. Duarte, "Study of enviromental sustainability: The case of Portuguese polluting industries," *Energy*, vol. 30, no. 8 SPEC. ISS., pp. 1247–1257, 2005, doi: 10.1016/j.energy.2004.02.006.

[2] M. Ahmed, C. Shuai, K. Abbas, F. U. Rehman, and W. M. Khoso, "Investigating health impacts of household air pollution on woman's pregnancy and sterilization: Empirical evidence from Pakistan, India, and Bangladesh," *Energy*, vol. 247, p. 123562, 2022, doi: 10.1016/j.energy.2022.123562.

[3] Z. Wang, W. Wei, and F. Zheng, "Effects of industrial air pollution on the technical efficiency of agricultural production: Evidence from China," *Environ. Impact Assess. Rev.*, vol. 83, no. May 2019, p. 106407, 2020, doi: 10.1016/j.eiar.2020.106407.

[4] C. Sun, S. Xu, M. Yang, and X. Gong, "Urban traffic regulation and air pollution: A case study of urban motor vehicle restriction policy," *Energy Policy*, vol. 163, no. February, p. 112819, 2022, doi: 10.1016/j.enpol.2022.112819.

[5] C. Protano, M. Manigrasso, P. Avino, and M. Vitali, "Second-hand smoke generated by combustion and electronic smoking devices used in real scenarios: Ultrafine particle pollution and age-related dose assessment," *Environ. Int.*, vol. 107, no. May, pp. 190–195, 2017, doi: 10.1016/j.envint.2017.07.014.

[6] M. Elheddad, C. Benjasak, R. Deljavan, M. Alharthi, and J. M. Almabrok, "The effect of the Fourth Industrial Revolution on the environment: The relationship between electronic finance and pollution in OECD countries," *Technol. Forecast. Soc. Change*, vol. 163, no. December 2020, p. 120485, 2021, doi: 10.1016/j.techfore.2020.120485.

[7] S. L. Paralovo *et al.*, "Observations of particulate matter, NO2, SO2, O3, H2S and selected VOCs at a semi-urban environment in the Amazon region," *Sci. Total Environ.*, vol. 650, no. 2, pp. 996–1006, 2019, doi: 10.1016/j.scitotenv.2018.09.073.

[8] A. Kerimray, A. Bakdolotov, Y. Sarbassov, V. Inglezakis, and S. Poulopoulos, "Air pollution in

Astana: Analysis of recent trends and air quality monitoring system," *Mater. Today Proc.*, vol. 5, no. 11, pp. 22749–22758, 2018, doi: 10.1016/j.matpr.2018.07.086.

[9] L. Aventaggiato, A. P. Colucci, G. Strisciullo, F. Favalli, and R. Gagliano-Candela, "Lethal Hydrogen Sulfide poisoning in open space: An atypical case of asphyxiation of two workers," *Forensic Sci. Int.*, vol. 308, p. 110122, 2020, doi: 10.1016/j.forsciint.2019.110122.

[10] T. Xu, L. Wang, X. Wang, T. Li, and X. Zhan, "Heavy metal pollution of oil-based drill cuttings at a shale gas drilling field in Chongqing, China: A human health risk assessment for the workers," *Ecotoxicol. Environ. Saf.*, vol. 165, no. May, pp. 160–163, 2018, doi: 10.1016/j.ecoenv.2018.08.104.

[11] F. E. Rahayu and Sunaryo, "Itm-04 : Analisis Reduksi Gas H 2 S Untuk Meningkatkan Kualitas Biogas Berbahan Baku Sampah Organik Buah- Buahan," *Pros. Semin. Nas. Fis.*, vol. 3, pp. 287–288, 2014, [Online]. Available: http://journal.unj.ac.id/unj/index.php/prosidingsnf/article/view/5535

[12] Z. Liang, D. Wu, G. Li, J. Sun, F. Jiang, and Y. Li, "Experimental and modeling investigations on the unexpected hydrogen sulfide rebound in a sewer receiving nitrate addition: Mechanism and solution," *J. Environ. Sci. (China)*, vol. 125, pp. 630–640, 2023, doi: 10.1016/j.jes.2021.12.038.

[13] H. Ohge and R. D. Madoff, "Studies of Toxicity and Odor Resulting from Hydrogen Sulfide Produced by the Fecal BacteriaInternational Fellowship #I-102," *Semin. Colon Rectal Surg.*, vol. 17, no. 4, pp. 177–181, 2006, doi: 10.1053/j.scrs.2006.10.006.

[14] B. Ren, N. Lyczko, Y. Zhao, and A. Nzihou, "Simultaneous hydrogen sulfide removal and wastewater purification in a novel alum sludge-based odor-gas aerated biofilter," *Chem. Eng. J.*, vol. 419, no. March, p. 129558, 2021, doi: 10.1016/j.cej.2021.129558.

[15] D. Czajczyńska, R. Krzyżyńska, and H. Jouhara, "Hydrogen sulfide removal from waste tyre pyrolysis gas by inorganics," *Int. J. Hydrogen Energy*, no. xxxx, 2022, doi: 10.1016/j.ijhydene.2022.03.082.

[16] C. Wang *et al.*, "Effective removal of hydrogen sulfide from landfill gases using a modified iron pentacarbonyl desulfurization agent and the desulfurization mechanism," *Sci. Total Environ.*, vol. 839, no. February, pp. 1–9, 2022, doi: 10.1016/j.scitotenv.2022.156160.

[17] Y. Wang, Z. Wang, J. Pan, and Y. Liu, "Removal of gaseous hydrogen sulfide using Fenton reagent in a spraying reactor," *Fuel*, vol. 239, no. October 2018, pp. 70–75, 2019, doi: 10.1016/j.fuel.2018.10.143.

[18] M. Lei, H. Ren, S. Luo, W. Yang, and Z. Gao, "Surface Science Analysis of the adsorption characteristics of gasification pollutants (HCl , COS , H_2S , NH_3 and HCN) on Ti-anchored graphene substrates," *Surf. Sci.*, vol. 725, no. February, p. 122148, 2022, doi: 10.1016/j.susc.2022.122148.

[19] L. Song *et al.*, "Screening of zeolites for H_2S adsorption in mixed gases: GCMC & DFT simulations," *Microporous Mesoporous Mater.*, vol. 328, no. July, p. 111495, 2021, doi: 10.1016/j.micromeso.2021.111495.

[20] D. Li, W. Li, and J. Zhang, "The adsorption behavior of H2S, SO2, CO, and O3 on non-noble metal doped BN: A theoretical study," *Surfaces and Interfaces*, vol. 30, no. March, p. 101958, 2022, doi: 10.1016/j.surfin.2022.101958.

[21] H. jian Yang *et al.*, "Modification of nanoscale polymeric adsorbent for the preparative separation and purification of tacrolimus from fermentation broth of Streptomyces tsukubaensis," *J. Chromatogr. A*, vol. 1675, p. 463180, 2022, doi: 10.1016/j.chroma.2022.463180.

[22] Shreya, A. K. Verma, A. K. Dash, P. Bhunia, and R. R. Dash, "Removal of surfactants in greywater using low-cost natural adsorbents: A review," *Surfaces and Interfaces*, vol. 27, no. October, p. 101532, 2021, doi: 10.1016/j.surfin.2021.101532.

[23] T. Owen *et al.*, "Materials Today : Proceedings Tensile properties of epoxy resin filled with activated carbon derived from coconut shell," *Mater. Today Proc.*, no. xxxx, 2022, doi:

10.1016/j.matpr.2022.06.568.
[24] Y. Yue, Y. Liu, W. Zhang, J. Guo, Y. Gong, and Y. Yu, "Amidoxime functionalized low-cost cellulose-based adsorbent derived from waste cigarette filters for efficient heavy metal removal," *J. Environ. Chem. Eng.*, vol. 10, no. 3, p. 107846, 2022, doi: 10.1016/j.jece.2022.107846.
[25] E. Wina, "Tanaman Pisang sebagai Pakan Ternak Ruminansia," *Wartazoa*, vol. 11, no. 1, pp. 20–27, 2001.
[26] "http://www.ajer.org/," vol. 2, no. 9.
[27] M. Hailu, T. Seyoum Workneh, and D. Belew, "Effect of packaging materials on shelf life and quality of banana cultivars (Musa spp.)," *J. Food Sci. Technol.*, vol. 51, no. 11, pp. 2947–2963, 2014, doi: 10.1007/s13197-012-0826-5.
[28] A. Mahmood, N. Ngah, and M. N. Omar, "Phytochemicals constituent and antioxidant activities in Musa x paradisiaca flower," *Eur. J. Sci. Res.*, vol. 66, no. 2, pp. 311–318, 2011.
[29] H. Mohd Zaini, J. Roslan, S. Saallah, E. Munsu, N. S. Sulaiman, and W. Pindi, "Banana peels as a bioactive ingredient and its potential application in the food industry," *J. Funct. Foods*, vol. 92, no. March, p. 105054, 2022, doi: 10.1016/j.jff.2022.105054.
[30] K. G. Akpomie and J. Conradie, "Banana peel as a biosorbent for the decontamination of water pollutants. A review," *Environ. Chem. Lett.*, vol. 18, no. 4, pp. 1085–1112, 2020, doi: 10.1007/s10311-020-00995-x.
[31] S. Sumiyati, H. S. Huboyo, and B. S. Ramadan, "Potential Use of Banana Plant (Musa spp.) as Bio-sorbent Materials for Controlling Gaseous Pollutants," *E3S Web Conf.*, vol. 125, no. 201 9, pp. 1–6, 2019, doi: 10.1051/e3sconf/201912503015.
[32] H. S. Huboyo, B. Zaman, B. S. Ramadan, and A. D. Prinaningrum, "Promising Adsorption of Sulfidic Acid Gases Using Wet Banana Plant Adsorbent (Musa spp.)," *J. Eng. Technol. Sci.*, vol. 54, no. 1, 2022, doi: 10.5614/j.eng.technol.sci.2022.54.1.10.
[33] N. Fahmi, I. Herdiana, and R. Rubiyanti, "PENGARUH METODE PENGERINGAN TERHADAP MUTU SIMPLISIA DAUN PULUTAN (Urena lobata L.)," *Media Inf.*, vol. 15, no. 2, pp. 165–169, 2020, doi: 10.37160/bmi.v15i2.433.
[34] W. D. Widodo, K. Suketi, and A. Fitriansyah, "Pemantapan Satuan Panas sebagai Kriteria Panen Pisang Raja Bulu," vol. 12, no. 2, pp. 99–107, 2021.
[35] W. D. Widodo, K. Suketi, and R. Rahardjo, "Evaluasi Kematangan Pascapanen Pisang Barangan untuk Menentukan Waktu Panen Terbaik Berdasarkan Akumulasi Satuan Panas," *Bul. Agrohorti*, vol. 7, no. 2, pp. 162–171, 2019, doi: 10.29244/agrob.7.2.162-171.
[36] D. Wirnas and M. Surahman, "Development of Selection Criteria in Banana (Musa sp.) Based on Path Analysis," *Indones. J. Agron.*, vol. 54, no. November, pp. 48–54, 2005, [Online]. Available: http://jamu.journal.ipb.ac.id/index.php/jurnalagronomi/article/view/1264/369
[37] N. Sitohang, "Pembiakan anakan (*sucker*) pisang Barangan (*Musa paradisiaca* L.) secara *in vitro*," *Biota*, vol. 13, no. 2, pp. 1–3, 2008.
[38] W. O. S. sariamanah A. Munir and Agriansyah Ahdiat, "Karakterisasi Morfologi Tanaman Pisang (Musa paradisiaca L.) di Kelurahan Tobimeita Kecamatan Abeli Kota Kendari," *J. AMPIBI*, vol. 1, no. 3, pp. 32–41, 2016, [Online]. Available: http://ojs.uho.ac.id/index.php/ampibi/article/download/5043/3766
[39] C. Y. K. Putri, F. S. Pranata, and Y. R. Swasti, "Kualitas Muffin Dengan Kombinasi Tepung Pisang Kepok Putih (Musa paradisiaca forma typica) Dan Tepung Labu Kuning (Cucurbita moschata)," *Biota J. Ilm. Ilmu-Ilmu Hayati*, vol. 4, no. April, pp. 50–62, 2019, doi: 10.24002/biota.v4i2.2471.
[40] dan Dedin Finatsiyatull Rosida, "EVALUASI NILAI GIZI TEPUNG PRA-MASAK PISANG TANDUK DAN PISANG RAJA NANGKA [Nutritional Evaluation of Pre-cooked 'Tanduk' and 'Raja Nangka' Plantain flour]," *Has. Penelit. J. Teknol. dan Ind. Pangan*, vol. XXII, no. 2, 2011.
[41] G. Byarugaba-Bazirake, "The effect of enzymatic processing on banana juice and wine," *Diss. Present. degree Dr. Philos.*, no. December, 2008, [Online]. Available:

https://scholar.sun.ac.za/handle/10019.1/1320
[42] E. Oksanen and S. Kontunen-Soppela, "Plants have different strategies to defend against air pollutants," *Curr. Opin. Environ. Sci. Heal.*, vol. 19, p. 100222, 2021, doi: 10.1016/j.coesh.2020.10.010.
[43] L. D. Emberson *et al.*, "Ozone effects on crops and consideration in crop models," *Eur. J. Agron.*, vol. 100, no. May, pp. 19–34, 2018, doi: 10.1016/j.eja.2018.06.002.
[44] P. A. Lin *et al.*, "Stomata-mediated interactions between plants, herbivores, and the environment," *Trends Plant Sci.*, vol. 27, no. 3, pp. 287–300, 2022, doi: 10.1016/j.tplants.2021.08.017.
[45] S. Kuppusamy, P. Thavamani, M. Megharaj, K. Venkateswarlu, Y. B. Lee, and R. Naidu, "Potential of Melaleuca diosmifolia leaf as a low-cost adsorbent for hexavalent chromium removal from contaminated water bodies," *Process Saf. Environ. Prot.*, vol. 100, pp. 173–182, 2016, doi: 10.1016/j.psep.2016.01.009.
[46] A. G. Adeniyi and J. O. Ighalo, "Biosorption of pollutants by plant leaves: An empirical review," *J. Environ. Chem. Eng.*, vol. 7, no. 3, p. 103100, 2019, doi: 10.1016/j.jece.2019.103100.
[47] A. Apriwandi, E. Taer, and R. Farma, "Analysis of Cyclic Voltammetry dan Galvanostatic Charge Discharge Electrode Supercapacitor based on activated carbon from Kepok Banana Leaf (Musa balbisiana)," *J. Aceh Phys. Soc.*, vol. 10, no. 4, pp. 94–101, 2021, doi: 10.24815/jacps.v10i4.19491.
[48] A. Gundogdu, C. Duran, H. B. Senturk, M. Soylak, M. Imamoglu, and Y. Onal, "Physicochemical characteristics of a novel activated carbon produced from tea industry waste," *J. Anal. Appl. Pyrolysis*, vol. 104, pp. 249–259, 2013, doi: 10.1016/j.jaap.2013.07.008.
[49] W. Zheng *et al.*, "Capture of fire smoke particles by leaves of Cunninghamia lanceolata and Schima superba, and importance of leaf characteristics," *Sci. Total Environ.*, vol. 841, no. April, p. 156772, 2022, doi: 10.1016/j.scitotenv.2022.156772.
[50] P. P. Gao *et al.*, "Contribution of PM2.5-Pb in atmospheric fallout to Pb accumulation in Chinese cabbage leaves via stomata," *J. Hazard. Mater.*, vol. 407, no. August 2020, p. 124356, 2021, doi: 10.1016/j.jhazmat.2020.124356.
[51] "Adsorption of chromium (Cr) from tannery wastewater using low-cost.pdf."
[52] C. Du, Y. Song, S. Shi, B. Jiang, J. Yang, and S. Xiao, "Preparation and characterization of a novel Fe3O4-graphene-biochar composite for crystal violet adsorption," *Sci. Total Environ.*, vol. 711, p. 134662, 2020, doi: 10.1016/j.scitotenv.2019.134662.
[53] E. R. K. Fernandes, C. Marangoni, O. Souza, and N. Sellin, "Thermochemical characterization of banana leaves as a potential energy source," *Energy Convers. Manag.*, vol. 75, pp. 603–608, 2013, doi: 10.1016/j.enconman.2013.08.008.
[54] X. Jin *et al.*, "International Journal of Biological Macromolecules High lignin , light-driven shape memory polymers with excellent mechanical performance," *Int. J. Biol. Macromol.*, vol. 219, no. July, pp. 44–52, 2022, doi: 10.1016/j.ijbiomac.2022.07.134.
[55] Y. Qin, D. Yang, W. Guo, and X. Qiu, "Investigation of grafted sulfonated alkali lignin polymer as dispersant in coal-water slurry," *J. Ind. Eng. Chem.*, vol. 27, pp. 192–200, 2015, doi: 10.1016/j.jiec.2014.12.034.
[56] Z. Rouzitalab, D. M. Maklavany, S. Jafarinejad, and A. Rashidi, "Lignocellulose-based adsorbents: A spotlight review of the effective parameters on carbon dioxide capture process," *Chemosphere*, vol. 246, p. 125756, 2020, doi: 10.1016/j.chemosphere.2019.125756.
[57] C. I. K. Diop, M. Tajvidi, M. A. Bilodeau, D. W. Bousfield, and J. F. Hunt, "Evaluation of the incorporation of lignocellulose nanofibrils as sustainable adhesive replacement in medium density fiberboards," *Ind. Crops Prod.*, vol. 109, no. April, pp. 27–36, 2017, doi: 10.1016/j.indcrop.2017.08.004.
[58] "A hydrogen bond based self-healing superhydrophobic octadecyltriethoxysilane–lignocellulose or silica coating.pdf."
[59] Y. Yang, X. Xiong, Y. Fan, Z. Lai, Z. Xu, and F. Luo, "Insight into volatile iodine uptake

properties of covalent organic frameworks with different conjugated structures," *J. Solid State Chem.*, vol. 279, no. August, p. 120979, 2019, doi: 10.1016/j.jssc.2019.120979.
[60] L. Shao *et al.*, "Facile preparation of oxygen-rich porous polymer microspheres from lignin-derived phenols for selective CO2 adsorption and iodine vapor capture," *Chemosphere*, vol. 288, no. P1, p. 132499, 2022, doi: 10.1016/j.chemosphere.2021.132499.
[61] Suhas, P. J. M. Carrott, and M. M. L. Ribeiro Carrott, "Lignin - from natural adsorbent to activated carbon: A review," *Bioresour. Technol.*, vol. 98, no. 12, pp. 2301–2312, 2007, doi: 10.1016/j.biortech.2006.08.008.
[62] Q. Zheng, H. Zhang, H. Mi, Z. Cai, Z. Ma, and S. Gong, "High-performance flexible piezoelectric nanogenerators consisting of porous cellulose nanofibril (CNF)/poly(dimethylsiloxane) (PDMS) aerogel films," *Nano Energy*, vol. 26, pp. 504–512, 2016, doi: 10.1016/j.nanoen.2016.06.009.
[63] C. Jin, S. Han, J. Li, and Q. Sun, "Fabrication of cellulose-based aerogels from waste newspaper without any pretreatment and their use for absorbents," *Carbohydr. Polym.*, vol. 123, pp. 150–156, 2015, doi: 10.1016/j.carbpol.2015.01.056.
[64] N. T. Cervin, C. Aulin, P. T. Larsson, and L. Wågberg, "Ultra porous nanocellulose aerogels as separation medium for mixtures of oil/water liquids," *Cellulose*, vol. 19, no. 2, pp. 401–410, 2012, doi: 10.1007/s10570-011-9629-5.
[65] L. Riva, A. Fiorati, and C. Punta, "Synthesis and application of cellulose-polyethyleneimine composites and nanocomposites: A concise review," *Materials (Basel).*, vol. 14, no. 3, pp. 1–22, 2021, doi: 10.3390/ma14030473.
[66] "High CO2 adsorption by amino-modifed bio-spherical cellulose.pdf."
[67] F. Arliandi, Yulianty, E. Nurcahyani, and S. Wahyuningsih, "UJI EFEKTIVITAS MULSA DAUN PISANG KEPOK (Musa paradisiaca L .) TERHADAP PERTUMBUHAN VEGETATIF TANAMAN TOMAT (Lycopersicum esculentum Mill .)," *Pros. Semin. Nas. Perhimpun. Biol. Indones. XXV*, pp. 1–2, 2019.
[68] R. L. Orwell, R. L. Wood, J. Tarran, F. Torpy, and M. D. Burchett, "Removal of benzene by the indoor plant/substrate microcosm and implications for air quality," *Water. Air. Soil Pollut.*, vol. 157, no. 1–4, pp. 193–207, 2004, doi: 10.1023/B:WATE.0000038896.55713.5b.
[69] C. Treesubsuntorn and P. Thiravetyan, "Removal of benzene from indoor air by Dracaena sanderiana: Effect of wax and stomata," *Atmos. Environ.*, vol. 57, pp. 317–321, 2012, doi: 10.1016/j.atmosenv.2012.04.016.
[70] C. Treesubsuntorn, P. Suksabye, S. Weangjun, F. Pawana, and P. Thiravetyan, "Benzene adsorption by plant leaf materials: Effect of quantity and composition of wax," *Water. Air. Soil Pollut.*, vol. 224, no. 10, 2013, doi: 10.1007/s11270-013-1736-5.

The seafarer's level of understanding on ballast water management convention

Iksiroh el Husna[1], Dian Wahdiana[1], Nasri[1], Anissofiah Azise Wijinurhayati[1], Widya Putri[1]

[1] Politeknik Ilmu Pelayaran Semarang Jl. Singosari Raya No.2A, Wonodri 50242 Kota Semarang, Central Java, Indonesia

Anissofiah1987@gmail.com

Abstract. Due to practical and economic considerations, the sea transportation is increasingly in demand as a means of transporting world trade activities. The ships without cargo will use ballast water as a stabilizer. Ballast water is often carried by various foreign microorganisms that harm the ecosystem. IMO promulgated the 8 September 2017 Ballast Water Management (BWM) Convention to prevent the spread of harmful pathogenic microbes. This study aims to determine the understanding of seafarers of the provisions of BWM. The descriptive-analytical method used a cross-sectional approach using questionnaires and in-depth interviews. The sample consisted of 47 seafarers; 3 samples were removed because they did not meet the requirements. The results show that most respondents do not understand the provisions of BWM correctly. None of the respondents knew when the BWM was enacted. In some questions, less than 10% of the answers are correct. However, on the importance of handling ballast water, there is 43.18% answered correctly, meaning that there is a concern for respondents about the dangers of it, and it is essential to address it immediately. We suggest that socialization among seafarers can be through seminars or special training on ballast water.

1. Introduction

The development of technology in transportation from time to time is made so that human life becomes easier. Of the three elements of transportation modes, sea, air and land are the most ideal for supporting the world of trade is the mode of sea transportation. The large carrying capacity makes sea transportation (ships) the main choice for export-import activities. Stated that sea transportation transports more than 80% of world trade commodities with high economic value [1][2]. As an archipelagic country with 17,504 islands, Indonesia is ideal for using ships as a means of transportation [3].

Mega tankers, bulk-cargo ships, will carry large amounts of ballast water. Ballast Water (Ballast Water) is water filled on ships that do not carry cargo to maintain the ship's balance; water is taken from coastal waters in an area after the ship has unloaded or is empty [4]. The ballast water will be discharged upon arrival at the port where the ship docks to load cargo. Ballast water is also useful for controlling the movement of ships while sailing or being hit by waves [5] [6]. The introduction of alien species through ships dates back 100 years, but it wasn't until the 1970s that the first biological samples were taken from ballast water samples. Since then, more than 1,000 species have been identified from ballast water, including human pathogenic bacteria; it is estimated that 3,000–7,000 species are transferred daily around the world [7]. Approximately 10 billion tons of ballast water are transported worldwide each year, the equivalent of about 4 million Olympic-sized swimming pools [8].

In 2012, Rivera examined organisms in ballast water tanks and found that living bacteria were 32% freer than plankton which was only 18% [9]. Dobbs conducted research in the waters of Singapore, Mexico and Virginia, United States of America, regarding the operation of ballast water on ships that can potentially spread Vibrio cholerae bacteria to all ports in the world [10]. The study also assessed antibiotic resistance in Vibrio cholerae bacteria found in ballast water. Altug examined the Turkish Marmara Sea from 2009 to 2010; the study aimed to study foreign species that entered the Turkish Marmara Sea from ballast water output. This study found at least 38 species of bacteria, 27 of which are pathogenic [11]. In 2017, El-Husna found pathogenic microbes (Vibrio cholerae, Escherichia coli and Enterococcus intestinal) from ballast water with no difference from pathogenic microbes on the coast of Cilacap [12]. In 2022 El-Husna found 9 (nine) types of microbial pathogens from ballast water discharged at the port of Tanjung Emas Semarang, containing V. cholerae (> 800 CFU/100 ml sample), E. coli (. 400 CFU/100 ml sample) and E [13]. intestinal (300 cfu/100 ml sample) which exceeds the threshold allowed by the Standard D-2 Ballast Water Management (BWM) Convention [14] [15].

Table 1. The D-2 BWM Convention Performance Standard

Microorganism category	Regulation D-2
Plankton, size > 50 μm	< 10 viable cells / m3
Plankton, size 10-50 μm	< 10 viable cells / mL
Toxicogenic *Vibrio cholerae*	< 1 Colony Forming Unit / 100 mL
Escherichia coli	< 250 Colony Forming Unit / 100 mL

Ballast Water Management (BWM) Convention is an international convention for controlling and managing ballast water and ship sediments. The BWM Convention is an international maritime treaty in 2004 which obliges signatory countries to the convention to ensure that ships flying their national flag comply with standards and procedures for managing and controlling ship ballast water and sediment. Damage to the marine environment due to the discharge of ballast water [14] [15].

2. Methodology

This study uses an analytical description method with a cross-sectional approach. They are using questionnaires and in-depth interviews to obtain data. The study was conducted on 47 seafarers who work on the deck or engine room, with qualifications of strata 2 and 1. Of the 47 respondents, 3 data were issued due to dishonest answers, thus reducing the validity of the data.
There is a list of questions for respondents in Table 2.

Table 2. List of questions

NO	QUESTION
1	Do you agree that Indonesia ratifies the BWM Convention by IMO?
2	Do you agree that the Indonesian government makes laws and regulations as a derivative of the BWM Convention?
3	Is it okay for vessels that do not exchange ballast water to be penalized?
4	Do you agree that vessels that do not treat ballast water before it is disposed of will be penalized?
5	Do you agree to meet the requirements of standard D-2, ballast water treatment facilities provided at the port?

Table 2 List of questions (Continued)

NO	QUESTION
6	When the BWM Convention was promulgated?
7	When did Indonesia ratify the BWM Convention?
8	What are laws and regulations of a derivative of BWM Convention?
9	What is the D-1 BWM Convention standard?
10	What is the D-2 BWM Convention standard?
11	How far is the ballast water discharge from the outer land?
12	How many meters depth of ballast water discharge?
13	How many kinds of ballast water treatments are on the ship before discharge?
14	What chemical compounds are used for ballast water treatment?
15	Is the treatment of ballast water before it is urgently carried out?

3. Results and discussion

There are respons of cluster 1 question. The results on whether respondents agree with the promulgation of the BWM Convention by IMO Indonesia ratified it 44 people (100%) stated strongly agree. On whether they agree if the Indonesian government makes laws and regulations as a derivative of the BWM Convention to protect Indonesian waters, 44 people (100%) agreed. When the question was asked whether it was agreed for ships that did not exchange ballast water, four people (9.09%) answered they did not agree, and 40 people (90.90%) answered they agreed. On whether it is agreed for ships that do not treat ballast water before being discharged, they are given a sanction, four people (9.09%) answered disagreeing, and 40 people (90.90%) answered agree. On whether they agree to comply with the standard requirements of the D-2 BWM Convention, ballast water treatment facilities are provided at the port; 44 people (100%) answered strongly agree. Seeing the answers given by respondents, we are quite optimistic that seafarers understand the importance of regulations governing shipping activities to protect the maritime environment from the invasion of foreign microorganisms that pose a risk to the environment and health from ballast water.

However there are still disagree to give sanction for the offense of D-1 and D-2 regulation of BMW convention. With the reason for domestic shipping is imposible for ballast water exchange at sea. Ballast water treatment on board, most of ship still dont have facilities for ballast treatment. To huild up the ballast water treatment facility require a lot of money, so that many ship owner not do it

There are respons of cluster 2 question . When asked about when the BWM Convention was promulgated, 0 people answered correctly, 24 (54.54%) answered incorrectly, 16 (36.36%) answered they did not know, and 4 (9.10%) forgot. On the question of when Indonesia ratified the BWM Convention, 4 (9.10%) people answered correctly, 23 (52.26%) answered they didn't know, 14 (31.82%) answered incorrectly, and 3 (6.82%) forgot. For the question of the number and year of the laws and regulations which are derivatives of the BWM Convention, 1 (2.27%) people answered correctly, 31 (70.46%) people answered they didn't know, 11 (25.00%) people answered incorrectly and 1 (2.27%) people answered forgot. On the question, what is known about the standard D-1 BWM Convention? 4 (9.10 %) people answered correctly, 8 (18.18%) people answered incorrectly, and 31 (70.45%) people answered they didn't know, and 1 (2.27 %) answered forgot. For the question of knowledge about the D-2 standard, the results are 6 (13.64%) people answered correctly, 26 (59.09%) people answered incorrectly, 11 (25.00%) people did not know, and 1 (2.27%) answered forgot. On the question of how far the ballast water discharge is from the outermost land, 6 (13.64%) people answered correctly, 22 (50.00%) people answered incorrectly, 14 (31.82%) did not know, and 2 (4.54%) answered forgot. On the question of how many meters the depth of ballast water discharge is, 9 (20.46%) people answered correctly, 21 (47.72 %) answered incorrectly, and 14 (31.82%) answered they did not know. On the question of how many kinds of ballast water treatment on the ship before being discharged, 3 (6.82%) people answered correctly, 12 (27.27%) people answered incorrectly, and 26 (59.09%) did not know, and 3 (6.82%) forget. On the question of what chemical compounds are used for ballast water treatment, 6 (13.64%) people answered correctly, 12 (27.27%) people answered incorrectly, 25 (56.82%) people answered they didn't know, and 1 (2, 27%) answered forgot. And for whether the ballast water treatment before being discharged is urgent, 19 (43.18%) people answered correctly, and 14 (31.82%) answered incorrectly. 6 (13.64%) people answered they didn't know, and 5 (11.36) %) people answered they forgot.

According the most reponden they are not understand with BMW convention. However about the promulgated BMW Convention none correc answer (0%), then the uncorrectly, unknow and forget answer are mayority for all question. Its mean the seaferers less literate and not awareness with their knowledge.

The results of respondents' answers vary greatly in their level of knowledge, depending on their experience while sailing. Seafarers sailing in international waters usually have experience in onboard ballast water management because countries that have ratified the BWM Convention will enforce the provisions of the BWM Convention so that it requires ships to meet the D-1 standard and the D-2 standard. For questions about the urgency of ballast water management before being discharged, there were 19 (43.18%) people who answered correctly, meaning that there was concern from seafarers to treat ballast water before it was disposed of so as not to pollute the environment.

There is a need for further socialization and education to seafarers about the dangers of ballast water and its management, as well as an understanding of the BWM Convention. Socialization can be through seminars, workshops or familiarization while on board by ship management.

References

[1] Fugazza, Marco., Hoffman, Jan., 2017. Liner Shipping conectivity as determinant of trade. Journal of Shipping and trade. 2. Article number 1 (2017). Published 02 March 2017.

[2] Erga, O.K.H., J.M. Andres, O. Enger, dan O. Vadstein. 2019. Microorganisms in ballast water: Disinfection, community dynamics, and implications for management. Science of The Total Environment. 657:704-716. DOI: https://doi.org/10.1016/j.scitotenv.2018.12.004.

[3] Lasabud, R., 2013. Pembangunan Wilayah Pesisir dan Lautan dalam Perspektif Negara Kepulauan Republik Indonesia (Regional Development in Coastal and Ocean in Archipelago Perspective of The Republic of Indonesia). Jurnal Ilmiah Platax Vol. I-2, Januari 2013 ISSN: 2302-3589

[4] Sayinli,B., Dong, y., Park, Y., Bhatnagat, A. 2021, Recent progress and Challenges facing ballast water treatment – A Review. Chemosphere 291 (72): 132776. Nopember 2021, DOI: 10.1016/j.chemosphere.2021.132776

[5] Gollasch, S., David, M., 2018. Ballast water: problems and management. In: World Seas: an Environmental Evaluation Volume III: Ecological Issues and Environmental

[6] Ardura, A., Martinez, J.L., Zaiko, A., Garcia-Vazquez, E., 2021a. Poorer diversity but tougher species in old ballast water: Biosecurity challenges explored from visual and molecular techniques. Mar Pollut Bull. 2021 Jul;168:112465. doi: 10.1016/j.marpolbul.2021.112465. Epub 2021 May 12. PMID: 33991987 DOI: 10.1016/j.marpolbul.2021.112465

[7] Gollasch, S., Minchin, D., David, M., 2014. The Transfer of Harmful Aquatic Organisms and Pathogens with Ballast Water and Their Impacts. Global Maritime Transport and Ballast Water Management pp 35–58. First Online: 10 October 2014

[8] IMO, 2019. Ballast water management – the control of harmful invasive species. https://www.imo.org/en/MediaCentre/HotTopics/Pages/BWM-default.aspx

[9] Rivera, I. N. G., Souza, K. M. C., Souza, C. P., & Lopes, R. M. (2012). Free-living and plankton-associated vibrios: Assessment in ballast water, Harbor areas, and coastal ecosystems in Brazil. Frontiers in Microbiology, 3(JAN), 1–8. https://doi.org/10.3389/fmicb.2012.00443

[10] Dobbs, F.C., Goodrich, A. L., Thomson III, F. K., Hynes, W., 2013. Pandemic Serotypes of Vibrio cholerae Isolated from Ships' Ballast Tanks and Coastal Waters: Assessment of Antibiotic Resistance and Virulence Genes (tcpA and ctxA). Microbiology of Aquatic Systems. Published: 30 January 2013.

[11] Altug, G., Gurun, S., Cardak, M., Ciftci, P. S., & Kalkan, S. (2012). The occurrence of pathogenic bacteria in some ships' ballast water incoming from various marine regions to the Sea of

Marmara, Turkey. Marine Environmental Research, 81, 35–42. https://doi.org/10.1016/j.marenvres.2012.08.005

[12] El-Husna, I., S. Anggoro, H.R. Sunoko, dan O. Setyani. 2017. Impact of Ballast Water on Environment Health. Advanced Science Letters, American Scientific Publishers. 23(3):2432-2434. DOI: https://doi.org/10.1166/asl.2017.8689.

[13] El-Husna, I., S.Anggoro., H.R.Sunoko., Subagiyo. 2022Bacteriological Study of Ballast Water at Tanjung Emas Port, Semarang. Indonesian Journal of Maritime Scences, March 2022. Vol 27 (1): 45-52

[14] IMO, 2009.Ballast Water Management Convention and the Guidlines for its implenetation, 2009 edition. London. ISBN: 978-92-801-1503-1

[15] IMO, 2017. Ballast Water Management how to do it, ratify/implement/enforce. 2017 edition. London. ISBN: 978-92-801-1681-6

Study of environmental carrying capacity in Pemalang Regency

A H Kahfi[1], B P Samadikun[1], A Sarminingsih[1],

[1]Department of Environmental Engineering, Faculty of Engineering, University of Diponegoro, Indonesia

Abstract. To achieve sustainable development goals, the environment is one of the aspects that is taken into account. Environmental problems such as population growth, changes in land use, climate change lead to a decrease in the carrying capacity of the environment for the community. Pemalang Regency is included in the administrative area of Central Java Province, Indonesia. This writing aims to determine the extent of the Environmental Carrying Capacity in Pemalang Regency in the sector of water resources and land availability, as well as the resilience of the Environmental Carrying Capacity to the projected population growth in Pemalang Regency. This study found that the water resources sector has deficit since the first year of projected population growth in 2020. While in the land availability sector, it was found that the surplus condition occurred in the first year of projected population growth from 2020 to 2022

Keywords: Environmental Carrying Capacity, Water Carrying Capacity, Land Carrying Capacity.

1. Introduction

In Law no. 32 of 2009 concerning Environmental Protection and Management, every regional spatial planning must be based on a Strategic Environmental Study and determined by taking into account the carrying capacity and capacity of the environment with KLHS, RPPLH and utilization of natural resources.

The carrying capacity of the environment includes the carrying capacity of land and the carrying capacity of water. The carrying capacity of land and carrying capacity of water according to Muta'ali (2014) is basically a comparison of availability and demand or demand and supply. Environmental carrying capacity of each region will have different capabilities that are influenced by the conditions and characteristics of the resources in the relevant expanse of space. While the capacity of natural resources depends on the ability, availability, and need for land and water, so that the determination of Environmental Carrying Capacity is determined based on 3 (three) approaches, namely:

a) Land capability for spatial use allocation.
b) Comparison between supply and demand for land.
c) Comparison between supply and demand for water.

If the supply > demand, the component's carrying capacity is declared surplus and if supply < demand, the component's carrying capacity is declared in deficit or exceeded.

Pemalang Regency is one of the areas in Central Java Province, Indonesia. Located between 8° 52' 30" and 7° 20' 11" South Latitude and between 109° 17' 30" - 109° 40' 30" East Longitude. Total area

± 113.885 Ha. Population growth and development in various sectors in Pemalang Regency is very likely to cause a decrease in the carrying capacity of the environment, so that it can threaten the sustainability of human life and other creatures. To avoid this, it is necessary to study the carrying capacity of the environment in Pemalang Regency, with the hope that it can be the basis for planning and utilization of natural resources and spatial planning in Pemalang Regency.

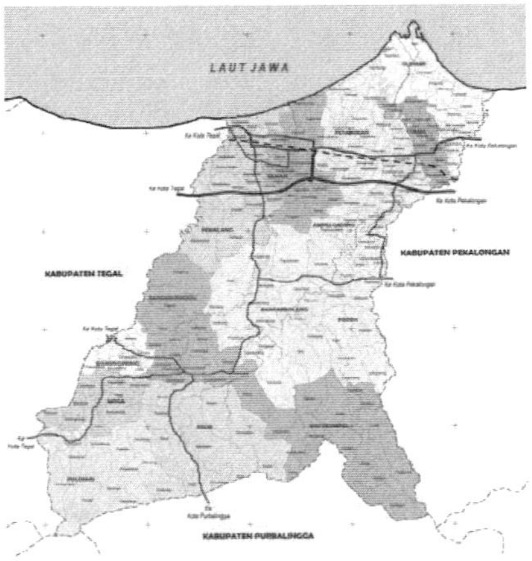

Figure 1. Pemalang Regency Map

This writing aims to determine the extent to which the carrying capacity of the environment in Pemalang Regency is in the water resources sector and land availability, as well as the resilience of the environmental carrying capacity of the projected population growth in Pemalang Regency.

2. Methodology

This research was conducted with the coverage of water and land resources in Pemalang Regency. By calculating the availability and demand for water and land resources, then comparing the value between the availability and the need for these two resources. So that it is known how much the carrying capacity value is in Pemalang Regency, as well as the resilience of the Environmental Carrying Capacity to the projected population growth in Pemalang Regency.

2.1. Water carrying capacity

Based on the guidebook for calculating the carrying capacity and carrying capacity of the environment, to determine the availability and demand for water are as follows:

Water Supply :

$C = \Sigma (Ci \times Ai) / \Sigma Ai$

$R = \Sigma Ri / m$

$SA = 10 \times C \times R \times A$

Where :

SA = Water supply (m3/year)
C = Weighted runoff coefficient
Ci = Land use coefficient
Ai = Land use area
R = average annual rainfall (mm/year)
Ri = Annual rainfall at the station i
m = Number of rain observation stations
A = Area (Ha)
10 = Conversion factor of mm.ha to m3.

Water Demand :
DA = N x KHLA

Where :
DA = Total water demand (m3/year)
N = Total population
KHLA = Water needs for a decent life (1600m3/person/year)

2.2. Land carrying capacity

Based on the guidebook for calculating the carrying capacity and carrying capacity of the environment, to determine the availability and demand for water are as follows:

Land Supply. :
SL = (Σ (Pi x Hi) / Hb)) x (1/Ptvb)

Where :
SL = Land supply (Ha)
Pi = Annual commodity productivity (ton)
Hi =Comodity unit price
Hb = Rice unit price / Kg
Pvtb = Rice productivity (ton/Ha)

Land Demand :
DL = N x KHLL

Where :
DL = Land Demand
N = Total population
KHLL = Land area for living per person s a decent living requirement per person divided by local rice productivity, a decent living requirement per person is assumed to be 1 ton of rice/person/year

3. Results

3.1. Runoff coefficient

In the Regulation of the Minister of the Environment No. 17 of 2009 concerning Guidelines for Determining Environmental Carrying Capacity in Spatial Planning, the value of the Runoff Coefficient (C) is determined in table 1

Table 1. Runoff Coefficient.

Surface Description	Runoff Coefficient
City, asphalt road, tile roof	0,7 – 0,9
Industrial area	0,5 – 0,9
Multi-unit settlements, shops	0,6 – 0,7
Housing complex	0,4 - 0,6
Villa	0,3 – 0,5
Garden, cemetary	0,1 – 0,3
Heavy earth yard :	
a. > 7 %	0,25 – 0,35
b. 2 - 7 %	0,18 – 0,22
c. < 2 %	0,13 – 0,17
light earth yard :	
d. > 7 %	0,15 – 0,20
e. 2 - 7 %	0,10 – 0,15
f. < 2 %	0,05 – 0,10
Heavy land	0,40
maedow	0,35
Agricultural cultivation land	0,30
Production forest	0,18

3.2. Land use

Pemalang Regency has an area of 113.885 Ha, consisting of 14 sub-districts and 222 sub-districts and villages. Pemalang Regency is an agricultural area with a percentage of 31.03% of paddy fields. Then the area of land in a row is 24.17% production forest, 24.08% garden, 11.87% settlement, 4.61% protected forest, 2.43% pond, 1.04% river, the rest is conservation land, industry, moor, field and sea. Land use in Pemalang Regency can be seen in table 2

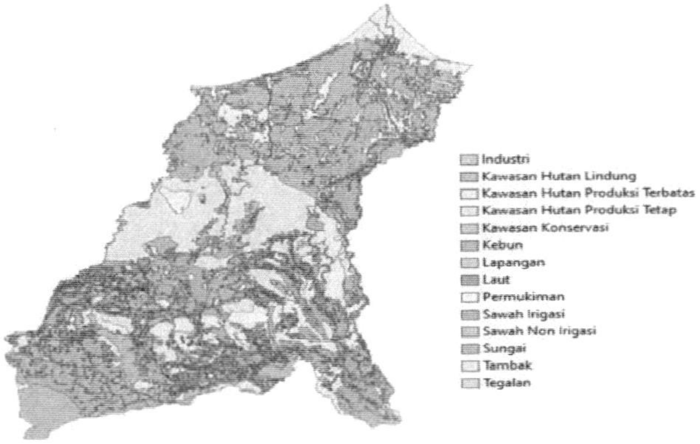

Figure 2. Pemalang Regency Land Use Map

Table 2. Land Use Area

Land Use	Land Area in Ha (Ai)	Run Off Coefficient (Ci)	Ai x Ci
Settlements, city, road	13.519,01	0,90	12.167,11
Industry	174,39	0,90	156,95
Protected forest	5.253,32	0,18	945,60
Production forest	27.529,80	0,18	4.955,36
Conservation land	57,34	0,18	10,32
Kebun	27.427,43	0,25	6.856,86
Tegalan	620,09	0,25	155,02
Irrigated rice field	34.369,31	0,30	10.310,79
Non-irrigated rice field	965,90	0,30	289,68
Field	2,30	0,35	0,81
Tambak	2.771,38	1,00	2.771,38
Rivers	1.189,08	1,00	1.189,08
Sea	5,84	1,00	5,84
	113.884,89		39.814,80
	Weighted runoff coefficient	**Σ (AixCi) / Σ Ai**	**0,3496**

3.3. Population number and projection

Population Projection can be calculated by arithmetic method as follows :

$$Pn = Po + Ka\,(Tn - To)$$
$$= Po + (P2 - P1) / (T2 - T1) \times (Tn - To)$$

Where :

Pn = projected population in year n

Po = total population data for the last year
Ka = the average number of population growth
Tn = year of the last population data
To = year of initial population data
P1 = total population data in the initial year
P2 = total population data in the last year
T1 = year of initial population data
T2 = year of the last population data

Based on data from the Central Statistics Agency, the population of Pemalang Regency in 2010 was 1,263,273 people and experienced population growth until 2019 of 1,203,814 people. Can be seen in table 3.

Table 3. Population Number And Projection

Population Data		Projection	
Year	Number of Population	Year	Projected Population
2010	1.263.273	2020	1.307.207
2011	1.269.078	2021	1.311.601
2012	1.277.437	2022	1.315.994
2013	1.279.597	2023	1.320.388
2014	1.284.238	2024	1.324.781
2015	1.288.566	2025	1.329.175
2016	1.292.573	2026	1.333.568
2017	1.296.272	2027	1.337.962
2018	1.299.433	2028	1.342.355
2019	1.302.814	2029	1.346.748

3.4. Rainfall

Pemalang Regency has 19 rainfall monitoring stations. The average amount of rainfall in Pemalang Regency in 2021 is 4,427.1 mm/year. Can be seen in table 4.

Table 4. Rainfall

No	Monitoring Station	Annual Rainfall
1	Kejene	3.516
2	Warungpring	4.166
3	Kecepit	4.400
4	Randudongkal	4.371
5	Nambo	4.434
6	Moga	6.217
7	Pulosari	6.780
8	Belik	7.343
9	Banjardawa	2.622
10	Sungapan	2.811
11	Karangsuci	2.889
12	Klareyan	2.871
13	Karangtengah	2.675
14	Sokowati	2.615
15	Bantarbolang	3.441
16	Pedagung	4.149
17	Watukumpul	6.342

18	Sipedang	5.733
19	Bongas	6.740
	Σ Ri	84.115
	R = Σ Ri	4.427,1

3.5. Comodity productivity

Pemalang Regency has a rice harvest area of 97,558 hectares, with a productivity of 591,584.3 tons/year. So that the productivity of rice is 6.060 kg/ha/year. Based on productivity data for 2021 at the Department of Agriculture. Productivity of other commodities from the highest in a row is pineapple honey 76,628.3 tons/year, corn 62,143 tons/year, mango 20,536 tons/year, broiler chicken 8,390 tons/year, ginger 5,756.2 tons/year, jasmine 5,010,2 tons/year, big chili 4,740 tons/year, shallot 2,432.4 tons/year. The complete productivity of Pemalang Regency commodities can be seen in table 5.

Table 5. ComodityPruductivity

No	Comodity	Production Quantity ton/th (Pi)	Price /ton (Hi)	Pi x Hi
1	Rice	591.584,300	5.000.000	2.957.921.500.000
2	Corn	62.143,000	4.800.000	298.286.400.000
3	Soya bean	365,600	14.000.000	5.118.400.000
4	Cassava	892,000	3.000.000	2.676.000.000
5	Sweet potato	1.763,000	6.000.000	10.578.000.000
6	Peanuts	245,000	28.000.000	6.860.000.000
7	Red onion	2.432,400	28.000.000	68.107.200.000
8	Big chili	4.740,000	46.000.000	218.040.000.000
9	Curly chili	384,000	30.000.000	11.520.000.000
10	Cayenne pepper	562,000	32.000.000	17.984.000.000
11	Durian	1.116,300	30.000.000	33.489.000.000
12	Mango	20.536,000	8.000.000	164.288.000.000
13	Honey pineapple	76.628,600	20.000.000	1.532.572.000.000
14	Beef	1.481,220	125.000.000	185.152.500.000
15	Buffalo meat	26,724	125.000.000	3.340.500.000
16	Lamb (kambing)	1.317,536	120.000.000	158.104.320.000
17	Lamb (domba)	351,216	120.000.000	42.145..920.000
18	Layer chicken	103,414	40.000.000	4.136.560.000
19	Broiler Chicken	8.390,154	40.000.000	335.606.160.000
20	Free-range chicken	2.166,118	80.000.000	173.289.440.000
21	Duck	297,359	60.000.000	17.841.540.000
22	Ginger	5.756,245	20.000.000	115.124.900.000
23	Lime	110,000	10.000.000	1.100.000.000
24	Cardamon	45,283	79.500.000	3.599.998.500
25	Aromatic ginger	26,672	24.000.000	640.128.000
26	Turmeric	188,554	12.000.000	2.262.648.000
27	Galangal	127,632	12.000.000	1.531.584.000
28	Lempuyang	32,578	15.000.000	488.670.000
29	Aloe vera	29,042	4.000.000	116.168.000
30	God crown	277,400	20.000.000	5.548.000.000
31	Noni / mengkudu	24,008	9.000.000	216.072.000
32	Sambiloto	18,000	90.000.000	1.620.000.000
33	Lemongrass	129,870	20.000.000	2.597.000.000

34	Temuireng	21,876	12.000.000	262.512.000
35	Temukunci	19,020	12.000.000	228.240.000
36	Curcuma	22,659	12.000.000	271.908.000
37	Jasmine	5.010,200	40.000.000	200.408.000.000
	Σ (Pi x Hi)			6.583.073.668.500

3.6. *Water carrying capacity*

Water Supply :

SA = 10 x C x R x A

Water Demand :

DA = N x KHLA

Population projections are calculated based on the latest available population data from 2010 to 2019. Population projection figures can be calculated starting in 2020 for 10 years until 2029. Based on formula, the water carrying capacity of Pemalang Regency based on population projections, it is found that the Pemalang Regency has experienced a deficit since the calculation of the population projection for 2020. Along with the projected population growth, the value of the deficit is getting bigger every year. Can be seen in table 6.

Table 6. Pemalang Regency Water Carrying Capacity Projection

Projection		Water Carrying Capacity		Description
Year	Projected Population	Supply	Demand	
2020	1.307.207	1.762.612.569	2.091.531.200	Defisit
2021	1.311.601	1.762.612.569	2.098.560.000	Defisit
2022	1.315.994	1.762.612.569	2.105.590.400	Defisit
2023	1.320.388	1.762.612.569	2.112.619.200	Defisit
2024	1.324.781	1.762.612.569	2.119.649.600	Defisit
2025	1.329.175	1.762.612.569	2.126.678.400	Defisit
2026	1.333.568	1.762.612.569	2.133.708.800	Defisit
2027	1.337.962	1.762.612.569	2.140.737.600	Defisit
2028	1.342.355	1.762.612.569	2.147.768.000	Defisit
2029	1.346.748	1.762.612.569	2.154.796.800	Defisit

3.7. *Land carrying capacity*

Land Supply :

SL = (Σ (Pi x Hi) / Hb)) x (1/Ptvb)

Land demand :

DL = N x KHLL

The land carrying capacity of Pemalang Regency based on population projections, it was found that Pemalang Regency experienced a surplus from 2020 to 2022. However, in 2023 the carrying capacity of the land will begin to experience a deficit in line with the projected population growth. And the deficit is increasing every year. Can be seen in table 7.

Tabel 7. ProyeksiDayaDukungLahanKabupatenPemalang

Projection		Land Carrying Capacity		Description
Year	Projected Population	Supply	Demand	
2020	1.307.207	217.263	215.711	Surplus
2021	1.311.601	217.263	216.436	Surplus

2022	1.315.994	217.263	217.161	Surplus
2023	1.320.388	217.263	217.886	Defisit
2024	1.324.781	217.263	218.611	Defisit
2025	1.329.175	217.263	219.336	Defisit
2026	1.333.568	217.263	220.061	Defisit
2027	1.337.962	217.263	220.786	Defisit
2028	1.342.355	217.263	221.511	Defisit
2029	1.346.748	217.263	222.236	Defisit

4. Conclusions
Based on the results of the study, it can be seen that the carrying capacity of the environment in Pemalang Regency based on production data in 2021, spatial planning in the 2018 RTRW and projections from population data until 2019 is known to have experienced a deficit in the water resources sector since the first year of projected population growth in 2020. Meanwhile, in the land availability sector, it was found that the surplus condition occurred in the first year of the projected population growth from 2020 to 2022, while for the following years there would be a deficit condition. These results can be used as a reference for the Pemalang Regency government in making efforts to manage water resources and management in the field of food availability to become even better

5. References
[1] Law No. 26 of 2007 concerning Spatial Planning.
[2] Minister of Environment Regulation No. 17 of 2009 concerning Guidelines for Determining Environmental Carrying Capacity in Spatial Planning
[3] Muta'ali, L (2011). Environmental Carrying Capacity Based on Spatial Planning. Indonesian Journal of Geography, 43 (2), 142-155.
[4] LawNo. 32 of 2009 concerning Environmental Protection and Management.
[5] Meilani (2013). Environmental Carrying Capacity of Rasau Jaya District Based on Land Availability and Needs. Journal of Wetland Environmental Technology, 1 (1).
[6] Guidelines for Determining the carrying capacity and carrying capacity of the environment (2014). Ministry of Environment Deputy 1 for Environmental Management
[7] Admadhani DN, Hajil AHS, Susanawati LD (2014). Analysis of Water Supply and Water Demand for Carrying Cpacity Assessment (Case STudy of Malang). Journal of Natural Resources and Environment, 1 (3), 13-20.
[8] A R Pramesty, Azwa N, A Asripin (2014). Calculation of Environmental Carrying Capacity Based on Water Availability and Land Productivity in TujuhBelas District, Bengkayang Regency. Journal of Wet Environmental Technology, 2(1), 1-10.
[9] Sumberdaya J, Fadilah N, Tunggul A, Haji S, Widiatmono R (2015). Water Balance Model for Carrying Capacity Assessment (Case Study of Batu). Journal of Natural Resources and Environment, 1 (2), 7-13
[10] Dian Hudawan Santoso (2015). Study of Water Carrying Capacity on Bintan Island, Riau Islands Province. Journal of Environmental Science & Technology, 7(1), 01-17.
[11] Pemalang Regency Regulation No. 1 of 2018 concerning the Pemalang Regency Spatial Plan for 2018 – 2038
[12] Ignatius Sriyana (2018). Evaluation of Watershed Carrying Capacity for Watershed Management (a Case Study on Bodri Watershed, Central Java, Indonesia). MATEC Web of Conference, 195, 05003.
[13] Ade Mulawarman (2019). Carrying Capacity of Water and Food Availability in Sukamaju District. LINEARS Journal, 2 (2), 92-99.

[14] Nasrudin (2019). Carrying Capacity of Grompol Watershed Settlements, Central Java. Earth and Environmental Science, 328 (2019) 012072.
[15] K M Maulana, F Lihawa, S Maryati (2020). Analysis of Water Carrying Capacity in Pulubala Sub-Watershed, Gorontalo Regency, Gorontalo Province. Earth and Environmental Science, 575 (2020) 012220.
[16] Muhammad Ma'arijHarfadli (2020). Study of environmental Carrying Capacity and Water Criticality Index Based on Availability Water in Balikpapan City. Presipitasi Journal, (17-3), 253-162.
[17] Central Bureau of Statistic(2022). Pemalangin Numbers
[18] Comodity Production Data at Pemalang Regency Agriculture Office in 2021 (2022).

IOP Conf. Series: Earth and Environmental Science 1098 (2022) 012066 doi:10.1088/1755-1315/1098/1/012066

Food waste humification: a process analysis

B Zaman[1]*, P Purwono[2]

[1]Department of Environmental Engineering, Faculty of Engineering, Diponegoro University, Semarang- Indonesia 50275
[2]Center for Science and Technology, Univeristas Islam Negeri Raden Mas Said Surakarta, Jl. Pandawa, Pucangan, Kartasura 57168, Indonesia

badruszaman2@gmail.com

Abstract. Food and Agricultural Organization estimates that roughly 1.3 billion kitchen trash are generated annually on a global scale. By using composting, we could convert our kitchen waste into a properly sanitized, humus-rich, relatively stable product that improves plant growth by conditioning soil Introduction. The main objective of this study is to conduct a review of fundamental of composting, humus, and enhanced humidification. Humus is created from degraded organic compound by the polymerization and condensation of organic components such polyphenols, reducing sugars, and amino acids. Humification can be increased through vermicomposting, the addition of biochar made from digestate, the addition of microbes to food waste compost, the addition of cow manure, multistage inoculation, solid state anaerobic digestion (SSAD), and composting hybrid processes.

1. Introduction

Food waste includes food that hasn't been eaten and food that has been prepared in places like restaurants, schools, and factories [1]. There will be less work and expense involved in transporting food if there is less loss and waste of food at home. Composting is an effective method for managing food waste because it produces fewer harmful byproducts and a more valuable end product than other disposal methods [2]. By using composting, we could convert our kitchen waste into a properly sanitized, humus-rich, relatively stable product that improves plant growth by conditioning soil [3]. Food waste composting is acceptable if it results in nontoxic compost (germination index value > 80%, thermophilic time > 10 days), high humification (humus content > 25%, E4/E6 ratio 4) and stable physicochemical qualities (pH value > 7, moisture content 5%) [4]. The main objective of this study is to conduct a mini review of fundamental of composting, humus, and enhanced humification.

2. Compost and humus

Compost is a type of natural fertilizer comprised of decayed organic matter. The end product of composting is a collection of partially decomposed organic materials known as humus [5]. Composting results in the production of humic compounds and a rise in the amount of humic acid-like organic-C, but it results in a decrease in the amount of fulvic acid (FA)-like organic-C and water-extractable organic-C due to microbial degradation [6]. Humification is the process of decomposing organic matter into humic compounds, which are mature organic matter [7]. Humin (molecular weight [MW] = 100,000–10,000,000 kDa), humic acids (HAs; MW = 10,000–100,000 kDa), and fulvic acids (FAs; MW =1000–10,000 kDa) are the three primary fractions that make up the components of humic (Pettit, 2004).

Humus is created from degraded organic compound by the polymerization and condensation of organic components such polyphenols, reducing sugars, and amino acids. The degree of polymerization (DP), which is a method that may be used to determine the maturity level of compost, is one of the humus indicators. This method is also one of the humus indicators. According to Ko et al. [8], if the compost has a DP value that is larger than 1.6, it has reached its mature state and can be utilized in agriculture and other fields.

3. Enhanced humidification

Cai et al., [9] have utilized vermicomposting to decompose food waste combined with soybean meal and sugarcane bagasse for five weeks. The amount of humic and humic acids went up because there was more sugarcane bagasse (SCB), but the amount of fulvic acid went down because the rate of humification went up. Vermicomposting with the addition of SCB led to an increase in humification [9], due to an increase in the activity of microorganisms responsible for humification [10]. Also, polysaccharides that are made when lignocellulosic materials break down in SCB are thought to be the main sources of H2S, along with the reduced carboxyl group content that happens during humification.

The humification could be enhanced by adding biochar produced from digestate [11]. The addition of 25% biochar is optimal for encouraging the degradation of organic matter, minimizing NH_3 emissions and increasing the formation of humic-acid-like matter, thereby increasing humification. Evaluation of degradation, emission, and humification was carried out on days 5, 11, 19, 27, 35, 43, and 50. The compost was manually turned over.

The process of adding microbes to compost made from food waste can serve as a bioaugmentation method for the synthesis of humic acid (HA) [12]. In order to demonstrate how bioaugmentation works, many methods like as statistical analysis, thermogravimetric analysis, structural equation modeling, and Fourier transform infrared spectroscopy (FTIR), among others, can be utilized. Bioaugmentation can hasten the production of humic acid and the decomposition of food waste by increasing the rate at which these processes occur.

According to Xu et al. [13], a multistage inoculation pattern can be used to increase both humification (with a high germination index of 120.9%) and humic substances (with a high humus index of 4.3 and a biological index of 1.4). Strains of Bacillus subtilis, Geobacillus thermoleovorans, and Lactobacillus amylophilus were used in the inoculum. Composting food waste through a solid state anaerobic digestion (SSAD) and composting hybrid process resulted in a rise of 28.4% and 17.6% in the amount of humic acid in the compost on day 15 and day 21, respectively. The rate at which the digestate's HA content grows is sped up dramatically when cow manure is added to the composting process [14].

4. Conclusion

Humification is the process of decomposing organic matter into humic compounds, which are mature organic matter. Humus is created from degraded organic compound by the polymerization and condensation of organic components such polyphenols, reducing sugars, and amino acids. One sign of humus is the degree of polymerization (DP), and a DP value of more than 1.6 means that the compost is mature enough to be used in agriculture. Vermicomposting, adding charcoal produced from digestate, adding microbes to food waste compost, adding cow manure, multistage inoculation, solid state anaerobic digestion (SSAD), and composting hybrid processes are all ways to promote humification. Vermicomposting is also known as worm composting.

Acknowledgments

The authors would like to acknowledge their gratitude to DRPM for providing funds for this research in the form of a PTUPT grant for financing year 2022 No. 187-72/UN7.6.1/PP/2022.

References

[1] Charkhestani A and Kebria D Y 2022 Investigation of effective microorganisms bioaugmentation in an on-site aerobic food waste composter *Int. J. Environ. Sci. Technol.*

[2] Wei Y, Li J, Shi D, Liu G, Zhao Y and Shimaoka T 2017 Environmental challenges impeding the composting of biodegradable municipal solid waste: A critical review *Resour. Conserv. Recycl.* **122** 51–65
[3] Suresh A, Alex R A and Abraham J 2021 Fungal Mediated Conversion of Food Waste to Compost *Curr. Trends Biotechnol. Pharm.* **15** 4–9
[4] Zhao Y, Weng Q and Hu B 2022 Microbial interaction promote the degradation rate of organic matter in thermophilic period *Waste Manag.* **144** 11–8
[5] Zakarya I A, Suhaimi N S and Kamaruddin A 2020 Comparative evaluation of compost quality, process and organic materials and adoptability potential to complement by Compost Quality Index (CQI) *IOP Conference Series: Earth and Environmental Science* vol 616, ed N N.M., R N.L., T S.S., Z I.A., Y S.Y., I T.N.T., A R.C. and I N.M. (School of Environmental Engineering, Universiti Malaysia Perlis, Perlis, Malaysia: IOP Publishing Ltd)
[6] Bernal M P, Alburquerque J A and Moral R 2009 Composting of animal manures and chemical criteria for compost maturity assessment. A review *Bioresour. Technol.* **100** 5444–53
[7] Flores R M 2014 Origin of Coal as Gas Source and Reservoir Rocks *Coal and Coalbed Gas* ed R M B T-C and C G Flores (Boston: Elsevier) pp 97–165
[8] Ko H J, Kim K Y, Kim H T, Kim C N and Umeda M 2008 Evaluation of maturity parameters and heavy metal contents in composts made from animal manure *Waste Manag.* **28** 813–20
[9] Cai L, Gong X, Ding H, Li S, Hao D, Yu K, Ma Q, Sun X and Muneer M A 2022 Vermicomposting with food processing waste mixtures of soybean meal and sugarcane bagasse *Environ. Technol. Innov.* **28** 102699
[10] Liu L, Guo X, Wang S, Li L, Zeng Y and Liu G 2018 Effects of wood vinegar on properties and mechanism of heavy metal competitive adsorption on secondary fermentation based composts *Ecotoxicol. Environ. Saf.* **150** 270–9
[11] Wang N, Huang D, Bai X, Lin Y, Miao Q, Shao M and Xu Q 2022 Mechanism of digestate-derived biochar on odorous gas emissions and humification in composting of digestate from food waste *J. Hazard. Mater.* **434** 128878
[12] Zhou X, Li J, Zhang J, Deng F, Chen Y, Zhou P and Li D 2022 Bioaugmentation mechanism on humic acid formation during composting of food waste *Sci. Total Environ.* **830** 154783
[13] Xu M, Yang M, Sun H, Meng J, Li Y, Gao M, Wang Q and Wu C 2022 Role of multistage inoculation on the co-composting of food waste and biogas residue *Bioresour. Technol.* **361**
[14] Bai S, Xi B, Li X, Wang Y, Yang J, Li S and Zhao X 2021 Anaerobic digestion of chicken manure: Sequences of chemical structures in dissolved organic matter and its effect on acetic acid production *J. Environ. Manage.* **296** 113245

The influence of drying time on pellet fuel moisture content

Yustina M Pusparizkita[1*], Alif F Hidayatullah[1], Norrisal F Anwar[1], Junaidi[1], Sudarno[1]

[1]Department of Environmental Engineering, Universitas Diponegoro, Jl. Prof. Soedharto, S.H. Tembalang, Semarang, Indonesia

yustinapusparizkita@gmail.com

Abstract. Pellets are solid fuels made from biomass. This work was aimed at finding out the effect of drying time on the moisture content in sawdust-made pellets using a moisture analyzer. The pellet was divided into 4 groups and each group was experimented with up to 5 times, namely: non-drying pellet and drying pellet at 100°C for 10, 15 and 20 minutes after the pellet comes out of the pelletizing machine. The results showed that the water content of the pellet was 31% for the non-drying pellet and 22%; 21%; 15% for the drying time of 10, 15, and 20 minutes. The water content was still slightly higher than the specified standard.

1. Introduction

The energy crisis that occurs in the world can occur due to an increase in the human population, which has an impact on increasing energy needs. Additionally, the continuous use of fossil energy can increase the occurrence of an energy crisis. Therefore, it is necessary to transition from fossil energy to alternative energy, one of which is the use of biodiesel, biomass, bio-pellets, biogas, and bio-briquettes [1]. According to the Indonesian Energy Outlook, land use in Indonesia produces 434,000 GW or equivalent to the production of 255 million barrels of oil per year. This value can play an important role in shifting the use of fossil fuels to alternative energy [2].

Pellets are solid fuels made from biomass which are converted using a cylindrical densification technique with a size smaller than the size of a briquette and have a length of approximately 6-25 mm with a diameter of 10-12 mm. Different uses of biomass have been considered for the production of pellet fuel, such as wood, agricultural waste, which contained cellulose. Furthermore, pellets are an environmentally friendly energy because pellets use organic waste in their production process and produce less emissions than wood. Pellet can burn immediately, which reduces the effectiveness of burning pellets [3]. Still, pellet is capable of producing high calories (heat) and environmentally friendly so that it can be used for household and small to medium scale industries, small and medium capacity power plants

Pellets have different characteristics depending on the materials. According to Indonesian standard SNI 8021 2014, the quality parameters of the pellets are: 12% moisture content, 1.5% ash content, 80% volatile matter, 14% fixed carbon, 400 Cal/g of caloric value and density is 0.8 g/cm^3. Moisture content is an important factor because it affects the strength and processing speed of pellet production. In addition, the moisture content is one of the most important factors that affects the efficiency, stability, and complete combustion of solid biofuels. Higher calories are provided by low-ash and moisture-containing pellets and are mechanically durable [4]. In addition, the high moisture content of wood pellets may have a negative impact on pellet performance, reducing net energy output in combustion, and generating large emissions of air pollutants [5]. Insufficient moisture content will tend to lead to a blockage in the molding die, which requires considerable repair works, such as electrical drilling to remove the blocked [6,7]. Therefore, the objective of this study was to investigate the effects of drying time on the moisture content after the pellet was discarded from the pelletizing process. This is crucial because it is related to the quality of the pellets produced.

2. Material and methods

The material used for this experiment was sawdust waste obtained from the Sawmill Center in Semarang, Indonesia. Before being further processed into pellets, the moisture content of the sawdust was analyzed. Sawdust was mixed with 10% by weight of tapioca starch as a binder and 10% by weight of water before pelletization. The sample was then inserted into the pelletizing machine.

A roller rotating pelletizing machine was used to make the pellets. The die inside the pelletizer was powered by an electric motor, where the die rotation will initiate the rotation of the rollers, which will force the compost materials and compress them through the die holes to form pellets. The pellets that emerge from the die holes were cut with cutting knives to obtain pellets of acceptable length (<40 mm) and the pellets were collected in a basin. Fresh pellets that leave the pelletizer were usually at high temperatures and were left to cool for 1 h.

To assess the effects of drying time, the pellet samples were placed on trays and dried in the oven. The oven capacity used is 53 L (Memmert UN 55, Germany). For the experiment, three variations of drying time were established, namely 10, 15 and 20 minutes, with a drying temperature of 100 °C. After drying, all samples were analyzed for moisture content using a moisture analyzer (Shimadzu MOC63u, Japan). The water content was measured with different samples (minimum of 5 replications).

3. Results and discussion

The moisture content of the raw materials used for the manufacture of pellets was approximately 27%. After the production process was completed, the water content measurement of all samples can be seen in Table 1. According to the results of the five pellet sample measurements, the highest moisture value for non-drying pellets was 33.3% and the average moisture value was 31%. From the results obtained, it can be proven that after the pelletizing process ended, the water content of the pellets was still relatively high.

Table 1. The moisture content of the pellets

Sample	**% moisture content**			
	Non-drying	***Drying***		
		10 minutes	**15 minutes**	**20 minutes**
1	32.19	24.57	20.81	16.5
2	32.53	20.71	21.87	19.74
3	27.7	24.27	24.07	14.57
4	29.29	20.55	18.48	13.28
5	33.3	23.08	21.93	13.04

The drying time will definitely affect the moisture content of the pellet. The moisture content for the 10-minute drying process produced a moisture content of approximately 23%. The highest sample moisture content for the 10-minute variation was 24%, while the lowest was 20%. A longer drying of 15 minutes produced pellets with an average moisture content of 21%. In addition, for the variation of the longest drying time in this study, the resulting pellet moisture content was 13-19%. On average, the moisture content of the pellet for 20 minutes has been reduced to 15%.

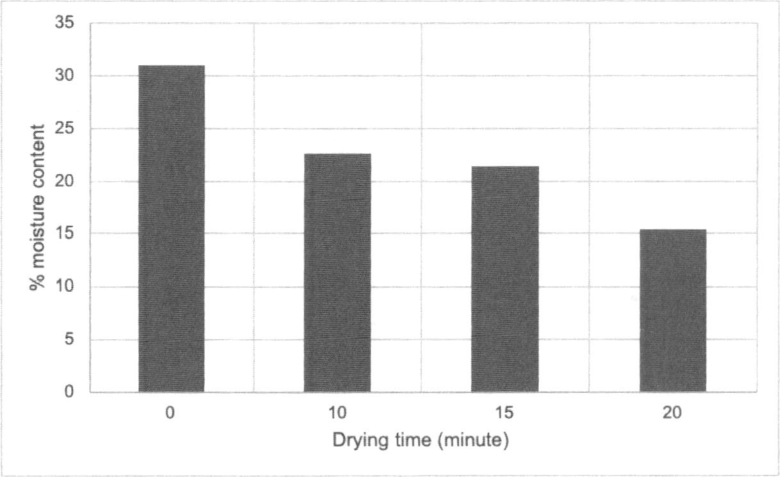

Figure 1. Average moisture content of pellets after drying different times

The general comparison can be clearly seen in Figure 1. The faster the drying time, the more water the pellet retains. However, because of the long drying times, more water accumulated in the pellet evaporated. However, efforts to reduce the water content are not optimal, as the water content remained above the specified standards. Generally, the optimal moisture content of the feedstock during palletization appears to be 8–13%[8].

4. Conclusion
The three variations of drying time for pellets produced products with average water content of approx. 15%-31%. Pellets with the lowest water content were produced with the longest drying process. However, large amounts of water-rich pellets do not meet the standard for the use of solid fuel. As a result, the drying temperature must be increased, and further studies need to be done.

5. Acknowledgement
The authors would like to express their special thanks to the Diponegoro University Strategic Research Faculty, RKAT Grants Funds 2022 (grant number 356/UN7.5.3.2/HK/2022), which has funded this research.

References
[1] Wahyono Y, Hadiyanto H, Zuli Pratiwi W and Dianratri I 2021 "Biopellet" as One of Future Promising Biomassbased Renewable Energy: a Review *E3S Web Conf.* **317**

[2] Munawar S S and Subiyanto B 2014 Characterization of Biomass Pellet Made from Solid Waste Oil Palm Industry *Procedia Environmental Sciences* **20** 336–41

[3] Muhammad Yerizam, Muhammad Zaman, Taufiq Jauhari, Nur Yuli, Riwen Setiawan, and Umaidella Affrilla 2021 Production of Bio-Pellet Briquettes From Coconut Shell Waste as Alternative Energy for Household Scale *Proceedings of the 4th Forum in Research, Science, and Technology (FIRST-T1-T2-2020)* (Atlantis Press) pp 57–61

[4] Proskurina S, Junginger M, Heinimö J, Tekinel B and Vakkilainen E 2019 Global biomass trade for energy— Part 2: Production and trade streams of wood pellets, liquid biofuels, charcoal, industrial roundwood and emerging energy biomass *Biofuels, Bioproducts and Biorefining* **13** 371–87

[5] Liu Z, Quek A and Balasubramanian R 2014 Preparation and characterization of fuel pellets from woody biomass, agro-residues and their corresponding hydrochars *Applied Energy* **113** 1315–22

[6] Zafari A and Kianmehr M H 2014 Factors affecting mechanical properties of biomass pellet from compost *null* **35** 478–86

[7] Mavaddati S, Kianmehr M H, Allahdadi I and Chegini G R 2010 Preparation of Pellets by Urban Waste Compost *International Journal of Environmental Research* **4** 665–72

[8] Sgarbossa A, Costa C, Menesatti P, Antonucci F, Pallottino F, Zanetti M, Grigolato S and Cavalli R 2014 Colorimetric patterns of wood pellets and their relations with quality and energy parameters *Fuel* **137** 70–6

IOP Conf. Series: Earth and Environmental Science 1098 (2022) 012068 doi:10.1088/1755-1315/1098/1/012068

Socio-ecological Risk Mitigation in the Construction of the Sicincin-Padang Toll Road, West Sumatra

Nora Eka Putri[1], Helmi[2], Melinda Noer[3], Yossyafra[4]
Universitas Negeri Padang[1], Andalas University[2,3,4]

*noraekaputri@fis.unp.ac.id

Abstract. Although infrastructure development aims to improve the welfare of the whole society, it poses adverse environmental effects in practice. One of such developments carried out by the Indonesian government is the trans-Sumatra toll road, which created social and environmental problems, particularly on the Sicincin-Padang toll road in West Sumatra. The affected communities in this province lost their homes, agricultural land and livelihoods to lifestyle changes. One of the villages experienced significant flooding due to large water discharge after clearing the land. Therefore, this research aims to explain the mitigation of socio-ecological risks in developing toll road infrastructure in West Sumatra. This research was carried out using the multi-method approach with qualitative and quantitative data analyzed by CATWOE analysis. The results explained that the construction of toll roads in West Sumatra has social and environmental impacts. Social and ecological risk mitigation is carried out by formulating policies through a soft system methodology (SSM) by involving many parties, especially affected communities, in decision-making.

1. Introduction

Infrastructure development aims to improve people's welfare [1]. One of such structures is the toll road because it is considered to provide easy access and mobilization of people from one place to another. The implementation of this development is carried out on land, which is a significant necessity. This is usually purchased by the government or agency responsible for the execution of the project. However, during its implementation, several obstacles are encountered, especially those related to lands owned by individuals or communities [2,3], state or private companies [4,5] and the indigenous peoples [6,7]. This also poses social and environmental risks, especially for affected communities. Among the risks posed by land acquisition and toll road construction are forced asset transfer [8,9,10,11], the emergence of poverty [12,13], natural disasters [14], weather and climate change [15], damage soil structure, ecosystem disturbance and biodiversity [16].

Social and environmental risks encountered during infrastructure development, especially in the construction of toll roads, also occur in Indonesia. This condition is widespread in certain areas, such as the Sicincin-Padang toll road in West Sumatra. These social and environmental risks include loss of land and long-term impacts. Although the government usually compensates the affected communities, their survival and future generations cannot be guaranteed. This also affected the social interaction of several communal communities in West Sumatra. Although people have strong kinship ties, with the construction of the toll road, the houses were separated from each other. Another risk factor is the change in livelihoods from agricultural practices to non-agricultural services due to the differences in skills, which is also likely to increase poverty. Its impact on the environment includes flooding, as seen during toll road construction activities in Nagari Sungai Buluh Selatan, which caused the forest to become barren and increased the discharge of river water.

Social and environmental risks must be on a priority scale whose negative impact can be overcome to realize sustainable infrastructure development. Preliminary studies have been analyzed from various points of view on the effects of toll road construction on society and the environment. Regardless of the fact that multiple efforts and policies such as the concept of "green space" [17,18,19] and Environmental Impact Problem Analysis [2,20,21] have been employed to overcome these risks, they have all proved abortive. Efforts to solve social problems have also been carried out with the BOT concept and CSR fund management [22,23,24,25], but still proved adamant. When social and environmental impacts are handled in a clear way, communities find it difficult to progress. It becomes necessary to employ combined efforts and solutions to the social and environmental risks caused by the construction of toll roads. This can be achieved by implementing a risk mitigation policy for these two aspects using the soft system method (SSM). The novelty involves finding a participatory mitigation policy model using a soft system methodology to handle the social and environmental risks encountered during toll road construction.

2. Research methods

This research adopted a multi-method involving a positivist paradigm and the combination of two or more approaches in the soft system methodology (SSM). Data were acquired by various means, such as interviews, mailing questionnaires to the affected communities, documentation, and observation [26]. Interviews were held with stakeholders involved in the acquisition of land to construct the Sicincin-Padang toll road, West Sumatra. The SSM data can be categorized into three, first, documentation from policies or regulations related to land acquisition on a macro and meso basis, then interviews and questionnaires as microdata. In addition, this software system methodology comprises several stages [26,27,28].

The first stage involves the identification of problematic situations at the meso and microlevels through interviews, distribution of questionnaires, documentation, and observations related to the social and environmental risks encountered during the construction of the Sicincin- Padang toll road in West Sumatra. The Head of Land Procurement Division of the National Land Agency (BPN) for the West Sumatra Region, the Commitment Making Officer (PPK) for Land Procurement of the Ministry of Public Works and Housing of Indonesia (PUPR RI), the Regent of Padang Pariaman, and the Governor of West Sumatra for the tenure 2015 to 2020 were interviewed. This also included the village head (walinagari) and the land acquisition committee. Meanwhile, questionnaires were distributed to 325 research samples from affected communities.

The second stage includes problematic situations detected from interviews and observations of experience-based organizations. In addition, the data were obtained based on participatory observation, with the output portraying a "rich picture." The following steps in the soft system methodology were proposed by Checland [26].

The third stage involves the creation of a complete root definition with CATWOE analysis (Customers, Actors, Transformations, Worldviews, Owners, and Environmental Constraints). Compares the research data to that of CATWOE. In the fourth stage, a conceptual model was developed using the following criteria: efficacy, efficiency, and effectiveness. The fifth stage involves comparing the conceptual model with the actual situation. In the sixth stage, recommendations for change and problem-solving actions based on comparisons and debates were compiled in the subsequent phase. Finally, the seventh stage is to propose a model or policy as a new and practical effort to be employed by policy makers and stakeholders concerned with the acquisition of land for toll road construction. Subsequently, a participatory-based socioecological mitigation policy model with respect to the construction of toll roads in west Sumatra was formulated.

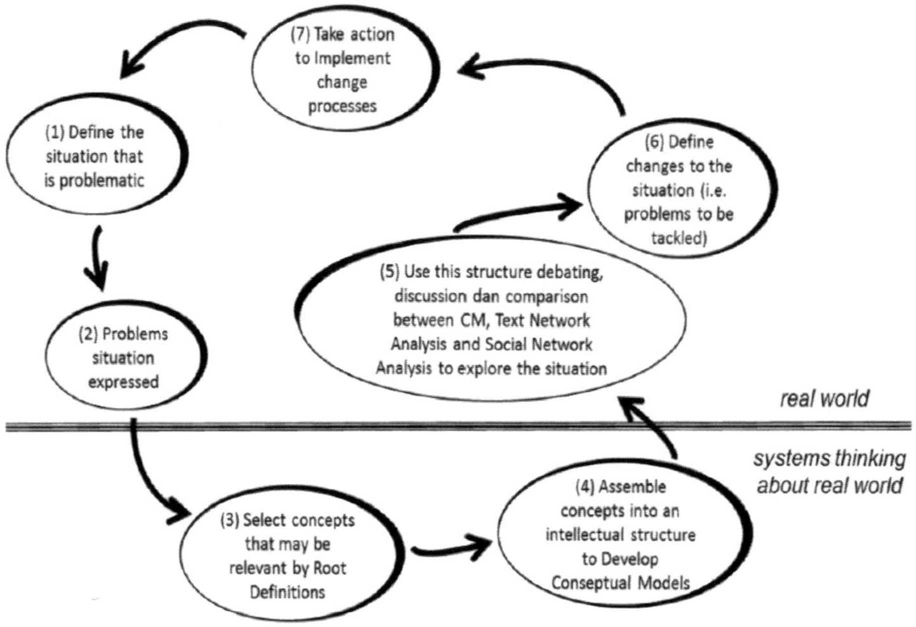

Figure 1. Multi-method-based SSM [26,29]

3. Research Findings

Toll road construction in West Sumatra, specifically in Sicincin-Padang, is the first of its kind in the region. Before the construction process, the government mapped out various development plans, such as acquiring land and compensating owners. In the plan to acquire land, the government, as the body in charge, issues a document that starts with carrying out feasibility research. This involves the technical, economic, environmental, and social explanation of the site. However, it turns out that land acquisition has triggered environmental and social problems. The research findings are explained through the soft system methodology stages described in the previous section.

3.1. Stage 1: Define the Problematic Situation

Based on the results of in-depth interviews with stakeholders, it was discovered that land acquisition planning does not involve the participation of affected communities. Furthermore, it was reported that the implementation of feasibility research to assess economic, environmental, and social likelihood does not yet involve the entire affected community. Therefore, the survey does not represent the opinions or expectations of these people. The Environmental Impact Problem Analysis (AMDAL) carried out by relevant agencies does not guarantee the sustainability of the environment. Socially, the construction of community affected by the toll roads loses land (assets) due to the procurement implemented procurement for the public interest. Despite their refusal, the government used their land, which had various implications, such as people being forcibly displaced, loss of livelihood, inability to interact with family and relatives socially, etc.

Table 1. Tabulation of Average Respondents' Statements on the Sicincin-Padang Toll Road Construction Process

No	Statement	Mean	Category
1	The land acquisition planning process involved all communities affected by the construction of the Sicincin-Padang toll road	2,9 (59%)	Slightly Low
2	All communities whose land was used for the construction of the Sicincin-Padang toll road attend every public consultation activity	3,6 (72%)	Currently
3	The community continues to interact and carry out social, economic, and customary activities with groups in their old places before they were moved to new areas	3,34 (66%)	Currently
4	The value of land compensation was changed after public consultation	3,0 (60%)	Slightly Low
5	The government and investors prioritize the need for the indigenous people of affected communities to be employed in the toll road construction	2,25 (45%)	Low
6	They prioritize humanitarian assistance, education, and health facilities at land acquisition locations	2,20 (44%)	Low
7	The government and investors must ensure environmental sustainability by engaging in regular pollution repairs after the land acquisition	2,21 (44,2%)	Low

Source: Data processed by researchers, 2022

The table shows the social and environmental risks encountered in constructing the Sicincin-Padang toll road and how people feel at that location. Based on the quantitative and descriptive data obtained from 325 respondents, it is a known fact that the statements submitted by the majority were in the category of medium-low social and environmental risks encountered during the construction of the Sicincin-Padang toll road.

Figure 2. Toll Road Path for Agricultural Land (source: PT Hutama Karya documentation, 2021)

This figure shows the real condition of the Sicincin-Padang toll road construction process that passes through the agricultural areas of the affected communities, thus impacting the livelihood patterns and the standard of living.

3.2. Stage 2: Problems Situation Expressed

Problems related to environmental and social aspects are richly illustrated in the following picture.

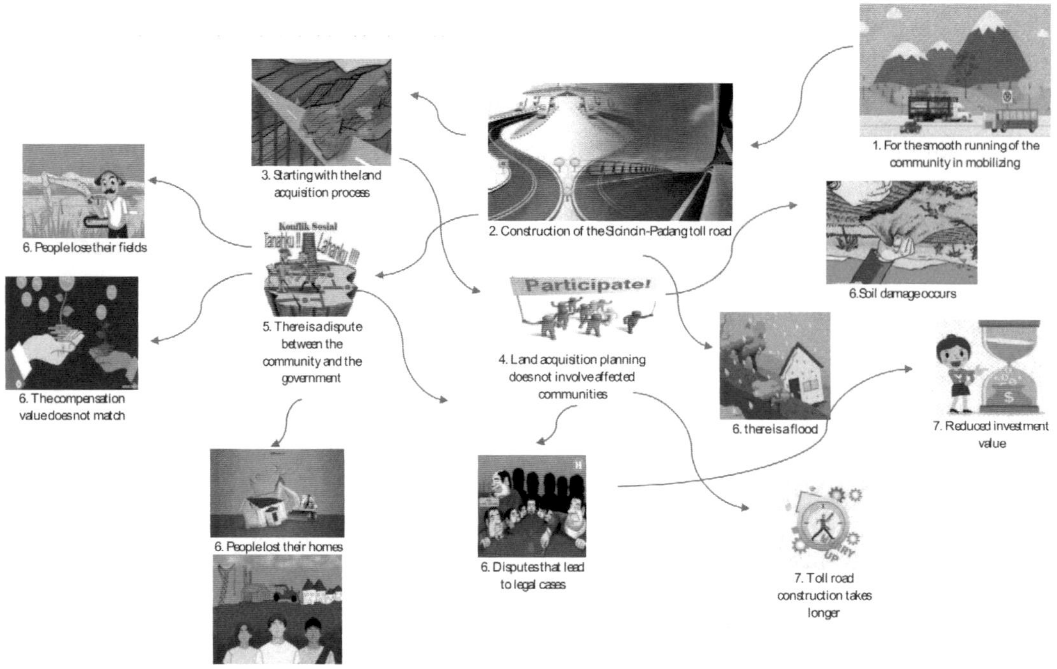

Figure 3. Socio-Ecological Risk Mitigation Problems

3.3. Stage 3: Select Concepts that may be Relevant to Root Definitions

The problem encountered is the construction of acquisition of land for the toll roads. This was done before the construction process began to guarantee the availability of land for the erection of the road. However, regardless of the feasibility of research during its implementation, certain problems have still encountered. The most prominent is determining the road route and the compensation received by the community. This also has an impact on the environmental and social problems experienced by the people. A CATWOE analysis is used to illustrate the relationship of this problem with respect to the interests of each party, and it is stated as follows.

Table 2. Root Definitions Problems with CATWOE Analysis

No	Component	Definition of Each Component
1	Customer/ Client	Communities who are entitled to receive compensation for land acquisition for the Sicincin-Padang toll road section, West Sumatra
2	Actors	Land Procurement Task Force, the West Sumatra Provincial Land Agency (BPN), PPK for the Land Procurement Section of the Sicincin-Padang Toll Road Ministry of Public Works and Housing of Indonesia (PUPR RI), Local Government (West Sumatra Province and Padang Pariaman Regency), NGOs, academics
3	Transformation Process	Realizing social and environmental risk mitigation systems for toll road construction that pay attention to aspects of community participation
4	Weltanschauung	The realization of a participatory risk mitigation policy or model that pays attention to environmental and social aspects and is equitable for affected communities
5	Owner	BPN, Ministry of Public Works and Housing of Indonesia (PUPR RI)

No	Component	Definition of Each Component
6	Environment Constrain	Understanding and implementing regulations governing participatory-based socio-environmental risk mitigation systems (taking into account the aspirations of affected communities)

Source: data processed by researchers, 2022

This was continued by developing a conceptual model involving the measurement of certain criteria.

3.4. Stage 4: Assemble concepts into an intellectual structure to develop conceptual models

The conceptual model describes the relationship between certain activities and the diverse role of each party in achieving their respective goals [28, 30]. It is measured based on three criteria, namely efficacy, efficiency, and effectiveness [26].

Table 3. Measurement Criteria for the Conceptual Model of the Socio-ecological Mitigation System Model

Efficacy	Increasing the frequency of community involvement in toll road construction sites and formulating a risk mitigation system related to the occurrence of social and environmental impacts.
Efficiency	The participation of affected communities in formulating a social risk mitigation system for toll road construction tends to minimize the occurrence of social and environmental impacts.
Effectiveness	The implementation of toll road construction triggers efficient community mobility and reduces social and environmental risks, especially for those who live in the location.

Based on these measurement criteria, a conceptual model for mitigating these social-environmental risks is developed, and outlined in the form of a chart, as follows:

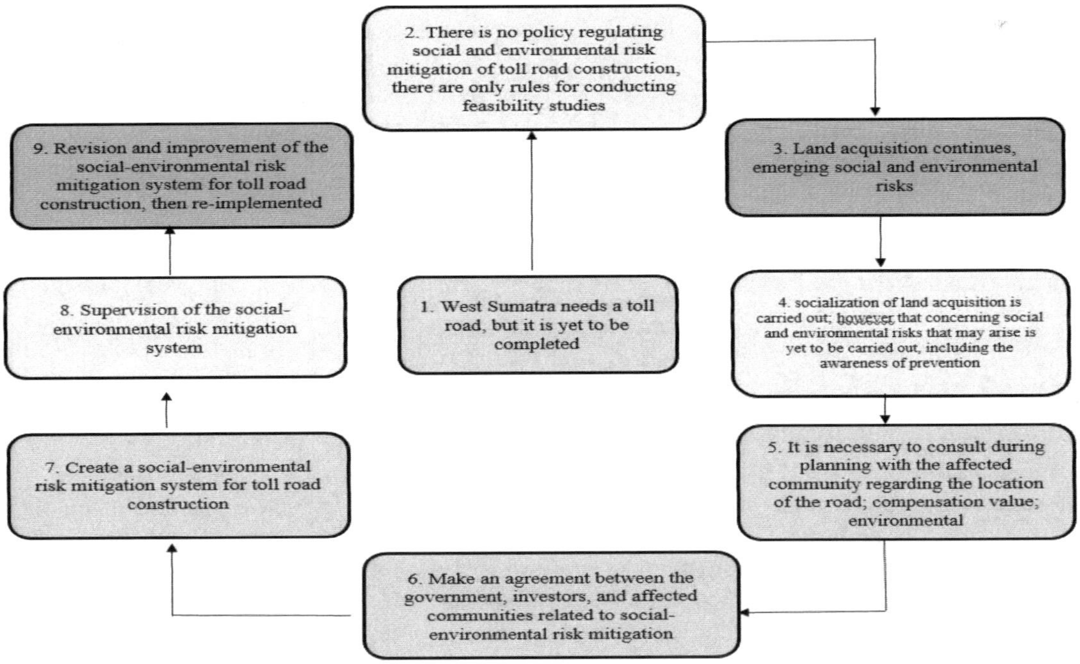

Figure 4. Conceptual Model of Socio-Ecological Risk Mitigation of Toll Road Construction

The resulting conceptual model tends to function effectively if all strategic assumptions are met. Furthermore, these are divided into three categories. The macro category mandates the need for regulations that govern the social-ecological risk mitigation system related to the development of toll roads. Assuming that it does not yet exist, there is a need to enact a policy on this matter. Based on the meso aspect, every stakeholder involved in the construction of toll roads must comply with the stipulated rules and tasks, irrespective of their diverse roles and functions. These are related to the community's implementation of socialization and public consultation. It is also related to the micro category's strategic assumptions that the condition of the affected community needs to be considered by policymakers in mitigating these socio-ecological risks. Stakeholders must discuss and deliberate with affected communities regarding the hopes and concerns that overshadow the construction of the toll road. Good communication between these parties tends to lead to participatory planning or decision-making processes. The three categories serve as a reference for formulating a socio-ecological disaster risk mitigation system for toll road development. This must be monitored and evaluated to ensure its stability and sustainability.

3.5. Stage 5: Comparison of models of the real world with the conceptual model

At this stage, the conceptual model is tested with real-world conditions. The evaluation was conducted through interviews with several experts to verify and validate the model. Based on this analysis, the social-ecological risk mitigation system is yet to be monitored by standard regulations. There is a need to carry out social and environmental feasibility research before the implementation of the toll road construction. However, this does not guarantee that there will be no risk during the development process. This is because to a greater extent, this project is a feasible physical development and the tendency to develop the affected community is only a formality.

The most important aspect is to establish effective communication between stakeholders and affected communities. To a greater extent, the toll road construction carried out by the government tends to be a sustainable business-oriented project. Although the essence is to develop the community, if the construction turns out to be poor, it is concluded that the development has not been successful. The comparison of real conditions with the conceptual model involves the affected community developing a sustainable social-ecological risk mitigation system.

The sixth and seventh stages are centered on developing a social-ecological risk mitigation system and providing policy recommendations. Based on this explanation, it was reported that the social-ecological mitigation system designed by stakeholders who formulated the policies has to pay attention to aspects of community participation, analyze their needs, mitigate planning based on the principles of ecological sustainability and the social risk index related to land acquisition.

4. Discussion

The socio-ecological mitigation scheme related to the implementation of toll road construction in West Sumatra was analyzed using a soft system methodology. The results obtained show that it is significant during the development process. Socio-ecological risk mitigation is an effort to overcome social and environmental problems and realize sustainable development. Elliot stated that it is realized through environmental and ecosystem management boosted by 10 ecological principles. These include social tolerance, ecologically sustainable, economically viable, technologically feasible, administratively permitted by law, politically wise, culturally inclusive, ethically defensible, and tend to be communicated effectively [31].

Based on the results, the most crucial principle that stakeholders lack is the implementation of effective communication between the parties involved. It is considered important because, through effective communication, risk mitigation efforts are carried out properly, including environmental problems arising from the construction of the toll road, such as flooding and other potential natural disasters. This is consistent with previous research that boosts infrastructure development, thereby overcoming natural disasters, including compliance with existing policies or regulations [32]. Furthermore, it is also related to the coordination of actors involved in the mitigation system, thereby achieving administrative achievements by maintaining effective communication with the community [33,34].

The next aspect relates to ecological sustainability, which aims to maintain an ideal ecosystem of the natural environment. The natural potential of the unspoiled Sicincin-Padang toll road construction site needs to be maintained and preserved. This is because ecological sustainability also involves preventing various diseases that are bound to infect humans [35]. Furthermore, the social-ecological risk mitigation system in toll road construction tends to pay attention to government funding because its lack causes the project to not function effectively [36]. This also includes climatic conditions because the resilience of this construction depends on the tools, materials, and technology used [37]. Socio-ecological mitigation systems that include climate variables in overcoming social and environmental risks have been used in infrastructure development in developed countries [34]. This needs broad attention in developing nations, including Indonesia, and the formulated policy needs to consider these aspects.

Based on the conceptual model and real conditions regarding the social-ecological risk mitigation system, the policy formulation must include environmental principles during development. In addition, it also adopted the concept of green infrastructure, which led to the participation of affected communities in formulating and supervising the policy, a participatory social-ecological risk mitigation system model associated with toll road construction.

5. Conclusion

Construction of toll roads in developing countries, such as the one erected in Sicincin-Padang, Indonesia, is relatively experiencing social and environmental constraints. The socio-ecological risk mitigation was carried out using a soft system methodology (SSM). This approach explains the problem more clearly using the CATWOE analysis to produce a conceptual model as the basis for policy formulation. Furthermore, it can be compared with real conditions, allowing stakeholders in designing a participatory social-ecological risk mitigation policy or model that pays attention to the communication between the parties involved. This refers to ecological principles and green infrastructure to realize sustainable toll road development.

6. References

[1] Putri, N. E., Helmi, Noer, M., & Yossyafra 2021 Risk Analysis and Potential Benefits of Infrastructure Development. *MIMBAR : Jurnal Sosial Dan Pembangunan*, ***37***(2), 236–248

[2] Athfal, I. F 2019 *Pengadaan Tanah untuk Pembangunan Jalan Tol di Kabupaten Pemalang*.

[3] Damayanti, E 2019 Pelaksanaan Tukar Guling (ruislag) Tanah Wakaf di Kota Semarang Untuk Pembangunan Jalan Tol Trans Jawa. In *Journal of Politic and Government Studies*. ejournal3.undip.ac.id.

[4] Lova, E. F., Fadli, M., & Susilo, H 2016 Pengadaan Tanah Dalam Pembangunan Infrastruktur Jalan Tol oleh Badan Usaha Milik Negara (Persero). *Jurnal Universitas Brawijaya*, 1–30

[5] Simamora, B 2017 Pelaksanaan Pengadaan Tanah Untuk Kepentingan Umum Jalan Tol Pekanbaru–Kandis–Dumai. *Jurnal Hukum Respublica*, ***17***(1), 170–188

[6] Dewi, A. R., Sutaryono, & Nurhikmahwati, A 2020 Pemetaan Masalah Pengadaan Tanah dengan Objek Tanah Ulayat (Kasus Jalan Tol Padang-Sicincin). *BHUMI: Jurnal Agraria Dan Pertanahan*, ***6***(2), 277–291

[7] Reba, W. H 2018 The Existance of Indigenous Legal Community Rights And The Position In Indonesia Legal System. In *Papua Law Journal*. pdfs.semanticscholar.org.

[8] Ding, C 2007 Policy and praxis of land acquisition in China. *Land Use Policy*, ***24***(1), 1–13

[9] Hall, J. D., & Savage, I 2019 Tolling Roads to Improve Reliability. *Journal of Urban Economics*, ***113***(July 2018), 103187

[10] Meadows, J., Annandale, M., & Ota, L 2019 Indigenous Peoples' participation in sustainability standards for extractives. *Land Use Policy*, ***88***(July), 104118

[11] Reyes-García, V., Fernández-Llamazares, Á., Bauchet, J., & Godoy, R 2020 Variety of indigenous

peoples' opinions of large infrastructure projects: The TIPNIS road in the Bolivian Amazon. *World Development*, ***127***.

[12] Cernea, M. M 2000 *Impoverishment Risks , Risk Management , and Reconstruction : A Model of Population Displacement and Resettlement*. 1–61.

[13] Eguavoen, I., & Tesfai, W 2012 Social impact and impoverishment risks of the Koga irrigation scheme, Blue Nile basin, Ethiopia. *Afrika Focus*, ***25***(1), 39–60

[14] Bensi, M., Kiureghian, A. Der, Asce, M., & Straub, D 2015 *Framework for Post-Earthquake Risk Assessment and Decision Making for Infrastructure Systems*. *Id*, 1–17.

[15] Vij, S., Moors, E., Ahmad, B., Md. Arfanuzzaman, Bhadwal, S., Biesbroek, R., Gioli, G., Groot, A., Mallick, D., Regmi, B., Saeed, B. A., Ishaq, S., Thapa, B., Werners, S. E., & Wester, P 2017 Climate adaptation approaches and key policy characteristics: Cases from South Asia. *Environmental Science & Policy*, ***78***, 58–65

[16] Kadarisman, M 2015 Implementasi Kebijakan Sistem Transportasi Darat dan Dampaknya terhadap Kesejahteraan Sosial di Jakarta. *Jurnal Manajemen TRansportasi Dan Logistik*, ***02***(01), 59–78

[17] Ammenberg, J., Anderberg, S., Lönnqvist, T., Grönkvist, S., & Sandberg, T 2018 Resources , Conservation & Recycling Biogas in the transport sector — actor and policy analysis focusing on the demand side in the Stockholm region. *Resources, Conservation & Recycling*, ***129***, 70–80

[18] Chi, S., & S, B. J. K 2016 Comparative Case Study on Cost- Benefit Analysis For Toll Road Projects. *27th ARRB Conference-Linking People, Places and Opportunities*.

[19] Connolly, C., Livy, M. R., Qiu, Y., Klaiber, H. A., Polytechnic, C., Obispo, S. L., & States, U. 2019 Landscape and Urban Planning Capitalization of interconnected active transportation infrastructure. *Landscape and Urban Planning*, ***182***, 67–78

[20] Novianti, T 2011 *Pemodelan Risiko Pendapatan Proyek Infrastruktur Jalan Tol dengan Pendekatan Fault Tree Analysis*. ***6***(2), 138–149.

[21] Sandhyavitri, A., & Saputra, N 2019 Analisis Risiko Jalan Tol Tahap Pra Konstruksi (Studi Kasus Jalan Tol Pekan Baru-Dumai). *Jurnal Teknik Sipil*, ***9***(1), 1–19

[22] Akimura, S 2015 Transportation statistics that can contribute to policies and social infrastructure development aimed at ensuring the healthy growth of cities and providing support for smooth economic activity. *IATSS Research*, ***39***(1), 9–18

[23] Chayandito, F 2006 Pembangunan Berkelanjutan, Ekonomi Dan Ekologi, Sustainability Communication Dan Sustainability Reporting. *Jurnal Fakultas Ekonomi Dan Bisnis LMFE*, *i*(022), 1–12.

[24] Gross, J., LeRoy, G., & Janis-aparicio, M 2002 *Community Benefits Agreements : Making Development Projects Accountable*. Good Jobs First.

[25] Short, J., & Kopp, A 2005 Transport infrastructure : Investment and planning . Policy and research aspects *. *Transport Policy*, ***12***, 360–367

[26] Muhammaditya, N., Hardjosoekarto, S., Herwantoko, O., Fany, Y. G., & Subangun, M. I 2021 Institutional Divergence of Digital Item Bank Management in Bureaucratic Hybridization: An Application of SSM Based Multi-Method. *Systemic Practice and Action Research*, *0123456789*

[27] Jassbi, J., Makvandi, P., Ataei, M., & Sousa, P. A. C 2011 Soft system modeling in transportation planning: Modeling trip flows based on the fuzzy inference system approach. *African Journal of Business Management*, ***5***(2), 505–514

[28] Petir Papilio, M. S. M 2015 Model kebijakan pengelolaan agroindustri bioenergi dalam perspektif kelestarian lingkungan (Soft System Methodology Sebagai Suatu Pendekatan). *Penelitian Dan Aplikasi Sistem Dan Teknik Industri (PASTI)*, ***IX***(1), 10–18.

[29] Burge, S 2015 An Overview of the Systems Methodology. In *Systems Thinking: Approaches and Methodologies* (pp. 1–14). Burge Hughes Walsh

[30] Irwanti, A., Hariyadi, P., & Handoko, L. T 2020 A Model of Government Intervention upon a Technology Transfer Program for the Product Innovation Improvement of Food Micro-Small Enterprises. *International Journal of Innovation, Creativity and Change*, ***14***(8).

[31] Barnard, S., & Elliott, M 2015 The 10-tenets of adaptive management and sustainability – Applying

an holistic framework for understanding and managing the socio-ecological system. Environmental Science & Policy, **51**, 181–191

[32] Maikhuri, R. K 2017 Socio-ecological vulnerability : Assessment and coping strategy to environmental disaster in Kedarnath valley , Uttarakhand , Indian Himalayan Region Machine Translated by Google International Journal of Disaster Risk Reduction Socio-ecological vulnerabil. *International Journal of Disaster Risk Reduction*, ***25***, 111–124

[33] Kolden, C. A 2019 *Case Report A Socio-Ecological Approach to Mitigating Wildfire Vulnerability in the Wildland Urban Interface : A Case Study from the 2017 Thomas Fire*. ***1***, 1–19

[34] Shi, L 2020 Beyond flood risk reduction : How can green infrastructure advance both social justice and regional impact ? *Socio-Ecological Practice Research*, ***2***(4), 311–320.

[35] Combs, M. A., Kache, P. A., Acker, M. C. Van, Gregory, N., Plimpton, L. D., Tufts, D. M., Fernandez, M. P., & Wasser, M. A. D 2021 ecological drivers of multiple zoonotic hazards in highly urbanized cities. *Global Change Biology*, *00*(February 2022), 1–20.

[36] Jellineka, S., Rumpffa, L., Driscollb, D. A., Parrisa, K. M., Wintlea, B. A., & Place, B. (n.d.). *Modelling the benefits of habitat restoration in socio-ecological systems*.

[37] Hayes, S., Desha, C., Burke, M., Gibbs, M., Hayes, S., Desha, C., Burke, M., Gibbs, M., Desha, C., & Burke, M 2019 Leveraging socio-ecological resilience theory to build climate resilience in transport infrastructure. *Transport Reviews*, ***39***(5), 677–699

Analysis of the application of injection wells in the Undip Tembalang campus area as an effort to reduce drainage flow loads

A Sarminingsih[1*], M A Budihardjo, N Damayanti[1], A P P Priyandoyo[1], N T Salsabila[1]

[1]Department of Environmental Engineering, Faculty of Engineering, Universitas Diponegoro, Semarang – Indonesia 50275

aniksarminingsih@lecturer.undip.ac.id

Abstract. Increased infrastructure development in the Undip Tembalang Campus Area causes an increase in the frequency of floods and inundation, as well as a decrease in groundwater availability Undip still uses groundwater wells as the primary source to meet clean water needs. One alternative to groundwater conservation is to inject rainwater using infiltration wells. The hydrology analysis was carried out using the EPA SWMM 5.2 software, while the hydraulic analysis was carried out by dredging the sediment along the channel. Calculation of planned infiltration wells is calculated according to SNI 03-2453-2002. The dimensions of this infiltration well are designed to be 1 m in diameter and 18 m deep. Based on the results of the analysis, it found that runoff in the Undip Campus Area was 66,467.97 litres/second. The calculation results show that the Undip Campus Area requires 1350 infiltration wells with a rainwater harvesting volume of 14148 m^3. The total runoff that can be reduced using infiltration wells is 45,978.74 litres/second or 69% of the total runoff. This plan is hoped to be an alternative water source in the Undip Tembalang Campus Area.

1. Introduction

In general, drainage can be defined as a technical action to reduce excess water, either from rainwater, seepage, or excess irrigation water from an area/land, so that the function of the area/land is not disturbed [1]. (Suripin, 2004). Drainage is to divert excess water (mainly rainwater) to the nearest water body. Excess water is immediately drained into drainage channels, rivers and the sea so as not to cause inundation or flooding [2]. (Hardjosuprapto, 1999). Several factors causing flooding include reduced river capacity, increased flood discharge due to changes in land use, damage to flood control buildings, and others[3]. The drainage system is a system designed to overcome the problem of excess water. Drainage buildings are water structures intended to control excess water, by absorbing, temporarily storing, and draining water to a place but without causing new negative impacts [4].

Diponegoro University (Undip) is one of the leading universities in Indonesia which is located in Tembalang, Semarang City.The available document regarding the development of the Undip campus is the Review of the Masterplan for the Undip campus development plan for 2013-2026 which was published in 2017 [5]. The master plan document also reviews the management plan for the drainage system and clean water in the Undip area. For the evaluation of the drainage system, rain data for the period 2003-2012 was used. Recommendations from the study include increasing the capacity of drainage channels in several locations and installing infiltration wells. Meanwhile, for the management of clean water, they still use drilled wells.

Along with the increase in development in the Undip area, in recent years there have been puddles in several places such as roads and parking lots that disrupt campus activities. Considering these problems, this study aims to evaluate the drainage system in the Undip area based on changes in land cover and changes in hydrological data until 2021. This is considered important considering that changes in land cover in the Undip area and changes in hydrological data will greatly affect surface runoff and will ultimately affect drainage capacity. The concept of handling drainage problems will use the concept of sustainable urban drainage system, where runoff water outside the Undip area is kept to a minimum. Efforts are being made to include rainwater into infiltration wells. Infiltration wells not only function to reduce surface runoff, they are also intended as a conservation effort to increase the groundwater level, so that groundwater as a source of clean water in the Undip area will be maintained.

2. Methodology

The methods for data collection are divided into two, namely primary data (survey and field documentation) and secondary data (literature study and copying data) [6]. In addition, there is data analysis carried out with the following stages:

2.1. Existing Condition Analysis

This analysis is carried out to determine the existing conditions of the study area, in the form of an overview, the existing drainage system and inundation areas that can be done in two ways, namely site surveys and analyzing physical conditions (contour, spatial detail plan and system flow direction).

2.2. Determination of Design Flood Discharge Analysis

In this analysis, the required data is rainfall data, calculation of concentration time, calculation of designed rainfall, and the area that becomes the flow area. The stages of calculating the design flood discharge are:

1. Determination of Rain Observation Station
2. Determination of Design Rainfall
3. Frequency Analysis
4. Determination of Design Rainfall Intensity
5. Flow Coefficient Calculation
6. Concentration Time Calculation
7. Design Flood Discharge Calculation .

2.3. Drainage System Redesign Analysis with SUDS Concept

In this stage, the calculation and design of drainage channels using hydrological and hydraulic analysis methods will be carried out, then an analysis will be carried out on the calculations that have been carried out with the EPA-SWMM 5.2 application modeling [7]. This simulation model checks the suitability of the drainage network capacity to accommodate the maximum runoff discharge in the design rainfall. After that, recommendations are given to apply the concept of Sustainable Urban Drainage System in accordance with the conditions of the planning area and the needs in the area.

3. Results and Discussions

3.1. Existing Condition Analysis

The existing drainage channel in the planning area uses a mixed drainage system because the drainage system not only drains rainwater, but also waste water from buildings. According to the results of the field survey, damaged drainage channels, several inundation points are still found when the rainfall is high, there is still garbage and sediment that clogs the flow of the drainage system, and the maintenance of drainage channels is not optimal. One of the worst inundation points is in the Widya Puraya Area where the inundation point reaches a height of 70 cm and lasts an average of 30 minutes [8].

Figure 1. Puddle of Water in the Widya Puraya Area

3.2. Hydrology Analysis

Hydrology analysis is needed to plan the drainage system. Hydrology analysis is the most important factor in planning the size of storage and drainage facilities. This analysis is necessary to overcome the surface runoff that occurs so as not to cause the inundation. The output of the hydrology analysis is to get the planned flood discharge in analyzed channel, then a hydraulic analysis is carried out using the EPA-SWMM 5.2 modeling application.

3.2.1. Regional Rainfall

Rainfall data needed to calculate the maximum daily rainfall from the nearest rain gauge station in the study area. The rain stations used in calculating the hydrology of the drainage system of the Undip Tembalang Campus Area are the Meteseh Rain Post and the Klipang Rain Post with an observation period of 10 years from 2012 to 2021 which were obtained from Semarang City Meteorology, Climatology and Geophysics Agency (BMKG). Rainfall data used in the calculation can be seen in Table 1.

Table 1. Maximum Rainfall Data for Meteseh Rain Station and Klipang Rain Station

No	**Year**	**Rain Data (mm)**	
		Meteseh Rain Post	**Klipang Rain Post**
1	2012	107	87
2	2013	109	93
3	2014	71	91
4	2015	110	157
5	2016	121	50
6	2017	90	95
7	2018	97	149
8	2019	120	95
9	2020	85	108
10	2021	83	104

3.2.2. Frequency Analysis

The purpose of frequency analysis of hydrological data is about the magnitude of extreme events related to their frequency of occurrence through the application of probability distributions. Frequency analysis is an analysis of the repetition of an event to determine the amount of rainfall or discharge for a certain return period using statistical calculation methods. Frequency analysis is based on statistics and available data to obtain the probabilities used.

Table 2. Frequency Analysis Results [9]

No	Distribution Type	Requirement	Calculation Results	Description
1	Gumbel	Cs ≈ 1.1396	-0.48	Not Qualified
		Ck ≈ 5.4002	3.35	
2	Log Normal	Cs ≈1.137	-0.96	Not Qualified
		Ck ≈ 5.383	4.08	
3	Log-Person type III	Cs ≠ 0	-0.9582	Qualified
		Cv ≈ 0.3	0.03	
4	Normal	Cs ≈ 0	-0.48	Not Qualified
		Ck ≈ 3	3.35	

The calculation of flood discharge for the return period T using the Log Pearson Type III method can be found using the formula [1]:

$$\text{Log XT} = \text{Log } \bar{X} + K.Sd$$

Where:
XT = Return Period Rainfall (mm)
X = Average Rainfall
K = K value for Log Pearson type III Distribution
Sd = Standard Deviation

The results of the calculation of the maximum daily rain using the Log Pearson Type III method are as follows:

1. The return period is 2 years, and the rainfall value is 99 mm
2. The return period is 5 years, and the rainfall value is 111.98 mm
3. The return period is 10 years, and the rainfall value is 119.44 mm
4. The return period is 20 years, and the rainfall value is 125.92 mm
5. The return period is 50 years, and the rainfall value is 133.72 mm

3.2.3. Rainfall Intensity Calculation
Rainfall intensity is the height of rainfall that occurs per unit time. Semarang City is a Metropolitan City with an area of 101-500 hectares. Based on the Minister of Public Works Regulation No. 12 of 2014, Semarang City uses a 5-10 year return period [10]. The results of the calculation of rainfall intensity with the Mononobe method are shown in Figure 2 with the calculation of the intensity of rainfall which is formulated as follows:

$$I = \frac{R_{24}}{24}\left[\frac{24}{T_c}\right]^{2/3}$$

Where:
I = Rainfall intensity (mm/hour)
R24 = Rainfall planned return period (mm)
Tc = Rainfall duration (hours)

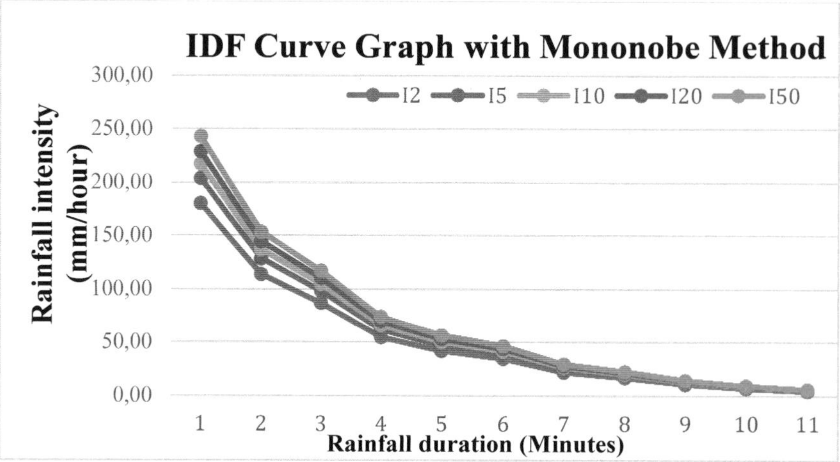

Figure 2. Graph of IDF Curve with Mononobe Method

3.3. modelling EPA SWMM 5.2 existing condition

Simulation of existing conditions using EPA SWMM 5.2 was carried out to determine the response of water flow which shows the occurrence of water runoff under existing conditions in the planning area[4]. The existing canal data was obtained from direct measurements of drainage channels in the planning area and the master plan of the University of Diponegoro in 2020-2029.

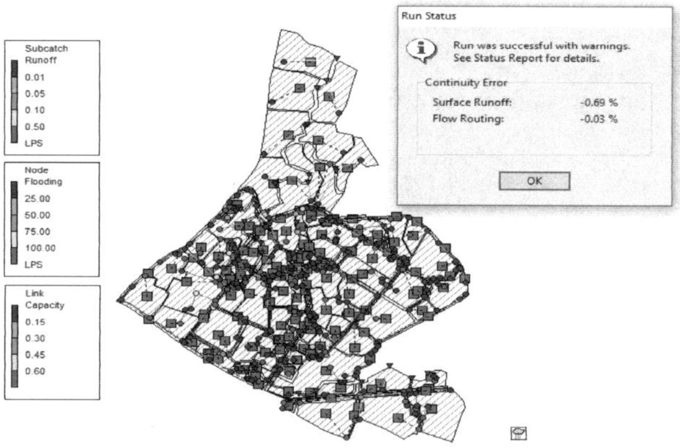

Figure 3. SWMM simulation results in existing conditions

The simulation results that have been carried out show the surface runoff value of -0.69% and the value of continuity error flow routine -0.03%. The analysis of the EPA SWMM modelling of the existing condition is focused on the sixth hour, where the peak hour occurs. The simulation found 128 overflow channels because the drainage capacity was not suitable or damaged, and there was a pile of sediment. For example, Channel I Taman Inspiration A, Jl. Prof. Eko Budihardjo Sipil, and Jl. Prof Soedarto Sipil-Geodesi

3.4. EPA SWMM modelling in Normalized condition

The SWMM modelling simulation in this normalized condition is carried out by carrying out a modelling scenario, namely dredging sedimentation in the channel. If there is still runoff in the normalization modelling simulation of sediment dredging, then this is the basis for applying the concept of sustainable drainage or SUDS. Sediment in the drainage channel of the planning area has a height of about 1 cm – 20 cm. after dredging the sediment, the overflow channel is reduced because the drainage capacity is increased.

Figure 4. SWMM simulation results in the planning area after sediment dredging

Normalization is only in the form of dredging sediment without adding the dimensions of the drainage channel because the condition of the planning area itself is no longer possible to add channel dimensions. From the simulation on normalization engineering, it is found that 64 channels are still overflowing, as in the Jl. Prof. Soedarto SAMWA 2. There is a channel that does not flood after normalization, for example channel I Taman Inspirasi A

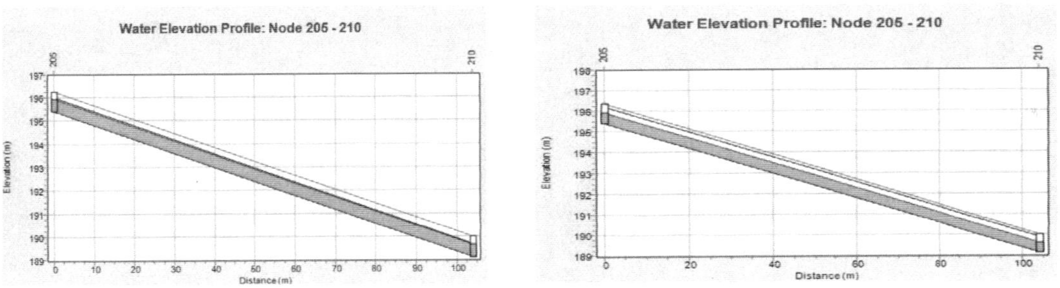

Before After

Figure 5. Example Channel Profile Jl. Prof. Soedarto SAMWA 2 Before and After Sediment Dredging

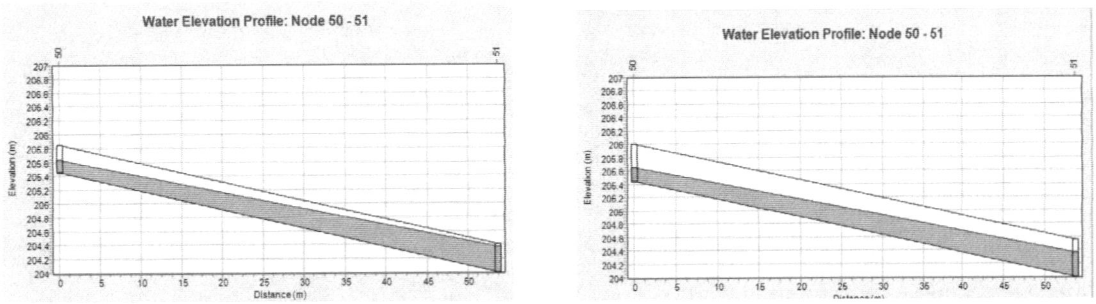

Before After

Figure 6. Example Channel Profile Taman Inspirasi A Before and After Sediment Dredging

3.5. Application of Injection Well

The concept of Sustainable Urban Drainage System (SUDS) is applied to the planning area of the Undip Tembalang Campus Area because it can manage runoff on a small scale starting from the source of the problem. The SUDS used in the planning is infiltration wells by draining rainwater from the roof of the building and the road leading to infiltration wells which will later be absorbed into the ground.

The planned injection well is 1 meter in diameter with a depth of 18 meters, which will be placed next to the drainage channel with a distance between injection wells of 6 meters. Later, water from the drainage channel will not flow to the outlet but will flow into the injection well to be absorbed into the groundwater surface.

Figure 7. SWMM simulation results for the application of injection well

Input data into the EPA SWMM 5.2 software from the concept of Sustainable Urban Drainage System, namely by opening a hydrology project on SUDS Controls, which is the application menu for implementing a sustainable drainage system consisting of a Rain Garden, Infiltration Trench, Rain Barrel, Permeable Pavement, Bioretention and so on [7].

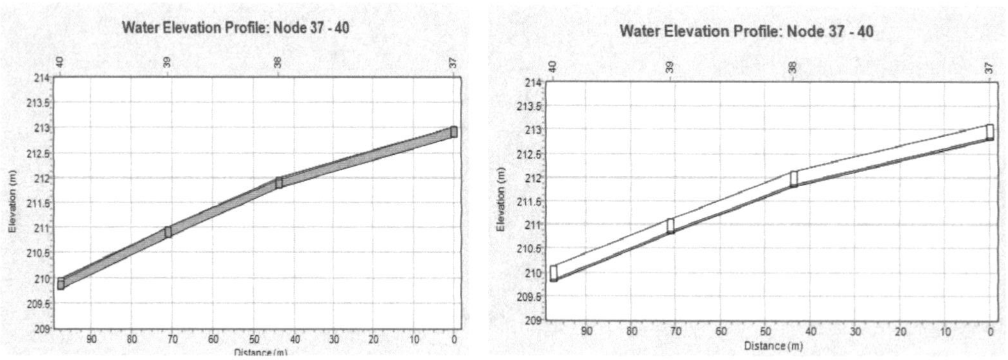

Before After

Figure 8. Example of Inspiration Garden Channel Profile A 2 Before and After the application of injection well

Calculation of planned infiltration wells is calculated according to SNI 03-2453-2002 [10]. The soil permeability coefficient in the planning area is 0.77 [11], with a porous layer height of 0.786. It is assumed that the average rain is 2 hours, so the total injection well capacity is 14148 m3.

Based on the results of the analysis, it found that runoff in the Undip Campus Area was 66,467.97 litres/second. Requires 1350 infiltration wells to reduce the runoff by 45,978.74 litres/second, or 69% of the total runoff.

4. Conclusion

Based on the results of the evaluation of the drainage system in the Undip area, it was concluded that there were several canal sections that were unable to accommodate the runoff discharge due to sedimentation or also because the dimensions of the channel were not sufficient.

It is necessary to maintain channel properly including sediment dredging and waste management so that it does not enter the canal. In order to anticipate the increase in runoff due to changes in land cover into built-up areas as well as the possibility of increasing rainfall intensity due to climate change, it is necessary to apply SUDS/LID by infiltrating runoff water using injection wells.

The number of injection wells required is 1350 units. The application of this injection well can reduce the total runoff value in each sub-watershed by 45,978.74 liters/second from 66,467.97 liters/second or the percentage decrease in the total runoff value of 69%. This injection well is also expected to be able to conserve groundwater, which is currently used as a source of clean water.

Acknowledgment

The author would like to express his deepest gratitude to the Department of Environmental Engineering, Faculty og Engineering Diponegoro University for the opportunity for strategic research and other relevant agencies who have helped the author provide references in the form of secondary data that are indispensable in this research..

References

[1] Suripin. 2004. Sistem Drainase Perkotaan Yang Berkelanjutan. Andi Offset. Yogyakarta
[2] Hardjosuprapto, M. M. 1999. Drainase perkotaan. Dep. Pekerjaan Umum: Jawa Barat
[3] A Sarminingsih and M Hadiwidodo (2021).Evaluation of flood risk as a step in determining the "Acceptable Risk" criteria. Case study: Dengkeng River, Klaten, Central Java, IOP Conf. Ser.: Earth Environ. Sci. 894 012045
[4] Kamiana (2011) 'Teknik Perhitungan Debit Rencana Bangunan Air Oleh : I Made Kamiana', (May).
[5] Yodya Karya, (2017). Penyempurnaan Review Masterplan Kampus Undip Tahun 2013-2026. Tim Swa Kelola Undip
[6] A Sarminingsih., M Hadiwidodo, A Rezagama, K S Sausan, Nurullah.(2021). Hydraulic analysis of Semarang River in supporting the drainage channel as water tourism. IOP Conf. Ser.: Earth Environ. Sci. 896 012010
[7] Rossman, A. Lewis, Storm Water Management Model Reference Manual, Water Quality III (2016).
[8] https://regional.kompas.com/read/2019/03/27/23273821/feb-undip-kebanjiran-hingga-sepaha-orang-dewasa-ini-penyebabnya
[9] Soewarno. 1995. Hidrologi aplikasi metode statistik untuk analisis data. Bandung. Penerbit Nova
[10] Badan Standarisasi Nasional. (2002). SNI 03-2453-2002
[11] Daru, K.M., Hamnda, A.W., Priyo, N.P., Dwi, K. (2013) Perencanaan Saluran Drainase Bersumur Resapan Di Lingkungan Kampus Universitas Diponegoro Tembalang-Semarang

IOP Conf. Series: Earth and Environmental Science 1098 (2022) 012070 doi:10.1088/1755-1315/1098/1/012070

Virtual Waste Community: Sustainable Living in Digital Era

B P Redyantanu[1], Y A Yatmo[1*], P Atmodiwirjo[1]

[1] Department of Architecture, Faculty of Engineering, Universitas Indonesia

yandiay@eng.ui.ac.id

Abstract. Waste is a problem in our society and still depends on the conventional management system. This paper discusses the interrelationship among virtual social media, waste management practice, and community-based economics in supporting sustainability and zero waste in the digital era. In particular, it attempts to identify the potential connections of social media in extending new environmental practices. Observation of two types of Indonesian Instagram accounts (personal and group) of waste-related illustrates that social media can act as an alternative platform to respond to waste issues. Qualitative case studies demonstrate the relationship between virtual social media, waste, and creative economic potential in various communities in Indonesia. This analysis focuses on independent efforts, especially regarding their potential related to environmental and economic factors. The findings show that virtual platforms related to urban waste management play an important role in three aspects. The personal aspect is related to building a sustainable lifestyle through informational and awareness content, the economic aspect is related to generating alternative circular creative economic for environmental communities, and environmental practice focuses on forming an independent and participatory waste management mechanism. Such connection in the virtual platforms brings new green digital infrastructure possibilities supporting sustainability.

1. Introduction

This paper explores the virtual relationship in social media Instagram that supports the expansion of waste management practices in urban areas. We often consider waste useless, but it plays an essential role in the life of the city environment and its society [1]. The observation of diverse waste management practices has the opportunity to open up various roles or multi-dimensional subjects, actions, and relationships between them [2]. The virtual concept is to see all relations between multiple actors and objects, which work as a system. [3]. Exploring the development of current waste management practices becomes relevant to see the future of the sustainability of a city.

Virtual communities and their digital relationships are part of our society today. Digital virtual relationships can act as a representation or a reflection of our environment and a response to societal changes. Social media, including Instagram, can be a representation of people's lives, including a reflection of their response to environmental sustainability. Still, they are becoming wider so that various new digital virtual connections based on the same attention to particular interests bring together multiple layers of society across dimensions and groups.

Instagram as a daily digital platform, has features and functions that allow the creation of connections that support various things in life [4]. Instagram users in Indonesia are very dominant in terms of number and intensity of usage [5]. Social media, a virtual connection connecting many people, can potentially solve problems in various ways, including allegations of urban waste management.

Conventional systems still dominate the majority of urban waste management. The waste management system in Indonesia is currently in a condition that needs improvement [6]. We can do many things to change this condition. Simple things can be started by increasing awareness of the environment, then reaching broader implications [7]. Activation of waste management efforts at the urban micro level has the opportunity to be more operational, contextual, and sustainable, rather than waiting for improvements on a large scale by the government.

This study aims to highlight the emergence of local Instagram accounts in Indonesia that focuses on waste awareness and management. This investigation seeks to discover the critical role of Instagram as a small to medium-level supporter for realizing a life with better municipal waste management. Digital activation is one of the keys to comprehending this improvement effort [8]. The connection between the expansion of information and the development of waste awareness has the opportunity to open alternative waste management practices and the creative economic potentials that arise from it.

2. Literature Review

Social media is an internet-based platform that brings together people from various backgrounds. The focus of this platform is user-generated content. Social media is not only a connecting society in terms of everyday use in today's digital era. However, social media has become a new form of the socialization process. Social media can simplify the flow of communication, transfer information quickly, and can help people with various needs. Social media can be a bridge for creating an instant and fast virtual relationship, which was not possible in the past.

As one of Indonesia's most significant social media, Instagram has become a representation and reflection of society's virtual and digital relationships and connections. Using Instagram has become a part of everyday life for Indonesian people. The ability of social media to connect various parties according to specific interests and concerns can carefully open multiple creative solutions to several types of urban problems.

In general, social media changes at least three things about society. New identities, new forms of social interaction, and new patterns of information distribution are the hallmarks of changes in this digital era. This digital social paradigm can be structured in various forms of integrated communication. According to Garcia [9], there are several models of interaction and communication in virtual digital relationships between users. The processes that occur in social media are interactive and multi-directional. The four social media connection categories are blogs, recommendations, bookmarking, and social networking. Various possibilities of virtual relationships that are formed lead to the potential use of social media to activate solutions for societal problems.

Social media, especially in Indonesia, is driven mainly by public figures and corporations on a large scale. Public figures and commercial media can be effective instruments for sharing information and building awareness in the community. This opportunity can evoke and direct urban lifestyles in more positive ways. Most people follow what they see on Instagram as an unconscious influence.

Regarding waste, an environmentally friendly lifestyle is starting to be encouraged, especially for young social media users [10]. Some of the lifestyle represented in social media is influenced by various social aspects that lead to environmental sustainability. This digital space can open opportunities spontaneously, broadly, and diversely for activating the movement based on environmental sustainability. Social media, especially Instagram, shows various accounts related to the interest based on awareness to waste as a movement that supports sustainability.

The commonality of attention to special things gives rise to various virtual communities on social media. Community support can influence individual behavior in environmental action [11]. Massively formed motivation in social media can generate significant movement. Social force is able to present a new perspective on reducing consumption and waste from daily activities [12]. Motivation delivered through social media becomes important in activating solutions to waste. Campaigns through social media lead to positive things, especially in waste production and management system [13]. The keyword community is essential here because there is a similar spirit, shared concern, and mutual support in realizing positive goals regarding waste reduction and management practice.

One of the concepts that emerged in social media is zero waste and circular economy. Individual waste management plays a crucial role since many types of waste begin at the household level [14,15]. The concept of zero waste does not mean actual zero waste but uses any material at an optimum level. Zero waste is the conservation of all resources for responsible production and consumption, which does not threaten the environment and human life [16]. The potential for zero waste does not only have an impact on improving the environment but even on economic potential [17][18]. The circular economy can replace a conventional economy based on reducing, reusing, recycling, and recovering materials in the production, distribution, and consumption phases [19–22]. Social media can be a virtual activation that becomes a concrete movement for environmental improvements.

This study investigates the connections between Instagram, waste management, and a community-based circular economy. Understanding virtual connections as a potential for better environmental and city management opens opportunities for other manifestations of green infrastructure in urban life.

3. Methodology

This study discusses awareness and practice of waste management based on virtual digital communities on Instagram social media. The study format is a qualitative case study. It is relevant in discussing this topic because it can be an actual reflection of our society today [23]. This study uses a particular search method to find suitable cases.

Instagram is a social media platform in this study [24]. Searches for community accounts are carried out using advance searches with the filter of follower number (more than 10k), location (Indonesia), and specific keywords [25], using *Inflact* search (https://inflact.com/tools/instagram-search/). The keywords are focused on *'zero-waste'* and *'sampah'* with combined results of around 150 accounts. In addition, other search methods are also carried out by opening polls on Instagram stories so that responses from considerate users can lead to recommended personal Instagram accounts that don't have any specific *'sampah'* description on their account. The third search method is through recommendations from several people who have implemented an environmentally friendly lifestyle related to waste, to find additional case studies of both group and personal influencer accounts.

Based on a limited *Inflact* account search focusing on waste, the first criterion is that they are still actively posting. The second criterion is that the user has a significant impact and interaction (more than 10k followers and active likes and comments). There are two categories: personal and group (media, communities, and organization). The three aspects that are the primary concern of this study include users, content, and interactions within them. Five to fifteen specific accounts selected as representatives focus on the impact of personal awareness and information access, an alternative circular economy, and creative aspects of waste management practices.

The virtual observation identifies the activities, visual content, and various other digital media (applications, group chat features, and so on) in the account. The visual analysis of the content shows the relationships that occur through diagrams to describe the virtual connections. This connection act as the basis for expanding different digital green infrastructures to support better waste management.

4. Result and Discussion

Exploration of Instagram's social media is focused on searching for users, content, and interactions that occur within them. The main concern of this study is to cover the topic of awareness and waste management, as well as its impact in three main respects, personal aspects, economic aspects, and aspects of waste management practices, as seen in Figure 1.

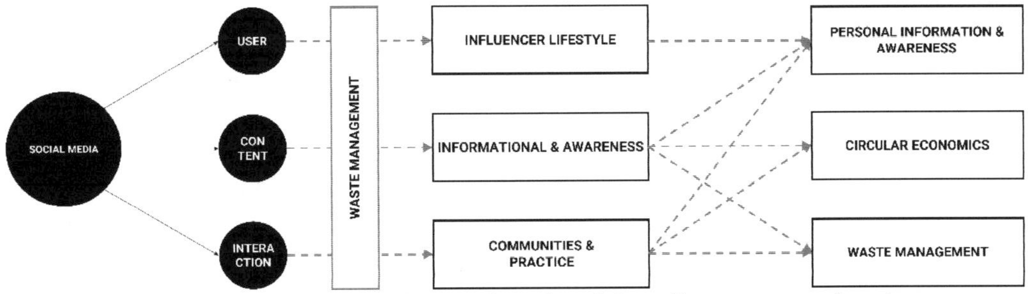

Figure 1 Study framework of waste-based social media account exploration

4.1. Personal: informational and awareness

Visual based content that leads to an understanding of waste problem can be found in personal-based accounts and media-based accounts, where both have opportunities to raise awareness and convey information [26]. A study says the emergence of a sense of necessity to live green and sustainable is more substantial if you have joined a community. Based on an exploration through search, polls, and recommendations, the personal account @atiit (Astri Puji Lestari) is the one that has quite a lot of influence on awareness of waste. In the highlight on her Instagram profile (Figure 2), Astri conveys her journey in awareness of waste through her anxiety about various packaged products used in daily life through a narrative stories. Astri invites her followers (60k followers) to participate and not burden the environment through better waste management, which is distributed to various parties who manage the waste. Messages and captions play an important role, where the proper caption can manipulate the perception of the reader [27]. The information conveyed is in the form of questions, invitations, and suggestions, thus enabling a more intense (two-way) information exchange interaction.

Figure 2 Personal Instagram account that shows the personal waste journey
Source: https://www.instagram.com/stories/highlights/17938369510093116/

Figure 3 Instagram post with visual and textual narrative format
Source: https://www.instagram.com/p/CgOD3zovL_z/

Media-based accounts convey messages, information, and invitations to manage waste better through informative and thematic visual content. Message delivery techniques are crucial in understanding the message [27]. One media-based account that is quite aggressive in voicing good waste management is @021Suarasampah. This account, with 26k followers, was initiated by Andhini Miranda, one of the zero waste practitioners. In the Instagram account @021Suarasampah, various information on various types of waste and efforts to managc it arc thc main conccrns in the content. For example, in one of his posts (Figure 3), this account shares messages through interesting visualizations and narration of factual information through narrative captions on posts. Stories about waste are not necessarily conveyed as instruction but through an exciting story fact. Most content posts want to share that through simple things; we can contribute quite a lot to reducing waste production, especially in community activities. Some other similar influencing accounts are @akademiminimsampah, @rethinkcampaign, @pilahsampah, @demibumi.id, @kotatanpasampah, @biomagg. Good waste management focuses not only on guilt but must be built from self-awareness slowly [28]. Information conveyed in a narrative format is more acceptable because we can digest it like experiencing a story.

Influencers are roles that can be relevant to specific relationships in content distribution, interaction and appearance on social media. The two accounts above are examples of how personal accounts and media-based accounts build awareness through cognitive and emotional aspects so that the invitation does not seem like an order. Information and content that is conveyed visually and textually, needs to appear authentic, attractive, and also personal and genuine, so that they can have a strong attachment to their followers. Through the exploration above, it can be conveyed that the impact on the personal aspect of social media's virtual relationship with waste is Instagram's ability to build information and awareness interestingly. Integrating content that appears unconsciously and also supported by Instagram algorithm, in social media consumption in everyday life raises awareness slowly and more deeply.

4.2. Economics: creative circular economic of environmental company

Organizational and community accounts can also be found on Instagram. One organization and business-based account that appears in polls and searches is @rekosistem. This account with 29k followers is a waste management company account. The ecosystem platform provides complete waste management information, from depositing it to its conversion into economic value in the form of vouchers and points.

The flow in the circular economy is to open up opportunities for sustainable transactions [19]. Social media platforms open up opportunities for various people to join in a socio-economic population that opens up opportunities for relationships with the environment. Rekosistem is an example of a platform initiated by an independent waste management company, where virtual digital technology such as applications can facilitate the occurrence of this economic mechanism. Figure 4 shows an example of how the flow that occurs in depositing waste becomes a monetary value through an application.

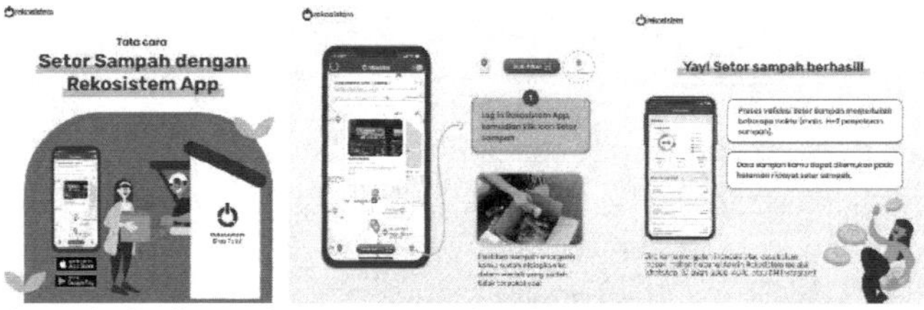

Figure 4 Application based platform that support circular economic based on waste management
Source: https://linktr.ee/rekosistem

Through the exploration above, the study can convey that social media has a role in connecting users to waste management companies. Waste management companies have various platforms such as applications, catalogues, etc. Social media, especially Instagram, has a role in being a virtual access and bridge on these multiple platforms. This fact is supported by more intense access to social media than other platforms (websites, applications, and so on). Some other accounts with similar character are @waste4change, @plasticbank, @siklusrefill, @jangjoindonesia, @universaleco.id, @evowareworld. More accessible access to information is accommodated by social media channels, leading to various other particular digital platforms (websites, group chat, informational catalogue, and so on).

4.3. Practices: community and participatory waste management practices

One of the most prominent accounts on Indonesian local Instagram related to waste is @zerowaste.id_official. This account is based on a non-profit community that invites people to minimize waste. The concept of zero waste does not mean zero waste but increases the use value of an item to the optimal point [16]. Waste does not always have to be leftovers but can be in the form of goods that can no longer function under certain conditions. One of the participatory efforts initiated by this community is the 'tukarbaju' activity, namely the management of clothing waste to be more beneficial to other people who are more suitable (Figure 5). The creative practice of changing clothes seems unrelated to waste management practices. However, with this activity, we can reduce the possibility of waste production from piles of old clothes.

Figure 5 Social media platform that connects particular communities
Source: https://www.instagram.com/tukarbaju_/

The Instagram platform allows for personal practice based on the exchange of information within the community. Sustainable communication focuses on the ecology and its people [29]. Zero waste concept that is closely related to the optimal use of goods [30], applied through a variety of other similar communities. Through Instagram also, recommendations for similar accounts can be intelligently generated through algorithms so that the possibility of access to various communities becomes more effective [31]. A sustainable community is a community that is initiated by the continuity of activities and togetherness of its members. Other similar communities based accounts are @re.pairs.id, @revup.id, @lyfewithless, @letsswap.id, @tri.cycle.id. The intense communication of exchange of information and ideas is more accommodated through this virtual platform.

Through the exploration above, this study can convey that social media has a role in connecting users with similar interests and focus. Through social media too, waste management practices are varied creatively. Activities such as changing clothes reflect the method of managing used goods according to the zero waste concept, where goods have an optimal level of usability. Social media, especially Instagram, accommodates the meeting of various parties who have the same concern so that the expansion of waste management practices becomes more widespread.

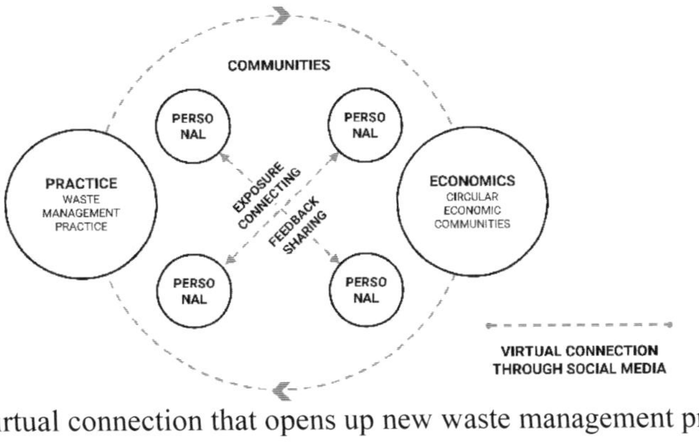

Figure 6 Virtual connection that opens up new waste management practice

In general, through the three types of research on the aspects above: personal aspects, economic aspects, and creative management practice, it is seen that social media has an important role. The virtual relationship formed through Instagram allows for various things that are difficult to imagine happening without going through this medium (Figure 6). Virtuality here is a connection made possible through a digital medium [3]. All activities, actors, and society are actual, while relationships, community, communication, and exchange of various information and awareness embodiments are the virtual potential to achieve a better environment.

5. Conclusion

This paper is an early exploration study that explores that the virtual connection formed by social media, especially Instagram, which has an essential role in realizing awareness through the expansion of information transfer and better management of urban waste. By observing the variety of accounts, content, and interactions in waste-based social media, it was revealed that there are at least three roles related to waste. The roles are building information and personal awareness, supporting a new alternative circular economy, and promoting independent management participation for sustainability.

Information dissemination and waste awareness are relevant on social media platforms because of their integration into everyday use. In addition, social media intelligent algorithm recommending various similar content according to particular interests is also an added value. The circular economy that arises from waste management is also possible through a community formed based on the same interests and concerns. Access of various waste management companies by the community becomes easy through the Instagram. Delivering information is the key to achieving an alternative waste management system. The last aspect is creativity that is formed in a participatory and independent method, where waste management is no longer on the principle of throwing away but becomes more than that. More optimum use of materials can occur when supported by particular communities.

Instagram has the potential to become a virtual digital-based green infrastructure related to independent, effective, and participatory waste management. This micro-level virtual infrastructure is not impossible to be a supporter and an alternative to standard conventional systems. Life in the digital era opens up opportunities for better environmental practices in the future. This analysis in this study is based on a limited number of cases; therefore, the result might not identify various forms of virtual connections. Expanding this study to include more cases to describe the virtual connection comprehensively becomes necessary.

Acknowledgments

The study presented in this paper is supported by Penelitian Dasar Unggulan Perguruan Tinggi (PDUPT), a research grant from Kementerian Pendidikan, Kebudayaan, Riset dan Teknologi Republik Indonesia (Ministry of Education, Culture, Research and Technology of Republic of Indonesia), under Grant Number: NKB-998/UN2.RST/HKP.05.00/2022.

References

[1] Lynch K 1990 The waste of place *Places* **6**
[2] Yatmo Y A, Atmodiwirjo P and Paramita K D 2013 Whose waste is it anyway? *J. Urban Des.* **18** 534–52
[3] Kalaga W 2003 The trouble with the virtual *Symplokē 11* **11** 96–103
[4] Hendricks D 2013 Complete history of social media: Then and now *Small Bus. trends* **8**
[5] Simon K 2022 Digital 2022: Indonesia — DataReportal – Global Digital Insights *Datareportal*
[6] Fatimah Y A, Widianto A and Hanafi M 2020 Cyber-physical system enabled in sustainable waste management 4.0: a smart waste collection system for Indonesian semi-urban cities *Procedia Manuf.* **43** 535–42
[7] Michael C and Utama D N 2021 Social media based decision support model to solve Indonesian waste management problem: An improved version
[8] Sivitanides M and Shah V 2011 The era of digital activism *Conference for Information Systems Applied Research* vol 4
[9] Garcia I 2011 Social media-integration: theory-model *Soc. Media Today*
[10] Pratiwi P Y, Handra T and Choirisa S F 2021 Determinants of Zero Waste Lifestyle Adoption Among Generation-Z *Conference Series* vol 3 pp 371–84
[11] Cole E J and Fieselman L 2013 A community-based social marketing campaign at Pacific

University Oregon: Recycling, paper reduction, and environmentally preferable purchasing *Int. J. Sustain. High. Educ.*
[12] Cecere G, Mancinelli S and Mazzanti M 2014 Waste prevention and social preferences: the role of intrinsic and extrinsic motivations *Ecol. Econ.* **107** 163–76
[13] Lumampauw A, Lestari R, Marta R F and Fernando J 2020 Awakening Revealed through Rose's Visual Analysis on\# LangkahHijau Commercial Ads *Avant Garde* **8** 121–33
[14] Filippides E 2019 Self Storage: A Contemporary Archaeology of Domestic Interiority *Interiority* **2** 5–23
[15] Marco E, Williams K and Oliveira S 2021 Prioritising storage practices: A new approach to housing design thinking *Interiority* **4** 223–48
[16] Săplăcan Z and Márton B 2019 Determinants of adopting a zero waste consumer lifestyle *Reg. Bus. Stud.* **11** 25–39
[17] Lehmann S 2011 Optimizing urban material flows and waste streams in urban development through principles of zero waste and sustainable consumption *Sustainability* **3** 155–83
[18] Zaman A U 2016 A comprehensive study of the environmental and economic benefits of resource recovery from global waste management systems *J. Clean. Prod.* **124** 41–50
[19] Kirchherr J, Reike D and Hekkert M 2017 Conceptualizing the circular economy: An analysis of 114 definitions *Resour. Conserv. Recycl.* **127** 221–32
[20] Shrivastava A, Jain G, Kamble S S and Belhadi A 2021 Sustainability through online renting clothing: Circular fashion fueled by instagram micro-celebrities *J. Clean. Prod.* **278** 123772
[21] Kale E 2021 *Influencer marketing in circular economy* (Masaryk University, Faculty of Economics and Administration)
[22] Happonen A and Ghoreishi M 2022 A mapping study of the current literature on digitalization and industry 4.0 technologies utilization for sustainability and circular economy in textile industries *Proceedings of Sixth International Congress on Information and Communication Technology* pp 697–711
[23] Paramita K D, Karimah A and Yatmo Y A 2021 Collective Strategies and Spatialities of Neighborhood Food Coproduction during COVID-19 Pandemic *Int. J. Technol.* **12** 1228–38
[24] Yang C and others 2021 Research in the Instagram Context: Approaches and Methods *J. Soc. Sci. Res.* **7** 15–21
[25] Ramjaun T A 2021 Exploring the\# zerowaste lifestyle trend on Instagram *Sustain. Debate Policies, Gend. Media* 205
[26] Lehbrink M 2020 *Being green on Instagram: A qualitative study on how green influencers are composing their messages and arguments of sustainability in their Instagram posts* (University of Twente)
[27] Cheng T, Woon D K and Lynes J K 2011 The use of message framing in the promotion of environmentally sustainable behaviors *Soc. Mar. Q.* **17** 48–62
[28] Koestner R, Houlfort N, Paquet S and Knight C 2001 On the Risks of Recycling Because of Guilt: An Examination of the Consequences of Introjection 1 *J. Appl. Soc. Psychol.* **31** 2545–60
[29] Kilbourne W E 2004 Sustainable communication and the dominant social paradigm: can they be integrated? *Mark. Theory* **4** 187–208
[30] Song Q, Li J and Zeng X 2015 Minimizing the increasing solid waste through zero waste strategy *J. Clean. Prod.* **104** 199–210
[31] Cheung C M K and Lee M K O 2009 Understanding the sustainability of a virtual community: model development and empirical test *J. Inf. Sci.* **35** 279–98

Managing Informal Economic Environment based on Daytime and Night-time Transition

F Gunawan[1], **Y A Yatmo** [1*]

[1] Department of Architecture, Universitas Indonesia, Depok 16424, West Java, Indonesia

Corresponding author (yandiay@eng.ui.ac.id)

Abstract. Informal economy is one of the examples of life cycle attributes contributing to the overall sustainable environment. Pasar Lama Tangerang in Banten, Indonesia is known as a market with continuous economic cycle from morning until night-time with different economic actors. This study investigates the management of the environment from daytime to night-time as part of informal and night-time economy. The discussion of night-time economy is currently limited and has been addressed separately from daytime economy. We argue that investigating the transition and management from daytime to night-time and its consequences of a night-time market is a crucial strategy of managing future informal economy. The methodology of this study was conducted qualitatively by capturing the transition by night-photography on multiple visits and the observation was done by night-walking and informal interviews, while tracing of spatial changes was done by differentiation in visiting hours. The study finds that differentiation of economic hours and implementation of spatial layering and zoning of all the economic actors, formal and informal, from an existing environment would be useful as environmental management strategy.

1. Introduction

This paper studies the strategy of managing an environment as part of informal and night-time economy. Until recently, such investigations tend to overlook the temporal continuity within the environmental management, separating the day time management from the night time. The transition, the temporal management and its consequences become crucial for the environmental management and, arguably contribute to the future sustainability of informal markets.

This study took place at the Pasar Lama (Tangerang Old Market), a market in Tangerang, Banten Province, Indonesia. The market operates almost 24 hours, starting from 6 AM to almost 10 PM. Our observations indicate that there are two different kinds of activities and actors involved in the market, namely the daytime market and the "nighttime" one. The daytime market functions as a traditional market that sells fresh vegetables, dairy, and household goods, operating from 6.00 AM until 2.30 PM, meanwhile the nighttime market starts, interestingly, from 3.30 PM and is constituted by the street vendors selling foods and beverages until around 10 PM. However, the operations are not strictly separated, they gradually shift from one mode to the other on a daily basis. The objective of this study is to discover ways of managing those operations and what contributes to the sustainability and livability of such environment.

2. Theoretical Framework

Night-time theories stemmed from the industrialization due to the expansion of electrical network from state to the domestic [1,2]. Quoting Melbin, Shaw proposed this phenomenon as "colonization of night by day". The consequences of this is the shift from task-oriented to clock-oriented temporalities, the rise of incessacy, and the creation of new subjects or actors with behaviors that are more relaxed and permissive [2]. The biggest impact due to electrification is the creation of nighttime economy based on its prolonged working hours or *incessacy*. Incessacy marked a shift where power was not reliant on an individual anymore but possibly displaced by tools of monitoring and surveillance, rhythms of power changed daily or weekly, and the invasion of bodies and domesticity [1]. One important factor that Shaw identified is the flow of alcohol with the production of subjectivity at night combined with the avoidance of surveillance could produce an opposite effect when compared with the activities of day, and this potentially could provide new experiences in these spaces [1]. Avoidance of surveillance is one indicator where night time behavior is a potential destructive force and also the potential loss of nocturnality [1]. This prompted Gwiazdzinski to propose four futures for the night: the first is colonization where a distinct timespace and a distinct population to provide traction as part of the city, the second is banalization where the night is completely colonized and no distinction between night-time city and day-time city, the third is explosion where existing tensions surrounding the night-time city continue to grow and the last one is harmonization where solidarity with day-time city is cultivated but the night-time city keeps its characteristics [1].

Night-time economy (NTE) was previously studied in the UK to "encourage the deregulation and development of the alcohol and leisure industries" [2]. However, the spread of NTE created new spaces of entertainment which extended to public nightlife on the streets [1]. Nocturnal activities were not bound to be empty, ritually set for primitive activities, or a quiet domestic life anymore. While Shaw acknowledged the contribution of Melbin's study regarding NTE, especially the reading of spatial and nocturnal frontier, he also cautioned that it was limited to the experience of the American Midwest [1].

Since Melbin's studies, various recent studies regarding daytime or night-time economy were done to discover how visible urban kioks and how they perform socially in a particular central market in Tel Aviv [3], night-time entertainment zones stemming from marginal buildings in downtown American Midwest [4], and perception of street vendors compared in daytime and night-time [5]. Aside from the last study, the first two studies highlight how daytime and night-time economy affected the spatiality and dynamics between their location, the users, and the experiences attained in the spatial negotiations.

Drawbacks of nocturnal experiences such as criminal behaviors, safety and comfort within darkness related to perception of night, and separation of gender activities. According to Melbin quoted by van Liempt, changes to night-time gave platform to "special atmospheres with particular activities, experiences, and possibilities, including criminal acts, non-conventional behaviors, or organizing rebellion" [2]. Van Liempt also added that all of the above include but not limited to trying to be someone else, meeting with questionable characters, or doing activities usually not permitted, contributing to violent crime, damage and anti-social behavior [2]. Perception of the night is always attributed to negative associations, according to Edensor, such as the 'Dark Ages', 'dark forces, 'dark deeds', the 'dark net' and the 'dark side' [6]. In addition, these associations are always connected to "what should and should not happen in the dark of night" [6]. In the global South, but in particular in rural India, darkness is used as a cover or *purda* to evade social stigma where women can defecate in the open and be separated from "other men", as well as facilitation of movement and empowerment [7].

While the existing knowledge of studies focused solely on daytime or nocturnal experiences, this study presents a transitional spatial dynamic that occurs from daytime to night-time market including the spatial management, negotiations, and consequences between the everyday routine of the street vendors in Pasar Lama Tangerang. The urgency of this study is propelled by the lack of references and understanding regarding "the all-pervasive aspect of human experience, geographical investigation of daylight, darkness, and illumination" [6].

IOP Conf. Series: Earth and Environmental Science 1098 (2022) 012071 doi:10.1088/1755-1315/1098/1/012071

3. Methodology

The location of Pasar Lama Tangerang (Tangerang Old Market) is in Banten province, Indonesia. The size of the research location is along Jalan Kisamaun and spans around 400 meters from the entrance to the end of the market. The entrance of the night market starts at the intersection of Jalan Kisamaun and Jalan A. Damyati, marked by a stone monument. The Kisamaun street is a one-way street between 7 AM and 2.30 PM where cars and motorcycles freely pass through. The street is barricaded on the east and west elevation by multi-story shophouses which house various retailers. The location of study and the typical and existing condition can be seen in Figure 1.

Figure 1. Location of research and typical condition before transition. *Source:* Google Maps and Authors

Variations in the survey hours were done intentionally to capture the differences in traffic users, flow of circulation between users, and the spatial changes in-between. Recording of existing and the transformation thereafter helped in providing an overall view of what transpired on a daily basis. The sample of research is limited to 15-20 street vendors as a partial segment of those lining up the street.

Informal and unstructured interviews were done in the second visit. The first interview was conducted with a parking attendant and each parking attendant was allowed to manage 10 cars on a particular stretch. The second interview was done to a street vendor attendant to inquire about a few existing street vendors already set-up on the western side of the street. The information gathered from the interview was that those vendors start in the morning and continue to operate through the evening. However, the interviewee revealed that the side-street vendors pay a lower rent compared to the ones starting at 2.30 PM due to the closeness of the eventual night-market pedestrian traffic. Information based on the interviews will be used as analytical points in the discussion below.

Night-walking was conducted on all the observations except on the second visit. The second visit observed the transition time when all the street vendors started to arrive and setup their carts between 2.30 PM to 3.30 PM.

To capture and record the observations, several equipments were used: a DSLR camera with a tripod, a smartphone on a gimbal, a camera attachment to the smartphone in the form of FLIR camera, and an app-based luminance capturing method. In addition to photos, videos were also recorded to capture certain aspects of the observation where limitations of night-photography might possibly hinder an in-depth understanding of the existing spatial management. Utilization of photos and videos has been underused as a research tool and research methodology [8]. However, tracing, diagramming and mapping became crucial in the analysis process to uncover certain attributes within the study case [9].

4. Analysis

The transformation and management analysis of Pasar Lama Tangerang in this section will be divided into three parts: the typical condition before 2.30 PM, the transition period between 2.30 – 3.30 PM, and the night-market period between 3.30 until closing around 9.30 PM.

4.1. Typical condition before 2.30 PM

During morning activities, the motorcycle parking is typically reserved on the western side of the street while cars and motorcycles occupy the eastern side. Some portion of the street has notiecable parking lines meant for cars. This parking arrangement leaves a space of around 3 to 4 meters for vehicles to pass through in-between. Pedestrians space is reserved in front of the shophouses but only with a width of roughly one meter.

Top view of a segment of Pasar Lama Tangerang where the typical condition occurs before 2.30 PM on a daily basis can be seen in Figure 2. Area A is the location of car parking along the western side of the street. The car parking area was invisibly divided per ten cars for one parking attendant. The boundary of this spatial management was done based on initial negotiation between the attendants. While typical, parallel parking was sometimes allowed within a ten-minute to thirty-minute parking time with the condition that no parking brake was practiced during. This arrangement would allow the parking attendant to push the parallel-parked car(s) when the inside car needed to leave. Area B is the typical traffic direction going north to south for cars and motorcycles, but numerous users would disregard this management and going the opposite way such as motorcyclists and pedicabs. While pedestrians could use the allotted sidewalk in front of the shophouses, the majority still used parts of Area B as their area of traffic going north and south, co-mingling with cars and other motor vehicles. Area B is typically the dominant area where continuous spatial negotiations and dynamics occurred. Area C is the typical area for motorcycle parking and throughout the day, various visitors would come and go using this mode of transportation. Area D is the sidewalk in front of the shophouses reserved for pedestrians. From our observations, this section was minimally used since the width was not adequate for two people side by side or passing by. In addition, some of the store fronts would use this space as an extension of their retail space, further minimizing usable traffic space for pedestrians. The consequence of this is the weaving circulation of pedestrians to navigate along the space.

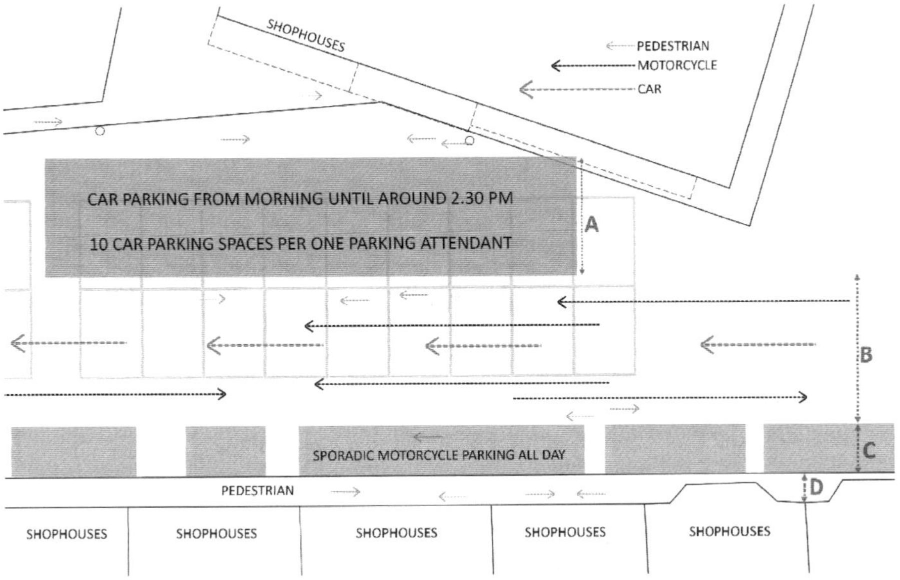

Figure 2 Typical spatial management of Pasar Lama Tangerang before 2.30 PM. *Source:* Authors

4.2. Transition period between 2.30 PM and 3.30 PM

The transition period occurred daily starting from 2.30 PM with a prior warning to visitors arriving late that they must vacant their parking spot due to arriving street vendors setting up shop. In Figure 4, some vendors arrived early and were already at their designated spot. The spots were determined by the scale of fees from the local precinct with the management and understanding that the closer they were to the pedestrian traffic, the higher the fees (Figure 3).

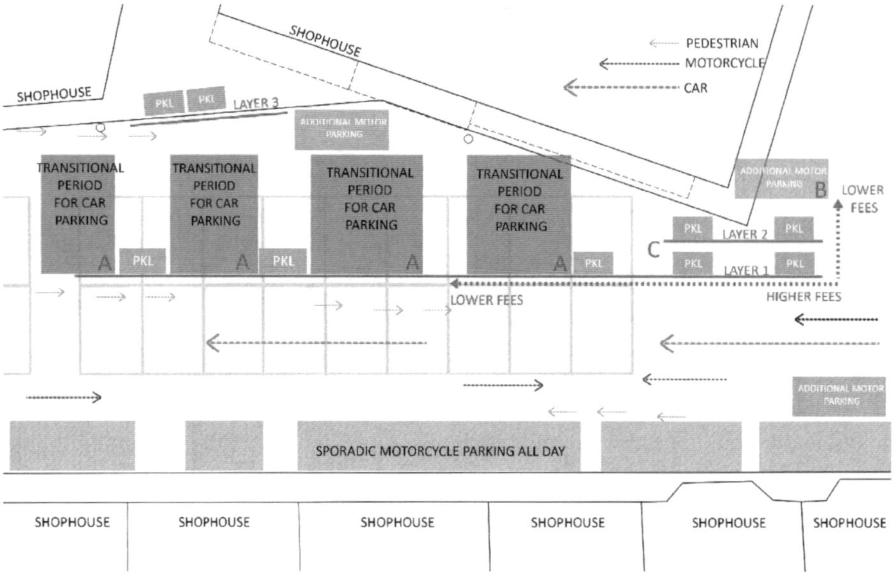

Figure 3 Layers of street vendor placement. *Source:* Authors

As seen in Figure 3, the Layer 1 vendors have the luxury of setting up their back-of-the-house equipment with variable extensions. The areas occupied by these were previously designated for car parks. (Figure 3, area A) Behind Layer 1 are Layer 2 vendors which leave little space in between and require pedestrians to navigate the "insertion" space due to the back-of-the-house setup of Layer 1 vendors (Figure 3, area C and Figure 4). In addition, Layer 2 vendors are squeezed around the additional motorcycle parking due to excess visitors. (Figure 3, area B). During the transition period, parking attendants working from morning had officially closed up for the day, replaced by different economic actors with its own spatial hierarchy and negotiations.

Figure 4 Insertion space between Layer 1 and Layer 2 vendors. *Source:* Authors

4.3. *Night-market period between 3.30 PM and 9.30 PM*

The spatial configuration and management had changed dramatically from the setup before 2.30 PM. The typical spatial configuration of the night-market consisted of all the Layer 1 street vendors on the western side of the street with their own back-of-the-house setups, creating a wall of storefronts with interspersed gaps in between. The typical section of this spatial arrangement can be seen in Figure 5.

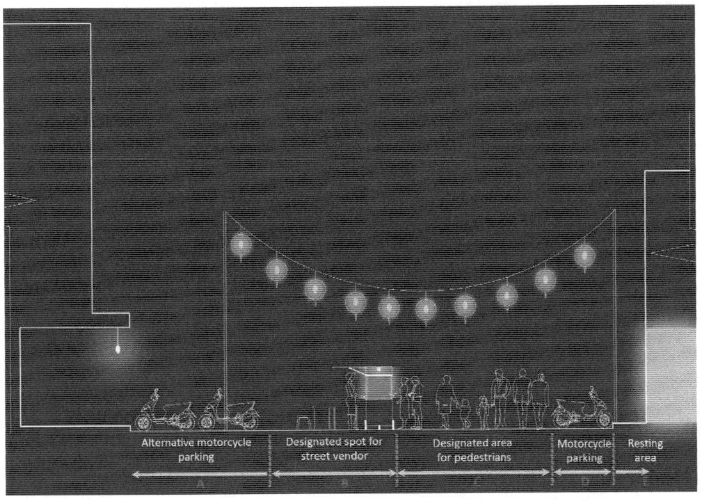

Figure 5. Typical section of night-market. *Source:* Authors

Figure 6 illustrates a dynamic spatial management where area A still functions as an operable shophouse with alternative motorcycle parking in front, area B is Layer 1 of street vendors with their typical setup where previously designated for carpark, area C is now heavily occupied by pedestrians going north and south while motorcyclists and pedicab drivers co-mingled within and jostled for economic survival, area D is the typical motorcycle parking but flexibly shared with other informal vendors, and Area E were now being used as a resting area for weary visitors where the shophouse is closed. The overall view of the research area is illustrated in Figure 6.

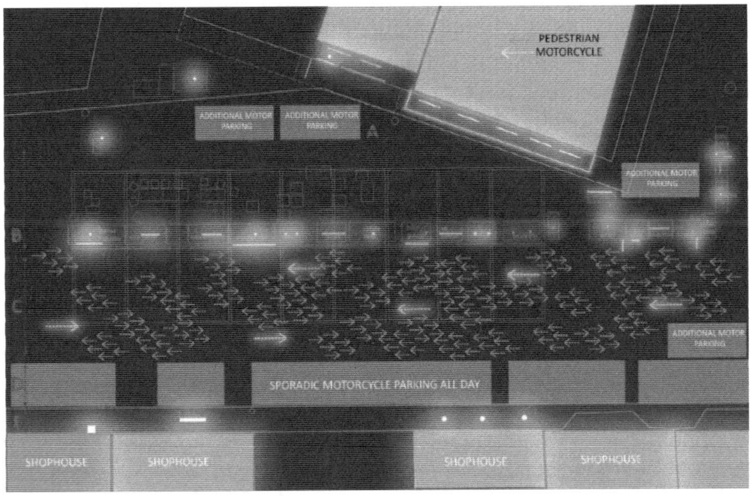

Figure 6 Overall plan view of typical night-market. *Source:* Authors

While not shown in this particular stretch of street, the eastern side of the street mainly occupied by motorcycle parking also provided spaces for "loose and sporadic" street vendors vying for economic value and subsistence. Interspersed on the eastern side, street performers and beggars also had their own spatial stake and claim for the same exact reason.

5. Discussion

Pasar Lama Tangerang is a case study where spatial management, changes, and negotiations revolve daily based on economic cycles between daytime and night-time including all the actors involved. The morning market persisted until around 2.30 PM before the transition period began. Early spatial boundaries such as car parking or motorcycle parking, visible or invisible, were replaced by different economic participants, staking claim to their designated yet invisible territories in order to utilise the time-space previously allotted for them. In the transition time starting at 2.30 PM, street vendors sporadically arrived at their invisible yet fixed spots for operational business hours. These spots created different layers of economic co-existence since their locations were determined by specific fees from the precinct. Our finding shows that Layer 1 is the closest to the nighttime pedestrian traffic with subsequent layers operating behind. A creation of "insertion space" sprang up between Layer 1 and Layer 2 as a consequence of limited space management (Figure 4). This space forced the visitors to compress their personal boundaries in order to arrive at their favourite street vendor located within, creating a different kind of passage of movement [10]. Layer 3 is allowed due to their prolonged operational hours from morning until nighttime. The location of Layer 3 along Pasar Lama Tangerang is sporadic based on their previous arrangement and necessity. In this study, Layer 3 is visible on the western side of the street on the pedestrian sidewalk.

At nighttime, Layer 1 street vendors created a wall of storefronts with interspersed gaps to accommodate visitors going through. This storefront is a barrier separating a front and a back. Back of the store is actually an extension of the front for some vendors when their offering is a full meal to be consumed with tables and chairs. After purchase, the visitors had the option to walk around the wall through the gap for their meal consumption. This arrangement revealed that the front and the back was interchange-able, depending upon the needs of the visitors and allowance given by the vendors. Another spatial management emerged when one or a few vendors have visitors lined up at the front of the store at the same time while waiting for their meal to be prepared. The sudden, stagnate, and intermittent bubble of concentrated bodies would create a traffic jam for all users.

This emergence required another spatial arrangement for the pedestrians, especially during peak hours at dinner time or weekend. Pedestrian visitors at night were constantly required to negotiate their spatial boundaries due to the traffic in both directions and interjected with motorcycles, pedicabs, and cars. The management of nighttime street vendors occupy and define informal and temporary storefronts. The storefronts co-exist horizontally, front and back, and also the right and left of their spot. Allowance must be given by the left and right of it when one or several storefronts are heavily waited upon by visitors.

In comparison with another similar setup in Semawis, Semarang, this study case is unique in itself since the transition starts at 2.30 PM as opposed to 5 PM at Semawis. Another important difference is the operating days; Pasar Lama Tangerang operates daily Monday to Sunday from 2.30 PM until 9.30 PM while Semawis only operates on the weekend from Friday to Sunday, from 5 PM until closing time at 11 PM. This indicates that vendors at Pasar Lama Tangerang rely heavily on the transition and continuous presence for their economic subsistence.

Pasar Lama Tangerang is another evidence that nighttime is an extension of public life on the streets as proposed by Shaw [1], however the manifestations are vastly different than previously studied. In this particular case, the street vendors are considered fully "in-place", not otherwise perceived [5], all are within next to each other without the risk of invisibility [3], and no perceived drawback of nighttime experiences [2], [6].

6. Conclusion

The findings of this study indicate that the understanding of the layers of territories, boundaries, mobile and stationary actors as key in managing urban environment based on daytime-nighttime transition. The added layers and spaces contributed to the malleability and organic shifting due to the continuous and overlapping activities between all actors. While we can perceive somewhat definitive boundaries and territories within each or several vendors, they are always negotiated without verbal allowance but always recognised. The findings also reveal that managing existing informal economic environment, such as Pasar Lama Tangerang, provides broader values, additional inclusivity, and multiple experiences at night-time.

This becomes an important knowledge that would guide our perspective in seeing that differentiation of daytime and night-time economy as environmental management does not require a distinct separation. However, transition and extension from daytime to night-time for an informal economy is an important strategy in creating and providing a sustainable and livability of an urban environment. In addition to that, this study raises further questions on the possibility of duplicating this management model at another location with similar economic demography.

Acknowledgements
The study presented in this paper is supported by Penelitian Disertasi Doktor (PDD) 2022 under Grant Number: NKB-1019/UN2.RST/HKP.05.00/2022, a research grant provided by Direktorat Riset, Teknologi, dan Pengabdian kepada Masyarakat, Direktorat Jenderal Pendidikan Tinggi, Riset, dan Teknologi (Directorate Research, Technology and Community Services, Directorate General of Higher Education, Research and Technology of Republic Indonesia).

References

[1] Shaw R 2021 *The Nocturnal City* (Routledge)
[2] van Liempt I, van Aalst I and Schwanen T 2015 Introduction: Geographies of the urban night *Urban Studies* **52** 407–21
[3] Natapov A and Grinshpun H 2020 Hidden in the most visible place: measuring visual accessibility and social performance of urban kiosks *Journal of Urban Design* **25** 412–32
[4] Campo D and Ryan B D 2008 The Entertainment Zone: Unplanned Nightlife and the Revitalization of the American Downtown *Journal of Urban Design* **13** 291–315
[5] Yatmo Y A 2009 Perception of street vendors as 'out of place' urban elements at day time and night time *Journal of Environmental Psychology* **29** 467–76
[6] Edensor T 2015 Introduction to geographies of darkness *cultural geographies* **22** 559–65
[7] Kumar A and Shaw R 2020 Transforming rural light and dark under planetary urbanisation: Comparing ordinary countrysides in India and the UK *Trans Inst Br Geogr* **45** 155–67
[8] Garrett B L 2011 Videographic geographies: Using digital video for geographic research *Progress in Human Geography* **35** 521–41
[9] Atmodiwirjo P, Johanes M and Yatmo Y A 2019 Mapping stories: representing urban everyday narratives and operations *Urban Des Int* **24** 225–40
[10] Paramita K D and Schneider T 2018 Passage Territories: Reframing Living Spaces in Contested Contexts *in* **1** 113–29

Synthesis of Geopolymer Concrete Using Flyash

S U Amjad[1*], M A Budihardjo[1], I B Priyambada[1], R Asghar[2]

[1]Department of Environmental Engineering, Faculty of Engineering, Diponegoro University, Indonesia
[2]College of Civil Engineering and Architecture, Shandong University of Science and Technology Shandong Province, 266590 P.R, China

syeduzairamjad@students.undip.ac.id

Abstract. Geopolymer concrete (GPC) reduces the environmental impacts of conventional Portland cement concrete, therefore, this research aims to investigate the compressive characteristics of GPC by utilizing the readily available industrial waste i.e., coal based Flyash in the Khyber Pakhtunkhwa province of Pakistan. For the preparation of GPC, Na_2SiO_3 was used as an activator. After the preparation of GPC specimens, their compressive characteristics were investigated at different ages of specimens, Flyash content, curing time, and curing conditions. Moreover, the relationship between compressive strength and modulus of elasticity of GPC was also evaluated. From the testing of GPC specimens in a compression testing machine (CTM), their compressive strength was found to be similar to that of conventional Portland cement concrete i.e., in the range of 20-40 MPa tested at the age of 7 and 28 days respectively. Furthermore, elevated temperature curing was found to increase the compressive strength of GPC significantly as compared to that of ambient curing. From the cost analysis, Flyash-based GPC was found to be cost effective as compared to conventional Portland cement concrete

1. Introduction

After water, concrete is the second most frequently utilized substance on the planet [1] Its constituents such as cement require a higher amount of energy to be prepared, this energy is obtained by burning fossil fuels. Along with this, the hydration reaction of cement with water during the preparation of concrete is an exothermic process and it releases energy in the form of heat into the environment while forming bonds. These two are the main reasons that contribute towards the great carbon footprint of concrete. About 9 percent of carbon dioxide in the environment is due to the manufacturing or use of concrete and its constituents. [2]

In recent decades, attempts are being made to produce greener concrete with the lowest carbon footprint and highest strength and sustainability. One of these attempts, is to make green concrete known as geopolymer concrete (GPC) using geopolymer materials available abundantly throughout the globe. Mainly the materials having properties the same as cement are used for this purpose, called supplementary cementitious materials (SCMs). The SCMs include Flyash (FA), Condensed Silica Fumes (CSF), Ground Granulated Blast Furnace Slag (GG, BFS), and various other industrial and agricultural wastes that are rich in silica and alumina. [3] GPC attains its strength on the basis of polymerization of inorganic minerals i.e., the geopolymers. These geopolymers belong to a group of three-dimensional aluminosilicate materials made from sources like clay, red mud, and Flyash which were discovered by Davidovits in 1979.[4] The primary characteristics of geopolymers include their rapid development of compressive strength, low permeability, improved chemical and thermal resistance, better mechanical properties under both normal and extreme conditions, and tendency to significantly reduce the mobility of the majority of heavy metal ions contained within the geopolymer structure. [5]

The Flyash based GPC can have advantageous and varied qualities, according to numerous studies. For instance, GPC has demonstrated good performance in high temperatures, high early age strength, and good resistance to acid and sulphate attacks. [6], [7] In recent years, it has been demonstrated that GPC may produce high early and ultimate compressive strengths under ambient curing conditions, as well as good workability when additives like Portland cement, calcium hydroxide, or powdered granulated blast furnace slag are used. [8] Most experts agree that GPC is less harmful to the environment than Portland cement concrete, however further research is needed to support this conclusion. Moreover, the literature available on the mechanical and structural properties of GPC is still not enough to make it an industrial material like ordinary Portland cement (OPC) concrete.

Therefore, this research paper aims to investigate the mechanical properties (i.e., the compressive characteristics in particular) of GPC by utilizing locally available Flyash. For this, the compressive strength of GPC manufactured from different concentrations of Flyash to OPC i.e., 0%, 10%, 20%, 30%, 40%, and 50% (by mass) was evaluated using Na_2SiO_3 as an activator. The compressive characteristics of GPC were investigated at different ages of specimens i.e., 1 day, 3 days, 7 days, 14 days, and 28 days, curing time and curing conditions. Furthermore, the comparison between elevated temperature and ambient temperature was done. The cost analysis among the ordinary Portland cement concrete and geopolymer concrete was also compared.

2. Methodology

2.1. Materials

2.1.1. Flyash

The Flyash which was used in this research was obtained from the coal power plants located in Khyber Pakhtunkhwa province of Pakistan.

2.1.2. Cement

In this research Ordinary Portland Cement prepared by "Bestway, Pakistan" was also used to manufacture cubes which were used in comparison with Geopolymer cubes. The physical properties and chemical properties of OPC are shown in table 1. The chemical composition of OPC was obtained from a research article by Amin and Ali. [9] However, the standard laboratory tests were conducted in accordance with the ASTM guidelines [10]–[13] to determine the physical properties of above mentioned OPC.

Table 1: Physical Properties and Chemical Composition of OPC

Cement Composition		Physical Properties of Cement	
Constituents	**Content %**	**Parameter**	**Value**
CaO	61.7	Specific Gravity	3.01
SiO_2	21.0	Consistency	28.4%
Al_2O_3	5.04	Initial Setting Time	1 h and 39 min
MgO	2.56	Final Setting Time	3 h and 47 min
SO_3	1.51	Specific Surface	319.5 m^2/kg
Fe_2O_3	3.24		
IR	0.65		
Free Lime	0.98		
LOI[3]	1.83		

2.1.3. Activator
In this research Sodium Silicate (Na_2SiO_3) prepared by a South Korean company named "DAEJUNG" was used as an activator in powdered form. The purpose of using an activator is that it helps to activate silica and alumina content present in Flyash.

2.1.4. Aggregates
Using ASTM C136, [14] the nominal maximum size of coarse aggregate (inches) was determined to be 0.7 inches down. The sand originates from a nearby source which is Lawrencepur Pakistan. The computed sand fineness modulus was 2.47 as per ASTM C136. [14] Using ASTM C127, [15] the Available Moisture Absorption was 1.42 percent, the Available Moisture Content was 1.05 percent, the Bulk Specific Gravity was 2.493 g/cc, and the Apparent Specific Gravity was 2.685 g/cc.

2.2. Sample Preparation
For the investigation of the compressive characteristics of GPC using locally available Flyash in Khyber-Pakhtunkhwa (Pakistan), 3 cylindrical specimens were prepared to achieve each type of experimental assessment. To achieve the desired research objectives, GPC specimens were prepared by replacing Flyash with cement in the ratio of 0%, 10%, 20%, 30%, 40%, and 50%. Moreover, GPC specimens were also prepared to investigate the effect of curing conditions (ambient or elevated temperature curing) and curing time on their compressive strength.

2.3. Curing of Specimens
Two curing conditions were used in this experiment. The first type of curing was elevated temperature curing at 70 °C for 24 h. The other curing condition was ambient curing where specimens were kept at in open atmospheric temperature of about 23°C till the day of testing. GPC specimens subjected to different curing conditions were tested separately to investigate the effect of the type of curing on their compressive strength.

3. Results and Discussions
3.1. Compressive Strength of Flyash Based GPC
For the determination of the compressive behavior of all the GPC specimens, a compression strength test was conducted using a compression testing machine (CTM) according to the manual of ASTM C39. [16] The experimental findings of the GPC compressive strength used in this investigation are compiled in table 2. Their compressive strength was found to be similar to that of conventional Portland cement concrete i.e., in the range of 20-40 MPa tested at the age of 7 and 28 days respectively. The highest strength was attained i.e., 39.98 MPa when OPC was replaced with 30% of Flyash. It implies that fly ash concrete can generate significant strength due to the pozzolanic reaction after 28 days.

Table 2: Compressive Strength of Flyash Based GPC

Flyash (%)	1 day	3 days	7 days	14 days	28 days
0	5.6	14.36	21.4	31.4	34.9
10	5.92	14.8	22.1	32.95	36.6
20	6.08	15.2	23.15	34.2	37.93
30	6.4	16.54	24.95	36.1	39.98
40	5.92	14.8	22.45	32.44	36.7
50	5.44	13.6	20.2	31.73	33.85

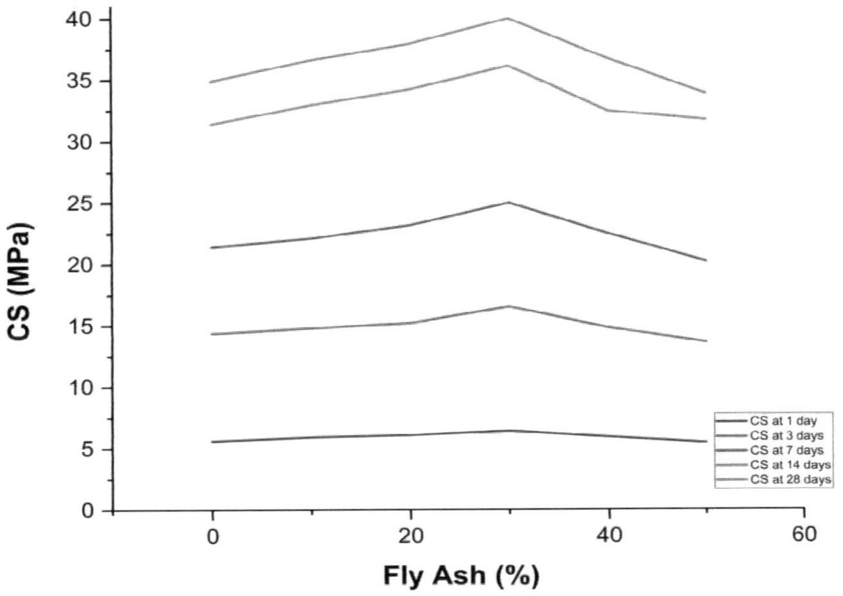

Figure 1: Comparison of FA and CS on different days *CS: Compressive strength

3.2. Compressive Strength of GPC and Curing Conditions

The GPC compressive strengths at 7 and 28 days for elevated temperature curing (70 °C) ranged from 10 to 45 MPa, whereas the compressive strengths of GPC at 7 and 28 days for ambient curing ranged from 05 to 40 MPa, respectively. Since the geo-polymerization process needed heating to speed up the interaction between the alkali activator and Flyash, it can be noted that the heat curing temperature had an impact on the early strength of GPC. When compared to ambient curing, the compressive strength of GPC that was cured at 70 °C increased by 35% after 7 days. At 28 days, the strengths of samples cured at elevated temperatures were 8% greater than those cured at room temperature.

Table 3: Compressive Strength of GPC Subjected to Ambient and Elevated Temperature Curing

Age of Specimens	Compressive Strength (MPa)	
	Ambient Curing	Elevated Temperature Curing
1	6.4	11.24
3	16.54	31.74
7	24.95	33.58
14	36.1	39.99
28	39.98	43.26

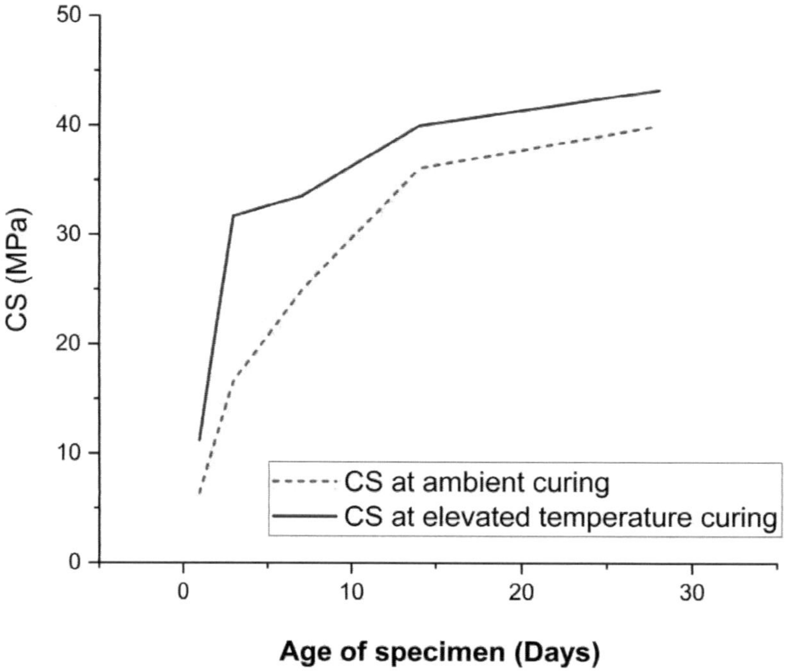

Figure 2: Ambient curing vs elevated temperature curing

3.3. Compressive Strength of GPC and Curing Duration

The compressive strength increased as curing time increased, as indicated in table 4. The results suggest that for the synthesis of geopolymers with higher compressive strengths, a longer period of low-temperature curing is preferred. This is due to the fact that getting a geopolymer without cracks depends a lot on the water that is contained in geopolymers and how it is removed when it evaporates. As a result, curing at a lower temperature is preferred rather than drying quickly during the process.

Table 4: Impact of Curing Time on GPC Compressive Strength

Curing Time (Days)	Compressive Strength (MPa)
5	17.86
7	24.95
9	28.96
12	34.21
15	37.36

3.4. Modulus of Elasticity of GPC

Specimens of different grades of concrete were cast and a compressive strength test was done on these specimens. The compressive strength test yields a stress-strain curve for each concrete grade. From the initial elastic portion of the stress strain curve, we determine the slope and this slope gives us the value of the modulus of elasticity. The compressive strength and modulus of elasticity of every grade of concrete were now determined. The ACI provides a relationship between the modulus of elasticity and the compressive strength of conventional concrete. However, as per the author's knowledge no such relationship exists for the GPC, therefore this research has worked to provide the relationship between the modulus of elasticity and the compressive strength of GPC. This relationship is presented in equation 1.

Table 5: Modulus of Elasticity of Various Grades of GPC

S. No	Compressive Strength (MPa)	Modulus of Elasticity (MPa)
1	16.37	19251
2	21.48	21435
3	26.89	21002
4	32.74	24244
5	36.84	26409
6	39.98	27113

$$E = 4312\sqrt{f_c'} \quad (1)$$

3.5. Cost Analysis of GPC

The ingredients in OPC Concrete and GPC Concrete are shown in table 6 and table 7 respectively along with their prices. The price of ingredients are local prices and varies from region to region. These prices were taken from a local material supplier in the district of Mansehra, Khyber Pakhtunkhwa province of Pakistan. From the cost analysis, GPC was found to be 11% cheaper as compared to OPC concrete. The GPC is cheaper as compared to OPC Concrete because Flyash is a byproduct and there is no usage of it, so it cost less as compared to OPC Concrete.

Table 6: Cost analysis of 1m^3 of OPC concrete

Material	Rates in (PKR)	Unit	Quantity in Kg	Amount in PKR
Cement	1080	50 Kg	350	7560
Fine Aggregate	1716	T	642	1102
Coarse Aggregate	1839	T	1270	2335
SuperPlastizer	300	Kg	7.5	2250
Total		PKR		13247

PKR: Pakistani Rupee, T: 1000 kg.

Table 7: Cost analysis of 1m^3 of GPC concrete

Material	Rates in (PKR)	Unit	Quantity in Kg	Amount in PKR
Flyash	300	50 Kg	378	2268
Fine Aggregate	1716	T	554	950
Coarse Aggregate	1839	T	1294	2379
Na_2SiO_3	50	Kg	124	6200
Total		PKR		11797

PKR: Pakistani Rupee, T: 1000 kg.

4. Conclusions

Based on the experimental investigation of compressive characteristics of GPC utilizing locally available Flyash from the coal power plants located in Khyber Pakhtunkhwa province of Pakistan, the subsequent conclusion may be drawn.

- The optimum concentration of Flyash was revealed to be 30% to achieve the maximum compressive strength of GPC.
- The elastic modulus was found to be 4312 times the square root of compressive strength of GPC.
- In comparison to ambient curing, elevated temperature curing was discovered to be increasing the compressive strength of GPC. Moreover, increasing the curing time was also found to increase the compressive strength of GPC.
- From the cost analysis, GPC was found to be cost effective than OPC concrete. It was found to be 11% cheaper as compared to OPC concrete.

References

[1] C. R. Gagg, "Cement and concrete as an engineering material: An historic appraisal and case study analysis," *Engineering Failure Analysis*, vol. 40, pp. 114–140, 2014.

[2] Y. Geng, Z. Wang, L. Shen, and J. Zhao, "Calculating of CO2 emission factors for Chinese cement production based on inorganic carbon and organic carbon," *Journal of Cleaner Production*, vol. 217, pp. 503–509, 2019.

[3] A. Mehta and R. Siddique, "An overview of geopolymers derived from industrial by-products," *Construction and Building Materials*, vol. 127, pp. 183–198, 2016.

[4] J. W. Wang and T. W. Cheng, "Production geopolymer materials by coal fly ash," in *Proceedings of the 7th International Symposium on East Asian Resources Recycling Technology*, 2003, vol. 263, p. 266.

[5] J. G. S. Van Jaarsveld, J. S. J. Van Deventer, and L. Lorenzen, "The potential use of geopolymeric materials to immobilise toxic metals: Part I. Theory and applications," *Minerals engineering*, vol. 10, no. 7, pp. 659–669, 1997.

[6] E. I. Diaz, E. N. Allouche, and S. Eklund, "Factors affecting the suitability of fly ash as source material for geopolymers," *Fuel*, vol. 89, no. 5, pp. 992–996, 2010.

[7] S. E. Wallah, “Creep behaviour of fly ash-based geopolymer concrete,” *Civil Engineering Dimension*, vol. 12, no. 2, pp. 73–78, 2010.
[8] P. Nath, P. K. Sarker, and V. B. Rangan, “Early age properties of low-calcium fly ash geopolymer concrete suitable for ambient curing,” *Procedia Engineering*, vol. 125, pp. 601–607, 2015.
[9] K. Ali, “Chemical Analysis and Comparison of Ordinary Portland Cement of Khyber Pakhtoon Khwa Pakistan,” *Chemical Engineering Research Bulletin*, vol. 14, no. 1, pp. 45–49, 2010.
[10] C. ASTM, “188 (2009),” *Standard Test Method for Density of Hydraulic Cement. ASTM International, West Conshohocken, PA, ASTM C*, pp. 188–09.
[11] C. ASTM, “191-08. Standard test methods for time of setting of hydraulic cement by vicat needle,” *West Conshohocken, PA: ASTM International*, 2008.
[12] C. ASTM, “Standard test method for amount of water required for normal consistency of hydraulic cement paste,” 2011.
[13] A. C204, “C204. Standard test methods for fineness of hydraulic cement by air-permeability apparatus,” *Annual Book of ASTM; American Society of Testing and Materials: West Conshohocken, PA, USA*, 2007.
[14] C. ASTM, “136,” *Standard Test Method for Sieve Analysis of Fine and Coarse Aggregates," ASTM International, West Conshohocken, PA*, 2002.
[15] C. ASTM, “128-15. Standard test method for density, relative density (specific gravity), and absorption of fine aggregate,” *ASTM, West Conshohocken, PA, USA*, 2015.
[16] “Standard Test Method for Compressive Strength of Cylindrical Concrete Specimens.” https://www.astm.org/c0039_c0039m-14.html (accessed Aug. 02, 2022).

Sustainable urban communities: Territoriality of waqf-based communal place in Kampong Cities

Agus Rochani[1] **, Nany Yuliastuti**[1,2] **, Budi Sudarwanto**[1,3] ,

[1,2,3] Doctoral Program in Architecture and Urban Science, Diponegoro University, Semarang, Indonesia
[1] Department of Urban and Regional Planning, Sultan Agung Islamic University (UNISSULA), Semarang, Indonesia
[2] Department of Urban and Regional Planning, Diponegoro University, Semarang, Indonesia
[3] Department of Architecture, Diponegoro University, Semarang, Indonesia

Correspondence email: n.yuliastuti@pwk.undip.ac.id

ABSTRACT: One of the problems in today's urban development is the scarcity of communal spaces in settlements. The communal space is a shared space where the community performs social interactions, establishes the community, and performs collective activities in the local community. One form of local wisdom that develops in forming communal spaces is the waqf land. Almost the entire Muslim community is involved in the running of the waqf. Likewise, in the city of Semarang, with its predominantly Muslim citizens, almost all kelurahan areas have waqf land. From this waqf land various communal spaces such as places of worship, education, and social, economic, and environmental infrastructure were developed. This study aims to formulate the territoriality of communal spaces that developed from the waqf process in Kauman Village as a representation of dense settlements in Semarang City. The research uses a holistic single case study method and inductive analysis to find the common thread of spatial territoriality. The study found that territoriality was developed through the enforcement of religious values, the legality of ownership status, institutional systems, and the development of complementary spatial functions according to communal needs. This system continues to serve the needs of the community in the long term.
Keywords: Territoriality of space, Land waqf, Urban settlements, Kauman Village

1. Introduction

The life of urban settlements requires communal spaces as a medium for citizen interaction and communication. Through interaction, communication will be formed between two or more people that form a reciprocal relationship that affects each other [1]. The formation of social relationships is strongly influenced by the existence of a space that serves as a forum for individual and group activities. The community space is a place for the meeting, interaction, and cohesion of collective actions. Through this process, humans will always seek harmony and create various spaces according to the needs of society.

As the population increases, urban settlements cannot more easily meet the need for communal space. Technology dependence is increasing, while togetherness and direct manual interaction are decreasing [2]. This impacts the decline in social life, the growth of individualistic attitudes, and the loss of social values. Meanwhile, social sustainability is primarily determined by their social bonds, so communal spaces as a place to gather, socialize, and interact with each other are indispensable.

Social sustainability is the ability of a city to provide a safe environment for social interaction, communication, and cultural development [3]. To realize this sustainability, efforts are needed to develop creative community energy in forming sustainable communal spaces as the ideals of the Sustainable Development Goals (SDGs), which at point 11 emphasizes the importance of developing direct community participation [4].

One of the systems for the development of communal spaces that have long developed in the community of the waqf community [5]. Waqf is Islamic philanthropy manifested in the form of surrendering assets (land or buildings) for goodness to fellow human beings [6]. Waqf space grows from the spirit of charity by forming a network of people, places, capital, and activities in a belief system [7]. This system creates communal spaces inclusive of various sectors of life sectors [8][9]. The purpose of land waqf is to form a communal space for common interests so that it is owned and operated by the community based on religious principles. The Waqf space is included in the semi-public category because its use requires a catalyst or filter for those who use it [9]. Therefore, communal spaces tend to have secondary territories because they are owned by many people who already know each other [10].

Territoriality is a system of territorial control by the community that manifests itself in the status of ownership [11], exclusive defense, symbolization of place, attitude of ownership, and arrangement of objects [12]. Territorial boundaries are formed by a unified group of people with the same interests or religious values [14] as described in meeting shared functional needs [14]. An understanding of territoriality can be built by observing group behavior in managing nonspatial and spatial aspects in the form of personalization (*defense*), placement (*occupancy)* and attachment [10][15]. Observing behavior settings shows how stable combinations of activities and places repeatedly occur to form a consistent relationship (synapomorphy) within a specific time [10]. This system in the context of sustainability can be observed according to social, economic, and environmental aspects [16]. The sustainability of communal space can also be seen from The perspective of time; the longer a space can last (across generations), the stronger the bond between the area and the existing social system will be [17].

So far, there have been many studies on communal spaces, which have observed aspects of the role of stakeholders [2] [18], the process of making a place [19], as well as spatial multilocality [20]. The waqf theme has also developed in terms of spatial distribution (GIS) [22], real estate development [23], and the architectural morphology of the city [24]. These kinds of research generally have not touched on space's meso- and micro-aspects, especially on territoriality. This is a theoretical gap that will be filled through this research.

The research aims to formulate the territoriality of communal spaces that developed from the waqf process in urban settlements. The location studied is Kauman Village, a dense settlement with a lot of waqf land in the central area of Semarang City. The Kauman village is known as the center of Islamic civilization, dubbed the village of the Qur'an. Many waqf spaces have developed in this area. This research will build an understanding of the territoriality of space formed in running the waqf system in *Kampung* Kauman Semarang.

2. Method

The research was developed in the constructivism paradigm of the behavior of a community's life. With a single case study holistic approach, the researcher raised the journey of the waqf land in Kauman Village as a unique case so that it can be studied in depth as a whole entity [25]. Case studies are developed from philosophical and psychological thinking in which researchers describe community life experiences and culminate in the essence of the experience. All aspects of the territoriality of waqf are investigated to find its identity in it [26]. The analysis was carried out inductively for obtain general themes to specific themes and interpret the meaning of the data into the research findings [27]. Participants' experiences are explored in stages to get a pure, original, and authentic experience. In this approach, theoretical propositions are used only as *background knowledge* [29] to frame the topic of territoriality in the context of waqf space in the community of Kauman Village.

3. Results and Discussion

Kauman village is a densely populated residential district inhabited by a community of santri and kyai [29]. Kauman village is located in the center of the city's trade and service area, next to the square and Johar Market, so it is characterized by a modern and capitalist urban [30]. Kauman has a very high population density of around 8,968 people / km ^{2} [31]. This village is included in the administrative area of Bangunharjo Village and Kauman Village, Central Semarang District, Semarang City. *Kampung* Kauman has an area of 13 ha, with physical boundaries to the north are Jalan Pemuda stores, to the east are Kanjengan stores, and Jalan Alun-Alun west, to the south are Jl. Wachid Hasyim (Kranggan); and to the west of the Indonesian Hajj Brotherhood complex (PHI) and the Main Bank. The majority of the population of Kauman is Muslim (99%), so Islamic ideology is firmly embedded in people's daily lives. The solid Islamic identity in this area has made Kauman dubbed the "*Kampung* Al-Qur'an," which has a variety of religious activities and waqf spaces that grow from the community's spirit in carrying out religious teachings. The orientation of the location of *Kampung* Kauman can be seen in Figure 1.

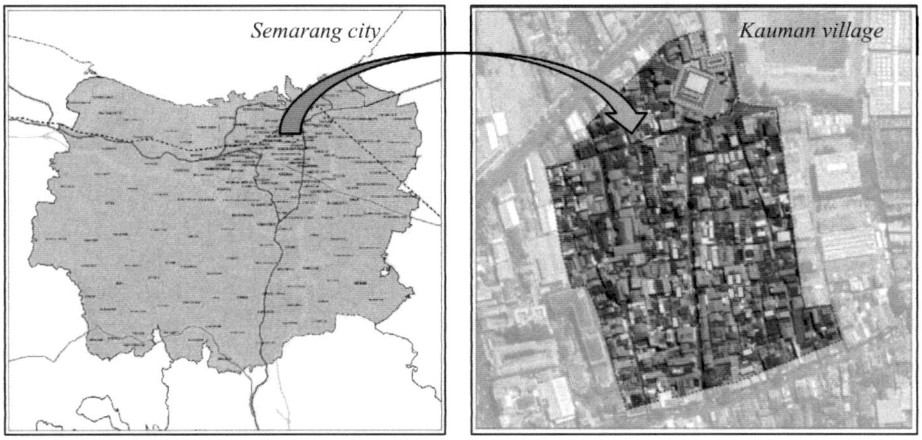

Figure 1. The location *Kampung* Kauman in the City of Semarang
Source: Secondary data processing, 2022

Various Islamic institutions that grew and managed waqf include the Great Mosque of Semarang (a relic of the Islamic kingdom of Demak); the Roudlotul Qur'an Islamic Boarding School in the 1950s; Waqf Board Madrasah in 1953, which eventually changed to Waqf Board Elementary School in 1962. Maarif Nahdatul Ulama (NU) Educational Institution in 1957, which was embodied in a Tsanawiyah Madrasah School, Hasanudin Junior High School (SMP) and Pungkuran Elementary School; Tahafidul Qur'an Islamic Boarding School in 1972. Furthermore, the Al-Iman Foundation was founded in 1983 by Kauman entrepreneurs, with services in the field of elementary education and Islamic symbols in Kauman; and the Roudlotul Quran Education Center (TPA) in 1987.

4. Space personalization

The personalization of the waqf space can be seen from the historical process that occurred. The oldest waqf in Kauman Village is the Great Mosque of Semarang, founded in the XVI century AD, during the Sultanate of Demak. The land of the Great Mosque was formed from the decree of the Dutch *staatsblad*, which stated that four mosques in the Dutch East Indies owned the *Bondo* Masjid land. The four mosques are the Great Mosque of Semarang, the Great Mosque of Demak, the Great Mosque of Kendal, and the Great Mosque of Kaliwungu. The land was a gift from the Kingdom of Demak to be used as a mosque's wealth. In the 1960s, along with the development of the PKI in Indonesia, the lands of the Bondo Masjid were looted. In response to this problem, the government, through the Ministry of Religion of the

Republic of Indonesia, issued a Decree of the Minister of Religion (KMA) number 92 of 1962, which contains the recognition of *Bondo land* as property of the mosque [32]. The ownership of the mosque on this land is intended for financing the mosque and managing the mosque's grave. According to the KMA letter, all land from the Great Mosque of Semarang is declared as waqf. Therefore, the Great Mosque was formed not naturally but grew from a royal gift that was secured by making it a waqf land [1]. Other waqfs in the settlements of *Kampung* Kauman settlements, such as the Assajiddin Mosque, Al-Iman Mosque, Al Ishlah Mosque, Ar Ridwan Mosque, and Muhyiddin Mosque, are formed naturally from individual waqfs. The visualization of the worship space in Kauman Village is shown in Figure 2.

Figure 2. Waqf room in Kauman Village
Source: Primary survey, 2022

Waqf of educational space in the form of schools and Islamic boarding schools was born from the thought of the importance of education for the younger generation of Islam. As happened with the Sultan Agung Waqf Foundation (YBWSA) in the 1950s, the community established a foundation based on awareness of the development of Islamic education in the city of Semarang. In the 1980s, Pondok Roudlotul Qur'an was also found to provide a place to recite the Koran for the mosque congregation in Kauman. The office space developed in Kauman Village is in the form of a shop on Jalan Kauman. The first room is an RQ drinking water business managed by the Roudlotul Qur'an Islamic Boarding School. The second room is the Marwa Shop, organized by the Sunday Morning Study Foundation (PAP) in 1939. The third room is a perfume shop in the courtyard next to the Great Mosque of Semarang. With this effort, the cottage, foundation, and takmir get business profits that can be used to support the operational costs of the institution.

People believe that waqf is a sustainable charity, a source of endless rewards. The community entrusts waqf to foundations and kyai because they are required to manage assets. The development of waqf is also supported by infaq and alms so that the function of the waqf room can run well. According to the informant, the dominant factors that encourage the occurrence of waqf in Kauman Village are the desire to do charities for those who can afford it. Because they do not have offspring, assets are considered valid if they are donated, the family's desire to make charities in a congregation, waqf is developed through cooperation in constructing buildings, carrying out parental orders, and waqf to solve land disputes. [2]. Waqf is carried out as a pledge to transfer assets from wakif to nadzir, registered with the Office of Religious Affairs (KUA) and Land Deed Making Officials (PPAT) so that it has strong legal provisions. The factors above show that waqf forms a strong bond between wakif and beneficiaries. The construction of dormitories is not carried out by the cottage, but is financed by the residents in cooperation. The cottage party is only a user of space, while for completeness and financing, it is fulfilled by residents and is supported by outside donors.

The description above shows that spatial personalization is realized through space's physical development according to waqf's purpose. In contrast, non-spatial personalization is carried out by clarifying land ownership status and enforcing Islamic principles in carrying out the waqf mandate. The

[1] Muhaimin, Secretary of the Great Mosque of Semarang, interview, 2022
[2] Muhaimin, Caretaker of Roudlotul Qur'an Islamic Boarding School, interview, 2022

Waqf in Kauman Village was based on a high sense of trust from the community to its managers. It can be seen from the community's involvement in donating their energy, thoughts, and assets to support the sustainability of the function of the waqf space. Each waqf is supported by a strong social system based on regional boundaries, ideological similarities, vision, mission, and goals of the institution (foundation), including family ties. This is related to Suprapti's research which states that waqf in Kauman Village is formed by the spirit of doing good deeds and maintaining family traditions [33].

5. Space placement

Space placement is the process of forming space toward the desired function. On the basis of observations, the waqf space of the Great Mosque has physical elements in the form of the main room, terrace, ablution area, management office, shop, tower, and courtyard. The Great Mosque of Semarang has a functional relationship with the surrounding spaces, including the city square of Semarang, the shopping area along Jalan Kauman and the village behind the mosque. The Great Mosque has a large area and a high tower, which gives it a monumental character as the center of Islamic civilization in Semarang City. This mosque is also a *tetenger* (marker) for the identity of the Qur'an Village in Semarang City. The physical setting of the mosque and other prayer rooms in *Kampung* Kauman is characterized by Islamic architectureThe physical setting of the educational space can be divided into the categories of formal and informal schools. Traditional schools in the form of SD Sultan Agung 1.3 and SD Pungkuran/SMP Hasanuddin 6 have physical settings in the form of classrooms, management offices, teachers' rooms, hallways and courtyards. At the same time, the informal school in the state of the Roudlotut Qur'an Islamic Boarding School has a physical setting in the form of a nanny's house, a place for reciting the Koran, a prayer room, and a residence for the students. The business space in Kauman village is in the form of a store on Jalan Kauman. This building has a 2-story form with the function of freedom as a store/trade storefront and a residential building on the top floor. Details of the character of space and activities formed in the waqf space can be seen in Table 1.

Table 1. Physical *settings*, behavior, and users of waqf space in Kauman Village

No	Building	Room	Activity
1	Mosque	The main room, the kholwat room, the terrace, the ablution area, the courtyard, the tower, and the shop.	• Rowatib prayer, Friday prayer, Eid prayer, reciting the Koran, i'tikaf, purification. • Commemoration of Islamic holidays, management of zakat, infaq and alms, takmir meetings, haul, congregational spirits, recitation, tahlilan, calling the call to prayer, administration of the takmir's office. • Buy and sell
2	Prayer room	The main room, terrace, ablution area, courtyard, Kindergarten	• Rowatib prayer, reciting, purification. • Commemoration of Islamic holidays, management of zakat, infaq and alms, takmir meetings, congregational spirits, recitations, tahlilan, recitations, calling the call to prayer. • Learning to teach Kindergarten students.
3	School	Classroom, teacher's room, library, schoolyard.	• Class learning, practicum, sports, scouts, teacher meetings, school administration. • Ceremonies, skills training, sports, extracurricular. • Village cultural activities.
4	Islamic boarding school	Recitation, ustad, administrative, prayer, student dormitory, Al-Qur'an Education Park (TPA), community hall, communal toilet, and bath (MCK).	• Memorization deposits, boarding meetings, recitations, living. • Recitation, mujahadah, sports, skills training, village cultural activities • Learning to teach TPA, community meetings, preparation for cooperation, a commemoration of national holidays, and public toilets.
5	Shops	Souvenir shop, foundation office.	• Buying and selling, foundation meetings, recitations.

Source: Primary survey, 2022

The description shows that each waqf building is developed according to the purpose of the waqf so it is trustworthy [34]. The primary function of space manifests the waqf. The relationship between the waqf space and the surrounding space forms a mutually reinforcing complementarity. The community can change the character of the waqf space as long as it does not violate the established manifestation function. This process forms a balance between individual interests and the interests of waqf as a shared space [35]. The reciprocal relationship is also seen from the funding aspect, where the construction of the waqf space comes from the community, while the community benefits from the waqf building. This mutually beneficial relationship forms a balance of welfare so that waqf can survive in the long term.

6. Place attachment

Kauman Village is a densely populated settlement full of Islamic nuances. The existence of mosques, prayer rooms, Islamic boarding schools, and mobile recitation forums signifies the solid Islamic culture in this environment. Although the natives of Kauman live in the middle of the city, they still know each other and are familiar with each other. They have been united by the unity in the mosque since childhood so that the values of togetherness and community still feel as strong as life in the village [3]. The community interprets waqf assets as assets that "legally stop." Every property waqf will stop changing ownership status, so it cannot be traded anymore. Waqf is a mandate that must be managed thoughtfully to benefit the environment. Through this effort, the charity of the founder of waqf (wakif) will continue to flow as a highly expected reward and will provide welfare to many people [4]. The high value of land property in Kauman Village demands that nadzir actually hold the mandate because waqf assets have the potential to be diverted into personal wealth [5].

The Waqf in Kauman village forms the emotional and functional bonds of the residents. Emotional attachments are formed from religious behavior that colors people's lives, thus creating a symbolic connection in waqf spaces. Meanwhile, functional extension is developed through the function and physical contact with the daily behavior [37] . In terms of managing the waqf space, they help each other to be able to carry out their functions as expected. To maintain the operation of mosques and prayer rooms, the takmir raises funds through infaq and alms and publicly announces financial positions. When in need of funds for the renovation of the building, the takmir reports it and gets sufficient support from the congregation. Likewise, in Pondok Roudlotul Qur'an, the learning is purely for memorizing the Qur'an, and the students are free of charge. To support the operational process, the cottage employs regular donors from inside and outside the village to run well. Islamic ties also seek to equalize welfare, such as providing allowances from the Great Mosque to teachers and religious teachers in Kauman Village. [6]. This process shows that the funding of waqf space involves infaq and alms to form a circulation of interrelated worship. The cyclical relationship between the waqf property and the charitable services created by the manager shows a natural structure that benefits each other and ensures the continuity of community service [24].

The sustainability of waqf can be seen in the community's decision to form spaces that complement each other and are maintained in the community's life cycle. The variety and variation of the existing waqf space show that waqf not only at the level of religious activity but also serves social, cultural and even economic functional needs [14]. The Waqf that continues from generation to the next indicates that this system has social, economic, and environmental sustainability [16]. Waqf philanthropy increases the sense of belonging to the environment, community in social activities, as well as community initiatives to care for and managing the built space [37].

[3] Fahmi, Chairman of the Al-Iman Mosque Takmir, interview, 2022
[4] Solikin, Chairman of the Assajiddin Mosque Takmir, interview, 2022
[5] Muhaimin, Caretaker of Roudlotul Qur'an Islamic Boarding School, interview, 2022
[6] Nur Istiqomah, Principal of Pungkuran Elementary School, interview, 2022

7. Conclusion

Waqf is a worship behavior, so the waqf space's territoriality is formed based on Islamic Shari'a law. The waqf space in Kauman Village was created from community donations and from changing the status of *Bondo* Mosque to waqf land. Personalization is realized through the legal status of the waqf land and the application of Islamic laws in the process of formation of the space. The waqf space is not only for religious activities, but also develops in social functions and environmental infrastructureThis indicates that waqf is carried out consistently and adaptively.

The Waqf that grows from individual awareness is institutionalized in a social system, thus forming solid and functional ties in society. The construction of mosques, Islamic boarding schools, and other social spaces is a unified system with the settlement environment. The community is the owner and guardian of the area, so it has a high persistence in maintaining the function of the space. The Waqf in Kauman Village is a symbol of local wisdom that can survive in the long term, so it must be preserved. Given the large number of waqf spaces and the magnitude of their role in society, future urban spatial planning policies must accommodate waqf spaces as a socially protected space to maintain sustainability and become a symbol of good deeds in people's lives.

8. Acknowledgment

We want to thank the Directorate General of Higher Education, Research and Technology of the Republic of Indonesia, Diponegoro University and Sultan Agung Islamic University who has supported this research. The authors also thank the people of *Kampung* Kauman who have provided information, observation data, and interviews in the field.

References

[1] N. M. Swanendri and G. A. M. Suartika, "Try and Find it - Understanding Public Space and Civil Society in Bali," *IOP Conf. Ser. Earth Environ. Sci.*, vol. 396, no. 1, pp. 0–11, 2019, doi: 10.1088/1755-1315/396/1/012030.
[2] S. R. Tamariska, A. A. Ronauly, M. S. Nugraha, T. Nurhasan, and A. Z. Fillah, "Studi Komparasi Perumahan Terencana Dan Perumahan Tidak Terencana (Perumahan Sukaluyu Dan Kampung Tubagus Ismail Bawah)," pp. 285–292, 2020.
[3] A. Putra G and Z. Zahro H, "Desain Partisipasi Dalam ruang Publik, Ruang komunal Sosial Dalam Kampung Heritage TawangsariSebagai Salah Satu Bentuk Aplikasi Unsur Keberlanjutan Sosial," *Pawon J. Arsit.*, vol. III, pp. 25–36, 2019.
[4] Bappenas, *Metadata Indikator: Pilar Pembangunan Lingkungan*. 2020.
[5] A. Lestari and A. Sari, "Upaya peningkatan kesalehan sosial dengan zakat dan wakaf," vol. 2, pp. 451–466, 2016.
[6] J. Wales, M. Aslam, S. Hine, S. Tawal, and L. Wild, "The role and impact of philanthropic and religious schools in developing countries." University of Birmingham, Institute of Education University of London, 2015.
[7] A. Singer, "Replace stasis with motion to fathom the persistence of Waqf: The complex histories and legacies of a Muslim institution," *Muslim World*, vol. 108, no. 4, pp. 702–716, 2018, doi: 10.1111/muwo.12270.
[8] S. Abdullah, "Potensi dan Kekuatan Modal Sosial dalam Suatu Komunitas," *SOCIUS J. Sosiol.*, vol. XII, pp. 15–21, 2013.
[9] Jon Lang, *Creating Architectural Theory*. New York: Van Nostrand Reinhold Company, 1987.
[10] Joyce Marcella Laurens, *Arsitektur dan perilaku manusia*. Surabaya: Gramedia Widia Sarana, 2004.
[11] I. Altman, *The Environment and Social Behavior*. Wadsworth: Monterey, CA, 1975.
[12] Pastalan, *Privacy, Territoriality, and Personal Space – Proxemic Theory, in Lang, Jon., 1987. Creating Architectural Theory*. New York: Van Nonstrand Reinhold, 1970.
[13] H. Kertajaya, *Arti komunitas*. Jakarta: Gramedia Pustaka, 2008.

[14] Soenarno, *Kekuatan Komunitas Sebagai Pilar Pembangunan Nasional*. Jakarta: Fakultas Teknik Universitas Muhammadiyah, 2002.
[15] B. S. dan Haryadi, *Arsitektur, lingkungan dan perilaku*. Yogyakarta: Gadjah Mada University Press, 2010.
[16] I. M. Yunita and I. N. S. S. Wijaya, "Tingkat keberlanjutan permukiman kampung warna-warni, kelurahan jodipan, kota malang," *Plan. Urban Reg. Environ.*, vol. 10, no. 1, pp. 145–156, 2021.
[17] K. Kusumastuti *et al.*, "Peran Kelompok Swadaya Masyarakat Dalam Mewujudkan Penataan Kampung Yang Berkelanjutan (Studi Kasus : Kampung Ngemplak, Jebres, Kota Surakarta)," *Desa-Kota*, vol. 3, no. 2, p. 171, 2021, doi: 10.20961/desa-kota.v3i2.45466.171-178.
[18] R. Sara, M. Jones, and L. Rice, "Austerity urbanism: connecting strategies and tactics for participatory placemaking," *CoDesign*, vol. 17, no. 4, pp. 493–509, 2021, doi: 10.1080/15710882.2020.1761985.
[19] T. Shwartz-Ziv and R. Strier, "Place-making: Toward a place-aware community practice agenda," *Br. J. Soc. Work*, vol. 52, no. 1, pp. 61–78, Jan. 2022, doi: 10.1093/bjsw/bcaa219.
[20] O. Scharbrodt, "A minority within a minority?: the complexity and multilocality of transnational Twelver Shia networks in Britain," *Contemp. Islam*, vol. 13, no. 3, pp. 287–305, 2019, doi: 10.1007/s11562-018-0431-0.
[21] A. Fahmi and E. Sugiarto, "Sistem Informasi Geografis E-Wakaf Sebagai Model Pendukung Keputusan Pemberdayaan Wakaf Produktif," *Ikra-ith Inform.*, vol. 1, no. 2, pp. 87–93, 2017, [Online]. Available: http://e-wakaf.info.
[22] F. Asni, M. A. Mahamud, and J. Sulong, "Socio-economics and management of Muslim cemetery waqf using istibdal and GIS method in Penang state," *J. Islam. Account. Bus. Res.*, vol. 11, no. 7, pp. 1343–1362, 2020, doi: 10.1108/JIABR-01-2019-0026.
[23] S. N. M. Ali, A. H. M. Noor, N. Johari, N. S. Fauzi, N. N. Chuweni, and N. R. P. Ismail, "Hungry for housing: Waqf real estate development - A social welfare alternative," *MATEC Web Conf.*, vol. 66, 2016, doi: 10.1051/matecconf/20166600068.
[24] M. G. Çöteli, "Islam Kentinde Vakiflarin Ticaret Bölgesinin Olusumu Üzerine Etkisi: Hayrat-Akar," *Found. J.*, 2016.
[25] R. E. Stake, *The Art of Case Study Research*. SAGE Publications Inc, 1995.
[26] J. W. Creswell, *Research Design, Pendekatan Metode Kualitatif, Kuantitatif, dan Campuran*, Cetakan IV. Yogyakarta: Pustaka Pelajar, 2019.
[27] Amedeo Giorgi, *The descriptive phenomenological method in phychology: A modified husserlian approach*. Pittsburgh: Duquesne University Press, 2009.
[28] R. k Yin, *Studi kasus, desain dan metode*. Depok: Rajawali Pers, 2002.
[29] N. A. Setiawati, "Manajemen Sejarah Berbasis Komunitas: Pengembangan Kawasan Kauman sebagai Living Museum," *Bakti Budaya*, vol. 1, no. 1, p. 101, 2018, doi: 10.22146/bb.37934.
[30] A. Suprapti, "Control of spatial protection in Kauman Semarang Journal of Archutecture and Urbanism," *J. Archutecture Urban.*, vol. 414 268, 2017.
[31] BPS Kota Semarang, "Statistik Daerah Semarang Tengah Dalam Angka Tahun 2021," p. 22, 2021.
[32] A. F. Yusuf, *Melacak Banda Masjid Yang Hilang*. Semarang: Penerbit Aneka Ilmu, 2000.
[33] A. Suprapti and D. Iswanto, "Mapping the socio-religious spaces of a kampong: A case of Kauman Semarang," *IOP Conf. Ser. Earth Environ. Sci.*, vol. 213, no. 1, 2018, doi: 10.1088/1755-1315/213/1/012048.
[34] S. Denoix, *A Mamluk Institution for Urbanization: The Waqf. In Behrens-Abouseif, Doris (Eds.) The Cairo Heritage: Essays in Honor of Layla Ali Ibrahim,*. 2000.
[35] H. Nour, "Reconsidering The Waqf: Traditional Mechanism of Urban Regeneration in Historic Muslim Cities," 2015.
[36] N. Ujang, *Place Attachment Towards Shopping District in Kuala Lumpur City Centre*. Universiti Peutra Malaysia. Unpublished Theses Degree of Doctor of Philosophy, 2008.

[37] A. Rochani, N. Yuliastuti, and B. Sudarwanto, “Philanthropy in Settlement Infrastructure Development,” *IOP Conf. Ser. Earth Environ. Sci.*, vol. 832, no. 1, 2021, doi: 10.1088/1755-1315/832/1/012010.

Business Model Innovation: Lesson Learned From EV Ecosystem In Indonesia

M T Amir[1], B Prabawani[2]

[1]Faculty of Economics and Social Science, Universitas Bakrie, Jl HR Rasuna Said, Kav C-22, Jakarta Selatan 12920, Indonesia
[2]Faculty of Social and Political Science, Universitas Diponegoro, Jalan Professor Soedarto SH, Tembalang, Kota Semarang, Jawa Tengah 50139

Abstract. Although relatively developed in advanced countries, the electric vehicle (EV) industry in Indonesia is still in the early stages of development. Designing a business model with an ecosystem perspective is critical for relevant stakeholders because there are still many uncertainties to be faced, including the unclear structure of the industry. This study examines the case of how an e-bus company creates a business model by taking into account the important role of stakeholders in their ecosystem. Top management of a manufacturing company and its parent company were in-depth interviewed. The roles and critical challenges of regulators, battery manufacturers, and assembly plants are explored. The practice of creating business models will be compared with applying business models in the literature. The discussion will focus on the need of the focal company in creating value through approaching regulators, collaborating with battery manufacturer, and creating communication as the innovations. The findings in this study are expected to be input for e-bus business actors for sustaining the development stage and stakeholders in the existing ecosystem, including the government or regulators.

.
Keywords: Electric vehicles; business model innovation; business ecosystem; regulators

1. Introduction

Although relatively developed in advanced countries, Indonesia's electric vehicle (EV) industry is still in the early stages of development. As a country that agrees to reduce gas emissions, and has issued regulations to accelerate electric-based motor vehicles, the progress experienced by every stakeholder in the electric vehicle ecosystem in Indonesia can still be slow. A study from the Institute for Essential Services Reform (IESR) found that adopting electric vehicles is indeed low compared to the government's target of 20% (around 400 thousand) of total vehicle production in 2025. [1]. Still to the IESR study, by mid-2020, the number of electric vehicles was 2,279. This slow adoption shows that the government's intention and commitment to reducing greenhouse gas emissions, which is 29-41 percent by 2030, has not demonstrated satisfactory progress.

On the other hand, the development of the global trend regarding the transition of electric vehicles indicates the existence of business opportunities in the field of transportation. Various manufacturing companies or sole agent of automotive brand holders also seem to have taken the initiative to develop the market. Recently there was a collaboration called the "EV Smart Mobility Joint project," which involved five major manufacturers from Japan agreeing to advance and establish a partnership for the electric car ecosystem[2].

Considering the ecosystem's development is essential, especially for companies whose industries are at the beginning of development. This need was experienced by PT X, a startup company that intends to become an electric bus manufacturer, which is now in the early stages of developing its business. PT X is a subsidiary of one automotive parts manufacturer in Indonesia for commercial and passenger vehicles. By making sustainability a future orientation, PT X needs to get around the design of its business model so that it can develop as expected.

The literature still diversely defines the electric vehicle ecosystem, and according to the IESR, it consists of the following elements: Charging infrastructure, Vehicle manufacturing/suppliers, User awareness and acceptance, Supply chains of batteries and electric vehicle components, incentives and supporting policies from the government, and telematics/industry [1].

With the situation's complexity, each stakeholder's interest in developing this electric vehicle ecosystem is different. This condition has implications for the complexity of designing an effective business model for a manufacturing company. Designing a business model with an ecosystem perspective is critical for relevant stakeholders because there are still many uncertainties, including the industry's unclear structure. As a stage of development, decisions about the business model become decisive. Therefore, the theme of changing or redesigning the business model plays a significant role. Sustainability and technological developments drive exponential changes and demand that companies be creative in determining business models [3].

This study examines the case of how an e-bus manufacturing company creates a business model by taking into account the critical role of stakeholders in the electronic vehicle ecosystem in Indonesia. The top management of a manufacturing company and its parent company were in-depth interviewed. The roles and critical challenges of regulators, battery manufacturers, smart operators, and assembly plants are explored. Who is considered necessary, and how to design the business model. The study contributed to several areas that overlap: Strategic management, managing innovation, and business development. The following section will review the study of business innovation and its relation to sustainability. Next the electric vehicle ecosystem, especially regarding electric buses, is reviewed. The results and discussion section shows how a supplier company sees the development of the ecosystem situation and adapts to the company's business model.

2. Literature review

2.1. Business model innovation

The business model formulation concept from Teece (2010), namely how the company conducts its business and explains how it creates, submits, and captures values, is still becoming a norm for scholars and practitioners. The business model innovation factory adds a business model also related to viability in its application. The business development model is characterized by a series of critical decisions that determine how a business earns revenue, incurs costs, and manages risks[4]. Thus, Innovation on business models also updates related decisions: what your offerings will be, when decisions are made, who makes them, and why. Its success will affect revenue, costs, and risks. According to Lindgard et al. (2009), business model innovation and innovation in terms of value proposition, revenue model, and operating model.

Adapting to the environment requires the leader to evaluate the effectiveness of his business model. The emergence of new business people who present technological disruption makes the existing rules of the game change. Osterwalder and colleagues [5] revealed that there is a demand for the organization to continuously reinvent itself before it is too late. Therefore, business model innovation becomes a strategic imperative. Companies need to explore the future while excelling and exploiting the present advantages. These companies simultaneously build a culture of innovation, expose the ongoing business, and can last in harmony. They compete with superior business models and transcend industry boundaries. Sustainability can be one of the drivers of innovation. Business model innovation is believed to be the key for companies that want to seize new opportunities and leaps in their progress [3].

2.2. Sustainability as an innovation driver

Seeing sustainability as the company's agenda and goals, can lead to the development of innovative business models. Although sustainability has been widely echoed, such as through efforts to compare it with corporate social responsibility (CSR), with the rapid changes in the field of sustainability, mediocre CSR practices have been left behind. Many studies show its development is relatively sluggish, and those who are first movers are perceived as overrated. Also, many are waiting or just moderately adapting. Companies need to take advantage of creative imagination, courage, and stamina to make other companies seem to move slowly.

The Sustainable Development Goals from the United Nations can drive the need for exponential business model innovation because it emphasizes economic, social, and environmental goals. Adams et al. [6] said that innovating in sustainability reflects important business capabilities, whether related to incremental, radical, or even disruptive innovation. Because it changes how businesses operate, the business model innovation emerged as a necessary mechanism to integrate sustainability into the business (Jolink and Niesten, 2015). Business model innovation is emerging as a potential mechanism to incorporate sustainability into business [7], [8].

Because the business model focuses on value creation, the shift in the perspective of economic value creation to psychology, sociology, and ecology becomes essential. The view of sustainability integrates everything until the meaning of its value becomes more holistic [7]. Evans et al. (2017) offer this holistic view, covering environmental, social, and economic values. The breakthrough business model commissioned by the Business and Sustainability development commission says that the SDGs tend to be more radical to carry [3]. The company moves from an incremental mindset and ambition to an exponential mind here. The business model needs to underline four key areas: 1)The measure of economic benefits in the SDGs environment demands transformation at the sector level. 2) A new business model that both insurgents will often initiate and radical incumbers who break traditions, 3) A new financial model that aligns with high social and social performance 3)A new social contract between business, government, and society.

As more stakeholder considerations become, as well as the demands of assessing their impact on the entire business network, it will bring higher complexity in its implementation. Previously, Teece [9] also assumed that the business model is a technological innovation and commercialization vehicle.

2.3. Electric vehicle ecosystem

Studies on the electric vehicle ecosystem in Indonesia tend to mention that this ecosystem has not been well built. As stated at the beginning, a study from the Institute for Essential Services Reform (IESR) notes five elements of the electric vehicle ecosystem a) charging infrastructure, b) models and supplies of electric vehicles; C) public awareness and acceptance, d) supply chain of batteries and electric vehicle components e)incentives and supporting policies from the government. A study by Abdelkafi et al. [10] said the technology needed to be considered in electric car systems and is a critical element of the ecosystem. This includes technologies from battery components and telematics in bus operation. Complementing the supply elements, Abldelkafi et al. [10] suggest the details of actors, consisting of manufacturing for automobile components and batteries, suppliers, service providers, and government as well as substitution goods.

As a new industry, ecosystem elements wait and see for the development of ecosystem effectiveness. The government is vital in enabling actors to initiate and accelerate initiatives. Fiscal incentives are determined to accelerate the transition by exemptions from VAT, registration taxes, and toll fees that can reach 55% percent of the price of vehicles [11]. Non-fiscal incentives are also expected to help accelerate adoption.

Government incentives largely determine the element of awareness and acceptance. Experience in pre-existing countries such as Norway, Sweden, America, or China must be considered for Indonesian government incentive policies. Packages provide incentives to make the value reach>50% of the vehicle price [1]. Without it, the cost of vehicles is still high, and the Chinese government even provides additional incentives in the form of subsidies. Especially for electric buses, government factors that at the same time act as users can be related to central government policies. Here business actors, in the form of traders, face two challenges: the first is convincing users or the market, and the second is convincing the government and associated elements which hold the regulatory authority in which it is interested. Certain countries encourage the development of electric vehicle manufacturing with research and development funds, financial incentives related to intellectual rights, and public service procurement for public services.

For models and supplies, it can be expected that it is also very dependent on government policies. Here business actors, in the form of traders, manufacturing has two challenges: the first is convincing users or the market, and the second is convincing the government and related elements which hold the regulatory authority in which it is interested. Certain countries encourage the development of electric vehicle manufacturing with research and development funds, financial incentives related to intellectual rights, and procurement for public services (IESR).

In terms of the electricity supply network, the government is again expected to be present and play an essential role. This network is necessary and should be considered a public service. Nevertheless, something that may not be too much of a problem for bus companies, especially those whose fleets already have regular travel lanes.

3. Method

This research uses a qualitative approach to case studies [12]. It tends towards the descriptive direction of how a company creates a business model based on an ecosystem perspective. For this, two top management, namely the CEO and COO of an electric bus supplying actor company, were interviewed in-depth for a total of 5 interviews. Three webinars with speakers from the chairman of the company's parent company were also observed. Some secondary data, such as one recording guest lecture on electric vehicles, and two other general webinars and materials from electronic media, were used as additional materials. The selection of respondents is based on the respondents' knowledge of developing an electric vehicle business from traditional businesses that BA has previously owned. Their long work experience can undoubtedly reflect the depth of strategic thinking needed in this study.

The concept of the business model from Teece [9] and ecosystem models from IESR [1] and Zulkarnain research institutes [13] was used as the lens and created interview guides. Data validation was carried out through follow-up interviews, triangulation with two energy transition research institute researchers for the most critical data, and comparing print media news.

4. Results and discussion

4.1. Brief profile and condition of the company

Company X is a subsidiary of PT Y, an auto parts manufacturing company. Company Y is part of a conglomerate in Indonesia engaged in mining & energy, property, electronic media, and metal and construction. The initial strategic step carried out by PT Y, as the parent company of PT X, is to cooperate with BYD Auto, the number one electric vehicle company in China and the world's number two since 2018. Then PT Y formed PT X and initially became an electric bus assembler completely knocked down through a partnership with a carousel company in Indonesia. With the capacity of PT Y, which now plays a leading role in terms of Manufacturer for auto components, general casting & Machining lines, and Precision machining for automotive industries, the company has long-term plans as an OEM manufacturer for electric buses.

Until mid-2022, 30 buses have been successfully marketed for DKI Jakarta province. Several provinces are now willing to try this bus and become potential buyers. The company's strategy studies the development of the transport electrification ecosystem and actively leads the direction of this ecosystem, especially with the government considered the key to success. When this report was prepared, PT X anticipated additional orders from DKI Jakarta Province and West Java province.

4.2. Dynamic of business model

PT X's position is considered a first mover or pioneer in supplying electric buses in Indonesia, so it is necessary to implement a more flexible business model in the early stages of this development. Top management feels that development is still quite dynamic because there is some uncertainty waiting for the readiness of the ecosystem. The company's position was temporary while still as a brand holder agent. But in the long run, the company intends to become an OEM for electric buses and a battery manufacturer. With its parent company's capacity to facilitate related parts production, this intention is quite reasonable.

4.3. The innovation in the business model

The business model is very dependent on the ecosystem, while the ecosystem in Indonesia is still immature and not yet established. Meanwhile, from the company's side, following the existing business model in more developed countries with its EV ecosystem is not necessarily an optimal choice. This is in line with the industry's embryonic concept, which suggests focusing on exploratory activities while understanding market needs, and making more frequent adaptations to business model strategies [14]. For this reason, the business model innovation applied here is to secure partners as critical resources that are important to move to the next stage so that the dynamics of the ecosystem can be more predicted. The company's current business model strategy is to strengthen partnerships: 1) Bus assembler, 2) Battery supplier, 3) Retrofit company. This strategy enhances the company's position to face other competitors and negotiate with stakeholders at a later stage of the business. This choice is considered to provide the most combination of values for the company at this stage.

4.4. Regulator/government as the huge market

The government plays a crucial role in the success of the company's short and long-term plans. The central government regulation regulates the acceleration of the use of electric vehicles, stating that at least eight parties need to be monitored continuously by the company if the development of the ecosystem is to be mastered. Among them are the Ministry of National Development Planning, Ministry of Stated-Owned Enterprise, Ministry of public works and housing, Ministry of Finance, and Ministry of energy and mineral resources. As regulators or route owners, some of the schemes offered will be the primary considerations of the company; 1)Bus rapid transfer, 2)Buy the services (from Ministry of Transportation), 3)Regional government, 4)Military, and 5)Private company.
The above schemes make the government-related institution the primary target market for the company, and the operator company is part of the regional company. The success of the mechanism of purchasing operators with the company as the brand holders agent in DKI Jakarta province provides excellent value because it builds a sense of trustworthiness for related parties.
Business related to government programs requires facilitating policies in the financial sector, especially budget policies. Because the economic scheme of the financial infrastructure has not yet supported the needs of electric buses, the value of which can be 3-5 times that of ordinary buses, policies on budgets and business actors need to observe their development closely. This arrangement can allow the bank to provide funding to the operator company, and the supplier's relationship can be smooth. For this, the company needs to master and approach critical persons. Among the strategies, PT X implemented is appointing several important figures as company advisors. A similar strategy is also applied for

obtaining partner work from abroad. This business model allows companies to be up-to-date and know the policy directions that regulators and foreign partners developed.

4.5. Become a battery technology supplier

The company realizes the importance of electrical technology to ensure electric buses' smooth operation and performance. The company is now directing its attention to becoming an electrical technology player by partnering with global players in the battery sector and investing in the ownership of related partner companies.

4.6. The end-of-life of the battery

How to manage the end-of-life of the battery needs to be a concern for being a battery supplier, in line with the increasingly advanced study of battery recycling [15]. Some studies found that EV batteries can be recycled and used to store electricity for building purposes. This practice has been proven to work well in the case of the use of used batteries in Amsterdam. In a further study, the recycling process was also found to be profitable, as indicated by the addition of the number of industries developing in several countries; many factories were established in Asia, Europe, and North America [16]. The focal company in the study needs to learn these two important things.

4.7. Partnership and alliance factor

The company is very aware of being able to adapt to the demands of every element in the ecosystem. At this stage, while waiting for preparations to become an OEM of an electric bus, PT X ensures cooperation is established in the area of brand holding, assembler, battery manufacturing, companies for retrofitting, universities for research, and global consulting institutions. The company adds value to each ecosystem element where one partnership gives the confidence to reach the other potential partners.
The company realizes the importance of electrical technology to ensure electric buses' smooth operation and performance. The company is now directing its attention to becoming an electrical technology player by partnering with global players in the battery sector and investing in the ownership of related partner companies.

4.8. Value creation through public relations

The company utilizes various communication channels and methods to build an image and trust as a pioneer. Success after success from the development stage can convince other elements of the ecosystem, especially the government as a market. For example, the company's success in getting orders from the DKI Jakarta government helps and becomes valuable in dealing with stakeholders. New change and transformation communication techniques to convince audiences and create an urgency utilized by companies.

5. Conclusions and suggestions

Being a bus supplier actor in the electric bus ecosystem gives two options; as a first mover or as a follower. These two options provide very different consequences in developing a business model. With the rapid changes in the ecosystem elements, the company needs to anticipate changes in business models.

As a pioneer, the focal company in this study needs to secure the decisions of elements of the ecosystem and make the basis for developing the business model. Among the ecosystem elements that must get priority to be approached are regulators/governments. Business actors must have access and resource persons to develop this regulatory decision. The best way to ensure this is: Master the regulations and be actively involved. Having access to the parties that determine the direction is significant. The company's next element of the ecosystem must prioritize battery supply and battery technology. Another value creation is good communication. Here, companies can consider various public relations strategies, including storytelling. With the need to change rapidly, including modifying the business model, companies must develop adaptive learning capacity for their employees. Furthermore, to be different from other competitors, the company needs to leverage existing businesses network in its internal parent company and utilize its strengths with a network that will benefit PT X.

References

[1] I. Marciano, "Mengembangkan Ekosistem Kendaraan Listrik di Indonesia: Pelajaran dari Pengalaman Amerika Serikat, Norwegia dan Cina," Jakarta, Indonesia, 2021.
[2] G. M. Finesso, "Jejak mobil listrik dan ikhtiar jepang demi langit biru," *Harian Kompas*, 2022.
[3] J. Elkington, J. Lim, and L. Smith, "Breakthrough Business Models Exponentially more social, lean, integrated and circularNo Title," 2016.
[4] K. Girotra and S. Netessine, "Four paths to business model innovation," *Harv. Bus. Rev.*, vol. 92, no. 7, pp. 96–103, 2014.
[5] A. Osterwalder, Y. Pigneur, A. Smith, and F. Etiemble, *The Invincible Company*. Wiley, 2020.
[6] R. Adams, S. Jeanrenaud, J. Bessant, D. Denyer, and P. Overy, "Sustainability-oriented Innovation: A Systematic Review," *Int. J. Manag. Rev.*, vol. 18, no. 2, pp. 180–205, Apr. 2016, doi: 10.1111/IJMR.12068.
[7] S. Schaltegger and M. Wagner, "Sustainable entrepreneurship and sustainability innovation: categories and interactions," *Bus. Strateg. Environ.*, vol. 20, no. 4, pp. 222–237, May 2011, doi: 10.1002/BSE.682.
[8] A. Jolink and E. Niesten, "Sustainable Development and Business Models of Entrepreneurs in the Organic Food Industry," *Bus. Strateg. Environ.*, vol. 24, no. 6, pp. 386–401, Sep. 2015, doi: 10.1002/BSE.1826.
[9] D. J. Teece, "Business Models, Business Strategy and Innovation," *Long Range Plann.*, vol. 43, no. 2–3, pp. 172–194, Apr. 2010, doi: 10.1016/J.LRP.2009.07.003.
[10] N. Abdelkafi, S. Makhotin, and T. Posselt, "Business model innovations for electric mobility: What can be learned from existing business model pattern?," *Int. J. Innov. Manag.*, vol. 17, no. 1, Mar. 2013, doi: 10.1142/S1363919613400033.
[11] W. K. Jaya, "Menunggu insentif dan ekosistem kendaraan listrik," *Harian Kompas*, p. 6, 2022.
[12] R. K. Yin, "Case study methods," in *Handbook of Complementary Methods in Education Research*, 2012.
[13] Zulkarnain, P. Leviäkangas, T. Kinnunen, and P. Kess, "The Electric Vehicles Ecosystem Model: Construct, Analysis and Identification of Key Challenges," *Manag. Glob. Transitions*, vol. 12, no. 3, pp. 253–277, 2014.
[14] D. Vittori, A. Natalicchio, U. Panniello, A. Messeni Petruzzelli, and F. Cupertino, "Business Model Innovation between the embryonic and growth stages of industry lifecycle," *Technovation*, p. 102592, Aug. 2022, doi: 10.1016/J.TECHNOVATION.2022.102592.
[15] M. Pagliaro and F. Meneguzzo, "Lithium battery reusing and recycling: A circular economy insight," *Heliyon*, vol. 5, no. 6, p. e01866, Jun. 2019, doi: 10.1016/J.HELIYON.2019.E01866.
[16] M. Pagliaro and F. Meneguzzo, "Electric Bus: A Critical Overview on the Dawn of Its Widespread Uptake," *Adv. Sustain. Syst.*, vol. 3, no. 6, p. 1800151, Jun. 2019, doi: 10.1002/ADSU.201800151.

Acknowledgments
Authors wishing to acknowledge the support of World Classs University Task Force of University of Diponegoro.

Effect of cow manure on biogas production based on rice husk waste in SSAD conditions

H H A Matin[1,4*], S Syafrudin[2], and S Suherman[3]

[1] Doctor Program of Environmental Science, School of Post Graduate Studies, Universitas Diponegoro, 50275 Semarang, Indonesia
[2] Department of Environmental Engineering, Faculty of Engineering, Universitas Diponegoro, 50275 Semarang, Indonesia
[3] Department of Chemical Engineering, Faculty of Engineering, Universitas Diponegoro, 50275 Semarang, Indonesia
[4] Study Program of Environmental Science, Faculty of Mathematic and Natural Science, Universitas Sebelas Maret, 57126 Surakarta, Indonesia

*hawalihashfi@staff.uns.ac.id

Abstract. The development of biogas technology is currently very fast along with the increasing need for energy and expensive costs used to buy fossil energy. Rice husk-based biogas has begun to be widely developed because of the high potential that can be utilized from the biomass. The aim of this study was to study the effect of the addition cow manure combined with rice husks for biogas production under SSAD conditions. Variable comparison of rice husk and cow manure as follows: 4:1, 3:1, 2:1, 1:1, 1:2, 1:3, and 1:4. The study was carried out in duplicate on laboratory scale with C/N ratio 35, 27% total solids and rumen fluid as a source of methanogenic bacteria. Biogas is observed every two days until its productivity decreases. It was found that the best biogas productivity was in the 2:1 variable range, namely 39.49 mL/grTS followed by 1:1 21.91 mL/grTS, 3:1 13.80 mL/grTS, 1:2 13.31 mL/grTS, 1:3 10.90 mL/grTS, 1:4 82 mL/grTS and 4:1 9.41 mL/grTS. After calculating the biogas production rate constant with Gompertz, the highest productivity was obtained at 1:1 variable, with maximum biogas (A): 65.69 mL/grTS, daily biogas production (U): 0.37 mL/grTS.day, and minimal time for biogas formation (λ): 25th day.

1. Introduction

Alternative energy continues to be a very interesting topic in Indonesia and even in the world. This is the result of rumors that always appear that the world's fossil energy reserves continue to run low. This is even confirmed by a press release from the Ministry of Energy and Mineral Resources number: 028.Pers/04/SJI/2021 it was stated that oil reserves in Indonesia will be available for the next 9.5 years, while the age of Indonesia's natural gas reserves reaches 19.9 years assuming no sources of oil and gas reserves are found. The information was issued in 2021 so that if no new oil and gas wells are found, then oil will run out in the 2030s and natural gas will run out in the 2040s. Therefore, many researchers both in Indonesia and the world are competing to find and develop new and renewable energy as an alternative to fossil energy which continues to run out.

As an alternative energy to replace fossil energy, biogas technology is a technology that is easy to operate. This is evidenced by its very development, such as in China and India where there are many agricultural and livestock areas [1]–[4]. This condition is similar to Indonesia so that biogas technology is very attractive to be developed in this agrarian country [5]. Biogas technology has been developed by many researchers using livestock waste as raw materials and is commonly referred to as first generation biogas technology. This first generation was very successful because it was so easy to convert livestock waste into biogas energy. As a result of the high agricultural waste, currently researchers are also developing biogas with agricultural waste as raw material and commonly referred to as second generation biogas technology [6]–[8]. The principle of biogas technology is to ferment organic material (waste) anaerobically without oxygen with the help of methanogenic bacteria. In general, biogas produces 60% methane gas (CH_4), 38% carbon dioxide (CO_2), and the rest is other supporting gases 2% [9]. Raw materials that can be used for biogas production include livestock waste such as cow dung, goats, horses, pigs, poultry, and agricultural waste such as corn cobs, sugarcane bagasse, rice straw, rice husks and other liquid waste which is degradable organic material [10]–[19].

The raw material for biogas technology that has been proven reliable is cow dung and currently many biogas technologies are being developed using agricultural waste (biomass) [6], [20]. One of the biomass is rice husk. Rice husks are found in Indonesia, especially during the rice harvest season. The rice harvest season in Indonesia reaches three times a year so that the production of rice husks in each rice mill house is very large. On the other hand, the use of rice husks is currently still low, such as for planting media, cooking fuel and fuel for making bricks. This activity has not been able to absorb the existing rice husk generation, so the other option is to burn it even though the activity of burning rice husk can harm health for both humans and the environment [21]. The high lignin contained in rice husk makes the material difficult to be degraded by microorganisms so that preliminary treatments such as physics, chemistry, and biology are needed to be easily converted into biogas [22].

Chemical pretreatment for biomass with lignocellulosic may increase biodegradation rate and yield biogas. He et al. [23] reported NaOH pretreatment of rice straw increase biogas production SS-AD with the highest yield at an organic loading rate of 50 g/L, compared to the untreated rice straw. Further, NaOH pretreatment can reduce barriers during the fermentation process so that it can increase efficiency and lower production costs. Millet et al. [24] stated that the level of fineness of hardwood treated with NaOH increased from 14% to 55%, it also caused a decrease in lignin levels from 24-55% to 20%. Chandra et al. [25] cultivate rice straw with L-AD system (5% TS) and pretreatment with NaOH 3%. The treatment NaOH 3% succeeded in raising the production of biogas as much as 132% rather than rice straw without pretreatment.

Biogas production using anaerobic digestion (AD) has two benefits, namely being able to process the waste and produce biogas which can be used as an alternative energy source in the form of renewable energy. In addition, other advantages was obtained such as reduce greenhouse gas emissions and produce fertilizers in the form of organic fertilizers which can increase the community's economy [8-9]. The AD process based on TS content is classified into L-AD and SS-AD. L-AD operates at TS less than 15%, whereas SS-AD TS content is higher than 15% [7], [26].

Many researchers have studied biogas production testing in SSAD systems with high lignocellulosic biomass feedstock [17], [18], [21], [27]–[30]. However, the production of biogas using a combination of rice husks and cow dung as raw materials has not been studied. So the focus of this research is to study the effectiveness of biogas production with these two raw materials at 27% SSAD condition and C/N ratio 35. Then proceed to study the kinetics of biogas production rate with Gompertz non-linear regression which adjusts the microorganism growth scheme [7], [31]–[34].

2. Materials and Methods

In this study, the main raw material is rice husk combined with cow dung. The rice husks were taken from the rice mill house in the Rowosari area, Tembalang, Semarang, while the cow dung was taken from the Slaughterhouse (RPH) in the Penggaron area, Pedurungan, Semarang. This study used rumen fluid which was also taken from the same abattoir with cow dung.

Prior to running the research, pre-treatment of rice husk was carried out as a delignification process so that it could be digested by anaerobic bacteria. Pretreatment was carried out chemically with 3% technical NaOH from local store. The rice husks were contacted with NaOH for 24 hours after which they were cleaned until the pH returned to neutral and continued with the drying process until the conditions were the same as when the rice husks were taken from the rice mill house [21].

This research was conducted in batches on a laboratory scale. The reactor used is 1 liter in size and made of airtight polyethylene so that anaerobic conditions can be formed. The reactor is equipped with a rubber valve that can be opened and closed for the biogas production measurement process. The measurement of biogas is carried out every 2 days with the help of a laboratory measuring cup with units of milliliters [5], [13]. The description of the reactor and the process of measuring biogas production can be seen in Figure 1 below.

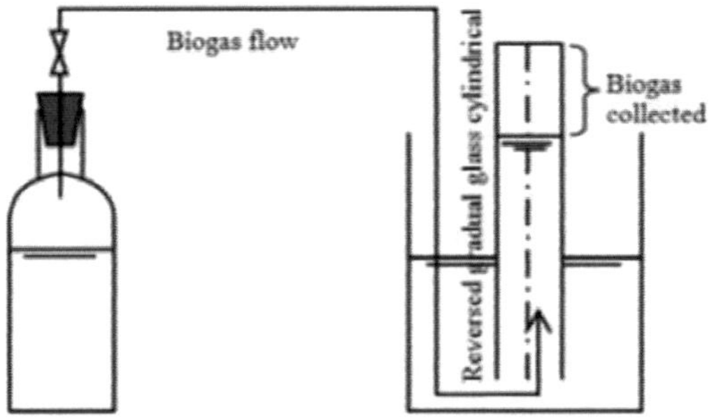

Figure 1. Research scheme of batch anaerobic digestion

The implementation procedure in this study begins with chemical pretreatment of rice husks. Furthermore, it was conditioned by SS-AD, with a TS of 27% and a C/N ratio of 35 [21]. The C/N ratio was regulated with the help of adding nutrients using technical urea [18]. Variables set to achieve the objectives of this study are: rice husk (RH) and cow manure (CM) as follows: 4:1, 3:1, 2:1, 1:1, 1:2, 1:3, and 1:4. Biogas production was measured until the gas no longer formed significantly.

In this study also studied the rate of biogas production kinetics. With the Gompertz equation which was analyzed using Microsoft excel software, the constant model of A, U, and λ was obtained. Gompertz equation can be seen as follows:

$$P = A.exp\left\{-exp\left[\frac{Ue}{A}(\lambda - t) + 1\right]\right\} \quad (1)$$

where:

P = Cumulative biogas production (mL/grTS)
A = Maximum biogas production (mL/grTS)
U = Constant Maximum biogas production rate (mL/grTS.day)
λ = Duration of lag phase (day)
t = time to review cumulative production of biogas (day)
e = Euler number (e = 2.71828 ...)

3. Result and Discussion

3.1. Study of Adding Cow Manure to Biogas Production

The purpose of this study was to examine the effect of the addition of cow dung on biogas production using rice husk as raw material and operated under SSAD conditions. In this study, observations were made in the laboratory for 90 days (until biogas productivity was no longer significant). Observations were made with variations in the variable of RH:CM as follows: 4:1, 3:1, 2:1, 1:1, 1:2, 1:3, and 1:4. Daily biogas production data can be seen in Figure 2.

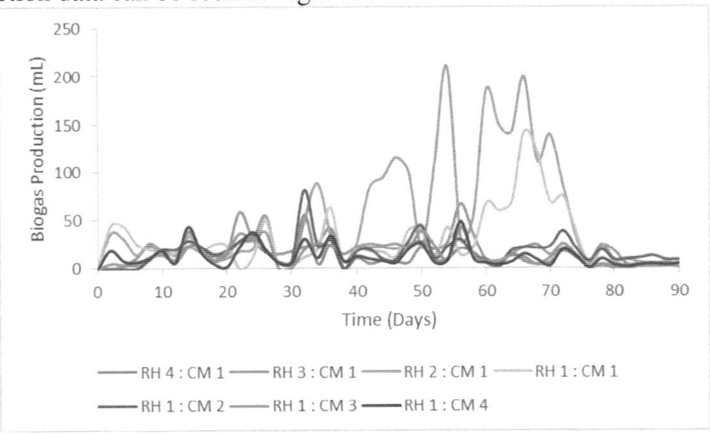

Figure 2. Daily biogas production observation data for 90 days

Figure 3 below shows the cumulative biogas observation data for 90 days. In part A is cumulative in milliliters and part B is per unit TS in one anaerobic reactor. Part A shows the cumulative results of 4:1, 3:1, 2:1:1:1, 1:2, 1:3, 1:4 respectively as follows: 512 mL, 856 mL, 2449.5 mL, 1359 mL, 826 mL, 676.5 mL, and 584 mL. Based on the cumulative data, it was found that the best biogas production was in the variable RH 2: CM 1, which reached 2449.5 mL. Furthermore, for part B, the cumulative results per TS unit are as follows: 8.25 mL/grTS, 13.80 mL/grTS, 39.49 mL/grTS, 21.91 mL/grTS, 13.31 mL/grTS, 10.90 mL/grTS, and 9.41 mL/grTS. In line with part A, the best biogas production in the variable RH 2: CM 1 is 39.49 mL/grTS.

The 2:1 variable produces higher biogas than the other variables under SSAD conditions. This variable contains twice as much rice husk as cow dung. Biogas with rice husk as raw material has the characteristics of gas production which is produced slowly but with a long continuity. This is due to the difficulty of processing rice husks by anaerobic bacteria in the reactor due to the lignin content even though chemical pretreatment using NaOH has been carried out [7], [8], [21]. Next on this variable is cow dung with half the content of rice husks. Cow dung used as raw material for biogas is very easily converted by anaerobic bacteria and has a fairly high gas production characteristic in the early stages of anaerobic processing [10], [13], [35]. The combination of the two between rice husks and cow dung gives positive results on biogas productivity.

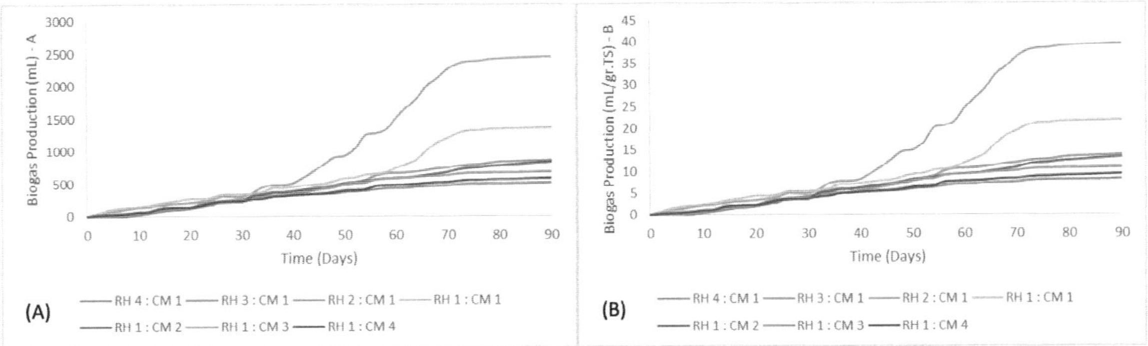

Figure 3. Biogas production observation data for 90 days: A. Cumulative (mL); B. Cumulative per unit TS (mL/grTS)

3.2. Study of Production Rate of Adding Cow Manure to Biogas Production

After getting the best biogas production variable, a non-linear regression mathematical model was calculated using the Gompertz equation to get the biogas production rate. In Table 1, the results of the model constants (A) are the maximum biogas production, (U) are the daily biogas production rate, and (λ) are the minimum time for biogas formation.

Table 1. Kinetic constant in biogas production RH 2 : CM 1

Constant Model	Unit	Value
A	mL/grTS	53.60
U	mL/grTS.day	0.80
λ	days	28.34

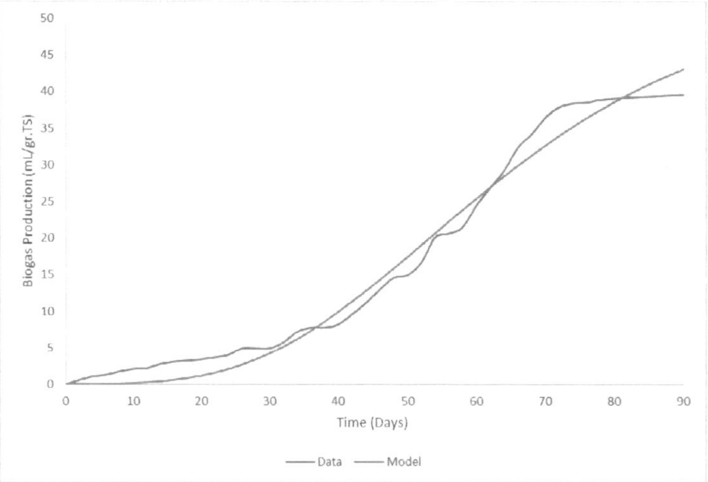

Figure 4. Relationship between research data and model calculation results on biogas production RH 2 : CM 1

Based on the calculation of the non-linear regression mathematical model using the Gompertz equation using Microsoft Excel software, the constants A, U, and λ were obtained respectively as follows: 53.60 mL/grTS, 0.80 mL/grTS.day and day 28, with r2 = 0.9844.

4. Conclusion

Based on the research we have done, cow dung has an effect on biogas production based on rice husk waste under SSAD conditions. This is indicated by the biogas productivity which increases with the addition of cow dung in each variable. It was found that the biogas productivity with the best RH:CM ratio was in the variable range of 2:1, namely 39.49 mL/grTS followed by 1:1 21.91 mL/grTS, 3:1 13.80 mL/grTS, 1:2 13.31 mL/grTS, 1:3 10.90 mL/grTS, 1:4 82 mL/grTS and 4:1 9.41 mL/grTS. After calculating the biogas production rate constant with Gompertz, maximum biogas (A): 53.60 mL/grTS, daily biogas production rate (U): 0.80 mL/grTS.day, and minimal time for biogas formation (λ): 28th day.

References

[1] Bhattacharya S C, Salam P A, Pham H L, and Ravindranath N H 2003 Sustainable biomass production for energy in selected Asian countries *Biomass Bioenergy* **25**(5) 471–482
[2] Cheng H and Hu Y 2010 Municipal solid waste (MSW) as a renewable source of energy: Current and future practices in China *Bioresour. Technol.* **101**(11) 3816–3824
[3] Mittal S, Ahlgren E O, and Shukla P R 2017 Barriers to biogas dissemination in India: A review

Energy Policy **112** 361–370
[4] Singh R and Setiawan A D 2013 Biomass energy policies and strategies: Harvesting potential in India and Indonesia *Renew. Sustain. Energy Rev.* **22** 332–345
[5] Putri D A, Saputro R R, and Budiyono B 2019 Biogas Production from Cow Manure *Int. J. Renew. Energy Dev.* **1**(2) 61
[6] Torrijos M 2016 State of Development of Biogas Production in Europe *Procedia Environ. Sci.*, **35** 881–889
[7] H Matin H H A, Syafrudin S, and Suherman S 2020 Solid State Anaerobic Digestion for Biogas Production from Rice Husk *E3S Web Conf.* **202** 08005.
[8] Syafrudin, Nugraha W D, Putri S A, Matin H H A, and Budiyono 2020 Enhancement of biogas production from rice husk using mechanical pretreatment (grinding) in Liquid Anaerobic Digestion (L-AD) *E3S Web Conf.* **202** 08003
[9] Al Seadi T. *et al.* 2008 *Biogas Handbook* (Denmark: University of Southern Denmark Esbjerg)
[10] Budiyono, Manthia F, Amalin N, Matin H H A, and Sumardiono S 2018 Production of Biogas from Organic Fruit Waste in Anaerobic Digester using Ruminant as The Inoculum *MATEC Web Conf.* **156** 03053
[11] Mönch-Tegeder M, Lemmer A, and Oechsner H 2014 Enhancement of methane production with horse manure supplement and pretreatment in a full-scale biogas process *Energy* **73** 523–530
[12] Ye J *et al.* 2013 Improved biogas production from rice straw by co-digestion with kitchen waste and pig manure *Waste Manag.* **33**(12) 2653–2658
[13] Budiyono B, Riyanta A B, Sumardiono S, Jos B, and Syaichurrozi I 2021 Optimization of parameters for biogas production from bagasse using taguchi method *Polish J. Environ. Stud.* **30**(5) 4453–4461
[14] Okonkwo U C, Onokpite E, and Onokwai A O 2018 Comparative study of the optimal ratio of biogas production from various organic wastes and weeds for digester/restarted digester *J. King Saud Univ. - Eng. Sci.* **30**(2) 123–129
[15] Purwoko T, Sari S L A, and Mahadjoeno E 2017 Bioethanol Production From Rice and Corn Husks After Enzymatic and Microbes Hydrolysis and Yeast Fermentation *Biotek.* **14** 19–23
[16] Sumardiono S, Riyanta A B, Matin H H A, Kusworo T D, Jos B, and Budiyono 2017 Increasing biogas production from sugar cane baggase by anaerobic co-digestion with animal manure *MATEC Web Conf.* **101** 02014
[17] Nugraha W D, Syafrudin, Keumala C F, Matin H H A, and Budiyono 2018 The Effect of Acid Pre-Treatment using Acetic Acid and Nitric Acid in The Production of Biogas from Rice Husk during Solid State Anaerobic Digestion (SS-AD) *E3S Web Conf.* **31** 01006
[18] Matin H H A and Hadiyanto 2018 Biogas Production from Rice Husk Waste by using Solid State Anaerobic Digestion (SSAD) Method *E3S Web Conf.* **31** 02007
[19] Carlini M, Castellucci S, and Moneti M 2015 Biogas production from poultry manure and cheese whey wastewater under mesophilic conditions in batch reactor *Energy Procedia* **82** 811–818
[20] Shehu B, Ibn U, and Ismail N 2012 Anaerobic Digestion of Cow Dung for Biogas Production A*RPN Journal of Engineering and Applied Sciences* 7(2) 169–172
[21] Matin H H A and Hadiyanto H 2018 Optimization of biogas production from rice husk waste by solid state anaerobic digestion (SSAD) using response surface methodology *J. Environ. Sci. Technol.* **11**(3) 147–156
[22] Bochmann G and Montgomery L F R 2013 Storage and pre-treatment of substrates for biogas production *Biogas Handb. Sci. Prod. Appl.*, 85–103
[23] He Y, Pang Y, Li X, Liu Y, Li R, and Zheng M 2009 Investigation on the Changes of Main Compositions and Extractives of Rice Straw Pretreated with Sodium Hydroxide for Biogas Production *Energy Fuels* **23**(4) 2220–2224
[24] Millett M A, Baker A J, and Satter L D 1976 Physical and chemical pretreatments for enhancing

cellulose saccharification *Biotechnol. Bioeng. Symp. United States* **6**
[25] Chandra R, Takeuchi H, and Hasegawa T 2012 Hydrothermal pretreatment of rice straw biomass: A potential and promising method for enhanced methane production *Appl. Energy*, **94** 129–140
[26] Brown D, Shi J, and Li Y 2012 Comparison of solid-state to liquid anaerobic digestion of lignocellulosic feedstocks for biogas production *Bioresour. Technol.* **124** 379–386.
[27] Mirmohamadsadeghi S, Karimi K, Zamani A, Amiri H, and Horváth I S 2014 Enhanced solid-state biogas production from lignocellulosic biomass by organosolv pretreatment *Biomed Res. Int.* **2014** 350414
[28] Ge X, Xu F, and Li Y 2016 Solid-state anaerobic digestion of lignocellulosic biomass: Recent progress and perspectives *Bioresour. Technol.* **205** 239–249
[29] Syafrudin, Nugraha W D, Agnesia S S, Matin H H A, and Budiyono 2016 Enhancement of Biogas Production from Rice Husk by NaOH and Enzyme Pretreatment *E3S Web Conf.* **31** 02002
[30] Syafrudin, Nugraha W D, Ardinata L G, Kencanawardhani I H, Matin H H A, and Budiyono 2017 The Influence of Total Solid (TS) Content to Biogas Production from Rice Husk Waste During Solid State Anaerobic Digestion (SS-AD) *Adv. Sci. Lett.* **23**(3) 2204–2206, 2017.
[31] Syaichurrozi B I and Sumardiono S 2017 Kinetic Model of Biogas Yield Production from Vinasse at Various Initial pH: Comparison between Modified Gompertz Model and First Order Kinetic Model *Res. J. Appl. Sci. Eng. Technol.* **7**(13) 2798–2805
[32] Sarlinda F, Sarto S, and Hidayat M 2018 Kinerja dan kinetika produksi biohidrogen secara batch dari sampah buah melon dalam reaktor tangki berpengaduk *J. Rekayasa Proses* **12**(1) 32 [in Indonesian Language].
[33] Chouaibi M, Ben Daoued K, Riguane K, Rouissi T, and Ferrari G 2020 Production of bioethanol from pumpkin peel wastes: Comparison between response surface methodology (RSM) and artificial neural networks (ANN) *Ind. Crops Prod.* **155**(4) 112822
[34] Budiyono B, Syaichurrozi I, Suhirman S, Hidayat T, and Jayanudin J 2021 Experiment and modeling to evaluate the effect of total solid on biogas production from the anaerobic co-digestion of Tofu liquid waste and rice straw *Polish J. Environ. Stud.* **30**(4) 3489–3496
[35] Kasisira L L and Muyiiya N D 2009 Assessment of the Effect of Mixing Pig and Cow Dung on Biogas Yield *CIGR Ejournal*, **XI**, **2003**, 1–7, 2009.

The Effect of Mewis Duct Energy Saving Device to Propeller Performance

A Trimulyono[1]*, I P Mulyatno[1], A F Rachmat[1]

[1]Department of Naval Architecture, Faculty of Engineering, Universitas Diponegoro, Semarang, Indonesia.

*Correspondence: anditrimulyono@lecturer.undip.ac.id

Abstract. Installation of ESD (Energy Saving Device) can improve ship propulsion performance. Mewis Duct is one type of ESD (Energy Saving Device) multi-component device that combines nozzle and fin into the nozzle. The structure can minimize losses due to small losses at the ship's stern and rotational losses or loss of thrust in the slipstreams area. Mewis Duct can reduce power by about 3-8% and increase thrust on the propeller by about 2-5%. This study aims to improve the performance of the INSEAN e779a propeller type using Mewis Duct. The modified Mewis ducts are the number of fins four, five, six asymmetrical and four symmetrical fins using the computational fluid dynamics (CFD) approach. The CFD code is based on the RANS (Reynolds - Averaged Navier Stokes) equation with the turbulent model is k-ε. This study found that installing Mewis duct as ESD in a ship increased the propeller thrust by 3-5% and the torque by 3-4%.

1. Introduction

The energy-saving device (ESD) is one of the ways to increase propeller performance by adding the device to the propeller or ship. ESD can minimize energy loss caused by propeller performance which can reduce energy loss imposed on a propeller performance, then improve ship performance. The study of ESD was performed by Trimulyono et al. for propeller B-series with the combination of ESD [1]. Later, The study was carried out with some variations in angle and fin diameter of PBCF [2]. The study of propeller B-series was conducted in open water schema using computational fluid dynamics (CFD) by Agung and Anita [3]. In addition, a comparison of blade area B-series propeller was conducted by Putra et al. [4]. Moreover, the study of propeller B-series in different advanced coefficients was performed by Fitriadi et al.[5]. Recently, a comparison study of propeller B-series with Kaplan using PBCF was conducted by Berlian et al.[6]. Furthermore, the study of the effect of the incline angle of PBCF using the B-series was performed by Akbar and Utama [7]. The previous research showed B-series propeller was commonly used to perform propeller studies because the empirical data was available compared to other propeller types. It was demonstrated many studies were performed for PBCF with B-series. However, there is a lot of ESD available in the market, such as Mewis duct, hull vane, vortex fin generator, etc.

The present study aims to improve propeller performance by installing the Mewis duct on the propeller. Propeller INSEAN e779a was used in this study, the study was carried out with CFD based on the volume of fluid method (VOF). The velocity advanced was based on the velocity speed service of the Kriso Container Ship (KCS). In addition, there are four combinations of Mewis duct, i.e., Mewis duct with four fins, five fins, six fins, and symmetrical four fins. The results showed that the Mewis duct increased the propeller performance by 5% by increasing propeller thrust.

2. Methods

This paper used a numerical computation of propeller performance in CFD based on Reynold Average Navier Stokes (RANS). Fig. 1 depicts propeller INSEAN e779a developed by Italian National Institute for Naval Architecture Studies and Testing (INSEAN)/ Italian Ship Model Basin [8]. The propeller itself was made for research aims. Fig. 2 shows Mewis duct installation in a stern hull ship; in this study, the mechanism of Mewis duct set up in the stern frame was ignored. In addition, only velocity of advance (v_a) was used based on before and after Mewis duct installation. After v_a is known, the next step is the value of v_a is used as inlet velocity in the computational propeller domain. It was used to simplify the geometry without a ship hull and decrease computation time. Fig. 3 depicts the numerical domain for fluid and propeller, consisting of stator and rotator. The stator is the static domain in which the fluid enters the tube. Then the rotator is the rotating domain that the propeller rotates likes in the ship. Table 1 indicates CFD's numerical setup for the fluid and propeller domain. The fluid domain remains static with a constant velocity of 0.61 m/s entering and rotating for the propeller domain. This study's propeller angular speed is 5.2 revolutions per second (RPS).

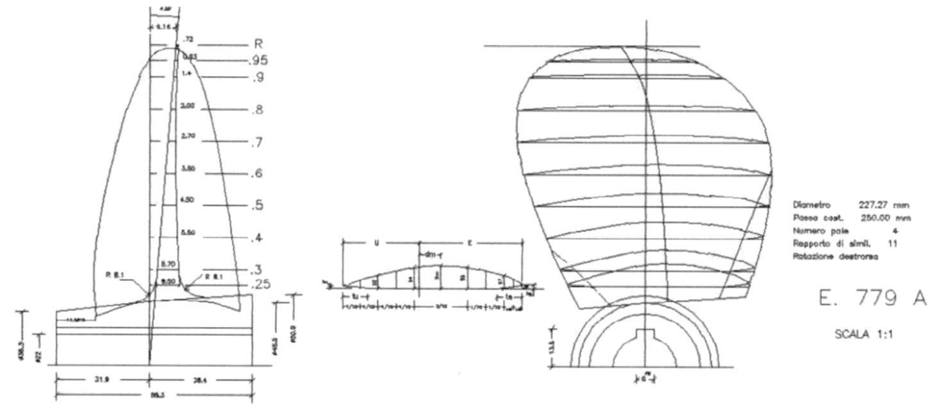

r/R	E	U	dm	te	tu	he	hu	r
.25	35.32	27.93	7.53	8.61	8.41	1.30	0.80	0.40
.3	39.40	31.37	7.00	8.20		1.08		0.20
.4	41.59	35.63	6.48	6.98		0.92		
.5	43.16	39.44	5.67	4.80		0.34		
.6	43.08	42.39	4.67	0.81		0.04		
.7	42.18	44.57	3.14					
.8	38.24	45.26	1.08					
.9	28.54	41.72	−2.30					
.95	14.25	30.95	−5.24					
1	0.00	0.00	−8.67					

r/R	S1	S2	S3	S4	S5	S6	S7	S8	Tm
.25	3.09	4.74	6.28	7.40	7.60	6.40	4.86	3.06	8.20
.3	2.26	3.92	5.30	6.18	6.22	5.22	3.80	2.46	6.70
.4	1.97	3.40	4.46	5.10	5.10	4.40	3.32	2.23	5.50
.5	1.50	2.53	3.43	4.12	4.10	3.40	2.60	1.80	4.50
.6	1.18	1.97	2.62	3.19	3.22	2.78	2.17	1.68	3.50
.7	0.96	1.52	1.98	2.40	2.47	2.10	1.60	1.24	2.70
.8	0.78	1.21	1.54	1.78	1.82	1.50	1.14	0.90	2.00
.9	0.60	0.81	1.03	1.20	1.22	1.00	0.77	0.60	1.40
.95	0.46	0.60	0.76	0.82	0.82	0.78	0.60	0.41	0.90
1									0.00

Figure 1. Propeller geometry of INSEAN e779a

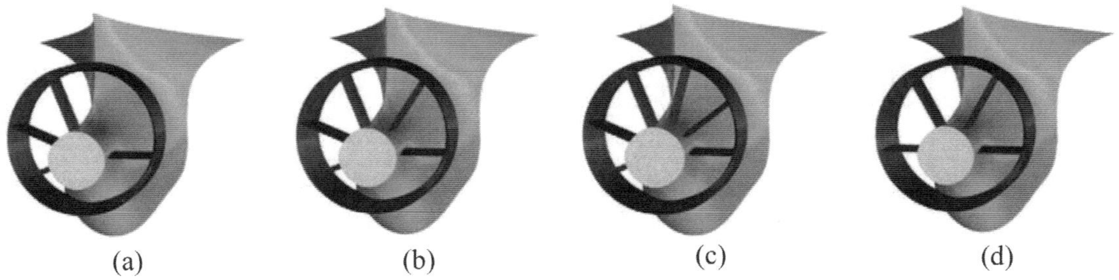

Figure 2. Mewis duct model with (a) 4 fin, (b) 5 fin, (c) 6 fin, (d) symmetrical 4 fin.

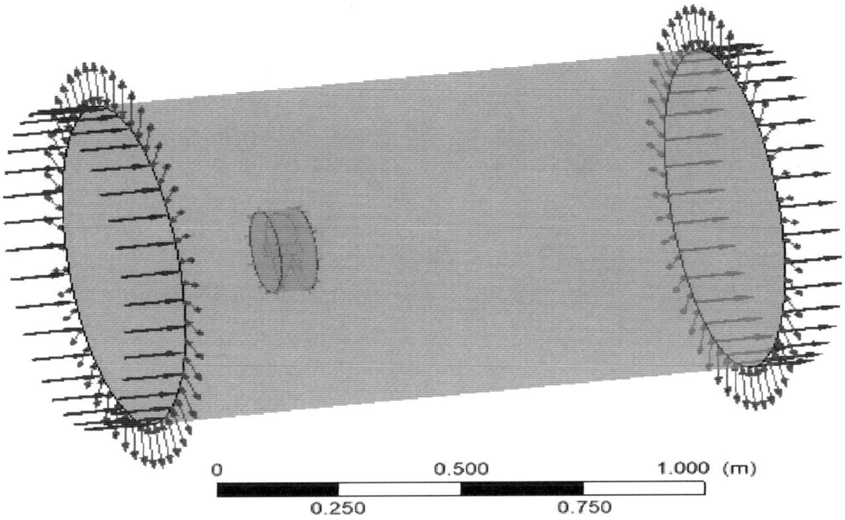

Figure 3. Computational domain of propeller

Table 1. Parameters setup for CFD computation

Domain	Fluid	Propeller
Type	Fluid	Fluid
Location	B1214	B38
Materials		
Water		
Fluid Definition	Material Library	
Morphology	Continuous Fluid	
Settings		
Buoyancy Model	Non Buoyant	
Domain Motion	Stationary	Rotating (5.2 RPS)
Reference Pressure	1 atm	

Heat Transfer Model	None
Turbulence Model	K-epsilon
Turbulent Wall Functions	Scalable

3. Results and Discussion

Mewis duct is one of the ESD to remedy stream flow in front of the propeller with a smaller nozzle diameter than the propeller diameter. In this study, several fin numbers and positions were carried out to increase propeller performance. Fig. 4 (a) shows the comparison of the propeller without the Mewis duct; the stream flow after passing the propeller is more divergent as the velocity decreases. Fig. 4 (b) indicates propeller with Mewis duct; the stream flow becomes convergent in front of the propeller; after passing the propeller, the stream flow shows a similar trend became convergent as a result, the velocity increases and then force acts in the propeller increases. Table 2 shows the validation results of propeller Wageningen B series for propeller thrust. It shows the CFD results has a good agreement with experiment, which the difference is below 3%. Table 3 indicates a comparison of ships without Mewis duct and with the installation of Mewis duct, which showed that Mewis duct in all variations improves the propeller performance by increasing the propeller thrust. The highest performance indicates by the Mewis duct using asymmetrical six fins inside the nozzle. The value of propeller thrust is similar for all cases in this study; some variations are needed to deepen the effect of advanced velocity and propeller revolution. Fig. 5 shows a similar trend to Fig. 4 that in all variations, the Mewis duct is improved the stream flow in front of the propeller. The velocity affected by the nozzle in front of the propeller that stream flow became convergent, and then the velocity increased; as a result, increasing the force act on the propeller.

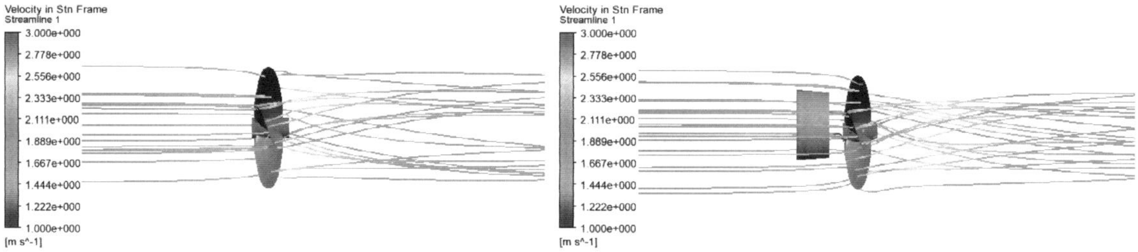

(a) (b)

Figure 4. Comparison of propeller without (a) and with Mewis duct (b).

Table 2. Validation results for propeller thrust and torque

Parameters	Exp.	*CFD*	Error (%)
	22,91	23,17	1,35
	34,45	35,03	1,67
Thrust	50,23	50,91	1,99
	66,04	67,49	2,15
	91,63	93,19	2,32

Table 3. Thrust and torque propeller with and without Mewis duct

No	Propeller	v_a	Thrust (N)	%
1	Ship w/o Md		22.4555	0
2	Ship +Md1		23.5007	4.68
3	Ship +Md2	0,61	23.5338	4.80
4	Ship +Md3		23.5816	5.01
5	Ship +Md4		23.5218	4.75

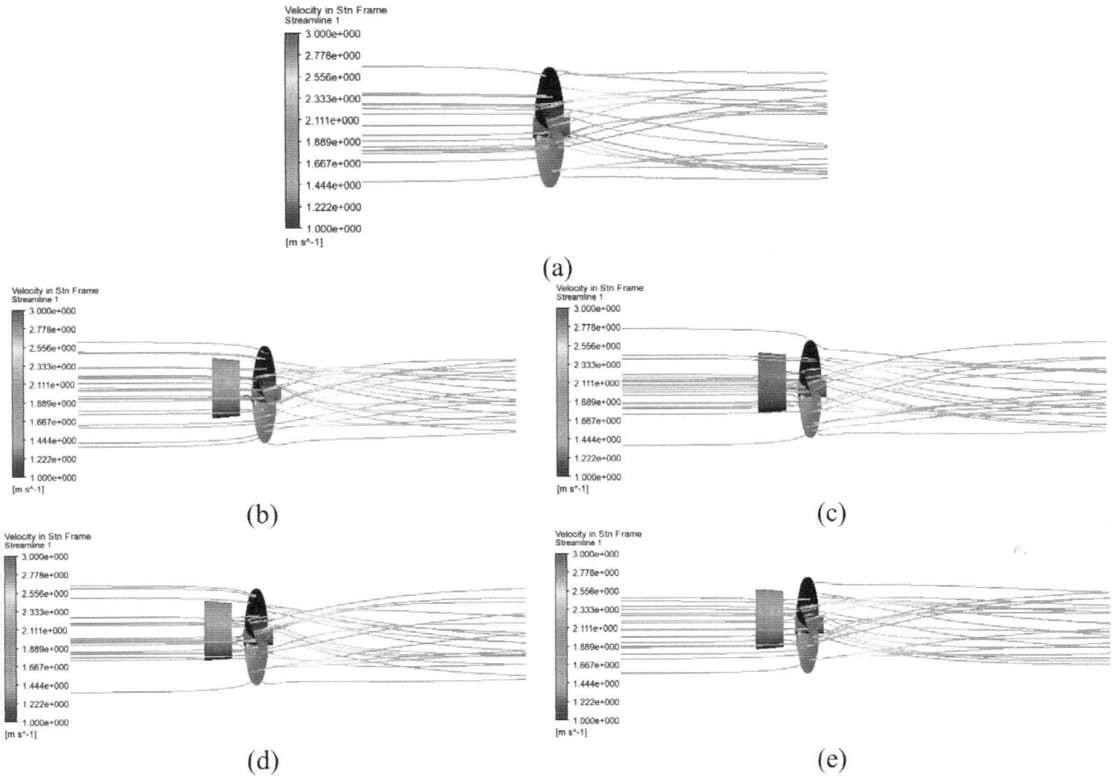

Figure 5. Comparison of propeller INSEAN e779a (a) and with Mewis duct (b).

4. Conclusion

Mewis duct as one of ESD for propeller can be one solution to improve propeller performance. The results show that the Mewis duct improved the propeller performance by 5% by increasing the thrust propeller. The best design of Mewis duct shows using six fins compares to other designs. Nevertheless, this study was preliminary for a combination of ESD such as PBCF or nozzles. Future work of optimization needs to carry out using a sophisticated method such Analytic Hierarchy Process (AHP) [9].

5. Acknowledgments

The authors are sincerely grateful to the Faculty of Engineering, Diponegoro University, for funding this work under scheme Penelitian Strategis with contract number No.3178/S/perkapalan/4/UN7.5.3.2/PP/2022.

References

[1] Trimulyono, A., Manik, P., Huda, N. Pengaruh Penggunaan Energy Saving Device Pada Propeller B4 55 Dengan Metode CFD. *Kapal:Jurnal Ilmu Pengetahuan dan Teknologi Kelautan* **2013**, *10*, 147–153.

[2] Trimulyono, A., Jatmiko, A.B., Mulyatno, I.P., Yudo, H. The Effect of Propeller Cap Angle and Fin Size of PBCF on Propeller Performance. *IOP Conf. Ser. Earth Environ. Sci.* **2022**, *972*, doi:10.1088/1755-1315/972/1/012045.

[3] Purwana, A., Hidayati, A. Analisa Karakteristik Baling-Baling B Series Di Air Terbuka Dengan CFD. *Kapal:Jurnal Ilmu Pengetahuan dan Teknologi Kelautan* **2014**, *11*, 21–25.

[4] Bangkit Setyabudi, P., Chrismianto, D., Rindo, G. Selam Midget 150m Dengan Variasi Skew Angle Dan Blade Area Ratio (AE/AO) Menggunakan Metode CFD. *Kapal:Jurnal Ilmu Pengetahuan dan Teknologi Kelautan* **2016**, *13*, 109–118, doi:10.14710/kpl.v13i3.12352.

[5] Fitriadhy, A., Adam, N.A., Quah, C.J., Koto, J., Mahmuddin, F. CFD Prediction of B-Series Propeller Performance in Open Water. *CFD Letters* **2020**, *12*, 58–68.

[6] Adietya, B.A., Utama, I.K.A.P., Aryawan, W.D., Sutiyo. CFD Analysis into the Effect of Using Propeller Boss Cap Fins (PBCF) on Open and Ducted Propellers, Case Study with Propeller B-Series and Kaplan-Series. *CFD Letters* **2022**, *14*, 32–42, doi:10.37934/cfdl.14.4.3242.

[7] Abar, I. A. C., Utama, I.K.A.P. Effect of the Incline Angle of Propeller Boss Cap Fins (PBCF) on Ship Propeller Performance. *International Journal of Technology.* **2019**, *10*, 291–319, doi:https://doi.org/10.14716/ijtech.v10i5.2256.

[8] Salvatore, F., Pereira, F.J.A., Felli, M., Calcagni, D., Di Felice, F. Description of the INSEAN E779A Propeller Experimental Dataset. **2006**.

[9] Muhammad, A.H., Thariq, A., Yusuf, Z., Yudo, H., Yasir, M. The Selection of Propeller and Primary Engine Matching of A 30 GT Fishing Vessel (Case Study KM Inka Mina 759). *Kapal:Jurnal Ilmu Pengetahuan dan Teknologi Kelautan* **2022**. doi: https://doi.org/10.14710/kapal.v19i2.46070.

Adaptation Strategies of Grobogan Regency Farmers in Face of Climate Change

L Esariti [1], R S Nida [1], W Handayani [1], and I Rudiarto [1]

[1] Department of Urban and Regional Planning, Faculty of Engineering, Diponegoro University, Indonesia

landungesariti@lecturer.undip.ac.id

Abstract. The advent of climate change in Grobogan Regency is marked by the lengthy dry season, prevalence of rat attacks and changes in the growing season due to weather instability. This article aims to explore adaptation strategies for farmers in dealing with the problem of declining crop productivity due to climate change. Based on observations, expert interviews, and questionnaires delivered to the population, it was concluded that the strategies implemented among the farmers were aimed at maintaining livelihoods. The survey results indicated that there were different agricultural patterns in the western and eastern parts of Grobogan Regency and, as such, the farmers employed different adaptation strategies respective to their locations. In West Grobogan, focus was put on managing farmland with irrigation sourced from the Glapan dam. Meanwhile, in East Grobogan, rainfed agriculture, particularly for corn and soybean commodities, was customary. Analysis of the exploratory research suggested that 2 adaptation strategies are implemented by the farmers. First, the farmers commonly migrate and seek supplementary jobs to ensure income stability. Second, the farmers use planting techniques and seeds with superior quality and short harvest periods, which they invented.

Keywords: adaptation, climate change, farmers, strategy

1. Introduction

Climate change is a phenomenon characterized by erratic changes in the world's long-term weather patterns. Climate change occurs due to continuous changes or shifts in climate variables, such as temperature and rainfall, that occur over a long period of time, between 50 to 100 years (Ministry of Environment, 2004). United Nations Framework Convention on Climate Change (UNFCCC) defines climate change as an occurrence caused either directly or indirectly by human activities which alter the composition of the global atmosphere and natural climate variability over comparable time periods. Climate change is also influenced by unstable weather conditions, for example, erratic rainfall, frequent storms, extreme air temperatures, and drastic changes in wind direction (Ratnaningayu, 2013), as in [1].

Study on climate change is relevant to the sustainability of livelihoods because climate change variables, i.e., environmental factors, impact activities in agriculture, fisheries and other sectors that concern livelihood. Likewise, climate change can also be inextricably linked to adaptive capacity. Losses attributed to climate change often prompt people, especially farmers, to respond and create strategies so that agricultural productions and livelihoods can be sustained and remain stable. In Asia, climate change has a particularly large impact on agricultural land conditions. Crop and Livestock

Diversification and Changes in Timing of Farm Operations [2] is one of the most frequently used adaptation strategies. Such an example of an adaptation strategy is developed by farmers as a response to a detrimental experience which has left them vulnerable, lest it repeat itself.

Adaptation strategies are provisional in nature. The strategies adopted will depend on the respective local conditions, type of climate change that occurs, as well as the capacity and quality of the human resources available in conjunction with the institutional environment that may accelerate the realization of these adaptation strategies. The accuracy and relevance of the adaptation strategies adopted by farmers depends on how sensitive the assessment of impacts and vulnerabilities is [3], [4]. According to [4], adaptation takes into consideration extreme environmental threats and is manifested as longer-term activities.

Indonesia as an agricultural country depends on the stability of its agricultural production as the main source of living for most rural communities. Climate change typically has a greater negative impact on poorer farming households because they have lower capacity to adapt to changing climatic conditions [5]. In such case, the adaptation strategies employed by these farming families in response to effects of climate change are oriented towards maintaining their level of livelihood. Short-term and long-term adaptation strategies aimed at maintaining the sustainability of agricultural production have also been developed and implemented world-wide. Amongst them are innovations in terms of irrigation by corn farmers in Switzerland [6], soil fertility enriching cultivation and road access improvement for the distribution of agricultural products in Eastern Himalayan villages [7], as well as adaptation in the form of migration and off-farm activities in several countries [8][9][10] [11].

Grobogan Regency as one of the largest corn producing area and the third largest producer of soybeans in Indonesia is, naturally, a region of priority in terms of the development of agricultural land [12]. Based on an expert interview from the Agriculture Office of Grobogan Regency, climate change factors that threaten the stability of agricultural productions includes floods, droughts, and rats. In addition to the emergence of these detrimental factors within the last 5 years, production has been further worsened by fluctuations in rainfall. These conditions thus prompt the inception of the adaptation strategies put into practice by Grobogan farmers. On As a matter of course, most of the locals migrate out of Grobogan to better ensure their livelihoods. Those who stay in Grobogan are family members of non-productive age that still depend on agricultural activities. This article discusses the extent of the adaptation strategies established by Grobogan farmers in an effort to secure family livelihoods in response to the threat of climate change that disrupts their agricultural productions. With dissimilar characteristics of agricultural lands in West and East Grobogan, this research will explore these two locations distinctly. This article will dive into 3 aspects, namely identification of agricultural activities, characteristics, and adaptation strategies conducted in West Grobogan and East Grobogan. Finally, in conclusion, the researchers summarize the types of adaptation strategies taking place and provide recommendations based on the implications of this research.

Praveen & Sharma [13] asserted that the five most important risk factors in the agriculture field are production risk, economic risk, marketing risk, legal and environmental risk and unskilled human resources. Production risk, in particular, is affected by climate change as it affects yields in crop and livestock. Among the production risks are the unpredictable weather conditions, disease incidence, and pests. The 5th IPPC Report defined the "*ability of systems, institutions, humans and other organisms to adjust to potential damage, to take advantage of opportunities, or to respond to consequence as adaptive capacity*" [14]. Many literatures argues that adaptation capacity encompasses internal and external factors. Among the important factors that influence farmers adaptation strategies and decisions frequently mentioned are education, farming experience, access to agricultural extension, weather forecasting, marketing information and agricultural income source [7], [15], [16], [14].

This article attempts to discern the adaptation strategies developed by farmers at Grobogan Regency in response to climate change. Climate Change disturbances to agricultural land in Grobogan are prevalent, varying in forms, from droughts to pest attacks. In a research on agricultural land at Pelem village, Batu City, [11] mentioned several adaptation strategies that were established by the locals. Based on the regression calculations conducted, 4 push factors can be identified to be quite prominent in the make-up and implementation of the climate change adaptation strategies. These factors are, first, the presence of farmer groups and activities involved within; second, crop yields; third, crop seed sources; and lastly, weather forecast information. [14] recommends that adaptive capacity analysis involves five parameters, namely: livelihoods, institutions and rights, knowledge and information, innovation and technology, and a flexible future looks at decision-making and governance to assess local adaptive capacity. In addition, [17] noted that acquiring technical education, access to credit and previous experience of droughts are the major determinants of farm-level adaptation measures.

Most of the losses related to climate change in Grobogan are caused by droughts. Droughts occur both in the West and East Regions of Grobogan. In a study similar to Grobogan's condition, [18] concluded that several adaptation strategies were in place including: 1) the building of a dam to rise water from the river; (2) the making of field wells; (3) fertilizers and pesticides dose adjustments, both chemical and plant based; (4) the use of short-duration rice varieties; (5) the shifting of planting period for some time and; (6) the pumping of water sourced from sewers.

In a research related on adaptation strategies of rice farmers in Bangladesh, [10] identified 2 important aspects that prompts the inceptions of the adaptations that were in practice, namely farming experience and farmer to farmer extension. The level of farming experience of a head of a household affects the carrying out of different adaptation strategies. In fact, experienced farmers have better knowledge and information about climate change. Indirectly, this experience is shared among farmers and play important role in the adaptations implemented at the farmer group / community level. In addition, farmer-to-farmer extension services provide a form of social capital and personal social network [19]. Information plays an important role in finding new agricultural strategies and adaptive technologies that can be implemented. With the knowledge of agricultural supervisors who understand local conditions, they provide easy access to the dissemination of this information. Therefore, adaptation strategy should be linked to the understanding of adaptative capacity.

Adaptive capacity is linked to how much insight and knowledge are gained from previous exposures to vulnerability to develop strategies to cope with future changes (Brooks and Adger 2004, Armitage 2005) in [16]. Adaptive capacity can be manifested in individual, household and organization/community levels. Therefore, adaptive capacity is sensitive to the extent of social circle. Adaptive capacity comprises level of understanding influenced by political, economic and social attributes within an individual, household and organization / community level. Knowledge and management ability also play influence on the level of adaptation capacity. [20] argues that adaptation capacity is determined by the social and human assets within frame of sustainability. In fact, several researchers also conclude that the combination of social and human assets create a social capital that could accelerate the proses of adaptation measures. Whereas, adaptation strategy is an action taken by farmers, as a subject of the matter, to minimize loss caused by climate change and maintain the stability of crop production. The end objective of an adaptation strategy is to guarantee that family livelihood is achieved and sustained.

2. Research Method

This study uses quantitative methods involving literature review as the framework for data analysis. The research locations were chosen based on the expert interviews from the Grobogan Agriculture Office (*Dinas Pertanian Kabupaten Grobogan*) and the Regional Development Planning Agency of Grobogan (BAPPEDA). The expert interviews indicated two distinct agricultural patterns in West and East Grobogan. The research locations that represented West Grobogan were Kedungjati and Tanggungharjo Districts. Whereas, in East Grobogan they were Pulokulon and Kradenan Dictricts.

Primary surveys were conducted by observations in both areas whereby photographs of the types of agricultural conditions were taken. In-depth interviews were also conducted with the community leaders to investigate research variables related to the physical conditions of the agricultural land, social and economic conditions of farmers, as well as the forms of adaptations adopted. Secondary survey data to further support findings were obtained from several related agencies such as Central Bureau Statistic, District Offices (*Kantor Kecamatan*) and other relevant data needed.

3. Findings

3.1. Adaptation Strategies in West Grobogan

Effects of climate change as experienced by several regions of the world today need to be anticipated, especially with respect to economic sectors that depend on and are susceptible to changes environmental conditions, such as agriculture. This bring about the foundation as to why farmers have to devise adaptation strategies to prevent and reduce losses attributable to climate change [18]. The current conditions demand that farmers become more competitive and resilient in dealing with climate change [8]. The adaptive response of farmers to environmental shocks will affect how sources of livelihood be managed and maintained. As such, the negative effects of climate change need to be measured and preventive measures need to be made. [21] confirmed that efforts to increase the welfare of farmers can be linked to the extent of adaptive capacity of a community. Adaptations to climate change have been formulated and implemented by farmer groups in Grobogan Regency. Grobogan Regency is the area with the largest agricultural land in Central Java Province, which covers approximately 83,833 hectares [12]. Statistically, most of the local population are farmers. Moreover, the agricultural conditions and practices found in Grobogan Regency are such that they can be satisfied by the physical characteristics of the land. The western part of Grobogan, thus, consists mostly of irrigated rice fields, whereas, the agricultural land of East Grobogan is dominated by dry lands.

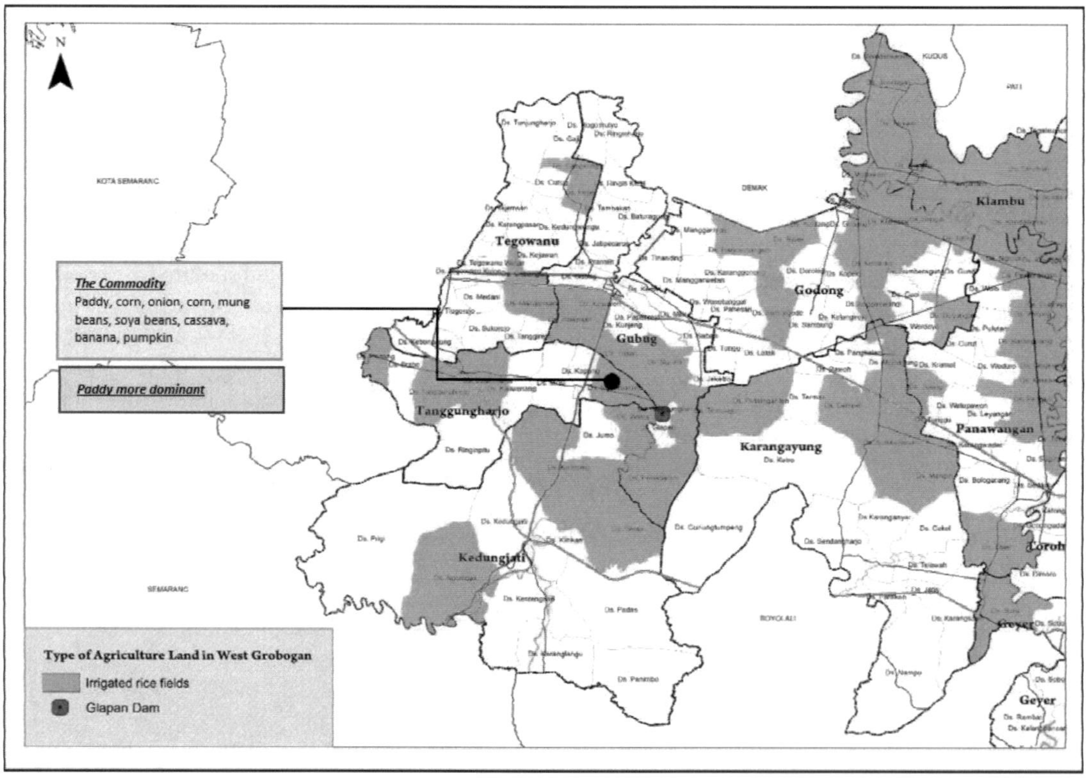

Figure 1. Map of West Grobogan Agricultural Land Use

The various conditions and agricultural problems that exist in Grobogan Regency encourage farmers to take on adaptation strategies. The adaptation strategies in effect were aimed to maintain adequate family income to provide for daily life essentials. In the western part of Grobogan, people choose to migrate to earn additional income. This is because, agricultural activities as a soul source of income are typically only able to provide income of approximately Rp 2,000,000- Rp 3,000,000 per harvest period.

If 3-4 months harvest period is assumed after the planting season, this would mean that farmers only get Rp 500,000-Rp 750,000 / month (far below the Regional Minimum Wage of around Rp 2,300,000 / month). Furthermore, after the harvest period, farmers still have to save their income as capital and compensation for the losses experienced during crop failing.

Another adaptation strategy existing was the shifting of the planting period based on estimation of the season and weather at the time. Some farmers prefer to shift their planting season or even decide not to plant on their farms when the weather conditions were uncertain, to avoid big losses lest their crops fail. It was also commonly found that farmers in western Grobogan practice ***oyotan*** leasing. ***Oyotan*** is a rental system per planting period per certain area. Oyotan rental price is around Rp 1,500,000 /planting season for a land area of 0.167 hectare for 2 months. It is a typical practice by farmers that either do not own their own land or have a small parcel of land to lend.

3.2. Adaptation Strategies in East Grobogan

The consequence of climate change on agricultural land was also felt in the eastern part of Grobogan, such as in Pulokulon and Kradenan Districts. These areas have agricultural land dominated by rainfed rice fields in which they often experience droughts. Accordingly, the local farmers would tend to opt for crops that are comparably drought resistant such as corn, soybeans, and green beans. The data found showed that the droughts that occurred in recent years have resulted in the decline of agricultural profits. As was in 2018, when the rainy season was delayed, the planting period, as a result, was set back for 2 months. This condition was further worsened by the prolonged dry season in 2019.

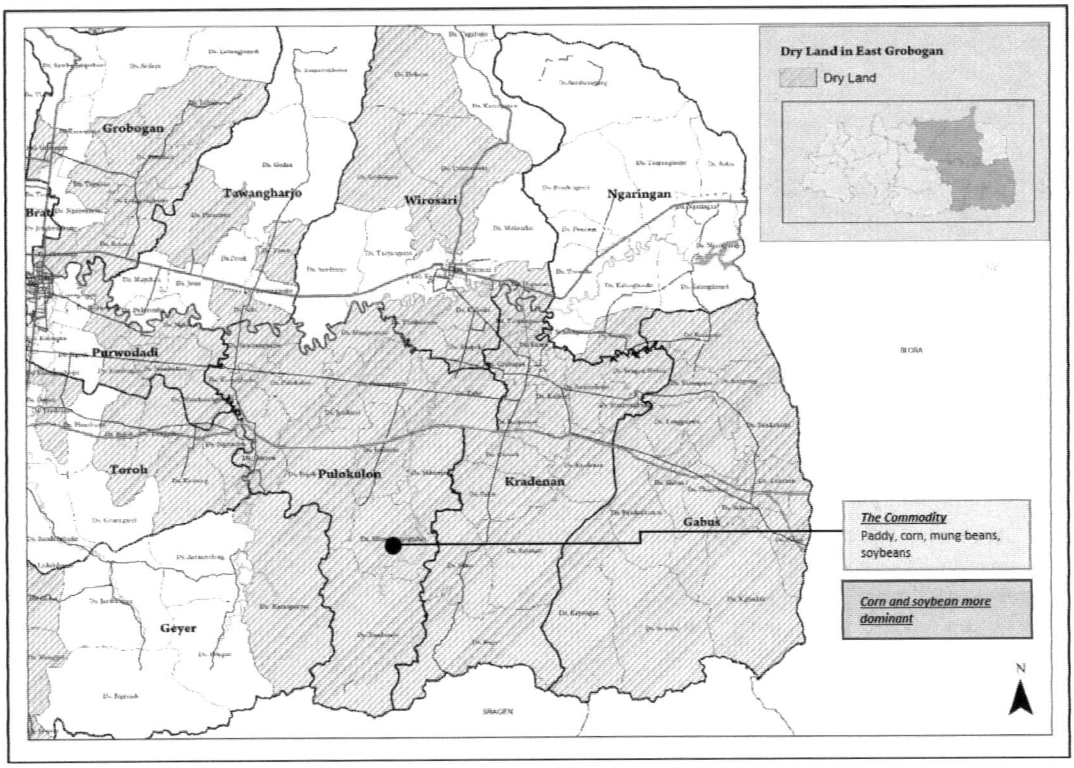

Figure 2. Map of East Grobogan Agricultural Land Type

When compared to the agricultural land in West Grobogan, the lands in East Grobogan tend to be drier as irrigation depend only on rainwater. As such, 4 adaptation strategies could be identified. First was the cultivation of drought resistant crop varieties such as corn, soybeans, and green beans. Secondly, farmers breed farm animals for supplementary income, such as cows. Cows provide a means of investment or saving. They can, later on, be sold when the farmers are in urgent need. Thirdly, farmers plant short-duration crops. [18] recommends planting drought-resistant varieties or short-duration crops as a strategy in adapting to drought disasters. By planting short-duration crops, farmers could have up to 4 harvests per year. The expert interview also indicated that short-duration crops have been developed with harvest requiring about 85-76 days. Lastly, farmers implement intercropping *(tumpangsari)* or ***methuk*** techniques. The intercropping *(tumpangsari)* technique is an alternative method. It is implemented by having more than one type of agricultural crops per planting period. In addition, ***methuk*** technique involves planting of the next set of crops before the harvest of the previous arrives. To illustrate, when soybeans are planted and its harvest is near, the green peas will then be planted in between or among the soybean plants. On a typical agricultural land in eastern Grobogan, green peas are often planted last. It is because green peas are more resistant to drought than other crops. These adaptation strategies made the areas of Pulokulon and Kradenan, despite often being hit by droughts, to still have good yields. [8] believed that farming performance is influenced by an effective resource management, the ability to utilize existing skills, and knowledge-based implementations. Likewise the adaptation strategies that exist have a positive impact on the local agricultural productivity [15][11].

Table 1. Current Climate Change Adaptation Strategy Practices

	Adaptation Method	*West Grobogan*	*East Grobogan*
1	Farming practices (type of plants cultivated)	Paddy, onion, corn, mung beans, soya beans, cassava, banana, pumpkin	Paddy, corn, mung beans, soybeans
2	Crop diversification (innovations in farming technique)	Shifting of planting period, *oyotan* leasing	Intercropping (*tumpangsari*), *methuk*, short-duration crop planting, *oyotan* leasing
3	Off-farm employment	Migration and work in the non-agricultural sector	Animal breeding

Source: Analysis Result, 2022

Based on the interviews from the Agriculture Office of Grobogan Regency, other adaptation strategies also exist to overcome causes of crop failure due to climate change. These strategies include the provision of water pump facilities, irrigation facilities and agricultural infrastructures, as well as cropping pattern regulation. In addition, the Agriculture Office of Grobogan Regency has suggested that farmers apply for agricultural insurance with affordable instalments and more sustainable benefits. The agricultural insurance would help the farmers reduce the problem of procuring farming capital and mitigate financial losses as a result of climate change [22]. In any case, education and information transfer on climate change understanding and awareness are critical and urgently needed. This should also be in conjunction with the estimation and calculation of impacts of climate change on local agriculture, which were still limited and less intensively looked into.

4. Conclusion: Future Policy Implications

It is clear that agricultural adaptation strategies found in Grobogan Regency are manifested into two distinct strategies. The first was the migration activities in search of additional jobs prevalent in the western part of Grobogan with the objective of maintaining family livelihood, wherein income from agricultural activities was less than sufficient. Meanwhile in East Grobogan, the adaptation strategy was more focused on the plantation techniques and methods.

These findings are in line with [4]. The research by Simoes advised three prerequisites in developing adaptation strategies in respond to climate change. These prerequisites are the building of awareness on current climate variability, future climate change, and their potential impacts, the building of capacity at the local (community) level in the implementations of innovative technologies as well as provisions in technical assistance, and lastly, improvement efforts in the quality of human resources for initiative and appropriate adaptation strategies. The implementation of these aspects in Grobogan Regency could be manifested in several means of development. First and foremost is through building an understanding of climate change among the farmers, so that they are aware of climate change and its impact on agriculture. Second is to create a database of knowledge and information. It is apparent that access to information, such as weather forecast and new plantation techniques, helps the farmers to implement effective adaptation measures. This should also be followed by trainings invocational skills that may increase the capacity of farmers in increasing crop productivity. This can take form as farmer-to-farmer information network lead, facilitated, assisted, and supervised by agricultural experts in from government.

References

[1] D. Nurhayati, Y. Dhokhikah, and M. Mandala, "Persepsi dan Strategi Adaptasi Masyarakat Terhadap Perubahan Iklim di Kawasan Asia Tenggara," *J. Prot. J. Lingkung. Berkelanjutan*, vol. 1, no. 1, pp. 39–44, 2020.
[2] T. W. B. E. Department, *Climate Change and Agriculture*. 2003.
[3] M. Rondhi, A. F. Khasan, Y. Mori, and T. Kondo, "Assessing the role of the perceived impact of climate change on national adaptation policy: The case of rice farming in Indonesia," *Land*, vol. 8, no. 5, 2019, doi: 10.3390/land8050081.
[4] A. F. Simões, D. C. Kligerman, E. L. La Rovere, M. R. Maroun, M. Barata, and M. Obermaier, "Enhancing adaptive capacity to climate change: The case of smallholder farmers in the Brazilian semi-arid region," *Environ. Sci. Policy*, vol. 13, no. 8, pp. 801–808, 2010, doi: 10.1016/j.envsci.2010.08.005.
[5] M. Bello, S. E. S., O. E. Galadima, and A. I., "Knowledge, Perception and Adaptation Strategies to Climate Change Among Farmers of Central State Nigeria," *Sustain. Agric. Res.*, vol. 2, no. 3, p. 107, 2013, doi: 10.5539/sar.v2n3p107.
[6] A. M. R. Fadina and D. Barjolle, "Farmers' adaptation strategies to climate change and their implications in the Zou department of South Benin," *Environ. - MDPI*, vol. 5, no. 1, pp. 1–17, 2018, doi: 10.3390/environments5010015.
[7] P. Datta and B. Behera, "Assessment of adaptive capacity and adaptation to climate change in the farming households of Eastern Himalayan foothills of West Bengal, India," *Environ. Challenges*, vol. 7, no. July 2021, p. 100462, 2022, doi: 10.1016/j.envc.2022.100462.
[8] Sumaryanto, "Strategi Peningkatan Kapasitas Adaptasi Petani Tanaman Pangan Menghadapi Perubahan Iklim," *Jurnal*, vol. Volume 30, no. Climate Change, pp. 73–89, 2012.
[9] A. . Fallis, "Impact of Climate Change On Food Production," *J. Chem. Inf. Model.*, vol. 53, no. 9, pp. 1689–1699, 2013.
[10] M. A. R. Sarker, K. Alam, and J. Gow, "Assessing the determinants of rice farmers' adaptation strategies to climate change in Bangladesh," *Int. J. Clim. Chang. Strateg. Manag.*, vol. 5, no. 4, pp. 382–403, 2013, doi: 10.1108/IJCCSM-06-2012-0033.
[11] M. Priyanto, H. Toiba, and R. Hartono, "Climate Change Adaptation Strategy : Affecting Factors and," *J. Ekon. Pertan. dan Agribisnis*, vol. 5, pp. 1169–1178, 2021.
[12] S. of G. Regency, *Grobogan Regency in Figures 2022*. 2022.
[13] B. Praveen and P. Sharma, "A review of literature on climate change and its impacts on agriculture productivity," *J. Public Aff.*, vol. 19, no. 4, pp. 1–15, 2019, doi: 10.1002/pa.1960.
[14] T. Matewos, "The state of local adaptive capacity to climate change in drought-prone districts of rural Sidama, southern Ethiopia," *Clim. Risk Manag.*, vol. 27, no. May 2019, p. 100209, 2020, doi: 10.1016/j.crm.2019.100209.
[15] M. Abid, U. A. Schneider, and J. Scheffran, "Adaptation to climate change and its impacts on food productivity and crop income: Perspectives of farmers in rural Pakistan," *J. Rural Stud.*, vol. 47, pp. 254–266, 2016, doi: 10.1016/j.jrurstud.2016.08.005.
[16] L. W. Welsh, J. Endter-Wada, R. Downard, and K. M. Kettenring, "Developing adaptive capacity to droughts: The rationality of locality," *Ecol. Soc.*, vol. 18, no. 2, 2013, doi: 10.5751/ES-05484-180207.

[17] U. Patnaik, P. K. Das, and C. S. Bahinipati, "Development interventions, adaptation decisions and farmers' well-being: evidence from drought-prone households in rural India," *Clim. Dev.*, vol. 11, no. 4, pp. 302–318, 2019, doi: 10.1080/17565529.2017.1410084.

[18] C. Muslim, "Mitigasi Perubahan Iklim dalam Mempertahankan Produktivitas Tanah Padi Sawah (Studi kasus di Kabupaten Indramayu) Climate Change Mitigation In Maintaining Land Productivity Rice Rice Fields (Cases ; Regency of Indramayu) Chairul Muslim Indonesian Cent," *J. Penelit. Pertan. Terap.*, vol. 13, no. 3, pp. 211–222, 2013.

[19] E. Saptutyningsih, D. Diswandi, and W. Jaung, "Does social capital matter in climate change adaptation? A lesson from agricultural sector in Yogyakarta, Indonesia," *Land use policy*, vol. 95, no. September 2018, p. 104189, 2020, doi: 10.1016/j.landusepol.2019.104189.

[20] E. J. Udoh, S. B. Akpan, and E. F. Uko, "Assessment of Sustainable Livelihood Assets of Farming Households in Akwa Ibom State, Nigeria," *J. Sustain. Dev.*, vol. 10, no. 4, p. 83, 2017, doi: 10.5539/jsd.v10n4p83.

[21] E. K. Austin, T. Handley, A. S. Kiem, J. L. Rich, D. Perkins, and B. Kelly, "Drought, wellbeing and adaptive capacity: Why do some people stay well?," *Int. J. Environ. Res. Public Health*, vol. 17, no. 19, pp. 1–17, 2020, doi: 10.3390/ijerph17197214.

[22] I. N. Hidayati and Suryanto, "Pengaruh Perubahan Iklim Terhadap Produksi," *J. Ekon. dan Stud. Pembang.*, vol. 16, no. 1, pp. 42–52, 2015.

Acknowledgement
This research was financially supported by The Faculty of Engineering, Diponegoro University, Indonesia via Strategic Research Grant 2022.

Determination of peak hour and maximum daily factors on domestic usage patterns in the Special Region of Yogyakarta

B Syahputra[1,*], **Nafiah**[1], **and H Poedjiastoeti**[1]

[1]Departement of Civil Engineering, Sultan Agung Islamic University, Semarang, Indonesia

*benny@unissula.ac.id

Abstract. Water use during peak hours and daily maximum are two interrelated terms in water usage patterns. Fluctuations in water use can be determined by planning standards, namely, estimates of peak hour and daily factors, so as to optimize air production and improve services. This study aims to calculate the maximum peak hour and daily factors. This research was carried out in the Special Region of Yogyakarta with a sampling technique carried out using a stratified proportional random sampling as many as 200 respondents. The results showed that the use of water at peak hours was 266.73 liters / day, which was in the morning between 06.00 and 08.00 am, while the maximum daily use of water is 774.09 litres/day which is on Sunday. The peak hour factor in the Special Region of Yogyakarta is 1.30 and the maximum daily factor is 1.26, when compared to the standard established by the Directorate General of Human Settlements, Ministry of Public Works, the peak hour factor is 1.5 while the daily maximum is 1.1. This means that the peak hour and maximum daily factors cannot be generalized in the Special Region of Yogyakarta.

1. Introduction

Fluctuations in water demand in a place are strongly influenced by population conditions, and in general, it is shown that the denser the population, the lower the peak load. Settlements in sub-urban areas (periphery) will provide a greater peak hour load than settlements in big cities. Fluctuations in water demand in an area are determined by the number of people who use water, the local environment, and the conditions of the water supply itself.

In fluctuations in water use, they cannot be separated from the so-called peak hours and daily maximum. The peak hour in water usage is the most water usage at certain hours during one day (24 hours), to determine the amount of water usage in one day, the peak hour factor is needed. While the maximum daily use is the most water usage on certain days of the week, to determine the amount of water use in one week, the maximum daily factor is needed. [1].

Taking into account the current conditions, the problem of water usage patterns always fluctuates, because people have different habits of using water [2]. The research, which was conducted in the Yogyakarta Special Region, was conducted to determine the pattern of water use in the area, which is used as an effort to expand the water distribution network.

This research generally aims to find the peak hour factor and maximum daily factor that can be used to determine the amount of water use at certain hours and the amount of water use on certain days, in addition to that it can also be used to predict community water needs, both now and in the future, and can also be used to calculate reservoir dimensions.

2. Methodology

In this study, the sample taken was 200 respondents. The sampling method was carried out using stratified proportional random sampling. Stratified proportional random sampling is used to accurately describe the characteristics of a heterogeneous population, the population concerned must be divided into uniform layers (strata) and each layer will be taken randomly. The stratified sample is the type of work of the head of the family, the level of family income and the education level of the head of the family. Data analyses were using Equation (1), (2), and (3).

$$Qri = \frac{Qh}{7} \tag{1}$$

where:
Qri: is the daily average discharge in a week (m^3/day)
Qh: is the flow rate every hour (m^3/day)
7 is the number of days in a week

$$f\ (peak\ hour) = \frac{Qhm}{Qri} \tag{2}$$

where:
F (peak hour): is the peak hour factor
Qhm: is the peak hour discharge in one day (m^3/day)
Qri : is the daily average discharge in a week (m^3/day)

$$f\ (max.day) = \frac{Qdm}{Qri} \tag{3}$$

where:
f (max.day) : is the maximum daily factor
Qdm: is the maximum debit day in one week (m^3/day)
Qri : is the daily average discharge in a week (m^3/day)

The category of city and domestic water use is often used as a standard in planning water distribution networks, this standard is set by the Ministry of Public Works, domestic water use determined by the Ministry of Public Works divides based on the number of residents as shown in Table 1.

Table 1. Water use by city category [4]

Categories	Total population	City category	Domestic water use (LPCD*)
I	> 1000000	Metropolitan	190
II	500000 - 1000000	Big	170
III	100000 - 500000	Currently	150
IV	20000 - 100000	Small	130
V	3000 - 20000	Sub-district capital	100
VI	< 3000	Village	60

*LPCD = liters per capita per day.

The domestic water use determined by the Ministry of Public Works in Table 1 is the result of an empirical approach, namely by comparing domestic water use in various regions with a certain population. However, the use of domestic water as determined by the Director General of Human Settlements of the Ministry of Public Works does not consider socio-economic aspects and household habits in using water, so that the use of domestic water will certainly differ from one area to another, including research conducted in the Special Region of Yogyakarta different from other regions, although the categories of cities in the two regions are the same.

As a comparison, the use of water by type of activity can be seen in table 2, this type of activity is found in areas that have 4 seasons.

Table 2. Types of water use activities [5]

No.	Types of activities	Water usage (%)
1.	Flushing the toilet	41
2.	Shower and wash	37
3.	Activities in the kitchen	6
4.	Drinking water	5
5.	Washing clothes	4
6.	House cleaning	3
7.	Watering plants	3
8.	Cleaning family furniture	1

The amount of water use is inseparable from the activities carried out daily, causing different patterns of water use at each time of the day. Based on the results of research conducted by Leaden et al. (1990) in the United States, there are variations in water use at certain times of the day (Table 3).

Table 3. Variations in water usage for one day [6]

No.	Descriptions	Water usage hours
1	Lowest water usage rate	11.00 pm - 5.00 am
2	Highest water usage rate	5.00 am - 12.00 am (peak usage at 07.00 am - 08.00 am
3	Medium water usage rate	12.00 am - 05.00 pm (calm water usage around 3.00 pm)
4	Increased night use	05.00 pm - 11.00 pm (peak at06.00 pm - 08.00 pm)

Table 3 above describes water use activities that are closely related to water use at peak hours and daily maximums; water use is certainly closely related to the routine activities of each household. The use of water in the Special Region of Yogyakarta is certainly very different as stated by Leiden et al. (1990) in the United States.

3. Results and discussion

3.1. Domestic water usage by type of activity

The use of domestic water based on this type of activity is a human activity in utilizing daily water which is carried out routinely, where the activity is carried out in the household, so that the use of domestic water carried out outside the household is not included in this study, even though the type of activity is the same as at home.

Household water use in the Special Region of Yogyakarta consists of types of activities that contribute to the amount of domestic water use, several types of activities that contribute to this are:

a. bath
b. washing clothes
c. cooking/drinking
d. wash kitchen utensils
e. wash the floor
f. ablution
g. washing the vehicle
h. watering plants
i. other water usage

The use of domestic water in each type of activity can be seen implicitly in Figure 1.

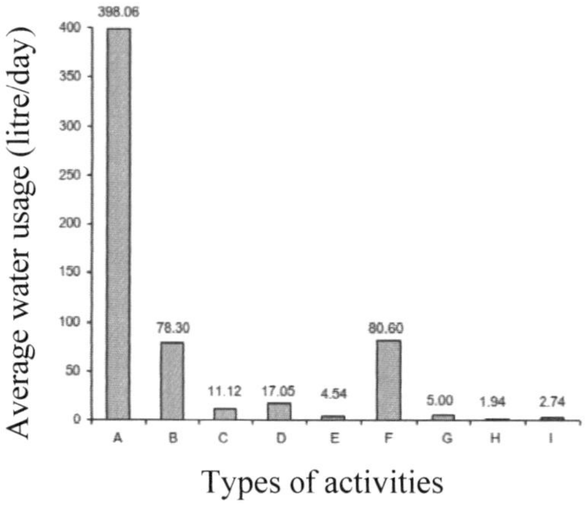

Figure 1. Domestic water usage by type of activity

3.1.1. Bath. The use of domestic water in each type of activity is dominated by the type of bathing activity, which is 39.06 litres/day, from the average number of family members as many as 5 people there are as many as 60.57% or as many as 3 people who carry out routine activities every day, such as going to work, going to school, or other activities that are carried out every day, this contributes greatly to the use of water in this type of bathing activity, because every morning activity it is always preceded by a shower. In addition to routine activities that contribute to the large use of domestic water for bathing activities, it was also found that the use of domestic water for this type of bathing activity is strongly influenced by a habit of the community in utilizing water, namely the habit of using water for bathing is influenced by the presence of a place for bathing. Water reservoirs (water containers), the use of water in the type of bathing activity in the household using a bath is relatively larger when compared to using a bucket, this is evident from the 200 respondents studied, there are 152 households that use the bath, and 48 households which uses a bucket as a shelter for bathing. The use in the type of bathing activity using a bathtub has an average water use of 413.37 liters/day, while using a bucket has an average water use of 349.58 liters/day. From the description above, it can be concluded that the use of water in the type of bathing activity using a bathtub tends to use more water compared to using a bucket [7].

3.1.2. Washing clothes. The average use of water to wash clothes is 78.30 liters/day. There are 96.5% or as many as 193 households doing laundry activities at home, while the remaining 3.5% or as many as 7 households do not do laundry activities at home, however, households that do not wash clothes are included as dividers in calculate the average use of water. This situation contributes to the small use of water for this type of work [8].

3.1.3. Cooking/drinking. The use of water for cooking/drinking activities is 11.12 liters/day. This activity ranks fifth in the most water use in the Special Region of Yogyakarta, the size of the use of water in this type of activity is strongly influenced by the number of family members and the daily mobility of family members. From as many as 200 respondents, it turns out that the average number of family members is 5 people and those who carry out routine activities are 3 people per household, but from the average number of family members who carry out routine activities, there are 60 households that have both parents. The number of working parents contributing to the use of water in the type of cooking/drinking activities is small, this is because often lunch or drinking activities are carried out outside the home [9].

3.1.4. Wash kitchen utensils. The type of activity to wash kitchen utensils ranks fourth most in water utilization, which is 17.05 liters/day. This type of activity is closely related to the type of cooking/drinking activity, so the reason for the size of the use of water in this type of activity is also the same as the type of cooking/drinking activity, in addition to the same reason as the type of cooking/drinking activity, there are also other reasons that cause the size of the activity [10]. The use of water in the type of activity to wash kitchen utensils, namely the frequency of washing kitchen utensils. In the Special Region of Yogyakarta, the frequency of kitchen utensils varies, from 200 respondents studied, it turns out that there are 13% or as many as 26 households who do this type of activity washing kitchen utensils with a frequency of 3 times a day, 43.5% or as many as 87 households with a frequency of 3 times a day. 2 times a day, 43% or as many as 86 households with a frequency of 1 time a day, and 1 household does not wash kitchen utensils.

3.1.5. Wash the floor. The washing of floors is a type of activity that ranks seventh out of nine types of activities that use water, which is 4.54 liters/day. The size of the use of water in this type of activity is closely related to the number of households that carry out floor washing activities and the frequency with which they are carried out in one week. Of the 200 respondents studied in the Special Region of Yogyakarta, it turned out that there were 56.5% or as many as 113 who carried out this type of floor washing activity, while the remaining 43.5% or as many as 87 households did not do this type of floor washing activity, this was caused by the household. It does not have a tile or ceramic floor, but only a dirt floor.

There were also variations in the frequency of floor washing; from 113 households that carried out this type of floor washing activity, there were 46.02% or as many as 52 households that carried out this type of floor washing activity with a frequency of every day; 17.70% or as many as 20 households with a frequency of once every two days; 19.47% or as many as 22 households with a frequency of once every 3 days; and 16.81% or as many as 19 households with a frequency of once a week.

3.1.6. Ablution. Of the 200 respondents studied in the Special Region of Yogyakarta, there are 167 Muslim households, while the remaining 31 are Catholic, 1 Christian household, and 1 Hindu household. The use of water in the type of ablution activity does not include non-Muslim religions as a divisor in calculating the average use of water. Based on the results of the study, it is known that the average use of water for ablution every day is 80.60 liters/day. Based on field data, it was found that the average every time for ablution is 5 liters, so the average use of water for ablution is the number of family members x the average each time ablution x prays 5 times, then the result is: 5 x 5 x 5 = 125 liters/day, which means that it can be concluded that 80.60: 25 = 3.224. This means that from the average number of family members as many as 5 people, it turns out that only 3 people pray, or in other words, the average household in the Special Region of Yogyakarta that does not pray is 2 people.

3.1.7. Washing vehicles. The type of vehicle washing activity ranks sixth in terms of water use, which is 5 liters/day. The size of the use of water in this type of activity is closely related to the number of households that carry out vehicle washing activities and the media used every time they wash their vehicles. Of the 200 respondents studied, 25% or as many as 50 households carried out this type of vehicle washing activity, while the remaining 75% or as many as 150 households did not carry out this type of vehicle washing activity with details for the following reasons: 1. As many as 23 households do not perform any type of vehicle washing activity because they do not have a vehicle, either in the form of a motorbike, car, or other form of vehicle; 2. Although up to 127 households do not carry out this type of vehicle washing activity because they always use a cloth every time they clean the vehicle, they do not need water.

Households who carry out this type of vehicle washing activity also experience variations in the media used every time they wash their vehicles, the media that are often used every time they wash their vehicles are two, namely buckets or water hoses. Of the 50 households that carried out this type of vehicle washing activity, there were 80% or as many as 40 households that carried out this type of vehicle washing activity using bucket media, while the remaining 20% or as many as 10 households carried out this type of vehicle washing activity using a water hose. Different media resulted in different average use of water in the type of vehicle washing activity; Based on the calculation results, it was found that the average use of water using bucket media was 9 liters/day, while the average use of water using water hose media was 64 liters/day.

3.1.8. Watering plants. The activity of Watering Plants uses the smallest amount of water compared to other types of activities, which is 1.94 liters/day. Of the 200 respondents studied, 7.5% or up to 15 households that used their home yards to grow plants, while the remaining 92.5% or 185 households did not use their home gardens as a place to plant plants, but were only left as a place to grow plants. for children's play. Based on the description above, it can be concluded that households in the Yogyakarta Special Region are not interested in growing plants in their yards, so the use of water for watering plants is also relatively small.

3.1.9. Other water usage. Other types of water use activities occupy the eighth highest rank in water utilization, which is 2.74 liters /day. Types of other water use activities include feeding livestock and filling fish tanks; these two types of activities include other types of water use activities and include contributing to the large use of domestic water, including these two types of activities in domestic water use. , because the water used is sourced from households. Of the 200 respondents, there were 3.5% or as many as 7 households that carried out other types of water use activities with details of 4 households carrying out livestock drinking activities with an average water use of 32.68 liters / day, and 3 households do this type of activity to fill fish tanks with an average water use of 139.27 liters / day, which means that 193 households in the Special Region of Yogyakarta do not carry out other types of water use activities, based on the results of field observations, it is known that 193 the household is not allowed to carry out other types of water use activities, due to the limited land available.

Each type of activity provides a different contribution to the use of domestic water; the implicit contribution of each type of activity is presented in Table 4.

Table 4. Percentage of water use in each type of activity

No.	Types of activities	Water usage (%)
1	Bath	66.42
2	washing clothes	13.06
3	cooking/drinking	1.86
4	wash kitchen utensils	2.84
5	wash the floor	0.76
6	ablution	13.45
7	washing the vehicle	0.83
8	watering plants	0.32
9	Other water usage	0.46
	Total	100

3.2. *Peak hours and maximum daily factors*

The use of water at peak hours and daily maximum is closely related to time, since use of water at peak hours is the largest amount of water used for domestic purposes at certain hours of the day, while the maximum daily amount is the highest amount of water used for domestic purposes at certain times of the day and certain days of the week.

Peak hour and maximum daily water usage play a very important role in determining the peak hour and maximum daily factors, based on the calculation results, the following values are obtained:

- maximum daily average = 774.09 liters/day (Sunday);
- average water usage in one week = 612.91 liters/day;
- average daily peak hours = 347.92 liters/day;

- Average peak hour water usage = 266.73 liters/day (morning).

From the data above, it can be determined the peak hour and maximum daily factors by using equations (4) and (5) as follows:

$$f\ (peak\ hour) = \frac{347.92}{266.73} = 1.30 \quad (4)$$

$$f\ (max.day) = \frac{774.09}{612.91} = 1.26 \quad (5)$$

According to the Directorate General of Human Settlements, Ministry of Public Works (1996), the peak hour factor is 1.5, while the maximum daily factor is 1.1. The standard set by the Directorate General of Human Settlements, Ministry of Public Works (1996) is very different from the results of research conducted in the Special Region of Yogyakarta, where the peak hour factor in the area is 1.30 and the maximum daily factor is 1.26, so that if in the Special Region of Yogyakarta, water distribution network will be carried out, so the standards set by the Directorate General of Human Settlements, Ministry of Public Works (1996) are not suitable, and it is recommended to use peak hours and maximum daily factors in the results of this study. It can be concluded that the peak hour and maximum daily factors set by the Directorate General of Human Settlements; Ministry of Public Works (1996) cannot be generalized to every region.

4. Conclusion

Per capita water use in the Special Region of Yogyakarta has matched the use of water in small towns set by the Directorate General of Human Settlements, Ministry of Public Works at 130 LPCD. The amount of water use can continue to increase along with the increase in population, but not only the use of water is not only due to population growth, but also because of the advancement of human life.

The peak hour of water use is 266.73 liters / day, which is in the morning between 06.00 and 08.00 am, while the maximum daily water use is 774.09 liters / day on Sunday. The peak hour factor in the Special Region of Yogyakarta is 1.30 and the maximum daily factor is 1.26, when compared to the standard established by the Directorate General of Human Settlements, Ministry of Public Works, the peak hour factor is 1.5 while the daily maximum is 1.1. This means that the peak hour and maximum daily factors set by the Directorate General of Human Settlements, Ministry of Public Works, cannot be generalized in the Special Region of Yogyakarta.

Acknowledgments

I express my deepest gratitude for the assistance from the Dean of the Faculty of Engineering, Sultan Agung Islamic University, and the Deputy Dean for their financial assistance and motivation. By the will of Allah this research can be completed properly.

References

[1] Syahputra B 2013 *Pro. Inter. Conf.on Infrastructure Develop.* p143-146
[2] Syahputra B 2011 *Pros. Semin. Nas. Sains & Teknologi* **1** B10-B14
[3] Syahputra B 2012 *J. Lingk. Sultan Agung* **1** 1–15
[4] Direktorat Jendral Cipta Karya 1996. *Petunjuk Teknis Air Bersih***.** Dept. P.U.
[5] Fair GM, Geyer JC and Okun DA 1971. *Elements of Water Supply and Wastewater Disposal.* Second Edition, (New York : John Wiley & Sons) p615
[6] Leiden VDF, Troise FL and Todd DK 1990 *The Water Encyclopedia* Second Edition, (USA : Lewis Publishers, Inc.)
[7] Syahputra B, Kiono BFT and Sudarno 2021 *IOP Conf. Ser.: Earth Environ. Sci.* **896** 012033
[8] Departemen Pekerjaan Umum. 2020. *Modul Proyeksi Kebutuhan Air Dan Identifikasi Pola Fluktuasi Pemakaian Air. Perencanaan Jaringan Pipa Transmisi Dan Distribusi Air Minum*. Jakarta
[9] Billings RB and Jones CV 2008. *Forecasting urban water demand.* (Denver: American Water Works Association)

[10] Syahputra B and Nafiah 2021 *IOP Conf. Ser.: Earth Environ. Sci.* **896** 012034

Creating Green Industries through Circular Economy

Sudharto P. Hadi

Faculty of Social and Political Science and Graduate Program of Environmental Sciences, Universitas Diponegoro, Semarang, Indonesia

sudhartophadi@yahoo.co.id

Abstract. The role of industrial sector is very important aligned with a program of economic transformation heading to competitive manufacture. However, industrial activities also cause adverse environmental and social impacts such as air and water pollution, odour, vibration, noise, traffic jam, land subsidence, flood, in turn, deteriorating people's health. One policy to deal with this problem is by encouraging industries to participate on business performance rating on environmental management called proper. To achieve the environmental excellent, companies must adopt life cycle assessment (LCA) and corporate social innovation. This paper observes the companies adopting LCA and practicing circular economy based on LCA. The type of research is a descriptive analysis. The data collection techniques consist of literature review, informal interview with relevant persons and webinar. The study showed that the companies adopting LCA benefit from reducing emission, energy used, the volume of water consumed and the waste being generated, in turn, reducing the cost of production. The benefit of practicing corporate social innovation includes reduction of waste heading to zero waste, consumer's loyalty and good relation with community. It is recommended for industries to formulate corporate social innovation at the initial stage of the program through social mapping.

1. Introduction

The contribution of industrial sector to Gross Domestic Product (GDP) is currently 17.3 percent [1]. Indonesian government expects the role of industrial sector in contributing economic growth more significant aligned with one of the priority programs called economic transformation from resource based economic to knowledge based economic heading to competitive manufacture [2]. The government plans to build twenty-seven industrials estates across Indonesia [3]. However, industrial activities cause adverse environmental and social impacts such as air and water pollution, huge solid waste, odor, vibration, noise, traffic jam, flood, land subsidence disrupting day to day activities of people and deteriorating people's health. Almost current industrial estate creates such environmental and social problems, hence the demands on business to realize a green industry are getting higher, especially through the corporates' innovation process [4]. In a study of literature articles published by Scopus indexed journals in the social, business, and economic fields within the last 5 years, the implementation of green industry mainly requires innovation in production systems, energy consumption, and emissions as well as sources and types of materials [5,6]. A change towards a more environmentally friendly industry in turn encourages efficiency, changes in mechanisms, and attracts investment [4,6]. However, the green industry in relation to the utilization of externalities from the production process in the form of ash and solid waste in the extractive industry has not been a major concern in previous studies.

China is the country with the largest energy consumption and top emitter in the world in 2021 and the US is in second place. Indonesia ranks 15th for energy consumption and 10th for emission contributions from all surveyed countries. [7] China as the largest economy with its manufacturing industry, uses an innovation-driven program, namely energy consumption and emission reduction (ECER) as a policy approach to overcome industrial problems [8]. This is essential because air pollution in China has caused 1.3 million premature deaths and cost 267 billion yuan [6], while in Indonesia, the government encourages industries to participate in a program called Proper that is a business performance rating on environmental management. The program is intended to create environmental compliance and environmental excellent [9]. Environmental compliance is characterized by industries managing its environmental well and given a blue color rank. Environmental excellent is divided by two categories. The first is the industries could manage its environmental impact proactively by adopting LCA (life cycle assessment) resulting environmentally friendly product, energy efficiency, reduction of water consumption, emission, waste generation, improving bio-diversity and implementing community development. They are awarded green color rank. The second is the industries could implement corporate social innovation explored from LCA creating self-reliance community and awarded with gold color rank. Industries categorized as gold color rank actually practice circular economy because they utilize its waste to be economic products creating people's prosperity. This paper observes two industries implementing circular economy based on LCA. The aim is to identify the benefits obtained and the problems encountered by industries in implementing circular economy through social innovation based on LCA.

2. Research Method

This is descriptive analysis using two industries (companies), gold Proper holder, as case studies. They are a subsidiary company of PLN (electric stated owned company), Steamed Power Plant (PLTU) in Tanjung Jati, Jepara Central Java and Badak LNG Enterprise, a subsidiary company Pertamina (state owned oil company) based in Bontang, East Kalimantan.

This study mainly used secondary data available at Proper Secretariat, Ministry of Environment and Forestry, Proper presentation by the two companies and followed by discussion, Webinar on Social and Innovation on July 12, 2021. This also utilized interview with relevant persons at the two companies as supporting data. The data is analyzed qualitatively using an explanation building in that the researcher creates an explanation about the case. The use of various data collection techniques is a form of source triangulation or called construct validity. [10].

3. Literature Review

Green industry is a method to attain sustainable economic growth and promote sustainable economies. It includes policy making, improved industrial production processes and resource-efficient productivity [11]. Life cycle assessment (LCA) is an instrument to achieve green industry. Life Cycle Assessment (LCA) is a method based on cradle to grave (analysis of cycle of production process from raw material, production, transportation, use and waste) utilized to identify the energy used, the water consumed, the emission raised, the waste generation and the environmental impacts caused [12]. By definition, innovation is the application of better solutions that meet the requirement, unarticulated needs, existing market. It can be said that innovation is more effective products, process, services, technologies or business that are readily available to markets, governments and society. Under the Ministry of Environment and Forestry Decree no 1 of 2021, proper participating companies determined to be gold proper candidate are obliged to adopt eco-innovation and social innovation [13]. Social innovation refers to the design and implementation of new solutions that imply conceptual, process, product or organizational change, which ultimately aim to improve the welfare and wellbeing of individual and communities. From the perspective of business, social innovation is an effort to integrate community problems with the policy, strategy and program of the company. The practice of circular economy for companies means conserving materials, extending a product's lifetime through repair and reuse, and ultimately recycling

A literature review done by Dionisio and Raupp de Vehas found that new paradigm could reinvent institutions for solving problems by giving business a new way to innovate [14]. Mirvis et al [15] explore how companies learn to engage in successful social innovation through the acquisition of tacit knowledge from external parties.

4. Results and Discussions

The two companies used as case studies are a subsidiary company of PLN (electric stated owned company), Steamed Power Plant (PLTU) in Tanjung Jati, Jepara Central Java and Badak LNG Enterprise, a subsidiary company Pertamina (state owned oil company) based in Bontang, East Kalimantan.

The crucial problem encountered by steamed power plant with its raw material of coal is its waste called FABA (fly ash bottom ash). As quoted from Suara Baru.id [16] the volume of fly ash reach to 361 ton per year and has been utilized only 30 percent for batching plant and cement factory. PLTU has to provide large area for storage of FABA for the rest of 70 percent and also manage it in such a way to prevent air pollution. Based on Government Regulation no 22 of 2021 on organizing of protection and environmental management determined that FABA is no longer considered hazardous waste and now can be utilized for useful products such as concrete bricks, paving, precast concrete and other construction materials. PLTU utilized FABA through community development scheme build a FABA's House producing these various products. In a month they produce 10,400 concrete bricks and 3,100 paving. Until the mid of August 2021, PLN Tanjung Jati B distributed 315,663 paving and 29,242 concrete bricks.

PT Badak NGL (Natural Gas Liquified) launch a program called Salin Swara (Sampah Keliling Swadaya Masyarakat or community self-help solid waste around) [17]. The idea is an investment based on solid waste management by self-help. Community formed solid waste bank to collect waste from community and solid non-hazardous waste from PT Badak NGL such as aluminum, polyurethane, calcium silicate, plastics, used drum, used cooking oil. Various solid waste supplied by company to be processed as useful products such as cool box, drum furniture, concrete brick, and bio-diesel. Cool box has been utilized for fish storage and cool drinks. Non-hazardous waste has been recycled to be concrete bricks result in 4.61 tons per year [18]. The two industries have adopted circular economy by utilizing waste for economic products. It meets criteria of social innovation in which these activities explored through LCA. In addition, it creates shared value. The two companies benefit having reduction and minimization of waste heading to zero waste. Community benefits of having job and income by involving with this activity.

5. Conclusions and Recommendations

As gold proper rank holder, the two companies adopted eco-innovation and social innovation. Through eco-innovation based on LCA, the two companies manage its environmental impacts from the stage of raw material, production, transportation, use and waste. They benefit from having energy efficiency, reduction of volume of water being consumed, reduction emission and waste generation. By practicing circular economy through social innovation both companies are successful minimizing its waste and creating the prosperity of community. The two companies are categorized as green industries because they are successful in synergizing profit, people and planet. It is recommended for industries to formulate corporate social innovation at the initial stage of the program through social mapping. By doing that the industries enable to empower people to be able to process industries waste for various economic products by themselves and not only receiving any products from the industries.

References

[1] Kementerian Investasi 2021 *Industri Manufaktur untuk Percepatan Pertumbuhan Ekonomi Indonesia* [Internet] Publikasi. 2021.
[2] Yuda TK 2021 *Ekonomi Berbasis Pengetahuan* Opini.
[3] Indraini A 2020 *Daftar 27 Kawasan Industri Baru* Industri.

[4] Grillitsch M, Hansen T 2019 *Green industry development in different types of regions* Eur Plan Stud. **27** 11 pp 2163–83
[5] Padilla-Lozano CP, Collazzo P 2022 *Corporate social responsibility, green innovation and competitiveness – causality in manufacturing* Compet Rev. **32** 7 pp 21–39
[6] Ma X, Talluri S, Ferguson M, Tiwari S 2022 *Strategic production and responsible sourcing decisions under an emissions trading scheme* Eur J Oper Res. **303** 3 pp 1429–43
[7] The World Bank 2019 *CO2 emissions (metric tons per capita)* [Internet] Data. 2019.
[8] Zhou Y, Zhuo C, Deng F 2021 *Can the rise of the manufacturing value chain be the driving force of energy conservation and emission reduction in China?* Energy Policy. **156** May pp 112408
[9] Nurbaya S 2022 *Inovasi Sosial untuk Indonesia Maju* Jakarta; 2022
[10] Conde CF 2021 *A Quick Guide to Case Studies* 2021
[11] UNIDO 2010 *A greener footprint for industry* Vienna; 2010
[12] Reliantoro S 2020 *Proper sebagai Platform Pengelolaan Lingkungan Perusahaan untuk Mencapai Environmental Excellence* Jakarta; 2020
[13] Ministry of Environment and Forestry Republic of Indonesia 2021 *Regulation of the Minister of the Environment Number 01 of 2021 concerning the Company Performance Rating Program in Environmental Management* Republic of Indonesia; 2021
[14] Dionisio M, de Vargas ER 2020 Apr *Corporate social innovation: A systematic literature review* Int Bus Rev. **29** 2 pp 101641
[15] Mirvis P, Herrera MEB, Googins B, Albareda L 2016 *Corporate social innovation: How firms learn to innovate for the greater good* J Bus Res. **69** 11 pp 5014–21
[16] Priyanto H 2021 Aug *Ubah Limbah Jadi Berkah, Ternyata Ini yang Dilakukan Rumah Produksi FABA PLTU Tanjung Jati B* Suarabaru.id. 2021.
[17] Pahalawan GI 2021 *Mewujudkan Komitmen Perusahaan Melalui Proper* Webinar Ready for Proper. 2021.
[18] Safruddin R 2021 *Best Practice Badak LNG dalam Proper dan Pencapaian SDGs* In: Webinar Nasional SDGs. SDGs Centre Diponeooro University; 2021.

Domestic waste briquetting as refuse-derived-fuel for power plant alternative energy (case study: Bali Province)

I M W Widyarsana[1] and D Saraswati[1]

[1]Department of Environmental Engineering, Faculty of Civil and Environmental Engineering, Bandung Institute of Technology, Jl. Ganesa 10, Bandung

imww.research@gmail.com

Abstract. Mobilization and population growth caused waste generation and energy supply increments. It requires more efficient waste management and treatment methods. On the other side, the availability of fossil fuel resources decreased, it urges alternative energy sources to take place. One of the waste-to-energy implementations is using domestic waste into briquettes as refuse-derived fuel (RDF) for gasification fuel as the solutions to overcome this problem. Bali Province is one of the targets of the government's acceleration program for the construction of waste processing plants into electric energy based on Presidential Regulation 35/2018. Waste-to-energy can be applied as co-firing in power plant. This study aims to determine the potential utilization of refuse-derived-fuel as power plant at Suwung Sarbagita Landfill and Bali Province as study case. Wastes are produced into briquette as refuse-derived-fuel. The proximate test results of briquette characteristics were analyzed in laboratory. Potential utilization calculation using data of waste generation in Bali Province, briquette characteristics especially calorific value, and optimal coal and briquette ratio of co-firing process. From the calculation with assuming using Integrated Gasification Combined Cycle (IGGC) system technology with efficiency of 45%, waste in Suwung Sarbagita Landfill, Bali Province has the potential to generate electricity of 101.6 MW.

1. Introduction

Based on data from the Central Statistics Agency of Indonesia, there is an increment in population growth of 1.2% per year (1). This condition resulted in the rising waste generation in Indonesia which is 67.8 million tons per year (2). This problem is faced by most developing countries including Indonesia. Some people in Indonesia still applied the conventional method of waste management. Waste is considered as something dirty, a source of odor and disease, reduces any comfort, and gives a bad impression. Waste management in Indonesia also still refers to the concept of 'collect-haul-dispose' and ended up in a landfill. The landfill has a limited-service period and cannot always accommodate waste generated from Indonesian society. Most landfills are operated by an open-dumping method which results in high polluting potential toward the environment. Several landfills are also not yet equipped with an appropriate facility such as drainage system, methane gas collection, and leachate management which indicate a highly probable risk of polluting the environment (3).Waste management improvement is needed to avoid environmental and social problems in later years (4).

On the other hand, the limited availability of the main energy source has resulted in the need for alternative energy sources development to meet the energy consumption. The implementation of waste to energy is expected to be able to overcome this problem. Waste to Energy (WtE) technology is a waste treatment to convert unrecycled waste or residue, so that it can produce energy. Waste-to-energy convert waste into energy or raw materials that are more beneficial, environmentally friendly (5) and significantly reduce waste volume. It is considered to solve waste problems quickly while at the same time producing energy resources to achieve a reduction in greenhouse gas emissions target by 29% in 2030 (6).

Most of waste-to-energy technologies require some method of waste processing to maximize the performance and potential of energy generation. Therefore, most of the WtE facilities are located side by side with a material recovery facility or landfill. Waste is separated from unprocessed or inert materials, and processed by specific technology or method based on its materials (7).

The biggest obstacles to the use of household waste as fuel to obtain energy are the high-water content, heterogeneous shapes and sizes, and the mixed conditions of organic and plastic waste makingit difficult to sort. Direct utilization of waste in thermal installations such as incineration is difficult because of the high-water content (5). Indonesian household waste has the characteristics of a fairlyhigh-water content, which is above 50%, so it requires briquetting as one of the pre-treatment processesto reduce its water content before it can be used as fuel (6). Pre-treatment in processing waste into refuse-derived fuel (RDF) aims to improve the quality of raw materials, in this case reducing the water content, in order to produce fuel with good quality and can meet the requirements or applicable quality standards. Plasma gasification is one of energy conversion method from waste to electricity. Plasma gasification system can convert waste briquette into syngas. This syngas has potential energy that canbe generated into electricity. In this study, the calculation of the thermal capacity of waste when it is generated into energy, as well as the power generated by a power generator. Potential utilization calculation using data of waste generation in Bali Province, briquette characteristics especially calorific value, and optimal coal and briquette ratio of co-firing process.

2. Objective

This study aims to determine the potential utilization of refuse-derived-fuel as power plant at three variation of scope area: Bali Province, Denpasar City and Suwung Sarbagita Landfill, with an assumption of using Integrated Gasification Combined Cycle (IGGC) system technology and economic feasibility calculation using *Net Present Value* (NPV) analysis.

3. Materials and Method

The research methodology carried out is shown in the flow chart in Figure 1.

3.1. Data Collecting

Data related to briquetting process, gasification system and waste-to-energy conversion method in this study were collected from published research papers and academic articles. In this study, data of waste composition using secondary data from previous research. Data of waste generation for each scope of area are gathered to support result analysis.

3.2. Waste Briquette Producing

Waste samples from research area are collected and dried to reduce the moisture content and then processed to be made into briquettes. Furthermore, the waste will be shredded and molded into briquettes. These briquettes are tested to the laboratory for calorific value calculation.

3.3. Potential Energy Calculation

The energy potential of waste is described as the thermal capacity contained in the waste that can be used as fuel for power generation. This calculation refers to the previous case study research on the calculation of potential waste to energy conducted by Maarif et al. (8), Ramachar (9) and Usmadiansyah

(10). Thermal capacity is calculated from the total of waste in Suwung Sarbagita Landfill, Denpasar City, and waste generation in Bali Province, then the weight of organic waste and the calorific value of waste are calculated, so that energy can be generated (kWh).

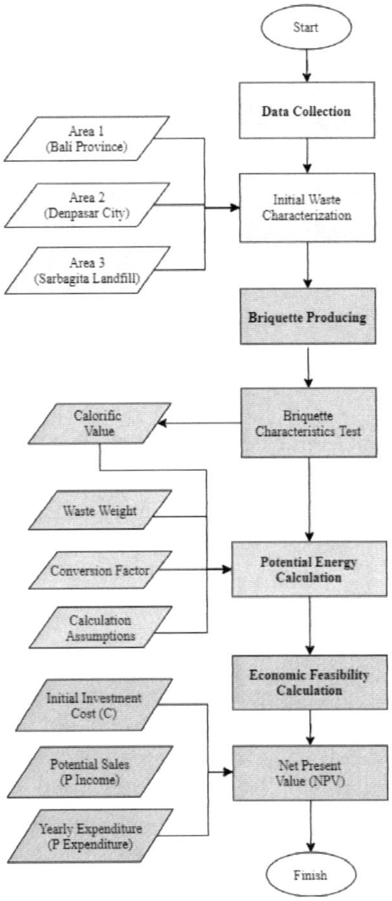

Figure 1. Research flow diagram

To calculate the potential for electrical energy from waste in Bali Province, the following steps arecarried out.

a. Determine the amount of weight of waste to be calculated;
b. Determine the calorific value of waste;
c. Determine the amount of energy output to heat input (1 kcal = 0.001163 kWh);
d. Determine the amount of energy produced through the plasma gasification process. It is assumed that the energy produced in the gasification reactor has an efficiency of 80% in converting waste energy into synthetic gas and has 10% losses from the gasification process (11);
e. Determine the amount of electrical energy generated from the process of generating electrical energyby an electrical energy generation system. It is assumed that the generating system used IntegratedGasification Combined Cycle (IGGC) technology which has an efficiency of 45% (12);
f. Determine the amount of electrical energy that can be used in the electricity system in the Provinceof Bali. It is assumed that the energy used for own use in supplying energy for this process is 5% ofthe total energy generated.

3.4. *Economic Feasibility Calculation*

In determining the economic feasibility of this installation, Net Present Value (NPV) analysis can be used as a tool for making financial decisions. If NPV > 0, the gasification installation can be said to be economically feasible. On the other hand, if NPV < 0, the installation is not economically feasible. Economic analysis is needed to determine the cost of producing electrical energy per kWh. Cost calculation refers to the Waste to Energy Power Plant Project for Bali Province in Indonesia feasibility study report by JFE Engineering Corporation Clean Authority of Tokyo (13), Calculation Guide with Waste Handling Cost Calculator (14) with waste service balance based on material flow analysis of the research results of Widyarsana et al (15). To calculate the economic feasibility, the following steps are carried out.

a. Calculate the amount of initial investment cost (C);
b. Calculate of potential sales of electrical energy (P income);
c. Calculate the amount of expenditure each year (P expenditure);
d. Calculate Net Present Value (NPV).

4. Result and Discussion

4.1. *Waste Composition*

In this study, data of waste composition using secondary data as described on Table 1 below.

Table 1. Waste composition variation

Waste Component	**Bali Province** (3)	**Denpasar City** (16)	**Sarbagita Landfill** (16)
Organic	65 %	60,5 %	63,5 %
Plastic	16 %	17 %	11 %
Paper	9 %	14 %	22 %

4.2. *Waste to Electricity Potential*

If the calorific value of waste in Suwung Sarbagita landfill is 27.72 MJ/kg or 6620.8 cal/gr (16), then the calculation of electrical energy that can be generated from 1 ton of waste are detailed at Appendix A. (10).

From the calculation, for 1 ton of waste from Suwung Sarbagita Landfill can be converted into electrical energy of 2370.061 kWh. With the identical steps, calculations were made for the average calorific value of total waste in Bali Province and waste generation in Denpasar City. From the calculation results, it is obtained that the energy potential that can be generated from 1 ton of waste produced in the Province of Bali with a calorific value of 5701.3 kcal/kg is 2040.9 kWh and for waste in Denpasar City with a heating value of 5538.8 kcal/kg is 1982.73 kWh.

Assuming a waste power plant will be built at the Suwung Sarbagita Landfill, the next calculation is estimating power capacity that can be generated by this power plant. Based on the calculation of the material flow analysis of the Province of Bali in 2020, the amount of waste that enters the landfill is 436,137 tons/year or 53.02% of the total waste generation in the Province of Bali. Meanwhile, the Suwung Sarbagita landfill received 77.66% of the waste that went into the landfill from Denpasar City, Badung Regency, Gianyar Regency and Tabanan Regency (15). With this number, the composition of waste that cannot be treated by thermal method is metal waste and other inert waste, which is only 4.15%. Therefore, 95.85% of waste has potential energy capacity that can be processed thermally (16). The amount of combustible waste generation at Suwung Sarbagita Landfill in 2020 is 324,647.8 tons/year. These calculations are detailed in Appendix A.

Therefore, the waste that enters the Suwung Sarbagita landfill has the potential to generate 769,434.9 MWh of electricity per year by processing waste using gasification and electricity generation with the IGCG system. From this energy potential, the capacity of electric power generated is 87.83 MW. With the same calculation, the power that can be generated from waste in the Province of Bali,

Denpasar City, and Suwung Sarbagita landfill is shown in Table 2 as follows.

Table 2. Calculation of waste power capacity in Bali Province

Scope of Area	Bali Province	Denpasar City	Suwung Sarbagita Landfill
Waste Generation (ton/year)	436.137,41	221.982,05	324.647,8
Calorific Value (kcal/kg dry mass)	5701,3	5538,8	6.620,8
Potential Energy (kWh/ton)	2040,90	1982,73	2370,05
Total Potential Energy (MWh/year)	890.113,02	440.130,91	769.433,75
Electric Power Potential (MW)	**101,6**	**50,2**	**87,8**

In comparison, Bali has around 1,274 MW of electricity from a number of suppliers. The electricity supply is obtained from 380 MW PLTU (steam power plant) Celukan Bawang, PLTG (gas power plant) Pesanggaran 344 MW, PLTG Gilimanuk 130 MW, and PLTU Pemaron 80 MW. A total of 340 MW comes from submarine cable transmission from Java to Bali. The peak electricity load in Bali was recorded at 902 MW. In 2021, it is predicted to reach 932 MW (17). Thus, the existence of waste power plant or *Pembangkit Listrik Tenaga Sampah* (PLTSa) in Bali Province which can process waste that goes to the landfill, can provide approximately 11% of the electricity needs in Bali Province.

4.3. Economic Feasibility Calculation

The gasification installation which has a capacity of 1000 tons per day is planned to operate for 10 years in 2021-2030. In calculating the current value of money, according to data from Bank of Indonesia, the interest rate (i) as of January 2020 (assuming normal conditions before the pandemic) is 5%, which is the average of interest rates from various banks in Indonesia, with a time period (n) of 25 year. From the waste generation data in Bali Province, the capacity of the gasification installation for PLTSa used is assumed to be 1000 tons per day using plasma gasification technology. The calculations of economic feasibility are detailed in Appendix B.

4.3.1. Initial investment cost (C) calculation.

The capital cost required to build a waste to energy installation (13) with a capacity of 1000 tons per day is USD 118,000,000. If the project is implemented in 2021, the currency value of 1 USD in 2021 is equal to IDR 14,042 and the increase factor is 1.1, then the investment cost of power plant capacity is IDR 1,656,956,000,000 (1.67 trillion rupiah).

4.3.2. Potential income (P income) calculation.

Based on the Regulation of the Minister of Energy andMineral Resources Number 27 of 2014 about the purchase of electricity for biomass, IDR 1150/kWh multiplied by F factor is the incentive constant according to the location of the purchase of electricity by State Electricity Company or PT PLN Persero if it is connected at medium voltage by PT PLN Persero. F factor for Bali Island of 1.50. From the calculation B.1, P income of this installation are IDR18,706,617.108,155.

4.3.3. Expenditure calculation (P expenditure).

The average annual cost required for the operation of awaste to energy installation *(13)* with a capacity of 1000 tons per day for 25 years is USD 7,000,000. Ifthe project is implemented in 2021, the currency value of 1 US dollar in 2021 is equal to IDR 14,042.00, so the estimated operational costs are IDR 98.294,000,000 (98.3 billion per year). Operational and maintenance costs per kwh from the calculation are IDR 114/kWh.

For PLTSa with plasma gasification technology, no fuel costs are required. However, the cost required is to transport the waste from temporary waste storage or TPS over Denpasar City, Badung Regency, Gianyar Regency and Tabanan Regency to the Suwung Sarbagita landfill area. Based on calculations using the Waste Handling Cost Calculator (14) with a total ideal number of trucks transporting waste to the landfill as many as 200 trucks per day, the transportation cost is IDR

38,576,978,048 (38.6 billion per year). Operational and maintenance costs per kwh from the calculation is IDR 45/kWh.

4.3.4. Calculating Net Present Value (NPV).
The NPV from the calculation are IDR 16.912.790.130.107. Because NPV>0, the project for the procurement of a waste treatment plant can besaid to be economically feasible. The recapitulation of the calculation of the gasification installation budget requirements can be concluded as follows.

- Initial investment cost : IDR 1.656.956.000.000
- Potential income per year : IDR 1.327.275.231.173
- Operational cost per year : IDR 98.294.000.000

5. Conclusion
From the research results, it is concluded that waste generation in Bali Province has the potential to generate 101.6 MW of electrical power, Denpasar City waste of 50.2 MW and Suwung Sarbagita landfill has the potential to generate electricity of 87.8 MW by processing waste using gasification and power generation with the Integrated Gasification Combine Circle (IGGC) system. The calculation of economic feasibility analysis resulted that this installation is feasible to be run and applied in the future.

References
[1] Indonesia Central Statistics Agency (BPS) 2020 *Hasil Sensus Penduduk* (Jakarta: IndonesiaCentral Statistics Agency)
[2] Ministry of Environment and Forestry of Indonesia (KLHK) 2020 *KLHK: Indonesia Memasuki Era Baru Pengelolaan Sampah* (Jakarta: Ministry of Environment and Forestry of Indonesia)
[3] Widyarsana IMW and Elprida A 2019 *Int. J. GEOMATE* 2019 **17** 164-71
[4] Widyarsana IMW and Rahayu AI 2021 *J. Mater. Cycles Waste Manag.* **23** 2051-64
[5] Confederation of European Waste-to-Energy Plants (CEWEP) 2020 *What is waste to energy* [Online] https://www.cewep.eu/ what-is-waste-to-energy/
[6] Ministry of Energy and Mineral Resources of Indonesia (ESDM) 2020 *Ini Capaian Kelistrikan2020* (Jakarta: Ministry of Energy and Mineral Resources of Indonesia)
[7] Triyono B, Prawisudha P, Pasek AD and Mardiyati 2018 *AIP Conf. Proc.* **1984** 020004
[8] Ma'arif S, Widyawidura W, Ariditio NM, Kurniasari HD and Kismurtono M 2019 *Int. J. Renew.Energy Res.* **9** 354-62
[9] Ramachar T, Rao GC, Umamahesh M and Nagamouli D. 2014 *Int. J. Chem. Sci.* **12** 1345-54
[10] Usmadiansyah 2017 *Jurnal ELKHA* **9** 28-34
[11] Janajreh IS, Raza S and Valmundsson AS 2013 *Energy Convers. Manag.* **65** 801-9.
[12] Anyaegbunam FNC 2013 *Am. J. Eng. Res.* **2** 65-75
[13] JFE Engineering Corporation Clean Authority of Tokyo 2017 *Feasibility Study of Joint CreditingMechanism Project by City-to-City Collaboration Waste to Energy Power Plant Project for Bali Province in Indonesia Final Report* (Tokyo: JFE Engineering Corporation Clean Authority of Tokyo)
[14] Ministry of Public Works and Housing Profile Indonesia (PUPR) 2021 *Calculation Guide with Waste Handling Cost Calculator* (Jakarta: Ministry of Public Works and Housing Profile Indonesia)
[15] Widyarsana IMW, Damanhuri E and Agustina E 2020 *J. Mater. Cycles Waste Manag.* **22** 405–15
[16] LAPI ITB 2019 *Study of Potential Use of Municipal Solid Waste to Energy in Sarbagita LandfillService Area: Final Report* (Bandung: LAPI ITB)
[17] Saputro W and Nurhayat W. Kumparan 2019 [cited 2021 March 1. Available from: https://kumparan.com/kumparanbisnis/bali-ingin-mandiri-listrik-apa-bisa-1rLzGRLEdwS/full.

Appendix A
Waste Energy Potential Calculation

- $Combustible\ waste\ generation\ at\ Suwung\ Sarbagita\ Landfill$
  $= 95.85\% \times 77.66\% \times 436{,}137 \frac{\text{tons}}{\text{year}} = 324{,}647.8 \frac{\text{tons}}{\text{year}}$
- $Electrical\ energy\ per\ year = Waste\ generation\ in\ landfill \times\ nett\ electrical\ energy$
  $= 324{,}647.8 \frac{\text{tons}}{\text{year}} \times\ 2370{,}061\ \text{kWh}$
  $= 769{,}434{,}916.6\ \text{kWh}$
  $= 769{,}434.92\ \text{MWh/year}$
- $Electrical\ power\ potential = \frac{Electrical\ energy\ per\ year}{365\ day \times 24\ hour} = \frac{769{,}434.92\ MWh}{365\ day \times 24\ hour} = 87.83\ MW$

Appendix B
Electrical Energy Calculation

- $Total\ waste\ calorific\ value = waste\ mass\ (1\ ton) \times waste\ calorific\ value$
  $= 1000\ kg\ \times 6620.8 \frac{kcal}{kg} = 6{,}620{,}810\ kcal$
- $Energy\ to\ heat\ input\ (unit\ conversion) = 6{,}620{,}810\ kcal \times 0.001163\ kWh/kcal$
  $= 7700\ kWh$
- $Energy\ generation\ from\ plasma\ gasification\ process$
  $= 80\% \times 90\%\ \times 7700\ kWh = 5544\ kWh$
- $Electrical\ energy\ generation\ from\ IGCC\ system$
  $= 45\% \times 5544\ kWh = 2494.8\ kWh$
- $Nett\ electrical\ energy\ (after\ self - consumption\ power\ reduced)$
  $= 2494.8\ kWh - (2494.8 - 5\%) = 2370.061\ kWh$

Appendix C
Economic Feasibility Calculation

C.1 Potential income (P income) calculation

- $Economic\ value\ of\ waste\ conversion = 2{,}108{,}040{,}87\ \times \text{IDR}\ 1150/\text{kWh} \times\ 1.5$
  $= \text{IDR}\ 3{,}636{,}370{,}496/\text{day} = \text{IDR}\ 1{,}327{,}275{,}231{,}173/\text{year}\ (1.33\ \text{trillion/year})$

The total amount of income obtained by the waste processing installation from sales each year can be found using the equation, as follows.

- $P\ income = A\ (P/A\ ; 5.0\%\ ; 25)$
  $= \text{IDR}\ 1{,}327{,}275{,}231{,}173\ (14{,}094) = \text{IDR}\ 18{,}706{,}617.108{,}155$

C.2 Potential expenditure (P expenditure) calculation

- $Transportation\ Cost = \frac{IDR\ 38{,}576{,}978.048}{365\ days \times 1000\ tons/day \times 2370{,}061\ kWh/tons} = IDR\ 45/kWh$
- $Total\ expenditure\ per\ year$
  $= Total\ operational\ cost + total\ transportation\ cost$
  $= \text{IDR}\ 98.294{,}000{,}000 + \text{IDR}\ 38{,}576{,}978{,}048 = \text{IDR}\ 136{,}870{,}978{,}048$

C.3 Net Present Value calculation

- $NPV = Rp18{,}706{,}617{,}108{,}155 - Rp136{,}870{,}978{,}048\ - Rp1{,}656{,}956{,}000{,}000$
  $= Rp.\ 16.912.790.130.107$

Does higher income lead to more renewable energy consumption? evidence from Indonesia

A Setyadharma[1]*, P E Prasetyo[1], S Oktavilia[1], B D Fortuna[1] and I F S Wahyuningrum[1]

[1] Faculty of Economics, Universitas Negeri Semarang, Semarang, Indonesia

andryan@mail.unnes.ac.id

Abstract. Indonesia is aimed to reach the use of renewable energy by up to 23 percent by 2025. Increasing society's income is considered as one of the effective ways to encourage society to shift to renewable energy consumption. Therefore, the aim of this study was to investigate the impact of income on renewable energy consumption in Indonesia. This study employed Engle Granger Error Correction Model and used hydroelectricity consumption as the proxy of the renewable energy consumption and real gross domestic product (GDP) per capita as the proxy of the income. The result of this study suggested that hydroelectricity consumption can be written as a quadratic function of real GDP per capita. It means that at the beginning, increasing real GDP per capita reduced the hydroelectricity consumption, and at a certain level it turns to opposite direction, where the rise of the real GDP per capita increased hydroelectricity consumption. In conclusion, higher income leads to higher renewable energy consumption. The policy implication for policymakers is straightforward, i.e., it is important to increase the income of the society, not only for the sake of the wealth of the society, but also for the protection of the environment through more clean energy consumption.

1. Introduction

Global primary energy demand raised by 5.8% in 2021, higher than 2019 levels by 1.3%. Unfortunately, the high demand was fulfilled by fuel energies that cover 82% of the consumption of total energy in 2021 [1]. However, fossil fuels are non-environmentally friendly energy sources. Burning fossil fuels will release enormous amounts of carbon dioxide (CO_2) into the air and leads to climate change. CO_2 emissions recorded a new all-time high level in 2019 and it was a fourth consecutive high CO_2 ever recorded [2]. In addition, 4.3 million deaths in 2012 occurred due to indoor air pollution [3]. The concern over the severe impacts of fossil fuels have sparked the campaign of the importance of use of renewable energy sources. Especially, the United Nations also concerned about this matter and announced Sustainable Development Goal {SDG) to promote the use of modern energy for all by 2030 [3].

About 75% of the growth of the world's energy consumption because higher consumption in China, followed by India and Indonesia [2]. Indonesia's primary energy supply is from fossil fuels (it accounts 88% of the national power capacity), but Indonesia is aimed to reach the use of renewable energy by up to 23% by 2025 and reach net-zero emissions by 2060. However, signs of progress in renewable energy projects have shown no promising development. For example, installed capacity of renewable energy sources in Indonesia remains low, as it does not meet to achieve the 23% target. Furthermore, investments on renewable energy in Indonesia are still low, the total investment is only 30% in the energy segment in 2021. Renewable energy projects only received USD 1.1 billion while fossil fuels received USD 2.5 billion at the same time [4].

Humans are largely to blame to the destruction of the environment [5]. Their economic activities led to the degradation of the environment. Therefore, increasing the consumption of clean energy sources is important for the purpose of changing fossil fuels consumption to clean energy consumption. Recently, Indonesia's renewable energy consumption is kept at 19%. Since there is a low level of the use of renewable energy in Indonesia, it is vital to increase the renewable energy use through changing human behaviour to shift from use of fossil fuels to use renewable energy.

Previous studies have investigated the factors of renewable energy in developed and developing countries (for examples: [6], [7], [8], [9], [10] and [11]). Those studies indicated that per capita Gross Domestic Product (GDP), as the proxy of income, was considered as one of the effective ways to encourage society to shift to renewable energy consumption. Based on Consumption Theory, there is a connection between income and consumption. Since the theory is in general term, one can establish the connection between income and renewable energy consumption. [11] indicated that the rise in per capita income at the household level encouraged individuals paid more money on clean energy services to get better comfort for their life. Moreover, since the demand for renewable energy increases, additional renewable energy production is needed to cover the demand.

Even previous studies indicated positive impact of per capita GDP on renewable energy spending, other study showed the opposite direction [12] and some other findings did not find the impact of income on renewable energy (for examples: [13] and [14]). Therefore, it is concluded that the empirical results on the effect of income on the renewable energy consumption are not certain and mixed. In addition, to the best of our knowledge, the investigation about the impact of income on the renewable energy consumption in Indonesia does not exist. So, it is vital to investigate empirically how income may affect renewable energy consumption in Indonesia.

On top of it, this study also offers a new insight that renewable energy consumption can be formed as a quadratic function of income to examine the possibility of the existence of U-shape connection between income and renewable energy consumption. The U-shape means that initially greater income level will reduce renewable energy consumption, until at a particular level, that the greater income level will increase sustainable energy consumption. To the best of our knowledge, no previous studies examine the likelihood of the presence of U-shape link between income and renewable energy consumption.

To sum up, this paper makes two important contributions. Firstly, this study adds to current literature by justifying the connection between income and clean energy consumption in Indonesia. Secondly, this study tries to examine the likelihood of the presence of U-shape relationship between income and renewable energy consumption in the case of Indonesia. This study offers new-found evidence to current literature.

The structure of this paper is arranged as follow: Part 1 explained introduction and concisely discussed at the background and literature review on the subject. Part 2 explains the empirical method. Part 3 described the analysis of the results, and the Part 4 is the conclusion.

2. Methodology

This research applied econometric procedures, i.e., Engle-Granger Error Correction Model (E-G ECM) to mainly investigate the impact of income on renewable energy consumption in Indonesia. The variables are summarized as follow:

Table 1. Variables in the Model

Variable	Proxy of	Source
hydroelectricity consumption (HEC)	the renewable energy consumption	British Petroleum
real GDP per capita (YCAP)	Income	World Bank
Oil Price (OIL)	Crude Oil Price	British Petroleum
CO_2 (CO2)	CO_2 Emission	British Petroleum

HEC was a dependent variable; YCAP, OIL and CO2 were the independent variables. OIL and CO2 were used as those variables were likely have impact on renewable energy consumption. The E-G ECM was applied for this study because the E-G ECM offered not only the dynamic short run but also long run among the cointegrating variables. As economists are concerned mainly in long-run relations, the concept of E-G ECM is very convenient. The E-G ECM consists of two models. The basic model with quadratic function of real GDP per capita, known as the long run model, is expressed as follow:

$$L(\text{HEC})_t = \mu_o + \mu_1 L(\text{YCAP})_t + \mu_2 L(\text{YCAP})^2_t + \mu_3 L(\text{OIL})_t + \mu_4 L(\text{CO2})_t + e_t \ldots\ldots\ldots\ldots \quad (1)$$

In the next step, the equation (1) was converted into ECM form, known as the short run model, as follow:

$$\Delta L(\text{HEC}))_t = \delta_o + \delta_1 \Delta(L(\text{YCAP}))_t + \delta_2 \Delta((L(\text{YCAP}))^2)_t + \delta_3 \Delta(L(\text{OIL}))_t + \delta_4 \Delta(L(\text{CO2}))_t + \delta_5 \text{ECT}_{t-1} + \varepsilon_t \ldots\ldots\ldots\ldots\ldots \quad (2)$$

Where, L symbolized the logarithm function, HEC was hydroelectricity consumption, YCAP was real GDP per capita, YCAP^2 was real GDP per capita squared, OIL was raw oil price and CO2 was CO_2 emissions, e and ε were the errors or disturbances, t was the time. μ_o, μ_1, μ_2, μ_3 and μ_4 were the coefficients for the long run model, δ_0, δ_1, δ_2, δ_3, δ_4 and δ_5 were the coefficients for the short run model, Δ is the difference between the N_t observation and N_{t-1} observation. ECT was the error-correction term and ECT was the most important variable in E-G ECM model. The coefficient of ECT, in this case δ_5, informed the speed of adjustment to equilibrium takes place in each period. The value of α_5 lies from 0 to 1. In summary, if $\alpha_5 = 1$ then 100% of the adjustment occurs within a given time, or the adjustment is instant and finish within a given period; If $\alpha_5 = 0.5$ then 50% of the adjustment occurs in each period and if $\alpha_5 = 0$ then there is no adjustment at all.

E-G ECM must fulfil some requirements before can be used for the analysis. The requirements are (1) dependent variable and independent variables must not stationary at level, I(0); (2) dependent variable and independent variables must stationary at 1st difference, I(1); (3) all variables in the form pf cointegration on their linear combination. The stationary tests for both dependent and independent variables were ran using Phillips-Perron (PP) unit root tests (for conditions no. 1 and 2). The cointegration test was applied using Engle-Granger Cointegration test for the errors (for condition no. 3). If all requirements are achieved, E-G ECM can be applied.

3. Results and Discussion

the first requirement of the non-stationary variables was tested using with Phillips-Perron unit root tests. There are two types of Phillips-Perron tests, i.e., the test includes intercept, and the test includes trend & intercept. The Phillips-Perron tests showed that all variables were not stationary at level, as indicated by the probabilities values of all variables were greater than 10% (or 0.10). One variable (L (HEC)) was not stationary based on intercept-based calculation, two variables (L (YCAP) and L $(\text{YCAP})^2$) were not stationary based on both intercept- and trend & intercept-based calculations, and two variables (L(OIL) and L(CO2)) were not stationary based on trend & intercept-based calculations. For the second requirement, the unit root tests using Phillips-Perron tests suggest that all variables are stationary at first (1st) difference. The probabilities values of all variables were more than 10% (or 0.10) as presented in table 2. Therefore, it is concluded that the test fulfilled the first and second conditions of E-G ECM.

Table 2. Phillips-Perron Unit Root Test for Level and 1st Difference

Variables	Unit root tests in Level		Unit root tests in 1st Difference	
	Intercept	Trend and Intercept	Intercept	Trend and Intercept
L (HEC)	-2.189563	-5.207817	-8.384798	-8.207224

	(0.2139)	(0.0011)***	(0.0000)***	(0.0000)***
L (YCAP)	-0.851637 (0.7894)	-2.019710 (0.5674)	-5.477878 (0.0001)***	-5.375418 (0.0008)***
L $(YCAP)^2$	-0.735759 (0.8225)	-1.919506 (0.6196)	-5.248356 (0.0002)***	-5.145675 (0.0014)***
L (OIL)	-3.519227 (0.0143)***	0.871194 (0.9996)	-3.205647 (0.0299)***	-4.376286 (0.0086)***
L (CO2)	-6.419936 (0.0000)***	-1.354779 (0.8536)	-4.387319 (0.0017)***	-9.057867 (0.0000)***

Note: values in Parentheses () are probability values
*** significance at p-value ≤ 0.01, means the variables are stationary

Next requirement is the cointegration. The test demonstrated that the errors were stationary at level, I(0), as viewed in Table 3. The result of the test implied that variables in the model have a long-run relationship, in which it is important for the analysis. All three tests indicated that the model satisfied the all three requirements of E-G ECM, and it was continued to E-G ECM.

Table 3. Cointegration Test

Variable	**Unit root tests in Level**	
	Intercept	**Trend and Intercept**
Residuals	-2.189563 (0.2139)	-5.207817 (0.0011)***

The outcome of the short run of E-G ECM was shown in Table 4. Table 4 demonstrated that the coefficient of YCAP had negative value and significantly influence HEC and the coefficient of YCAP squared has positive value and it significantly influenced HEC. The results suggested that, in short run, in the beginning, if the real GDP per capita increased by 1%, the hydroelectricity consumption declined by 3.702%, and at certain point, the further increase of the real GDP per capita will led to the increase of the hydroelectricity consumption by 0.277% in with assumption that other variables are not changed, or *ceteris paribus*. This result suggested the existence of a quadratic function of real GDP per capita on hydroelectricity consumption. It means that at the beginning, increasing real GDP per capita reduced the hydroelectricity consumption because the Indonesian society preferred to fulfil the basic needs first without considering the renewable energy as one of the basic needs. It is important to note that renewable energy is not cheap. However, at a certain level, the more rise in the real GDP per capita increased hydroelectricity consumption because at this point, the society already fulfilled their basic needs, and they still have more money and immediately shift to spend their money on the renewable energy. In conclusion, higher income leads to higher renewable energy consumption, and it reduce the pressure on the environment.

Table 4. Outputs of Short Run ECM

Explained Variable: Δ(Logarithm of Hydroelectricity Consumption (HEC))	
Constant	0.016 (0.449)
Δ(LOG (YCAP))	-3.702 (-2.068)**
Δ(LOG (YCAP)^2)	0.277 (2.053)*
Δ (LOG (OIL))	-0.064 (-0.530)
Δ(LOG (CO2))	-0.090 (-0.169)
ECT_{t-1}	-1.058 (-5.198)***
Adjusted R^2	0.541

Note: *** significance at p-value ≤ 0.01; ** significance at p-value ≤ 0.05; * significance at p-value ≤ 0.10; t-statistics are indicated in parentheses.

Additionally, this study also included two variables that were likely to have influence on renewable energy consumption. Nevertheless, OIL and CO2 were insignificant in the short run. The insignificant results indicated that the changes in oil price and CO_2 emission will not affect the renewable energy consumption in the short run. Coefficient of ECT was negative and statistically significant which indicated there was evidence of the convergence from short run to long run in the model.

Table 5 showed the result of the long run of E-G ECM. Similar to short run result, the result in table 5 showed that the coefficient of YCAP had negative value and significantly affect HEC and the coefficient of YCAP squared had positive value and it significantly affected HEC. The results suggested that, in long run, if the real GDP per capita rose, the hydroelectricity consumption drops until at certain point, where the additional increase the real GDP per capita led to the increase of the hydroelectricity consumption in with assumption that other variables are not changed, or *ceteris paribus*. Again, there was evidence of the existence of a quadratic function of real GDP per capita on hydroelectricity consumption in long run. Therefore, this study supported previous studies, such as [6], [7], [8], [9], [10] and [11]. In the long run, the change of behaviour occurs when at initial stage, society seems does not care about the protection of the environment as little are spent on the clean energy consumption, but at certain level, society allocates their money on the renewable energy consumption as they understand the importance of the protection of the environment. As mentioned before, economists are more concern on the long run results and this study validates the long run relationship between income and renewable energy consumption. It is concluded that income become one of the vital ways to the transition from fossil fuels to clean energy.

Table 5. Outputs of Long Run ECM

Explained Variable: Logaritm of Hydroelectricity Consumption (HEC)	
Constant	4.158 (1.164)
LOG (YCAP)	-2.546 (-2.723)***

LOG (YCAP)^2	0.175 (2.657)***
LOG (OIL)	-0.323 (-0.829)
LOG (CO2)	-0.702 (-2.370)**
Adjusted R^2	0.751

Note: *** significance at p-value ≤ 0.01; ** significance at p-value ≤ 0.05; t-statistics are indicated in parentheses.

The results of two additional independent variables as follows. There was not enough evidence to support the claim that oil price affects hydroelectricity consumption. This result does not support the previous study by [10]. Indonesia still relies on oil for its production of gasoline, so the oil price changes seem do not affect any effort to shift to renewable energy consumption. Finally, the increase of CO_2 emission concentration level by 1% leads to the higher hydroelectricity consumption by 0.702%, with assumption other variables being equal. This study contradicts the previous study by [10]. When CO_2 emission concentration level increases, it begins to harm the environment. So, in this case, the society acknowledges the importance of moving toward the renewable energy consumption

4. Conclusion

The awareness of the significance of the clean energy consumption to protect of the environment has become a main objective in this study. The objective of this study was to examine the probability of the impact of income on the renewable energy consumption. This study used Engle Granger Error Correction Model and used hydroelectricity consumption as the proxy of the renewable energy consumption and real GDP per capita as the proxy of the income. The result of this study suggested that hydroelectricity consumption can be expressed as a quadratic function of real GDP per capita. It means that at the beginning, increasing real GDP per capita reduced the hydroelectricity consumption, and at a certain level it turned to reverse direction, where the rise of the real GDP per capita increased hydroelectricity consumption. In conclusion, income is undeniably vital keys to shift toward the transition from traditional to renewable energy sources. The policy implication for policymakers is straightforward, i.e., it is important to increase the income of the society, not only for the sake of the prosperity of the society, but also for the protection of the environment through more clean energy consumption.

References

[1] British Petroleum 2022 *bp Statistical Review of World Energy* https://www.bp.com/content/dam/bp/business-sites/en/global/corporate/pdfs/energy-economics/statistical-review/bp-stats-review-2022-full-report.pdf
[2] Rapier R 2020 *Fossil Fuels Still Supply 84 Percent Of World Energy — And Other Eye Openers From BP's Annual Review* https://www.forbes.com/sites/rrapier/2020/06/20/bp-review-new-highs-in-global-energy-consumption-and-carbon-emissions-in-2019/?sh=16b32b9566a1
[3] United Nations 2022 *Affordable and clean energy* https://www.un.org/sustainabledevelopment/energy/
[4] IESR 2021 *Indonesia Energy Transition Outlook 2022. Tracking Progress of Energy Transition in Indonesia : Aiming for Net-Zero Emissions by 2050*. Jakarta: Institute for Essential Services Reform (IESR). https://iesr.or.id/wp-content/uploads/2022/01/Indonesia-Energy-Transition-Outlook-2022-IESR-Digital-Version-.pdf
[5] Choudhary M P, Chauhan G S and Kushwah Y K 2015 *Environmental Degradation: Causes, Impacts and Mitigation* (NSRAPEM-2015), Maharishi Arvind College of Engineering and Technology, Kota, Rajasthan, India
[6] Apergis N and Payne JE 2014 *Appl Econ* **46** pp. 4519 – 4525
[7] Salim R A and Rafiq S 2012 *Energy Econ* **34** pp. 1051–1057

[8] Ackah I and Kizys R 2015 *Renew Sust Energ Rev* **50** pp. 1157–1166
[9] da Silva P P, Cerqueira P A and Ogbe W 2018 *Energy* **156** pp. 45–54
[10] Karacan R, Mukhtarov S, Barış I and İşleyen A 2021 *Energies* 14 p. 2947
[11] Ouedraogo N S 2013 *Energy Econ* **36** pp. 637–647
[12] Ergun S J, Owusu P A and Rivas M F 2019 *Environ Sci Pollut R*es **26** pp. 15390–15405
[13] Attiaoui I, Toumi H, Ammouri B, Gargouri I 2017 *Environ Sci Pollut Res* **24** pp. 13036–13048
[14] Marques A C and Fuinhas J A 2011 *Renew Sust Energ Rev* **15** pp. 1601–1608

IOP Conf. Series: Earth and Environmental Science 1098 (2022) 012082 doi:10.1088/1755-1315/1098/1/012082

Utilization of waste oil palm empty fruit bunches for bioethanol through chemical-enzymatic integration process technology and towards concept the implementation of circular economy

N Harihastuti[1*], F N Purnamastuti[2]

[1]Research Centre for Environmental and Clean Technology, National Research and Innovation Agency, Puspitek Serpong, Tangerang Selatan, West of Java, Indonesia
[2]Research Centre for Industrial Machinery and Manufacture Technology, National Research and Innovation Agency, Puspitek Serpong, Tangerang Selatan, West of Java, Indonesia

nanisoeharto@yahoo.com

Abstract. result obtained from the application of this integration technology is the novelty of the process to make bioethanol more efficient and environmentally friendly. The innovation technology of this can be concluded to have potential in the application of circular economy to process PEFB into the growth of palm plantation area in Indonesia continues to increase with an estimated 11.67 million Ha, which can produce 33.50 million tons of CPO. For 1 ton of palm oil, 230-250 kgs of PEFB will be produced. PEFB consists of cellulose, hemicellulose, and lignin which have great potential to be reused after being converted into bioethanol. The purpose of this research is utilization of PEFB for bioethanol through an innovative chemical enzymatic integration process. The method of the research that PEFB biomass waste into bioethanol through a hydrolysis process with ammonia. The saccharification process was carried out enzymatically using the cellulase enzyme obtained on-site from T. viride and N. crassa. The WWTP unit that processes liquid waste from the bioethanol unit disain to result in biogas. The generator units that process biogas energy can be used in unit operation. The bioethanol products and produce biogas energy and can be reused for bioethanol and WWT processes.

1. Introduction

The growth of the oil palm plantation area in Indonesia continues to increase, in 2014 the area of approximately 10.75 million hectares in 2016 reached 11.67 million hectares with a production of 33.50 million tons [1]. For 1 ton of palm oil, 120-200 kg of CPO will be produced, 230-250 kg of EFB, 130-150 kg of fiber, and 60-65 kg of shells, as well as 0.7 m3 of liquid waste [2] PEFB waste is the largest component of waste, namely 23-25%. Currently, the utilization of PEFB waste has only reached 20%, including for PEFB biorefinery products, including biovanillin with alkaline pretreatment [3] Xylose [4], Nanofiber [5] bio-oil [6], cellulose phosphate [7], Polymer [8], Lactic acid [9] biogas [10]

PEFB pretreatment has been widely carried out, in various ways, including using ammonia which produces higher sugar [11], dilute acid [12], Ionic liquid [13], Hydrothermal pre-treatment [14], alkaline pretreatment followed by hydrolysis using dilute acid [15], pretreatment with fungi [16] , Bacterial consortium [17], Microwave and acid pretreatment [18], Pyrolysis and hydrothermal [19] [20] states that solid waste in palm oil mills can be used as electrical energy with an electrical energy potential of 110 KW/ton FFB/hour. Bioconversion of PEFB to ethanol with on-site cellulase enzymes [21] Several microbes have been used in lignocellulosic biorefinery, namely Trichoderma asperellum UPM1 and Aspergillus fumigatus UPM2 [22], Neurospora crassa S1 [23] Saccharomyces cerevisiae [24]. Saccharification research for bioethanol production from lignocellulosic waste using on-site cellulase-based trichoderma viride [25]. Lignocellulosic biomass is used for the production of cellulase by bacteria [26]. [27] in their research stated that the processing of POME with an integrated system, that is, using an anaerobic UASF reactor and with membrane separation. According to [28] biogas from liquid waste from the palm oil industry can be utilized using methane capture technology, which is then converted into electrical energy. Bioethanol production can be carried out using microwave and enzymatic integration [29], Rhizopus oryzae [30], Ligninolytic enzymes [31], Fermentation and chemical heating [32], [33].

The results of studies from previous studies or those that have been carried out above, there has not been a single research conducted, which is related to research studies of bioethanol production based on PEFB biomass waste through chemical-enzymatic integration process innovations. The study approach consists of three mutually supportive processes, namely the bioethanol process unit which processes PEFB biomass waste into bioethanol, the WWTP process unit which processes liquid waste from the bioethanol unit into biogas, and the steam generator unit which converts biogas and lignin into energy, are used in the bioethanol unit and the WWTP unit.

The purpose of this research is the utilization of PEFB waste into bioethanol through a chemical-enzymatic integration process, which further examines the application of circular economy in a sustainable process, which has the impact of preventing environmental pollution, saving energy use, preventing greenhouse gas emissions, and global warming.

2. Methodology

This research method is carried out on the basis of: literature study in the form of studies or research results that have been carried out from journals both national and international journals, especially those related to the topic (oil palm, PEFB, bioethanol, biogas, bioenergy, etc.). Then an evaluation of the results of the research is carried out, for further innovation of an integrated research concept that combines chemical and enzymatic processes to obtain bioethanol based on Palm Empty Fruit Bunches optimally. In addition , biogas energy is also obtained from the treatment of wastewater.

The integration of technology in the form of developing a pretreatment process for raw materials (biomass), using an enzymatic chemical process simultaneously will increase the efficiency of bioethanol production. The process of preparing PEFB raw materials is carried out through a hydrolysis process and a saccharification process, followed by a fermentation process and a distillation process. For the preparation of treatment of PEFB raw materials, it is done by chopping it using a chopping machine and sorting it with a sieving machine. Then proceed with the process of hydrolysis and saccharification: Heating at 100°C for 3 hours with 12.5% Ammonium hydroxide. Alkanolamine for 10 minutes (at a temperature of 80-150°C). No chemicals: first heating at 140°C-180°C, 1-2 hours, 105-150 psig, and second heating at 190°C-210°C, 2-10 hours, 167-262 psig.

The products of the best hydrolysis (ammonia, alkanolamine, and no chemicals) are fed into the saccharification reactor. The cellulase enzyme solution obtained on site from T. viride and N. crassa was then added using PEFB as the substrate. The research parameters on on-site cellulase production include humidity and substrate size, which greatly affect the quantity of cellulase. The fermentation process used S

cerevisiae and Z. mobilis at 30°C, pH 4, for 72 hours with 10% v/v input. Then the distillation process was continued through multilevel distillation and the dehydration process was continued using a synthetic Na zeolite membrane through a vaporizer, molecular sieve, and vacuum regenerant process.
Integrated technology in the biodigestion process of wastewater from the distillation process in bioethanol production, will produce biogas through an anaerobic process in a digester that is integrated with a bioethanol process unit and a biogas-based electricity generator unit.

3. Results and discussion

Palm Empty Fruit Bunches as show in fig. 1, which is a solid waste of oil palm consisting of cellulose and hemicellulose is also called holocellulose. These two components occupy 70% of the total composition of PEFB and the lignin content is 17%. PEFB is currently mostly used for various purposes, such as alternative energy materials, mulch, paper filling materials, etc. Several studies have been developed on the use of PEFB, including composting, pulping, charcoal briquettes, polypots for oil palm and rubber nurseries, plastic fillers that can be degraded, and lightweight chipboard.

Figure 1. Palm Empty Fruit Bunches [3]

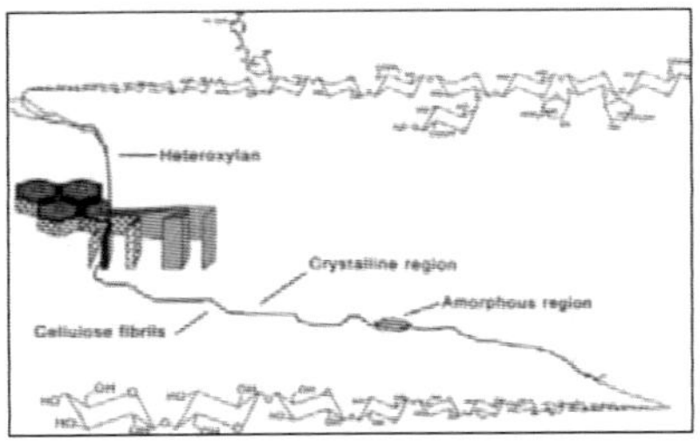

Figure 2. Lignocellulose [26]

Lignocellulose is the main component of plants that describes the amount of renewable sources of organic material. Lignocellulose is mainly composed of lignin, cellulose, and hemicellulose as shown in Figure 2 Lignin (Fig. 3) is the hardest non-carbohydrate component in plant cell walls. The lignin content in wood is about 25-40%, depending on the type of wood. Lignin functions as a binder between cells and strengthens wood cell walls. Lignin is a polymer of phenylpropane units that are cross-linked with each other so that they are not soluble in water as shown in Fig.3 [34].

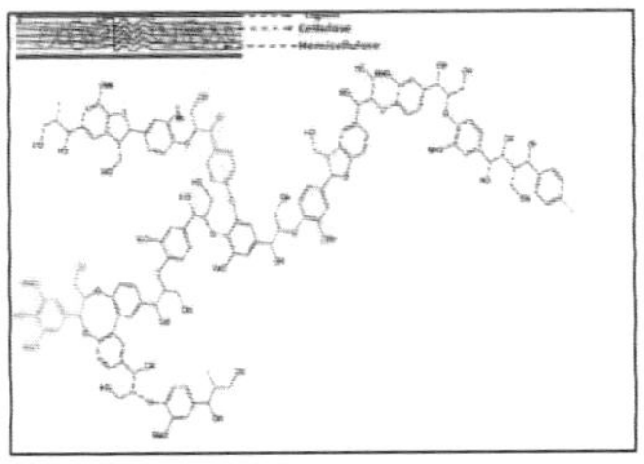

Figure 3. lignin [34]

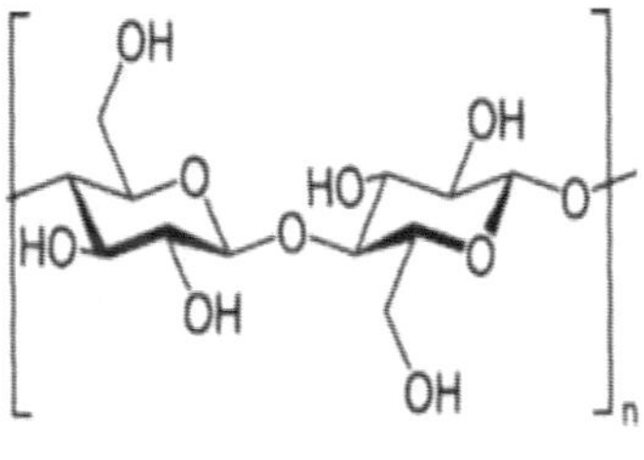

Figure 4. Cellulose [35]

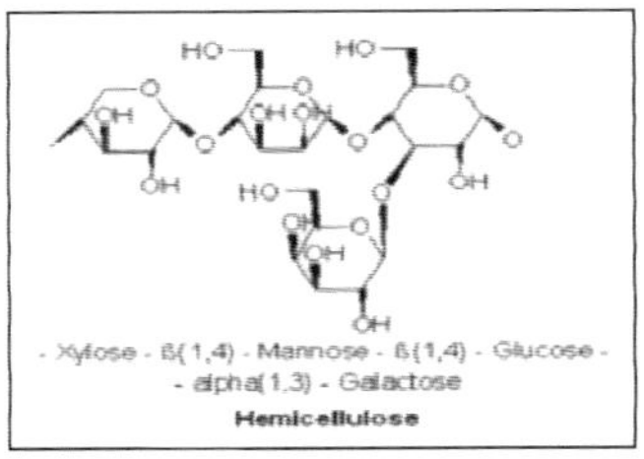

Figure 5. Hemicellulose [35]

Cellulose is the main component of plant cell walls. The cellulose content in higher plants is 35-50% of the dry weight of the plant. Cellulose is made up of unbranched glucose polymers, as shown in Figure 4. Hemicellulose is a branched heteropolysaccharide that is noncrystalline, is not fibrous, and easily expands. In contrast to cellulose which is only composed of glucose, hemicellulose is composed of several different sugar monomers, such as D-xylose, D-galactose, and D-mannose as shown in Figure 5. The hemicellulose content is approximately 25-30%. Hemicellulose polymer chains are short and amorphous. Hemicellulose has a polymerization degree of 50-200. Hemicellulose functions as a support for cell walls and acts as an adhesive between single cells found in plant stems.

From the composition of PEFB, hemicellulose is a very potential material to be used because of the amount of xylose content. Before being used, hemicellulose must first be broken down into monomers, and one way is by using the hydrolysis process. Hydrolysis is the process of breaking down complex molecules into smaller molecules with the help of acids or enzymes as catalysts. An example of a hydrolysis process is the breakdown of hemicellulose into xylose using sulfuric acid or enzymes derived from microbes [35].

Enzymatic hydrolysis of hemicellulose resulted in a more specific hydrolysis product and no further hydrolysis of the sugar formed. Xylanase enzymes can be used to hydrolyze xylan, which is a polymer of xylose and xylooligosaccharides. Enzymatic hydrolysis has several disadvantages: less xylose is produced than hydrolysis with acid, it takes longer, the price of the enzyme is more expensive than the price of acid, and the action of the enzyme is inhibited by the product. Therefore, hydrolysis with acid is more applicable because the production cost is lower than hydrolysis with enzymes.

The saccharification stage is the stage of breaking down complex sugars into simple sugars carried out in a tube on a series of equipment for the production of bioethanol. Saccharification involves the following processes:

- Cooling the slurry to the optimal temperature for the saccharification enzymes to work
- Optimum pH regulation of enzymes
- Appropriate addition of enzymes (glucoamylase)
- Maintain pH and temperature in the range of 50° to $60^{o}C$, until the saccharification process is complete (done by testing the resulting simple sugars) [36].

The fermentation process stage will run for a few hours after all the ingredients are put in the fermenter. With a translucent fermenter (from glass for example), it will appear small air bubbles will appear from inside the fermenter. These air bubbles are CO2 gas produced during the fermentation process. During this fermentation process, try to keep the temperature not exceeding 36oC and the pH is maintained at 4.5-5. The fermentation process runs for approximately 72 hours or approximately 2.5 days. One sign that fermentation is complete is that there are no more air bubbles to be seen. The ethanol content in the fermented liquid is around 7% – 10%.

The next stage is the process of distillation and dehydration. After the fermentation process is completed, put the fermented liquid in the evaporator or boiler. Heat maintain evaporator and the temperature between 79-81°C. At this temperature, the ethanol has evaporated, but the water has not evaporated. The ethanol vapor flows into the distillator. Bioethanol will come out of the distillator discharge pipe. At the first distillation, the ethanol content is usually still below 95%. If the ethanol content is still below 95%, the distillation needs to be repeated again (reflux) until the ethanol content is 95%. When the ethanol content is 95%, dehydration or water removal is performed. To remove water, you can use quicklime or synthetic zeolite. Add quicklime to ethanol. Leave it overnight. After that it is distilled again until the water content is approximately 99.5% (Jeon et al, 2014) dehydration can also be done with a NaA membrane [37].

The utilization of palm oil effluent in electrical energy can provide multiple benefits, both from an environmental and an economic perspective. From the environmental side, efforts to capture methane gas emissions from liquid waste will reduce the amount of methane gas (CH4) emissions that have the potential to cause global warming. For an economic point of view, the utilization of palm oil effluent can provide financial benefits, the first from the Clean Development Mechanism (CDM) program in the form of selling carbon or Certified Emission Reduction (CER) and secondly selling electrical energy.

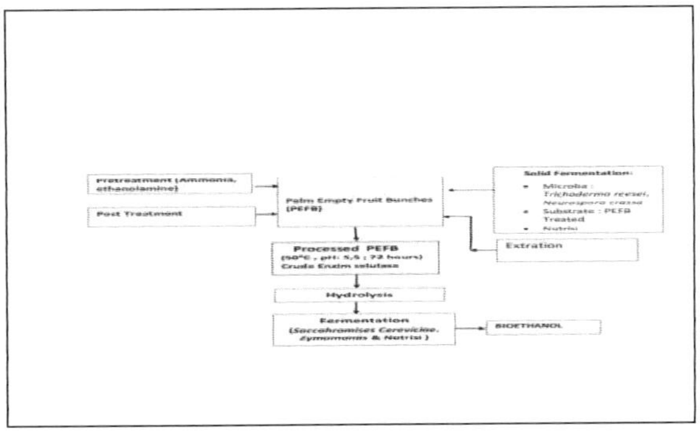

Figure 6. Bioethanol production scheme with on-site cellulase enzyme

The bioethanol unit aims to process PEFB biomass waste into bioethanol, as shown in Fig.6. The liquid waste is then processed in the WWTP unit into treated clean water, which will be reused in the bioethanol unit as shown in Fig. 7 and Fig. 8 below. The lignin that has been separated becomes the input of the generator unit in the supply of hot steam, as shown in Fig.7. Integration of technology in the form of the development of biomass pretreatment processes, using local microbe isolates, and enzymatic chemical processes will simultaneously increase the efficiency of bioethanol production, as shown in Fig. 9.

The bioethanol production unit with PEFB waste as a raw distillation unit. The pretreatment unit in the form of PEFB is chopped using a chopping machine and material, consists of a raw material pretreatment unit (hydrolysis and saccharification), a fermentation unit, and a sorted by a sieving machine as shown in Fig.6 above. Furthermore, the process of hydrolysis and saccharification. Products from the best hydrolysis (ammonia, alkanolamine, and no chemicals) are feed into the saccharification reactor. Then, the cellulase enzyme solution obtained on-site from T. viride and N. crassa was added using PEFB as substrate. The research parameters on on-site cellulase production include humidity and substrate size which greatly affect the cellulase quantity, as shown in Fig.9.

The fermentation unit used S cerevisiae and Z. mobilis at 50°C, pH 5,5, for 72 hours with 10% v/v input. Distillation unit with multilevel distillation followed by dehydration of bioethanol with synthetic Na zeolite membrane with vaporizer, molecular sieve, and vacuum regenerant components. Below is the design of each recommended process flow diagram to achieve optimal circular economy as follows on Fig.7, Fig.8, and Fig.9.

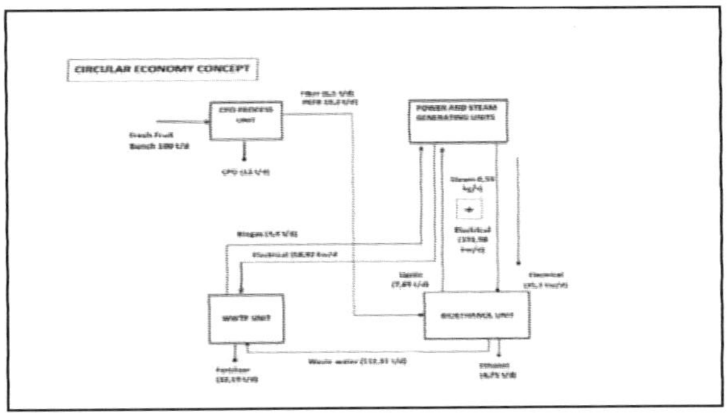

Figure 7. Chemical-enzymatic technology integration process flow diagram

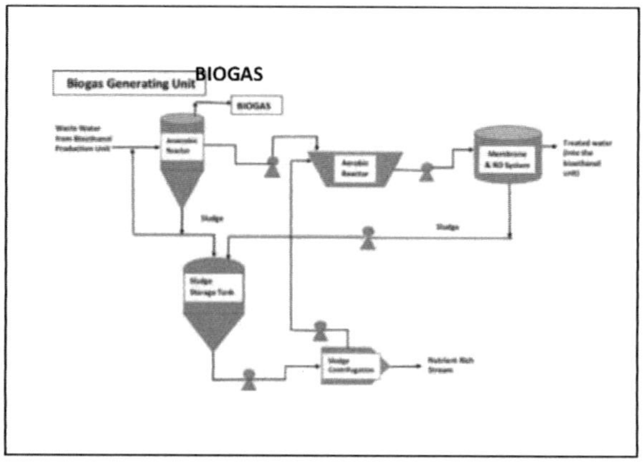

Figure 8. Design of WWTP generating biogas unit from bioethanol waste

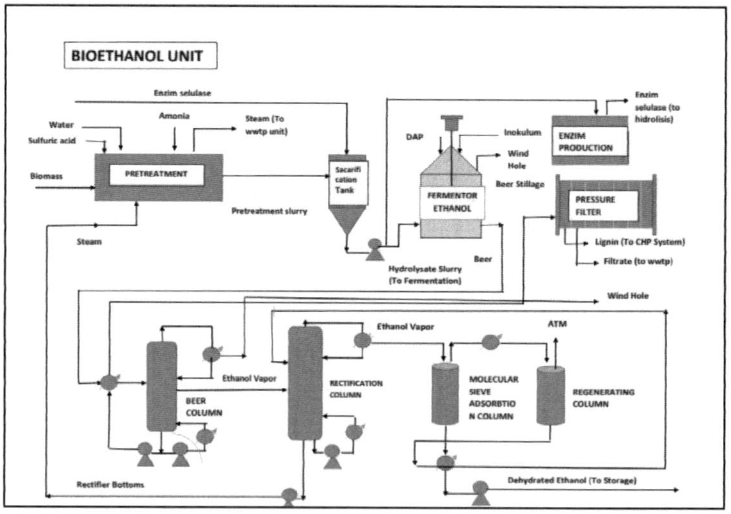

Figure 9. Design of Process flow diagram intregrated of bioethanol production

4. Conclusion

From the results of the evaluation and study of this research, it can be said that the palm oil industry,which is Palm Empty fruits Bunches into bioethanol and produces biogas, can save costs in energy supply and environmental management. The use of bioethanol and biogas is very influential in reducing the effects of greenhouse gases and global warming, so that it is friendly to the environment, gaining added value and income. The palm oil industry will increase by obtaining bioethanol and biogas as well as additional income from CER (Certified Emission Reduction). The use of treated clean water and fertilizer materials from the WWTP unit will reduce operating costs for the industry. Overall, it is a concept of applying circular economy optimal to achieve in industry.

Acknowledgments

The authors express their gratitude to the Research Center for Enviromental and Clean Technology, the National Research and Innovation Agency, for supporting this research study.

References

[1] Ditjen Perkebunan RI. 2015. Statistik Perkebunan Indonesia Komoditas Kelapa Sawit 2014-2016. Kementan.

[2] Mahajoeno. E.. Lay. B.W.. Sutjahjo. S.H.. Siswanto. 2008. Potensi Limbah. Cair Pabrik Minyak Kelapa Sawit Produksi Biogas. Jurnal Biodiversitas 9(1):48.

[3] A. Bahrin. E. K.. Ramli. N.. Phang. L. Y.. & Abd-Aziz. S. (2016). Alkaline Hydrolysate of Oil Palm Empty Fruit Bunch as Potential Substrate for Biovanillin Production via Two-Step Bioconversion. Waste and Biomass Valorization. 1–11.

[4] Zhang. Z.. Vancov. T.. Mackintosh. S.. Basu. B.. Lali. A.. Qian. G.. Doherty. W. O. S. (2016). Assessing dilute acid pretreatment of different lignocellulosic biomasses for enhanced sugar production. Cellulose. 23(6). 3771–3783.

[5] Solikhin. A.. Hadi. Y. S.. Massijaya. M. Y.. & Nikmatin. S. (2016). Novel Isolation of Empty Fruit Bunch Lignocellulose Nanofibers Using Different Vibration Milling Times-Assisted Multimechanical Stages. Waste and Biomass Valorization.

[6] Sarwono. R.. Pusfitasari. E. D.. & Ghozali. M. (2016). Hydrothermal liquefaction of palm oil empty fruit bunch (EFB) into bio-oil in different organic solvents. 60015. 60015.

[7] Rohaizu. R.. & Wanrosli. W. D. (2015). Production of cellulose phosphate from oil palm empty fruit bunch: Effect of chemical ratio. Journal of Physics: Conference Series.

[8] Faizi. M. K.. Shahriman. A. B.. Majid. M. S. A.. Shamsul. B. M. T.. & Ng. Y. G. (2017). An overview of the Oil Palm Empty Fruit Bunch (OPEFB) potential as reinforcing fibre in polymer composite for energy absorption applications.

[9] Chin. S. X.. Tasirin. S. M.. Chan. H.. & Chia. H. (2016). Catalytic Conversion of Empty Fruit Bunch (PEFB) Fibres into Lactic Acid by Lead (II) ions. Bioresources.Com. 11((1)). 2186.

[10] Ariffin. M. A.. Wan Mahmood. W. M. F.. Harun. Z.. & Mohamed. R. (2016). Medium-scale gasification of oil palm empty fruit bunch for power generation. Journal of Material Cycles and Waste Management. 1–9.

[11] Zulkiple. N.. Maskat. M. Y.. & Hassan. O. (2016). Pretreatment of Oil Palm Empty Fruit Fiber (PEFB) with Aquaeous Ammonia for High Production of Sugar. Procedia Chemistry. 18(Mcls 2015). 155–161.

[12] Zhang. D.. Ong. Y. L.. Li. Z.. & Wu. J. C. (2012). Optimization of dilute acid-catalyzed hydrolysis of oil palm empty fruit bunch for high yield production of xylose. Chemical Engineering Journal. 181–182. 636–642.

[13] Yoricya. G.. Aisyah. S.. Dalimunthe. P.. Manurung. R.. & Bangun. N. (2016). Kelapa Sawit dalam Sistem Cairan Ionik. Jurnal Teknik Kimia. 5(1). 1–7.

[14] Muhd Ali. M. D.. Tamunaidu. P.. Nor Aslan. A. K. H.. Morad. N. A.. Sugiura. N.. Goto. M.. & Zhang. Z. (2016). Hydrothermal pre-treatment of oil palm empty fruit bunch into fermentable sugars. IOP Conference Series: Earth and Environmental Science. 36(July).

[15] Loow. Y. L.. Wu. T. Y.. Jahim. J. M.. Mohammad. A. W.. & Teoh. W. H. (2016). Typical conversion of lignocellulosic biomass into reducing sugars using dilute acid hydrolysis and alkaline pretreatment. Cellulose. 23(3). 1491–1520.

[16] Kamcharoen. A.. Champreda. V.. Eurwilaichitr. L.. & Boonsawang. P. (2014). Screening and optimization of parameters affecting fungal pretreatment of oil palm empty fruit bunch (EFB) by experimental design. International Journal of Energy and Environmental Engineering. 5(4)..

[17] Hazaimeh. M.. Abd Mutalib. S.. Abdullah. P. S.. Kok Kee. W.. & Surif. S. (2014). Enhanced crude oil hydrocarbon degradation by self-immobilized bacterial consortium culture on sawdust and oil palm empty fruit bunch. Annals of Microbiology. 64(4). 1769–1777.

[18] Fatriasari. W.. Anita. S. H.. & Risanto. L. (2016). Microwave Assisted Acid Pretreatment of Oil Palm Empty Fruit Bunches (EFB) to Enhance Its Fermentable Sugar Production. Waste and Biomass Valorization..

[19] Chan. Y. H.. Tan. R. R.. Yusup. S.. Lam. H. L.. & Quitain. A. T. (2016). Comparative life cycle assessment (LCA) of bio-oil production from fast pyrolysis and hydrothermal liquefaction of oil palm empty fruit bunch (EFB). Clean Technologies and Environmental Policy. 18(6). 1759.

[20] Tiyono. tiyono (2011) Potensi Pembangkit Listrik Tenaga Biomassa Sawit PKS Ngabang Sebesar 110 KW/Ton Sawit per jam. Jurnal Penelitian Teknik Elektro. 4 (2). pp. 1-6. ISSN 1979-343X.

[21] Zhu. Y.. Xin. F.. Zhao. Y.. & Chang. Y. (2014). An integrative process of bioconversion of oil palm empty fruit bunch fiber to ethanol with on-site cellulase production. Bioprocess and Biosystems Engineering. 37(11). 2317 2324.

[22] Zanirun. Z.. Bahrin. E. K.. Lai-Yee. P.. Hassan. M. A.. & Abd-Aziz. S. (2014). Effect of physical and chemical properties of oil palm empty fruit bunch. decanter cake and sago pith residue on cellulases production by Trichoderma asperellum UPM1 and Aspergillus fumigatus UPM2. Applied Biochemistry and Biotechnology. 172(1). 423–435.
[23] Li. Q.. Ng. W. T.. & Wu. J. C. (2014). Isolation. Characterization and Application of a Cellulose-degradingStrain Neurospora crassa S1 from Oil Palm Empty Fruit Bunch. Microbial Cell Factories. 13(157). 1–8.
[24] Chauhan. M. (2016). Bioethanol Production Using Saccharomyces cerevisiae with Different Perspectives : Substrates . Growth Variables . Inhibitor Reduction and. 5(2). 2–5.
[25] Rame. Nani Harihastuti. Silvy Djayanti. 2016. Optimalisasi Proses Sakarifikasi Untuk Produksi Bioetanol Dari Limbah Padat Industri Pati Aren Menggunakan Trichoderma Viride Berbasis Enzim Selulase On-Site. Prosiding Seminar Nasional Hasil-Hasil Penelitian Pascasarjana. SPS UNDIP Semarang.
[26] Oke. M. A.. Suffian. M.. & Annuar. M. (2016). Mixed Lignocellulosic Biomass Degradation and Utilization for Bacterial Cellulase Production. Waste and Biomass Valorization.
[27] Ahmed. Y.. Yaakob. Z.. Akhtar. P.. Sopian. K.. 2015. Production of Biogas and Performance Evaluation of Existing Treatment Processes in Palm Oil Mill Effluent (POME). Renewable and Sustainable Energy Reviews. Volume 42. pp. 1260-1278.
[28] Febijanto. I. 2010. Potensi Penangkapan Gas Metana dan Pemanfaatannya Sebagai Bahan Bakar PembangkitListrik di PTPN VI Jambi. Jurnal Ilmu Teknologi Energi. Vol. 1 No. 10 Februari 2010: 30-47. Pusat Teknologi Sumberdaya Energi. BPPT. Jakarta.
[29] Sudha. A.. Sivakumar. V.. Sangeetha. V.. & Devi. K. S. P. (2015). Enhancing fermentable sugar yield from cassava pulp for bioethanol production: Microwave-coupled enzymatic hydrolysis approach. Bioprocess and Biosystems Engineering. 38(8). 1509–1515.
[30] Sahlan. M.. & Sudiyani. Y. (2012). Simultaneous Saccharification and Fermentation of Oil Palm Empty Fruit Bunch for Bioethanol Production by Rhizopus oryzae. International Journal of Environment and Bioenergy International Journal of Environment and Bioenergy.
[31] Plácido. J.. & Capareda. S. (2015). Ligninolytic enzymes: a biotechnological alternative for bioethanol production. Bioresources and Bioprocessing. 2(1). 23.
[32] Nurul Adela. B.. Nasrin. A. B.. Loh. S. K.. & Choo. Y. M. (2014). Bioethanol production by fermentation of oil palm empty fruit bunches pretreated with combined chemicals. Journal Applied Environmental and Biological Sciences. 4(10). 234–242.
[33] Christia. A.. Setiowati. A. D.. Millati. R.. Karimi. K.. Cahyanto. M. N.. Niklasson. C.. & Taherzadeh. M. J. (2016). Ethanol production from alkali-pretreated oil palm empty fruit bunch by simultaneous saccharification and fermentation with mucor indicus. International Journal of Green Energy. 13(6).
[34] Suhartati. S.. Puspito. R.. Rizali. F.. & Anggraini. D. (2016). Analisis Sifat Fisika dan Kimia Lignin Tandan Kosong Kelapa Sawit asal Desa Sape.Kabupaten Sanggau. Kalimantan Barat. Jurnal Kimia Valensi :Jurnal Penelitian Dan Pengembangan Ilmu Kimia. 2(1). 24–29.
[35] Harsono. H.. Putra. A. S.. Maryana. R.. Rizaluddin. A. T.. H'ng. Y. Y.. Nakagawa-izumi. A.. & Ohi. H. (2016). Preparation of dissolving pulp from oil palm empty fruit bunch by prehydrolysis soda-anthraquinone cooking method. Journal of Wood Science. 62(1). 65–73.
[36] Muktham. R.. Bhargava. S. K.. Bankupalli. S.. & Ball. A. S. (2016). A Review on 1 st and 2 nd Generation Bioethanol Production-Recent Progress. (September). 72–92.
[37] Wei. H. J.. Tsai. T. Y.. & Weng. Y. H. (2016). Pilot-scale study on dehydration of synthetic and lignocellulosic ethanol by NaA membrane. Korean Journal of Chemical Engineering. 33(4). 1362–1368.

AUTHOR INDEX